Capillary Electrophoresis

THEORY and PRACTICE

Second Edition

NEW DIRECTIONS in ORGANIC and BIOLOGICAL CHEMISTRY

Series Editor: C.W. Rees, CBE, FRS
Imperial College of Science, Technology and Medicine, London, UK

Advisory Editor: Alan R. Katritzky, FRS
University of Florida, Gainesville, Florida

Published Titles

Activated Metals in Organic Synthesis
Pedro Cintas

The Anomeric Effect
Eusebio Juaristi and Gabriel Cuevas

Aromatic Fluorination
James H. Clark, David Wails, and Tony W. Bastock

Asymmetric Synthetic Methodology
David J. Ager and Michael B. East

Capillary Electrophoresis: Theory and Practice
Patrick Camilleri

C-Glycoside Synthesis
Maarten Postema

Chemical Approaches to the Synthesis of Peptides and Proteins
Paul Lloyd-Williams, Fernando Albericio, and Ernest Giralt

Chiral Sulfur Reagents
M. Mikołajczyk, J. Drabowicz, and P. Kiełbasiński

Chirality and the Biological Activity of Drugs
Roger J. Crossley

Cyclization Reactions
C. Thebtaranonth and Y. Thebtaranonth

Dianion Chemistry in Organic Synthesis
Charles M. Thompson

Lewis Acids and Selectivity in Organic Synthesis
M. Santelli and J.-M. Pons

Mannich Bases: Chemistry and Uses
Maurilio Tramontini and Luigi Angiolini

Organozinc Reagents in Organic Synthesis
Ender Erdik

Synthesis Using Vilsmeier Reagents
C. M. Marson and P. R. Giles

Vicarious Nucleophilic Substitution and Related Processes in Organic Synthesis
Mieczyslaw Makosza

Capillary Electrophoresis

THEORY and PRACTICE

Second Edition

EDITED BY

Patrick Camilleri

SmithKline Beecham Pharmaceuticals
Harlow, Essex, United Kingdom

CRC Press
Boca Raton   Boston   New York   Washington, D.C.   London

Library of Congress Cataloging-in-Publication Data

Capillary Electrophoresis / edited by Patrick Camilleri. — 2nd. ed.
    p. cm. — (New directions in organic and biological chemistry)
    Includes bibliographical references and index.
    ISBN 0-8493-9127-X (alk. paper)
    1. Capillary electrophoresis. I. Camilleri, Patrick. II. Series.
    [DNLM: 1. Hepatitis B virus. QW 710 G289h]
  QP519.9.C36C373 1997
  5436'.0871—dc21

                                                     97-22491
                                                      CIP

© 1998 by CRC Press LLC

No claim to original U.S. Government works
International Standard Book Number 0-8493-9127-X
Library of Congress Card Number 97-22491
Printed in the United States of America 2 3 4 5 6 7 8 9 0
Printed on acid-free paper

The Editor

Patrick Camilleri was born in Malta where he received his secondary and graduate education. His Ph.D. was in the area of gas kinetics, under the supervision of the late Professor Howard Purnell, at The University College of Swansea in Wales. This was followed with a postdoctoral fellowship with Professor Anthony Kirby at The University Chemical Laboratory in Cambridge.

After having lectured for three years in physical chemistry at The Royal Universtiy of Malta, Dr. Camilleri joined Shell Research, working on the physicochemical characterization of a range of agrochemicals. He remained at Shell for 10 years, then joined SmithKline Beecham, where he currently holds the position of Director of Separation Sciences, heading a department specializing in biological and chemical separations.

In 1995 he was made Visiting Professor at Imperial College, and in 1996 was awarded the Jubilee Silver Medal by the Chromatographic Society (UK), in recognition of distinguished contributions to the development of separation science, in particular, capillary electrophoresis.

Dr. Camilleri has given a number of public lectures during his scientific career on a wide range of toipics, which include photosynthesis, quantitative structure-activity correlations, chiral discrimination, and carbohydrate analysis. To date, he has published over 135 publications in refereed international journals.

Contributors

Patrick Camilleri, Ph.D.
SmithKline Beecham Pharmaceuticals
Harlow, Essex, United Kingdom

Norman J. Dovichi
Department of Chemistry
University of Alberta
Edmonton, Alberta, Canada

Joe P. Foley, Ph.D.
Department of Chemistry
Villanova University
Villanova, Pennsylvania

James S. Fritz, Ph.D.
Department of Chemistry and Ames
  Laboratory
U.S. Department of Energy
Iowa State University
Ames, Iowa

András Guttman, Dr.Sc., Ph.D.
Genetic BioSystems
San Diego, California

Werner G. Kuhr, Ph.D.
Department of Chemistry
University of California
Riverside, California

Kurt R. Nielsen, Ph.D.
Energy Biosystems
The Woodlands, Texas

Yehia S. Mechref, Ph.D.
Department of Chemistry
Oklahoma State University
Stillwater, Oklahoma

George N. Okafo, Ph.D.
SmithKline Beecham Pharmaceuticals
Harlow, Essex, United Kingdom

David Perrett
Department of Medicine
St. Bartholamew's Hospital Medical College
West Smithfield, London, United Kingdom

Ziad El Rassi, Ph.D.
Department of Chemistry
Oklahoma State University
Stillwater, Oklahoma

Herbert E. Schwartz, Ph.D.
FzioMed Inc.
Redwood City, California

Pierre Thibault, Ph.D.
Institute for Biological Sciences
National Research Council/ Canada
Ottawa, Ontario, Canada

Anders Vinther, Ph.D.
Novo Nordisk
Gentofte, Denmark

Contents

Preface

The importance of capillary electrophoresis (CE) as an analytical tool has increased dramatically over the last ten years. In addition to the continued effort in free-zone or open tubular CE, there has been tremendous impetus in the development of a number of other modes of operation — micellar electrokinetic chromatography (MEKC), isoelectric focusing (IEF), and gel and affinity electrophoresis. It is becoming increasingly appreciated that, unlike other separation techniques, changing from one mode of CE to another is very often simple and only involves a change of the buffer solution. Moreover, CE is a welcome technique when the amount of material to be analyzed is sparse.

The complementarity of CE to other wet techniques, especially high performance liquid chromatography and slab gel electrophoresis, has made its presence essential in a modern analytical laboratory. Judging from my experience in the pharmaceutical industry, CE appears to be entering a new phase. It is undergoing a change from being an exploratory technique, of mainly academic interest, to one which is being applied to solve "real" analytical problems. It is being used effectively to monitor purity or impurity profiles of drug substances during the research, development, and manufacturing phases.

This book outlines the basic theoretical aspects of the separation and detection methodology of CE. Throughout the ten chapters the applicability of this technique is demonstrated for the analysis of a range of analytes differing widely in physicochemical properties, such as size, charge, and hydrophobicity. Chapters dealing with the CE analysis of inorganic ions and carbohydrates have also been included in this edition. The extensive and detailed coverage of the literature for each of the topics covered makes this book an excellent reference work. As in the first edition, three appendices have been included to enhance the practical aspects of the book. Appendix I gives a list of CE suppliers, and Appendix II tabulates a number of buffer solutions suitable for CE. In Appendix III, experiments have been designed that are suitable for university or college students, or anyone who wishes to become familiar with CE.

As was the case in the first edition, there will undoubtedly be some errors, although it is hoped that these will be infrequent and will not diminish the scientific quality of this book.

I am very grateful to the authors for their commitment, patience, and cooperation. Manuscripts submitted were found to be highly organized and of excellent quality so that only a limited amount of editing was necessary.

Patrick Camilleri
Harlow, Essex, United Kingdom
August 1997

CHAPTER 1

History and Development of Capillary Electrophoresis

Patrick Camilleri

CONTENTS

I. INTRODUCTION

In 1833 and 1834 Michael Faraday studied a number of electrode reactions quantitatively. He summarized his findings by formulating his Laws of Electrolysis. This was followed by considerable activity in the remaining years of the 19th century, when prominent scientists, in particular Hittorf, Helmholtz, and Kohlrauch, performed landmark experiments in the area of the electrophoresis of small inorganic ions. They outlined theories for the mechanism of electrophoresis, discovered the role of electroosmosis, and studied the mobility of ions in an electric field to determine physical parameters such as ionic radii and transport numbers.

Tiselius was one of the first to carry out electrophoresis experiments on proteins. His thesis titled, "The Moving Boundary Method to Study the Electrophoresis of Proteins," was published in 1930. Tiselius utilized the electric charge carried by these macromolecules to achieve some pioneering separations of blood plasma proteins in free solution. He developed a "moving boundary" method by which he was able to separate albumin from α-, β-, and γ-globulin. Detection of these proteins was carried out by recording changes in the refractivity of their boundaries or their characteristic ultraviolet absorbance on a photographic film. Tiselius also demonstrated that the electrophoretic mobility of proteins can be related to their molecular weight, assuming an idealized molecular shape. In recognition of his pioneering work in electrophoresis, Tiselius was awarded the Nobel Prize in 1948.

One of the early problems encountered in free-solution electrophoresis is that of band broadening due to thermal effects caused by electrical heating. This effect was partly resolved by Tiselius, who introduced efficient cooling of the electrophoresis cell by fast circulation of water maintained at 4°C. However, an important solution to this problem has been to carry out electrophoresis in buffer utilizing supporting media such as paper, cellulose acetate, starch, agarose, and polyacrylamide. These media not only function to restrict diffusion of bands, but can also act as molecular sieves, especially in the case of polyacrylamide gels.

Capillary electrophoresis (CE) is the most recently introduced electrophoretic technique in which, unlike conventional electrophoresis, the use of a supporting medium is not necessary and analysis can be carried out in free aqueous solution. It will be shown later in this chapter and more extensively in the other chapters of this book that capillary electrophoresis can be used for the separation of a variety of analytes of different sizes and charge: simple molecules, organic and inorganic ions, peptides, proteins, carbohydrates, and nucleic acids. In some cases, the same capillary can be used for a variety of structural compounds by performing consecutive electrophoresis runs under different buffer conditions, varying the buffer composition and/or adjusting pH.

Both size and charge are important factors in obtaining resolution of biomolecules in capillary gel electrophoresis, one of the modes of operation of capillary electrophoresis, a situation similar to that found in conventional gel electrophoresis. However, small differences in charge contribute largely to resolution achieved in open tubular capillary electrophoresis, the most common mode of operation of CE. Neutral molecules are efficiently separated using micellar electrokinetic chromatography (MEKC).

For a greater appreciation of some of the advantages of capillary electrophoresis over conventional paper and gel electrophoresis, an account is given in the following section about the separation of charged biological molecules (proteins and nucleic acids) using the latter electrophoresis techniques. This is followed by a brief historical account and an overview of some of the developments that have taken place in the field of capillary electrophoresis over the past 10 to 20 years. Special emphasis will be given to the wide and ever-growing range of applications of this technique.

II. PAPER AND GEL ELECTROPHORESIS

The use of filter paper as a supporting medium for electrophoresis has been widely used since the late 1940s. This separation technique has been most successful for the resolution of ionized compounds of relatively small molecular weight such as amino acids, lipids, nucleotides, and charged sugars. A mixture of analytes is applied at the center of a long strip of filter paper, the ends of which are placed in close contact with the cathode and anode. Typically a voltage of about 100 V/cm is applied across the strip. The rate of migration of these organic ions depends on their size and the magnitude of the applied current, whereas the direction of migration is related to the nature of the net charge (positive or negative) which they carry at a particular pH of the buffer medium. This methodology has been used in the case of amino acids and short peptides to empirically determine their isoelectric point at a known ionic strength by measuring the pH at which these analytes become electrophoretically immobile or electrically neutral. Detection of analytes is usually carried out by appropriate staining methods.

The outstanding success of paper electrophoresis in its early introduction was due mainly to good resolution, the possibility of accurate quantitative and qualitative analysis, and the requirement that the amount of analytes could be analyzed in less than 1-mg quantities. Moreover, the use of paper as the supporting medium for electrophoresis allowed suitable variations in the methodology described above. Enhanced resolution could be obtained by carrying out electrophoresis at a particular pH in one dimension followed by a second electrophoresis at another pH in a second dimension. Analytes were also separated by paper chromatography followed by electrophoresis. These techniques were widely used for the amino acid analysis of proteins and peptides.

The use of gelatin and agar gels as supports in electrophoresis has been recognized for at least a hundred years. A major advancement in gel electrophoresis took place about 25 years ago when it was recognized that electrophoresis of proteins can be carried out with high resolution on polyacrylamide gels under denaturing conditions using urea or, more commonly, a detergent such as sodium dodecyl sulfate (SDS).[2] This molecule has both polar (negatively charged sulfate) and nonpolar (hydrocarbon) moieties, allowing it to bind to most proteins. In boiling aqueous media containing SDS, proteins are denatured and bind to the detergent in a constant weight ratio. Under these conditions, the net charge of the polypeptide becomes insignificant compared to the negative charges due to the bound SDS. These SDS-protein complexes cannot be differentiated from their respective molecule-charge densities and will only migrate differently from one another in polyacrylamide gel electrophoresis according to their differences in size. This sieving effect can be altered by changing the ratio of bisacrylamide crosslinked to acrylamide. After separation, peptides and proteins are located on the supporting medium by staining with a dye such as coomassie brilliant blue R250 dissolved in a mixture of methanol and acetic acid. This process is then followed by destaining and washing in the same solvent mixture.

SDS polyacrylamide gel electrophoresis is useful because it can give a rough estimate of the molecular weight of a protein from the measurement of its mobility. Figure 1.1 shows the separation of four proteins on an SDS polyacrylamide gel system. A plot of mobility of these proteins with $\log_{10}$ (molecular weight) gives a good linear relationship.

Most SDS-polyacrylamide gel systems have insufficient resolving power for polypeptides with a molecular weight lower than about 10,000. Resolution of these molecules has been found to improve considerably when electrophoresis is carried out in the presence of SDS,

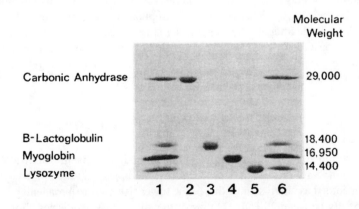

Figure 1.1 Polyacrylamide gel electrophoresis of four proteins: lane 2, carbonic anhydrase; lane 3, P-lactoglobulin; lane 4, myoglobin; lane 5, lysozyme; lanes 1 and 6, mixture of the four proteins.

a high ratio of bisacrylamide crosslinked to acrylamide, and a high concentration of urea (8 M).[3,4] For peptides with molecular weights lower than about 3000, large deviations from the good linear relationship between the $\log_{10}$ of the molecular weight of oligopeptides and mobility are often observed.[3] This behavior is expected because of the greater significance of net charge and conformation on mobility at the smaller molecular weights.

Polyacrylamide gels are also used for the electrophoresis of nucleic acids.[5] Gel materials that have been used successfully for this purpose include agar, agarose, and mixtures of agarose and polyacrylamide. The latter gels are especially suitable for the resolution of high molecular weight RNA. SDS is usually added to the buffer solution when running a denaturing gel. The effect of this detergent is to stop nucleation of RNA, the occurrence of which can lead to inaccurate estimates of molecular weight.

Several methods are available for the detection of nucleic acids. Gels can be soaked overnight in a 0.1 to 0.2% solution of a dye such as methylene blue. Direct scanning of gels under ultraviolet light (wavelength, 254 nm) is fast and reliable. Incorporation of ethidium bromide (5 µg/mL) in the gel and the buffer allows the nucleic acids to be visualized by their fluorescence when the gel is viewed under ultraviolet light. ^{32}P-labeled RNA can be detected by autoradiography.

Another mode of operation of gel electrophoresis is isoelectric focusing, commonly abbreviated IEF. Svensson[6] published the theory behind this technique in 1961. Polyacrylamide or agarose gels are impregnated with carrier ampholytes of a wide range of pK_a values to produce a pH gradient on the application of a high voltage across the gel. These pH gradients allow resolution of proteins in relation to their respective isoelectric point. Thus although proteins of close molecular weight cannot be separated by SDS gel electrophoresis, IEF allows separation if their isoelectric points differ by as little as 0.01 of a pH unit.[7] Even higher resolution of complex mixtures of proteins can be obtained by two-dimensional gel electrophoresis. IEF is performed in the first dimension in a gel rod. This rod is then placed at one end of an SDS polyacrylamide slab gel and electrophoresis is carried out in the second dimension.[7]

Gel electrophoresis has been rarely used for the separation and identification of small charged molecules of molecular weight less than about 1000. Fixing these molecules by techniques adequate for larger molecules is very difficult primarily because of diffusion. Consequently, such molecules are easily washed out of the gel during staining and visualization procedures. Laborious techniques[8] involving freeze-drying of the polyacrylamide gel, followed by lyophilization of the gel and final staining by ninhydrin, have been found to be suitable for the analysis of amino acids and oligopeptides. Detection limits range from 0.1 to 5 µg.

One of the disadvantages of gel electrophoresis is the lack of complete automation. In many laboratories, gels have to be cast before electrophoresis. On completion of a separation, detection is only possible after at times lengthy staining and destaining procedures. The toxic properties of gel constituents such as polyacrylamide can also pose health hazards. As shall be discussed in Sections III and IV, some of the problems encountered in gel electrophoresis can be overcome using capillary electrophoresis technology.

III. CAPILLARY ELECTROPHORESIS

Over the last 5 to 20 years CE has proved to be a rapid and versatile analytical technique that combines simplicity with high reproducibility. Complex mixtures of analytes can be resolved and recorded as sharp signals due to the lower risk of zone broadening. The narrow diameter (normally between 20 and 100 µm) of the silica glass capillaries allows the application of high voltages and ensures rapid heat dissipation. In its various modes of operation (free zone, micellar, gel, isotacophoresis, and isoelectric focussing), CE can be utilized to analyze a wide variety of charged and uncharged species. Sizes range from that of small analytes such as metal ions and low molecular weight alcohols to larger molecules, oligosaccharides, proteins, and nucleic acids. Besides being complementary to other well-established electrophoresis techniques, CE is also orthogonal to other solution analytical methods, in particular reverse-phase high performance liquid chromatography (HPLC). In our laboratory we have met a number of cases where the quality of separation of polar ionic molecules by CE is substantially superior to that obtained by HPLC analysis. It is common practice to check the homogeneity of peptides from enzymic digests of proteins by preparative HPLC followed by CE analysis. In fact the availability of CE instrumentation in a modern laboratory is essential because it increases the opportunity for analysts to solve more analytical problems of ever-increasing complexity.

A. History

In 1967 Hjerten[9] recognized that carrying out electrophoresis in narrow diameter tubes reduces thermal effects. This author also showed that it was possible to carry out electrophoretic separations using capillaries of 300 μm internal diameter and to detect separated bands by ultraviolet absorbance. In 1970 Neuhoff and co-workers[10] used polyacrylamide gel-filled glass tubes (dimensions: 450 mm internal diameter and 3 cm in length) to analyze for minute quantities of protein. In this case, analytes were detected by staining and autoradiography techniques. However, one of the earliest demonstrations of the advantages of using small diameter Pyrex® tubing (internal diameter 200 to 500 μm) for free-zone electrophoresis was provided in 1974 by R. Virtanen of the Helsinki University of Technology in Finland. In his thesis[11] for the degree of Doctor of Technology this author describes a potentiometric detection method for zone electrophoresis applied to the quantitative analysis of the alkali cations, Li^+, Na^+, and K^+. Electropherograms obtained were analyzed theoretically, "taking into account longitudinal diffusion, dispersive factors, and the uneven potential gradient at a zone." Virtanen also mentioned the importance of electroosmotic flow in influencing electrophoretic behavior of an analyte.

In 1979 Mikkers, Everaerts, and Verheggen[12] performed free-zone electrophoresis in Teflon® tubing of 200 μm internal diameter in order to reduce convection problems. A number of both inorganic (chloride, sulfate, chromate, and chlorate) and organic (adipate, acetate, glutainate, naphthalene-2-monosulfate, and benzyl-DL-aspartate) anions were resolved over a short period of time (less than 10 minutes), using both UV (254 nm) and conductometric detection. Although these authors discovered that the application of very small amounts of analytes achieves symmetric peak shapes, they were not able to show the high separation efficiency of CE due to poor detector sensitivity and large injection volumes.

Jorgenson and Lukacs[13,14] were the first to demonstrate that the use of even narrower capillaries (< 100 μm internal diameter) produced highly efficient electrophoretic separations. Voltages as high as 30 kV and fluorescence detection were used. These authors gave a brief description of some of the theoretical aspects of CE and showed that electroosmosis played an important part in determining the mobility of ionic species, affecting both resolution and analysis time. Separation efficiencies in excess of 4×10^5 theoretical plates were demonstrated by Jorgenson and Lukacs, who also used CE in open-tubular capillaries to resolve mixtures of dansyl and fluorescamine derivatives of amino acids, dipeptides, and simple amines.[13,14]

The publications of Jorgenson and Lukacs[13,14] drew the attention of a number of scientists from various scientific disciplines (analysts, physical chemists, and biochemists) and were effectively the beginning of a yearly rise in the number of publications and reviews on CE, published from 1981 to date[15] (Figure 1.2). The introduction of commercial CE instrumentation from late 1988 also enhanced the speed of development and application of this tech-

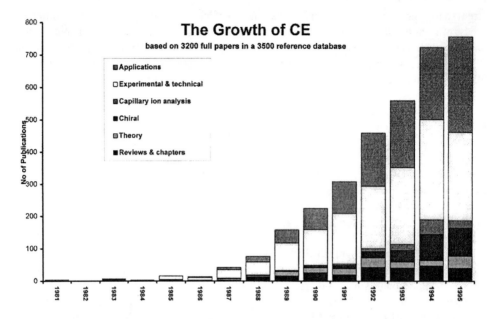

Figure 1.2   The annual growth of publications on capillary electrophoresis from 1981 to 1995. (Data from Perrett, D., *Database on Capillary Electrophoresis*, with entries to December 1995.)

nique. Variations in capillary design and the discovery of a number of modes of operation of CE have enabled the continued success and application of this separation technique over the last ten years.

The considerable activity in this area of separation science was analyzed in 1991 by Braun and Negydiosi-Rozsa[16] who attempted to predict the future prospects for the growth of CE by using a logistic equation of the following form:

$$g(t) = b/\{1 + a\ exp\ (-kbt)\}$$

where g is the cumulative number of publications or citations on CE and t is the time in years. The value of g is considered to reach the limit b as CE has passed the "maturity" stage. The terms a and k in the above equation are constants. Assuming that the development of CE were to follow a sigmoidal time curve, it was thought that this technique would reach the point of inflection (that is, $b/2$) around 1993. In the first edition of this book we showed the growth of publications in CE until 1991 when the substantial yearly rise in the number of publications related to CE was clearly evident. As shown in Figure 1.2 this growth has shown signs of plateauing after 1994, although the rate of application of CE is on the increase, compared to pure, theoretical studies.

B. Instrumentation

The experimental setup for CE is simple. A schematic diagram of a typical CE system is given in Figure 2.1 (Chapter 2). The high-voltage direct current (dc) power supply normally delivers up to 35 W and is connected to two platinum electrodes extending into buffer solution. The ends of the fused-silica capillary tube (20 to 200 μm i.d.) are dipped in the two buffer containers. The flexibility of the capillary tube is assured by the use of polyimide-coated capillaries. At some distance (about 150 to 200 mm) from the negative end of the capillary tube a window is created by etching off or burning about 4 mm of the polyimide coating. This optical window is positioned in a detector assembly, where most commonly, an ultra violet (UV) light beam is focused radially into the center of the capillary. Signals are electronically collected, stored, and visually displayed on a recorder or printer.

The dimensions of a home-built or commercial CE system are of the same order as those of an HPLC system. For home-built systems greater care has to be exercised when the high voltage is switched on; gadgets such as perspex casing and safety locks should be necessary features of these instruments. The availability of a wide choice of commercial CE systems (see Appendix I) means it is no longer necessary to build one's own instrumentation. Besides safety, a major advantage of commercial instruments is automated injection using a number of modes, in particular, pressure and electrophoretic injections. Sample trays containing a large number of samples are also commercially available allowing unattended analytical measurements. However, some precautions have to be exercised when running a sequence of several samples. A number of electrophoretic runs using the same cathodic and anodic buffer solutions will eventually cause a change in pH, especially when the volume of buffer solutions is less than 1 mL. Thus, steps have to be taken to change buffer solutions during a sequence of analytical measurements. It is also good practice to include an internal standard to avoid errors in lack of reproducibility in injection.

Currently the large majority of CE studies have been carried out using sensitive UV detectors that provide wavelength selection from 190 to 700 nm. At least four commercial instruments equipped with a diode array facility have also been marketed; the quality of spectra obtained from these instruments is generally good and, in most cases, is comparable to that generated by diode array detection instruments for HPLC analysis. Other sensitive but less common detection systems such as fluorescence,[17,18] mass spectroscopy,[19,20,] Raman,[21] amperometry,[22,23] conductivity,[24] and radiochemical[25] have also been used with CE. The commercial availability of an argon laser induced fluorescence (LIF) detector and a variety of mass spectrometers have enhanced the use of these two detection systems with CE separation.

Laser excited fluorescence is playing an important part in the detection of trace levels of analytes in biological studies. For example, a recent study has shown the use of CE/LIF in the analysis of primary amine compounds, labeled with a fluorophore, in human cerebrospinal fluid.[18] CE/LIF is also being applied to great effect in DNA sequencing,[26,27] using a confocal fluorescence capillary array scanner. In this system, the laser excitation beam and the emitted fluorescence are usually perpendicular to each other. This allows the detection of analytes traveling in an array of capillary columns with high sensitivity. The use of a number of capillaries allows the simultaneous electrophoresis and processing of a larger number of samples over a shorter period of time. It is envisioned that capillary array electrophoresis and LIF detection will be used extensively for disease diagnostics, genetic typing, and genome mapping.

It has long been recognized that mass spectrometric detection should add an important capability to CE analysis, that is, it will allow chemical structure determination and/or identity. Electrospray ionization (ESI),[19,28] fast atom bombardment (FAB),[29] and matrix-assisted laser desorption-ionization time-of-flight (MALDI-TOF)[20] are the three mass analyzers that have been successfully coupled to CE separation instrumentation. Until recently CE/ESI has been the method of choice for the analysis of a variety of molecules[19,28] in a wide range of molecular weights (including proteins), whereas CE/FAB has been used mostly to analyze molecules of moderate size, such as simple organic molecules and small peptides.[29]

Coupling of CE with ESI has been relatively difficult to implement, especially because in some cases the buffer/salt conditions in the CE separation procedure have been found to be incompatible with the mass spectrometry ionization process. Unlike ESI, MALDI-TOF mass spectrometry combines tolerance of diverse analytical conditions with excellent sensitivity. Direct coupling of the latter technique with CE is not essential. CE fractions can be deposited directly on a MALDI probe as a single spot and the molecular weight determined. This methodology has been successfully applied for the analysis of mixtures of proteins and peptide digests.[20,30] The further development of CE-MS may provide an important tool for determining not only the molecular mass of polypeptides and proteins but also is providing primary structure information from samples containing nanomolar concentration of analytes contained in volumes less than 1 µl.

Fraction collection after CE separation has been used not only for mass spectrometric analysis, as described above, but also as a method for the purification of peptides in nanogram quantities prior to Edman sequencing. To collect peptides, we have used both dynamic elution,[31] employing a pressure once zones are close to the end of a capillary, or electroelution[32] by stopping the current and moving the capillary into a microvolume of buffer at an appropriate time. More recently, a high-precision fraction collector has been described by Karger et al.[33] This device detects analytes close to the end of the capillary and automatically collects up to 60 µl fractions (see Chapter 4) without interruption of the electric field.

C. Modes of Operation

Several modes of CE operation have been developed over the past ten years. The most common are (a) open tubular or capillary zone electrophoresis (CZE), (b) micellar electrokinetic chromatography (MEKC), (c) capillary gel electrophoresis, (d) capillary isoelectric focusing, and (e) capillary isotacophoresis. To date, CZE has been the most popular CE mode, and studies using this technique account for over 60% of the publications in the open literature. Reports have been published regarding the analysis of a wide range of charged simple organic molecules, inorganic ions, peptides, and proteins.

MEKC was introduced by Terabe[34] in 1984. The use of this mode of operation of CE is increasing substantially as, unlike CZE, it allows the separation of both charged and uncharged molecules in an electroosmotically driven system. The resolution of neutral molecules requires the addition of surfactant to the buffer solution at a concentration above its critical micelle concentration (cmc). Differences in migration of uncharged species is mainly due to variations in the partitioning characteristics of these analytes across micelles. MEKC is the CE mode that is closest to reverse-phase HPLC, and the retention of an analyte is similarly described in terms of a capacity factor, k' (see Chapter 4). Sodium dodecyl sulfate and bile salts, in particular taurodeoxycholate, are widely used surfactants. The development of new micellar phases is a very active area of research. Recently Terabe[35] introduced the use of a double-chain (commonly known as *gemini*) surfactant for MEKC. Due to the low cmc of this surfactant (~0.8 mM) excellent resolution of mixtures of analytes is achieved even when used at concentrations between 2.5 and 10 mM.

Gels and sieving polymers have been widely reported for the separation of biopolymers, peptides, proteins, oligonucleotides, and DNA restriction fragments. Polymers such as dextran or polyethyleneglycol, transparent to UV, are preferred to crosslinked polyacrylamide.[36] As in the case of conventional gel electrophoresis (Section II), SDS is added both to the separation polymer and to the protein mixture to be analyzed. The approximate molecular weight of an unknown protein can be determined from a linear semilog plot of the molecular weight of standards versus the inverse of relative migration times. As fluorescence detection is normally the method of choice in the detection of oligonucleotides, liquid linear polyacrylamide matrices have been used very successfully to resolve oligonucleotides according to size.[37]

To date, capillary isotacophoresis and capillary isoelectric focusing (CIF) have been the least used modes of operation of CE. The former technique has been applied to great effect as a preconcentration method for dilute mixtures of analytes.[38] In CIF a pH gradient under the influence of an electric field is set up using ampholytes. After focusing, bands are slowly pushed across the detection window of the capillary to minimize band broadening. One important application of CIF is the rapid determination of the isoelectric point (pI) of a protein.

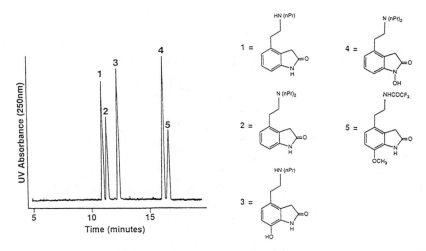

Figure 1.3 Capillary electrophoresis of the drug ropinirole (peak 2) and a number of related molecules. Separation conditions: 20 mM sodium phosphate buffer (pH 7.82); voltage 10 kV; separation length of capillary, 50 cm; radius of capillary, 50 pm; temperature 25°C.

IV. APPLICATIONS OF CAPILLARY ELECTROPHORESIS

The large number of applications of CE is given in considerable detail in Chapters 2 through 10. The following sections briefly outline the usefulness of CE as an analytical tool that complements other more established techniques, in particular HPLC.

A. Resolution of Small Molecules and Inorganic Ions by Free-Zone Capillary Electrophoresis

Figure 1.3 shows the resolution achieved in the analysis of the anti-Parkinson drug ropinirole (2) and a number of related molecules using CZE. This example demonstrates the principle of the mechanism for resolution in free-zone CE. Assuming that these analytes have a similar size, small differences in their pK_a values result in different charge densities, in turn resulting in differences in migration times and excellent resolution over a short period of analysis. The pK_a values of secondary aliphatic amines are usually lower than those of tertiary amines. For instance, the pK_as of di-n-propyl and tri-n-propyl amines are 11.00 and 10.65, respectively. At a pH close to 8, a fractionally higher proportion of compound (2) will be less positively charged than compound (1). Hence the latter molecule will move slightly faster than (2), giving the excellent resolution shown in Figure 1.3. The pK_a for a phenolic hydroxy group is around 10. Thus the net positive charge on (3) will be less than that of either (1) or (2). The hydroxy group in compound (4) has a much lower pK_a (about 5) than that in (3), resulting in the zwitterionic character of (4). In compound (5) the electron-withdrawing nature of the — COCF3 group makes the amide slightly acidic and negatively charged at pH 7.93. As a consequence, this molecule travels the slowest of the five in an electric field.

A number of studies have been reported on the determination of inorganic ions by CE. Complexation of metal ions with 8-hydoxyquinoline-5 sulfonic acid has been used with direct UV detection.[39] Alternatively, the problem of the lack of UV or fluorescence shown by analytes is overcome utilizing indirect UV or fluorimetric detection. Foret et al.[40] separated several metal ions by CE using a background electrolyte containing hydroxyisobutyric acid as a complexing counter ion and creatine as a UV absorbing co-ion for indirect detection. Complete resolution and analysis was achieved in less than five minutes. These authors applied this method to the analysis of a sample of gas lighter flint alloy and were able to detect the presence of magnesium, potassium, and four lanthanide elements. Indirect detection is also useful in the analysis of anionic species (inorganic and organic). In this case an absorbing anionic compound, such as chromate or 1,2,4,5-benzetetracarboxylic acid, is included in the separation buffer.[41]

B. Micellar Electrokinetic Chromatography (MEKC)

As has already been mentioned in Section III.C, resolution of a mixture of nonionic analytes is achieved in MEKC only because of a combination of kinetic and thermodynamic phenomena. The use of charged micelles is essential in MEKC. These micelles migrate under the influence of their electrophoretic mobility and the electroosmotic mobility of the bulk electrolyte. Uncharged (neutral) molecules will have free energies of partitioning between the moving micellar phase and the aqueous phase, depending mainly on their hydrophobic characteristics, in a manner similar to reverse-phase HPLC. The more hydrophobic solutes interact more strongly with the hydrophobic moiety of the micelle and will be slowed down by this process. SDS and the bile salts (deoxycholic acid and taurodeoxycholic acid) are the most common anionic surfactants.[42,43] An example of a cationic surfactant that has been used in MEKC is dodecyltrimethylammonium bromide (DTAB).[44]

MEKC has been applied to the separation of a variety of molecules which include aromatic hydrocarbons, nucleic acids, vitamins, antibiotics, barbiturates, and porphyrins. The separation of 36 illicit drug substances (opium alkaloids, amphetamines, hallucinogens, barbiturates, benzodiazepines, and cannabinoids) by MEKC was found[45] to be both faster and more selective than HPLC. In fact, it was suggested that MEKC is "well suited for drug screening."[45] An MEKC method with indirect UV detection in the anionic mode has also been reported for the analysis of aminoglycoside antibiotics.[46]

C. Chiral Discrimination

As shown in Figure 1.2, CE is increasingly becoming a useful method for the analysis of racemic mixtures and for the enantiomeric purity determination of a number of pharmaceuticals. Cyclodextrins have been the most popular chiral selectors,[47,48] although other naturally occurring discriminating agents, e.g., proteins, have been added to the separation buffer. Chiral surfactants such as the bile salts,[43] amino acids,[49,50] and carbohydrates[51] have also been used to directly resolve the enantiomers of racemic analytes. In all cases of chiral resolution transient complexes are formed between one or both of the enantiomers and the chiral additive, and differences in the free energy of formation of these complexes leads to enantiomeric separation.

The addition of cyclodextrin to the micellar electrolyte solution can also give highly efficient separations. Figures 1.4 (a), (b), and (c) illustrate the resolution of the 1-cyano-2-substituted-benz [f] isoindole (CBI) derivatives of a number of amino acids. The excellent resolution shown was obtained using a mixture of taurodeoxycholic acid (50 mM) and β-cyclodextrin (20 mM) at pH 7.

D. Separation of Peptides and Proteins

Mixtures of peptides are relatively easy to resolve. Analysis of enzymic digests by CZE is straightforward, and optimization in resolution is achieved by changing buffer type and concentration, pH, and ionic strength. Enzymic digests are often analyzed by reverse-phase HPLC before collection of peaks is carried out for internal sequencing of the parent protein. It is now becoming common practice to analyze collected peaks by CE before sequencing is

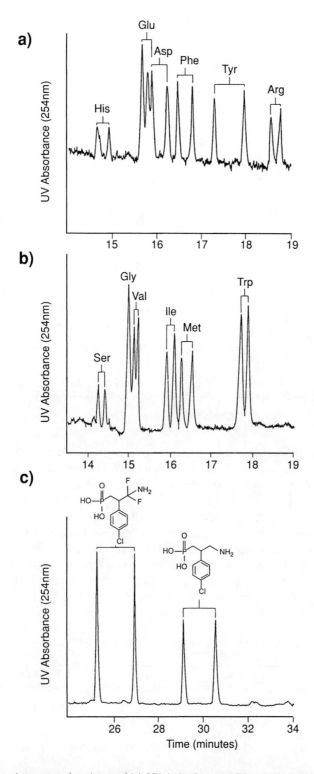

Figure 1.4 Electropherogram of a mixture of (a) CBI derivatives and (b) DL-amino acids and (c) phaclofen and difluorophaclofen. Separation conditions: 20 mM β-cyclodextrin, 50 mM taurodeoxycholic acid, 30 mM sodium phosphate, and 10 mM boric acid buffer (pH 7.0); voltage 9 kV for (a) and (b) and 10 kV for (c); separation length of capillary 50 cm; radius of capillary 50 μm; temperature 27°C.

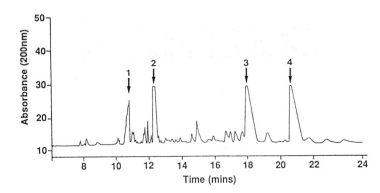

Figure 1.5 Capillary electrophoresis of the tryptic fragments of elcatonin for the micropreparation of peptide fragments 1 to 4. Edman sequencing yields obtained from fragments 1 to 3 are shown in Table 1.1.

pursued. The complementarity of the two orthogonal techniques can often eliminate the possibility of not resolving fairly similar peptides.

CE has also been successfully used for the micropreparation of peptides after an enzymic digest.[32] Figure 1.5 shows the electropherogram of peptides from a tryptic digestion of elcatonin, a protein closely related to the calcium-regulating hormone, calcitonin. The shape

$$\text{—CO(CH}_2)_5\text{—}$$
Ser-Asn-Leu-Ser-Thr-NHCHCO-Val-Leu-Gly-*Lys*-Leu-Ser-Gln-Glu-Leu-His-
Lys-Leu-Gln-Thr-Tyr-Pro-*Arg*-Thr-Asp-Val-Gly-Ala-Gly-Thr-Pro-NH₂

Elcatonin

of the peaks is not symmetrical because these tryptic fragments were undiluted from a starting concentration of elcatonin of about 10^{-4} molar. Peptides from this one experiment were collected and each was subjected to Edman sequencing. Table 1.1 shows the yields obtained from three of the peptides. As expected, no sequence was possible from fraction 4 (residues 1 to 11), which is a peptide derived from the blocked N-terminus of elcatonin.

The separation of proteins by free-zone CE is more difficult than for peptides. Proteins tend to adsorb onto the untreated silica capillary walls and can cause peak tailing and loss of resolution. Thus the electrophoresis of proteins is preferably carried out using capillaries where the inside wall is chemically modified (covalently or dynamically) to reduce electroosmotic flow and protein adsorption.[52] An alternative[53] is to adjust the pH of the buffer in the range of 8 to 11 where both the uncoated silica surface and the proteins are negatively charged;

Table 1.1   Sequencing Yields (pmol) for Peptide Fractions 1 through 3

Cycle	Fraction 1 (residues 12–18)	Fraction 2 (residues 19–34)	Fraction 3 (residues 25–32)
1	Leu (24)	Leu (13)	Thr (12)
2	Ser (8)	Gln (13)	Asp (13)
3	Gln (13)	Thr (5)	Val (9)
4	Glu (8)	Tyr (4)	Gly (20)
5	Leu (15)	Pro (9)	Ala (15)
6	His (2)	Arg (2)	Gly (16)
7	Lys (2)		Thr (5)
8			Pro (3)

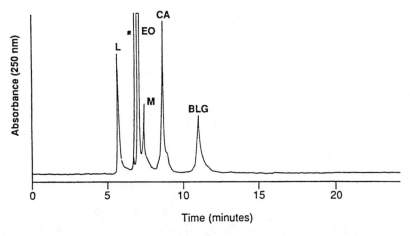

Figure 1.6   Capillary electrophoresis of the four proteins separated by polyacrylamide gel electrophoresis shown in Figure 1.1. L, lysozyme; EO, electroosmotic flow; M, myoglobin; CA, carbonic anhydrase; BLG, P-lactoglobulin. Separation conditions: 250 mM phosphate buffer (pH 6.5); voltage 5 kV; separation length of capillary 20 cm; radius of capillary 50 μm.

electrostatic repulsion decreases adsorption phenomena. A low pH also has the same effect on decreasing adsorption. However, analyzing proteins at these high pH values may have detrimental effects on the stability of proteins, especially if CE is being used in a micro-preparative mode, and care has to be exercised to retain the enzymic activity of a collected protein. An effective method for decreasing protein adsorption is by ion-pairing with a polyanionic species as shown in Chapter 8.

Good resolution of proteins at physiological pH is possible if a high concentration (100 to 200 mM) of the buffer constituents is used.[54] Figure 1.6 shows the separation of four proteins by this method. These proteins are the same as those separated by SDS PAGE in Figure 1.1. Whereas resolution by the latter technique was achieved on the basis of size, separation by CE is mainly dependent on differences in the ionization *and* size, and therefore the charge density of these proteins.

It has already been mentioned in Section III.C that acrylamide gel-filled capillaries have also been used to separate proteins. Both denaturing,[55] and nondenaturing conditions[56] can be used.

E. Carbohydrates and Oligosaccharides

The absence of chromophores in most carbohydrates does not allow direct UV detection. The pK_a of simple sugars is about 12, so these molecules are only ionized at highly alkaline pHs. Thus until recently, analysis of simple sugars has been normally carried out by anion-exchange HPLC at pHs between 12 and 14, followed by pulsed amperometric detection. Buffer solutions at pHs greater than 12 have also been used in the analysis of carbohydrates by free-zone-CE, again using amperometric detection at a constant potential.[57] However, the most common methodology for the detection of sugars has involved derivatization prior to CE analysis. Such reactions usually involve a Schiff interaction of a highly UV absorbent or fluorescent aromatic amine with the aldehyde form of the sugar, followed by reduction to stable products. Reagents that have been used include 2-aminopyridine,[58] 6-amino-quino-line,[59] and aminonaphthalensulfonic acids.[60] As these reagents can be ionized over a wide pH range, separation of the derivatized carbohydrates is by free-zone CE, very often in the presence of borate buffer. Recently we used MEKC in the analysis of a number of carbohy-drates, derivatized with 2-aminoacridone.[61] This fluorophore does not carry a charge even at pHs close to 9 so that it does not interfere with analysis as excess reagent is trapped in the slow migrating micelles.

The increasing importance of the biological function of the oligosaccharide moiety in glycoproteins has demanded the development of rapid and sensitive analytical techniques. The reductive amination reactions mentioned in the preceding paragraph have also been used in the CE fingerprinting of complex and branched polysaccharides, after release from the glycoprotein.[62] Levels as low as 1 pmol can be analyzed. A more comprehensive account of the analysis of oligosaccharides by CE is given in Chapter 7.

F. Nucleosides, Nucleotides, and Nucleic Acids

The ionization of nucleic acid-related molecules over a wide pH range makes them ideal candidates for analysis by CE. Recently, in collaboration with Professor A.J. Kirby at Cambridge University (U.K.), we have been investigating the mechanism of the hydrolysis of ribonucleic acid (RNA). As a model of this reaction (see Figure 1.7) we have studied the hydrolysis of 3′,5′-uridyluridine (6); intermediates and products include, 2′,3′-cyclic phosphate (9), uridine (8), uridine-2-monophosphate (10), uridine-3-monophosphate (11) and the isomeric form of (6), namely 2′,5″-uridyluridine (7). As shown in Figure 1.7, separation of all seven compounds in about ten minutes allows the acquisition of kinetic data under various conditions over a relatively short period of time.

Capillary gel electrophoresis has also been a popular technique for the analysis of oligonucleotides.[63,64] Polyacrylamide gel-filled capillaries have been used for this purpose. Polysiloxane-coated capillaries in conjunction with methylcellulose buffer additives (to reduce electroosmotic flow) give good resolution separations of DNA fragments. Separations with these coated capillaries are performed at negative polarity under constant voltage. The addition of ethidium bromide to the buffer considerably increases resolution of the fragments.[64] In these studies migration time has been found to increase relative to the number of base pairs. Separations are again carried out at negative polarity. This method has wide applications, ranging from the rapid separation of small synthetic oligonucleotides to the resolution of large DNA restriction fragments

Much effort is now being devoted to sequencing deoxyribonucleic acid (DNA) in relation to the Human Genome Project. If this ominous task is to be completed over a relatively short time, a sequencing method faster than the automated conventional slab-gel method is necessary. It is expected that the recently developed capillary array electrophoresis methodology will play an important role in the high-speed sequencing of DNA fragments.[65] As many as

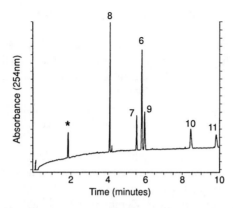

Figure 1.7 Electropherogram showing the separation of 3′,5′-uridyluridine from its products of hydrolysis. Separation conditions: 340 m*M* phosphate/ borate buffer (pH 9.1); voltage 20 kV; separation length of capillary 20 cm; radius of capillary 50 μm. The peak denoted by an asterisk marks the migration time of the electroosmotic front.

25 gel-filled capillaries are used simultaneously, and DNA restriction fragments are sequenced using two or four fluorescently labeled dye primers, followed by multicolor fluorescence detection. The commercialization of such a sequencer will obviously usher in a new era in high-throughput DNA analysis.

G. Pharmaceutical and Clinical Applications

The analysis of pharmaceutical formulations and biofluids by CE is of great interest, especially the potential complementarity of this method to HPLC. The reported applications of CE in clinical and pharmaceutical analysis are on the increase compared to three years ago. Methodology, using either free-zone CE or MEKC, has been developed for the analysis of several compounds such as salycylate (aspirin), acetaminophen (paracetamol), cimetidine, hypoglycemic drugs, antiepileptic drugs, morphine, and cocaine in urine or blood plasma from drug-treated mammals.

CE analysis is most appropriate when it is necessary to confirm the presence of drugs or metabolites in biological fluids without precise quantification. The acquisition of accurate quantitative data requires that a number of precautionary measures be taken. For instance, the effect of the biological matrix on the migration time of an analyte can be pronounced in CE so that the design of suitable internal standard procedures is essential when either formulations (for quality control) or biofluids are directly analyzed by this method.[66] Alternatively, as is the case for other analytical methods, cleaning of the sample prior to CE analysis may be desirable. Moreover, a careful choice of the loading procedure (that is, hydrodynamic vs. electrokinetic injection) can be used to obtain quantitative analytical information from CE.

V. IMPROVEMENTS AND OTHER APPLICATIONS OF CE

Advancements in instrument and capillary technology, together with improvements in CE methodology, are occurring at a fast rate, ensuring that the number of applications of this technique is steadily on the increase. Solutions to existing difficulties in analysis of samples by CE are being discovered and in some cases readily commercialized. The availability of diode array and laser induced fluorescence (LIF), in addition to single wavelength detection has made CE a more versatile technique. Improvements have also occurred in sensitivity by making use of sample preconcentration either by HPLC prior to CE or by isotacophoresis methods.

Modification of the structure of the capillary has lead to improvements in sensitivity. For instance, Tsuda et al.[67] have reported that replacing ordinary cylindrical capillaries with rectangular ones can lead to a 20-fold gain in sensitivity, without any loss in the efficiency of resolution. Z-shaped cells of 3 mm pathlength have also been used with a gain in sensitivity of up to 14-fold being reported.[68] To achieve the same enhanced sensitivity, capillaries containing a bubble cell at the point of detection have been produced by one of the CE instrument manufacturers (see Appendix I).

Improvements in capillary technology have also had an impact on the quality of analysis by CE. For instance, the analysis of basic proteins using ordinary capillaries has often been impossible due to electrostatic attraction between the negatively charged silanol groups and positive regions of these macromolecules. Commercially available coated capillaries can now be used to considerably reduce adsorptive effects. Alternatively, the addition of polyanionic species such as the dodecasodium salt of phytic acid can allow the analysis of proteins with pI values in excess of 9 (see Chapter 8).

Although CE has been used mainly as an analytical tool, the simplicity and reliability of this technique has excited many workers from a number of other scientific disciplines — physical chemists, biologists, and biochemists. Several studies have appeared in which this technique has been used successfully to determine physicochemical constants such as pK_a and pI values, binding constants, octanol-water partition, and diffusion coefficients. CE is especially attractive when the amount of available material is low.

Ionization or dissociation constants (pK_a values) of weak acids and bases can be conveniently determined by CE.[69] Thus, for an equilibrium between the free base B and its protonated form BH+, the variation of the net electrophoretic mobility (μ_e) with pH is given by the relationship:

$$pH = pK_a - \log_{10} [\mu_e/\mu_{BH+} - \mu_e)]$$

where μ_{BH+} is the mobility of the fully protonated form. An analogous relationship can be derived for the ionization of a weak acid of the type HA.

$$pH = pK_a - \log_{10} [\mu_e/\mu_{A-} - \mu_e)]$$

where μ_{A-} is the electrophoretic mobility of the deprotonated species, A-.

In the same way that the pK_a of simple molecules is determined, the pI value of peptides and proteins can be determined by CE. This is a relatively simple procedure. The analyte and a neutral marker such as mesityl oxide are studied at different pH values. The pH at which the electrophoretic mobilities of the marker and a substrate coincide is equivalent to the pI value. Using this method, more than one determination can be carried out at the same time, using a mixture of analytes.

Complexation constants between cyclodextrins and guest molecules are usually determined by spectroscopic or solubility measurements. Because a change in charge density occurs on complexation, CE is an appropriate technique to determine inclusion constants. The experimental procedure involves monitoring electrophoretic mobility of an analyte under the influence of increasing amounts of cyclodextrin in the separation buffer. For a simple 1:1 complex between cyclodextrin (CD) and a guest molecule, Gareil et al.[70] determined inclusion constants (K′) from a plot of electrophoretic mobility versus –log[CD]. This plot gives a sigmoidal curve, and log K′ is obtained from the point of inflection. In a similar manner, Francois et al.[71] determined stability constants of 18-crown-6 complexes with K+, Ba2+, Sr2+, and NH4+. Stability constants determined in this manner were found to be in agreement with literature values.

A CE study of the binding of the broad spectrum antibiotic vancomycin to the mucopeptide mimic N-acetyl-D-Ala-D-Ala was reported[72] by us in early 1992. In this study we found
that binding constants determined by CE were in good agreement with values reported in the
literature using either differential UV spectrometry or nuclear magnetic resonance. These
studies were found to be useful not only as an alternative means of determining binding
constants but also in providing information about the net charge characteristics of the complexes formed. A number of other reports have now appeared in the literature in which
capillary affinity electrophoresis has been used to study protein–ligand interactions.[73-75]

Using MEKC, it is possible to evaluate surfactant critical micelle concentrations (cmc).[76]
The retention factor (k′) of an analyte in MEKC at low micelle concentrations is directly
proportional to the micelle concentration:

$$k' = P_{mw} \, V \, \{[S] - cmc\}$$

where P_{mw} is the partition coefficient of solute between the micelle and aqueous phases, V
is the molar volume of the surfactant, and [S] is the total concentration of the surfactant. A
plot of k′ vs. [S] for several analytes gives a value of the cmc extrapolated to the point
where k′ is zero. The cmc of a surfactant can also be obtained by measuring the change in
current with increasing concentration of surfactant.[51] A break in the linear relationship
between current and concentration occurs at the cmc of a surfactant due to the difference
in the conducting properties of its monomeric and aggregated (micellar) forms. The cmc
values obtained by both of the above CE methods are in good agreement with values
determined by viscosity measurements.

The use of SDS in MEKC for the rapid estimation of log P has been reported by us and
by other authors.[77,78] Recently we have shown[79] that capacity factors of a number of structurally unrelated standard compounds and drug substances can be adequately related to log
P when deoxycholic acid micelles are used in MEKC. These micelles differ from those formed
by SDS both in shape and in the number of monomers forming each micelle. In the case of
SDS, about 80 monomers form a spherical micelle, compared to 3 to 4 forming cylindrical
micelles in the case of deoxycholic acid. Figure 1.8a shows an electropherogram obtained
for the separation of a mixture of drugs. A linear plot of the logarithm of capacity factors of
32 analytes, including these drugs, against log P is given in Figure 1.8b. Capacity factors
were measured in relation to the migration of halofantrine, an antimalarial drug, which
completely associates with the micelles. This methodology for estimating log P cannot be
satisfactorily applied to negatively charged molecules. Such compounds will be electrostatically repelled from the micelle and will migrate under the free-zone mode. To avoid this
problem Ishihama et al.[80] developed a microemulsion electrochromatography method to determine the migration index (MI) of a number of neutral and anionic solutes, calculated from
the measured capacity factors relative to the capacity factors of reference solutes.

VI. CONCLUSION AND FUTURE PROSPECTS

Although CE has only been available as a technique for the past 10 to 20 years (commercial instruments appeared in 1988), advances in this area of separation technology have
been taking place at a progressive rate. Moreover, it is now widely accepted that the complementarity of this technique to other more established wet techniques, in particular HPLC,
will make it an essential and valuable tool in a modern analytical laboratory. Hyphenated
techniques, in particular CE/MS and HPLC/CE, are also expected to increase in their popularity. The use of CE to determine the physicochemical and biochemical properties of a variety
of molecules should make this technique attractive to scientists from other disciplines. Such

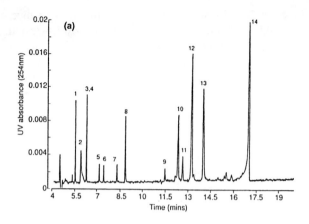

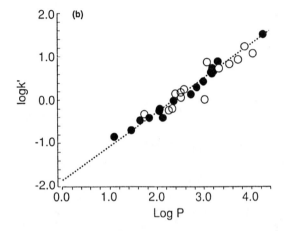

Figure 1.8 (a) Electropherogram of 14 drugs using MEKC. Separation conditions: buffer, 50 mM boric acid adjusted to pH 9 with potassium hydroxide and containing 40 mM of the sodium salt of deoxycholic acid: voltage, 22 kV; uncoated capillary, separation length, 50 cm and 50 μm i.d.; temperature, 25°C: (1) acetaminophen, (2) mephenytoin, (3) lidocaine, (4) chromatin, (5) tolbutamide, (6) warfarin, (7) dexamethasone, (8) ropinirole, (9) testosterone, (10) nabumetone, (11) prazepam, (12) fluoxetine, (13) progesterone, and (14) halofantrine. (b) Correlation of log P with log k′ for a number of standard compounds (●) and drug substances (○).

measurements can be carried out on a very small amount of sample (about 5 to 10 μl) and can provide valuable information suitable for the design of biologically active molecules.

Micromachined capillary electrophoresis systems may offer a new dimension in the future prospects of CE. These miniaturized devices can lead to notable enhancements in separation speed and can allow multichannel analysis, increasing sample throughput. Microchip electrophoresis separations have been reported using free-zone,[81] gel,[82] and MEKC[83] modes of capillary electrophoresis. Important applications of CE microchip instrumentation are expected to be in the areas of rapid DNA sequencing, bioassay, and combinatorial technology.[84]

REFERENCES

1. Tiselius, A., *The moving boundary method of studying the electrophoresis of proteins (Thesis)*, Nova Acta Regiae Societatis Scientiarum Upsaliensis, 1930, Ser IV, Vol. 7, Number 4.
2. Swank, R. T. and Munkers, K. D., Molecular weight analysis of oligopeptides by electrophoresis in polyacrylamide gel with sodium dodecyl sulphate, *Anal. Biochem.*, 39, 462, 1971.

3. Hames, B. D., in *Gel Electrophoresis of Proteins — A practical approach*, Hames, B. D. and Rickwood, D., Eds., IRL Press. Oxford, 1986, 17.

4. Douglas. M., Finkelstein, D., and Butow, R. A., Analysis of products of mitochondrial protein synthesis in yeast: genetic and biochemical aspects, in *Methods in Enzymology*, Fleischer, S. and Packer, L., Eds., Academic Press, New York, 1979, Vol. 56, 58.

5. Grierson, D., *Gel Electrophoresis of RNA in Gel Electrophoresis of Nucleic Acids — A practical approach*, Rickwood, D. and Hames, B. D., Eds., IRL Press, Oxford, 1987, I.

6. Svensson, H., Isoelectric fractionation, analysis, and characterization of ampholytes in natural pH gradients. I. The differential equation of solute concentrations at a steady state and its solution for simple cases, *Acta Chem. Scan.*, 15, 325, 1961.

7. Vesterberg, O., A short history of electrophoretic methods, *Electrophoresis*, 14, 1243, 1993.

8. Nishizawa, H., Suzuki, K., and Abe, Y., Fixation of amino acids and oligopeptides by freeze-drying in polyacrylamide gel electrophoresis, *Electrophoresis*, 10, 498, 1989.

9. Hjerten, S., Free zone electrophoresis, *Chromatogr. Rev.*, 9, 122, 1967.

10. Neuhoff, V., Schill, W.B., and Sternbach, H., Micro-analysis of pure deoxyribonucleic acid-dependent ribonucleic acid polymerase from *Escherichia Coli*, *Biochem. J.*, 117, 623, 1970.

11. Virtanen, R., Zone electrophoresis in a narrow-bore tube employing potentiometric detection — a theoretical and experimental study, Vol 123 *Acta Polytechnica Scandinavia* (Helsinki), 1974.

12. Mikkers, F. E. P., Everaerts, F. M., and Verheggen, T. P. E. M., High performance zone electrophoresis, *J. Chromatogr.*, 169, 11, 1979.

13. Jorgenson, J. and Lukacs, K. D., Zone electrophoresis in open-tubular glass capillaries, *Anal. Chem.*, 53, 1298, 1981.

14. Jorgenson, J. and Lukacs, K. D., Capillary zone electrophoresis, *Science*, 222, 266, 1983.

15. Perrett, D., *Database on Capillary Electrophoresis*, with entries to Dec. 1995.

16. Braun, T. and Negydiosi-Rozsa, S., Capillary electrophoresis: prospects for growth, *Trends in Anal. Chem.*, 10 (9), 266, 1991.

17. Kuhr, W. G. and Yeung, E. S., Indirect fluorescence detection of native amino acids in capillary zone electrophoresis, *Anal. Chem.*, 60, 1832, 1988.

18. Bergquist, J., Douglass Gilman, S., Ewing, A. G., and Ekman, R., Analysis of human cerebrospinal fluid by capillary electrophoresis with laser-induced fluorescence detection, *Anal. Chem.*, 66, 3512, 1994.

19. Garcia, F. and Henion, J. D., Gel-filled capillary electrophoresis/mass spectrometry using a liquid junction-ionspray interface, *Anal. Chem.*, 64, 985, 1992.

20. Walker, K. L., Chiu, R. W., Monnig, C. A., and Wilkins, C. L., Off-line coupling of capillary electrophoresis and matrix-assisted laser desorption/ionization time-of-flight mass spectrometry, *Anal. Chem.*, 67, 4197, 1995.

21. Chan, C. and Morris, M. D., On-line multi-channel Raman spectroscopic detection system for capillary zone electrophoresis, *J. Chromatogr.*, 540, 355, 1991.

22. Wallingford, R. A. and Ewing, A. G., Capillary zone electrophoresis with electrochemical detection, *Anal. Chem.*, 59, 1762, 1987.

23. Wallingford, R. A. and Ewing, A. G., Capillary zone electrophoresis with electrochemical detection in 12.7 µm diameter columns, *Anal. Chem.*, 60, 1972, 1988.

24. Huang, X. and Zare, R. N., Improved end-column conductivity detector for capillary zone electrophoresis, *Anal. Chem.*, 63, 2193, 1991.

25. Pentony, J. S. L., Zare, R., and Quint, J., On-line radioisotope detection for capillary electrophoresis, *Anal. Chem.*, 61, 1642, 1989.

26. Drossman, H., Luckey, J. A., Kostichka, A. J., D'Cunha, J., and Smith, L., High-speed separations of DNA sequencing reactions by capillary electrophoresis, *Anal. Chem.*, 62, 900, 1990.

27. Takahashi, S., Murakami, K., Anazawa, T., and Kambara, H., Multiple sheath-flow gel capillary-array electrophoresis for multi-color fluorescent DNA detection, *Anal. Chem.*, 66, 1021, 1994.

28. Smith, R. D., Olivares, J. A., Nguyen, N. T., and Udseth, H. R., Capillary zone electrophoresis — mass spectrometry using an electrospray ionization interface, *Anal. Chem.*, 60, 436, 1988.

29. Mosley, M. A., Deterding, L. J., Tomer, K. B., and Jorgenson, J. W., Determination of bioactive peptides using capillary zone electrophoresis-mass spectrometry, *Anal. Chem.*, 63, 109, 1991.

30. Licklider, L., Kuhr, W. G., Lacey, M. P., Keough, T., Purdon, M. P., and Takigiku, R., On-line microreactors/capillary electrophoresis/mass spectrometry for the analysis of proteins and peptides, *Anal. Chem.*, 67, 4170, 1995.

31. Camilleri, P., Okafo, G. N., and Southan, C., Separation by capillary electrophoresis followed by dynamic elution, *Anal. Biochem.*, 178, 196, 1991.

32. Camilleri, P., Okafo, G. N., Southan, C., and Brown, R., Analytical and micropreparative capillary electrophoresis of the peptides from calcitonin, *Anal. Biochem.*, 36, 198, 1992.

33. Miller, O., Foret, F., and Karger, B. L., Design of a high-precision fraction collector for capillary electrophoresis, *Anal. Chem.*, 67, 2974, 1995.

34. Terabe, S., Otsuka, K., Ichikawa, K., Tsuchuya, A., and Ando, T., Electrokinetic separations with micellar solutions and open-tubular capillaries, *Anal. Chem.*, 56, 111, 1984.

35. Tanaka, M., Ishida, T., Araki, T., Masuyama, A., Nakatsuji, Y., Okahara, M., and Terabe, S., Double-chain surfactant as a new and useful micelle-forming reagent for micellar electrokinetic chromatography, *J. Chromatogr.*, 469, 648, 1993.

36. Fung, E. N. and Yeung, E. S., High speed DNA sequencing by using mixed poly(ethylene oxide) solutions in uncoated capillary columns, *Anal. Chem.*, 67, 1913, 1995.

37. Guttman, A., Nelson, R. J., and Cooke, N., Prediction of migration behaviour of oligonucleotides in capillary gel electrophoresis, *J. Chromatogr.*, 593, 2971, 1992.

38. Foret, F., Szoko, E., and Karger, B. L., On-column transient and coupled column isotacophoretic preconcentration of protein samples in capillary zone electrophoresis, *J. Chromatogr.*, 608, 3, 1992.

39. Timerbaev, A., Semenova, O., and Bonn, G., Metal ion capillary zone electrophoresis with direct UV detection: comparison of different migration modes for negatively charged chelates, *Chromatographia*, 37, 497, 1993.

40. Foret, E., Fanall, S., Nardi, A., and Bocek, P., Capillary electrophoresis of rare earth metals with indirect UV absorbance detection, *Electrophoresis*, 11, 780, 1990.

41. Cousins, S. M., Haddad, P. R., and Buchberger, W., Evaluation of carrier electrolytes for capillary zone electrophoresis of low molecular-mass anions with indirect UV detection, *J. Chromatogr.*, 671, 397, 1994.

42. Vindevogel, J. and Sandra, P., Resolution optimization in micellar electrokinetic chromatography: use of Plackett-Burman statistical design for the analysis of testosterone esters, *Anal. Chem.*, 63, 1530, 1991.

43. Cole, R. O., Sepaniak, M. J., Hinze, W. L., Gorse, J., and Oldiges, K., Bile acid salt surfactants in micellar electrokinetic capillary chromatography: application to hydrophobic molecule separations, *J. Chromatogr.*, 557, 115, 1991.

44. Kaneta, T., Tanaka, S., Taga, M., and Yoshida, H., Migration behaviour of inorganic anions in micellar electrokinetic capillary chromatography using a cationic surfactant, *Anal. Chem.*, 64, 798, 1992.

45. Weinberger, R. and Lude, I. S., Micellar electrokinetic capillary chromatography of illicit drug substances, *Anal. Chem.*, 63, 823, 1991.

46. Ackermans, M. T., Everaerts, F. M., and Beckers, J. L., Determination of aminoglycoside antibiotics in pharmaceuticals by capillary zone electrophoresis with indirect UV detection coupled with micellar electrokinetic capillary chromatography, *J. Chromatogr.*, 606, 229, 1992.

47. Ueda, T., Kitamura, F., Metcalf, T., Kuwana, T., and Nakamoto, A., Separation of naphthalene-2,3-dicarboxaldehyde-labelled amino acid enantiomers by cyclodextrin modified micellar electrokinetic chromatography with laser-induced fluorescence detection, *Anal. Chem.*, 63, 2979, 1991.

48. Nardi, A., Ossicini, L., and Fanali, S., Use of cyclodextrins in capillary zone electrophoresis for the separation of optical isomers: resolution of racemic tryptophan derivatives, *Chirality*, 4, 56, 1992.

49. Otsuka, K. and Terabe, S., Enantiomeric resolution by micellar electrokinetic chromatography with chiral surfactants, *J. Chromatogr.*, 515, 221, 1990.

50. Swartz, M. E., Mazzeo, J. R., Grover, E. R., and Brown, P. R., Separation of aminoacid enantiomers by micellar electrokinetic capillary chromatography using synthetic chiral surfactants, *Anal. Biochem.*, 231, 65, 1995.

51. Tickle, D., Jones, R., Okafo, G. N., Camilleri, P., and Kirby, A. J., Glucopyranoside-based surfactants as pseudostationary phases for chiral separations in capillary electrophoresis, *Anal. Chem.*, 66, 4121, 1994.

52. Regnier, F. E. and Towns, J. K., Capillary electrokinetic separations of proteins using nonionic surfactant coatings, *Anal. Chem.*, 63, 1126, 1991.

53. Laurer, H. H. and McManigill, D., Capillary zone electrophoresis of proteins in untreated fused silica tubings, *Anal. Chem.*, 58, 166, 1986.

54. Green, L. and Jorgenson, J., Minimising adsorption of proteins on fused silica in capillary zone electrophoresis by the addition of alkali metal salts to the buffer, *J. Chromatogr.*, 478, 63, 1989.

55. Cohen, A. and Karger, B., High performance sodium dodecyl sulphate polyacrylamide gel capillary electrophoresis of peptides and proteins, *J. Chromatogr.*, 397, 409, 1987.

56. Tsuji, K., High performance capillary electrophoresis of proteins: sodium dodecyl sulphate-polyacrylamide gel filled capillary column for the determination of recombinant biotechnology derived proteins, *J. Chromatogr.*, 550, 823, 1991.

57. Colon, L. A., Dadoo, R., and Zare, R. N., Determination of carbohydrates by capillary zone electrophoresis with amperometric detection at a copper electrode, *Anal. Chem.*, 65. 476, 1993.

58. Honda, S., Iwase, S., Makino, A., and Fujiwara, S., Simultaneous determination of reducing monosaccharides by capillary zone electrophoresis as the borate complexes of N-2-pyridylglycamines, *Anal. Biochem.*, 176, 72, 1989.

59. Nashabeh, W. and El Rassi, Z., Capillary zone electrophoresis of linear and branched oligosaccharides, *J. Chromatogr.*, 600, 279, 1992.

60. Stefansson, M. and Novotny, M., Electrophoretic resolution of monosaccharide enantiomers in borate-oligosaccharide complexation media, *J. Am. Chem. Soc.*, 115, 11573, 1993.

61. Greenaway, M., Okafo, G. N., Camilleri, P., and Dhanak, D., A sensitive and selective method for the analysis of complex mixtures of sugars and linear oligosaccharides, *J. Chem. Soc. Chem. Commun.*, 1691, 1994.

62. Camilleri, P., Harland, G.B., and Okafo, G. N., High resolution and rapid analysis of branched oligosaccharides by capillary electrophoresis, *Anal. Biochem.*, 230, 115, 1995.

63. Guttman, A. and Cooke, N., Capillary gel affinity electrophoresis of DNA fragments, *Anal. Chem.*, 63, 2038, 1991.

64. Guttman, A., Nelson, R. J., and Cooke, N., Prediction of migration behaviour of oligonucleotides in capillary gel electrophoresis, *J. Chromatogr.*, 593, 2971, 1992.

65. Huang, X. C., Quesada, M. A., and Mathies, R. A., DNA sequencing using capillary array electrophoresis, *Anal. Chem.*, 64, 2149, 1992.

66. Guzman, N. A., Berck, C. M., Hernandez, L., and Advis, J. P., Capillary electrophoresis as a diagnostic tool: determination of biological constituents present in urine of normal and pathological individuals, *J. Liquid Chromatogr.*, 13, 3833, 1990.

67. Tsuda, T., Swedler, J. V., and Zare, R. N., Rectangular capillaries for capillary zone electrophoresis, *Anal. Chem.*, 62, 2149, 1990.

68. Chervet, J. P., Van Soest, R. E. L, and Ursem, M., Z-shaped flow cell for UV detection in capillary electrophoresis, *J. Chromatogr.*, 543, 439, 1991.

69. Beckers, J. L., Everaerts, F. M., and Ackermans, M. T., Determination of absolute mobility, pk values and separation numbers by capillary zone electrophoresis: effective mobility as a parameter for screening, *J. Chromatogr.*, 537, 407, 1991.

70. Gareil, P., Pernin, D., Gramond, J., and Guyon, F., Free solution capillary electrophoresis as a new method for determining inclusion constants for the complexes between cyclodextrins and guest molecules, *J. High Res. Chromatogr.*, 16, 195, 1993.

71. Francois, C., Morrin, Ph., and Dreux, M., Effect of the concentration of 18-crown-6 added to the electrolyte upon the separation of ammonium, alkali and alkaline-earth cations by capillary electrophoresis, *J. Chromatogr.*, 706, 535, 1995.

72. Carpenter, J. L., Camilleri, P., Dhanak, D., and Goodall, D., A study of the binding properties of vancomycin to dipeptides using capillary electrophoresis, *J. Chem. Soc. Chem. Commun.*, 804, 1992.

73. Gomez, F. A., Avila, L. Z., Chu, Y., and Whitesides, G. M., Determination of binding constants of ligands to proteins by affinity capillary electrophoresis: compensation for electroosmotic flow, *Anal. Chem.*, 66, 1785, 1994.

74. Kuhn, R., Frei, R., and Christen, M., Use of capillary affinity electrophoresis for the determination of lectin-sugar interactions, *Anal. Biochem.*, 218, 131, 1994.
75. Mammen, M., Gomez, F. A., and Whitesides, G. M., Determination of the binding of ligands containing N-2,4-dinitrophenyl group to bivalent monoclonal rat anti-DNP antibody using affinity capillary electrophoresis, *Anal. Chem.*, 67, 3526, 1995.
76. Khaledi, M. G., Smith, S. C., and Strasters, J. K., Micellar electrokinetic capillary chromatography of acidic solutes: migration behaviour and optimization strategies, *Anal. Chem.*, 63, 1820, 1991.
77. Greenaway, M., Okafo, G., Mannallack, D., and Camilleri, P., Micellar electrokinetic chromatographic studies in D_2O based buffer solutions, *Electrophoresis*, 15, 1284, 1994.
78. Herbert, B. J. and Dorsey, J. G., n-Octanol-water partition coefficient estimation by micellar electrokinetic capillary chromatography, *Anal. Chem.*, 68, 744, 1995.
79. Adlard, M., Okafo, G., Meenan, E., and Camilleri, P., Rapid estimation of octanol-water partition coefficients using deoxycholate micelles in capillary electrophoresis, *J. Chem. Soc., Chem. Commun.*, 2241, 1995.
80. Ishihama, Y., Oda, Y., and Asakawa, N., A hydrophobicity scale based on the migration index from microemulsion electrokinetic chromatography of anionic solutes, *Anal. Chem.*, 68, 1028, 1996.
81. Jacobson, S. C., Koutny, L. B., Hergenroder, R., Moore, A. W., and Ramsey, J. M., Microchip capillary electrophoresis with an integrated postcolumn reactor, *Anal. Chem.*, 66, 3472, 1994.
82. Effenhauser, C. S., Paulus, A., Manz, A., and Widmer, H. M., High-speed separation of antisense oligonucleotides on a micromachined capillary electrophoresis device, *Anal. Chem.*, 66, 2949, 1994.
83. Moore, A. W, Jacobson, S. C., and Ramsey, J. M., Microchip separations of neutral species via micellar electrokinetic capillary chromatography, *Anal. Chem.*, 67, 4184, 1995.
84. Chu, Y., Kirby, D. P., and Karger, B. L., Free solution identification of candidate peptides from combinatorial libraries by affinity capillary electrophoresis/mass spectrometry, *J. Am. Chem. Soc.*, 117, 5419, 1995.

CHAPTER 2

General Instrumentation and Detection Systems Including Mass Spectrometric Interfaces

Pierre Thibault and Norman J. Dovichi

CONTENTS

0-8493-9127-X/98/$0.00+$.50
© 1998 by CRC Press LLC

I. INTRODUCTION

An overview of the important historical developments of capillary electrophoresis, pre-sented in Chapter 1, emphasized the pivotal role of Arne Tiselius in the development of modern electrophoresis.[1] Over the two decades that followed this original contribution, major efforts focused on improving the performance of this separation format by using anticonvec-tive media. The recognition that the high surface-to-volume ratio of capillary columns pro-vides excellent heat transfer properties, thereby allowing use of high electric fields for fast and efficient separations, was at the base of the expansion of this technique. During the early development of capillary electrophoresis (CE), researchers also realized the potential of this technique for trace level analysis. The work of Edstrom in 1953 was one of the early demonstrations of CE, in which hundreds of picograms of RNA were sampled from a single cell using a micromanipulator.[2] Detection of the separated nucleic acids involved use of ultraviolet absorbance as revealed by photomicrographs taken at 257 nm and 275 nm.

A number of other important developments have marked the progress of CE. Early investigations by Neuhoff and co-workers on the analysis of proteins in polyacrylamide gel-filled tubes of capillary dimensions represented one of the first demonstrations of capillary gel electrophoresis.[3,4] Protein bands isolated from an isoelectric focusing gel were subse-quently applied to a microslab gel for two-dimensional electrophoresis using only minute amounts of analyte. This technology remains important today in miniaturized two-dimen-sional polyacrylamide gel electrophoresis for protein analysis. Similarly, the pioneering work of Hjerten and later Everaerts and co-workers provided essential understanding of free-zone electrophoresis and pointed to the necessity of using small diameter capillaries to minimize convection effects.[5-7] However, the wider acceptance of CE as a successful analytical tech-nique required significant technological advances in the production of inexpensive capillaries of small diameters (< 200 μm i.d.) and the development of high sensitivity optical detectors.

The modern era of CE, which led to current instrumentation, is considered by many to have started with the seminal publications of Jorgenson and Lukacs in the early 1980s.[8,9] These investigations presented simple instrumentation using fused-silica capillaries of 75 to 100 μm i.d. dimensions similar to those used in capillary gas chromatography. Separations were effected with field strengths of several hundred V/cm, and detection of the analyte bands was achieved using modified optics taken from HPLC detectors. The simplicity of the instrumentation and the extraordinary separation efficiencies obtainable on such electro-phoretic systems were partly responsible for the explosion of interest that rapidly followed.

CE has now become a mature analytical technique providing practical and innovative solutions to challenging separation problems. The versatility and range of applications of CE derive from its unique characteristics and advantages compared to other complementary analytical techniques. The current five formats of CE — capillary zone electrophoresis (CZE), capillary isotachophoresis (CITP), capillary gel electrophoresis (CGE), capillary isoelectric focusing (CIEF), and micellar electrokinetic chromatography (MEKC) — provide different principles of separation and extend the application of CE to different classes of analytes, from small organic and inorganic ions to complex biopolymers, such as oligosaccharides, peptides, proteins, and DNA.

Another attribute of CE is its high mass sensitivity and small injection volumes. This comes as a result of the high separation efficiencies and high sensitivity optics available on most CE instruments. In terms of mass detection, analysis at sub-picomole levels is typically achievable on most commercial instruments using UV detection. In view of the relatively small sample loadings (pL to nL volumes) characteristic of zone electrophoresis, such sensitivity appears modest when expressed in terms of concentration detection limits. However, this limitation can be alleviated using sample preconcentration techniques, such as sample stacking or transient isotachophoresis, and enhancement of concentration detection limits by at least two orders of magnitude have been reported.[10-12] Further improvements in signal detectability compared to conventional incoherent light sources can be obtained using detectors based on laser-induced fluorescence.

These remarkable advances have potential benefits in the field of molecular biology to simplify the standard assay used in the analysis of gene promoter activity, to determine point mutations in genetic sequence, or to investigate DNA–DNA or DNA–protein interactions. The possibility of separating and characterizing small quantities of material also provides unique opportunities for the development of CE-based enzyme assays or for the determination of binding constants of protein–protein or drug–protein interactions. CE can also be used as an on-line microreactor where small amounts of substrate can be digested by immobilized enzyme contained within a segment of the capillary. The number of CE applications is far too large to be cited conveniently in the present chapter. Rather, the reader is referred to fundamental reviews that have appeared regularly in selected issues of *Analytical Chemistry*.[13,14]

Improvements in detection technology have always played a fundamental role in the wider acceptance of CE. While the remainder of the instrumentation is simple, particularly for free-zone electrophoresis, the development of rugged, simple, and high-sensitivity detection technology remains a significant and important issue in CE. This chapter reviews the current status and basic instrumentation involved in CE, with particular emphasis on detection systems.

II. GENERAL INSTRUMENTATION

CE is a remarkably simple analytical separation technology. The basic instrumentation consists of four parts as indicated in Figure 2.1. A capillary is required for the separation, a high-voltage power supply is required to drive the separation, a detector is required to determine the presence and amount of analyte, and a safety interlock-equipped enclosure is used to protect the operator from the high voltage. The entire operation, including capillary conditioning, sample analysis, data acquisition and processing, can be automated, as is the case for most commercial CE instruments.

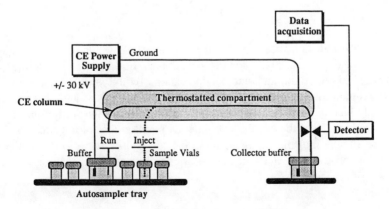

Figure 2.1 Schematic representation of a basic capillary electrophoresis instrument.

While not strictly necessary, most commercial instruments provide some mechanism for temperature control in the capillary. This temperature control serves two purposes. By use of a thermal bath with high heat capacity, the capillary can be cooled and its temperature can be actively controlled. Active cooling does not improve the separation in most electrophoresis experiments because thermal gradients across the capillary, and not absolute temperature rise, limit the efficiency of the separation. Because the temperature gradient across the interior of the capillary is independent (to a first approximation) of the temperature of the outside wall of the capillary, the separation efficiency is not improved by actively cooling the capillary.[15,16] Constant temperature is most important in obtaining reproducible retention times for analytes. The viscosity of aqueous solutions decreases by about 2.5% for each degree of temperature rise.[17] Because electrophoretic mobility is inversely proportional to viscosity, variations in temperature lead to variations in separation times for analytes. This situation is difficult to control, particularly in summer if no temperature regulation is available in the laboratory.

A. Sampling

1. Sample Introduction

Capillary electrophoresis is inherently a microchemical technique; capillaries used in electrophoresis typically have a total volume of a few nanoliters to a few microliters. In order to avoid excessive band broadening only a small fraction of the capillary can contain sample, so that only a minute amount of sample can be loaded onto the capillary.

The volume of a capillary, V_{column}, is given by the volume of a cylinder of length L and inner diameter d

$$V_{column} = \frac{\pi d^2 L}{4} \qquad (1)$$

For example, the volume contained within a 50-cm-long capillary of 50 μm i.d., is about 1 μL.

The maximum allowable injection (and detection) volume is related to the column volume and the number of theoretical plates; to preserve high separation efficiency, only a minute sample volume can be injected onto the capillary. The maximum allowed injection volume to avoid noticeable peak broadening may be estimated as

$$V_{inject} = \frac{\Theta V_{column}}{\sqrt{N}} \qquad (2)$$

where V_{inject} is the injection volume, Θ^2 is the fraction of allowable peak broadening, and N is the number of theoretical plates for the separation.[18] For a separation that provides one million theoretical plates and allowing 5% broadening by the injection, i.e., $\Theta = \sqrt{0.05} = 0.22$, the maximum injection volume is only 225 pL. While this volume is certainly microscopic, the 225-picoliter sample occupies a 100-μm-long slug in the 50-μm-diameter column, or approximately 0.2% of the capillary volume. Different modes of injection can be used to introduce the sample into the capillary: these include hydrostatic, electrokinetic, hydrodynamic (pressure or vacuum), and split flow. A brief outline of these injection modes follows.

a. Hydrostatic

Injection may be made based on a pressure difference by raising the sample reservoir above the detection reservoir. The pressure difference induced between the two vessels drives analyte onto the capillary. To terminate the injection, the capillary is removed from the sample and placed in a vial containing running buffer. The injection volume produced in hydrostatic injection is given by

$$V_{inject} = \frac{\rho g \pi r^4 \Delta h t_{inject}}{8 \eta L} \tag{3}$$

where ρ is the buffer density, g is the gravitational force constant, Δh is the height difference between collection and injection reservoirs, and η is the buffer viscosity.[19] Subtle difficulties are associated with hydrostatic injection. In particular, volume must be corrected for the finite time necessary to raise and lower the injection reservoir.

b. Electrokinetic

Instead of pressure-driven injection, electrokinetic injection may be used. The capillary tip is dipped into the sample for a moment, potential is applied for a few seconds, and the capillary is removed to the separation buffer. Notice that the number of moles of analyte injected onto the capillary depends on the electroosmotic flow (EOF) and electrophoretic mobility of the analyte. As pointed out by Jorgenson, analyte discrimination can occur, and sample components with the highest electrophoretic mobilities will be preferentially introduced over those with lower mobilities.[9] The electrokinetic injection volume, V_{inject}, is estimated by

$$V_{inject} = \frac{E_{inject}}{E_{separate}} \times \frac{t_{inject}}{t_{retention}} \times V_{capillary} \tag{4}$$

where E_{inject} is the injection potential, $E_{separate}$ is the separation potential, t_{inject} is the time of injection, and $t_{retention}$ is the retention time for the analyte. This calculation assumes that the conductivity of the sample is the same as that of the separation buffer.[20] As an example, consider an analyte that requires 5 minutes to elute at 30 kV potential in the 50-μm, 50-cm-long capillary. An applied potential of 1 kV for 2 seconds will deliver 225 pL sample volume. High accuracy injection is produced by carefully monitoring the injection voltage and current.[21] The presence of hydrostatic flow can introduce subtle systematic errors in electrokinetic injections, and care must be taken to avoid pressure-driven flow during the separation. The injection and detection buffers should be maintained at equal height.

c. Hydrodynamic

There are several ways in which a sample can be dynamically introduced onto the CE column, namely vacuum, pressure, or split flow. In all cases, the amount injected will depend on the dimensions of the column and on the viscosity of the solution.

Several commercial CE systems utilize vacuum injection. The advantage for the instrument manufacturer in using this approach over pressure injection is that it is only the vials at the detector end of the capillary that require a gas-tight seal and not the (more numerous) sample vials. Such a sample introduction technique is, however, difficult to implement when the detection system has inherent physical constraints such as for a mass spectrometer, or when the collector buffer is not easily accessible. In such cases, more elaborate procedures have to be developed for column conditioning. This type of injection procedure yields reproducible sample loadings with typically less than 2% relative standard deviation.

The sample can also be introduced by pressurizing the vial ($\approx$0.5 psi) for a set time, forcing a small volume of the solution into the CE column. As noted above, this does require special individual septa for sample vials. Commercial CE instruments can provide fixed pressure settings (0.5 psi for the actual injection and 20 psi for capillary conditioning) or variable pressure through the use of compressed gases. This type of sample introduction is particularly attractive and convenient when used in combination with mass spectrometric detection.

d. Split Flow

A rotary injection valve has been reported for capillary electrophoresis.[22] Use of a metal valve is unacceptable because electrolysis of the separation buffer leads to formation of bubbles in the capillary, destroying the separation. Instead, a ceramic injector was developed for applications in capillary electrophoresis. The injector produces a relatively large injection volume, 350 nL, so a split injection system is required for high efficiency separations in capillary electrophoresis. This split flow injection is similar in principle to the corresponding injection technique in GC, in which only a portion of the analyte solution is actually introduced onto the column. An HPLC syringe (10 μL) is used to inject the sample into one port of a three-way injection block.[23] The tip of the CE column seats into the top port by means of a septum seal, and the third (split-vent) port, opposite the syringe port, is connected with capillary tubing of varying lengths (3 to 5 cm) and/or i.d. (0.005 to 0.020 in), which provides the split ratio. Thus, to inject 10 nL of a sample solution onto the CE column would require a 1-μL injection with a split ratio of 100. Obvious advantages of this injection system are its simplicity, low cost, and ease of use. Although reproducibility is dependent on the accurate dispensing of small volumes ($\approx$1 to 5 μL), obtaining and maintaining the correct split ratio has been found in practice to be the major problem associated with this injection technique.

e. Ubiquitous Injection

The phenomenon called ubiquitous injection is surprisingly important if the capillary is in contact with the sample for more than a few seconds, and is independent of the applied potential and hydrostatic pressure.[24] During the contact time, analyte diffuses into the capillary. For example, an analyte molecule of modest size has a diffusion coefficient $D = 5 \times 10^{-10}$ m^2 s^{-1} in water at 25°C. An analyte will diffuse a distance, $L_{diffusion}$, given by

$$L_{diffusion} = \sqrt{2\,Dt} \qquad (5)$$

where t is the contact time between the sample and capillary. A 15-second contact time will allow the analyte to diffuse about 125-μm onto the capillary, producing an effective injection volume ($V_{diffusion}$) of

$$V_{diffusion} = \sqrt{2\,Dt}\,\pi r^2 \qquad (6)$$

or 225 pL. That is, a 15-second contact time with the sample (with no applied potential) will introduce as much analyte as a 1 kV injection for 2 seconds. Because ubiquitous injection depends on the diffusion of analyte onto the capillary, the phenomenon is most important for low molecular weight analytes.

Ubiquitous injection can also be observed for large molecules, given both adequate contact time and detection sensitivity. Figure 2.2 presents an electropherogram of fluorescently labeled DNA loaded onto a capillary by simply placing the capillary in contact with the sample for 30 seconds. During the contact time, the analyte diffuses into the capillary. The

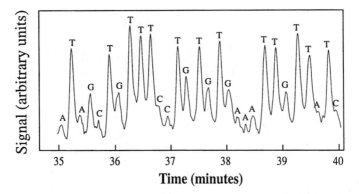

Figure 2.2 Example of ubiquitous injection of DNA sequencing fragments in capillary gel electrophoresis.

fragments that generated the peaks in Figure 2.2 range from 92 to 120 bases in length and are separated with a 50-cm-long, 5% acrylamide, 6% crosslinker gel-filled capillary at an electric field of 200 V/cm. The Richardson–Tabor peak height encoded sequencing technique was used to generate the sequencing samples; the peaks are ordered in amplitude as T > G > C > A. The low diffusivity of the relatively large sequencing fragments, along with the high viscosity of the sequencing gel, suggests that only a minute amount of analyte diffused onto the capillary. In this case, only ~10^{-20} mole of analyte is contained within each peak.

f. Photobleaching

Injection becomes problematic when dealing with short capillaries that are used for rapid separations. Because of the small volume of the short capillary, the amount of analyte injected must be very small to minimize band broadening. Jorgenson has used laser-induced photobleaching to develop a very high speed/small volume injection technique.[25] The sample contained fluorescently labeled amino acids. The labeled analyte underwent photobleaching under intense illumination from a moderate power laser beam. Brief interruption of the laser beam allowed a small slug of unbleached analyte to enter the capillary for separation. A lower power beam from the bleaching laser was used to excite fluorescence at the other end of the capillary. Injection times as short as 10 ms in a 1.5-cm-long capillary were described.

2. Preconcentration Techniques

As mentioned above, the injection size is typically limited to 1 to 2% of the capillary volume to minimize zone broadening in CZE. The constraints imposed on the sample loadings and the small dimensions of the typical flow cell are such that practical concentration detection limits in CZE-UV are in the order of 1 μg/mL. In order to improve these detection limits different approaches for on-line sample concentration have been developed. These preconcentration methods have provided concentration detection limits approaching those achievable by HPLC and have been crucial for the wider acceptance of CZE as a sensitive analytical technique.

a. Sample Stacking

One of the simplest methods of increasing sample loading is by exploiting the ionic strength differences between the sample buffer and the separation electrolyte. If the ionic strength of the analyte is less than that of the buffer, the phenomenon called stacking occurs. This phenomenon is based on isotachophoresis, where the low ionic strength sample and the high

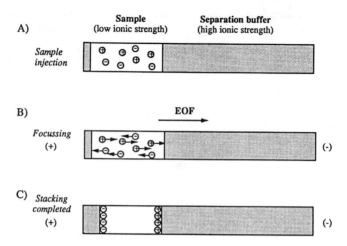

Figure 2.3 Enhancement of sample loading using stacking. (A) Injection of sample on an uncoated capillary, (B) migration of analytes during focusing step, and (C) completion of stacking. (Data courtesy of D. Y. Chen, University of Alberta.)

ionic strength separation buffer produce a voltage divider, with most of the potential applied across the lower conductivity sample. As a result, the ions in the sample are concentrated into the capillary until the ionic strength of the sample equals that of the separation buffer. A number of papers by Chien and Burgi consider this stacking phenomenon in great detail.[10,11,26] An example of sample stacking is presented in Figure 2.3. A plug of low ionic strength (low conductivity) sample is first introduced into the capillary by hydrodynamic, hydrostatic, or electrokinetic injection (Figure 2.3 A). In order to prevent dramatic differences in electroosmotic flow (EOF) between the sample and the separation buffer, and to minimize analyte denaturation upon the application of high field strengths, the sample is usually dissolved in 10- to 50-fold diluted buffer. Upon introduction of a potential across the capillary (Figure 2.3 B), the electric field will depend on the conductivity of the electrolyte at a given point. Regions with low conductivity (and high resistance) act as voltage dividers, with most of the voltage dropped across the low conductivity sample. As a result, the electric field is highest in regions where the sample buffer has lowest ionic strength. The high electric field drives the ionic components of the sample plug toward the higher ionic strength separation buffer that fills the majority of the capillary. After the stacking condition is completed (Figure 2.3 C), the components of the sample plug have been concentrated at the boundaries with the separation buffer so that the conductivity and electric field within the plug equal those of the separation buffer. The components of the analyte then undergo normal electrophoretic migration and separation. In this arrangement, the cations and anions concentrate in opposite directions, and a small plug of separation buffer is usually introduced after the sample to prevent any analyte loss. Such stacking procedures can yield a tenfold improvement in sample concentration.

It is also possible to stack larger sample volumes when the sample buffer can be removed following the stacking of the analyte bands at the buffer boundary. This technique is based on the principle that the electrophoretic velocity of the analytes is significantly faster than the electroosmotic velocity of the buffer.[26] If a high negative voltage is applied on the injection end of an uncoated capillary, the positive analyte ions will exit the column followed by the neutral sample buffer. When the sample buffer is almost completely ejected, the polarity of the electrodes is reversed to enable migration of the negatively-charged analytes in normal CZE mode. The separation of positively charged analyte can be effected by incorporating a buffer modifier such as tetradecyltrimethyl ammonium bromide (TTAB) to reverse the direction of the EOF. These large sample loadings can lead to sensitivity enhancements of several hundred-fold compared to conventional zone electrophoresis.

b. Combined Isotachophoretic Methods

Improvement in sample loadings can also be achieved using discontinuous buffer systems in a CITP capillary of suppressed EOF. The CITP preconcentration step is performed prior to the CZE separation using single or coupled column arrangements.[12,27] The single-column CITP-CZE configuration is currently the most interesting approach in view of its simplicity and ease of operation. The subject of this technique has been reviewed recently by Tjaden and co-workers.[28] In this method, the discontinuous buffer system is used to create isotachophoretic conditions at the beginning of the analysis. After the steady-state conditions have been reached, the diluted sample can be stacked in order of electrophoretic mobility with concentrations approaching that of the leading buffer.[12]

The steps involved in this procedure are illustrated in Figure 2.4. The capillary is first filled with a background electrolyte (BGE) whose mobility (μ_{BGE}) is greater than that of any ions in the sample of interest (μ_S). Following sample injection (Figure 2.4 A), a terminating electrolyte (T) of mobility less than that of the analyte is placed at the injection end, and the voltage is applied. As a result of the discontinuous buffer system, the field strength varies along the capillary. As the components separate, the field strength within individual zones changes, and once a steady state has been reached, all bands migrate at the same velocity (Figure 2.4 B). The isotachophoretic preconcentration is then stopped, the terminating buffer replaced with the leading electrolyte, and the voltage reapplied. Resolution of analyte bands through zone electrophoretic separation is established as soon as the leading ions of the electrolyte reach the focused sample and the sharp differences in electric strength are suppressed (Figure 2.4 C). In cases where the mobility of the analyte ion is higher than that of the leading electrolyte, the sample itself can be supplemented by a leading ion (L) such as NH_4^+, K^+, or Na^+ for cationic solutes. When proper electrolyte compositions are selected, the single-column arrangement can provide sample preconcentration of up to two orders of magnitude.[12,29] Although the loadability can be extended up to 1000-fold using a coupled

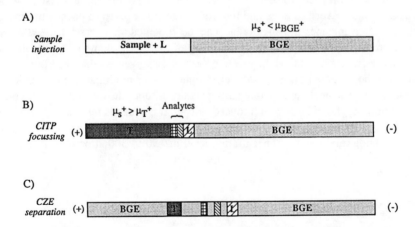

Figure 2.4 Capillary isotachophoresis/capillary zone electrophoresis for increased sample loading. A capillary with suppressed EOF is filled with the background electrolyte (BGE) followed by the sample containing a leading ion, L (A); the isotachophoretic focusing is performed using a terminator (T) of low mobility at the injection end (B); once the focusing is completed, the terminator is replaced with the background electrolyte and the separation is allowed to proceed by zone electrophoresis (C).

column configuration, the procedure is difficult to automate, requires monitoring of the separation currents, and additional equipment is often necessary.[12,27,28,30] Three different approaches to forming a discontinuous buffer system have been proposed by Schwer and Lottspeich, and offer valuable prerequisite conditions for the separation of a wide range of peptides.[31] The preceding discussion emphasizes that proper selection of electrolytes with appropriate electrophoretic mobilities is essential to the success of the isotachophoretic separation. The compositions, mobilities, and dissociation constants of many electrolytes suitable for CITP focusing can be found in the literature.[32]

c. Chromatographic Concentration

On-line trace enrichment can also be obtained by loading large volumes of sample using microcolumns containing adsorptive media followed by elution or electromigration onto a CZE column. These methods can use a single capillary arrangement where the adsorptive material is secured at one end of the column, or alternatively the chromatographic precon-centration can be effected in a dual-column arrangement separate from the CZE capillary. A review of different chromatographic preconcentrators has been presented recently.[33] Tandem columns comprising open tubular concentrators with surface-bound octadecyl were described by Cai and El Rassi for the analysis of different herbicides.[34] Enhancement of sample loadings was also reported for the enrichment of different analytes using a small bed of stationary phase inserted in an HPLC injection rotor or at the inlet of the CZE capillary.[35-38] Although the solid-phase extraction allowed a 100-fold preconcentration, such procedures have not gained wide acceptance due to the degradation of CE performance and the complexity of the instrumentation. Significant losses in separation efficiencies due to modification of the EOF, band broadening, and column overloading often compromise the separation.[33,39] Different attempts, including the reduction of stationary phase volume or the use of discontinuous buffer prior to zone electrophoresis, have been made to alleviate these difficulties.[33]

The most promising approach demonstrated thus far for chromatographic preconcentration involves the use of impregnated adsorptive membranes.[39] Polymeric styrenedivinyl benzene membranes have a high capacity for analyte adsorption, and elution of retained components can be effected with minimal volume of elution solvent. These membranes also have defined pore structure, smaller surface area, and reduced volume compared to packed microcapillary columns. For convenience, the membrane can be mounted in a cartridge system which facilitates its removal for independent conditioning of the CZE column or for off-line sample preparation. Special attention must be given to the elution solvent so that stacking or focusing conditions can be obtained prior to zone electrophoresis to minimize the dispersion of the analyte band. For basic analytes separated using acidic buffers, a leading electrolyte containing a dilute solution of NH_4OH can be introduced prior to the aqueous methanol elution solvent. The effectiveness of this approach was reported recently for the injection of approximately 50 μL of a peptide mixture at a concentration level of 150 to 300 attomoles/mL.[33]

B. Columns

1. Composition and Dimensions

Fused-silica capillary is almost universally used today in capillary electrophoresis, although both glass and Teflon® tubing were investigated by early workers.[7,8] The tubing is usually supplied with a thin outer coating of polyimide, which provides great strength and flexibility. The tubing is prepared from the same high purity silicon dioxide that is used for optical fibers, and is drawn from a larger preform in a furnace. The ratio of inner diameter to wall thickness is determined from the preform. The actual inner diameter is controlled by the rate at which the heated tubing is drawn from the preform; careful control of the furnace temperature and pulling rate results in very reproducible tubing dimensions. A typical capillary will vary in inner diameter by less than a micrometer from the start to the end of a 100-meter spool.

Tubing as thin as 2-μm i.d. is commercially available; most workers use 25- to 75-μm i.d. capillary. Since the cost of the tubing is proportional to the weight of the tubing, thinner-walled tubing is less expensive than thicker-walled tubing. One of the inherent advantages of using larger diameter capillary is the enhancement in sensitivity gained through longer optical pathlength. However, the increase in diameter is also accompanied by an increase in Joule heating due to lower surface-to-volume ratios and, more important, peak broadening. In optimizing the separation conditions (sample and separation buffers, field strength, etc.), it is also recommended that shorter capillary be used to speed up the analysis time. Once optimal conditions have been selected, the separation can be effected using longer capillaries with appropriate adjustment of field strength.

Absorbance and fluorescence measurements are often made directly on the capillary. Different brands of commercially available quartz capillaries have been investigated with respect to optical and surface properties.[40] The polyimide coating is removed from the capillary to provide an optical window for such detection. The coating can be removed with a gentle flame from a butane cigarette lighter although the use of an open flame makes the capillary wall quite brittle. Electric heating apparently removes the coating without damaging the capillary wall. However, heating at high temperatures will damage any coating on the interior of the capillary. The outer coating of the capillary can be removed without damage to the inner coating through use of a concentrated sulfuric acid bath at 100°C; this procedure must be performed with care in a fume hood.

While almost all workers use cylindrical fused-silica capillaries, rectangular capillaries can be used to provide a longer optical pathlength for transmission measurements.[41] Unfortunately, most rectangular capillaries are made of glass, have mediocre optical properties, and are extremely fragile.

Different alternatives to fused silica have been described. A hollow polypropylene fiber with significantly low EOF has been utilized as a column for CZE separation.[42] The hydrophobic nature of the fiber allows the surface to be coated with convenient surfactants so as to manipulate the magnitude and direction of the EOF. Capillaries made of polyfluorocarbon, polyethylene, and polyvinyl chloride have also been investigated.[43] Significant EOF was found with these capillaries, the magnitude of which was determined to be a function of electrolyte pH, organic modifier, and ionic strength. The EOF associated with these capillaries was attributed to the presence of carboxylic functionalities at the capillary surface.

2. Coatings and Buffer Additives

In particular applications of CE a capillary coating is essential for maintenance of the separation performance or for the reduction of analyte–surface interactions. The separation of proteins or analytes with basic functionalities can be severely compromised when the analysis is conducted on bare fused-silica capillaries. In other applications such as combined CITP-CZE separation or CIEF, the use of capillary with reduced EOF is a prerequisite for success of the analysis. A number of methodologies have been described in the literature for coating the internal surface of the capillary. The inner surface of the capillary wall can be modified by covalent attachment of coating agents or by the incorporation of surfactants and/or buffer additives to the separation buffer. These approaches not only reduce adsorption problems, but can also offer a means of manipulating the magnitude and direction of the EOF. Furthermore, peak resolution can be improved in situations where the electrophoretic mobility of the analyte is in the direction opposite to that of the EOF.[44]

The physical characteristics of the capillary surface directly influence the success of the CE analysis. The selection of modified capillary surfaces must be influenced by consideration

Table 2.1 Surface Modifications for Capillary Zone Electrophoresis Applications

Surface modification	Surface characteristics	Plates/m	pH range	Ref.
Coatings				
Polyethylene glycol	Weakly hydrophobic, reduced EOF	700–800 k	3–8	47
Polyethylene-propylene glycol	Weakly hydrophilic, reduced EOF	900–1000 k	3–7	48
Polyacrylamide	Hydrophilic, reduced EOF	200–300 k	2–10	48, 49
Polyvinylmethylsiloxanediol	Hydrophobic, reduced EOF	500–700 k	4–9	50
Polysiloxane (sulfonic acid)	Cathodal and constant EOF	≈200 k	3–9	51
Polysiloxane (quaternary amine)	Anodal EOF decreasing with pH	200–400 k	3–9	51
Aminopropyltrimethoxy silane	Anodal EOF decreasing with pH	200–400 k	<4	52
Buffer additives				
Non-ionic surfactants	Hydrophilic, reduced EOF	150–250 k	4–11	53
Cationic polymers	Anodal EOF decreasing with pH	200–500 k	<7	44, 54
Diaminoalkanes	Reduced EOF	≈300 k	6–8	55
Polyvinylalcohol	Reduced EOF	800–1000 k	3–10	56
Praestol (cationic polymer)	Anodal EOF decreasing with pH	≈200–300 k	3–10	56

of factors such as separation efficiencies, analyte recovery, variation of EOF, and influence of pH on the stability of the coating. The characteristics of several types of commercially available coated capillaries have been reported by Dougherty and Schure,[45] and an excellent review of different covalent coatings has been presented by Schomburg.[46] For convenience, a list of different surface modifications for CZE applications described for proteins and peptides is presented in Table 2.1. This list is obviously not exhaustive and is meant only to provide the reader with a description of different approaches taken to suppress or modify the EOF while simultaneously preventing analyte adsorption.

C. Sensitivity Requirements

To preserve separation efficiency, not only must the injection volume be held to a small fraction of the capillary volume, but the analyte concentration must also be significantly less than the ionic strength of the separation buffer.

As pointed out by Everaerts in an early capillary zone electrophoresis paper, the ionic strength of the sample must be much less (<1%) than the ionic strength of the separation buffer to avoid distortion of the band shape.[7] This distortion is associated with isotachophoresis; the high ionic strength analyte produces a region of high conductivity and low electric field, and the resulting distortion of the electric field leads to distortion of the peak shape. The extent of distortion of the peak shape depends on the relative mobilities of the analyte and the buffer components. Analytes that have higher mobility than the separation buffer ions will generate peaks that demonstrate fronting (peak A in Figure 2.5). Analytes that have the same ionic strength as the buffer will produce symmetrically broadened peaks (peak B in Figure 2.5). Analytes with lower mobility than that of the buffer components will demonstrate tailing (peak C in Figure 2.5). Typically, the ionic strength of the separation buffer is 0.01 M, which forces the analyte concentration to be less than 10^{-4} M to avoid band broadening. Indirect absorbance and fluorescence detection often use very dilute buffers, 100 μM or less. As a result, only extremely dilute analytes can be introduced onto the capillary without excessive band broadening.

The combination of small injection volume and low analyte concentration produces heroic requirements for detection sensitivity. For example, if the sample injection volume is 225 pL and the analyte concentration is less than 10^{-4} M, then the *maximum* amount of sample introduced onto the capillary is 25 fmol; larger amounts will produce a degradation in separation efficiency. If the separation is to have three orders of magnitude linear dynamic range, then the system must provide 25 amol (1 attomole = 1 amol = 10^{-18} mol) detection limit. Note that stacking conditions do not change this calculation. Relatively large amounts of dilute

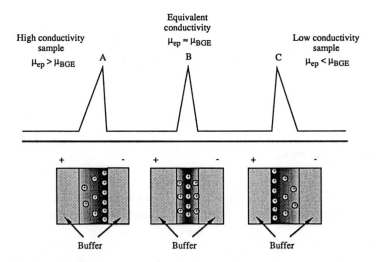

Figure 2.5 Peak asymmetry in capillary electrophoresis as a result of mobility differences between the analyte and the separation buffer.

sample can be concentrated by stacking; however the product of injection volume and sample ionic strength is limited to roughly 25 femtomoles before overloading effects are observed.

There are only a few reports where sufficiently dilute analyte was injected in a sufficiently small sample volume to produce peak broadening dominated by longitudinal diffusion. Smith et al. have published a separation in which 2.7 million theoretical plates were observed for dansyl-glycine using fluorescence detection.[57] Similarly, Cheng et al. have reported 2.7 million theoretical plates for FTC amino acids.[58]

Most capillary electrophoresis instruments rely on ultraviolet absorbance detection. These detectors, as we will see below, produce detection limits corresponding to ~5 femtomoles of analyte. As a result, most capillary electrophoresis data are generated under conditions that degrade the separation efficiency. Most often, relatively large injection volumes are used for the separation. The use of the large injection volume produces excess peak width and degraded separation efficiency. However, the peak height calibration curve remains linear. On the other hand, the use of high concentrations of analyte produces peak broadening that is related to the concentration injected. As a result, a calibration curve based on peak height will be nonlinear; and peak area must be measured to produce linear calibration curves for concentrated analytes.

Detection techniques not only must have high mass sensitivity, but the detection volume must also be small to minimize peak spread. To a first approximation, the limitations on detector volume are similar to those on injection volume; the detector must function with a sub-nanoliter probe volume to minimize extra-column band broadening.

III. DETECTORS

A. Optical Absorbance Detection

1. Conventional Transmission Measurement

All commercial electrophoresis instruments are provided with an ultraviolet absorbance detector. These detectors are based on a number of designs, but all are constrained to operate within the minute dimensions of the capillary. Figure 2.6 presents the basic design of an absorbance detector.[59] Light from a lamp is collected and imaged onto the capillary, and

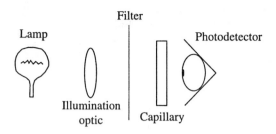

Figure 2.6 Absorbance detector.

either a monochromator or a filter is used to isolate a specific wavelength band for the absorbance measurement. The transmitted intensity is detected with a photomultiplier tube or silicon photodiode, and electronics are used to convert the transmitted intensity to an absorbance reading.

A number of different light sources can be used in the absorbance detector. Broad-band light can be produced from a xenon arc lamp, whereas special line sources can be used to generate light at specific wavelengths. Light from the source is spectrally filtered and imaged onto the capillary. There are fundamental physical limits to the performance of transmission detectors in capillary electrophoresis: only a small amount of light can be transmitted through the small capillary volume. Shot noise in this transmitted light determines the best possible detection limit. Other noise sources degrade the detection limit further.

Consider a black-body radiator at 5600 K. The amount of light produced by the hot lamp is described by Planck's black-body radiation equation

$$E_T(\lambda)d\lambda = \frac{2C^2h}{\lambda^5}\frac{d\lambda}{e^{hc/\lambda kT}-1} = \frac{1.19\times10^{-16}\,W\cdot m^2\,/\,ster}{\lambda^5}\frac{d\lambda}{e^{1.44\times10^{-2}(K\cdot m)/\lambda T}-1} \tag{7}$$

where E is the spectral radiance of a surface in units of W m^{-2} steradian^{-1} over the spectral region dλ, λ is wavelength in m, C is the speed of light, K is Boltzmann's constant, h is Planck's constant, and T is absolute temperature.[60] The spectral radiance, which describes the intrinsic brightness of the lamp, is given across a 1-nm spectral bandwidth for a 5600 K black body in Table 2.2.

Light from the lamp must be imaged onto the sample stream with an optical system. From the second law of thermodynamics, the radiance at the image of the lamp will be less than or equal to the radiance of the lamp itself; that is, the brightness of the image is less than the brightness of the lamp itself. The thermodynamic argument is that a perfect absorber would be heated by light from the lamp; the temperature (and spectral radiancy) of the object must be less than the temperature (and spectral radiance) of the lamp. The optical power that can be imaged onto the capillary is given by

$$P = E_t \times \Delta\lambda \times A \times \Phi_{collection} \times T_{filter} \tag{8}$$

where A is the illuminated area of the sample, $\Delta\lambda$ is the spectral bandwidth, $\Phi_{collection}$ is the collection efficiency of the illumination optics, and T_{filter} is the transmission of a filter or monochromator.

**Table 2.2 Spectral Radiance of a 5600 K Black Body in a
1 nm Bandwidth**

λ (nm)	250	300	400	500	600	700
E (Wm^{2}str^{-1})	4220	9390	19,000	22,500	21,500	18,500

Table 2.3 Power Delivered to Sample Through Optical System

λ (nm)	250	300	400	500	600	700
P (μW)	0.25	0.55	1.1	1.4	1.3	1.1

As noted above, the detector volume must be less than 225 pL to minimize extra-column peak broadening. As a result, only a small area of the capillary can be illuminated. The cross-sectional area of that volume in a 50-μm diameter capillary is 5×10^{-9} m² (100-μm × 50-μm). A high-efficiency illumination optic might collect 25% of the light from the lamp and direct it to the capillary. A monochromator system might transmit 5% of the incident light in the spectral band. The power passing through the capillary will be no higher than 6×10^{-11} times the spectral radiance of the lamp. Under these conditions, the optical power delivered to the sample is given in Table 2.3.

Thus, with a reasonable optical system, only a microwatt of optical power is incident on the capillary from a black-body light source. This limit is set by thermodynamics and can be increased only by using a hotter light source, by use of a wider spectral bandwidth, or by illumination of a larger cross-section of the capillary. Practically speaking, it is difficult to design an optical system with high collection efficiency. Typically, only a nanowatt of optical power can be transmitted through the capillary to the photodetector.[58]

Use of a high power lamp is usually of no value in capillary electrophoresis; 50 or 75 watt arc lamps will generate as much spectral radiancy as 500 watt lamps. The power of the lamp is usually increased by increasing the surface area of the hot discharge. As long as the hot discharge is larger than the illuminated volume of the capillary, an increase in the power of the lamp simply increases the area illuminated on the capillary. If a mask is used to restrict the illuminated area of the capillary, then increasing the power of the lamp does nothing to improve the amount of light passing through the capillary.

Much work has been devoted to maximizing the incident intensity onto the capillary. For example, ball lenses (small glass or sapphire spheres) have been designed to butt against the capillary (see Figure 2.7).[61] If the appropriate material and radius are chosen, a system with efficient illumination efficiency is produced. Sophisticated optical ray tracing programs are used in the design of the optical system. In every case, however, the illumination intensity produced in the capillary is no higher than that of the incoherent light source.

Shot noise in the number of detected photons limits the precision with which the transmitted light may be measured. The number of detected photons is given by

$$\text{Photons} = \frac{P \times \lambda}{1.986 \times 10^{-16}} \times \Phi_{\text{detection}} \times t_c \qquad (9)$$

where λ is wavelength in nanometers, P is in watts, $\Phi_{\text{detection}}$ is the detector quantum yield, and t_c is the detector time constant. The shot-noise in the transmitted signal is given by the square

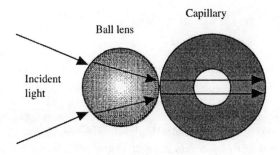

Figure 2.7 Use of a ball lens to illuminate capillary.

Table 2.4   Shot-Noise Limited Absorbance Detection Limit

λ (nm)	250	300	400	500	600	700
A_{min}	2×10^{-5}	1.4×10^{-5}	9×10^{-6}	7×10^{-6}	7×10^{-6}	6×10^{-6}

root of the number of detected photons. Table 2.4 presents the absorbance detection limit (3 s) due to shot noise for a 0.1-second time constant and 10% quantum efficiency photodetector.

The shot-noise limit can only be reached after heroic efforts to reduce source flicker noise, detector dark current, amplifier noise, and refractive index effects in the sample. In practice, absorbance detection limits are one to two orders of magnitude poorer than the shot-noise limit.

Both filters and monochromators have been used to isolate specific wavelength bands to illuminate the sample. Filters have the advantage of higher throughput and simple optical alignment and design. Monochromators offer the advantage of wavelength selection. On the basis of the specification from commercial electrophoresis instruments, it appears that use of a black-body light source and a monochromator leads to a factor of ten degradation in absorbance detection limit compared with use of a line source and a filter. Because a large number of intense spectral lines are available from simple discharge lamps, there often is little advantage in use of a continuum spectral source and monochromator.

These absorbance detection limits are measured across the 50-μm diameter capillary. Rays that travel through the center of the capillary have a longer optical path than rays that travel near the edge. Integrating the path over the cross-section of the capillary yields an average optical path that is about 80% of the inner diameter. This nonuniform path length across the capillary leads to negative deviations from Beer's law at absorbances greater than about 0.1. The deviations from Beer's law can be minimized by illuminating the center of the capillary.[61] However, any restriction in the illuminated area of the capillary produces a concomitant decrease in transmitted intensity and an increase in shot-noise.

For analytes with molar absorptivity of 10,000 Lmol^{-1}cm^{-1}, concentration detection limits are typically ~5×10^{-6} M. Given a 250 pL injection volume, the mass detection limit is then about 5 femtomoles. These detection limits are only a factor of 5 smaller than the upper limit in sample loading. However, dynamic range suffers in absorbance detection. Dynamic range can be extended by measurement of peak area rather than peak height, since high concentration samples increase the peak width rather than height. The issue of dynamic range is important, particularly in the pharmaceutical industry, where impurities must be certified at the part-per-thousand level. There is great advantage in having the most dilute components and the major components of the sample give rise to signals that are within the linear range of the detector.

Improved sensitivity is produced by measurement of absorbance at wavelengths shorter than 200 nm. Careful design is required to avoid excessive production of ozone by photolysis of oxygen, and a short path length is required between the lamp, capillary, and photodetector. Commercial instruments, based on the mercury 185 nm line, typically demonstrate a factor of 5 improvement in signal-to-noise ratio compared with measurements at 214 nm.

2. Increased Path Length Measurement

Because of the poor detection limits produced by conventional transmission measurements perpendicular to the capillary, there have been efforts to increase the sensitivity of transmission measurements. For example, a small bubble can be blown in the capillary to produce a region with increased diameter (see Figure 2.8). As analyte zones pass through the bubble, absorbance measurements can be made with increased pathlength. Preservation of the narrow electrophoresis zone in the expanded tube region may prove problematic with this arrangement.

Alternatively, a number of workers have described absorbance measurements along the capillary axis. If the sample bandwidth is significantly longer than the capillary axis, then

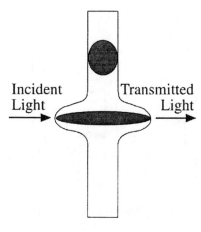

Figure 2.8 Increased pathlength using a bubble capillary cell.

improved absorbance sensitivity is produced. For example, an analyte that generates one million theoretical plates in a 50-cm-long capillary will occupy roughly a 500-μm section of the capillary. As a result, absorbance measurements made along the bandlength will generate sensitivity an order of magnitude higher compared with absorbance measurements made across the capillary axis.

One approach in axial-direct absorbance measurements uses a "Z" bend in the capillary,[61] with absorbance measurements along the long axis of the tube (Figure 2.9). As the analyte fills the bent arm of the capillary, the increased optical path leads to increased absorbance sensitivity. However, because the absorbance is integrated along the arm of the bent capillary, band broadening becomes important. If a second band enters the bend before the first band leaves, the peaks will overlap, degrading the separation efficiency. To preserve the separation efficiency of the capillary, the volume of the bend must be kept less than 225 pL; as a result, the Z bend approach to absorbance detection is difficult to implement in electrophoresis without seriously degrading the separation efficiency.

A different approach to axial absorbance detection has been reported by Yeung's group.[62] Rather than measuring absorbance perpendicularly to the capillary axis, enhanced pathlength is generated by using the capillary axis for the measurement (see Figure 2.10). In this system, the capillary is held rigid and straight. Light from either a lamp or laser is directed down the axis of the capillary. Precise alignment is important to maximize the amount of light reaching the detector and to minimize shot noise. A rather complicated optical system is required to illuminate the capillary effectively. If the refractive index of the capillary is less than that of the separation buffer, total internal reflection acts to guide the light down the capillary axis.

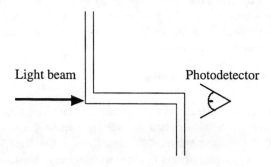

Figure 2.9 Increased pathlength with illumination of a "Z"-bend capillary.

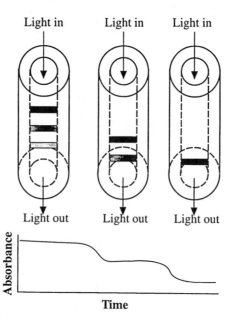

Figure 2.10 Axial illumination of capillary.

Pure dimethyl sulfoxide was used as the separation medium rather than aqueous buffer since the high refractive index of this nonaqueous solvent satisfied the total internal reflection conditions. The need to use this rather exotic separation buffer certainly is problematic. The development of low refractive index Teflon capillaries will be useful in this context because aqueous solvent can be used for the separations. The authors reported a sevenfold improvement in sensitivity compared to axial illumination of the capillary. This enhancement in sensitivity is exactly as predicted by the ratio of the electrophoretic bandwidth to the capillary diameter.

In this detector, the absorbance, which is integrated along the axis of the capillary, includes a contribution from all the components in the sample. Early in the separation, all analytes are present in the capillary and the absorbance is large. As each component of the sample elutes from the capillary, the absorbance decreases in a stepwise manner; baseline is observed when all analytes have eluted from the capillary.

The signal is the integral of the analyte absorbance along the capillary. To produce a plot that resembles conventional electrophoresis data, it is necessary to differentiate the absorbance signal numerically. This differentiation step introduces noise into the measurement.

Dynamic range is also an issue in the axial illumination method. The signal at the start of the separation is the sum of the absorbance of all components in the sample. The presence of a large number of components means that the absorbance along the capillary can attenuate the intensity of light significantly. The amount of analyte necessary to produce similar attenuation in a conventional transmission detector will be much larger, allowing greater dynamic range in conventional measurements.

3. *Laser-Based Absorbance Detection*

The spatial coherence of the laser allows the entire beam to be focused on a small spot. As a result, high power can be transmitted through the capillary and shot-noise is seldom significant. However, the limited wavelength tunability of the laser has resulted in few applications to absorbance measurements in capillary electrophoresis.

Hartwick's group reported a clever approach to increasing the optical pathlength.[63] After removing the polyimide coating of the capillary, they coated a portion of the capillary wall

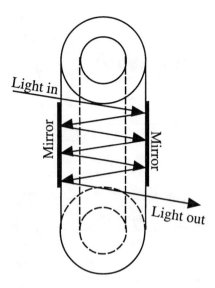

Figure 2.11 Multiple reflection in capillary.

with a layer of metallic silver to form a mirror, as indicated in Figure 2.11. A helium-neon laser beam (λ = 632.8 nm) was directed into the capillary at a shallow angle. The beam was reflected from the opposite wall and refocused by the curved surface into the center of the capillary. This reflection process was repeated many times before the laser beam exited from the capillary. By adjusting the angle of incidence and the spacing between the entrance and exit windows, as many as forty reflections were produced, which generates a concomitant increase in optical pathlength and a proportional increase in sensitivity. The maximum useful spacing between the entrance and exit windows is limited by band spread. To minimize band spread, this distance should be less than a few hundred micrometers. There is another limitation to the number of allowable reflections. Attenuation in the beam intensity with each reflection steadily decreases the transmitted intensity of the laser beam; if there is significant attenuation due to the imperfect reflections, the loss in beam intensity will result in an increase in shot-noise and a degradation in detection limit. Because of the low reflectivity of silver in the ultraviolet portion of the spectrum, it is not clear if a significant number of reflections could be used before attenuation of the beam becomes significant.

It does not appear likely that this multiple reflection cell would be useful with an incoherent light source. Incoherent light sources are characterized by large beam divergence. As a result, the beam tends to spread as it passes through the capillary. While the curved surface of the capillary refocuses the beam in the plane perpendicular to the capillary axis, the beam will diverge in the direction along the beam axis. For a well-collimated beam, light reaches the exit window after undergoing N reflections. In the presence of significant divergence, some light rays will reach the exit window after undergoing N-1 reflections. The light rays, originating from the same spot in the light source, can therefore undergo interference. Because the optical phase of the light rays depends on subtle refractive index variations in the sample, the interference of rays that have made a different number of reflections can lead to enhanced sensitivity.

4. Indirect Absorbance Detection

There is a need for universal detection in capillary electrophoresis. Indirect absorbance (and fluorescence) detection has been developed for use as a universal detector.[5,64] Transparent analyte can be detected in capillary electrophoresis if a component of the buffer absorbs light

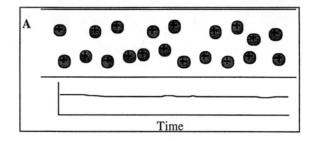

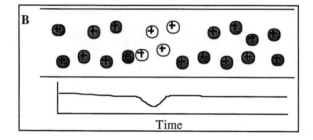

Figure 2.12 Indirect absorbance detection.

(see Figure 2.12). In the absence of analyte, the continually eluting buffer generates a high background absorbance level (Figure 2.12A). In the absence of analyte, the absorbance is constant. Because of the condition of constant electric field, the presence of ionic analyte in the capillary leads to the displacement of the buffer ions; the presence of analyte is accompanied by a decrease in the concentration of absorbing buffer. As a result, elution of analyte is accompanied by a decrease in the absorbance of the buffer (Figure 2.12B). This decrease in absorbance, or negative peak, is proportional to the amount of analyte present.

The detection limit produced by indirect absorbance is related to the absorbance of the buffer and the stability of the transmitted light intensity. In general, the background absorbance is stable to roughly 1%. The minimum detectable analyte concentration will be roughly 1% of the buffer concentration. Column overload effects are common in indirect detection experiments; recall that the analyte concentration must be less than 1% of the buffer concentration to avoid band-broadening effects. Band broadening can be acceptable if peak area rather than peak height is used to determine the amount of analyte present.

The use of a low concentration buffer leads to enhanced concentration detection limits; small amounts of analyte lead to a proportionally larger change in background absorbance. This concept has been taken to extreme values in fluorescence-based indirect detection, where micromolar buffers have been employed.[65] Of course, care must be taken not to exceed the buffering capacity of the separation medium.

The concept of indirect absorbance detection was described by Hjerten in the 1960s for zone electrophoresis,[5] and has been developed commercially to a high art by Millipore, particularly for the analysis of cations. Bocek in the Czech Republic has reported an impressive separation of rare earth metals; an eighteen-component mixture was resolved in slightly more than four minutes.[64]

B. Refractive Index and Refractive Index Gradient Detection

Refractive index (RI) measurements provide another approach to universal detection in capillary electrophoresis. Because most analytes have a refractive index different from that of the solvent, a high sensitivity monitor of refractive index may be used to screen a complex

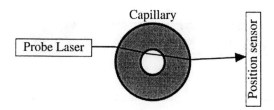

Figure 2.13 On-line refractive index detection in capillary electrophoresis.

sample for all components that are present in the sample. In reality, the low sensitivity of the detector to analyte and its high sensitivity to environmental fluctuations lead to mediocre performance for the refractive index detector. There have been very few applications of refractive index measurements to capillary electrophoresis.

A very simple and inexpensive refractive index detector for capillary-based separations was reported in 1986.[66] In the instrument, Figure 2.13, a laser beam is focused into a capillary. If the beam passes off-axis through the capillary, the beam's path is deflected by an amount given by Snell's law. When the beam intersects the capillary at glancing incidence, interference between the rays that pass through the liquid and the rays that only pass through the capillary wall produces a series of fringes. The position of these fringes changes as the refractive index of the solution changes. The position of the fringes may be monitored with several simple photodetectors. A particularly simple, high-sensitivity, small-volume refractive index detector is constructed by placing a photodiode on the edge of a fringe and observing the signal change as the fringe moves. Detection limits of $\Delta(RI) = 6 \times 10^{-7}$ were reported for a 500-μm-diameter capillary. Bruno et al. from Ciba–Geigy in Switzerland have reported detection limits of $\Delta(RI) = 3 \times 10^{-8}$ for capillary electrophoresis separation of underivatized carbohydrates in a 50 μm inner diameter capillary.[67] The primary noise source in refractive index measurements is associated with temperature fluctuations. Careful design of the capillary holder is required to maximize thermal stability of the capillary.

The refractive index detector measures the instantaneous refractive index of the solution. Because of the low-frequency response of the detector, the system is quite sensitive to thermally induced drift. Alternate approaches have been demonstrated in order to shift the refractive index signal to higher frequencies, thereby minimizing susceptibility to low frequency noise.[68,69] Modulation of the refractive index signal is produced by applying a 400-Hz modulation to the separation potential.[68] The modulated separation voltage produces a modulation in the analyte velocity and, as a result, in the refractive index signal. A lock-in amplifier is used to demodulate the refractive index signal and to reject low frequency drift in the measurement. Detection limits were about $\Delta(RI) = 5 \times 10^{-8}$. This velocity modulation technique should be valuable for most detection techniques. By introduction of a modulation on the analytical signal, low frequency drift should be minimized.

Instead of detecting the refractive index directly, it is possible to detect the refractive index gradient (the change in refractive index with distance along the capillary axis) with high sensitivity.[70-73] The refractive index gradient is proportional to the concentration gradient along the capillary axis. Because of the high separation efficiency produced in capillary electrophoresis, high concentration gradients, and hence refractive index gradients, can be observed. Consider three bands eluting from the capillary, as shown schematically in Figure 2.14. Peak A is compact, generates a large change in concentration with distance, and produces a large refractive index gradient. Peak B is broader than A. While it has the same maximum concentration as A, the concentration and refractive index gradients are less than those of A because the peak is more diffuse. Peak C suffers from the greatest dispersion and generates a very small gradient signal. Measurement of concentration gradient produces the highest

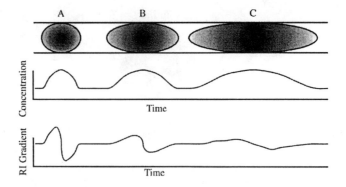

Figure 2.14 Refractive index gradient profiles.

signal for the sharpest bands. Because capillary electrophoresis is quite efficient, narrow peaks are produced and the concentration gradients are high.

The gradient detection is useful because it discriminates efficiently against slow variations in refractive index due to temperature or buffer impurities. An interesting application of the detector is in isoelectric focusing of proteins. The high molecular weight proteins have low diffusion and produce particularly sharp bands. On the other hand, the low molecular weight ampholyte components in the buffer have large diffusion coefficients, and they produce broad bands. The refractive index gradient detector effectively discriminates against the background concentration of ampholyte and instead produces high sensitivity for the tightly focused proteins.

In a simple instrument, Figure 2.15, a laser beam illuminates the capillary. The refractive index gradient associated with the concentration gradient leads to deflection of the beam along the capillary axis. As the analyte passes the laser beam, the beam is deflected by an amount proportional to the concentration gradient. A position sensor measures the beam deflection, producing a signal that is proportional to the concentration gradient. To convert the concentration gradient signal into a signal proportional to analyte concentration, the gradient signal is integrated with respect to time. Because integration is inherently a smoothing process, the resulting concentration signal is of low noise and high sensitivity.

Pawliszyn at the University of Waterloo reported detection limits corresponding to $\Delta(RI)$ = 2×10^{-6} for analytes that elute from the column in narrow peaks.[71] In the simple system,

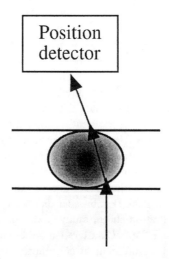

Figure 2.15 Refractive index gradient detection system.

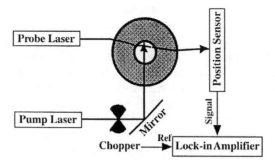

Figure 2.16 Thermooptical absorbance detection.

it is necessary for an analyte to migrate past the stationary detector. In isoelectric focusing, it is desirable to minimize the flow of analyte. Conventionally, a detector would be scanned along the capillary axis, detecting the presence of the analyte. Instead, Pawliszyn has illuminated the entire capillary with a collimated laser beam.[73] The beam profile is measured after the capillary with a photodiode array. The diode array generates the measurement of the refractive index gradient at 1024 points along the capillary axis without moving either the detector or the analyte. This stationary imaging system is powerful, particularly for the analysis of proteins by isoelectric focusing.

C. Thermo-Optical Absorbance

The major noise source in refractive index measurements is temperature-induced refractive index changes. These changes can be quite small, of the order of $\Delta T \sim 10^{-3}$ K. The sensitivity of refractive index measurements to temperature change has been used to construct a high-sensitivity absorbance detector (see Figure 2.16). The capillary is illuminated with a modulated pump laser beam at right angles to the refractive index probe laser. A modulated temperature rise (and refractive index change) is induced in the sample following absorbance of the modulated pump laser beam. The probe laser beam deflects in phase with the excitation frequency of the pump laser beam. A phase-sensitive detector produces low noise measurement of the probe deflection, which is proportional to the analyte absorbance. Because the temperature rise is linearly related to the laser power, improved detection limits are obtained by use of higher power lasers. In one approach a 2 mW pump laser beam with a modulation frequency of about 100 Hz was used to obtain absorbance detection limits of 6×10^{-6} measured across a 50-μm capillary.[74,75] The instrument has been applied as a detector for capillary zone electrophoresis separation of amino acids labeled with a chromophore which matched the wavelength of the excitation laser. Detection limits of 40 attomoles of the dimethylamino-azobenzene sulfonyl chloride derivative of methionine were obtained with a helium-cadmium pump laser beam operating at 442 nm.[75]

Most lasers operate in the visible portion of the spectrum and as a result are useful pump lasers only for highly colored analytes. More universal detection would be provided by use of lasers that operate in the ultraviolet portion of the spectrum. There are two examples in the literature of ultraviolet laser pumped thermo-optical detectors for capillary electrophoresis. Bornhop and co-workers reported the first application of ultraviolet laser-based thermo-optical absorbance detection in capillary electrophoresis.[76] They used a frequency-doubled argon ion laser (λ = 257 nm) to determine three dansyl-amino acids by high-speed capillary electrophoresis, with detection limits in the low 10^{-4} M range. The frequency-doubled argon ion laser is quite expensive and rather temperamental to operate, and it is not likely to see routine use in the analytical laboratory.

A low-power waveguide excimer laser for thermo-optical absorbance detection in capillary electrophoresis was also investigated for the analysis of phenylthiohydantoin (PTH) amino acid derivatives.[77] The laser produces quasi-cw output in the form of pulses of 100 W peak power, 100-ns duration, 5-mW average power, and repetition rates as high as 2 kHz. This waveguide laser is significantly less expensive than more conventional ultraviolet lasers, it is air cooled, and it operates from a standard 110 V line. The KrF laser of this type operates at 248 nm, which matches well the absorbance of the PTH-amino acids. With this laser, the thermo-optical absorbance detector produced detection limits of 4×10^{-6} absorbance units for PTH-glycine, two orders of magnitude better than those achieved using conventional detectors. A lock-in amplifier was used to demodulate the thermo-optical absorbance signal. Usually, a box-car averager would be used for this purpose. However, the thermal decay time matched the 610 Hz pump laser pulse repetition rate. As a result, the thermo-optical signal was nearly sawtooth in shape, and the lock-in amplifier provided excellent demodulation.

Figure 2.17 presents the separation of 20 PTH-amino acids by micellar capillary electrophoresis and thermo-optical absorbance detection. A 1.1 mV baseline signal is due to absorbance of the excimer laser beam by trace impurities in the separation buffer. The baseline disturbance at 4.5 minutes is due to the elution of trace amounts of acetonitrile, added to the analyte to effect dissolution. Acetonitrile produces a refractive index change that perturbs the optical alignment, producing the baseline disturbance. The PTH-amino acids are nearly baseline resolved, with slight overlap of alanine and glutamic acid and histidine and tryptophan. Detection limits (3 σ) range from 0.3 to 1 femtomole for the PTH amino acids injected onto the capillary, two orders of magnitude better than the detection limits produced by commercial capillary electrophoresis instruments using UV detection. Linear dynamic range in peak height extended for a factor of 3000 in analyte concentration.

The background signal in Figure 2.17 emphasizes the limitation of high sensitivity absorbance detection in capillary electrophoresis. In the ultraviolet portion of the spectrum, solvent impurities can produce significant background absorbance. As in indirect detection techniques, high sensitivity measurements in the presence of a high background signal limit the smallest change in absorbance to about 1% of the background level. As a result, high sensitivity absorbance measurements are not particularly useful with high background measurements. Either extremely clean buffers must be used for the measurement or more selective detection is required. One way of producing enhanced selectivity is by use of derivatization techniques to produce analytes that absorb in the visible portion of the spectrum, where the background is usually quite low. The other way of enhancing selectivity is by use of derivatizing techniques to produce highly fluorescent analytes.

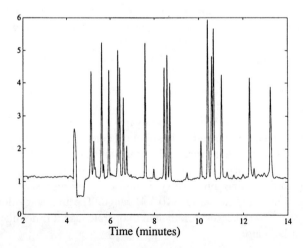

Figure 2.17 Separation of 45 fmol of PTH amino acids.

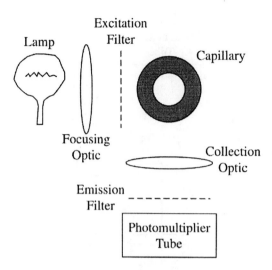

Figure 2.18 Filter fluorometer.

D. Fluorescence Detection

Fluorescence detection can provide both high sensitivity and high selectivity for capillary electrophoresis. Jorgenson reported the first application of fluorescence detection in capillary electrophoresis in 1981.[8] A typical fluorescence detector, Figure 2.18, has an excitation source, a wavelength selector, and focusing optics to image the excitation beam onto the capillary. The capillary is treated to remove the polyimide coating to provide an on-column detection window. Fluorescence is collected at right angles, spectrally filtered, and detected with a photomultiplier tube. Because of the low intensities produced by incoherent lamps, poor collection efficiency associated with most optical designs, and the high levels of scattered light produced by the capillary, fluorescence detection with conventional designs produces only an order of magnitude improvement over absorbance detection. For investigators who wish to construct a simple fluorescence detector, the use of an epi-illumination microscope is quite attractive. Guzman and co-workers reported sub-femtomole detection limits with a simple commercial microscope.[78]

Laser-induced fluorescence produces remarkable improvements in detection limits compared with the use of conventional incoherent light sources. In 1985, Zare's group reported the first application of lasers to fluorescence detection for capillary electrophoresis.[79] A 5-mW helium-cadmium laser beam, $\lambda = 325$ nm, produced the excitation beam. The fluorescence from the sample was collected with optical fibers, spectrally filtered, and detected with a photomultiplier tube. Femtomole amounts of dansyl-amino acids were detected with the system.

A number of improvements in fluorescence detection have been made in the past few years. Several groups have developed fluorescence detectors that produce detection limits of 10^{-19} moles of analyte introduced onto the capillary. These systems all rely on careful imaging of the collected fluorescence to block scattered laser light. Hernandez and co-workers have developed an elegant fluorescence detector based on epi-illumination design (see Figure 2.19).[80] In their system, a low-power argon ion laser beam is reflected from a dichroic filter and focused onto the capillary. Fluorescence is collected by the same optics, transmitted through the dichroic filter, spectrally filtered, and detected with a photomultiplier tube. The use of the same high-numerical aperture optics for illumination and collection results in a small depth of focus. As a result, light scattered at the capillary walls does not pass through the optical train and a low background signal is produced. As Hernandez pointed out, the epi-illumination detection system is useful in illuminating several capillaries to monitor several separations simultaneously. Mathies' group has demonstrated the use of epi-illumi-

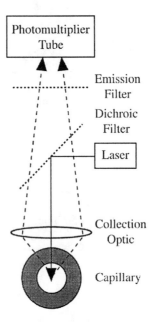

Figure 2.19 Epi-illumination fluorescence detection.

nation to monitor several separations. They placed a ribbon of capillaries in the microscope.[81] The capillaries were mounted on a motorized translation stage, and the fluorescence signal was monitored as the capillary array was scanned across the fixed laser beam. Up to 24 capillaries could be monitored simultaneously.

On-column fluorescence detection suffers from a relatively large amount of light scattering, which generates a large amount of background signal due to reflection and refraction at the capillary walls.[82] Ideally, then, fluorescence measurements should occur in a flow chamber with good optical quality and, in particular, with flat windows. While a square capillary could be used for electrophoresis, it is not possible to generate long segments of small-dimension capillary with flat windows. Instead, it is necessary to use a post-column fluorescence detection chamber to achieve extremely high-sensitivity fluorescence detection.

The design of a post-column detector requires care to avoid unacceptable band broadening due to dead volume in the plumbing used to transfer the sample from the capillary to the detection chamber. A post-column fluorescence detector based on the sheath flow cuvette has been developed and a schematic representation is shown in Figure 2.20.[83-89] In this system, the sample is transferred from the capillary to a flow chamber of exceptional optical quality with no loss of separation efficiency and very high detection sensitivity. The flow chamber is made from optically flat quartz, has 2-mm-thick windows, and has a 200-μm square hole in the center. The exit of the capillary is inserted into the flow chamber. Sheath fluid, identical to the separation buffer, is introduced into the cuvette at a flow rate of a few microliters per hour. This flow can be provided by a high precision syringe pump, or a simple alternative is to form a siphon by holding the sheath fluid in a vial that is raised a few centimeters above the waste collection vial. To complete the electrical circuit for electrophoresis, the stainless steel plumbing associated with the sheath stream is held at ground potential. Under the very low flow rate produced by electroosmosis, the sample stream travels as a ~10-μm-diameter stream through the center of the cuvette. Fluorescence is excited by a low-power laser beam focused on a 10-μm spot about 0.2 mm downstream from the capillary exit. Fluorescence is collected at right angles to both the sample stream and laser beam with a microscope objective. Spectral filters are used to reject scattered laser light and a 200-μm-radius pinhole is located in the reticule position of the microscope objective to restrict the field of view of a photo-

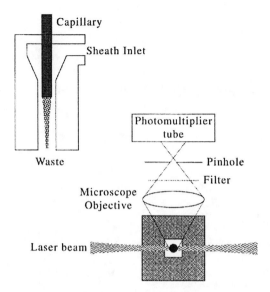

Figure 2.20 Sheath flow cuvette for fluorescence detection.

multiplier to the illuminated sample stream. For the highest-sensitivity measurements, the photomultiplier tube is cooled to −15°C to minimize detector dark current. The choice of laser is based on the spectral properties of the analyte. In general, lower background signals are observed for long wavelength excitation. The decreased background signal is due both to the decrease in scattered light intensity which scales roughly as λ^{-4}, and to the fact that there are relatively few impurity molecules that absorb long wavelength light to generate background fluorescence signals.

The sheath-flow cuvette has produced outstanding detection limits that approach the single molecule level. For example, Figure 2.21 presents the electropherogram generated when 625 yoctomoles (6.0×10^{-20} mol or 375 molecules) of rhodamine 6G (15 pL of a 4.1×10^{-11} M solution) were injected onto a 41-cm-long, 10-μm inner diameter capillary. The peak observed at 3.1 minutes corresponds to the elution of the analyte. Panel B presents a close-up of the region near the peak. The background signal of 575 photon counts per 100-ms integration time appears to be dominated by Raman scatter from the aqueous buffer. Panel C presents the peak itself. After subtraction of the baseline, the area under the peak contains 2360 excess photon counts. If there are no analyte molecules lost due to adsorption to the capillary walls, then each analyte molecule generates an average of 6.3 photon counts. In this system, the light collection efficiency is 15% (0.70 numerical aperture), while the filter transmission integrated over the analyte emission spectrum is about 25%, and the photomultiplier tube has quantum efficiency of about 15%. Detection of 6.3 photons per analyte molecule therefore implies that the analyte molecule emitted roughly 1000 photons while traversing the laser beam.

The number of photons emitted by an analyte molecule passing through a focused laser beam can be calculated as follows:

$$\text{Photons} = \int_{-\infty}^{\infty} I(t)\sigma\Phi dt = \frac{2\,P\sigma\Phi}{\sqrt{2\,\pi}\omega v} =$$

$$= \frac{2 \times (3 \times 10^{15}\,\text{photons/sec}) \times (1 \times 10^{-16}\,\text{cm}^2) \times 0.5}{\sqrt{2\pi} \times 10^{-3}\,\text{cm} \times 0.1\,\text{cm/sec}} \tag{10}$$

$$= 1000\,\text{photons}$$

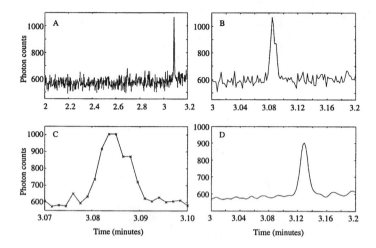

Figure 2.21 Injection of 375 analyte molecules. (A–C) raw data, (D) filtered data. (Data courtesy of D. Y. Chen, University of Alberta.)

where $I(t)$ is the irradiance of the laser beam in photon $cm^{-2}sec^{-1}$, σ is the absorbance cross-section in cm^2 ($\epsilon = 25,000$ L mol^{-1} cm^{-1} at 545 nm), Φ is the fluorescence quantum yield (0.5), P is the laser power (1 mW), ω is the beam spot-size (10 µm), and v is the linear velocity of the analyte molecule through the laser beam (volumetric flow ≈ 100 µL/hr). The agreement between the observed photon count per molecule and the calculated value is remarkable, particularly given the uncertainty in a number of the instrumental parameters.

Panel D of Figure 2.21 presents the result of passing the data of panel B through an optimum digital filter. The detection limit (3 s) calculated for this filtered data is 25 analyte molecules (40 ymol); minor improvements to the detection system should result in single molecule detection. The combination of the extraordinary sensitivity of fluorescence detection with the remarkable separation efficiency of capillary electrophoresis results in an analytical system with unparalleled performance.

The primary applications of the high-sensitivity detection system are found in biological analysis. For example, in DNA sequencing, the total amount of sequencing sample introduced onto the capillary often totals one attomole. The amount of each fragment averages perhaps a zeptomole. Novel peak-height encoded sequencing techniques require high signal-to-noise ratios, and detection limits of a few yoctomoles are mandatory.

Sweedler et al. have reported a particularly elegant fluorescence detection system for capillary electrophoresis (see Figure 2.22).[90] In their system, axial illumination was used to excite fluorescence; because of the long pathlength, high sensitivity was produced with a very low-power laser. To avoid band broadening produced by the long pathlength, they took advantage of the time-integration properties of a charged-coupled device (CCD) photodetector. The CCD, which consists of an array of silicon photocells, was first used for fluorescence detection in capillary electrophoresis by Cheng et al.[91] It is possible to move the charge accumulated in an array element to the next array element. In Sweedler's experiment, fluorescence from the zone was imaged onto the CCD.[90] As the zone moved down the capillary, its image swept across the CCD. Signal was integrated by passing the charge down the array at the same rate at which the image of the zone moved down the array. When the image of the zone passed the bottom of the array, the integrated charge was read off and stored in a computer. The result was a long-term integration of the fluorescence signal without band broadening. The signal from several bands can move down the array simultaneously, providing negligible dead-time in the measurement. In practice, a two-dimensional array was used so that the fluorescence signal was dispersed in the second dimension. Each column of the array

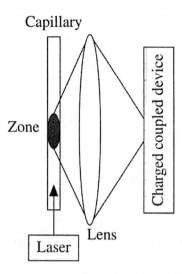

Figure 2.22 Time-integrated detection with charge-coupled device.

integrated the fluorescence intensity at a particular wavelength. At the end of the integration time, the spectrum was recorded. Detection limits for fluorescein-labeled amino acids fell in the range of 2 to 8×10^{-20} moles injected onto the capillary.

The fundamental limit to fluorescence detection is often associated with photobleaching of the analyte under the intense laser beam.[87] For example, a fluorescein molecule will undergo an average of 7800 cycles from the ground state to the excited state before undergoing photobleaching. In axial illumination systems, analyte is illuminated for the entire length of the capillary and, because of the long illumination time, photodestruction can occur before the analyte reaches the detection region. In Sweedler and Zare's work,[90] the illumination time was minimized by introducing a bend into the capillary immediately upstream from the detection region. The laser beam could not propagate beyond the bend, and photobleaching effects were minimized.

E. Derivatization for Fluorescence Detection

Proteins and peptides that contain tryptophan residues are the major examples of naturally fluorescent analytes. Such peptides are excited at about 280 nm but, because there are no inexpensive lasers available in that spectral region, high-sensitivity detection of native fluorescence is unlikely to become routine.

Rather than detection of native fluorescence, derivatization of analytes is required for high-sensitivity detection. As pointed out by Novotny, it is appropriate to choose a derivatizing reagent with spectral properties that match those of readily available lasers.[92] Table 2.5 presents a list of wavelengths produced by lasers that cost less than US$15,000, along with representative powers available at each wavelength. The new generation of helium-neon lasers is particularly valuable because of their low cost (<US$2,000), long life (many years), and interesting wavelengths. Because of the long life and modest cost of diode lasers, there is great anticipation of the development of new dyes that absorb strongly in the red and emit in the near infrared.

Lasers equipped with hard-seal mirrors usually have much longer lifetimes than lasers with interchangeable mirrors. For example, helium-cadmium lasers are available that allow access to either the 442 or 325 nm line simply by switching mirrors, but these lasers suffer from mediocre lifetime. Instead, lasers equipped with fixed, hard-seal mirrors can operate at only one wavelength but have excellent working lifetimes.

Table 2.5   Low-Cost Lasers

Laser	Wavelength (nm)	Power (mW)
Krypton-fluoride	248	10[a]
Helium-cadmium	325	10
Helium-cadmium	442	25
Argon ion	488	50
Argon ion	514.5	50
Helium-neon	543.5	1.5
Helium-neon	594	5
Helium-neon	633	10
Diode laser	650–700	50

[a] The krypton fluoride laser is pulsed up to 2000 times per second; average power is listed.

The overall cost of laser repair is so high that it appears to be cost effective to purchase several lasers, one for each wavelength desired. Similarly, fixed wavelength argon ion lasers are certain to have much longer lifetimes than lasers that contain a prism to scan a number of wavelengths. If it is anticipated that a number of lines from the argon ion laser will be useful for the particular work to be undertaken, the laser can be purchased with a hard-seal, broad-band mirror. The laser emits a few milliwatts at several discrete wavelengths across the blue and green portion of the spectrum. A single laser line can be isolated through use of an interference filter. The multiline laser tends to be more noisy than discrete wavelength lasers.

Fluorophores with strong absorbance in the range from 440 to 600 nm are most useful and common. Molecular Probes Inc. of Eugene, Oregon is an excellent source of fluorescent reagents.[93] A small sampling of readily available reagents for labeling amines is included in Table 2.6. Most of the strongest fluorophores are based on the structure of fluorescein or rhodamine. Such dyes have maximum molar absorptivities that approach 100,000 L mol^{-1} cm^{-1}. The dyes listed in Table 2.6 are all from the Molecular Probes catalog, except for the last one, which has been developed by Novotny's group.[94,95]

Isothiocyanates have received significant attention. They form derivatives with amines and offer the possibility of use as modified Edman degradation reagents. Unfortunately, the sluggish reactivity and large size of most of the derivatives have frustrated application to protein sequencing. Instead, the sulfonyl halide derivative is most useful for labeling amines.

A number of these dyes are similar to those used for labeling primers in DNA sequencing. Fluorescein, Texas red, tetramethyl rhodamine, and a modified fluorescein are the basis of

Table 2.6   Amine-Selective Labels

Label	Absorbance maximum (nm)	Molar absorptivity (L mol^{-1} cm^{-1})	Emission maximum (nm)
Fluorescein isothiocyanate	495	72,000	520
Tetramethyl rhodamine isothiocyanate	540	100,000	566
Rhodamine X isothiocyanate	580	80,000	605
Eosine-5-isothiocyanate	525	100,000	550
Erythrosine-5-isothiocyanate	535	100,000	569
Carboxyfluorescein, succinimidyl ester	490	66,000	520
Carboxytetramethylrhodamine, succinimidyl ester	545	60,000	575
7-Diethylaminocoumarin-3-carboxylic acid, succinimidyl ester	430	69,000	470
Sulfo-rhodamine sulfonyl chloride (Texas red)	596	85,000	615
Lissamine rhodamine B sulfonyl chloride	565	80,000	585
5-Dimethylaminonaphthalene-1-sulfonyl-chloride (dansyl-chloride)	340	4,500	580
p-Toluidinyl-naphthalene sulfonyl chloride	325	25,000	470
Carboxybenzoyl-quinolinecarboxaldehyde	450	?	550

the four fluorescently labeled primers used by Applied Biosystems in their sequencing protocol. The fluorescent dyes are conveniently excited with the 488 nm argon ion laser, whereas the rhodamine and Texas red dyes can be excited with either the 543.5 nm helium-neon laser or the 514.5 nm argon ion laser.

There is a strong effort to develop dyes that absorb in the red portion of the spectrum. These dyes may be excited by either the helium-neon laser at 632.8 nm or the solid-state diode lasers. An experimental, very low-power frequency-doubled diode laser ($\lambda = 415$ nm) has been used to excite fluorescein-labeled amino acids.[96]

There is one significant difficulty in labeling proteins and peptides. Because of the presence of primary amine groups on lysine side-chains, multiple labeling products are usually formed. For example, if a peptide contains M primary amino groups (M − 1 lysines and 1 N-terminal amino group), then a total of 2^M-1 possible products can be produced. For large proteins, the result of labeling will be a huge number of products, each producing a separate peak when analyzed by capillary electrophoresis. One solution to this multiple labeling problem is first to protect all primary amino groups by treatment with phenyl isothiocyanate to produce the phenyl thiocarbamyl derivative.[97] If the peptide is then treated with acid, the N-terminal amino acid is cleaved from the peptide, producing a free N-terminal amino group, which is available for labeling with an appropriate reagent. This approach replaces a positively charged ε amino group with a positively charged phenyl carbamyl group; the overall charge of the protein remains the same, and solubility is not expected to be significantly affected. As an alternative, the underivatized analyte can be separated by capillary electrophoresis and derivatized in a post-column reaction cell. Sheath flow designs have been used for the post-column detectors and an on-column "T" has been constructed to introduce derivatizing reagent during the separation.[98-100] Inevitably, the slow rate of reaction and post-column band broadening discourage application of post-column derivatization.

F. Electrochemical Detection

1. Conductivity Detection

Conductivity detection is an indirect detection technique that produces a universal response. Several groups have reported on conductivity detection, including Everaerts' early report on capillary electrophoresis.[6,7] Detection limits were in the sub-picomole range for simple anions, and a 16-component mixture was resolved in nine minutes. Another on-column conductivity detector used a casting technique to make a detection cell.[101]

Conductivity detection is difficult in capillary electrophoresis because of the presence of high potentials. The high-potential gradient along the capillary axis results in formation of a large DC potential between the two electrodes in contact with the capillary. This potential difference between the electrodes disappears if the two electrodes are carefully aligned to be opposite each other across the capillary, i.e., at precisely the same point along the capillary length. Zare's group has used a carbon dioxide laser to drill small holes in 50-μm inner diameter capillary.[102] Fine platinum wires were inserted into the holes, providing electrical contact with the buffer solution (Figure 2.23). The current flowing between the electrodes in response to a small AC voltage is proportional to the conductivity of the solution. Detection limits were in the sub-femtomole range for lithium, with a three order of magnitude linear dynamic range. Simplifications to the system have resulted in excellent performance in a simple, low-cost detection system.[103,104]

This detector measures small changes in the conductivity of the sample. It is therefore advantageous to use a low ionic strength, poorly ionized buffer system to produce a low conductivity background. However, construction of routine, simple conductivity detectors appears to be quite challenging because of the small dimensions of the capillary and the high separation field.

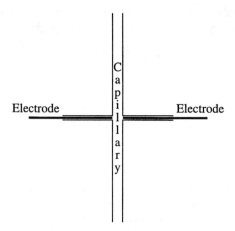

Figure 2.23 On-column conductivity detection.

2. Amperometric Detection

Electrochemically active analyte can be detected with high sensitivity through use of amperometric detection. At first glance, these measurements would appear to be impossible since the high electric field associated with the separation would be expected to interfere with the electrochemical measurement of minute currents. However, clever electrical isolation techniques have been applied to allow electrochemical measurements at ground potential.

Ewing's group has used a fractured capillary to create a section of field-free capillary (see Figure 2.24).[105,106] The fracture is covered with a piece of porous glass tubing, which is placed in a buffer solution held at ground potential. The section of capillary after the fracture is held at ground potential, and the flow of analyte is driven by electroosmosis. A short piece of field-free capillary is necessary to avoid excessive band broadening due to laminar flow, and with care, plate counts approaching 10^6 have been reported. A fine carbon-fiber working electrode is introduced into the field-free capillary using a micromanipulator to serve as the amperometric detector. Production of the porous-glass junction apparently is quite tedious. More recently, other workers have inserted the working electrode immediately at the exit of the capillary. Flow of analyte from the capillary tip then allows measurement in the field-free region in the buffer surrounding the capillary tip.

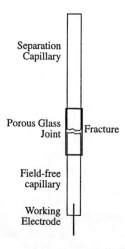

Figure 2.24 Amperometric detection.

Detection limits for the system are in the order of one attomole of electrochemically active analyte. As with fluorescence, electrochemical detection provides an important degree of specificity, which is particularly valuable when analyzing natural products, including the content of individual cells.

G. Radioactivity Detection

Several groups have described detection of radioactively labeled analytes that have been separated by capillary electrophoresis. Pentoney et al. used both a cadmium telluride solid-state detector and a plastic scintillator placed next to the capillary wall.[107] Decay from ^{32}P-labeled analytes, with sub-femtomole detection limits, were reported. This experiment is technically difficult because of the long decay time of phosphorus compared with the very short transit time past the detector window, so that relatively few decays are expected during that short time window. To increase the collection time, the electric field was stopped when analytes of interest were present in the detection chamber. After sufficient decay occurred, the electric field was reapplied and the separation continued until the next analyte of interest eluted. Clearly, this approach requires use of standard compounds with well-known retention times or a predetection sensor upstream from the detection chamber. Altria et al. have reported a similar system based on a TlI doped NaI scintillation counter for the very sensitive detection of compounds labeled with ^{99m}Tc.[108]

H. Mass Spectral Detection

The possibility of analyzing sub-picomole quantities of complex biomolecules, of molecular masses in excess of 200,000 Da, using efficient ionization/desorption techniques such as matrix-assisted laser desorption ionization (MALDI) and electrospray ionization (ESI), has greatly expanded the potential of mass spectrometry into the fields of recombinant technology, immunology, and protein chemistry. Continuing improvements in the sensitivity, speed of analysis, mass accuracy, and structural information obtainable, render the mass spectrometer as one of the most flexible and powerful detectors available for CE. In view of the increasing sample complexity and the requirement for obtaining structural information from limited amounts of material, mass spectral detection offers unparalleled advantages compared to single parameter detection systems. The potential of combined capillary electrophoresis-mass spectrometry (CE-MS) was quickly realized following the first reports on the analysis of vitamins, amino acids, dipeptides, and more complex peptides.[57,109,110]

In the past five years, there have been significant developments in CE-MS both in instrumentation and applications. An impressive number of publications have reported the use of CE-MS for the analysis of a wide variety of analytes ranging from inorganic ions to complex biomolecules such as proteins and nucleic acids. Several papers[111-113] and book chapters,[114,115] including the first edition of this book, have reviewed the coupling of CE and MS. Although different ionization techniques have been described in CE-MS applications, ESI remains the preferred method for direct coupling. However, the electrophoretic conditions providing adequate analyte resolution are sometimes incompatible with the operation of ESI mass spectrometry. Involatile buffers that are commonly used in CE separations, such as phosphate and borate, contribute significantly to chemical background problems. off-line methods which involve analyte collection and subsequent mass spectral analysis of the individual fractions thus offer a means of alleviating these difficulties. The merits and disadvantages of the off-line CE-MS coupling approach will be presented in Section J. The present section describes the different interfaces used for direct coupling of CE to mass spectrometry. Instrumental developments which led to electrophoretic conditions compatible with the use of different CE formats will also be presented.

Capillary electrophoresis by its nature is particularly well suited to the analysis of polar analytes with ionizable functionalities. These types of molecules have traditionally presented problems to the mass spectroscopist using conventional ionization techniques such as electron impact, whose range of application to organic compounds is limited by sample vaporization and thermal degradation of labile analytes. Over the last decade, significant developments in ionization techniques have enabled the formation of ionized molecules with minimum thermal damage. These methods have utilized two general processes: (1) formation of ions in the sample matrix prior to evaporation or desorption, and (2) rapid vaporization, e.g., by very rapid heating or sputtering by energetic particles, prior to ionization.[116] These ionization techniques include plasma desorption mass spectrometry (PDMS), matrix-assisted laser desorption mass spectrometry (MALDI), fast atom bombardment (FAB), thermospray (TSP), and atmospheric pressure ionization (API) techniques such as electrospray (ESI) and atmospheric pressure chemical ionization (APCI). All of these ionization techniques except TSP make use of ionization in, or from, the condensed phase and rely on the transport of sample ions or ion molecule complexes into the mass analyzer without exposure to high temperatures. Of these ionization techniques, continuous flow FAB (CF-FAB) and ESI have been used most successfully in direct CE-MS analysis, although some reports using APCI have been presented recently.[117] The various ionization techniques are briefly reviewed below.

1. Ionization Techniques

a. Electrospray and Ionspray Ionization

The analytical use of ESI was first described more than 20 years ago by Dole and co-workers.[118] However, the major developments of ESI were made by the research group of Fenn at Yale,[119,120] whose report of the formation of multiply charged ions using ESI first alerted the mass spectrometry community to the ability of ESI to dramatically extend the effective mass range of current mass spectrometers. In the case of ESI, a phenomenon was actually described by Zeleny in 1917[121] in which nebulization takes place by the exposure of a liquid surface to a high electric field. In an ESI source the analyte solution is introduced into dry air or nitrogen at atmospheric pressure through a metal capillary tube that is held at a potential of several kV relative to the walls of the ion source. The buildup of charges at the liquid surface creates such an instability in the liquid that coulombic repulsion forces are sufficient to overcome the surface tension, so that small (<1 μm) charged droplets separate from the liquid emerging from the capillary tube.[122] The technique works best with flow rates in the order of 1 to 10 μL/min. At flow rates higher then 10 μL/min the droplets become less uniform in size and the spray becomes unstable, leading to loss of ionization efficiency. In addition, the dispersion of liquids by electrical forces alone, as in the case of ESI, becomes more difficult if the percentage of water in the eluate is high. An efficient means of accommodating higher flow rates of up to 200 μL/min was achieved using a sheath nebulizer gas,[123] based on the earlier work of Iribarne and Thomson.[124,125] This nebulizer-assisted electrospray interface is often referred to as ionspray,[123] and for simplicity the acronym ESI will be used in this section to describe electrospray ionization with or without assisted nebulization. An excellent review on ESI has been presented by Kebarle and Tang.[126] Irrespective of how the electrically charged aerosol is formed, evaporation of solvents from the small charged droplets then takes place. As the size of the droplets decreases, the electric field at the liquid surface of the droplets increases, and gaseous analyte ions are produced from preformed ionic species in solution. In positive ion mode, this ionization technique gives rise to multiply protonated molecules $[M + nH]^{n+}$ where n assumes several integral values giving a series of related ions, or to adduct ions obtained from attachment of sodium or ammonium ions. Deprotonated molecules $[M - nH]^{n-}$ are typically observed in negative ion mode.[127]

b. Continuous Flow Fast Atom Bombardment

The application of FAB to the analysis of biomolecules was first demonstrated by Barber and co-workers.[128] The sample is dissolved in a viscous organic liquid matrix (e.g., glycerol) which is subsequently bombarded by primary beams of neutral atoms (Xe) or ions (Cs[+]), having a translational energy of up to 30 keV. This produces an intense thermal spike whose energy is dissipated through the outer layers of the sample lattice. Molecules are detached from these surface layers as a dense gas which is then ionized in the plasma just above the sample surface to produce mostly protonated or deprotonated molecules ([M + H][+] and [M − H][−]), or adduct ions of alkali salts or matrix. In order to reduce the chemical background contribution characteristic of standard or "static" FAB and to provide an alternative LC-MS interface, the ionization technique known as continuous flow fast atom bombardment (CF-FAB) was introduced.[129-132] Basically, in CF-FAB a fused-silica capillary transfer line is used to introduce a constant stream of solvent directly into the mass spectrometer ion source. The permissible flow rate of the solvent/matrix stream entering the source is determined by the vacuum system of the mass spectrometer and the proportion of the matrix. Flow rates of 1 to 5 μL/min are typical for solvent systems containing between 2 and 25% of a matrix such as glycerol. The target is generally heated (~40 to 50°C) to prevent evaporative freezing and to provide stability to the ion current. A wick can also be incorporated to remove excess buildup of matrix on the CF-FAB probe tip.[132] The matrix is usually delivered via a post-column arrangement, and both the liquid junction and the coaxial CE-MS interfaces (see following section) have been used with this ionization technique. Spectra generated by CF-FAB are simpler than those obtained by static FAB, and usually provide improved sensitivity.

2. CE–MS Interfaces

Reliable CE-MS interfaces, with either CF-FAB or ESI, usually require a "make-up" buffer to maintain optimal ionization efficiency and electrical contact at the end of the CE column, and to supplement the low electroosmotic flow for efficient transport of the analyte zone to the ion source. The buffer/solvent system used for the make-up flow should be compatible with the CE buffer and should minimize detrimental effects on the electroosmotic flow. Unfortunately, many of the involatile buffers that are commonly used in CE separations, such as phosphate and borate, are generally not suitable for CE-MS and often suppress the ionization of the analyte, yielding poor mass spectral sensitivity.

A major consideration in the design of a CE-MS interface based on a conventional low-pressure ion source is the minimization of vacuum effects associated with pressure differences between the ends of the capillary. This is important not only to minimize the volumetric flow entering the ion source, but also to retain the high separation efficiency of CE. Pressure-driven parabolic flow profiles, which can be obtained when isobaric conditions are not maintained (i.e., when low pressure sources or high nebulizer gas flow rates are used) can be a major source of zone broadening and loss of separation efficiencies.[133]

Three main approaches to interfacing CE to mass spectrometry have emerged in recent years. The first uses a coaxial capillary configuration with the CE column inside a sheath capillary through which the make-up buffer is added concentrically. In a second approach, the make-up solution is added through a "liquid junction," where the CE column is connected via a tee- or cross-junction to a transfer line which is in turn coupled to the ion source of the mass spectrometer. A make-up flow of a suitable buffer (containing a matrix such as glycerol in the case of CF-FAB) is added at the liquid junction. More recent investigations have presented a third interface termed "sheathless," where no liquid is supplemented to the CE flow and the electrical continuity is maintained through the use of metalized CE tips. The design and application of the various interfaces employing these approaches will be discussed in more detail below.

a. Coaxial Configuration

Electrospray Ionization — The pioneering work in the coupling of CE with mass spectrometry was performed by Smith and his co-workers,[57,134] who coupled CE with a quadrupole mass spectrometer using an ESI interface based on the LC-MS interface described by Whitehouse et al.[135] This first generation of CE-MS interface relied on the electroosmotic flow to provide adequate electrospray conditions.[134] Although the authors reported good separation efficiencies for amenable compounds (620,000 plates for a tetrabutylammonium ion), they clearly identified the limitations of the approach used with respect to the minimum flow required to maintain a stable electrospray ionization and the influence of the buffer composition and concentration on the intensity of the chemical background and mass spectral sensitivity.[57]

Having recognized the drawbacks of the earlier designs, Smith and co-workers developed a coaxial interface in which the metal contact at the CE terminus was replaced by a thin sheath of flowing liquid.[136] The CE column was introduced inside the stainless steel capillary and terminated at the probe tip. The sheath liquid flows inside a concentric tubing arrangement in the interspace between the CE column and the electrospray needle. The introduction of this sheath liquid ensures good electrical contact for efficient electrospray ionization while maintaining continuity of the voltage gradient across the CE capillary. Acetonitrile, methanol, or 2-propanol, containing small amounts of the CE buffer, were used as the sheath liquid at flow rates between 5 and 10 µL/min. Using this interface this group has published a series of papers[136-143] demonstrating its improved utility, including the first coupling of capillary isotachophoresis with mass spectrometry.[140,141]

The construction of such ESI probes was elaborate and comprised several layers of concentric tubings. Simpler and more flexible designs have been introduced which allow both the coaxial and liquid-junction approaches to be utilized.[144] A schematic representation of a simplified coaxial capillary arrangement for ESI and nebulizer-assisted ESI (ionspray) is presented in Figure 2.25. Briefly, the interface comprises two tees; the front one is used to introduce the nebulizer gas at a flow rate of typically 2 L/min, while the back tee provides a continuous stream of solvent to the interface. The stainless steel electrospray needle (27 gauge for 180 µm o.d. capillary) traverses the front tee and is inserted in the back tee in such a way that the sheath liquid can be introduced directly into the needle. The independent delivery of the make-up flow facilitates the optimization of the CE-MS interface, and provides a convenient means of calibrating the mass spectrometer and the injection volume. More recent coaxial interface designs have incorporated a retractable shutter arm activated by contact relay.[54] The controllable action of this arm avoids the accumulation of involatile residues from column washings and helps to maintain mass spectral sensitivity over the longer term.

Several important experimental parameters such as sheath buffer composition and flow rates, capillary dimensions, and CE electrolytes can influence the sensitivity of the coaxial CE-MS interface. The effects of capillary and electrospray needle dimensions on the stability and sensitivity of the coaxial CE-MS interface was investigated recently by Tetler and co-workers.[145] An improvement of 20 to 30% in ion current signal was obtained between CE columns of 180 µm and 360 µm outer diameters (75 µm i.d.), although these capillary configurations led to significant changes in ease of operation and optimization. The use of smaller i.d. capillary can also result in higher absolute signal intensity, even if smaller sample loadings are required to maintain comparable separation efficiencies. For capillary diameters ranging from 10 to 100 µm i.d., Wahl et al. found a decrease of only a factor of 2 to 4 in absolute signal intensities when the sample size was decreased by two orders of magnitude.[146]

The sheath buffer is usually composed of aqueous solutions of organic solvent with acidic or basic buffer modifiers to promote ionization in positive or negative ion modes. In the author's laboratory, a sheath buffer of 25% methanol and 0.2% formic acid in water has been found to be of general application in positive ion mode detection, while an aqueous solution of 10 mM ammonium formate in 25% methanol pH 9 is typically used for negative ion ESI.

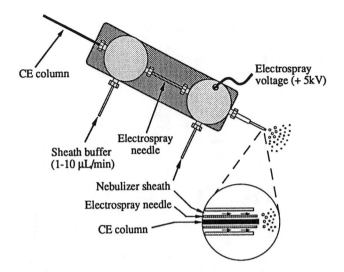

Figure 2.25 Schematic representation of the capillary electrophoresis–electrospray mass spectrometry interface using a coaxial sheath flow.

However, the sheath composition can have detrimental effects on the separation performance and analyte migration order, especially in situations where the electroosmotic flow is reduced. In positive ion mode, when the electrospray needle acts as the cathode, both the liquid sheath and the CE electrolyte anions migrate toward the anode. A moving boundary can be established if the sheath and CE electrolyte anions differ, and the migration rates of the analytes can change when they come in contact with this boundary.[147]

A large number of applications of the coaxial CE-MS interface using ESI have been presented in the past five years, and probably reflect its wider acceptance and ease of operation compared to other interface configurations. Most of these applications have used CZE separation formats, although recent developments have described the use of CIEF and MEKC in combination with ESI mass spectrometry. These applications include the analysis of small inorganic ions,[148,149] dyes,[105,151] toxins,[152-155] pharmaceuticals and drug metabolites,[155-160] chemical warfare agents,[161] peptides,[137-141,145,146,162-167] proteins,[138,141,147,168-171] drug conjugates,[172-174] complex carbohydrates,[175,176] and nucleotides.[177] An example of a CZE-MS separation using the coaxial interface is presented in Figure 2.26 for the analysis of CNBr peptides from ovalbumin.[54] This glycoprotein has one glycosylation site at Asn_{292}, but comprises a mixture of high mannose and hybrid oligosaccharides appended to the protein core. Selective identification of the glycopeptides from this complex digest is achieved by promoting the formation of oxonium fragment ions (m/z 163 and 204), characteristic of N-linked carbohydrates, using higher orifice voltage.[178] The microheterogeneity of the glycan composition is observed in the selected ion monitoring (SIM) profile (Figure 2.26 A) by a series of closely spaced peaks migrating between 16 and 18 min. The CZE-MS analysis obtained under full scan acquisition (Figure 2.26 B) enabled detection of all peptides for an injection of only 2 pmol of ovalbumin digest. An example of an extracted mass spectrum is presented in Figure 2.26 C for the glycopeptide observed at 17.2 min in Figure 2.26 B. The ESI mass spectrum shows abundant multiply protonated ions at m/z 887 and 1329 and the calculated molecular weight (Mr 2658) is consistent with the glycopeptide $CB_{288-298}$ having the high mannose structure $GlcNAc_2Man_6$. This example illustrates the advantages of combined CZE-MS analysis for the resolution of individual glycoforms and selective identification and characterization of peptides and glycopeptides present in complex digests.

The application of CZE-MS has also been demonstrated in several quantitative studies.[154,160] Henion et al. have demonstrated the use of a benchtop ion trap for the determination

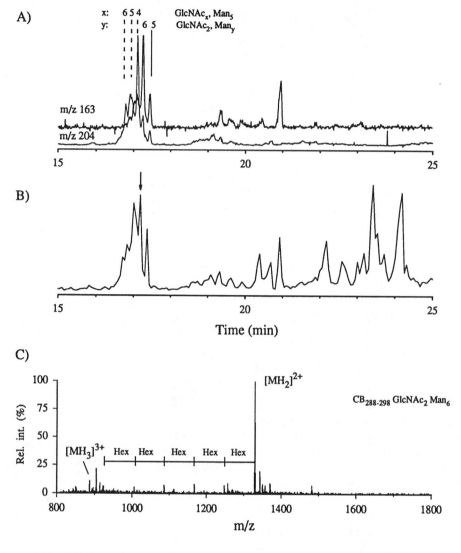

Figure 2.26 CZE-MS analysis of peptides arising from CNBr cleavage of ovalbumin using a coaxial interface. (A) Analysis performed under SIR mode using an orifice voltage of 100 V. (B) Total ion electropherogram for a narrow scan acquisition from m/z 800 to 1800 (orifice voltage 50 V). (C) Extracted mass spectrum taken at 17.2 min peak indicated by arrow in B. Conditions: 1.1 m × 50 μm i.d. capillary coated with 5% polybrene, 2% ethylene glycol, electrolyte: 2.0 *M* formic acid, polarity reversal with −30 kV, 2 pmol injection.

of alkaloids from tree barks and Chinese herbal medicines.[160] Full mass scan spectra were obtained for injected quantities as low as 370 attomoles of quinoline alkaloids, and internal standardization curves were demonstrated for injection levels of 0.6 to 16 pg with excellent linearity. Similarly, Bateman et al. compared the application of LC-MS and CZE-MS for the quantitation of microcystins in toxic strains of *Microcystis aeruginosa*.[154] Detection limits of 0.2 μg/mL (4 fmol injection) were obtained using CZE-MS, whereas the LC-MS method yielded a concentration detection limit of 0.05 μg/mL (500 fmol injection). A marked advantage of CZE over LC was its ability to resolve different desmethyl microcystin-LR positional isomers.[154] Perkins et al.[179] have compared the performance of packed-microcapillary column liquid chromatography (μLC) and CZE coupled with ESI, using a quadrupole mass spectrometer, for the analysis of sulfonamide antibiotics. The authors concluded that, although the narrower peaks in CZE-MS provided better detection limits than the μLC system, this was compensated for by the reduced sample loading capabilities of the latter.[179]

Continuous Flow Fast Atom Bombardment — The coupling of CE to CF-FAB mass spectrometry using a sheath flow CZE-MS interface was first described by Tomer and co-workers[52,110,167,180] and was based on earlier designs using packed-microcapillary (<75 μm i.d.) and open tubular (<10 μm i.d.) liquid chromatography (OTLC).[181,182] However, the requirements for CZE-MS have been found to be more stringent. With a 1-m-long, 75 μm i.d. CE capillary a vacuum induced flow of approximately 4 μL/min was reported,[133] and reduction of the column diameter to 12 to 15 μm was necessary to minimize the effects on separation efficiencies. The design of this CE-MS interface is similar to that described in Figure 2.25, except that the back tee was mounted at the end of the CF-FAB probe and was used to deliver the FAB matrix at a flow rate of 2 to 5 μL/min. The FAB matrix used in these studies contained either heptafluorobutyric acid (pH 3.5) to aid the detection of positive ions, or ammonium hydroxide (pH 9.0) for negative ion mode.

The possibility of independently varying the composition of the sheath buffer without affecting the separation has practical implications for the analysis of peptides and proteins. The amphoteric nature of these compounds dictates that they can be separated either in their anionic or cationic forms depending on the electrophoretic buffer chosen. Although good S/N ratios can be obtained in SIM mode for loadings of 100 fmoles of selected tripeptides, the negative ion sensitivity was approximately one order of magnitude less than that observed for positive ion detection.[52,110] In this case, peptides separated as anions using basic buffers were detected in positive ion mode using a FAB matrix composed of 25% aqueous glycerol pH 3.5. In SIM acquisition mode, efficiencies ranging from 220,000 to 480,000 plates were obtained for loadings of 60 to 75 fmol per component.

An integral probe for CZE-CF-FAB that incorporates features of both the coaxial and liquid-junction interfaces was presented by Suter and Caprioli for the analysis of synthetic peptides.[183] The sensitivity and performance of the coaxial CZE-MS interface were compared using ESI and CF-FAB ionization techniques.[184] The combination of CZE-MS with ESI was found to be superior to CF-FAB in view of the flexibility of using larger diameter capillaries (>15 μm i.d.) and the possibility of detecting higher molecular weight proteins via the observation of multiply charged ions.[184] In a separate study using the liquid junction interface (see next section), both ESI and CF-FAB yielded similar sensitivities for the analysis of aromatic sulfonates, quaternary ammonium compounds, and peptides, although the CF-FAB system resulted in increased peak broadening and was less amenable to routine use.[185]

b. Liquid Junction

Electrospray Ionization — The use of a liquid-junction in conjunction with ESI has been described by Henion and co-workers at Cornell.[185-191] A schematic diagram similar to that reported by this group is shown in Figure 2.27. The most significant change in design compared to the sheath flow CE-MS interface (Figure 2.25) is the configuration of the back tee. In this arrangement, a larger hole has been bored out from a standard 1/16-inch stainless steel tube, and the tee is sealed with a glass lens, thereby allowing adjustment of the capillaries inside the junction. The CE column terminates inside the liquid junction tee, approximately 10 to 20 μm from the transfer line or electrospray needle. The make-up fluid is supplied from a small reservoir and is drawn into the junction by a combination of gravity and Venturi vacuum due to the nebulizer gas introduced from the front syringe tee. Alternatively, a syringe pump can be used to introduce the make-up solution at 1 to 2 μL/min. An alignment tubing can be incorporated into the junction to minimize post-column peak broadening. The transfer

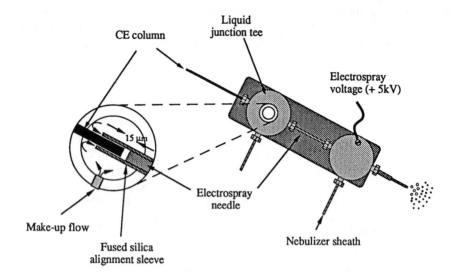

Figure 2.27 Schematic representation of the capillary electrophoresis–electrospray mass spectrometry interface using a liquid-junction.

line dimensions must be selected to avoid pressurization of the junction which could lead to a counter hydrodynamic flow in the CE column. A pressure of 20 to 40 mbar at the inlet capillary can be necessary in some configurations to counterbalance this effect.

This interface was initially applied to the separation of leucine enkephalin and other dynorphins[187] and has also been used to examine mixtures of synthetic[189] and tryptic peptides.[186] However, Henion's group has also been one of the most active in applying CE-MS not only to the detection of small molecules,[186,188,190] but also to their quantitative determination in biological and environmental matrices.[188,190] Mass spectral detection was achieved in either positive or negative ion modes. Separation efficiencies as high as 300,000 theoretical plates were reported for peptides using the liquid junction arrangement. In what could be a major advantage of the liquid-junction over the coaxial design, Garcia and Henion[191] demonstrated separations of dansylated amino acids, polyacrylic acids, aromatic sulfonates, and anionic surfactants using polyacrylamide gel-filled capillaries. Although modest separation efficiencies (50,000 to 100,000 plates) were achieved even in the presence of high concentrations of urea and Tris borate buffers, this application represents thus far the only demonstration of CGE in combination with mass spectrometry.

In spite of the distinct advantages of the liquid junction, this interface has not received a wide acceptance among CE-MS practitioners. One of the major limitations of this design compared to the sheath flow is the contamination of the liquid junction through the column washings required to regenerate the CE capillary inner surface. This obviously has implications for the overall sensitivity of the analysis. In a critical comparison of the coaxial and liquid junction interfaces, Pleasance et al. evaluated the ruggedness, ease of use, sensitivity, and electrophoretic performance of these two CE-MS interfaces.[144] A comparison of electropherograms obtained using these two interfaces is presented in Figure 2.28 for the analysis of five antibiotics.[144] The analyses were performed on the same day, under identical CZE conditions. The most obvious difference between the two CZE-MS techniques is the consistently longer migration times in the case of the liquid junction (Figure 2.28 B). This arises as a result of the additional length of transfer capillary (8 cm) and the discontinuity in the composition of the separation and liquid junction buffers. Further differences are also observed in terms of separation efficiency and sensitivity. The calculated theoretical plates for these antibiotics, analyzed using the liquid junction (Figure 2.28 B), were typically 60 to 120% lower than

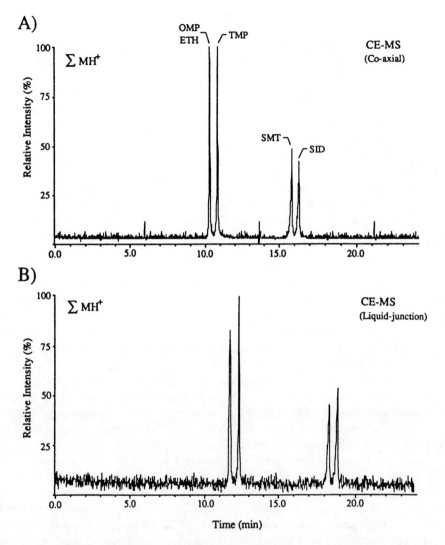

Figure 2.28 Separation of antibacterial drugs by CZE-MS using (A) coaxial and (B) liquid junction interface configurations. Conditions: 0.9 m × 50 µm i.d. bare fused silica, electrolyte: Tris pH 7.2, 25 kV, 40 nL injection. ETH, erythromycin, OMP, ometoprim, TMP, trimethoprim, SMT, sulfamethazine, SID, sulfaisomidine (From Pleasance, S., Thibault, P., and Kelly, J. F., *J. Chromatogr.*, 591, 332, 1992. With permission.)

those observed for the sheath flow interface (Figure 2.28 A). Consideration of the expected additional band broadening with the inclusion of the liquid-junction transfer capillary does not account for all the observed loss in plate number, and this was attributed to the junction itself.[144] This observation was supported by other workers using fluorescence detection, who reported a tenfold loss in plate number with the inclusion of a similar liquid junction.[192]

Continuous-Flow Fast Atom Bombardment Mass Spectrometry — The use of a liquid-junction interface for CE-MS was first reported by Minard et al.[193] using CF-FAB. Based on an earlier design for coupling other post-column detectors with CE, this group constructed the flow make-up interface by carefully cementing the ends of the CE capillary and a CF-FAB transfer line to a glass slide, leaving a gap of approximately 20 µm where the FAB matrix (aqueous methanol or acetonitrile containing 4 to 20% glycerol) was introduced. This

CF-FAB liquid junction was subsequently modified and refined by the same authors[194] and other groups.[192,195,196] The main changes in the design of the junction involved easier alignment of the two capillaries and improvement of its robustness. The design utilized by Reinhoud et al.[192,196] was machined from a Plexiglas® block for convenient viewing and allowed adjustment of the spacing between the CE capillary and the transfer line by means of retaining nuts and ferrules. The make-up liquid was introduced at a flow rate of 5 to 8 μL/min to the CF-FAB probe and consisted of 10 to 16% glycerol in water with 0.25% trifluoroacetic acid to improve conductivity. The same authors evaluated the degree of separation efficiency before and after the liquid junction and found that the plate counts were on average ten times lower after the junction. This was illustrated with o-phthalaldehyde derivatives of amino acids for which plate numbers of 300,000 were obtained before the liquid junction, versus 30,000 post-junction. This loss of performance was attributed to diffusion of the narrow analyte band during its transport to the CF-FAB probe. The separation of 25 pmol of β-endorphins was demonstrated using a double-focusing mass spectrometer using a narrow mass scan.[192]

Caprioli and co-workers[195] have developed a liquid-junction interface that incorporates some of the features of a coaxial design. The Plexiglas interface consists of a cross where a stainless steel cathode is introduced on top, while the lower arm comprises a PTFE tube which supplies the CF-FAB matrix solution (5% glycerol, 3% acetonitrile). A 10-mL syringe is connected to the cathode tubing and is used for injection by drawing the sample solution through the column (the authors indicate a 2 mL draw would introduce a 15-mm slug of sample corresponding to about 30 nL). The junction itself has a much lower dead volume than the design by Reinhoud et al.,[192] and no significant band broadening was reported. Suter and Caprioli have described an integrated probe for CE-MS using CF-FAB ionization in which a capillary may be either positioned at the entrance of a transfer capillary (liquid junction) or inserted into a larger sheath capillary (coaxial).[183] Using this integrated probe, these workers undertook a comparison of the two interfaces using a simple peptide mixture. Better performance was obtained using the coaxial CF-FAB approaches due primarily to a 75-fold reduction in dead volume. However, they indicated that the liquid junction provides several advantages over the coaxial CF-FAB interface, including lower vacuum-induced flow and use of larger and less brittle capillaries for on-line UV detection.[183] The application of the liquid junction was also demonstrated for the analysis of DNA adducts formed by covalent bonding of polycyclic aromatic hydrocarbon metabolites.[197] More recently, Wolf and Vouros incorporated sample stacking techniques to the same CF-FAB liquid-junction interface to improve concentration detection limits of the acetylaminofluorene deoxyguanosine 5′ mono-phosphate adduct by up to 3 orders of magnitude over that obtainable using a conventional CZE format.[198] The subject of enhanced sample loadings for CZE-MS experiments will be described in Section 3 below.

c. Sheathless Interface with Microelectrospray Ionization

The addition of make-up or sheath buffer for the liquid junction and the coaxial interface described above is required not only for efficient transport of the analyte to the ionization source, but also to maintain good electrical contact between the CE capillary and the ESI or CF-FAB probe tips. In view of the fact that the analyte band is diluted in a larger volume, and often dispersed as a divergent ion beam in the ionization source, these effects can lead to significant decreases in sensitivity. Previous investigations have reported improvement in sensitivity for the coaxial configuration using smaller CZE columns of 5 to 10 μm i.d., which also provided substantial reduction of the chemical background contribution.[146,171] However, recent advances in the fabrication of micro ESI tips have enabled the reduction of the ESI flow rate to the sub-microliter/min regime, and provided significant gains in sensitivity and increases in the signal-to-noise ratio.[199-202] Under these conditions, a focused Taylor cone is obtained for flow rates of 25 to 100 nL/min using a microelectrospray ion source. A theoretical description of the electrospray dispersion was presented in an excellent paper by Wilm and Mann.[202] In their model, the radius of the zone at the tip of the Taylor cone from which droplets are ejected, r_e, is defined as

$$ r_e \left(\frac{\rho}{4\pi^2\gamma \tan\left(\frac{\pi}{2}-\alpha\right)\left[\left(\frac{U_a}{U_t}\right)^2 - 1\right]} \right)^{1/3} \times \left(\frac{dV}{dt}\right)^{2/3} \tag{11}$$

where γ is the surface tension of the liquid, α the liquid cone angle, ρ the density of the liquid, U_t and U_a the threshold and applied voltages, and dV/dt the flow rate. For a given capillary dimension and solvent system, the emission diameter is proportional to the 2/3 power of the flow rate dV/dt. For an aqueous methanol solution flowing at 25 nL/min they calculated that r_e was 88 nm. The ionization/transmission efficiency for a peptide of 1800 Da introduced to the electrospray source at this flow rate was calculated to be 8×10^{-4}.[202] The focused plume of ions emerging from the Taylor cone results in a significant gain in sensitivity compared to that achieved with the conventional ESI source, and low attomole/μL sensitivity is typically achievable with these microsprayers.[201] In order to maximize mass spectral sensitivity, minimization of the droplet size to 200 nm or below is desirable. This is achieved by selecting capillaries of small diameters (<5 μm i.d.), which also allow reduction of the onset voltage U_t.

The combination of microelectrospray with CE is further complicated by the electrophoretic requirements for obtaining the desired separation. There is therefore an interplay between the operational parameters regulating the direction and magnitude of the electroosmotic flow and the capillary dimensions required for obtaining stable and efficient electrospray emitters. The development of a sheathless CE-MS interface using microelectrospray has been presented by different groups using tapered capillary tips coated with gold[203,204] or conductive epoxy adhesives.[205,206] An alternate approach to using microsprayers was proposed by Fang et al., in which the electrical contact was made by inserting a gold wire of 20 μm o.d. inside the CE capillary.[207] A schematic representation of these two sheathless CE-MS designs is presented in Figure 2.29. The sheathless interface using microelectrospray tips (Figure 2.29 A) is usually prepared from a single piece of fused-silica capillary, of which one of the ends has been tapered to give an outer diameter of 100 to 150 μm and an inner diameter of 10 to 30 μm. The surface of the tapered end is coated with conductive adhesives[205,206] or by gold sputtering deposition.[203,204] The reagent (3-mercaptopropyl)trimethoxysilane can be applied to the fused silica prior to gold deposition to extend the durability of the gold coating.[203] A variation of this design was proposed by Wahl and Smith,[206] in which a small fracture (pin hole) covered with gold adhesive was introduced at the electrospray tip. In the second sheathless interface (Figure 2.29 B), the polyimide coating of the fused-silica capillary is removed, and the tip is tapered to a diameter sufficient to introduce a 20 μm o.d. gold electrode.[207]

A comparison of the sheathless and coaxial interfaces revealed an improvement in sensitivity of up to tenfold using the sheathless design.[204,208] This improvement was associated with increased analyte signal intensity and reduction of background noise. For a mixture of peptide standards ranging from 500 to 3500 Da, the limit of detection was typically in the mid to low attomole range for SIR acquisition mode and low femtomole range for full mass scan acquisition.[204] While the coaxial capillary interface provides ease of operation and

A)

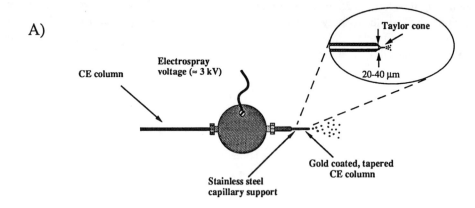

B)

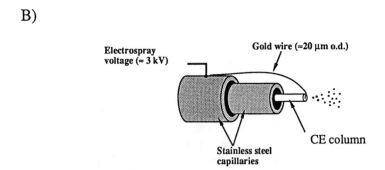

Figure 2.29 Schematic representation of the sheathless capillary electrophoresis–electrospray mass spectrometry interface using gold-coated capillary tips (A), and a thin gold wire electrode (B). (Adapted from Fang et al., *Anal. Chem.*, 66, 3696, 1994.)

simplicity of capillary preparation, future developments will likely focus on the sheathless design in view of its improved sensitivity.

The application of such a CE-MS interface was demonstrated for the analysis of complex proteolytic digests of cytochrome c using a 10 μm i.d. capillary.[205] This is illustrated in Figure 2.30 for the analysis of tryptic peptides from bovine, *Candida krusei*, and equine cyto-chrome c. The total ion electropherogram for the corresponding digests is presented in Figures 2.30 A–C, and extracted ion electropherograms taken from the same analyses are shown in Figures 2.30 D–F, respectively. These separations were obtained for the injection of approx-imately 30 fmol of the original protein. The CZE-MS analyses were conducted using a 50 cm × 10 μm i.d. capillary that was previously treated with 3-aminopropyltrimethoxy silane (APS). The CZE separation was conducted using polarity reversal conditions with a 10 mM aqueous ammonium acetate buffer pH 4.4 The electrophoretic conditions yielded fast electroosmotic flow and each analysis was completed in less than 6 min, although peak resolution was not achieved in many cases. However, individual tryptic peptides can be extracted from the relevant traces presented in Figures 2.30 D–F, and these confirmed the presence of a tryptic fragment common to all YIPGTK at m/z 678, along with the tryptic peptide EDLIAYLK specific to bovine and equine cytochrome c. It is noteworthy that most of the sheathless CZE-MS analyses presented thus far have used APS-coated capillaries, and higher separation efficiencies can be obtained using more acidic buffers and capillaries coated with linear polyacrylamide-bearing quaternary amine groups.[204]

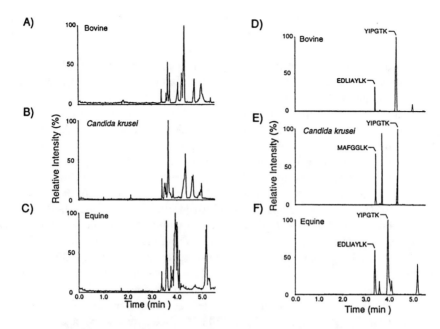

Figure 2.30 CZE-MS separations of tryptic digest of bovine, *Candida krusei,* and equine cytochrome c. Total ion electropherograms for narrow mass scan of m/z 600 to 1200 (A–C), summed extracted ion electropherograms for m/z 964, 678, and 723 for tryptic digest of bovine (D), *Candida krusei* (E), and equine (F) cytochrome c. Conditions: 0.5 m × 10 μm i.d. aminopropyltrimethoxy silane-coated capillary, 10 mM ammonium acetate-acetic acid pH 4.4, polarity reversal with −300 V · cm⁻¹, injection for 5 s at −1 kV. (From Wahl et al., *J. Chromatogr.,* 659, 217, 1994. With permission.)

3. Instrumental Developments

Over the past five years, significant progress has been made in the construction of more rugged CE-MS interfaces and in their applications to a wide variety of biological problems. While the previous section described different approaches for efficient CE-MS coupling, the present section will focus on emerging techniques which should broaden the use of mass spectrometry for on-line CE detection. In particular, the following discussion will present recent applications of CE-MS where special electrophoretic conditions were developed to facilitate the coupling of various separation formats to the mass spectrometer, or to enhance sensitivity by using on-line preconcentration techniques. Although a large number of applications have used ESI or CF-FAB ionization techniques, recent reports have described novel approaches that enabled coupling of CE to APCI or inductively coupled plasma mass spectrometry.

a. Separation Formats

In view of its simplicity and range of buffers available, CZE has been one of the most popular formats for on-line CE-MS coupling, as indicated by the number of publications reported thus far (in excess of 80% of the total CE-MS reports). A small sample of these applications was described in the preceding section. More recently, CZE was used in combination with microreactors to provide *in situ* digestion of proteins inside the capillary prior to mass spectral analysis.[209] Short capillary microreactors (25 or 50 cm length, 50 μm i.d.) were prepared using immobilized trypsin and carboxypeptidase Y, and the protein sample was introduced inside the microreactor where digestion was completed within one hour. An example of the use of an on-line microreactor coupled to CZE-MS with full mass scan acquisition was demonstrated for the digestion of approximately 500 pmol of oxidized insulin chain B.[209] The use of immobilized enzyme capillary microreactors was described previously by Amankwa and Kuhr[210] using UV detection, and shown to provide remarkable advantages over off-line digestion strategies in terms of sample requirements (picoliter to nanoliter injections), longer enzyme lifetimes, higher stability, and the possibility of reusing enzyme microreactors on multiple samples.

The application of CZE-MS to the monitoring of noncovalent associations in solution using neutral pH buffers has been described.[111,173,174] The noncovalent complex ribonuclease S, which comprises a 13.7 kDa protein plus a 2.1 kDa peptide, was studied at different pH values using CZE-MS with electrospray ionization.[111] Similarly, Henion and co-workers have investigated the noncovalent binding of FK506 and rapamycin to FKBP using CZE-MS.[173] Good agreement was found between binding affinity ratios determined for these two ligands using the CZE-MS technique compared to published affinity constants. The formation of noncovalent complexes of a sulfated pentasaccharide with the glycoprotein antithrombin III (Mr: 57882 Da) was described by Uzabiaga et al.[174] The CZE-MS analysis enabled observation of the native antithrombin III protein and its complex, whereas partial dissociation was observed when the sample was introduced to the ESI mass spectrometer by infusion.[174] The use of CZE-MS has also been described in applications to combinatorial chemistry for the selection of peptidyl ligands to vancomycin.[211] Electrophoretic conditions used buffers containing 20 mM Tris acetate pH 8.1 and 50 to 120 μM vancomycin, with coated capillaries to suppress the EOF. These conditions provided a direct measurement of binding constants of ligands in libraries, and the differences in mobilities with and without the receptor were used to identify potential peptide candidates.[211]

The use of buffer modifiers such as those exploited in MEKC separations was also investigated using CE-MS.[212-215] Organic solvents composed of zwitterionic surfactants such as N-dodecyl-N,N-dimethyl-3-amino-1-propanesulfate, at or near their critical micelle concentrations, have been used effectively to separate a series of closely related peptide variants of molecular masses of 7 to 8 kDa.[212] The CE-MS separation of a mixture of simple pentapeptides, differing by the substitution of only one aliphatic amino acid, was facilitated using an electrolyte containing a nonionic detergent, Genapol® C-100 (polyethylene glycol lauryl ether), and 1.25 mM SDS.[213] The characterization of enantiomers of the chiral drugs terbutaline and ephedrine was achieved by CE-MS using low pH electrolytes containing 0.5 mM β-cyclodextrin, which enabled the analysis of 1% of the (–) ephedrine in the presence of the (+) ephedrine enantiomer.[214] The direct coupling of MEKC to mass spectrometry can lead to significant chemical background noise due to the presence of nonvolatile surfactants or adduct ions. In order to alleviate these difficulties, a coupled capillary system providing voltage switching, and buffer renewal was designed and evaluated for the analysis of mepenzolate and its methyl analog pipenzolate, present at low μg/mL concentrations.[215] Recently, on-line coupling of MEKC to MS was demonstrated for a modified ion source consisting of a combination of an electrospray-type nebulizer and an APCI source.[216] This combined source enabled the analysis of aromatic amines (aniline, p-nitroaniline, and 1-naphthylamine) using separation buffers containing 20 mM phosphate with up to 50 mM SDS.[216]

The coupling of on-line CIEF to ESI mass spectrometry was presented by Tang et al. for the analysis of a simple protein mixture.[217] The polyacrylamide-coated capillary (20 cm × 50 μm i.d.) was entirely filled with the sample diluted in carrier ampholyte (2% pharmalyte) and focusing was performed at 10 kV with 20 mM NaOH at the cathode (located inside the ESI housing), while the anode was 20 mM phosphoric acid. Cathodic mobilization was effected by changing the buffer to a sheath solution containing 1% acetic acid in aqueous methanol, while maintaining a constant electric field of 500 V/cm. Alternatively pressure mobilization can be achieved using cathodic buffers compatible with ESI conditions. The focusing effect of CIEF enabled detection of proteins such as myoglobin, carbonic anhydrase, and cytochrome c at concentrations of 10^{-7} M.[217] The enhanced sample loadings provided by CIEF permit analysis of very dilute samples compared to those accessible by conventional CZE separation. CITP also enables larger injection volumes, although analyte band resolution is poor in this technique. The subject of sample preconcentration will be presented in Section b below.

The possibility of using the EOF as a means of transporting solvent through an LC column has been described previously by Pretorius and co-workers,[218] and this concept has led to the introduction of capillary electrochromatography (CEC).[219] The application of this hybrid separation technique coupled on-line to mass spectrometry was demonstrated recently by Gordon et al. for the analysis of low pmol levels of steroids using CF-FAB ionization.[220] This technique offers a viable alternative to MEKC for the separation of neutral compounds, and provides improved sample loadings, while separation efficiencies are typically much higher than those obtainable using LC. Other applications of on-line CEC-MS have been reported for the analysis of textile dyes[221] and pharmaceuticals[222] using ESI mass spectrometry. The combination of EOF with pressure-driven electrochromatography, referred to as pseudoelectrochromatography, was also described by Tjaden and co-workers.[223-224] The characteristic features of pseudoelectrochromatography, in terms of retention behavior and enhanced separation efficiencies, were compared to those of microcapillary LC for the separation of a series of sulfonamides.[224] This pressure-assisted variant of CEC originally presented by Tsuda was developed to suppress bubble formation.[225] The application of pseudoelectrochromatography was also reported by Schmeer et al. for the analysis of methylated peptides using 100 µm i.d. capillary columns packed with 1.5 µm i.d. reverse-phase silica gel.[226]

b. Enhancement of Sample Loadings

Different approaches to sample preconcentration were described in Section A.2 of this chapter, and the following discussion will focus only on the implications of such procedures when combined with mass spectrometric detection. In order to improve concentration detection limits in CZE-MS experiments, two main approaches have been developed — on-line isotachophoretic and chromatographic preconcentration techniques.

The coupling of CITP as a preconcentration step prior to CZE-MS analysis was first described by Tinke et al. for the analysis of anthracyclines, using a dual-column arrangement where the sample was first focused in a CITP column and subsequently introduced into the CZE column via a liquid junction.[227] The same authors have also described a theoretical equation for the calculation of the required split ratio of CITP zones into the CZE-MS system.[228] Karger and co-workers have developed a transient CITP procedure conducted in a single-capillary arrangement, in which the migration mode gradually changes from isotachophoresis to zone electrophoresis.[12] The focusing and separation steps of this approach have been described schematically in Figure 2.4. The application of on-line transient CITP-CZE to the enrichment of simple protein mixtures was demonstrated using ESI mass spectrometry and enabled a 200-fold preconcentration over that achievable using CZE.[229] Co-ions having a high electrophoretic mobility (NH_4^+ or H^+ for cationic separation) act as leading ions during the CITP focusing step. This approach also requires that the background electrolyte used for CZE separation contain a co-ion with low electrophoretic mobility that can serve as a terminating ion during the transient CITP focusing. The analysis of submicromolar solutions of protein standards of pI > 6 was presented by the same authors, using 20 mM 6-aminohexanoic acid/acetic acid pH 4.5 buffer as a background electrolyte, while the sample was supplemented with a leading co-ion (5 mM CH_3COONH_4).[229]

A similar approach was described by Lamoree et al. and enabled the injection of up to 1 µL (50% capillary volume) of trace levels of β-agonists in calf urine extracts.[230] The CE-MS analysis of paralytic shellfish poisoning toxins present at low nM levels in contaminated scallop tissues was achieved using a similar transient CITP preconcentration technique.[231] In this case, separations were conducted using polyacrylamide-coated columns with a universal terminator composed of 10 mM HCOOH and a background electrolyte of 50 mM morpholine pH 5. Unambiguous identification of individual toxins was facilitated using CITP-CZE combined with tandem mass spectrometry, and separation efficiencies in excess of 150,000 plates were reported.[231] More recently, electroextraction techniques have been combined with CITP preconcentration for further enhancement of sample loadings in CZE-MS experiments.[232] High field strengths were achieved in low conductivity organic solvents such as ethyl acetate and facilitated the rapid extraction of analyte ions into small buffer volumes. Detection of 2 to 5 nanomolar concentrations of clenbuterol and related drugs were reported using the combined electroextraction-CITP preconcentration procedures.[232] Although most of the above applications have used a coaxial capillary configuration, sample stacking techniques have also been developed using a liquid-junction interface in CE-MS experiments using CF-FAB ionization.[199] In this case the liquid junction provided an efficient means of conducting off-line field amplification stacking for the selective preconcentration of anionic adducts of deoxynucleotides and provided detection limits in the 10^{-8} M range for acetylaminofluorene deoxyguanosine monophosphate.[199]

An example of application of CITP preconcentration prior to CZE-MS analysis is presented in Figure 2.31 for the separation of a simple mixture of seven peptides. The analysis was conducted on a 95 cm × 50 μm i.d. linear polyacrylamide-coated capillary using 35 mM β-alanine/formate pH 3 as a background electrolyte and 10 mM formic acid as a terminator during the CITP step. A total of 2 ng of each component was injected on the column (0.5 μL injection of a 5 μg/mL solution), and all peptides were clearly observed from the total ion electropherogram (m/z 500 to 1500) shown in Figure 2.31A. The enhanced sample loading also facilitated the acquisition of MS-MS spectra for peptides present at low concentrations. In order to alleviate difficulties associated with the narrow peak width of CZE and the instrumental constraints of quadrupole instruments in terms of scanning speed, the CE field strength was lowered just before the peptide exited the capillary. This reduced migration speed strategy, previously described by Goodlett et al.,[143] allows sampling of a larger number of data points across the peak without loss of resolution. The electropherogram obtained by stepping the field strength from 30 kV to 10 kV at 25 min is presented in Figure 2.31B for the first four peptides. The MS-MS spectrum of bradykinin obtained under these electrophoretic conditions is shown in Figure 2.31C for product ions of $[M + 2H]^{2+}$ at m/z 531. As indicated, the MS-MS spectrum of this peptide was dominated by a large number of fragment ions corresponding to cleavage of the peptide bond, thus providing meaningful sequence information for the injection of only 2 pmoles of bradykinin.

Enhancement of sample loading prior to CZE-MS analysis has also been successfully accomplished using on-line chromatographic preconcentration techniques. Tomlinson and co-workers have been one of the most active groups reporting the successful coupling of on-line chromatographic preconcentrators prior to CZE-MS analysis.[33,38,39,233-236] Preliminary investigations of on-line analyte preconcentration[38] used a cartridge containing a small quantity of C-18 reverse-phase packing, similar to that described previously by other groups.[34-37] The separation of small hydrophobic peptides and mixtures of neuroleptic drugs, present at low pmol/μL concentrations, was demonstrated with this technique.[38] One of the major limitations of this chromatographic preconcentration approach was the significant peak broadening and loss of resolution attributed to large sample loading (≈5 μL), increased analyte-wall interactions, and reduction in the EOF.[38] Reduction of the size of the bed of the stationary phase or the use of discontinuous buffers prior to zone electrophoresis were proposed to alleviate these difficulties.[33]

Improvements in separation performance, when using chromatographic preconcentration procedures, were obtained by substituting the stationary phase with impregnated adsorptive membranes.[235] Best results have been obtained with polymeric styrenedivinyl benzene membranes which have a high capacity for analyte adsorption, smaller surface area, and reduced volume compared to packed microcapillary columns. In addition, elution of adsorbed com-

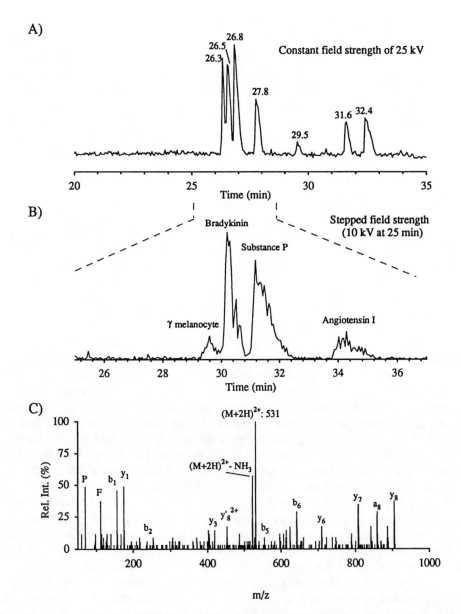

Figure 2.31 CITP-CZE-MS analysis of a mixture of seven peptides. Total ion electropherogram (m/z 500 to 1500) conducted under a constant field strength of 25 kV (A), and a stepped field strength (B). Product ion spectrum of m/z 531 obtained from combined CITP-CZE-MS-MS analysis. Conditions: 0.95 m × 50 μm i.d. capillary coated with linear polyacrylamide, 25 kV, background electrolyte: 35 mM β-alanine-formate pH 3.5, terminator: 10 mM formic acid, All separations were obtained for the injection of 0.5 μL of a 5 μg/mL of each peptide. (Fragment ion nomenclature according to Roepstorff, P. and Fohlman, J., *Biomed. Env. Mass Spectrom.*, 11, 601, 1984.)

ponents requires only minimal amounts of solvent. The application of membrane preconcentration with on-line CZE-MS, referred to as mPC-CE-MS, was demonstrated for the analysis of small peptides (Mr <1200 Da) present at μM concentrations.[234] The concurrent application of a low pressure (0.5 psi) differential together with the separation voltage provided shorter analysis times. Sample focusing using transient CITP immediately after sample elution improved separation efficiencies and peak resolution. The sample was eluted between a short zone (200 nL) of a leading stacking buffer composed of 0.1% NH_4OH and a terminating buffer of 1% acetic acid. Although most examples presented used uncoated capillaries, such an approach was also demonstrated using polybrene-coated capillaries for the analysis of low pmol levels of insulin-like growth factors and Bence-Jones proteins in patient urine.[236] More recent applications of the mPC-CE-MS technique used tandem mass spectrometry for the analysis of biologically derived major histocompatibility complex (MHC) class I peptides, where approximately 100 μL of the active HPLC fraction was introduced to the membrane cartridge.[33] The sequence of the active peptide was derived from the interpretation of the MS-MS spectrum using a database searching routine and subsequently synthesized to confirm the desired biological activity.

c. Coupling to Other Ionization Techniques

As mentioned in the preceding discussion, most CE-MS applications have been demonstrated using ESI and CF-FAB ionization techniques. Despite the fact that these techniques were found to be of practical use for a wide variety of ionic compounds, there are some limitations to the capability of ESI and CF-FAB for the analysis of analytes devoid of ionizable functionalities. Furthermore, significant compromises are required when attempting to couple MEKC to these ionization techniques. This problem is exacerbated when buffers composed of involatile salts are required to achieve the desired selectivity.

Takada et al. have recently described an alternate CE-MS interface using a combined nebulizer-assisted electrospray/atmospheric pressure chemical ionization (APCI) source using corona discharge.[117] In this system, the solution is first nebulized by electrospray, and the vaporized analytes are then ionized by ion–molecule reactions in the APCI source. The application of this new CE-MS interface to the analysis of 2 pmol of caffeine was demonstrated using a 20 mM sodium phosphate buffer. The intensity of the protonated molecule using APCI was practically unaffected by the presence of up to 40 mM sodium phosphate, whereas no detectable signal was obtained in the corresponding ESI experiment.[117] More recently, the same authors have demonstrated the utility of this combined ESI/APCI source for the MEKC separation of aromatic amines using SDS and phosphate buffers.[216]

The coupling of CE to inductively coupled plasma mass spectrometry (ICP-MS) for rapid metal speciation has been reported by different groups.[237-239] Olesik et al. have described a novel CE-ICP-MS interface based on a sheathless design, in which the electrical connection at the end of the CE column was made using conducting silver paint.[237] These authors reported detection limits as low as 0.06 ppb for Sr (90 amol injection), with analysis times of less than 2 min. One of the drawbacks of this interface was the extent of peak broadening associated with the use of high nebulizer gas flow rates (1 L/min), giving an effective aspiration rate from the CE column of 2 μL/min. Concentration detection limits were generally 60 times higher than those obtainable with ICP-MS without speciation, and this difference was accounted for by the smaller sample loadings of CE.[237] In two separate studies, a coaxial sheath flow interface was developed for the coupling of CE to ICP-MS, and provided efficient electrical contact with no noticeable suction or back pressure during the CE separation.[238,239] This CE-ICP-MS interface was evaluated using test samples containing selected alkali and heavy metal ions, including also Se and As species, and provided detection limits of 7 to 1000 pg/mL.[238] The elements Tl, As (V), and Mn yielded the best response, with detection limits of 7, 20, and 30 pg/mL, respectively, whereas a detection limit of 1000 pg/mL was obtained for potassium. The application of this CE-ICP-MS method was also demonstrated for the analysis of ferritin and metallothionein isoforms and enabled detection of Fe and Cd present at levels of 1.5 and 0.011 ppb, respectively.[239] Although these concentration detection limits are much higher than those observed with conventional ICP-MS, the incorporation of CITP preconcentration procedures should provide considerable improvements in sensitivity.

d. Mass Analyzers

The type of mass spectrometer employed also plays an important role in the design of the CE-MS interface. Quadrupole mass spectrometers have until recently been favored in CE-MS applications simply because of the technical difficulties in the introduction of liquids (leading to high pressure within the source) into high-voltage sources. Modern high resolution instruments, such as Fourier transform ion cyclotron resonance (FTICR) or sector instruments, require more elaborate ion guiding lenses and pumping systems in the ionization source assembly to maintain the required operational conditions and to prevent space charging effects. The very high separation efficiencies achievable by CE, and the resulting narrow peaks, can give rise to problems with the limited scanning speeds of sector or quadrupole instruments. Spectral acquisition over a large m/z range is often necessary to define the multiply charged envelope in the analysis of proteins by ESI-MS, for example, and can be problematic when fast cycle times are required. Alternatively, once the peaks of interest have been identified, the mass spectrometer can be set to monitor selected ions in a time-sharing fashion, using the selected ion monitoring (SIM) acquisition mode, such that only the ion current of these selected m/z values will be detected during the analysis. Under SIM acquisition, detection limits are typically two to three orders of magnitude lower than those obtainable under mass-scan acquisition mode. This improvement reflects the fact that in the SIM mode the mass spectrometer does not waste time (and sample) in examining regions of the m/z range that yield no useful information on the target analytes. Alternatively, scanning array detectors available on sector instruments can provide simultaneous ion detection over a larger mass range, and thus offer significant gains in sensitivity (10- to 100-fold) compared to the corresponding single point detector. The beneficial use of array detectors for the detection of trace level analytes in CE-MS experiments has been demonstrated by different groups.[197,240]

Improvement in ion utilization efficiency can be obtained using TOF or ion-trap mass spectrometers (FTICR and quadrupole ion trap instruments). In that respect, the TOF mass spectrometer offers significant advantages over sector or quadrupole instruments because the entire mass spectrum can be acquired with efficient ion transmission and fast duty cycles (up to 100 μsec/scan). This type of mass spectrometer is also attractive because of its simpler mass analyzer design, although fast data collection interfaces and elaborate ion optics for enhanced resolution are desirable features. To date, only a few reports have documented the use of TOF mass spectrometers as detectors for CE.[207,241,242] In one of the first reports of CE-MS using a TOF mass spectrometer, Fang et al. reported concentration detection limits in the low micromolar range for small peptides from the sea mollusk *Aplysia californica*.[207] Smith and co-workers have also demonstrated the use of sequential covariance data analysis to enhance S/N ratios in CE-MS experiments using a TOF mass spectrometer, thus allowing more immediate peak identification.[241]

Ion-trap mass spectrometers also offer significant gains in transmission efficiencies since all ions can be stored simultaneously and analyzed sequentially. The coupling of CE to a quadrupole ion-trap mass spectrometer has been accomplished by different groups using home-built or highly modified benchtop systems.[160,203,208] One interesting application of such ion-trap instruments is the combined use of broad-band collisional activation and resonance ejection to reduce the chemical and background noise present in the electropherogram.[243] Similarly, the FTICR instrument also offers excellent sensitivity, but has gained significant recognition for its high resolution (>100,000 with low ppm mass measurement error) and the possibility of conducting higher order tandem mass spectrometry experiments (MS^n). Smith's group at Battelle was one of the first to demonstrate the successful coupling of CE to FTICR-MS instruments using an external ESI source with ion injection into the magnetic field through either radio frequency ion guides or electrostatic lenses.[244,245] Although ion ejection and trapping parameters were not optimized, preliminary CE-MS experiments enabled the separation of simple peptides and proteins at femtomole levels with a resolution of 50,000.[245-246] More recent reports of CE-MS using FTICR mass spectrometers have incorporated CITP preconcentration techniques for the identification of intact and damaged oligonucleotides following X-ray irradiation.[247] On-line CE-MS with FTICR was also demonstrated for the analysis of α and β chains of hemoglobin extracted from human erythrocytes, and meaningful mass spectra were obtained for the injection of 4.5 fmol of hemoglobins corresponding to the sampling of approximately 10 erythrocytes.[248] Partial sequence information on these high molecular weight proteins was obtained using a combination of quadrupole axialization and sustained off-resonance irradiation (SORI) to yield MS-MS spectra dominated mainly by multiply charged fragment ions corresponding to cleavage of the peptide bonds.[248] The prospect of conducting CE-MS experiments using high resolution MS-MS opens new avenues in the identification of modification sites in altered proteins where only a limited amount of biological sample is available.

I. Nuclear Magnetic Resonance Detection

Nuclear magnetic resonance (NMR) is one of the most valuable tools for structural elucidation of unknown analytes and for the study of molecular structure in solution. However, the low sensitivity of NMR with respect to other analytical techniques and its inherent requirement for purified material have precluded the incorporation of this structural tool in practical applications where valuable biological extracts are available only in very limited quantities. Recent advances in NMR spectroscopy have included the development of microprobes for the analysis of low μg levels of sample present in only 1 to 10 μL volumes,[249] and the on-line coupling to HPLC.[250-253] However, previous investigations describing LC-NMR have used detector cells of 25 to 200 μL volume, which are naturally unacceptable for CE separations. One of the more significant contributions to the coupling of microseparation techniques with NMR was the recent design of a nanoliter sample cell for ¹H NMR spectroscopy by Sweedler and co-workers.[254] In this recent report the authors described a microcoil wrapped directly around the fused-silica CE column, providing an effective cell volume of 5 nL with detection limits in the nanogram range for short (<1 min) NMR acquisitions.[254] The hydrostatic injection was performed outside the magnet on a 48 cm × 75 μm i.d. CE column, and multiple NMR spectra were acquired (eight scans at a 2-s repetition rate) for each band. The application of on-line CE-NMR was demonstrated for the analysis of arginine, glycine, and cysteine present at levels of 0.7 M, and the concentration detection limit was estimated to be 35 mM.[254] The uniformity and configuration of the microcoil, capillary, and impedance matching circuitry relative to the sample volume caused significant band broadening and also resulted in line widths of 7 Hz at best for the 300 MHz NMR spectrometer equipped with a 7.05 T magnet. Further improvements in sensitivity by more than two orders of magnitude are expected from the use of superconducting microcoils.

It is important to keep in mind that recent progress made in the miniaturization of the NMR probe has had a significant impact on the resolution and sensitivity obtainable using this technique. As pointed out recently by Albert, the detection limits currently available using these refined probes (≈35 ng) restrict the type of application achievable by CE-NMR.[255] NMR spectroscopic investigations of biopolymers, which are currently performed in the stopped-flow mode, would not be directly applicable unless convenient CE field strength gradients are developed. Significant improvements in terms of concentration detection limits, possibly through sample preconcentration procedures, will be required before CE-NMR becomes a practical tool for obtaining stereochemical and other structural information from complex mixtures.[255] To this end the coupling of LC to NMR currently provides a more viable alternative in view of the present limitations in sensitivity.[256,257]

J. Off-Line Methods

Several reports have described fraction collection devices for CE,[258-272] and numerous commercial CE instruments now offer this capability. The micropreparative capability of CE has been demonstrated for a wide range of biochemistry applications, including the separation of oligonucleotides from gel electrophoresis,[258] the isolation of enzyme-active fractions,[259] and the collection of peptides prior to microsequencing.[260,261] Fraction collection was also applied to the isolation of peptides and proteins prior to mass spectral analysis, and a brief description of this application is provided below.

The isolation of analyte bands from the CE column is particularly challenging in view of the low EOF (< 300 nL/min) and the unavoidable interruption of electrical contact during capillary transfer which in turn causes temporal uncertainty in the collection process. Another limitation comes from the relatively small sample loadings that can be injected into the capillary column if one wishes to maintain adequate band resolution and separation efficiency. If a single injection and collection are made, this limitation imposes considerable demands on the sensitivity of the post-column detection system, in addition to the inherent difficulty of efficiently transferring ng and sub-ng sample sizes into a few μL of solution. Despite these problems, a number of different approaches have been described for micropreparative CE, and some of these fraction collection interfaces are described in Figure 2.32. One means of collecting fractions involves the formation of a fracture or "porous glass joint" in the capillary wall to establish electrical contact a few cm away from the capillary outlet (Figure 2.32 A).[262-264] CZE fractions can be collected directly onto a nitrocellulose-coated aluminum foil thus preventing any transfer losses prior to mass spectral analysis using PDMS.[263] Hydrodynamic forces can also be used to push the analyte zone out of the capillary while simultaneously maintaining the electrical connection. An alternate approach to individual fraction collection has also been developed in which a moving membrane is continuously passed beneath the capillary outlet and the collected stream supplemented by a sheath flow of matrix solution for adequate detection using MALDI-TOF.[265]

Other methods for the collection of the CE effluent have used a grounded capillary outlet. In one approach, a rotating moist membrane placed at the capillary end was used to continuously collect the analyte band and provided efficient electrical contact without interruption of separation current.[266-268] More recently, different coaxial interfaces were described which enable collection of sizable fractions of peptides and proteins with minimal memory effects.[269-272] In this approach (Figure 2.32 B), the CE column is supplemented with sheath flow rates of 1 to 25 μL/min, which provide both electrical contact and sizable volumes for collection. The sheath flow rate and capillary sizes must be carefully selected to prevent back streaming in the CE column and degradation of the separation performance. The coaxial

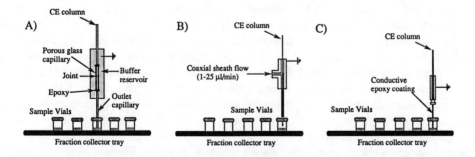

Figure 2.32 Schematic representation of fraction collection device using a porous glass capillary joint (A), a coaxial sheath flow (B), and a metallized capillary tip (C). See text for further description.

collection interface was investigated using post-column mass spectrometric analysis with MALDI and ESI. Detection limits in the order of 80 fmol for small peptides and less than 1 pmol for apolipoprotein AII were reported.[269] A similar coaxial column flow arrangement, with microscale fraction collection, was also used to deposit each isolated fraction for subsequent analysis on a Fourier transform mass spectrometer (FTMS).[270]

Walker et al. recently described a conductive capillary interface (Figure 2.32 C) and compared its performance to that of coaxial sheath flow fraction collection.[271,272] The capillary outlet was first tapered by abrasion, cleaned, and coated with gold epoxy. Although silver conductive adhesives provided suitable electrical contact, capillaries coated with silver were more prone to oxidation, and loss of coating resulted in sample contamination. By using this interface, CE fractions could be deposited directly on the MALDI probes and provided direct correlation of peak identity. The application of this interface was demonstrated in proteins and peptides arising from proteolytic digests, and good quality mass spectra were obtained from separations of as little as 25 fmol of protein.[272] The possibility of independently optimizing the CE and mass spectral conditions provides a clear advantage compared to direct CE-MS analysis. Indeed, involatile buffers common to CE separations (Tris, phosphate, citrate, borate, etc.) can be used in off-line experiments because adduct formation can be suppressed by the addition of strong acids to the MALDI matrix,[272] or, alternatively, fractions can be desalted prior to mass spectrometric analysis.

IV. CONCLUSIONS

Capillary electrophoresis is a valuable analytical tool which offers a number of different separation formats suited to the analysis of charged and neutral analytes. Its advantages in terms of separation formats and selectivity complement those of other separation techniques such as liquid chromatography. Very small volumes of sample are required (typically several nL are used per analysis), and a number of different separation conditions can be investigated using only 5 to 10 µL of the same sample. Improvement in sample loadings can also be achieved using chromatographic preconcentration or sample stacking. High-sensitivity detection systems are required to take advantage of the full benefits of CE. All commercial systems currently offer UV detection and in some cases a diode array detector is also available. While optical absorbance detectors rely on the presence of suitable analyte chromophores, refractive index and refractive index gradient detection are also attractive because of their universal response. On the other hand, when convenient reagents are available, the use of laser-induced fluorescence detection provides unparalleled sensitivity with detection limits approaching the single-molecule level. In view of the human genome initiative, such a gain in sensitivity opens new avenues in DNA sequencing, and this topic will be reviewed in Chapter 9. Most detectors used with CE offer single-parameter detection, and peak assignment often relies on the reproducibility of migration times or the use of mobility markers. There is thus a need for high sensitivity detectors that can provide structural information on the nature of the analyte. To this end, the coupling of CE with mass spectrometry provides a powerful structural tool for unambiguous peak identification, with sensitivity in the low femtomole range. Recent advances in the development of microcoils have permitted the coupling of CE to nuclear magnetic resonance spectroscopy. However, significant improvements in sensitivity will be required before the full benefits of this technique can be utilized in conjunction with electrophoretic separations. Future improvements in existing hardware and development of novel detectors should allow more sensitive and/or selective detection. Furthermore, the miniaturization of CE, as demonstrated for microfabricated channels on silicon wafers, is particularly attractive and could lead to important advances in decreasing analysis times and increasing sample throughput.

ACKNOWLEDGMENTS

N. Dovichi gratefully acknowledges a Steacie Fellowship from the Natural Sciences and Engineering Research Council of Canada. The data in Figures 2.3 and 2.21 were kindly supplied by D.Y. Chen of the University of Alberta. The assistance of J.F. Kelly, P. Blay, and S. Locke is gratefully acknowledged. We also thank Dr. S. Pleasance for his important contribution in the first edition of this chapter, and Drs. R.K. Boyd and A.W. McCulloch for valuable discussions and constructive criticisms during the review of the present chapter.

REFERENCES

1. Tiselius, A., A new apparatus for electrophoretic analysis of colloidal mixtures, *Trans. Faraday Soc.*, 33, 524, 1937.
2. Edstrom, J.E., Nucleotide analysis on the cyto-scale, *Nature*, 172, 908, 1953.
3. Neuhoff, V., Schill, W.B., and Sternbach, H., Micro-analysis of pure deoxyribonucleic acid-dependent ribonucleic acid polymerase from Escherichia coli, *Biochem. J.*, 117, 623, 1970.
4. Poehling, H.M. and Neuhoff, V., One and two-dimensional electrophoresis in micro-slab gels, *Electrophoresis*, 1, 90, 1980.
5. Hjertén, S., Free zone electrophoresis, *Chromatog. Rev.*, 9, 122, 1967.
6. Mikkers, F.E.P., Everaerts, F.M., and Verheggen, T.P.E.M., Concentration distributions in free zone electrophoresis, *J. Chromatogr.*, 169, 1, 1979.
7. Mikkers, F.E.P., Everaerts, F.M., and Verheggen, T.P.E.M., High-performance zone electrophoresis, *J. Chromatogr.*, 169, 11, 1979.
8. Jorgenson, J.W. and Lukacs, K.D., Zone electrophoresis in open-tubular glass capillaries, *Anal. Chem.*, 53, 1298, 1981.
9. Jorgenson, J.W. and Lukacs, K.D., Capillary zone electrophoresis, *Science*, 222, 266, 1983.
10. Chien, R.L. and Burgi, D.S., Sample stacking of an extremely large volume in high performance capillary electrophoresis, *Anal. Chem.*, 64, 1046, 1992.
11. Burgi, D.S. and Chien, R.L., Optimization in sample stacking for high performance capillary electrophoresis, *Anal. Chem.*, 63, 2042, 1991.
12. Foret, F., Szoko, E., and Karger, B.L., On-column transient and coupled column isotachophoretic preconcentration of protein samples in capillary zone electrophoresis, *J. Chromatogr.*, 608, 3, 1992.
13. Kuhr, W.G. and Monnig, C.A., Capillary electrophoresis, *Anal. Chem.*, 64, 389R, 1992.
14. Monnig, C.A. and Kennedy, R.T., Capillary electrophoresis, *Anal. Chem.*, 66, 280R, 1994.
15. Knox, J.H., Thermal effects and band spreading in capillary electro-separation, *Chromatographia*, 26, 329, 1989.
16. Gobie, W.A. and Ivory, C.F., Thermal model of capillary electrophoresis and a method for counteracting thermal band broadening, *J. Chromatogr.*, 516, 191, 1990.
17. Lux, J.A., Yin, H.F., and Schomburg, G., Construction, evaluation, and analytical operation of a modular capillary electrophoresis instrument, *Chromatographia*, 30, 7, 1990.
18. Meyer, V.R., High-performance liquid chromatography theory for the practitioner, *J. Chromatogr.*, 334, 197, 1985.
19. Rose, D.J. and Jorgenson, J.W., Characterization and automation of sample introduction methods for capillary zone electrophoresis, *Anal. Chem.*, 60, 642, 1988.
20. Huang, X., Gordon, M.J., and Zare, R.N., Bias in quantitative capillary zone electrophoresis caused by electrokinetic sample injection, *Anal. Chem.*, 60, 375, 1988.
21. Lee, T.T. and Yeung, E.S., Compensation for instrumental and sampling biases accompanying electrokinetic injection in capillary zone electrophoresis, *Anal. Chem.*, 64, 1226, 1992.
22. Tsuda, T., Mizuno, T., and Akiyama, J., Rotary-type injector for capillary zone electrophoresis, *Anal. Chem.*, 59, 799, 1987.
23. Tehrani, J., Macomber, R., and Day, L., *J. High Resolut. Chromatogr.*, 14, 10, 1991.
24. Fishman, H.A., Amudi, N.M., Lee, T.T., Scheller, R.H., and Zare, R.N., Spontaneous injection in microcolumn separations, *Anal. Chem.*, 66, 2318, 1994.

25. Monnig, C.A. and Jorgenson, J.W., On-column sample gating for high-speed capillary zone electrophoresis, *Anal. Chem.*, 63, 802, 1991.
26. Chien, R.L. and Burgi, D.S., On-column sample concentration using field amplification in CZE, *Anal. Chem.*, 64, 489A, 1992.
27. Foret, F., Sustàcek, V., and Bocek, P., On-line isotachophoretic sample preconcentration for enhancement of zone detectability in capillary zone electrophoresis, *J. Microcol. Sep.*, 2, 127, 1990.
28. Mazereeuw, M., Tjaden, U.R., and Reinhoud, N.J., Single capillary isotachophoresis-zone electrophoresis: current practice and prospects, a review, *J. Chromatogr. Sc.*, 33, 686, 1995.
29. Reinhoud, N.J., Tjaden, U.R., and van der Greef, J., Strategy for setting up single-capillary isotachophoresis-zone electrophoresis, *J. Chromatogr.*, 653, 303, 1993.
30. Stegehuis, D.S., Tjaden, U.R., and van der Greef, J., Analyte focusing in capillary electrophoresis using on-line isotachophoresis, *J. Chromatogr.*, 591, 341, 1992.
31. Schwer, C. and Lottspeich, F., Analytical and micropreparative separation of peptides by capillary zone electrophoresis using discontinuous buffer systems, *J. Chromatogr.*, 623, 345, 1992.
32. Pospichal, J., Gebauer, P., and Bocek, P., Measurement of mobilities and dissociation constants by capillary isotachophoresis, *Chem. Rev.*, 89, 419, 1989.
33. Tomlinson, A.J., Guzman, N.A., and Naylor S., Enhancement of concentration limits of detection in CE and CE-MS: A review of on-line sample extraction, cleanup, analyte preconcentration, and microreactor technology, *J. Cap. Elect.*, 6, 247, 1995.
34. Cai, J. and El Rassi, Z., Selective on-line preconcentration of triazine herbicides with tandem octadecyl capillaries-capillary zone electrophoresis, *J. Liq. Chromatogr.*, 15, 1179, 1992.
35. Debets, A.J.J., Mazereeuw, M., Voogt, W.H., van Iperen, D.J., Lingeman, H., Hupe, K.P., and Brinkman, U.A.T., Switching valve with internal micro precolumn for on-line sample enrichment in capillary zone electrophoresis, *J. Chromatogr.*, 608, 151, 1992.
36. Swartz, M.E. and Merion, M., On-line sample preconcentration on a packed-inlet capillary for improving the sensitivity of capillary electrophoretic analysis of pharmaceuticals, *J. Chromatogr.*, 632, 209, 1993.
37. Hoyt, A.M., Beale, S.C., Larmann, J.P. Jr., and Jorgenson J.W., Preparation and evaluation of an on-line preconcentrator for capillary electrophoresis, *J. Microcol. Sep.*, 5, 325, 1993.
38. Tomlinson, A.J., Benson, L.M., Braddock, W.D., Oda, R.P., and Naylor, S., On-line preconcentration-capillary electrophoresis-mass spectrometry (PC-CE-MS), *J. High Resol. Chromatogr.*, 17, 729, 1994.
39. Tomlinson, A.J. and Naylor, S., Systematic development of on-line membrane preconcentration-capillary electrophoresis-mass spectrometry for the analysis of peptide mixtures, *J. Cap. Elec.*, 5, 225, 1995.
40. Kohr, J. and Engelhardt, H., Characterization of quartz capillaries for capillary electrophoresis, *J. Chromatogr.*, 652, 309, 1993.
41. Tsuda, T., Sweedler, J.V., and Zare, R.N., Rectangular capillaries for capillary zone electrophoresis, *Anal. Chem.*, 62, 2149, 1990.
42. Nielen, N.M., Capillary zone electrophoresis using a hollow polypropylene fiber, *J. High Resolut. Chromatogr.*, 16, 62, 1993.
43. Schuetzner, W. and Kenndler, E., Electrophoresis in synthetic organic polymer capillaries: Variation of electroosmotic velocity and ζ potential with pH and solvent composition, *Anal. Chem.*, 64, 1991, 1992.
44. Wiktorowicz, J.E. and Colburn, J.C., Separation of cationic proteins via charge reversal in capillary electrophoresis, *Electrophoresis*, 11, 769, 1990.
45. Dougherty, A.M. and Schure, M.R., Covalent surface modification for capillary electrophoresis: characterization and effects of nonionic bondings on separations in capillary electrophoresis, in *Capillary Electrophoresis Technology*, Chromatographic Science Series, Vol. 64, Marcel Dekker, New York, 1993.
46. Schomburg, G., Technology of separation capillaries for capillary zone electrophoresis and capillary gel electrophoresis: The chemistry of surface modification and formation of gels, in *Capillary Electrophoresis Technology*, Chromatographic Science Series, Vol. 64, Marcel Dekker, New York, 1993.

47. Zhao, Z., Malik, A., and Lee, M.L., Separation of proteins and proteolitic digests of proteins by capillary electrophoresis on superox-coated open tubular columns, *J. Microcol. Sep.*, 4, 411, 1992.
48. Zhao, Z, Malik, A., and Lee, M.L., Solute adsorption on polymer-coated fused silica capillary electrophoresis columns using selected protein and peptide standards, *Anal. Chem.*, 65, 2747, 1993.
49. Cobb, K.A., Dolnik, V., and Novotny, M., Electrophoretic separations of proteins in capillaries with hydrolytically stable surface structures, *Anal. Chem.*, 62, 2478, 1990.
50. Schmalzing, D., Piggee, C.A., Foret, F., Carrilho, E., and Karger, B.L., Characterization and performance of a neutral hydrophilic coating for the capillary electrophoretic separation of biopolymers, *J. Chromatogr.*, 652, 149, 1993.
51. Huang, M., Yi, G., Bradshaw, J.S., and Lee, M.L., Charged surface coatings for capillary electrophoresis, *J. Microcol. Sep.*, 5, 199, 1993.
52. Moseley, M.A., Deterding, L.J., Tomer, K.B., and Jorgenson, J.W., Determination of bioactive peptides using capillary zone electrophoresis/mass spectrometry, *Anal. Chem.*, 63, 109, 1991.
53. Tows, J.K. and Regnier, F.E., Capillary electrophoretic separations of proteins using nonionic surfactant coatings, *Anal. Chem.*, 63, 1126, 1991.
54. Kelly, J.F., Locke, S.J., Ramaley, L., and Thibault, P., Development of electrophoretic conditions for the characterization of protein glycoforms by capillary electrophoresis-electrospray mass spectrometry, *J. Chromatogr.*, 720, 409, 1996.
55. Landers, J.P., Oda, R.P., Madden, B.J., and Spelsberg, T.C., High performance capillary electrophoresis of glycoproteins: the use of modifiers of electroosmotic flow for the analysis of microheterogeneity, *Anal. Biochem.*, 205, 115, 1992.
56. Schomburg, G., Belder, D., Gilges, M., and Motsch, S., Ionic and nonionic polymers as wall modifiers in capillary electrophoresis, *J. Cap. Elect.*, 3, 219, 1994.
57. Smith, R.D., Olivares, J.A., Nguyen, N.T., and Udseth, H.R., Capillary zone electrophoresis-mass spectrometry using an electrospray ionization interface, *Anal. Chem.*, 60, 436, 1988.
58. Cheng, Y.F., Wu, S., Chen, D.Y., and Dovichi, N.J., The interaction between capillary zone electrophoresis and a sheath flow cuvette detector, *Anal. Chem.*, 62, 496 1990.
59. Green, J.S. and Jorgenson, J.W., Design of a variable wavelength UV absorbance detector for on-column detection in capillary electrophoresis and comparison of its performance to a fixed wavelength UV absorbance detector, *J. Liquid. Chromatogr.*, 12, 2527, 1989.
60. Stair, R., Johnston, R.G., and Halbach, E.W., Standard of spectral radiance for the region of 0.25 to 2.6 microns, *J. Res. Natl. Bur. Stand. Sect. A*, 64A, 291, 1960.
61. Bruin, G.J.M., Stegeman, G., Van Asten, A.C., Xu, X., Kraak, J.C., and Poppe, H., Optimization and evaluation of the performance of arrangements for UV detection in high-resolution separations using fused-silica capillaries, *J. Chromatogr.*, 559, 163, 1991.
62. Xi, X. and Yeung, E.S., Axial beam absorption detection for capillary electrophoresis with a conventional light source, *Appl. Spectrosc.*, 45, 1199, 1991.
63. Wang, T., Aiken, J.H., Huie, C.W., and Hartwick, R.A., Nanoliter-scale multireflection cell for absorption detection in capillary electrophoresis, *Anal. Chem.*, 63, 1372, 1991.
64. Foret, F., Fanali, S., Nardi, A., and Bocek, P., Capillary zone electrophoresis of rare earth metal ions with indirect UV absorbance detection, *Electrophoresis*, 11, 780, 1990.
65. Kuhr, W.G. and Yeung, E.S., Optimization of sensitivity and separation in capillary zone electrophoresis with indirect fluorescence detection, *Anal. Chem.* 60, 2642, 1988.
66. Bornhop, D.J. and Dovichi, N.J., Nanoliter refractive index detector, *Anal. Chem.*, 58, 504, 1986.
67. Bruno, A.E., Krattiger, B., Maystre, F., and Widmer, H.M., On column laser-based refractive index detector for capillary electrophoresis, *Anal. Chem.*, 63, 2689, 1991.
68. Chen, C.Y., Demann, T., Huang, S.D., and Morris, M.D., Capillary zone electrophoresis with analyte velocity modulation. Application to refractive index detection, *Anal. Chem.*, 61, 1590, 1989.
69. Demana, T., Guhathakurta, U., and Morris, M.D., Effects of analyte velocity modulation on the electroosmotic flow in capillary electrophoresis, *Anal. Chem.*, 64, 390, 1992.
70. Hjertén, S., The history of the development of electrophoresis at Uppsala, *Electrophoresis*, 9, 3, 1988.

71. Pawliszyn, J., Nanoliter volume sequential differential concentration gradient detector, *Anal. Chem.*, 60, 2796, 1988.
72. Pawliszyn, J. and Wo, J., Moving boundary capillary electrophoresis with concentration gradient detection, *J. Chromatogr.*, 559, 111, 1991.
73. Wu, J. and Pawliszyn, J., Universal detection for capillary isoelectric focusing without mobilization using a concentration gradient imaging system, *Anal. Chem.*, 64, 224, 1992.
74. Yu, M. and Dovichi, N.J., Sub-femtomole determination of DABSYL amino acids using capillary zone electrophoresis and laser-based thermo-optical absorbance detection, *Mikrochim. Acta*, 3, 27 1988.
75. Yu, M. and Dovichi, N.J., Attomole amino acid analysis by capillary zone electrophoresis with thermo-optical absorbance detection, *Anal. Chem.*, 61, 37, 1989.
76. Bruno, A.E., Paulus, A., and Bornhop, D.J., Thermo-optical absorption detection in 25-μm-i.d. capillaries: capillary electrophoresis of Dansyl-amino acids mixtures, *Appl. Spectrosc.*, 45, 462, 1991.
77. Waldron, K.C. and Dovichi, N.J., Sub-femtomole determination of phenylthiohydantoin-amino acids: capillary electrophoresis and thermo-optical detection, *Anal. Chem.*, 64, 1396,1992.
78. Guzman, N.A., Trebilcock, M.A., and Advis, J.P., Increased sensitivity to analyze brain tissue constituents: use of capillary electrophoresis coupled to fluorescence microscopy, in *Techniques in Protein Chemistry II*, Villafranca, J.J., Ed., Academic Press, New York, 1991, 37.
79. Gassmann, E., Kuo, J.E., and Zare, R.N., Electrokinetic separation of chiral compounds, *Science*, 230, 813, 1985.
80. Hernandez, L., Escalona, J., Joshi, N., and Guzman, N., Laser-induced fluorescence and fluorescence microscopy for capillary electrophoresis zone detection, *J. Chromatogr.*, 559, 183, 1991.
81. Huang, X.C., Quesada, M.A., and Mathies, R.A., Capillary array electrophoresis using laser-excited confocal fluorescence detection, *Anal. Chem.*, 64, 967, 1992.
82. Lyons, J.W. and Faulkner, L.R., Optimization of flow cells for fluorescence detection in liquid chromatography, *Anal. Chem.*, 54, 1960, 1982.
83. Cheng, Y.F. and Dovichi, N.J., Sub-attomole amino acid analysis by capillary zone electrophoresis and laser induced fluorescence, *Science*, 242, 562 1988.
84. Wu, S. and N.J. Dovichi, High sensitivity fluorescence detector for fluorescein isothiocyanate derivatives of amino acids separated by capillary zone electrophoresis, *J. Chromatogr.*, 480, 141, 1989.
85. Cheng, Y.F., Wu, S., Chen. D.Y., and Dovichi, N.J., The interaction between capillary zone electrophoresis and a sheath flow cuvette detector, *Anal. Chem.*, 62, 496, 1990.
86. Swerdlow, H., Zhang, J.Z., Chen. D.Y., Harke, H.R., Grey, R., Wu, S., Fuller, C., and Dovichi, N.J., Three DNA sequencing methods based on capillary gel electrophoresis and laser-induced fluorescence, *Anal. Chem.*, 63, 2835, 1991.
87. Wu, S. and Dovichi, N., Capillary zone electrophoresis separation and laser induced fluorescence detection of fluorescein thiohydantoin derivatives of amino acids, *Talanta*, 39, 173, 1992.
88. Chen, D.Y., Swerdlow, H.P., Harke, H.R., Zhang, J.Z., and Dovichi, N.J., Low-cost, high sensitivity laser-induced fluorescence detection for DNA sequencing by capillary gel electrophoresis, *J. Chromatogr.*, 559, 237 1991.
89. Zhang, J.Z., Chen. D.Y., Wu, S., Harke, H.R., and Dovichi, N.J., High sensitivity laser-induced fluorescence detection for capillary electrophoresis, *Clinical Chem.*, 37, 1492 1991.
90. Sweedler, J.V., Shear, J.B., Fishman, H.A., Zare, R.N., and Scheller, R.H., Fluorescence detection in capillary zone electrophoresis using a charged coupled devise with time-delayed integration, *Anal. Chem.*, 63, 496, 1991.
91. Cheng, Y.F., Piccard, R.D., and Vo-Dinh, T., Charged-coupled device fluorescence detection for capillary zone electrophoresis (CCD-CZE), *Appl. Spectro.*, 44, 755, 1990.
92. Novotny, M., Recent advances in microcolumn liquid chromatography, *Anal. Chem.*, 60, 500A, 1988.
93. Haugland, R.P., *Handbook of Fluorescent Probes and Research Chemicals*, Molecular Probes, Eugene, Oregon 1989.

94. Liu, J., Hsieh, Y.Z., Wiesler, D., and Novotny, M., Design of 3-(4-carboxybenzoyl)-2-quinoli-necarboxaldehyde as a reagent for ultrasensitive determination of primary amines by capillary electrophoresis using laser fluorescence detection, *Anal. Chem.*, 63, 408, 1991.

95. Liu, J., Shirota, O., and Novotny, M., Capillary electrophoresis of amino sugars with laser-induced fluorescence detection, *Anal. Chem.*, 63, 413, 1991.

96. Higashijima, T., Fuchigami, T., Imasaka, T., and Ishibashi, N., Determination of amino acids by capillary zone electrophoresis based on semiconductor laser fluorescence detection, *Anal. Chem.*, 64, 711, 1992.

97. Zhao, J.Y., Waldron, K.C., Miller, J., Zhang, J.Z., Harke, H.R., and Dovichi, N.J., Attachment of a single fluorescent label to peptides for determination by capillary zone electrophoresis, *J. Chromatogr.*, 608, 239, 1992.

98. Tsuda, T., Kobayashi, Y., Hori, A., Matsumoto, T., and Suzuki, O., Post-column detection for capillary zone electrophoresis, *J. Chromatogr.*, 456, 375, 1988.

99. Rose, D.J. and Jorgenson, J.W., Post-column fluorescence detection in capillary zone electrophoresis using o-phthaldialdehyde, *J. Chromatogr.*, 447, 117, 1988.

100. Pentoney, S.L., Huang, X., Burgi, D.S., and Zare, R.N., On-line connector for microcolumns: applications to the on-column o-phthaldialdehyde derivatization of amino acids separated by capillary zone electrophoresis, *Anal. Chem.*, 60, 2625, 1988.

101. Foret, F., Deml, M., Kahle, V., and Bocek, P., On-line fiber optic UV detection cell and conductivity cell for capillary zone electrophoresis, *Electrophoresis*, 7, 430, 1986.

102. Huang, X., Pang, T.K.J., Gordon, M.J., and Zare, R.N., On-column conductivity detector for capillary zone electrophoresis, *Anal. Chem.*, 59, 2747, 1987.

103. Huang, X., Zare, R.N., Sloss, S., and Ewing, A.G., End column detection for capillary zone electrophoresis, *Anal. Chem.*, 63, 189, 1991.

104. Huang, X. and Zare, R.N., Improved end-column conductivity detector for capillary zone electrophoresis, *Anal. Chem.*, 63, 2193, 1991.

105. Wallingford, R.A. and Ewing, A.G., Capillary zone electrophoresis with electrochemcial detection, *Anal. Chem.*, 59, 1762, 1987.

106. Wallingford, R.A. and Ewing, A.G., Capillary zone electrophoresis with electrochemical detection in 12.7 μm diameter columns, *Anal. Chem.*, 60, 1972, 1988.

107. Pentoney, S.L., Zare, R.N., and Quint, J.F., On-line radioisotope detection for capillary electrophoresis, *Anal. Chem.*, 611, 1642, 1989.

108. Altria, K.D., Simpson, C.F., Bharij, A.K., and Theobald, A.E., A gamma-ray detector for capillary zone electrophoresis and its use in the analysis of some radiopharmaceuticals, *Electrophoresis*, 11, 732, 1990.

109. Smith, R. D., Barinaga, C. J., and Udseth, H. R., Improved electrospray ionization interface for capillary zone electrophoresis-mass spectrometry, *Anal. Chem.*, 60, 1948, 1988.

110. Moseley, M. A., Deterding, L. J., Tomer, K. B., and Jorgenson, J. W., Capillary zone electrophoresis-mass spectrometry using a coaxial continuous-flow fast atom bombardment interface, *J. Chromatogr.*, 516, 167, 1990.

111. Smith, R.D., Wahl, J.H., Goodlett, D.R., and Hofstadler, S.A., Capillary electrophoresis-mass spectrometry, *Anal. Chem.*, 65, 574A, 1993.

112. Niessen, W.M.A., Tjaden, U.R., and van der Greef, J., Capillary electrophoresis-mass spectrometry, *J. Chromatogr.*, 636, 3, 1993.

113. Cai, J. and Henion, J.D., Capillary electrophoresis-mass spectrometry, *J. Chromatogr.*, 703, 667, 1995.

114. Smith, R.D. and Udseth, H.R., Capillary electrophoresis mass spectrometry, in *Handbook of Capillary Electrophoresis*, Landers, J.P., Ed., CRC Press, Boca Raton, 1994, Chap. 8.

115. Tomer, K.B., Capillary zone electrophoresis-mass spectrometry: Continuous flow fast atom bombardment and electrospray ionization, in *Capillary Electrophoresis Technology*, Chromatographic Science Series, Vol. 64, Marcel Dekker, New York, 1993.

116. Chapman, J. R., *Practical Organic Mass Spectrometry*, John Wiley & Sons, Chichester, 1986, Chap. 5.

117. Takada, Y., Nakayama, K., and Yoshida, M., Atmospheric pressure chemical ionization interface for capillary electrophoresis/mass spectrometry, *Anal. Chem.*, 67, 1474, 1995.

118. Dole, M., Hines, R. L., Mack, L. L., Mobley, R. C., Ferguson, L. D., and Alice, M. B., *J. Chem. Phys.*, 49, 2240, 1968.

119. Yamashita, M. and Fenn, J. B., Electrospray ion source. Another variation of the free-jet theme, *J. Phys. Chem.*, 88, 4451, 1984.

120. Yamashita, M. and Fenn, J. B., Negative ion production with the electrospray ion source, *J. Phys. Chem.*, 88, 4671, 1984.

121. Zeleny, J., Instability of electrified liquid surfaces, *Phys. Rev.*, 10, 1, 1917.

122. Bruins, A. P., Mass spectrometry with ion sources operating at atmospheric pressure, *Mass Spectrom. Rev.*, 10, 53 1991.

123. Bruins, A. P., Covey, T. R., and Henion, J. D., Ionspray interface for combined liquid chromatography/atmospheric pressure ionization mass spectrometry, *Anal. Chem.*, 59, 2642, 1987.

124. Iribarne, J. V. and Thomson, B. A., On the evaporation of small ions from charged droplets, *J. Chem. Phys.*, 64, 2287, 1976.

125. Thomson, B. A., Iribarne, J. V., and Dziedzic, P. J., Liquid evaporation/mass spectrometry/mass spectrometry for the detection of polar and labile molecules, *Anal. Chem.*, 54, 2219, 1982.

126. Kebarle, P. and Tang, L., From ions in solution to ions in the gas phase, the mechanism of electrospray mass spectrometry, *Anal. Chem.*, 65, 972A, 1993.

127. Covey, T. R., Bonner, R. F., Shushan, B. I., and Henion. J. D., The determination of protein, oligonucleotide and peptide molecular weights by ionspray mass spectrometry, *Rapid Commun. Mass Spectrom.*, 2, 11, 1988.

128. Barber, M., Bordoli, R. S., Elliott, G. J., Sedgwick, R. D., and Tyler, A. N., Fast atom bombardment mass spectrometry, *Anal. Chem.*, 54, 645A, 1982.

129. Caprioli, R. M., Continuous-flow fast atom bombardment mass spectrometry, *Anal. Chem.*, 62, 447A, 1990.

130. Ito, Y., Takeuchi, T., Ishii, D., and Goto, M. J., Direct coupling of micro high performance liquid chromatography with fast atom bombardment mass spectrometry, *J. Chromatogr.*, 346, 161, 1985.

131. Caprioli, R. M., Fan, T., and Cottrell, J. S., Continuous-flow sample probe for fast atom bombardment mass spectrometry, *Anal. Chem.*, 58, 2949, 1986.

132. Caprioli, R. M., *Continuous-Flow Fast Atom Bombardment Mass Spectrometry,* John Wiley & Sons, Chichester, 1990.

133. Moseley, M. A., Deterding, L. J., Tomer, K. B., and Jorgenson, J. W., Capillary zone electrophoresis-fast atom bombardment mass spectrometry: design of an on-line coaxial continuous-flow interface, *Rapid Commun. Mass Spectrom.*, 3, 87, 1989.

134. Olivares, J. A., Nguyen, H. T., Yonker, C. R., and Smith, R. D., On-line mass spectrometric detection for capillary zone electrophoresis, *Anal. Chem.*, 59, 1230, 1987.

135. Whitehouse, C. M., Dreyer, R. N., Yamashita, M., and Fenn, J. B., Electrospray interface for liquid chromatographs and mass spectrometers, *Anal. Chem.*, 57, 675, 1985.

136. Smith, R. D., Barinaga, C. J., and Udseth, H. R., Improved electrospray ionization interface for capillary zone electrophoresis-mass spectrometry, *Anal. Chem.*, 60, 1948, 1988.

137. Loo, J. A., Udseth, H. R., and Smith, R. D., Peptide and protein analysis by electrospray ionization-mass spectrometry and capillary electrophoresis-mass spectrometry, *Anal. Biochem.*, 179, 404, 1989.

138. Loo, J. A., Jones, H. K., Udseth, H. R., and Smith, R. D., Capillary zone electrophoresis-mass spectrometry with electrospray ionization of peptides and proteins, *J. Microcol. Sep.*, 1, 223, 1989.

139. Edmonds, C. G., Loo, J. A., Barinaga, C. J., Udseth, H. R., and Smith, R. D., Capillary electrophoresis-electrospray ionization-mass spectrometry, *J. Chromatogr.*, 474, 21, 1989.

140. Udseth, H. R., Loo, J. A., and Smith, R. D., Capillary isotachophoresis-mass spectrometry, *Anal. Chem.*, 61, 228, 1989.

141. Smith, R. D., Loo, J. A., Edmonds, C. G., Barinaga, C. J., and Udseth, H. R., Capillary zone electrophoresis and isotachophoresis-mass spectrometry of polypeptides and proteins based upon an electrospray ionization interface, *J. Chromatogr.*, 480, 211, 1989.

142. Smith, R. D., Loo, J. A., Edmonds, C. G., Barinaga, C. J., and Udseth, H. R., Sensitivity considerations for large molecule detection by capillary electrophoresis-electrospray ionization mass spectrometry, *J. Chromatogr.*, 516, 157, 1990.

143. Goodlett, D.R., Wahl, J.H., Udseth, H.R., and Smith, R.D., Reduced elution speed for capillary electrophoresis/mass spectrometry, *J. Microcol. Sep.*, 5, 57, 1993.

144. Pleasance, S., Thibault, P., and Kelly, J.F., Comparison of liquid-junction and coaxial interfaces for capillary electrophoresis-mass spectrometry with application to compounds of concern to the aquaculture industry, *J. Chromatogr.*, 591, 325, 1992.

145. Tetler, L.W., Cooper, P.A., and Powell, B., Influence of capillary dimensions on the performance of a coaxial capillary electrophoresis-electrospray mass spectrometry interface, *J. Chromatogr.*, 700, 21, 1995.

146. Wahl, J.H., Goodlett, D.R., Udseth, H.R., and Smith, R.D., Use of small-diameter capillary for increasing peptide and protein detection sensitivity in capillary electrophoresis-mass spectrometry, *Electrophoresis*, 14, 448, 1993.

147. Foret, F., Thompson, T.J., Vouros, P., Karger, B.L., Gebauer, P., and Bocek, P., Liquid sheath effects on the separation of proteins in capillary electrophoresis/electrospray mass spectrometry, *Anal. Chem.*, 66, 4450, 1994.

148. Huggins, T.G. and Henion, J.D., Capillary electrophoresis/mass spectrometry determination of inorganic ions using an ionspray-sheath flow interface, *Electrophoresis*, 14, 531, 1993.

149. Corr, J.J., Covey, T.R., and Anacleto, J.F., IC-MS and CE-MS for elemental analysis using an ionspray interface, in *Proc. 42nd ASMS Conf. on Mass Spectrom. and Allied Topics*, Chicago, IL, May 29–June 3, 1994, pp. 340.

150. Tetler, L.W., Cooper, P.A., and Carr, C.M., The application of capillary electrophoresis/mass spectrometry using negative-ion electrospray ionization to areas of importance in the textile industry, *Rapid Commun. Mass Spectrom.*, 8, 179, 1994.

151. Varghese, J. and Cole, R.B., Optimization of capillary zone electrophoresis-electrospray mass spectrometry for cationic and anionic laser dye analysis employing opposite polarities at the injector and interface, *J. Chromatogr.*, 639, 303, 1993.

152. Pleasance, S., Ayer, S.W., Laycock, M.V., and Thibault, P., Ionspray mass spectrometry of marine toxins. III. Analysis of paralytic shellfish poisoning toxins by flow-injection analysis, liquid chromatography/mass spectrometry, and capillary electrophoresis mass spectrometry, *Rapid Commun. Mass Spectrom.*, 6, 14, 1992.

153. Buzy, A., Thibault, P., and Laycock, M.V., Development of a capillary electrophoresis method for the characterization of enzymatic products arising from the carbamoylase digestion of paralytic shellfish poisoning toxins, *J. Chromatogr.*, 688, 301, 1994.

154. Bateman, K.P., Thibault, P., Douglas, D.J., and White, R.L., Mass spectral analyses of micro-cystins from toxic cyanobacteria using on-line chromatographic and electrophoretic separations, *J. Chromatogr.*, 712, 253, 1995.

155. Hines, H.B., Brueggemann, E.E., Holcomb, M., and Holder, C.L., Fumonisin B1 analysis by capillary electrophoresis-electrospray ionization mass spectrometry, *Rapid Commun. Mass Spectrom.*, 9, 519, 1995.

156. Naylor, S., Tomlinson, A.J., Benson, L.M., and Gorrod, J.W., Capillary electrophoresis and capillary electrophoresis-mass spectrometry in drug and metabolite analysis, *Eur. J., Drug Metab. Pharmocokinet.*, 19, 235, 1994.

157. Tomlinson, A.J., Benson, L.M., and Naylor, S., On-line capillary electrophoresis-mass spectrometry for the analysis of drug metabolite mixtures; practical considerations, *J. Cap. Elect.*, 1, 127, 1994.

158. Tomlinson, A.J., Benson, L.M., Johnson, K.L., and Naylor, S., Investigation of the metabolic fate of the neuroleptic drug haloperidol by capillary electrophoresis-electrospray ionization mass spectrometry, *J. Chromatogr.*, 621, 239, 1993.

159. Tomlinson, A.J., Benson, L.M., Gorrod, J.W., and Naylor, S., Investigation of the in vitro metabolism of the H2-antagonist mifentidine by on-line capillary electrophoresis-mass spectrometry using non-aqueous separation conditions, *J. Chromatogr. B*, 657, 373, 1994.

160. Henion, J.D., Mordehai, A.V., and Cai, J., Quantitative capillary electrophoresis-ionspray mass spectrometry on a benchtop ion trap for the determination of isoquinoline alkaloids, *Anal. Chem.*, 66, 2103, 1994.

161. Kostiainen, R., Bruins, A.P., and Hakkinen, V.M.A., Identification of degradation products of some chemical warfare agents by capillary-electrophoresis-ionspray mass spectrometry, *J. Chromatogr.*, 634, 113, 1993.

162. Johansson, J.M., Huang, E.C., Henion, J.D., and Zweigenbaum, J., Capillary electrophoresis-atmospheric pressure ionization mass spectrometry for the characterization of peptides. Instrumental considerations for mass spectrometric detection, *J. Chromatogr.*, 554, 311, 1991.

163. Rosnack, K.J., Stroh, J.G., Singleton, D.H., Guarino, B.C., and Andrews, G.C., Use of capillary electrophoresis-electrospray ionization mass spectrometry in the analysis of synthetic peptides, *J. Chromatogr.*, 675, 219, 1994.

164. Hsieh, F.Y., Cai, J., and Henion, J.D., Determination of trace impurities of peptides and alkaloids by capillary electrophoresis-ionspray mass spectrometry, *J. Chromatogr.*, 679, 206, 1994.

165. Kostiainen, R., Lasonder, E., Bloemhoff, W., van Veelen, P.A., Welling, G.W., and Bruins, A.P., Characterization of a synthetic 37-residue fragment of a monoclonal antibody against herpes virus by capillary electrophoresis/electrospray mass spectrometry and ^{252}Cf plasma desorption mass spectrometry, *Biol. Mass Spectrom.*, 23, 346, 1994.

166. Perkins, J.R., Parker, C.E., and Tomer, K.B., The characterization of snake venoms using capillary electrophoresis in conjunction with electrospray mass spectrometry: Black Mambas, *Electrophoresis*, 14, 458, 1993.

167. Moseley, M.A., Jorgenson, J.W., Shabanowitz, Hunt, D.F., and Tomer, K.B., Optimization of capillary zone electrophoresis/electrospray ionization parameters for the mass spectrometry and tandem mass spectrometry analysis of peptides, *J. Am. Soc., Mass Spectrom.*, 3, 289, 1992.

168. Tsuji, K., Baczynskyj, L., and Bronson, G.E., Capillary electrophoresis-electrospray mass spectrometry for the analysis of recombinant bovine and porcine somatotropins, *Anal. Chem.*, 64, 1864, 1992.

169. Thibault, P., Paris, C., and Pleasance, S., Analysis of peptides and proteins by capillary electrophoresis/mass spectrometry using acidic buffers and coated capillaries, *Rapid Commun. Mass Spectrom.*, 5, 484, 1991.

170. Cole, R.B., Varghese, J., McCormick, R.M., and Kadlecek, D., Evaluation of a novel hydrophilic derivatized capillary for protein analysis by capillary electrophoresis-electrospray mass spectrometry, *J. Chromatogr.*, 680, 363, 1994.

171. Wahl, J.H., Goodlett, D.R., Udseth, H.R., and Smith, R.D., Attomole level capillary electrophoresis-mass spectrometric protein analysis using 5 μm i.d. capillaries, *Anal. Chem.*, 64, 3194, 1992.

172. Kostiainen, R., Franssen E.J.F., and Bruins, A.P., Capillary zone electrophoresis-ion spray mass spectrometry of a synthetic drug-protein conjugate mixture, *J. Chromatogr.*, 647, 361, 1993.

173. Hsieh, Y.L., Cai, J, Li, Y.T., Henion, J.D., and Ganem, B., Detection of noncovalent FKBP-FK506 and FKBP-rapamycin complexes by capillary electrophoresis-mass spectrometry and capillary electrophoresis-tandem mass spectrometry, *J. Am. Soc. Mass Spectrom.*, 6, 85, 1995.

174. Tuong, A., Uzabiaga, F., Petitou, M., Lormeau, J.C., and Picard, C., Direct observation of the non-covalent complex between human antithrombin III and its heparin binding sequence by capillary electrophoresis and electrospray mass spectrometry, *Carbohydr. Lett.*, 1, 55, 1994.

175. Kelly, J.F., Masoud, H., Perry, M.B., Richards, J.C., and Thibault, P., Separation and characterization of O-deacylated lipooligosaccharides and glycans derived from Moraxella catarrhalis using capillary electrophoresis-electrospray mass spectrometry and tandem mass spectrometry, *Anal. Biochem.*, 233, 15, 1996.

176. Auriola, S., Thibault, P., Sadovskaya, I., Altman, E., Masoud, H., and Richards, J.C., Structural characterization of lipopolysaccharides from *Pseudomonas aeruginosa* using capillary electrophoresis-electrospray mass spectrometry and tandem mass spectrometry, Snyder P., Ed., ACS symposium series, Vol. 619, pp. 149, 1996.

177. Zhao, Z, Wahl, J., Udseth, H.R., Hofstadler, S.A., Fuciarelli, A.F., and Smith, R.D., On-line capillary electrophoresis-electrospray ionization mass spectrometry of nucleotides, *Electrophoresis*, 16, 389, 1995.

178. Carr, S.A., Huddleston, M.J., and Bean, M.F., Selective identification and differentiation of N- and O-linked oligosaccharides in glycoproteins by liquid chromatography-mass spectrometry, *Protein Sci.*, 2, 183, 1993.

179. Perkins, J.R., Parker, C.E., and Tomer, K.B., Nanoscale separations combined with electrospray ionisation mass spectrometry: Sulfonamide determinations, *J. Am. Soc. Mass Spectrom.*, 3, 139, 1992.

180. Moseley, M. A., Deterding, L. J., Tomer, K. B., and Jorgenson, J. W., Coupling of capillary zone electrophoresis and capillary liquid chromatography with coaxial continuous-flow fast atom bombardment tandem sector mass spectrometry, *J. Chromatogr.*, 480, 197, 1989.

181. deWit, J. S. M., Deterding, L. J., Mosely, M. A., Tomer, K. B., and Jorgenson, J. W., Design of a coaxial continuous flow fast atom bombardment probe, *Rapid Commun. Mass Spectrom.* 2, 100, 1988.

182. de Wit, J.S.M., Parker, C.E., Tomer, K.B., and Jorgenson, J.W., Separation and identification of trifluralin metabolites by open-tubular liquid-chromatography negative chemical ionization mass spectrometry, *Biomed. Environ. Mass Spectrom.*, 17, 47, 1988.

183. Suter, M.J. and Caprioli, R.M., An integral probe for capillary zone electrophoresis/ continuous flow fast atom bombardment mass spectrometry, *J. Am. Soc. Mass Spectrom.*, 3, 198, 1992.

184. Deterding, L.J., Parker, C.E., Perkins, J.R., Moseley, M.A., Jorgenson, J.W., and Tomer, K.B., Capillary liquid chromatography-mass spectrometry and capillary zone electrophoresis-mass spectrometry for the determination of peptides and proteins, *J. Chromatogr.*, 554, 329, 1991.

185. Nichols, W., Zweigenbaum, J., Garcia, F., Johansson, M., and Henion, J., CE-MS for industrial applications using a liquid junction with ion-spray and CF-FAB mass spectrometry, *LC-GC*, 10, 676, 1992.

186. Lee, E. D., Muck, W., Henion, J. D., and Covey, T. R., Liquid-junction coupling for capillary zone electrophoresis-ionspray mass spectrometry, *Biomed. Environ. Mass Spectrom.*, 18, 844, 1989.

187. Lee, E. D., Muck, W., Henion, J. D., and Covey, T. R., On-line capillary zone electrophoresis-ionspray tandem mass spectrometry for the determination of dynorphins, *J. Chromatogr.*, 458, 313, 1989.

188. Lee, E. D., Muck, W., Henion, J. D., and Covey, T. R., Capillary zone electrophoresis-tandem mass spectrometry for the determination of sulfonated azo dyes, *Biomed. Environ. Mass Spectrom.*, 18, 253, 1989.

189. Johansson, I. M., Huang, E. C., Henion, J. D., and Zweigenbaum, J., Capillary electrophoresis-atmospheric pressure ionization mass spectrometry for the characterization of peptides: Instrumental considerations for mass spectral detection, *J. Chromatogr.*, 554, 311, 1991.

190. Johansson, I. M., Pavelka, R., and Henion, J. D., Determination of small drug molecules by capillary electrophoresis-atmospheric pressure ionization mass spectrometry, *J. Chromatogr.*, 559, 515, 1991.

191. Garcia, F. and Henion, J. D., Gel-filled capillary electrophoresis/mass spectrometry using a liquid junction-ionspray interface, *Anal. Chem.*, 64, 985, 1992.

192. Reinhoud, N. J., Niessen, W. M. A., Tjaden, U. R., Gramberg, L. G., Verheij, E. R., and van der Greef, J., Performance of a liquid-junction interface for capillary electrophoresis mass spectrometry using continuous-flow fast atom bombardment, *Rapid Commun. Mass Spectrom.*, 3, 348, 1989.

193. Minard, R. D., Chin-Fatt, D., Curry, P., and Ewing, A. G., *Proc. 36th ASMS Conference on Mass Spectrometry and Allied Topics,* San Francisco, CA, June 5-10, 1988, pp. 950.

194. Minard, R.D., Luckenbill, D., Curry, P., and Ewing, A.G., Capillary electrophoresis/flow FAB/MS, *Adv. in Mass Spectrom.*, 11, 436, 1989.

195. Caprioli, R. M., Moore, W. T., Martin, M., DaGue, B. B., Wilson, K., and Moring, S., Coupling capillary zone electrophoresis and continuous-flow fast atom bombardment mass spectrometry for the analysis of peptide mixtures, *J. Chromatogr.*, 480, 247, 1989.

196. Reinhoud, N. J., Schroder, E., Tjaden, U. R., Niessen, W. M. A., Ten Noever de Brauw, M. C., and van der Greef, J., Static and scanning array detection in capillary electrophoresis-mass spectrometry, *J. Chromatogr.*, 516, 147, 1990.

197. Wolf, S.M., Vouros, P., Norwood, C., and Jackim, E., Identification of deoxynucleoside-polyaromatic hydrocarbon adducts by capillary zone electrophoresis-continuous flow-fast atom mass spectrometry, *J. Am. Soc. Mass Spectrom.*, 3, 757, 1992.

198. Wolf, S.M. and Vouros, P., Incorporation of sample stacking techniques into capillary electrophoresis CF-FAB mass spectrometric analysis of DNA adducts, *Anal. Chem.*, 67, 891, 1995.

199. Dale, D.C. and Smith, R.D., Small volume and low flow rate electrospray ionization mass spectrometry of aqueous samples, *Rapid Commun. Mass Spectrom.*, 7, 1017, 1993.

200. Emmett, M.R. and Caprioli, R.M., Micro-electrospray mass spectrometry: Ultra-high sensitivity analysis of peptides and proteins, *J. Am. Soc. Mass Spectrom.*, 5, 605, 1994.

201. Andren, P.E., Emmett, M.R., and Caprioli, R.M., Micro-electrospray: Zeptomole/attomole per microliter sensitivity for peptides, *J. Am. Soc. Mass Spectrom.*, 5, 867, 1994.

202. Wilm, M.S. and Mann, M., Electrospray and Taylor-cone theory, Dole's beam of macromolecules at last?, *Int. J. Mass Spectrom. Ion Proc.*, 136, 167, 1994.

203. Kriger, M.S., Cook, K.D., and Ramsey, R.S., Durable gold-coated fused silica capillaries for use in electrospray mass spectrometry, *Anal. Chem.*, 67, 385, 1995.

204. Kelly, J.F., Thibault, P., and Ramaley, L.R., CZE-ESMS of peptides and proteins using a sheathless interface, *Proc. 43rd ASMS Conference on Mass Spectrometry and Allied Topics,* Atlanta, GA, May 21-26, 1995, pp. 1314.

205. Wahl, J.H., Gale, D.C., and Smith, R.D., Sheathless capillary electrophoresis-electrospray ionization mass spectrometry using 10 μm i.d. capillaries: Analyses of tryptic digests of cytochrome c, *J. Chromatogr.*, 659, 217, 1994.

206. Wahl, J.H. and Smith, R.D., Comparison of buffer systems and interface designs for capillary electrophoresis-mass spectrometry, *J. Cap. Elect.*, 1, 62, 1994.

207. Fang, L., Zhang, R., Williams, E.R., and Zare, R.N., On-line time-of-flight mass spectrometric analysis of peptides separated by capillary electrophoresis, *Anal. Chem.*, 66, 3696, 1994.

208. Ramsey, R.S. and McLuckey, S.A., Capillary electrophoresis/electrospray ionization ion trap mass spectrometry using a sheathless interface, *J. Microcol. Sep.*, 7, 461, 1995.

209. Licklider, L., Kuhr, W.G., Lacey, M.P., Keough, T., Purdon, M.P., and Takigiku, R., On-line microreactors/capillary electrophoresis/mass spectrometry for the analysis of proteins and peptides, *Anal. Chem.*, 67, 4170, 1995.

210. Amankwa, L.A. and Kuhr, W.G., Trypsin-modified fused-silica capillary microreactor for peptide mapping by capillary zone electrophoresis, *Anal. Chem.*, 64, 1610, 1992.

211. Chu, Y., Kirby, D.P. and Karger, B.L., Solution identification of candidate peptides from combinatorial libraries by affinity capillary electrophoresis/mass spectrometry, *J. Am. Chem. Soc.*, 117, 5419, 1995.

212. Nashabeh, W., Greve, K.F., Kirby, D., Foret, F., Karger, B.L., Reifsnyder, D.H., and Builder, S.E., Incorporation of hydrophobic selectivity in capillary electrophoresis: Analysis of recombinant insulin-like growth factor I variants, *Anal. Chem.*, 66, 2148, 1994.

213. Kirby, D., Greve, K.F., Foret, F., Vouros, P., Karger, B.L., and Nashabeh, W., Capillary electrophoresis-electrospray ionization mass spectrometry utilizing electrolytes containing surfactants, *Proc. 42nd ASMS Conference on Mass Spectrometry and Allied Topics,* Chicago, IL, May 29-June 3, 1994, pp. 1014.

214. Sheppard, R.L., Cai, J., and Henion, J.D., Chiral separation and detection of terbutaline and ephedrine by capillary electrophoresis coupled to ion spray mass spectrometry, *Anal. Chem.*, 67, 2054, 1995.

215. Lamoree, M.H., Tjaden, U.R., and van der Greef, J., On-line coupling of micellar electrokinetic chromatography to electrospray mass spectrometry, *J. Chromatogr.*, 712, 219, 1995.

216. Takada, Y., Sakairi, M., and Koizumi, H., On-line combination of micellar electrokinetic chromatography and mass spectrometry using an electrospray-chemical ionization interface, *Rapid Commun. Mass Spectrom.*, 9, 488, 1995.

217. Tang, Q., Harrata, K., and Lee, C.S., Capillary isoelectric focussing-electrospray mass spectrometry for protein analysis, *Anal. Chem.*, 67, 3515, 1995.

218. Pretorius, V., Hopkins, B.J., and Scheke, J.D., A new concept for high-speed liquid chromatography, *J. Chromatogr.*, 99, 23, 1974.

219. Dittmann, M.M., Wienand, K., Bek, F., and Rozing, G.P., Theory and practice of capillary electrochromatography, *LC-GC*, 13, 800, 1995.

220. Gordon, D.B., Lord, G.A., and Jones, D.S., Development of packed capillary column electrochromatography/mass spectrometry, *Rapid Commun. Mass Spectrom.*, 8, 544, 1994.

221. Lord, G.A., Gordon, D.B., Tetler, L.W., and Carr, C.M., Electrochromatography-electrospray mass spectrometry of textile dyes, *J. Chromatogr.*, 700, 27, 1995.

222. Lane, S.J., Boughtflower, Paterson, C., and Underwood, T., Capillary electrochromatography/mass spectrometry: Principles and potential for application in the pharmaceutical industry, *Rapid Commun. Mass Spectrom.*, 9, 1283, 1995.

223. Verheij, E.R., Tjaden, U.R., Niessen, W.M.A., and van der Greef, J., Pseudo-electrochromatography-mass spectrometry: a new alternative, *J. Chromatogr.*, 554, 339, 1991.
224. Dekkers, S.E.G., Tjaden, U.R., and van der Greef, J., Development of an instrumental configuration for pseudo-electrochromatography-electrospray mass spectrometry, *J. Chromatogr.*, 712, 201, 1995.
225. Tsuda, T., Electrochromatography using high applied voltage, *Anal. Chem.*, 59, 521, 1987.
226. Schmeer, K., Behnke, B., and Bayer, E., Capillary electrochromatography-electrospray mass spectrometry: A microanalysis technique, *Anal. Chem.*, 67, 3656, 1995.
227. Tinke, A.P., Reinhoud, N.J., Niessen, W.M.A., Tjaden, U.R., and van der Greef, J., On-line isotachophoretic analyte focusing for improvement of detection limits in capillary electrophoresis/electrospray mass spectrometry, *Rapid Commun. Mass Spectrom.*, 6, 560, 1992.
228. Reinhoud, N.J., Tinke, A.P., Tjaden, U.R., Niessen, W.M.A., and van der Greef, J., Capillary isotachophoretic analyte focusing for capillary electrophoresis with mass spectrometric detection using electrospray ionization, *J. Chromatogr.*, 627, 263, 1992.
229. Thompson, T.J., Foret, F., Vouros, P., and Karger, B.L., Capillary electrophoresis/ electrospray ionization mass spectrometry: Improvement of protein detection limits using on-column transient isotachophoretic sample preconcentration, *Anal. Chem.*, 65, 900, 1993.
230. Lamoree, M.H., Reinhoud, N.J., Tjaden, U.R., Niessen, W.M.A., and van der Greef, J., On-capillary isotachophoresis for loadability enhancement in capillary zone electrophoresis/mass spectrometry of β-agonists, *Biol. Mass Spectrom.*, 23, 339, 1994.
231. Locke, S.J. and Thibault, P., Improvement of detection limits for the determination of paralytic shellfish poisoning toxins in shellfish tissues using capillary electrophoresis/electrospray mass spectrometry and discontinuous buffer systems, *Anal. Chem.*, 66, 3436, 1994.
232. van der Vlis, E., Mazereeuw, M., Tjaden, U.R., Irth, H., and van der Greef, J., Combined liquid-liquid electroextraction-isotachophoresis for loadability enhancement in capillary zone electrophoresis-mass spectrometry, *J. Chromatogr.*, 712, 227, 1995.
233. Tomlinson, A.J., Braddock, W.D., Benson, L.M., Oda, R.P., and Naylor, S., Preliminary investigations of preconcentration-capillary electrophoresis-mass spectrometry, *J. Chromatogr. B*, 669, 67, 1995.
234. Tomlinson, A.J. and Naylor, S., Enhanced performance membrane preconcentration-capillary electrophoresis-mass spectrometry (mPC-CE-MS) in conjunction with transient isotachophoresis for analysis of peptide mixtures, *J. High Resol. Chromatogr.*, 18, 384, 1995.
235. Tomlinson, A.J., Benson, L.M., Braddock, W.D., Oda, R.P., and Naylor, S., Improved on-line membrane preconcentration-capillary electrophoresis (mPC-CE), *J. High Resol. Chromatogr.*, 18, 381 1995.
236. Tomlinson, A.J., Benson, L.M., Oda, R.P., Braddock, W.D., Riggs, B.L., Katzmann, J.A., and Naylor, S., Novel modifications and clinical applications of preconcentration-capillary electrophoresis-mass spectrometry, *J. Cap. Elect.*, 2, 97, 1995.
237. Olesik, J.W., Kinzer, J.A., and Olesik, S.V., Capillary electrophoresis inductively coupled plasma spectrometry for rapid elemental speciation, *Anal. Chem.*, 67, 1, 1995.
238. Lu, Q., Bird, S.M., and Barnes, R.M., Interface for capillary electrophoresis and inductively coupled plasma mass spectrometry, *Anal. Chem.*, 67, 2949, 1995.
239. Liu, Y., Lopez-Avila, V., Zhu, J.J., Wiederin, D.R., and Beckert, W.F., Capillary electrophoresis coupled on-line with inductively coupled plasma mass spectrometry for elemental speciation, *Anal. Chem.*, 67, 2020, 1995.
240. Tomlinson, A.J., Benson, L.M., Johnson, K.L., and Naylor, S., Investigation of drug metabolism using capillary electrophoresis with photodiode array detection and on-line mass spectrometry equipped with an array detection, *Electrophoresis.*, 15, 62, 1994.
241. Muddiman, D.C., Rockwood, A.L., Gao, Q., Severs, J.C., Udseth, H.R., Smith, R.D., and Proctor, A., Application of sequential paired covariance to capillary electrophoresis electrospray ionization time-of-flight mass spectrometry: Unravelling the signal from the noise in the electropherogram, *Anal. Chem.*, 67, 4371, 1995.
242. Andrien, B.A., Banks, J.F., Boyle, J.G., Dresch, T., Haren, P.J., and Whitehouse, C.M., A mass analyzer for high speed liquid chromatography, *Proc. 43rd ASMS Conference on Mass Spectrometry and Allied Topics*, Atlanta, GA, May 21-26, 1995, pp. 999

243. Ramsey, R.S., Goeringer, D.E., and McLuckey, S.A., Active chemical background and noise reduction in capillary electrophoresis/ion trap mass spectrometry, *Anal. Chem.*, 65, 3521, 1993.

244. Hofstadler, S.A., Wahl, J.H., Bruce, J.E., and Smith, R.D., On-line capillary electrophoresis with Fourier transform ion cyclotron resonance mass spectrometry, *J. Am. Chem. Soc.*, 115, 6983, 1993.

245. Wahl, J.H., Hofstadler, S.A., Zhao, Z., Gale, D.C., and Smith, R.D., On-line microseparations with Fourier transform ion cyclotron resonance mass spectrometry, *Proc. 42nd ASMS Conference on Mass Spectrometry and Allied Topics,* Chicago, IL, May 29-June 3, 1994, pp. 1019.

246. Johnson, P.J., Gross, D.S., Schnier, P.D., and Williams, E.R., CE/MS using a novel external ion source Fourier-transform mass spectrometer, *Proc. 42nd ASMS Conference on Mass Spectrometry and Allied Topics,* Chicago, IL, May 29-June 3, 1994, pp. 235.

247. Hofstadler, S.A., Zhao, Z., Smith, R.D., Budzinski, E., Dawidski, J., Gobey, J., and Box, H., CITP ESI-FTICR for the study of X-irradiation damaged tetranucleotides, *Proc. 43rd ASMS Conference on Mass Spectrometry and Allied Topics,* Atlanta, GA, May 21-26, 1995, pp. 448

248. Hofstadler, S.A., Swanek, F.D., Gale, D.C., Ewing, A.G., and Smith, R.D., Capillary electrophoresis-electrospray ionization Fourier transform ion cyclotron resonance mass spectrometry for direct analysis of cellular proteins, *Anal. Chem.*, 67, 1477, 1995.

249. Nano-NMR probe, literature from Varian Associates, Palo Alto, CA, 1994.

250. Bayer, E., Albert, K., Nieder, M., Grom, E., Wolff, G., and Rindlisbacher, M., On-line coupling of liquid chromatography and high field nuclear magnetic resonance spectroscopy, *Anal. Chem.*, 54, 1747, 1982.

251. Dorn, H.C., ¹H-Nuclear magnetic resonance: a new detector for liquid chromatography, *Anal. Chem.*, 56, 747A, 1984.

252. Laude, D.A., Lee, W.-K., and Wilkins, C.L., Reverse-phase high performance liquid chromatography/nuclear magnetic resonance spectrometry separations of biomolecules with 1-1 hard pulse solvent suppression, *Anal. Chem.*, 57, 1464, 1985.

253. Albert, K, Kunst, M., and Bayer, E.J., Reverse-phase high performance liquid chromatography/nuclear magnetic resonance on-line coupling with solvent non-excitation, *J. Chromatogr.*, 463, 355, 1989.

254. Wu, N., Peck, T.L., Webb, A.G., Magin, R.L., and Sweedler, J.V., Nanoliter volume sample cells for ¹H NMR: Application to on-line detection in capillary electrophoresis, *J. Am. Chem. Soc.*, 116, 7929, 1994.

255. Albert, K., Direct on-line coupling of capillary electrophoresis and ¹H NMR spectroscopy, *Angew. Chem. Int. Ed. Engl.*, 34, 641, 1995.

256. Wu, N., Peck, T.L., Webb, A.G., and Sweedler, J.V., On-line NMR detection of amino acids and peptides in microbore LC, *Anal. Chem.*, 67, 3101, 1995.

257. Sidelmann, U.G., Gavaghan, C., Carless, H.A.J., Spraul, M., Hofman, M., Lindon, J.C., Wilson, I.D., and Nicholson, J.K., 750-MHz directly coupled HPLC-NMR: Application to the sequential characterization of the positional isomers and anomers of 2-, 3-, and 4-fluorobenzoic acid glucuronides in equilibrium mixtures, *Anal. Chem.*, 67, 4441, 1995.

258. Cohen, A.S., Najarian, D.R., Paulus, A., Guttman, A., Smith, J.A., and Karger, B.L., Rapid separation and purification of oligonucleotides by high performance capillary gel electrophoresis, *Proc. Natl., Acad. Sci. U.S.A.*, 85, 9660, 1988.

259. Banke, N., Hansen, K., and Diers, I., Detection of enzyme activity in fractions collected from free solution capillary electrophoresis of complex samples, *J. Chromatogr.*, 559, 325, 1991.

260. Camilleri, P., Okafo, G.N., Southan, C., and Brown, R., Analytical and micropreparative capillary electrophoresis of the peptides from calcitonin, *Anal. Biochem.*, 198, 36, 1992.

261. Gagnon, J. and Goeltz, P., Sample preparation for sequence analysis: use of pressure mobilization with P/ACE 2000, Beckman application note DS-801, 1991.

262. Huang, X. and Zane, R.N., Use of on-column frit in capillary zone electrophoresis: sample collection, *Anal. Chem.*, 62, 443, 1990.

263. Takigiku, R., Keough, T., Lacey, M. P., and Schneider, R. E., Capillary-zone electrophoresis with fraction collection for desorption mass spectrometry, *Rapid Commun. Mass Spectrom.*, 4, 24, 1990.

264. Keough, T., Takigiku, R., Lacey, M. P., and Purdon, M., Matrix-assisted laser desorption mass spectrometry of proteins isolated by capillary zone electrophoresis, *Anal. Chem.*, 64, 1594, 1992.

265. van Veelen, P.A., Tjaden, U.R., and van der Greef, J., Off-line coupling of capillary electrophoresis with matrix-assisted laser desorption mass spectrometry, *J. Chromatogr.*, 647, 367, 1993.

266. Ericksson, K.O., Palm, A., and Hjertén, S., Preparative capillary electrophoresis based on adsorption of the solutes (proteins) onto a moving blotting membrane as they migrate out the capillary, *Anal. Biochem.*, 201, 211, 1992.

267. Cheng, Y., Fuchs, M., Andrews, D., and Carson, W., Membrane fraction collection for capillary electrophoresis, *J. Chromatogr.*, 608, 109, 1992.

268. Warren, W., Cheng, Y.-F., and Fuchs, M., Protein immunodetection using capillary electrophoresis with membrane fraction collection, *LC-GC*, 12, 12, 1994.

269. Weinmann, W., Parker, C.E., Deterding, L.J., Papac, D.I., Hoyes, J., Przybylski, M., and Tomer, K.B., Capillary electrophoresis-matrix-assisted laser desorption ionization mass spectrometry of proteins, *J. Chromatogr.*, 680, 353, 1994.

270. Castoro, J.A., Chiu, R.W., Monnig, C.A., and Wilkins, C.L., Matrix-assisted laser desorption ionization of capillary electrophoresis effluents by Fourier transform mass spectrometry, *J. Am. Chem. Soc.* 114, 7571, 1992.

271. Chiu., R.W., Walker, K.L., Hagen, J.J., Monnig, C.A., and Wilkins, C.L., Coaxial capillary and conductive capillary interfaces for collection of fractions isolated by capillary electrophoresis, *Anal. Chem.*, 67, 4190, 1995.

272. Walker, K.L., Chiu, R.W., Monnig, C.A., and Wilkins, C.L., Off-line coupling of capillary electrophoresis and matrix-assisted laser desorption/ionization time-of-flight mass spectrometry, *Anal. Chem.*, 67, 4197, 1995.

CHAPTER 3

Separation of Small Organic Molecules

Werner G. Kuhr

CONTENTS

I. PRINCIPLES OF ELECTROPHORETIC SEPARATIONS

The most fundamental mechanism of separation in electrophoresis utilizes the migration of charged molecules in an applied electric field. The generation of selectivity in most electrophoretic techniques involves a simple variation of the molecule's environment to either change the charge on the molecule (i.e., isoelectric focusing) or to physically retard the molecule's movement in the electric field (i.e., polyacrylamide gel electrophoresis, PAGE). This gives the experimenter a tremendous variety of approaches to enhance the selectivity of the separation process. The utility of the technique is such that it will separate anything ranging from inorganic ions to proteins to whole cells.

0-8493-9127-X/98/$0.00+$.50

Separations are accomplished through the movement of ions in an applied electric field. This movement is governed by the ion's electrophoretic *mobility* (μ), where this mobility is defined as the average velocity with which an ion moves under the influence of an applied potential field (normalized to the electric field strength, units of V cm^{-1}). The two primary factors that affect mobility include the applied electric force (F_{EF}), which depends on the charge of the ion (q) and the electric field strength (E (V/cm); Equation 1) and the frictional force (F_{FR}) that a molecule feels as it moves through a solution (Equation 2). The simplest estimation of the frictional force (which neglects the effect of other ions in solution, and merely assumes that the total retarding force is equal to the frictional drag determined by Stokes Law)

$$\overline{F}_{EF} = q\overline{E} \tag{1}$$

depends on a number of parameters, including the viscosity of the buffer (η), the velocity of the molecule (v; cm/s), and the size of the molecule (related to the radius of a "spherical" molecule, r).

$$\overline{F}_{FR} = 6\pi\eta r v \tag{2}$$

At equilibrium, these two forces are directly balanced, and the electrophoretic velocity (v; Equation 3) depends on all parameters involving both the analyte and the separation

$$\overline{v} = \frac{q\overline{E}}{6\pi\eta r} \tag{3}$$

column. As mentioned previously, most people concern themselves with the electrophoretic mobility (μ), which is the electrophoretic velocity normalized to the electric field strength (Equation 4).

$$\overline{\mu} = \frac{\overline{v}}{\overline{E}} = \frac{q}{6\pi\eta r} \tag{4}$$

The net effect of these balanced forces results in the following properties for the electrophoretic mobility of an ion:

1. Mobility is directly proportional to the charge of the ion
2. Mobility is inversely proportional to the viscosity of the solvent
3. Mobility is inversely proportional to the radius of the particle (represented by the diffusion coefficient)

It is important to note that all of these parameters are vector quantities, not scalar numbers. This merely indicates that each force or velocity has a directional component (either aligned with or against the electric field). This vector representation will be even more important when we consider the mobilities found in capillary electrophoresis.

A. Types of Electrophoretic Separations

There are three major categories of electrophoretic separations that are pertinent to capillary separations: moving boundary electrophoresis, zone electrophoresis, and steady-state electrophoresis. Each relies on different mechanisms for the introduction and separation of species in an applied electric field.

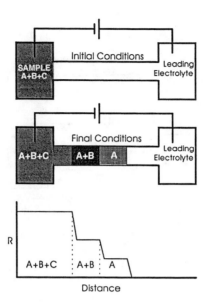

Figure 3.1 Moving boundary electrophoresis.

Moving boundary electrophoresis was the earliest electrophoretic method proposed by Tiselius. The experiment is initiated with sample as the only constituent in one reservoir (Figure 3.1). This reservoir is then connected via a long tube to another reservoir containing an electrolyte with a very high mobility. An appropriate potential is applied between these reservoirs, and the sample ions begin to migrate in the applied electric field. For example, if one were interested in separating anionic species, the sample reservoir would be set at a negative potential (the cathode), and analytes will migrate toward the positive pole (the anode). The fastest ions (the leading electrolyte, which, by definition, has the highest mobility) migrates along the tube, followed by successively slower components in the sample. As time goes on, the fastest ion in the sample emerges from the bulk of the sample at the same concentration as the leading electrolyte. Any decrease in the ionic strength of any segment of the tube produces an increase in the resistance of that section of solution, which in turn produces a voltage gradient that increases the velocity of the slower molecule in that segment (the same type of voltage gradient will be used to advantage in isotachophoresis). Successive zones contain mixtures of all faster ions and the next slower ion, until the bulk sample solution itself appears in the tube.

Obviously, the main problem with this approach is that only the leading component is separated from all other components in the sample. Detection of the different zones requires visualization of the entire separation medium, and the accuracy of the determination of each successive component is diminished, because one is effectively trying to measure a small signal on top of an ever-increasing background.

In *zone electrophoresis*, the most commonly used form in the capillary format, the entire separation matrix is filled with an excess of buffered electrolyte (Figure 3.2). An infinitely thin zone of sample is introduced at one end of the system, and a potential is applied. Each component migrates differentially along the length of the separation matrix, on which the velocity of this migration is determined by the mobility of each ion. Thus, each sample produces a discrete sample zone, similar to a chromatographic separation, and can be detected as it migrates past a detector placed at the end of the separation column (if that component is moving in the right direction!). This is the most popular format for analytical electrophoresis, since each component is physically separated from all other components, allowing more reliable quantitation and also isolation of individual components from a complex matrix. The basic tenets of electrophoretic separations assume that it is easy to modify the separation

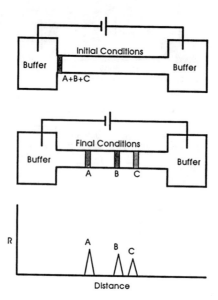

Figure 3.2 Zone electrophoresis.

conditions to allow optimization of any given separation through modification of the composition of the separation buffer, which, in turn, modifies the mobility of the analyte ion.

The simplest method for altering the electrophoretic mobility of an organic molecule involves changing the relative charge of the molecule by varying the pH of the buffer. Since most analytes have either acidic or basic groups, the extent of dissociation of these weakly acidic (e.g., carboxylate, sulfate, phosphate) or basic (e.g., amine, amide) groups is determined by the Henderson–Hasselbach Equation (Equation 5), which relates the pH of the solution to the pK_a of a weak acid or base and its degree of dissociation (α).

$$pH = pK_a + \log(1/\alpha_i - 1) \tag{5}$$

The *effective mobility* of an ion (μ_i) is simply the sum of all of the mobilities of all dissociable groups (Equation 6), each of which is determined independently by multiplying the degree of dissociation of each group (α_i) by the absolute mobility of each ion (μ_i^o). This assumes that the dissociation equilibrium kinetics are rapid and independent of one another.

$$\bar{\mu}_i = \sum_i \alpha_i \times \bar{\mu}_i^o \tag{6}$$

The resistance to movement for an ion can also change if its transport is hindered through the addition of a sieving agent if the molecules vary substantially in molecular size in free solution. Polyacrylamide gel electrophoresis is by far the most popular form of electrophoresis for the separation of large molecules (proteins, RNA/DNA fragments up to 200,000 MS). The degree of crosslinking of the polymer matrix determines pore size and has proven to be one of the most important mechanisms available to adjust the mobility of large molecules. Even larger molecules must be run in a matrix free of crosslinking, i.e., agarose gels are used for the separation of large strands of DNA and RNA, even for the separation of viruses and small bacteria.

**Effect of Cross-Linker Concentration
in Gel Electrophoresis**

% BIS	MW Range
5%	50,000–300,000
10%	10,000–100,000
15%	10,000–60,000

In steady-state electrophoretic separations, we create a gradient in the system so that each ion will migrate to a point where it has zero net mobility. In *isotachophoresis*, a voltage gradient is created by using special leading and terminating buffers (Figure 3.3). The leading buffer is chosen to contain an ion that moves faster than all sample ions (as described under moving boundary electrophoresis). Similarly, the terminating buffer contains an ion that moves slower than all sample ions. No buffer is added to the sample itself, so that when the potential is applied, zones separate and remain adjacent, since leading buffer will move fastest, fastest analyte second, etc., until the terminating buffer is reached. There is no buffer in the analyte to carry charge. Thus as zones spread out, the concentration of ions decreases (R increases). Since the current is constant, the electric field has to increase in that region, increasing the velocity of the molecules in that zone (without changing the mobility of the ion). Why don't zones move apart from one another? $E = IR$! The leading and trailing edges of each zone are constantly "refocused" by the changing electric fields. There is a net voltage gradient determined by the leading and terminating buffers (where the voltage gradient in each zone is determined by the mobility of its principal component), and the total ionic strength of each zone will be the same as the leading buffer. This has several important ramifications. Quantitation is very different from zone electrophoresis, because the amount of sample is determined by the length of the sample zone, and because the ionic strength, and hence concentration of each sample, must be the same throughout a given separation.

Isoelectric focusing is a technique that uses a pH gradient to change the mobility of the analyte ions (Figure 3.4). All large peptides and proteins have a pH in which their net mobility

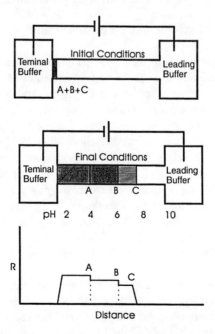

Figure 3.3 Isotachophoresis.

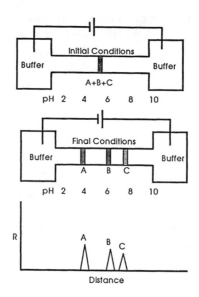

Figure 3.4 Isoelectric focusing.

is zero (i.e., the dissociation of positively and negatively charged groups balance, resulting in a net zero charge). This pH is termed the molecule's pI, or point of zero net charge. In isoelectric focusing, the sample is placed in a buffer containing a stationary pH gradient. When the electric field is applied, samples will move toward the region where the buffer pH equals their pI. At that point, their net charge equals zero, and the sample will no longer migrate. If an analyte molecule diffuses away from this point, it gains charge (due to the dissociation of either an acidic or basic group) and is rapidly migrated back to the pI region. This procedure actually focuses the samples into very tight zones, because any movement results in a change in charge, which results in migration back into the focused zone.

B. Problems with Standard-Format Electrophoresis

There are two major problems in the implementation of standard electrophoretic techniques for analytical determinations. The worst is detection of the separated species. In general, detection must be accomplished by visualizing the analyte in the separation medium (which in most cases contains a polyacrylamide or agarose gel). This means that one must scan the entire length of the tube or slab, and more important, that this solid support interferes with common detection methods used in other chromatographic separations (e.g., spectroscopic, conductivity, etc.). Therefore, detection methods commonly used in electrophoretic separations involve the use of autoradiography or staining and blotting techniques, which makes quantitation virtually impossible. With autoradiography, radioisotopes (e.g., C^{14}, P^{32}, S^{35}, I^{125}) must first be incorporated in the sample. Once the separation is complete, the gel (or slices of gel-filled tube) are placed in contact with X-ray photo paper for periods ranging from hours to weeks (depending on the level of radioactivity incorporated). Quantitation can be accomplished after the film is developed, but resolution is dependent on the grain size of the photographic emulsion.

Staining and blotting methods are even more difficult to quantitate. The separation gel is placed in a staining solution (e.g., amido black 10, coomassie blue, or fast green) for a few hours so the dye can diffuse into the gel completely and bind selectively to the analyte. The most difficult part of the procedure is destaining, in which residual dye is removed from the gel by passive diffusion for periods of up to 24 hours. Often, fixation of the sample within the gel is required to keep the sample from diffusing longitudinally in the gel during this step. The alternative to fixation, called blotting, involves the transfer of the sample protein from gel to a sensitized paper or membrane matrix (e.g., nitrocellulose). The trick is the quantitative transfer of the sample to the blotting matrix, which can be done either through diffusion, fluid flow, electromigration, or a combination of these techniques. The variety in blotting protocols (e.g., DNA diffusion = Southern blot; DBM paper/solvent flow = Northern blot; electromigration = Western blot) is an indication of the difficulty of transferring a protocol from one sample to another. Needless to say, it will be difficult to attain the level of quantitative precision expected in modern analytical chemistry with any technique involving sample blotting.

The second major difficulty in the standard electrophoretic experiment is the separation process itself. When the only mode of separation is migration, every other mode of mass transport introduces zone broadening. Heat produced through the passage of current through the separation buffer (a resistive medium, $P = I^2R$) increases the temperature of the buffer. The temperature is generated unevenly across the bed of the separation, with the greatest increase at the deepest point internally. Additionally, the main source of heat dissipation is thermal convection from the outside surfaces of the container, thus a large thermal gradient is generated between the interior and exterior surfaces of the electrophoretic bed. This thermal gradient, which can easily reach tens of degrees, causes turbulence and fluid convection, which tends to mix together any bands that may have been separated. The only way to minimize this type of convection is to reduce power production, which means decreasing the separation potential. For this reason, most commercial HV power supplies run at a potential less than 1000 V (using a pathlength of 20 cm, this limits the voltage gradient to 50 V/cm), and often in a constant power mode. This ensures that the thermal gradient does not exceed a tolerable threshold even if the solution resistance changes during the run. Additionally, increased temperatures also result in increased radial diffusion. The diffusion coefficient (D_o) increases roughly 2% for every 1°C gained. These effects become significant when the separation time exceeds one or two hours, a very common occurrence in conventional electrophoresis. This has completely eliminated the use of free-zone electrophoresis under these conditions and has severely limited the ultimate separation power available even in large, gel-filled media.

II. THEORY OF FREE-SOLUTION CAPILLARY ZONE ELECTROPHORESIS

Now we must consider what differentiates capillary electrophoresis from traditional electrophoretic separations. Basically, it is a matter of minimizing the generation of heat and maximizing heat transfer from the separation column. By minimizing the cross-sectional area of the separation medium (by about three to six orders of magnitude), one limits the amount of current flowing (for the same applied voltage) and also increases the amount of heat that can be effectively dissipated. Essentially, the amount of current that flows through the tube is a function of the cross-sectional area of the tube. For example, by decreasing the radius of a column by a factor of ten, you decrease the cross-sectional area by one hundred ($A = \pi r^2$), thereby decreasing the current a hundredfold. The lower power levels translate to lower temperature gradients, and the heat that is generated is more efficiently dissipated through the surface of the tube. The surface area of a capillary ($A_{cyl} = 2\pi rL$) becomes very large compared to its volume ($V_{cyl} = \pi r^2L$) as the radius of the capillary decreases because the ratio of surface area to volume is proportional to $1/r$. These two effects combine to make the temperature gradient between the center and wall of the capillary fall roughly with the square of the column radius. The size of the capillary, as well as the ionic strength of the buffer, determines the amount of power produced during an electrophoretic separation. For this reason, most capillaries used in CE have a diameter less than 100 μm and average around

50 μm. Other advantages of narrow-diameter capillaries will become apparent after a brief discussion of the efficiency of separation in CE.

A. Effect of Power Dissipation on Separation Efficiency

Many of the early theoretical discussions centered on the problems of power dissipation in larger i.d. capillaries. Hjerten developed much of the theoretical framework for minimizing zone broadening due to thermally induced density gradients when he realized that separation efficiency was limited by heat production and the generation of thermal gradients. His initial solution was to rotate a fairly large capillary (3 mm diameter) around its longitudinal axis, thereby homogenizing the distribution of heat throughout the column.[1] While this was not a practical instrumental arrangement, it demonstrated the effect of heat on zone broadening. The earliest reports of zone electrophoresis in relatively narrow capillaries was in instrumentation adapted from isotachophoresis by Mikkers et al.[2] They illustrated the separation of organic and inorganic anions in 200 μm i.d. Teflon® capillaries with UV or conductivity detection. They were unable to obtain high separation efficiencies because of the large injection volumes used. These were necessary because of poor detector sensitivity. They performed a theoretical evaluation of the effect of electrophoretic migration on concentration distributions in free-zone electrophoresis under these conditions.[3] Zones were found to be unsymmetrical when the concentration gradients induced by differential migration of different solutes produced inhomogeneities in the electric field. These calculations were confirmed by Thormann et al.,[4,5] who demonstrated nonsymmetrical broadening in overloaded separations due to the variation of electrophoretic velocities due to variation in electric field strength.

The realization that these problems can be avoided simply by using smaller i.d. capillaries was demonstrated by Lukacs and Jorgenson, who found that Joule heating could be virtually eliminated through the use of capillaries with an inner diameter of less than 100 μm.[6,7] Once thermal effects were eliminated, they quickly realized that the ultimate separation efficiency obtainable was only limited by diffusion along the radial axis. They developed the primary relationships governing separation efficiency and related the separation voltage, column diameter, length, and the relationship between solute and buffer concentration to resolution and separation efficiency.[8] It is important to realize that they were able to demonstrate these phenomena because they had used on-column fluorescence detection in this work. This meant that they could use very low sample concentrations and avoid many of the problems associated with sample overloading. Other workers, most notably, Tsuda et al., also demonstrated the current-voltage relationship to separation efficiency in the separation of several cations (pyridinium salts, metal ions) and anions (sulfonic acids) in CE with UV detection, but could not get as close to the theoretical maximum separation efficiency because of limitations in detector sensitivity.[9] The fundamental paradigm that CE works best when the sample concentration is only a small fraction of the buffer concentration will be stressed again and again in this chapter, because it is one of the most important concepts in understanding the practical implementation of capillary electrophoresis.

The effect of buffer type and concentration on analyte mobility, selectivity, and separation efficiency has been examined by a number of authors.[10-12] Low conductivity buffers and/or a large ratio of buffer concentration to analyte concentration were found to improve column efficiency and resolution. Rasmussen and McNair demonstrated that the relative velocity difference of two zones increased with increasing buffer concentration, but remained constant for a given concentration regardless of electrical field strength.[13] They also noted that increased dispersion due to heating from the increased buffer concentration counteracted any gains derived from the enhanced velocity differences. The effect of heat generation in CE on band broadening and separation efficiency was investigated by Grushka et al.[14] Their model compared ionic strengths of 0.1 M and 0.01 M and predicted significant loss of efficiency at high ionic strength and high field in columns larger than 75 μm i.d. because of temperature gradients across the column diameter. Maximum capillary diameters for efficient separations were calculated on the basis of buffer concentration and field strength. There are a number of ways to calculate the radial temperature gradient in capillaries. The radial temperature profile assumed earlier[14] was found to be parabolic at conventional power levels (up to 5 W), but was found to underestimate that observed at higher power levels.[14-16] Davis developed a numerical algorithm to calculate the radial profiles of temperature and ion mobility in an electrophoretic capillary from a steady-state equation of heat conduction.[17] A calculation of plate height employing this theory suggests that significant zone broadening only occurs when the difference in temperature between the capillary center and wall exceeds 5°C.

The internal temperature, the buffer viscosity, and the efficiency of heat removal from a silica capillary can be calculated by measuring the differential electroosmotic mobility at a low and a high voltage. Burgi et al. compared calculated capillary temperatures with empirically determined values. They found a linear relationship between these values when the power dissipation in the capillary was less than 3.0 W.[18] At low power, the efficiency of heat removal was found to be dictated by the heat transfer between the outer wall of the column and the surrounding environment, and not by the inside diameter of the column (as long as the column diameter was less than 100 μm). Vinther and Soeeberg calculated the temperature of zones in the capillary based on changes in viscosity, relative permittivity, zeta potential, and specific conductivity of buffer solutions.[19] From these data, they were able to determine heat transfer coefficients for liquid and air-cooled capillaries.

Joule heating is still a major problem in the separation of large molecules (i.e., proteins and oligonucleotides), because most buffer systems taken from conventional electrophoretic separations have very high ionic strength. Cohen et al. demonstrated the importance of efficient dissipation of Joule heat in PAGE, due to the high ionic strength of the buffers used.[20] They found that separation efficiency could be improved dramatically when thermoelectric (Peltier) devices are used to control the air temperature around the CE capillary.[21] The effect of air temperature variation was demonstrated by monitoring CE current. Thermal breakdown was found to occur at high power levels, which are needed in some protein separations.

B. Separation Efficiency in Free-Zone Capillary Electrophoresis

It is important to look more closely at all factors which can ultimately limit separation efficiency in CE. Such things as hydrodynamic flow (thermal convection, electroosmosis, and pressure-gradient induced flow), electrokinetic dispersion (resulting from variation in the electric field strength along the column), sample concentration, and the width of the sample zone (as well as the detection window) can limit the overall efficiency of separation. The minimization of these nondiffusive effects will be discussed after we have considered how efficient we can make these separations. What is the ultimate separation power of capillary electrophoresis if we have optimized all other aspects?

The maximum separation efficiency that can be obtained in the most general case, in which an analyte is introduced at one end of a tube, separated along the length of the tube, and detected as it leaves the tube, was first discussed by Giddings in 1969 and reviewed more recently by the same author.[22] The theoretical approach to this is quite simple, because most of the terms associated with band broadening are absent. In fact, the only term that remains is longitudinal diffusion. The efficiency of the separation can be represented as a number of theoretical plates (N), identical in definition to that used in chromatography. Jorgenson and Lukacs[6] demonstrated that N depends only on the total electrophoretic mobility (μ_{TOTAL}, units: cm^2 V^{-1} s^{-1}), the applied voltage (V, volts), and the diffusion coefficient of the molecule (D_o, unit: cm^2 s^{-1}), as shown in Equation 7.

$$N = \frac{\overline{\mu}_{TOTAL} V}{2D_o} \qquad (7)$$

The major complication for electrophoresis in small capillaries arises from the presence of electrically induced flow, or electroosmosis. This flow, discussed in greater detail below, is the result of the presence of a double-layer at the silica/water interface, which results in a slight excess of charge in the region adjacent to the capillary surface. These charged ions migrate toward the oppositely charged pole. (For silica/water, there is a net excess of positive ions, resulting in a net flow toward the cathode.) The magnitude of electroosmotic flow is quite large, because of the large potential gradient that can be applied in CE. Fortunately, it is quite easy to account for the contributions to electroosmotic flow with the theory used here. The flow rate gives an additional velocity component to the ions migrating in the electric field, such that the total velocity (v_{TOTAL}) of each molecule is the vector sum of the electrophoretic velocity (v) and the velocity imparted by electroosmotic flow (v_{EO}; Equation 8).

$$\overline{v}_{TOTAL} = \overline{v} + \overline{v}_{EO} \qquad (8)$$

Similarly, the total electrophoretic mobility is simply the vector sum of the electrophoretic mobility of the sample (μ) and the effective mobility due to electroosmotic flow (μ_{EO}), as shown in Equation 9 and Figure 3.5.

$$\overline{\mu}_{TOTAL} = \overline{\mu} + \overline{\mu}_{EO} \qquad (9)$$

It is important to realize that these are vector sums in that they are directional (aligned with or opposed to the electric field). This can be used to great advantage in CE, since all molecules that have a net velocity toward the detector will eventually be observed. Thus, the presence of electroosmotic flow allows the separation of both negatively and positively charged species in the same run. Additionally, it simplifies the theory, in that the total number of theoretical plates is described by Equation 10, which is just a combination of Equations 8 and 9. This implies that the highest efficiency is obtained when species are migrating at the fastest velocity (i.e., have the largest value of μ_{TOTAL}) and that we can use electroosmotic flow to help speed up the separation, and thereby increase the separation efficiency.

$$N = \frac{(\overline{\mu} + \overline{\mu}_{EO}) V}{2D_o} \qquad (10)$$

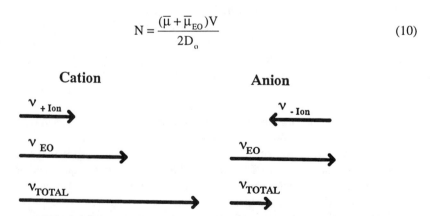

Figure 3.5 Electrophoretic mobility.

The ramifications of this relationship become more apparent when one realizes the effect this has on the time required for separation. The time required for the migration of a charged species from a separation capillary is simply the length of the capillary (L) divided by the linear velocity of the molecule (v_{TOTAL}; Equation 11). If we use the electrophoretic mobility (μ) instead of linear velocity in this calculation, we see that the elution time is minimized with a high mobility and that the application of a large electric field helps us in two ways.

$$t = \frac{L}{v_{TOTAL}} = \frac{L^2}{\overline{\mu}_{TOTAL} V} \tag{11}$$

First, we saw that the separation efficiency (N) is directly proportional to the applied voltage (V), since a molecule that moves through the column quickly does not have much time to be spread out by longitudinal diffusion. Now we see that this also results in a very fast separation, and we get the best efficiency in the shortest analysis time. This was illustrated recently by Monnig and Jorgenson,[23] who described a high-speed zone electrophoresis in a 1- to 2-cm-long fused-silica capillary. A very high electric field strength (>1,000 V cm^{-1}) and a short capillary length allowed the separation of a mixture of FTIC-labeled amino acids in 1.5 s. Formation of the analyte zone at the head of the capillary was controlled by laser-induced photolysis of a tagging reagent. This gating procedure, which produced a sample zone with a width of less than 100 ms (corresponding to a plug of sample approximately 20 μm long), was required to maintain the separation efficiency of the system. Ultimately, Joule heating of the buffer limited the speed and efficiency of the separation.

But is this the point at which we get the best separation? To answer that question, we must look at the parameter that expresses the quality of the separation, and that is *resolution* (R), or the ability of the system to separate two closely eluting species. If our separation efficiency is diffusion-limited (as we have assumed above), then the best resolution we can obtain is given by Equation 12,[6] which simply states that the resolution between two adjacent eluting species is directly proportional to the difference in their mobilities and inversely proportional to the square root of their average *total* mobility. Here we see that resolution increases with the square root of the applied voltage, again indicating that the best performance is observed at the highest electric field.

$$R = 0.177(\overline{\mu}_{i1} - \overline{\mu}_{i2})\left[\frac{V}{D_o(\overline{\mu}_{i(avg)} + \overline{\mu}_{EO})}\right]^{1/2} \tag{12}$$

This equation also points out the limitations imposed on the electrophoretic separation. As the speed of migration increases ($\mu_{AVG} + \mu_{eo}$ large), resolution decreases, simply because there is not enough time for the components to physically separate from one another. Thus, there are physical limitations on how fast we can perform the separation and maintain the same level of resolution. We will also see that other parameters in the experiment will limit the performance of the system as this speed increases. Zare and co-workers found that the size of the injection and detection zone were often the limiting quantities in the determination of separation efficiency.[24] They found that the principal limitation was usually the length of injection. Diffusion was found to be limiting only with an injection zone length of less than 3 mm for a separation length of 1 meter, when thermal and electrokinetic contributions were neglected. Let us now look at some of the other parameters characteristic of free-zone capillary electrophoresis that significantly affect separation efficiency.

C. Electroosmotic Flow

One of the most distinguishing features of capillary electrophoresis is electroosmotic flow. This bulk movement of solvent is caused by the small zeta potential (ζ) at the silica/water interface, which induces a minute excess of anionic charge at the surface of silica (Figure 3.6). This charge, presumably due to the presence of dissociated –SiOH groups on the surface of the capillary, is dependent on the pH of the electrophoretic buffer in the static diffuse double layer. Titration of the –SiOH $\Leftrightarrow$ –SiO$^-$ + H$^+$ equilibrium alters the degree of disso-ciation of the silanol (which has a pK_a of 6 to 7), thereby altering the zeta potential. This is directly related to the velocity of electroosmotic flow as shown in Equation 13 (ε_0 is the dielectric constant; η is solution viscosity; and E is electric field strength, V cm^{-1}).

$$\bar{\mu}_{EO} = \left(\frac{\varepsilon_o}{4\pi\eta} \right) E\zeta \qquad (13)$$

For the silica surface, an excess of cationic species in bulk solution migrate toward the cathode, producing net flow from anode to cathode. This means that the driving force of the flow stream originates at the walls of the capillary, producing a plug-like flow that has a flat velocity distribution across the capillary diameter, deviating only within a few nanometers of the capillary surface.[6,7] This flat flow profile and its effect on net mobility (i.e., elution of positively and negatively charged species in electrophoresis) and its use as a pump in packed-column capillary chromatography was originally demonstrated by Lukacs and Jorgenson.[8] Since that time, a number of groups have performed very thorough theoretical examinations of the parameters that govern the relationship between zeta potential and electroosmotic flow.[25-28] The effect of pH on electroosmosis in Pyrex®, silica, and Teflon capillaries using phenol as a neutral marker is shown in Figure 3.7.[8] Lambert and Middleton observed that

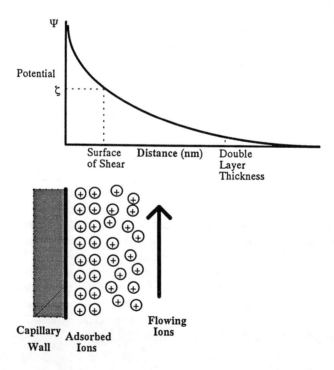

Figure 3.6 Electroosmotic flow.

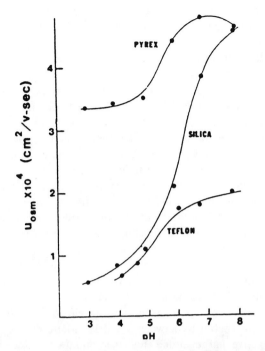

Figure 3.7 Effect of pH on electroosmosis at constant ionic strength. (From Lukacs, K. D. and Jorgenson, J. W., *J. High Resolut. Chromatogr. Chromatogr. Commun.*, 8, 407, 1985. With permission.)

the electroosmotic flow was dependent on the preconditioning of the fused-silica column. The flow velocity in a column previously exposed to acidic conditions was consistently lower than the flow from columns previously exposed to alkaline solutions.[29] For binary buffer mixtures containing water and a protic or aprotic dipolar solvent, pH of the solution was found to influence the electroosmotic flow.[30]

1. Effect of pH on Electroosmosis at Constant Ionic Strength

Most methods for the measurement of electroosmotic flow involve the addition of a small amount of a *neutral marker*, an uncharged molecule whose velocity is assumed to be completely derived from electroosmotic flow.[8, 9] The migration time for the elution of the marker is measured, and the velocity is the distance traveled divided by the migration time. This method does assume that there are no interactions between the marker molecule and anything else in the system. Another very simple method for measuring the rate of electroosmotic flow was introduced by Altria and Simpson,[31,32] who measured the change in weight in one buffer reservoir during the course of the experiment. By weighing the solution emerging from the capillary directly onto an analytical microbalance, they avoided problems with adsorption of an "unretained" marker. Flow rate was found to be inversely proportional to ionic strength, independent of column diameter, and decreased by methanol. The effect of pH was examined between 2 and 12. pH and applied voltage were found to increase electroosmotic flow rate linearly. Huang et al.[33] described another simple method for measuring electroosmotic flow, in which they measured the change in electrophoresis current when a buffer with a different ionic strength is introduced. Everaerts and co-workers described several methods for the measurement and control of electroosmotic flow using a post-column detector.[34] Van de Goor et al.[35] described two methods for the measurement of zeta potential to determine electroosmosis. The effluent could be weighed (as described above[31,32]) or the streaming potential could be measured as solvent was pumped through the column. They utilized both approaches in determining the zeta potential and electroosmotic flow of Teflon capillaries as a function of pH.

D. Alteration Of Electroosmotic Flow

A great deal of work has been done to manipulate electroosmotic flow. Fujiwara and Honda showed that the addition of NaCl reduced electroosmotic flow by decreasing double-layer thickness.[36] The addition of organic solvents increases the pK values of the surface silanol groups and produces a corresponding decrease in the electroosmotic velocity. Methanol decreases electroosmotic flow, while acetonitrile increases it, though not as dramatically.[37] Van Orman et al. observed that equivalent electroosmotic flows were obtained with different buffers if the ionic strength of the buffers was equivalent.[38] Control of the electrolyte composition, alone[39] or with the viscosity of the buffer,[40] can produce a wide range of electroosmotic flow velocities.[41] Salomon et al.[42] developed a model that accounts for the change in electroosmotic flow as the buffer concentration is changed. Important parameters for this model include the initial charge per unit area at the silica capillary wall, the thickness of the double layer adjacent to the capillary wall, and the equilibrium constant between the cations in the buffer and adsorption sites on the silica capillary. When compared with empirically derived data, this model was found to accurately predict system flow characteristics. Even the separation of some proteins in free-zone electrophoresis can be accomplished by adjusting the pH of the buffer to between 8 and 11, where the capillary wall and many proteins are electronegative and repel one another to minimize surface interactions.[43]

In addition to simple manipulation of the ionic composition of the buffer, several chemicals have been added to the buffer to alter the zeta potential developed across the capillary/water interface. The chemical additives are chosen for their ability to bind to the capillary wall and change the structure of the silica/water interface. For example, the direction of flow is reversed through the addition of a cationic surfactant, cetyltrimethyl-ammonium bromide (CTAB[31]) or tetradecyl-trimethyl-ammonium bromide (TTAB[44]). Putrescine was used to reduce electroosmotic flow,[43] and zero flow was found when s-benzyl thiouronium chloride is added to the buffer.[32] Triton-X was found to completely eliminate electroosmosis at high ionic strength.[45] Alternatively, polymeric resins can be employed to reduce electroosmosis. Various coatings have been evaluated for their ability to change the electroosmotic flow velocities of glass and polystyrene microspheres.[46] Covalently bound polyethylene glycol (MW > 5,000) was found most effective in reducing electroosmotic flow, was stable over the pH range 3.5 to 7.0, and was more effective than dextran, methylcellulose, amino- or diol-silanized columns.

Chemical derivatization of the capillary wall has been done extensively to change the properties of the silica/water interface. Silanization of the silica surface with trimethylchlorosilane (TMCS), used extensively in the preparation of deactivated columns for gas chromatography, was used early on to reduce electroosmotic flow.[7] McCormick studied the separation of proteins at low pH (3 to 5) in columns modified with various silating reagents.[47] The interaction of phosphate with the capillary surface was studied, and it was found to bind strongly to the silica surface. Modification reduced electroosmotic flow and shielded the surface from protein adsorption. Synthetic octapeptides with single amino acid substitutions were separated, as were larger proteins up to 77 k Daltons. Voltage programming was suggested to improve separation efficiency. One of the most effective reagents for derivatization of the surface seems to be γ-methacryloxypropyltrimethoxysilane, already used in the preparation of glass surfaces for gel electrophoresis. Hjerten prepared the capillary by silating with γ-methacryloxypropyltrimethoxysilane followed by crosslinking the surface-bound methacryl groups with polyacrylamide.[48] This preparation eliminated electroosmosis and minimized adsorption of analyte onto surface. The effectiveness of this treatment was illustrated with aromatic carboxylic acids and a hemoglobin sample, which was partially resolved in a 200 µm i.d. tube using UV detection. Most recently, Novotny and co-workers[49,50] produced hydrolytically stable surfaces of cellulose-derivative coatings for the separation of peptides, proteins, and glycoconjugates. Grignard reactions were used by Pesek and co-workers to form a direct carbon–silicon bond which produced a surface stable over the entire pH spectrum, and they found that these stable, hydrophilic poly(acryloylaminoethoxyethanol)-coated columns minimized adsorption of proteins and stabilized electroosmotic flow.[51]

A number of groups proposed instrumental procedures for direct control of electroosmotic flow in a fused-silica capillary.[52-55] In all these designs, a radial electric field of several kV (across the surface of the capillary relative to the inner solution) is produced. This alters the zeta potential at the inner wall and thus directly influences the direction and magnitude of electroosmotic flow. Although this technology is still in its infancy, it holds great promise for improving the speed and resolution of CE separations. Lee et al.[56] demonstrated direct control of electroosmosis by varying the zeta potential at the capillary/solution interface when this potential is small. A simple capacitor model predicts the effectiveness of the external electric field for controlling electroosmosis at different electrolyte concentrations and capillary dimensions. The direct control of the direction and velocity of electroosmotic flow was confirmed.[57] The effect of solution composition and capillary dimension on the control of the electroosmotic flow were also measured and analyzed in detail with a more elaborate capacitor model created by Hayes and Ewing.[55]

III. COLUMN PARAMETERS

Many of the available physical parameters associated with a capillary electrophoretic separation have been extensively investigated. These parameters include column material (e.g., fused silica, glass, Teflon), diameter (5 µm to 150 µm), length (1 cm to 100 cm) of the separation column, the electric field strength (10 to 1000 V/cm; which equals the applied voltage divided by the length of the column), and the most important factor, the composition of the separation buffer. Briefly stated, almost any combination of these parameters that produces a low power load (remember, $P = IE$) will work well. Additionally, it is important to realize that significant compromise of any one of these parameters may result in a dramatic reduction in overall performance.

Lukacs and Jorgenson examined many of the most important column parameters in their earliest work[8] and picked fused-silica columns primarily for their availability and ease of handing. It is important to realize that the material used for the column may play an important role in extending the utility of the technique. For example, the zeta potentials of glass, fused silica, and Teflon are different, and vary differentially with pH (Figure 3.7).[8] Additionally, the efficiency of heat removal from the separation column can be limited by the rate of heat transfer between the outer wall of the column and the surrounding environment, and not just by the inside diameter of the column.[18] Burgi et al. compared calculated capillary temperatures with empirically determined values.[18] They found a linear relationship between theoretical and experimental values when the power dissipation in the capillary was less than 3.0 W. A greater capacity for heat dissipation may allow the use of larger i.d. capillaries or higher ionic strength buffers. It is therefore likely that materials other than fused silica may eventually be used for column fabrication, even though almost all separations done today are accomplished in fused-silica columns.

The effect of the length and inner diameter of the fused-silica column (as well as the ionic strength of the buffer) on separation efficiency is much more straightforward. If our goal is to minimize thermal effects in the column (i.e., keep the radial thermal gradient to less than 5°C), then we must use conditions which minimize the electrophoretic current. In the simplest case, this can be accomplished by decreasing the diameter of the capillary (Figure 3.8), increasing the length of the capillary (Figure 3.9), or decreasing the ionic strength of

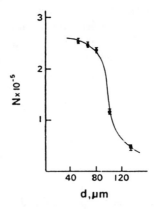

Figure 3.8 Separation efficiency as a function of column inner diameter. (From Lukacs, K. D. and Jorgenson, J. W., *J. High Resolut. Chromatogr. Chromatogr. Commun.*, 8, 407, 1985. With permission.)

the separation buffer (Figure 3.10). The effect of column diameter on separation efficiency is extremely dramatic. As shown in Figure 3.8, separation efficiency is essentially constant at column diameters less than 80 μm (where there is adequate heat dissipation and the separation efficiency is diffusion-limited), but it decreases precipitously as the column diameter is increased above 100 μm (where enough current passes to create a large [>5°C] radial temperature gradient), where all experiments were done with a 1-meter capillary filled with constant ionic strength buffer at 0.060 *M* at an applied voltage of 15 kV. This, of course, is due to increased heat production, since the surface area of the column decreases relative to its cross-sectional area as the radius increases. (Surface area is proportional to 2πr, while conductivity is proportional to the cross-sectional area, or πr².) The effect' of the length of the capillary is shown in Figure 3.9. Again, the separation efficiency is relatively constant at long lengths (>80 cm) but falls rapidly as the column is shortened. Since the resistance of the column is directly proportional to its length, we simply increase the current passing through the column, and again generate power, which generates heat. These data, taken together, indicate the maximum dimensions of the capillary as a function of the ionic strength of the separation buffer used. This does not mean that we cannot operate with columns >80

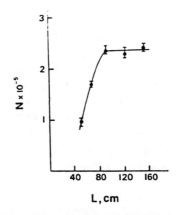

Figure 3.9 Separation efficiency as a function of column length. (From Lukacs, K. D. and Jorgenson, J. W., *J. High Resolut. Chromatogr. Chromatogr. Commun.*, 8, 407, 1985. With permission.)

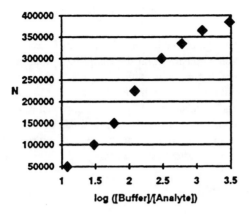

Figure 3.10 Effect of buffer concentration on separation efficiency.

μm i.d. or <80 cm in length, but that we must either reduce the applied voltage or decrease the ionic strength (hence the conductivity) of the buffer to compensate.

One of the most critical parameters governing the efficiency of separation is the choice of the separation buffer, and the most important property of that buffer is its ionic strength. This is important for two reasons. As the ionic strength of the separation buffer increases, the conductivity of the solution increases (increasing current flow and the power generated), resulting in increased temperature gradients and a reduction in separation efficiency similar to that found when the diameter of the column is increased (Figure 3.8). If this were our only concern, then one would want to use as dilute a buffer as possible. Unfortunately, the relative concentration of the buffer to the concentration of analyte will determine the homogeneity of the electric field in the analyte zone. This means that we must have a fairly large excess of buffer ions in solution to make sure that the sample ions do not change the local conductivity of the buffer (and thereby the electric field strength). This is illustrated in Figure 3.10, where the efficiency of separation is plotted as a function of the ratio of sample to background buffer.

When the sample is dilute compared to the buffer (<1% of the buffer concentration), the separation is very efficient and no distortion is observed. As the sample gets more and more concentrated, the ionic strength of the sample zone changes and the magnitude of the electric field decreases within the sample zone. This leads to a triangular shaped peak, as a result of electrophoretic "fronting," which severely degrades the separation efficiency. The rule of thumb to remember is that the sample concentration should never exceed 1% of the buffer concentration. This will lead to problems when relatively insensitive detectors are used in CE (such as UV absorbance) and will require a compromise between the sensitivity of the analysis and the efficiency of the separation.

The physical properties of the buffer solution can be used to advantage in the optimization of an electrophoretic separation. Camilleri and Okafo[58] demonstrated that CE in D_2O-based buffer solutions provided enhanced resolution compared to electrophoresis carried out in a water solution of the same acidity. These effects are thought to result from a lowering of electroosmotic flow due to the higher viscosity of D_2O and to a reduction of the zeta potential. They also found that the use of D_2O-based buffers lowers Joule heating and that pK_a differences can result in enhanced resolution.[59] Electrophoresis in the D_2O-based solution gave information complementary to that obtained in H_2O-based electrolytes of the same acidity.[60] For the analysis of the tryptic digest of salmon calcitonin and elcatonin, buffer solutions prepared with D_2O proved superior to buffers prepared in H_2O.[61] From a single separation of an elcatonin digest, three pure cleavage peptides were recovered in sufficient quantity to determine the amino acid sequence.

IV. SAMPLE INJECTION

The combination of the high efficiency of separation and the short analysis times possible with CE produces an extreme constraint on the sampling system used in CE. Samples must be introduced on-column with minimum volume (e.g., for a 50 µm i.d. column and a 2-mm-long injection plug, the total injection volume is 4 nL) in order to preserve this high separation efficiency.[24] Fortunately, there are several very simple approaches to sampling that can come close to satisfying these constraints. The simplest methods of sample introduction are direct electromigration of the sample onto the column[6,62] and hydrodynamic injection via the formation of a small pressure gradient between inlet and outlet of the column.[37,44,63]

A. Electrokinetic Injections

The easiest way to introduce the sample onto the electrophoresis column is via electromigration. The column is simply placed into the sample solution and a voltage is applied for a few seconds (i.e., 30 kV for 1 second) to introduce the sample into the column because of its net electrophoretic mobility (remember, this includes the mobility derived from electroosmosis) in the applied field. Following this, the column is moved back to the separation buffer and the voltage is turned on, beginning the separation. Burton et al. examined the effect of time and voltage of electromigration injection on separation efficiency in the MEKC of substituted purine nucleotides.[64] Row et al. reported high efficiencies in the separation of nucleotides and derivatives with MEKC when short electrokinetic injections were used.[65] An important factor to remember about electrokinetic injections is that the sample is brought onto the column via its own migration velocity. Since this is different for different ions (otherwise we wouldn't get a separation), this means that the amount of material introduced depends on the migration velocity of the ion. Huang et al.[66] quantified this relationship and found that electrokinetic injection induces a bias proportional to the total mobility of each ion. Since this is a systematic bias, peaks can be corrected for differential mobilities by normalization with their retention time. This problem can be avoided through the use of hydrostatic injection.[44] Peak area was linear with concentration of several low molecular weight (C1 to C6) carboxylic acids.[44] Kuhr and Yeung[67] monitored the shape of the injection plug with laser-induced fluorescence at a spot 5 mm from the inlet of the capillary. Addition of a 50 Mohm, 100-W load resistor parallel to the capillary reduced the width of the injection and improved the reproducibility of the electrokinetic injection. Huang et al. corrected separation efficiency for the differential velocities of analytes and the size of the detection zone.[24] After this was accomplished, the principle limitation of efficiency was found to be the width of injection. Diffusion was only limiting with an injection zone length of less than 3 mm, neglecting thermal and electrokinetic contributions.

B. Hydrodynamic Injections

There are three main ways to introduce a sample onto the head of the electrophoresis column via hydrodynamic flow: gravity flow (also known as hydrostatic sampling), application of a positive pressure at the sample reservoir, and application of a negative pressure (vacuum) at the detector end. Honda et al. automated a siphoning sampler for CE, produced via a hydrostatic pressure gradient from a difference in the height of the sample and injection buffers.[68] Sample injection time was linear with area, but not peak height, with a precision of about 1%, improved from 8% in manual mode. Tsuda et al. demonstrated that large sample volumes decreased separation efficiency in the separation of cations (pyridinium salts, metal ions) and anions (sulfonic acids) in free solution capillary electrophoresis using hydrodynamic injection procedures.[9] Rose and Jorgenson characterized both electrokinetic and hydrodynamic injection procedures, done manually or under computer control.[63] They calculated sample volumes from both procedures and found that the reproducibility of automated procedures is far superior to manual control. Schwartz et al. characterized an automated sample introduction system using electrokinetic and hydrodynamic flow.[69] The precision and linearity of each technique was demonstrated. A replenishment system, in which the running and detection buffer could be replaced, was found to improve the reproducibility of injection and separation procedures. Grushka and McCormick provided an excellent discussion of the importance of sample introduction technique in band broadening in CE.[70] They showed that mere insertion of the capillary into the sample solution can introduce sample into the capillary and thereby limit separation efficiency. They calculated the optimum zone length for a short separation (10 min) to be 0.4 mm, indicating the importance of sampling in CE.

C. Stacking

There are very simple procedures for concentrating dilute mixtures of ionic species before an electrophoretic separation. Typically, these procedures exploit the field enhancement that occurs in zones of low conductivity. Thus, ions in a large plug of low conductivity buffer are "stacked" at the interface between this zone and the high conductivity buffer which precedes it. Burgi and Chien have looked at many of the aspects concerning stacking as a means of sample preconcentration on-line in CE.[71-73] They investigated the effect of electroosmotic flow on the field strength in the stacking zone.[74] A model was developed to estimate the number of ions which could be injected under field amplified conditions. Electroosmotic flow in a fused-silica capillary column with a concentration step gradient was found to be the weighted average of the electroosmotic velocities of the pure buffers.[75] The mismatch between electroosmotic velocities of the two zones was found to enhance zone-broadening mechanisms. In a second paper, this model was further developed to predict the conditions for optimized zone variance under stacking conditions.[76] They also found that sample stacking in conjunction with gravity injection can provide more than an order of magnitude improvement in detection limits with little if any loss of resolution.[77] Judicious selection of field amplification conditions can permit simultaneous injection and concentration of both positive and negative ions.[78] Injection of a short plug of water before sample introduction permits a several hundred-fold enhancement in the amount of material injected.[79]

D. Other Injection Strategies

Several other methods have also been proposed for introducing small amounts of sample onto the CE capillary. Tsuda et al. described a rotary loop injector made of Teflon for CE with an injection volume of 350 nL.[80] The loop injection procedure was very reproducible, but the injection volume is still two orders of magnitude too large for efficient separations. Several investigators have utilized the principle of electrical splitting to put a fraction of a sample onto a column. Demyl et al. reported both the theory and practice of electrical splitting of sample in large-bore capillaries (200 µm[81]). These procedures gave split ratios of 40:1 from microliter samples. Verheggen et al. also described a sampling device for large diameter (250 µm i.d.) capillaries.[82] The sample is introduced through a feeder capillary and excess sample is eluted from a drain capillary, both of which are perpendicular to the separation capillary. The distance between these determines the sample volume. Sample zone lengths varied between 13 and 81 mm, with sample volumes of 1 to 5 µL. Finally, electrical splitting was used to vary the composition of the buffer during electrophoresis by Pospichal et al.[83] They provided the theory and apparatus for the simultaneous migration of different buffers into the injection end of the capillary to allow the control of buffer composition as a function of the current ratio. This three-pole design allows changes in buffer pH or electrolyte composition in isotachophoresis without physical manipulation of buffer reservoirs. While most investigators initially introduced buffer into the capillary by suction, Rohlicek and Deyl described an interface for filling small capillaries with a syringe, even under higher pressures.[84]

V. OPTIMIZATION OF SEPARATION IN FREE SOLUTION ZONE ELECTROPHORESIS

In the simplest case, the separation observed in free-zone electrophoresis should be strictly dependent on the effective mobility (μ_i) of each analyte. When considering only the electrophoretic mobilities of small molecules that are approximately the same size, the best separation is obtained at the pH where there is a maximum difference in the degree of dissociation between different molecules. This pH is easy to find — it is simply the average of the pK_a values for all molecules in the sample. Mobilities for more complicated molecules can be estimated based on literature values for their absolute mobilities (μ_i^o), pK_as, the pH of the buffer solution, and the Henderson–Hasselbach equation[67] (Equations 5, 6, and 12). Using tabulated literature values for pK_a and absolute mobilities, one can construct a series of simultaneous equations for the effective mobility of each molecule in the sample as a function of pH. These, in turn, can be used to calculate the optimum resolution between adjacent components. This optimum pH can be found by solving for the best total resolution (sum of resolution between each pair of components) or by specifying a minimum resolution for any pair.

When even larger molecules are involved (e.g., peptides, proteins), the best starting point is still the average of their respective isoelectric point (pI) values. While this will provide the best separation based on electrophoretic mobilities, one must also consider the effect of the buffer on the zeta potential of the capillary (which controls electroosmotic flow) and the effect of pH and buffer composition on sample solubility and adsorption of the sample onto the walls of the capillary (see Chapter 8 on protein separations). A good example of buffer optimization in the separation of peptides was provided in the analysis of the tryptic digest of human growth hormone (hGH) by Nielsen and Rickard.[85, 86] They used the amino acid composition to generate the molecular weight, pI, and hydrophilicity of each peptide (Table 3.1). While the pK_as of most carboxylate groups of amino acids are between 2 and 4 and the amine is deprotonated above pH 8 to 10, very different separations are obtained for peptides in these two regions. At low pH, (pH = 2.4) the separation is based on the dissociation of the carboxylate (since all amine groups are protonated at this pH), and most peptides have a net positive charge (since pH < pI), but the zeta potential is small, and electroosmotic flow is very low. As a result, the separation is slow, and there is considerable overlap between several components (Figure 3.11). At pH 8.1, most species are negatively charged, migrating against electroosmotic flow, which is very fast. The result is a fairly rapid separation (Figure 3.12), but again, all peptides could not be separated under these conditions. It is important to realize that different components overlap in each case, such that the two conditions together can separate all components.

Addition of other modifiers (mobile phase additives) can fine-tune the separation but are only truly useful once the basic conditions have been established. The separation buffer must have sufficient ionic strength to avoid overloading (this is important when using UV detection with concentrated samples). Since zwitterionic buffers (i.e., tricine, glycine, etc.) have low conductivity, one might think an obvious solution is to add an inert salt (e.g., NaCl or KCl) to increase conductivity. As shown in Figure 3.13, this doesn't work because the current increases proportionately, causing thermal broadening due to the higher current passed through the column (hence increased power levels and the production of a radial thermal

Table 3.1   Properties of Peptides Generated from Enzymatic Digest of hGH

Fragment number	Isoelectric point	Hydrophobicity	Molecular weight	Amino acid residues	Amino acid sequence
1	10.1	18.1	930	8	FPTIPLSR
2	5.8	17.5	978	8	LFDNAMLR
3	10.4	−12.3	382	3	AHR
4	4.2	45.5	2343	19	LHQLAFDTYQEFEEAYIPK
5	6.4	−15.7	404	3	EQK
6–16	7.3	74.8	3763	32	6: YSFLQNPQTSLCFSESIPTPSNR 16: NYGLLYCFR
7	4.5	−19.0	762	6	EETQQK
8	5.9	9.4	844	7	SNLELLR
9	6.4	72.1	2056	17	ISLLLIQSWLEPVQFLR
10	3.5	33.5	2263	21	SVFANSLVYGASDSNVYDLLK
11	4.0	13.0	1362	12	DLEEGIQTLMGR
12	4.0	−3.4	773	7	LEDGSPR
13	9.2	0.4	693	6	TGQIFK
14	9.0	−14.2	626	5	QTYSK
15	3.8	0.9	1490	13	FDTNSIINDDALLK
17	9.0	−13.9	146	1	K
18	6.1	10.5	1382	11	KDMDKVETFLR
19	4.0	12.1	1253	10	DMDKVETFLR
20–21	5.9	19.9	1401	13	20: IVQCR 21: SVEGSCGF

(The header row **Fragments from Enzymatic Digest of Human Growth Hormone** spans the full table.)

gradient, which causes mixing and zone broadening). Finally, additives that change electroosmotic flow can be used (e.g., morpholine, which is a cation that adsorbs to the fused-silica wall,[87] thus reversing the zeta potential and electroosmotic flow). While control of electroosmotic flow is imperative, the other effect that these modifiers have is to minimize the interaction between the sample and the capillary wall. Some improvement in separation efficiency can be realized this way, though there is still a great deal of black art involved in this type of optimization.

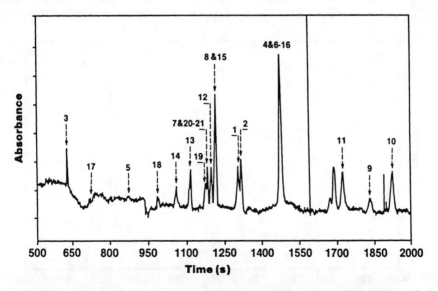

Figure 3.11   Electropherogram of hGH digest in 0.1 *M* glycine buffer, pH 2.4. (From Nielsen, R. G. and Rickard, E. C., *J. Chromatogr.*, 516, 99, 1990. With permission.)

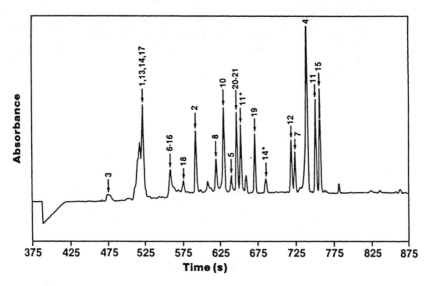

Figure 3.12 Electropherogram of hGH digest in pH 8.1 tricine buffer, 0.1 *M* tricine and 0.02 *M* morpholine (condition N). (From Nielsen, R. G. and Rickard, E. C., *J. Chromatogr.*, 516, 99, 1990. With permission.)

A. Other Peptide Separations

Miller and co-workers synthesized a 53-peptide mixture and assayed its purity by capillary-zone electrophoresis and mass spectrometry.[88] Rivier et al. utilized capillary-zone electrophoresis to determine the purity of a synthetic peptide mixture.[89] Chen et al. separated the anticoagulant peptide from its deletion by-products by free-zone electrophoresis.[90] Prusik et al. utilized CE and continuous free-flow zone electrophoresis to analyze and prepare pure fractions of synthetic growth hormone releasing peptide (GHRP).[91] Guarino and Phillips used capillary electrophoresis to assay for the purity of peptides.[92] Liu et al. derivatized amino acids and peptides from both standard solutions and biological samples with 3-(4-carboxy-benzoyl)-2-quinolinecarboxaldehyde at low sample concentration to form highly fluorescent

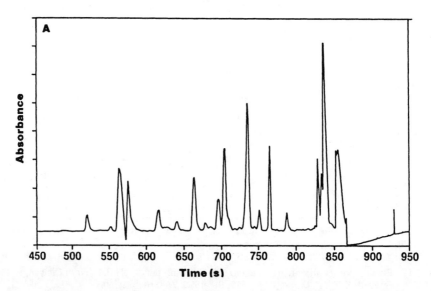

Figure 3.13 Effect of ionic strength on the separation of peptides at high pH. (From Nielsen, R. G. and Rickard, E. C., *J. Chromatogr.*, 516, 99, 1990. With permission.)

isoindole derivatives.[93] Minimum detectable quantities with CE and laser-induced fluorescence detection were in the low attomole (10^{-18} mol) range. Hortin and co-workers separated sulfated and nonsulfated forms of peptides by CE.[94] Kruegar et al. utilized CE and reverse-phase HPLC to determine the specificity and rate of cleavage of ACTH peptide bonds by endoproteinase Arg C (I).[95] Tran et al. derivatized samples of amino acids and peptides and analyzed these mixtures by free-zone and micellar CE.[96] The use of MEKC in combination with L and D-Marfey's reagent offered unequivocal means to confirm the presence of D-amino acid in an unknown sample. Bullock[97] developed a series of buffers encompassing the pH range from 3.5 to 9.0 for free solution capillary electrophoretic (CE) analysis of basic proteins in uncoated fused-silica capillaries. Separations of model proteins possessing isoelectric points between 9.1 and 11.0 have been achieved with efficiencies ranging from 95,000 to 690,000 theoretical plates. In each case, the modifier 1,3-diaminopropane was incorporated into the operating buffers at concentrations of 30 to 60 mM along with moderate levels of alkali metal salts to suppress protein-capillary wall interactions. The combination of these buffer additives allows the analyses to be performed at pH values below the protein isoelectric points.

Florance separated 14 of the 24 peptides that are fragments of the protein motilin. The influence of charge, hydrophobicity, secondary structure, and length of the motilin fragments on migration time was investigated.[98] Buoen and co-workers discussed the effect of slow conformational inversion of prolylproline residues on the separation of proline-rich peptides by CZE and HPLC.[99] The analysis of synthetic branched peptides and multiple antigen peptides by CZE was described by Pessi et al.[100] Wheat and co-workers examined tryptic peptides of cytochrome c by CE and HPLC.[101] Differences of a single residue between homologous proteins from different species could be observed. Yannoukakos et al. analyzed tryptic digest of proteins to determine sites of phosphorylation.[102] Young and Merion investigated species variations in the tryptic maps of cytochrome c by CE.[103] Ferranti et al. generated tryptic maps for each globin in human Hbs, then compared these maps to those for normal samples to identify abnormal globin.[104] Bushey and Jorgenson employed a two-dimensional reverse-phase HPLC–capillary electrophoresis instrument to identify differences in the tryptic digest fingerprints of horse heart cytochrome c and bovine heart cytochrome c.[105]

B. Complexation

A slightly more complicated way to modify the selectivity of the electrophoretic separation involves complexing the analyte with a suitable agent that will change either the net mass or total charge of the complexed ion. Snopek et al. reviewed separation strategies for electrokinetic chromatography, including micellar, inclusion, and complexation effects in modifying retention.[106] In the simplest case, metal ions can be complexed with inorganic anions to selectively alter the electrophoretic mobilities of the ions. This has been demonstrated in isotachophoresis by Gebauer et al.,[107] who separated nitrite, chloride, and sulfate as the Cd^{2+} complexes with 10 pmol detection limits. Drinking water was then assayed for content of these anions with conductivity detection. Aguilar et al. separated the iron, zinc, and copper cyanide complexes in electroplating solutions, detected with UV absorbance in free-zone CE.[108] Zare and co-workers reported the first chiral separations in CE,[109] where the separation was based on complexation of d,l-amino acids with Cu(II) complex of l-histidine. This resolved dansyl-derivatives of tyrosine, phenylalanine, aspartate, and glutamate. A more complete characterization of chiral separations will be found in Chapter 5 of this book.

Organic complexing reagents have been used to increase the selectivity of several separations. Walbroehl and Jorgenson investigated the use of solvophobic association in the separation of hydrophobic solutes.[110] Weakly hydrophobic analytes formed complexes with the tetrahexylammonium ion, allowing the separation of polycyclic aromatics in a water–acetonitrile solvent system. Carbohydrates, also normally neutral, were complexed with borate ion to form charged complexes after derivatization with 2-aminopyridine, whose UV absorbance was measured.[111] The derivatization was optimized and 12 sugars were separated at pH 10.5, and the mobilities of the sugars were related to their carbohydrate structure. Wallingford and Ewing improved the separation of serotonin and catecholamines in 9 μm i.d. columns[112] through complexation of the catechol with an inorganic anion, borate ion. Severe tailing of amines was minimized by formation of borate complexes and separation by MEKC or via addition of 20% isopropanol to the buffer. They also examined the effect of SDS and SOS concentration on the migration velocity of the borate-complexed catechols independently and in a mixed micelle solution.[113] Cohen et al. used Cu(II) complexes of nucleosides, nucleotides, and oligonucleotides in MEKC using SDS micelles.[114] Manipulation of the complexation constant by changing pH, metal, or surfactant concentrations produced differences in the selectivity of separation. Eighteen different oligonucleotides (8 bases each) were separated within 30 min. Similarly, Dolnik et al. found that the separation of oligonucleotides (polycytidines) was independent of pH or ionic strength, but could be influenced by complexation with spermine, with and without SDS.[115] Nucleotides were separated by free-zone CE with UV detection in silanized capillaries. Many chiral separations are based on the addition of cyclodextrins to the electrophoresis buffer to form inclusion complexes to induce separation of isomers in capillary electrophoresis (see Chapter 4).

C. Application of CE to Small Molecules

Inorganic Ions. A number of the parameters which influence CE separations of inorganic ions have been examined methodologically,[116] including choice of electrolyte anion, electrolyte pH, and the addition of an electroosmotic flow modifier. Optimized conditions were established for the separation of inorganic anions, organic acids, and alkylsulfonates, and the technique was applied to the determination of a variety of anionic solutes in several complex sample matrices (e.g., kraft black liquor). Several authors have used complexation to enhance either the separation or the detection of inorganic ions. Fourteen lanthanides were separated by CE in a background electrolyte containing hydroxyisobutyric acid as a complexing counterion and creatinine as a UV-absorbing co-ion for indirect detection (Figure 3.14).[117] Complete separation of the lanthanides was achieved in less than 5 minutes. Anions can also be

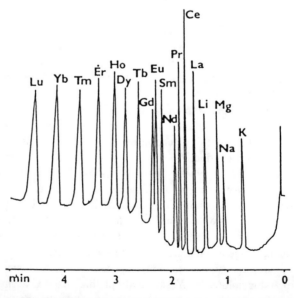

Figure 3.14 Separation of inorganic ions by capillary electrophoresis. (From Foret, F., Fanali, S., Nardi, A., and Bocek, P., *Electrophoresis*, 11, 780, 1990. With permission.)

separated by CE. For example, Honda et al. performed the simultaneous determination of iodate and periodate by capillary zone electrophoresis.[118] Migration of iodate and periodate was influenced by pH and salt concentration; detection was accomplished with UV absorbance with picomole detection limits.

Metal chelates were separated in SDS micellar solution with α, β, γ, δ-tetrakis(4- carboxyphenyl)porphine as chelating reagent.[119] The method shows great promise for the ultratrace detection of metal chelates.[120] MEKC was used to determine the solute-micelle binding constants (K_{mw}) of several different compounds.[121] Saitoh et al.[122] separated neutral complexes of chromium(III), cobalt(III), rhodium(III), and platinum(II) (or palladium(II)) by MEKC. The distribution coefficient of each metal complex was calculated from the capacity factor. The linear log-log relationship between the distribution and the partition coefficient was used for prediction of both the distribution coefficients and the migration times of other metal complexes, such as palladium(II) acetylacetonate and chromium(III) 3-methylacetylacetonate.

Fukushi and Hiro studied the effects of crown ethers[123] and α-, β- and γ-cyclodextrins[124] on the effective mobilities of various metal ions in capillary isotachophoresis (ITP). Twenty metal ions were separated by ITP and the zones were fractionated and analyzed off-line by particle-induced X-ray emission. The recovery, migration order, and separation efficiency were studied as a function of the concentration of a complexing agent, a-hydroxyisobutyric acid. The recovery of the metal cations was 100% with both electrolyte systems, except for Fe(II) and Zr(IV)O.[125] Krivankova and Bocek used ITP to determine Fe(III)-EDTA and free EDTA in a scrubbing liquid for desulfurization of waste gases from brown-coal gasification.[126] Swaile and Sepaniak[127] used laser-excited fluorescence to detect metal cations after complexation with 8-hydroxyquinoline-5-sulfonic acid (HQS−). By controlling mobile-phase parameters that affect the complexation reaction, the observed electrophoretic mobility of the metal was manipulated. Ca(II), Mg(II), and Zn(II) were detected at ppb levels and in complex sample matrices (i.e., blood serum).

Organic Acids. Karovicova et al.[128] determined formic, acetic, sorbic, and benzoic acids in mustards, jams, and syrups by dissolving in water, adjusting the pH to 10, and performing ITP analysis. Foret et al. separated triazine herbicides and their solvolytic products by CE in mixed water-ethanol buffers.[129] Jokl and Petrzelkova separated propranolol, metipranolol (I), and desacetylmetipranolol (II) with ITP.[130] Karovicova et al. determined nitrates in nine vegetables by ITP.[131] Kenney determined several organic acids in food samples.[132] Kopacek et al.[133] separated complex mixtures of humic substances by ITP after the addition of polyvinylpyrrolidone (PVP) to the leading electrolyte. Interaction of PVP with humic substances was found to differentiate humic and fulvic acids. Watarai developed a variation of MEKC which uses oil-in-water microemulsions as the electrophoretic media for the separation of ionic and nonionic samples containing aromatic compounds.[134]

Chadwick and Hsieh [135] used CE to separate *cis* and *trans* double-bond isomers of ionic species. The separation of fumaric acid and maleic acid in both the dianion and monoanion forms was demonstrated. Because these two isomers carry the same charge under the basic conditions of the running buffer, the separation is attributed to the difference in their hydrodynamic radii. Measurement of the electrophoretic mobilities allowed for quantitation of this difference. All-*trans*-retinoic acid and 13-*cis*-retinoic acid were also baseline separated.

Karovicova et al. used capillary ITP to determine citric, malic, oxalic, lactic, and ascorbic acids in fruit homogenates of *S. nigra* and *S. ebulus*,[136] and organic acids and H_3PO_4 during lactic fermentation of cabbage during sauerkraut manufacture and during the maturation of red wine.[137] Krivankova and Bocek[138] used an ITP method for the determination of pyruvate, acetoacetate, lactate, and 3-hydroxybutyrate in 1 to 10 μL of the untreated heparin plasma of patients with diabetes mellitus. Lutonska found that the determination of several organic acids (lactic, acetic, and butyric) on silages by capillary ITP was more accurate than an officially recognized classical method.[139] Vindevogel et al. found that CZE and MEKC, where all possible sources of metal ions (silica gel, injector, frits) were excluded, allowed the separation of hop bitter acids in beer production.[140] Oefner et al. evaluated the operating conditions for the isotachophoretic separation of organic acids.[141] The time of analysis was observed to be a function of the concentration of the leading electrolyte.

Amino Acids. The samples most frequently separated by CE have been the amino acids.[7,47,63,109,142-168] Since most small organic acids have very few intrinsic physical characteristics that can be used for detection (e.g., fluorescence, redox processes, even UV absorbance), precolumn derivatization of the sample has extended the utility of CE to many analytes. Many derivatization strategies have been employed, including dansyl (DNS)[7,109,142,143,153,154,159,163] o-pthalaldehyde (OPA),[63,151,154,161,162] napthalenedialdehyde (NDA),[155,162] fluorescein isothiocyanate (FITC),[144,145,156,169-172] fluorescamine,[152,154] and phenylthiohydantoin derivatives (PTH).[167,173-177] Additionally, derivatization has allowed the determination of n-alkyl amines as 4-chloro-7-nitrobenzofuran (NBD),[178,179] and carbohydrates as 2-aminopyridine derivatives.[111] Nickerson and Jorgenson[169] compared the relative sensitivities of OPA-, FITC-, and NDA derivatization procedures using amino acids with laser-induced fluorescence detection. An extremely rapid and efficient separation (75 sec) of 8 NDA-labeled amino acids (<0.5 s peak width) in short capillaries (35 cm, 10 µm i.d.) with high field (30 kV) was reported.[155] Detection with laser-induced fluorescence gave detection limits of 0.4 attomole.

Several reports of on-column or post-column reactor designs have also been reported. Pentoney et al. described an on-column reactor that utilized laser-drilled holes in the separation capillary to allow connection to a cross-flow reagent solution with minimal dead volume.[151] Amino acids were derivatized on-column with OPA with sub-femtomole detection limits in 75 µm i.d. capillaries. A somewhat simpler design was produced by Tsuda et al., who described a post-column reactor for derivatization using a four-way connector and fluorescamine derivatization of polyamines.[180] The flow rate of the derivatizing solution was optimized for sensitivity using putrescine as an analyte. Rose and Jorgenson reported the design of a post-column reactor for derivatization consisting of a sheath flow of reagent around an etched separation capillary.[173] This leads to sub-femtomole detection limits of amino acids with OPA derivatization and fluorescence detection with minimal band broadening. Linearity was over 3 orders of magnitude for amino acids and at low concentrations (> 0.1% (w/v)) for proteins. Nickerson and Jorgenson modified the post-column reactor design to optimize sensitivity and separation efficiency.[162] Laser-induced fluorescence detection of OPA- and NDA-derivatized amino acids used larger capillaries (a 50 µm i.d. separation capillary and 100 µm i.d., 375 µm o.d. reaction capillary). The reactor was characterized and optimized for reagent flow rate, generating detection limits of 44 amol myoglobin with a separation efficiency of 600,000 theoretical plates.

Several authors used derivatized amino acids to characterize their systems. As shown in Figure 3.15, Waldron et al. used fluorescence detection of FITC and dimethylaminoazobenzene (DAAB) derivatives of amino acids, and demonstrated much higher sensitivity for the fluorescein derivative (10^{-21} mole) than for the DAAB derivative (10^{-16} mole).[171] Nielsen et al. used positional isomers of substituted benzoic acids as model compounds to study the affect of system parameters (i.e., pH, electrolyte, ionic strength, addition of alcohols, counter ion, and temperature) on the electrophoretic separation. pH was again found to be the most effective parameter in optimizing resolution. However, thermostating of the capillary at elevated temperatures increased resolution and decreased analysis time.[181] Nishi et al.[182] resolved isomers of 2,3,4,6-tetra-O-acetyl-β-D-glucopyranosyl isothiocyanate (GITC)-derivatized DL-amino acids using micellar electrokinetic chromatography with SDS solutions. Of 21 DL-amino acids, optical resolution was achieved in 19 with neutral and alkaline conditions; aspartic acid and glutamic acid have an additional carboxyl group. Simultaneous resolution of more than ten GITC-derivatized DL-amino acids was achieved within 40 min in 0.2 *M* SDS. Tanaka et al. used cyclodextrin derivatives to separate dansyl-amino acids by CE.[183]

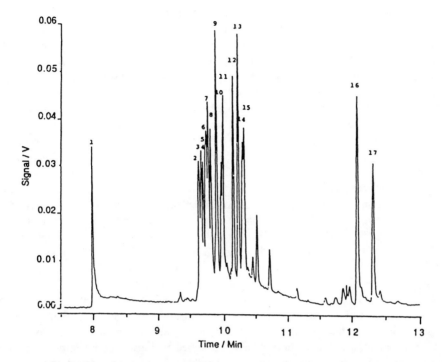

Figure 3.15 Separation of fluorescein isothiocyanate-labeled amino acids by capillary electrophore-sis/laser-induced fluorescence detection. (From Waldron, K. C., Wu, S., Earle, C. W., Harke, H. R., and Dovichi, N. J., *Electrophoresis*, 11, 777, 1990. With permission.)

Ong et al.[184] separated fifteen dansyl (DNS)-amino acids by MEKC. All the amino acids were separated within 26 minutes in a 40 mM SDS /phosphate-borate buffer (pH 7.56). Otsuka et al. used sodium N-dodecanoyl-L-valinate (SDVal) for chiral separations of PTH derivatives of six DL-amino acids (serine, a-aminobutyric acid, norvaline, valine, tryptophan, and nor-leucine).[177] Otsuka and Terabe[185] examined the effects of methanol and urea on enantiomeric resolution of PTH-DL-amino acids by MEKC with sodium N-dodecanoyl-L-valinate (SDVal). A mixture of four PTH-DL-amino acids was separated and each pair of enantiomers was also optically resolved. Otsuka and Terabe[175] obtained enantiomeric resolution of PTH-amino acids by MEKC using a chiral surfactant (digitonin) and an anionic chiral surfactant, sodium N-dodecanoyl-L-valinate under neutral conditions.

Catecholamines. The most sensitive analyses of catecholamines utilize amperometric detection with CZE.[152] The detection volume was minimized by placing the carbon-fiber working electrode directly into the end of the capillary. The electrochemical detector is decoupled from the separation capillary by a small break in the capillary surrounded by a porous glass joint, such that the applied potential drops across the joint while electroosmotic flow forces the buffer and analyte past the joint to the carbon-fiber working electrode posi-tioned inside the end of the capillary. This eliminates cross talk between the two electrical systems and allows sensitive amperometric detection of catechols and fluorescamine-amino acids with 100,000 plates. The performance of the electrochemical detector was optimized by miniaturizing both the separation column and the electrochemical detector.[87] A 5 μm o.d. carbon fiber was inserted into a 12.7 μm i.d. capillary, minimizing the annular flow region between the two surfaces and increasing the coulometric yield. Attomole detection limits were obtained for catechols (10 to 9 M concentration), and the apparatus was connected to a modified microinjector for direct sampling of the cytoplasm of a nerve cell.

Electrochemical detection of catecholamines was also demonstrated with MEKC.[186] Addi-tion of SDS micelles enhanced the separation efficiency to over 400,000 plates, and a smaller capillary diameter (26 μm) provided better detection limits (0.2 fmol) due to higher coulo-metric efficiency. Some attenuation in the amperometric signal was observed at high SDS concentrations, and several competing equilibria were suggested as a possible explanation. MEKC with SDS micelles was used to improve the separation of several ionic species (catecholamines), suggesting an ion-pairing reaction.[187] Electrochemical detection demon-strated enhanced resolution for cationic as well as nonionic species. Electrochemical detection of serotonin and catecholamines was enhanced by further miniaturization, where separation and detection was accomplished in 9 μm i.d. capillaries.[112] This decreased detection limits to attomole levels (24 nM concentration of 5-HT). Severe tailing of amines was minimized by the formation of borate complexes and separation by MEKC or via addition of 20% isopropanol to the buffer. Additionally, a Nafion-coated electrode was used to enhance sen-sitivity and selectivity. The separation of borate-complexed catechols was accomplished in micellar electrokinetic chromatography with electrochemical detection.[113] The effect of SDS and SOS concentration on the migration velocity of the borate-complexed catechols was determined independently and in a mixed micelle solution. Kaneta et al.[188] improved the resolution of catecholamines by controlling both their electroosmotic and their electrophoretic mobilities. The former was controlled by the addition of borate ion and a change in pH, resulting in a separation of ten catecholamines. Ong et al.[189] investigated the migration behavior of selected catechols and catecholamines in MEKC at different concentrations of micellar solutions and at different pH values for the electrophoretic media. The results successfully demonstrated the use of MEKC for the separation of a mixture of two catechols and six catecholamines. Tanaka et al.[190] also separated various catecholamines using com-plexation with boric acid. A short review by Lee and Heo looked at the detection of catechols by CE.[191] Tanaka et al. performed an isotachophoretic separation of catecholamines based on the inclusion complex formation with b-cyclodextrin. Separability was improved with increas-ing concentration of b-CD in the leading electrolyte. Six catecholamines were separated by using complex formation with b-CD.[192]

Olefirowicz and Ewing used CE with electrochemical detection to separate and detect attomole levels of neurotransmitter in picoliter volumes of cytoplasm withdrawn from single neurons of the pond snail *Planorbis corneus*.[193] This work demonstrated, for the first time, the direct detection of the cytoplasmic concentration of dopamine in single, intact neurons. They further optimized the use of CE with electrochemical detection by decreasing the inner diameter of the separation/sampling capillary to 2 μm (Figure 3.16).[194] Sample volumes as low as 270 fL were injected into the electrophoresis capillary with sub-attomole detection limits for easily oxidized species. Sampling of the cytoplasm is accomplished by inserting one end of the electrophoresis capillary directly into a single nerve cell. The high-voltage end of the electrophoresis capillary was etched with hydrofluoric acid to form a microinjector. Chien et al. compared results obtained with voltammetric electrodes and CE for dynamic and static monitoring of dopamine in the somal cytoplasm of the giant dopamine neuron of *P. corneus*.[195] The current status of the authors' microCE technique for the sampling and determination of neurotransmitters/catecholamines in single nerve cells was reviewed by Olefirowicz and Ewing.[196]

Other Biological Samples. Guzman et al. had a brief symposium report describing the analysis of brain tissue constituents with CE with laser-induced fluorescence detection.[197] Further work in this area by Hernandez et al.[198] examined the effects of cocaine and two other local anesthetics applied directly to the nucleus accumbens for 20 min by diffusion from a 4 mm microdialysis probe in freely moving rats. Cocaine (7.3 mM) was found to increase the extracellular concentration of dopamine. Liu and Chan analyzed ganglioside micelles by CE in uncoated fused-silica capillaries.[199] The mass sensitivity using UV absorp-tion (195 nm) was 10^{-11} mol. Ganglioside micelles including GM1, GD1b, and GT1b were resolved into separate peaks by CE. Prolonged incubation caused the ganglioside peaks to merge into a single species.

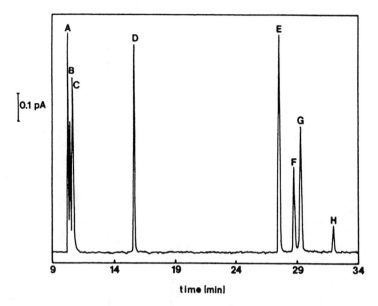

Figure 3.16 Separation of catecholamines by capillary electrophoresis with electrochemical detection. (From Olefirowicz, T. M. and Ewing, A. G., *Anal. Chem.*, 62, 1872, 1990. With permission.)

Nguyen et al.[200] monitored domoic acid in mussel homogenates by CE. UV absorbance detection at 242 nm provided a detection limit in wet tissues of 10 ppm. Nguyen et al. also used CE to quantitate nucleotide degradation in fish tissues and to provide a basis for determining the K value, an indicator of fish freshness.[201] The values obtained for the three compounds of interest, inosine monophosphate, inosine, and hypoxanthine, correlated very well with those obtained by enzymic assays. Thibault et al.[202] described a CE method with UV detection for the separation and determination of underivatized toxins associated with paralytic shellfish poisoning. Confirmation of the electrophoretic peaks was facilitated by mass spectrometric detection using an ion-spray CE–MS interface and by HPLC with fluorescence detection. Mereish et al.[203] used CE and HPLC methods for the analysis of palytoxin. The detection limit of the HPLC method was 125 ng/injection, while the CE method was 0.5 pg/injection.

Yeo et al. proposed the use of a systematic optimization scheme for the CE separation of nine plant growth regulators with a mixed carrier system consisting of three cyclodextrin modifiers.[204] The scheme utilizes the overlapping resolution mapping procedure and the interpretive optimization scheme to predict the optimum cyclodextrin composition for the separation of the plant growth regulators. Tsuda et al. determined free polyamines in rat tissues by CE with fluorescence detection.[205] After precipitation of proteins with perchloric acid, the sample solution was derivatized with fluorescamine using a microscale procedure. Ethylenediamine was added to the medium to avoid adsorption of polyamines onto the capillary wall. Matsumoto et al.[206] also described a CE method for polyamine analysis and compared it to HPLC. Huang et al. determined the nucleotide composition of base-hydrolyzed bulk RNA in five minutes by capillary zone electrophoresis with UV absorbance detection.[207]

Carbohydrates. Liu et al.[208] described the use of 3-(4-carboxybenzoyl)-2-quinolinecarboxaldehyde as a precolumn derivatization agent for amino sugars analysis. This procedure was used to analyze amino sugars in various biological mixtures. Low attomole (10^{-18} mol) limits of detection were reported for laser-induced fluorescence detection. This procedure was found to be useful for the analysis of reducing monosaccharides and oligosaccharides after transformation to the amino sugar via reductive amination.[209] Under optimized conditions, the minimum detectable quantities for monosaccharide solutes was determined to be

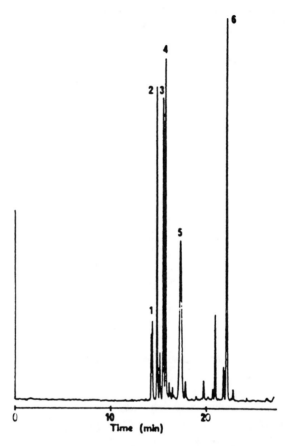

Figure 3.17   Separation of derivatized saccharides by capillary electrophoresis. (From Liu, J., Shirota, O., and Novotny, M., *Anal. Chem.,* 63, 413, 1991. With permission.)

at low attomole levels (0.5 amol for the CBCA derivative of galactose). Complex oligosaccharides, isolated from bovine fetuin by hydrazinolysis, were successfully mapped using this procedure (Figure 3.17).

Honda et al. derivatized various aldoses to their 3-methyl-1-phenyl-2-pyrazolin-5-one derivatives for UV detection and separated as borate complexes.[210] This system also provided good separation of derivatives of homologous oligoglucans having different interglycosidic linkages. Garner and Yeung[211] used indirect fluorescence detection and CE to detect mixtures of monosaccharides. The mass detection limit of fructose was 2 fmol using a 5-μm i.d. capillary with an efficiency of >600,000 theoretical plates. Deyl et al.[212] used CE to separate the products arising from the Maillard reaction of free amino acids (glycine, alanine, and isoleucine) with aldehydic sugars (glucose or ribose). The products of this reaction were separated without derivatization (UV detection at 220 nm), as phenylthiocarbamyl derivatives, and after derivatization with 2,4-dinitrophenylhydrazine. These separations were compared with those obtained by high performance liquid and thin-layer chromatography.

Honda et al.[213] analyzed 1-phenyl-3-methyl-5-pyrazolone (PMP) derivatives of reducing carbohydrates by CE with UV detection using an electrolyte containing alkaline earth metal salts. The PMP derivatives of isomeric aldopentoses were completely separated from each other by the interaction with these metal ions. The order of mobility for these derivatives was different from that observed in borate buffer, suggesting formation of different types of complexes. The extension of this procedure to the analysis of other monosaccharides and several oligosaccharides was also discussed. Al Hakim and Linhardt used CE for the analysis of non-, mono-, di-, and trisulfated disaccharides derived from chondroitin sulfate, dermatan sulfate, and hyaluronic acid.[214] Quantitation of disaccharides derived from chondroitin sulfate using chondroitin ABC lyase and mixtures of unsaturated disaccharide standards required only picogram quantities of sample. Ampofo et al.[215] separated eight disaccharide standards prepared from heparin, heparan sulfate, and derivatized heparins. Two of the standard heparin/heparan sulfate disaccharides, having an identical charge of −2, were not fully resolved in the sodium borate/boric acid buffer. The structure and purity of each of the eight disaccharides were confirmed using fast atom bombardment mass spectrometry and high-field 1H-NMR spectroscopy. Heparin and heparan sulfate were then depolymerized using heparainase, heparin lyase II, heparinitase, and a combination of all three enzymes. CE analysis of the products provided the disaccharide composition of each glycosaminoglycan.

Pharmaceutical Compounds. Salomon et al.[216] optimized the separation of seven tricyclic antidepressants (protriptyline, desipramine, nortriptyline, nordoxepin, imipramine, amitriptyline, and doxepin). Optimal resolution of this mixture was achieved by the addition of methanol to the buffer to decrease both the electroosmotic flow and the electrophoretic mobilities of the samples. Chmela and Stransky[217] described the isotachophoretic behavior and separation of similar drugs (amitriptyline, dosulepine, chlorpromazine, chlorprothixene, chlorothepin, levopromazine, nortriptyline, oxyprothepine, prochlorperzine, metipranolol, tranylcypromine) and their determination in pharmaceuticals (dosulepin, chlorpromazine, clorotepine). Snopek et al. used α−, β−, γ− and heptakis(2,6-di-O-methyl)-β-cyclodextrin as stereospecific selectors or electrolyte modifiers, both in capillary zone electrophoresis and isotachophoresis.[218] Several model isomeric compounds (including optical isomers of pharmaceutical interest) were resolved. Soluble alkylhydroxyalkyl-cellulose derivatives were added to the cyclodextrin-modified background electrolytes, and their presence was found to improve enantioselectivity and separation efficiency. Ong et al. used CE with UV detection at 214 nm to separate a group of nine antihistamines.[219]

Altria and Smith used CE and MEKC for the separation of the antidepressant GR50360A from potential manufacturing impurities.[220] Arrowwood and Hoyt developed a method for the determination of cimetidine (the active ingredient in the ulcer medication Tagamet®) in commercial preparations. Analysis of over 60 samples from common available formulations gave relative standard deviations of 1.9 to 6.4%.[221] Kenndler et al. determined arbutin in a crude drug preparation using UV detection at 214 nm.[222] Pietta et al.[223] used MEKC for the determination of flavonol-3-O-glycosides. These results were compared with those obtained by reverse-phase HPLC. Swedberg[224] used nonionic and zwitterionic surfactants to enhance separations of desipramine/nortriptylene and angiotensin III/Val₄-angiotensin III. The surfactants are only effective at or above the critical micelle concentration.

Krivankova et al.[225] used a combination of capillary ITP and CE to determine the coccidiocidic drug halofuginone in feed concentrates. The high load capacity of the isotachophoretic step and high sensitivity of the zone electrophoretic step enabled analysis of up to 25 μL of sample solution containing as little as 10^{-8} M halofuginone with excellent reproducibility (1% RSD). Meier and Thormann[226] quantitated thiopental in human serum and plasma by MEKC. These data were compared with equivalent data obtained by HPLC. Steuer et al. compared the utility of HPLC, SFC, and CZE for drug analysis. Factors considered in this analysis included separation efficiency, performance, sensitivity, optimization parameters, method development time, sample preparation, technical difficulties, orthogonality of the information obtained, and the possible application to various substance groups.[227] Swartz discussed various applications of CE to the pharmaceutical lab.[228]

Lloyd et al.[229] developed a CE assay for the antileukemic agent cytosine-b-D-arabinoside (ara-C). Solid-phase extraction and on-capillary peak concentration are used to improve the detection limit. Guzman et al. used CE for the determination of a recombinant cytokine in a pharmaceutical dosage form.[230] Tsikas et al. developed capillary ITP methods for the analysis of β-lactam antibiotics, i.e., penicillins and cephalosporins, in chemical and pharmaceutical preparations.[231] A leading electrolyte with a pH of 7.0 allowed the sensitive detection of several β-lactam antibiotics independent of the chemical structure of the side-chain of the penicillin or cephalosporin nucleus. Yeo et al. reported the separation of six antibiotics by CE with UV photodiode-array detection.[232]

Other Small Organic Molecules. Kenndler et al.[233] used CE for the determination of impurities in riboflavin 5'-phosphate (I). Tanaka et al.[234] separated the niacin derivatives of weak bases by capillary ITP. Yik et al. used electrochemical detection with CE for a mixture of B6 vitamins on a 50 μm i.d. column.[235] A calibration plot was linear over two orders of magnitude with a lower limit of detection of approximately 4 fmol. Zweigenbaum separated anionic components of Triton 770 by CE and MEKC.[236] Stover reported applications of CE developed for detergent, food additive, herbicide, animal nutrition, and biotechnology samples.[237] Wainright reviewed small molecule separations in uncoated and coated capillaries.[238]

Yik et al. investigated the application of CE to the analysis of environmental pollutants.[239] The separation of selected groups of compounds, including substituted phenols, polycyclic aromatic hydrocarbons, and phthalate esters, were discussed. The effects of relevant parameters, such as pH of the electrophoretic media, surfactant concentration, capillary length and applied voltage on separation efficiency were also considered. Northrop et al. used MEKC for separation and detection of organic gunshot and explosive constituents.[240] Twenty-six of these constituents were separated in <10 minutes with efficiencies in excess of 200,000 plates. The presence of gunshot residues in spent ammunition casings and the composition of six reloading powders and four plastic explosives were determined.

Sepaniak et al.[241] examined the role of the mobile phase in CE and MEKC for determining separation performance. The influences of ionic salt, surfactant, and organic solvent mobile phase additives on separation efficiency, retention, and elution range were discussed and demonstrated. Ghowsi and Gale[242] discussed CE with smaller capillary diameters for use as chemical sensors, electrokinetic field devices based on capillary field effect electroosmosis, and potential application for separation-based sensors. Takigiku and Schneider[243] used CE for the separation and quantitation of ribonucleoside and deoxyribonucleoside triphosphates. Capillaries were treated to reduce electroosmotic flow, and capillary zone electrophoresis was performed with negative voltage. Karovicova and Polonsky analyzed colorants by capillary isotachophoresis.[244]

Cole et al.[245] optimized binaphthyl enantiomer separation by CE. Bile salts were used in the mobile phases instead of conventional sodium dodecyl sulfate to provided a lower k' value and an optimal resolution for moderately hydrophobic compounds. Fanali and Bocek[246] separated enantiomers of tryptophan by CE, using α-cyclodextrin as a chiral active component in the background electrolyte. The separation of (–) and (+)-epinephrine was achieved by supplementing the background electrolyte with heptakis(2,6-di-O-methyl-β-cyclodextrin). As a practical application of the method, the quantitative analysis of (–) and (+) enantiomers in common pharmaceutical solutions of adrenaline is shown. Leopold and Gouesclou[247] separated chiral amino acids and peptides derivatized with Marfey's reagent.

REFERENCES

1. Hjerten, S., Free zone electrophoresis, *Chromatogr. Rev.,* 9, 122, 1967.
2. Mikkers, F. E. P., Everaerts, F. M., and Verheggen, T. P. E. M., High performance zone electrophoresis, *J. Chromatogr.,* 169, 11, 1979.
3. Mikkers, F. E. P., Everaerts, F. M., and Verheggen, T. P. E. M., Concentration distribution in free zone electrophoresis, *J. Chromatogr.,* 169, 1, 1979.

4. Thormann, W., Michaud, J. P., and Mosher, R. A., Theoretical and experimental separation dynamics in capillary zone electrophoresis, *Electrophor. '86, Proc. Meet. Int. Electrophor. Soc.*, 5th, 267, 1986.

5. Thormann, W., Description and detection in moving sample zones in zone electrophoresis: zone spreading due to the sample as a natural discontinuous element, *Electrophoresis*, 4, 383, 1983.

6. Jorgenson, J. W. and Lukacs, K. D., Zone electrophoresis in open-tubular glass capillaries, *Anal. Chem.*, 53, 1298, 1981.

7. Jorgenson, J. W. and Lukacs, K. D., Zone electrophoresis in open-tubular glass capillaries: preliminary data on performance, *J. High Resolut. Chromatogr. Chromatogr. Commun.*, 4, 230, 1981.

8. Lukacs, K. D. and Jorgenson, J. W., Capillary zone electrophoresis: effect of physical parameters on separation efficiency and quantitation, *J. High Resolut. Chromatogr. Chromatogr. Commun.*, 8, 407, 1985.

9. Tsuda, T., Nomura, K., and Nakagawa, G., Separation of organic and metal ions by high-voltage capillary electrophoresis, *J. Chromatogr.*, 264, 385, 1983.

10. Issaq, H. J., Atamna, I. Z., Metral, C. J., and Muschik, G. M., Factors that influence mobility, resolution, and selectivity in capillary zone electrophoresis. I. Sodium phosphate vs. potassium phosphate, *J. Liq. Chromatogr.*, 13, 1247, 1990.

11. Issaq, H. J., Atamna, I. Z., Muschik, G. M., and Janini, G. M., The effect of electric field strength, buffer type and concentration on separation parameters in capillary zone electrophoresis, *Chromatographia*, 32, 155, 1991.

12. Jones, W. R. and Jandik, P., Controlled changes of selectivity in the separation of ions by capillary electrophoresis, *J. Chromatogr.*, 546, 445, 1991.

13. Rasmussen, H. T. and McNair, H. M., Influence of buffer concentration, capillary internal diameter and forced convection on resolution in capillary zone electrophoresis, *J. Chromatogr.*, 516, 223, 1990.

14. Grushka, E., McCormick, R. M., and Kirkland, J. J., Effect of temperature gradients on the efficiency of capillary zone electrophoresis separations, *Anal. Chem.*, 61(3), 241, 1989.

15. Jones, A. E. and Grushka, E., Nature of temperature gradients in capillary zone electrophoresis, *J. Chromatogr.*, 466, 219, 1989.

16. Knox, J. H., Thermal effects and band spreading in capillary electro-separations, *Chromatographia*, 26, 329, 1989.

17. Davis, J. M., Influence of thermal variation of diffusion coefficient on nonequilibrium plate height in capillary zone electrophoresis, *J. Chromatogr.*, 517, 521, 1990.

18. Burgi, D. S., Salomon, K., and Chien, R. L., Methods for calculating the internal temperature of capillary columns during capillary electrophoresis, *J. Liq. Chromatogr.*, 14, 847, 1991.

19. Vinther, A. and Soeeberg, H., Temperature elevations of the sample zone in free solution capillary electrophoresis under stacking conditions, *J. Chromatogr.*, 559, 27, 1991.

20. Cohen, A.S., Paulus, A., and Karger, B.L., High performance capillary electrophoresis using open tubes and gels, *Chromatographia*, 24, 15, 1987.

21. Nelson, R. J., Paulus, A., Cohen, A. S., Guttman, A., and Karger, B. L., Use of Peltier thermoelectric devices to control column temperature in high-performance capillary electrophoresis, *J. Chromatogr.*, 480, 111, 1989.

22. Giddings, J. C., Harnessing electrical forces for separation. Capillary zone electrophoresis, isoelectric focusing, field flow fractionation, split-flow thin-cell continuous-separation and other techniques, *J. Chromatogr.*, 480, 21, 1989.

23. Monnig, C. A. and Jorgenson, J. W., On-column sample gating for high-speed capillary zone electrophoresis, *Anal. Chem.*, 63, 802, 1991.

24. Huang, X., Coleman, W. F., and Zare, R. N., Analysis of factors causing peak broadening in capillary zone electrophoresis, *J. Chromatogr.*, 480, 95, 1989.

25. Poppe, H., Cifuentes, A., and Kok, W. T., Theoretical description of the influence of external radial fields on the electroosmotic flow in capillary electrophoresis, *Analy. Chem.*, 68, 888, 1996.

26. Tavares, M. F. M. and McGuffin, V. L., Theoretical model of electroosmotic flow for capillary zone electrophoresis, *Anal. Chem.*, 67, 3687, 1995.

27. Gas, B., Stedry, M., and Kenndler, E., Contribution of the electroosmotic flow to peak broadening in capillary zone electrophoresis with uniform zeta potential, *J. Chromatogr.,* 709, 63, 1995.
28. Cohen, N. and Grushka, E., Controlling electroosmotic flow in capillary zone electrophoresis, *J. Chromatogr.,* 678, 167, 1994.
29. Lambert, W. J. and Middleton, D. L., pH hysteresis effect with silica capillaries in capillary zone electrophoresis, *Anal. Chem.,* 62, 1585, 1990.
30. Schwer, C. and Kenndler, E., Electrophoresis in fused-silica capillaries: the influence of organic solvents on the electroosmotic velocity and the zeta potential, *Anal. Chem.,* 63, 1801, 1991.
31. Altria, K. D. and Simpson, C. F., Measurement of electroosmotic flows in high-voltage capillary zone electrophoresis, *Anal. Proc.,* 23, 453, 1986.
32. Altria, K. D. and Simpson, C. F., High-voltage capillary zone electrophoresis: operating parameter effects on electroosmotic flows and electrophoretic mobilities, *Chromatographia,* 24, 527, 1987.
33. Huang, X., Gordon, M. J., and Zare, R. A., Bias in quantitative capillary zone electrophoresis caused by electrokinetic sample injection, *Anal. Chem.,* 60, 375, 1988.
34. Wanders, B. J., Van, de Goor, A. A. A. M., and Everaerts, F. M., Methods for on-line determination and control of electroosmosis in capillary electrochromatography and electrophoresis, *J. Chromatogr.,* 470, 89, 1989.
35. Van de Goor, A. A. A. M., Wanders, B. J., and Everaerts, F. M., Modified methods for off- and on-line determination of electroosmosis in capillary electrophoretic separations, *J. Chromatogr.,* 470, 95, 1989.
36. Fujiwara, S. and Honda, S., Determination of cinnamic acid and its analogs by electrophoresis in a fused silica capillary, *Anal. Chem.,* 58, 1811, 1986.
37. Fujiwara, S. and Honda, S., Effects of addition of organic solvent on the separation of positional isomers in high-voltage capillary zone electrophoresis, *Anal. Chem.,* 59, 487, 1987.
38. VanOrman, B. B., Liversidge, G. G., McIntire, G. L., Olefirowicz, T. M., and Ewing, A. G., Effects of buffer composition on electroosmotic flow in capillary electrophoresis, *J. Microcolumn,* Sep 2, 176, 1990.
39. Atamna, I. Z., Metral, C. J., Muschik, G. M., and Issaq, H. J., Factors that influence mobility, resolution, and selectivity in capillary zone electrophoresis. III. The role of the buffers' anion, *J. Liq. Chromatogr.,* 13, 3201, 1990.
40. Camilleri, P. and Okafo, G. N., Replacement of water by deuterium oxide in capillary zone electrophoresis can increase resolution of peptides and proteins, *J. Chem. Soc., Chem. Commun.,* 3, 196, 1991.
41. Bauer, H., Gruebler, G., and Wolf, B., Application of flow gradient techniques to high performance capillary zone electrophoresis (HPCE) of peptides and proteins, *Peptides,* 329, 1990.
42. Salomon, K., Burgi, D. S., and Helmer, J. C., Evaluation of fundamental properties of a silica capillary used for capillary electrophoresis, *J. Chromatogr.,* 559, 69, 1991.
43. Lauer, H. H. and McManigill, D., Capillary zone electrophoresis of proteins in untreated fused silica tubing, *Anal. Chem.,* 587, 166, 1986.
44. Huang, X., Luckey, J. A., Gordon, M. J., and Zare, R. N., Quantitative determination of low molecular weight carboxylic acids by capillary zone electrophoresis/conductivity detection, *Anal. Chem.,* 61(7), 766, 1989.
45. Foret, F. and Bocek, P., Use of optical fibers for the adaptation of UV detector LCD 254 for online detection in capillary columns, *Chem. Listy,* 83(2), 191, 1989.
46. Herrin, B. J., Shafer, S. G., van, A. J., Harris, J. M., and Snyder, R. S., Control of electroendoosmosis in coated quartz capillaries, *J. Colloid. Interfac. Sci.,* 115, 46, 1987.
47. McCormick, R. M., Capillary zone electrophoretic separation of peptides and proteins using low pH buffers in modified silica capillaries, *Anal. Chem.,* 60, 2322, 1988.
48. Hjerten, S., High-performance electrophoresis: elimination of electroendosmosis and solute adsorption, *J. Chromatogr.,* 347, 191, 1985.
49. Huang, M. X., Dubrovcakovaschneiderman, E., and Novotny, M. V., Self-assembled alkylsilane monolayers for the preparation of stable and efficient coatings in capillary electrophoresis, *J. Microcolumn Sep.,* 6, 571, 1994.

50. Huang, M. X., Plocek, J., and Novotny, M. V., Hydrolytically stable cellulose-derivative coatings for capillary electrophoresis of peptides, proteins and glycoconjugates, *Electrophoresis,* 16, 396, 1995.
51. Chiari, M., Nesi, M., Sandoval, J. E., and Pesek, J. J., Capillary electrophoretic separation of proteins using stable, hydrophilic poly(acryloylaminoethoxyethanol)-coated columns, *J. Chromatogr.,* 717, 1, 1995.
52. Ghowsi, K. and Gale, R. J., Field effect electroosmosis, *J. Chromatogr.,* 559, 95, 1991.
53. Lee, C. S., Blanchard, W. C., and Wu, C. T., Direct control of the electroosmosis in capillary zone electrophoresis by using an external electric field, *Anal. Chem.,* 62, 1550, 1990.
54. Wu, C. T., Lopes, T., Patel, B., and Lee, C. S., Effect of direct control of electroosmosis on peptide and protein separations in capillary electrophoresis, *Anal. Chem.,* 64, 886, 1992.
55. Hayes, M. A. and Ewing, A. G., Electroosmotic flow control and monitoring with an applied radial voltage for capillary zone electrophoresis, *Anal. Chem.,* 64, 512, 1992.
56. Lee, C. S., Wu, C. T., Lopes, T., and Patel, B., Analysis of separation efficiency in capillary electrophoresis with direct control of electroosmosis by using an external electric field, *J. Chromatogr.,* 559, 133, 1991.
57. Lee, C. S., McManigill, D., Wu, C. T., and Patel, B., Factors affecting direct control of electroosmosis using an external electric field in capillary electrophoresis, *Anal. Chem.,* 63, 1519, 1991.
58. Camilleri, P. and Okafo, G., Capillary electrophoresis in [2H]water solution, *J. Chromatogr.,* 541, 489, 1991.
59. Okafo, G. N., Brown, R., and Camilleri, P., Some physicochemical characteristics that make deuterium oxide-based buffer solutions useful media for capillary electrophoresis, *J. Chem. Soc., Chem. Commun.,* 13, 864, 1991.
60. Okafo, G. N. and Camilleri, P., Capillary electrophoretic separation in both water- and heavy-water-based electrolytes can provide more information on tryptic digests, *J. Chromatogr.,* 547, 551, 1991.
61. Camilleri, P., Okafo, G. N., Southan, C., and Brown, R., Analytical and micropreparative capillary electrophoresis of the peptides from calcitonin, *Anal. Biochem.,* 198, 36, 1991.
62. Jorgenson, J. W. and Lukacs, K. D., High-resolution separations based on electrophoresis and electroosmosis, *J. Chromatogr.,* 218, 209, 1981.
63. Rose, D. J. and Jorgenson, J. W., Characterization and automation of sample introduction methods for capillary zone electrophoresis, *Anal. Chem.,* 60, 642, 1988.
64. Burton, D. E., Sepaniak, M. J., and Maskarinec, M. P., The effect of injection procedures on efficiency in micellar electrokinetic capillary chromatography, *Chromatographia,* 21, 583, 1986.
65. Row, K. H., Griest, W. H., and Maskarinec, M. P., Separation of modified nucleic acid constituents by micellar electrokinetic capillary chromatography, *J. Chromatogr.,* 409, 193, 1987.
66. Huang, X., Gordon, M. J., and Zare, R. N., Current-monitoring method for measuring the electroosmotic flow rate in capillary zone electrophoresis, *Anal. Chem.,* 60, 1837, 1988.
67. Kuhr, W. G. and Yeung, E. S., Optimization of sensitivity and separation in capillary zone electrophoresis with indirect fluorescence detection, *Anal. Chem.,* 60, 2642, 1988.
68. Honda, S., Iwase, S., and Fujiwara, S., Evaluation of an automated siphonic sampler for capillary zone electrophoresis, *J. Chromatogr.,* 404, 313, 1987.
69. Schwartz, H. E., Melera, M., and Brownlee, R. G., Performance of an automated injection and replenishment system for capillary electrophoresis, *J. Chromatogr.,* 480, 129, 1989.
70. Grushka, E. and McCormick, R. M., Zone broadening due to sample injection in capillary zone electrophoresis, *J. Chromatogr.,* 471, 421, 1989.
71. Burgi, D. S., and Chien, R. L., Improvement in the method of sample stacking for gravity injection in capillary zone electrophoresis, *Anal. Biochem.,* 202, 306, 1992.
72. Chien, R. L. and Burgi, D. S., Sample stacking of an extremely large injection volume in high-performance capillary electrophoresis, *Anal. Chem.,* 64, 1046, 1992.
73. Chien, R.-L. and Burgi, D. S., On-column sample concentration using field-amplified CZE., *Anal. Chem.,* 64(8), 489A, 1992.
74. Chien, R. L., Mathematical modeling of field-amplified sample injection in high-performance capillary electrophoresis, *Anal. Chem.,* 63, 2866, 1991.

75. Chien, R. L. and Helmer, J. C., Electroosmotic properties and peak broadening in field-amplified capillary electrophoresis, *Anal. Chem.*, 63, 1354, 1991.

76. Burgi, D. S. and Chien, R. L., Optimization in sample stacking for high-performance capillary electrophoresis, *Anal. Chem.*, 63, 2042, 1991.

77. Burgi, D. S. and Chien, R.-L., Application of Sample Stacking to Gravity Injection in Capillary Electrophoresis, *J. Microcol. Sep.*, 3, 199, 1991.

78. Chien, R. L. and Burgi, D. S., Field-amplified polarity-switching sample injection in high-performance capillary electrophoresis, *J. Chromatogr.*, 559, 153, 1991.

79. Chien, R. L. and Burgi, D. S., Field amplified sample injection in high-performance capillary electrophoresis, *J. Chromatogr.*, 559, 141, 1991.

80. Tsuda, T., Mizuno, T., and Akiyama, J., Rotary-type injector for capillary zone electrophoresis, *Anal. Chem.*, 59, 799, 1987.

81. Demyl, M., Foret, F., and Bocek, P., Electric sample splitter for capillary zone electrophoresis, *J. Chromatogr.*, 320, 159, 1985.

82. Verheggen, T. P. E. M., Beckers, J. L., and Everaerts, F. M., Simple sampling device for capillary isotachophoresis and capillary zone electrophoresis, *J. Chromatogr.*, 452, 615, 1988.

83. Pospichal, J., Deml, M., Gebauer, P., and Bocek, P., Generation of operational electrolytes for isotachophoresis and capillary zone electrophoresis in a three-pole column, *J. Chromatogr.*, 470, 43, 1989.

84. Rohlicek, V. and Deyl, Z., Simple device for flushing capillaries in capillary zone electrophoresis, *J. Chromatogr.*, 480, 289, 1989.

85. Nielsen, R. G. and Rickard, E. C., Applications of capillary zone electrophoresis to quality control, *ACS Symp. Ser.*, 434, 36, 1990.

86. Nielsen, R. G. and Rickard, E. C., Method optimization in capillary zone electrophoretic analysis of hGH tryptic digest fragments, *J. Chromatogr.*, 516, 99, 1990.

87. Wallingford, R. and Ewing, A. G., Capillary zone electrophoresis with electrochemical detection in 12.7 µm diameter columns, *Anal. Chem.*, 60, 1972, 1988.

88. Miller, C., Hernandez, J. F., Craig, A. G., Dykert, J., and Rivier, J., Synthesis, purification and characterization of rat histone H2A (1-53)-NH2, *Anal. Chim. Acta*, 249, 215, 1991.

89. Rivier, J. E., Miller, C. L., Tuchscherer, G., Craig, A., Hernandez, J. F., Dykert, J., Raschdorf, F., and Mutter, M., Chromatographic characterization of template assembled synthetic proteins (TASP) and other large peptides using narrow-bore IEC, RPHPLC, CZE and MS analysis, *Peptides*, 80-86, 1990.

90. Chen, T. M., George, R. C., and Payne, M. H., Separation of anticoagulant peptide MDL 28,050 from its deletion by-products by capillary zone electrophoresis, *J. High Resolut. Chromatogr.*, 13, 782, 1990.

91. Prusik, Z., Kasicka, V., Mudra, P., Stepanek, J., Smekal, O., and Hlavacek, J., Correlation of capillary zone electrophoresis with continuous free-flow zone electrophoresis: application to the analysis and purification of synthetic growth hormone releasing peptide, *Electrophoresis*, 11, 932, 1990.

92. Guarino, B. C. and Phillips, D., High-performance ion exchange chromatography as a preparative adjunct to capillary electrophoresis, *Am. Lab.*, 23, 68, 1991.

93. Liu, J., Hsieh, Y. Z., Wiesler, D., and Novotny, M., Design of 3-(4-carboxybenzoyl)-2-quinolinecarboxaldehyde as a reagent for ultrasensitive determination of primary amines by capillary electrophoresis using laser fluorescence detection, *Anal. Chem.*, 63, 408, 1991.

94. Hortin, G. L., Griest, T., and Benutto, B. M., Separations of sulfated and nonsulfated forms of peptides by capillary electrophoresis: comparison with reversed-phase HPLC, *BioChromatography*, 5, 118, 1990.

95. Kruegar, R. J., Hobbs, T. R., Mihal, K. A., Tehrani, J., and Zeece, M. G., Analysis of endoproteinase Arg C action on adrenocorticotrophic hormone by capillary electrophoresis and reversed-phase high-performance liquid chromatography, *J. Chromatogr.*, 543, 451, 1991.

96. Tran, A. D., Blanc, T., and Leopold, E. J., Free solution capillary electrophoresis and micellar electrokinetic resolution of amino acid enantiomers and peptide isomers with L- and D-Marfey's reagents, *J. Chromatogr. 516*, 241, 1990.

97. Bullock, J. A. Y., L.C., Free solution capillary electrophoresis of basic proteins in uncoated fused silica capillary tubing, *J. Microcol. Sep.*, 3, 241, 1991.

98. Florance, J. R., Konteatis, Z. D., Macielag, M. J., Lessor, R. A., and Galdes, A., Capillary zone electrophoresis studies of motilin peptides. Effects of charge, hydrophobicity, secondary structure and length, *J. Chromatogr.,* 559, 391, 1991.

99. Buoen, S., Eriksen, J. A., Revheim, H., and Schanche, J. S., Conformational isomerism of a proline-proline containing peptide: verification by CE, HPLC and AAA, *Peptides,* 331, 1990.

100. Pessi, A., Bianchi, E., Chiappinelli, L., Nardi, A., and Fanali, S., Application of capillary zone electrophoresis to the characterization of multiple antigen peptides, *J. Chromatogr.,* 557, 307, 1991.

101. Wheat, T. E., Young, P. M., and Astephen, N. E., Use of capillary electrophoresis for the detection of single-residue substitutions in peptide mapping, *J. Liq. Chromatogr.,* 14, 987, 1991.

102. Yannoukakos, D., Meyer, H. E., Vasseur, C., Driancourt, C., Wajcman, H., and Bursaux, E., Three regions of erythrocyte band 3 protein are phosphorylated on tyrosines: characterization of the phosphorylation sites by solid phase sequencing combined with capillary electrophoresis, *Biochim. Biophys. Acta,* 1066, 70, 1991.

103. Young, P. M. and Merion, M., Capillary electrophoresis analysis of species variations in the tryptic maps of cytochrome c, *Curr. Res. Protein Chem.,* 3rd, 217, 1989.

104. Ferranti, P., Malorni, A., Pucci, P., Fanali, S., Nardi, A., and Ossicini, L., Capillary zone electrophoresis and mass spectrometry for the characterization of genetic variants of human hemoglobin, *Anal. Biochem.,* 194, 1, 1991.

105. Bushey, M. M. and Jorgenson, J. W., A comparison of tryptic digests of bovine and equine cytochrome C by comprehensive reversed-phase HPLC-CE, *J. Microcol. Sep.,* 2, 293, 1990.

106. Snopek, J., Jelinek, I., and Smolkova, K. E., Micellar inclusion and metal-complex enantioselective pseudophases in high-performance electromigration methods, *J. Chromatogr.,* 452, 571, 1988.

107. Gebauer, P., Deml, M., Bocek, P., and Janak, J., Determination of nitrate, chloride and sulfate in drinking water by capillary free-zone electrophoresis, *J. Chromatogr.,* 267, 455, 1983.

108. Aguilar, M., Huang, X., and Zare, R. N., Determination of metal ion complexes in electroplating solutions using capillary zone electrophoresis with UV detection, *J. Chromatogr.,* 480, 427, 1989.

109. Gassmann, E., Kuo, J. C., and Zare, R. N., Electrokinetic separation of chiral compounds, *Science,* 813, 1985.

110. Walbroehl, Y. and Jorgenson, J. W., Capillary zone electrophoresis of neutral organic molecules by solvophobic association with tetraalkylammonium ion, *Anal. Chem.,* 58, 479, 1986.

111. Honda, S., Suzuki, K., and Kakehi, K., Simultaneous determination of iodate and periodate by capillary zone electrophoresis: application to carbohydrate analysis, *Anal. Biochem.,* 177(1), 62, 1989.

112. Wallingford, R. A. and Ewing, A. G., Separation of serotonin from catechols by capillary zone electrophoresis with electrochemical detection, *Anal. Chem.,* 61(2), 98, 1989.

113. Wallingford, R. A., Curry, P. D. J., and Ewing, A. G., Retention of catechols in capillary electrophoresis with micellar and mixed micellar solutions, *J. Microcol. Sep.,* 1(1), 23, 1989.

114. Cohen, A. S., Terabe, S., Smith, J. A., and Karger, B. L., High performance capillary electrophoretic separation of bases, nucleosides and oligonucleotides, *Anal. Chem.,* 59, 1021, 1987.

115. Dolnik, V., Liu, J., Banks, J. J. F., Novotny, M. V., and Bocek, P., Capillary zone electrophoresis of oligonucleotides. Factors affecting separation, *J. Chromatogr.,* 480, 321, 1989.

116. Romano, J., Jandik, P., Jones, W. R., and Jackson, P. E., Optimization of inorganic capillary electrophoresis for the analysis for anionic solutes in real samples, *J. Chromatogr.,* 546, 411, 1991.

117. Foret, F., Fanali, S., Nardi, A., and Bocek, P., Capillary zone electrophoresis of rare earth metals with indirect UV absorbance detection, Electrophoresis (Weinheim, Fed. Repub. Ger., 11, 780, 1990.

118. Honda, S., Iwase, S., Makino, A., and Fujiwara, S., Simultaneous determination of reducing monosaccharides by capillary zone electrophoresis as the borate complexes of N-2-pyridylglycamines, *Anal. Biochem.,* 176(1), 72, 1989.

119. Saitoh, T., Hoshino, H., and Yotsuyanagi, T., Micellar electrokinetic capillary chromatography of porphinato chelates as a spectrophotometric approach to sub-femtomole detection of metal chelates, *Anal. Sci.,* 7, 495, 1991.

120. Saitoh, K., Chromatography using micelle solution, *Kagaku,* (Kyoto), 45, 884, 1990.
121. Kord, A. S., Strasters, J. K., and Khaledi, M. G., Comparative study of the determination of solute-micelle binding constants by micellar liquid chromatography and micellar electrokinetic capillary chromatography, *Anal. Chim. Acta,* 246, 131, 1991.
122. Saitoh, K., Kiyohara, C., and Suzuki, N., Mobilities of metal.beta.-diketonato complexes in micellar electrokinetic chromatography, *J. High Resolut. Chromatogr.,* 14, 245, 1991.
123. Fukushi, K. and Hiiro, K., Use of crown ethers in the isotachophoretic determination of metal ions, *J. Chromatogr.,* 523, 281, 1990.
124. Fukushi, K. and Hiro, K., Use of cyclodextrins in the isotachophoretic determination of various inorganic anions, *J. Chromatogr.,* 518, 189, 1990.
125. Hirokawa, T., Hu, J. Y., Eguchi, S., Nishiyama, F., and Kiso, Y., Study of isotachophoretic separation behavior of metal cations by means of particle-induced x-ray emission. I. Separation of twenty metal cations using an acetic acid buffer system and alpha-hydroxyisobutyric acid complex-forming system, *J. Chromatogr.,* 538, 413, 1991.
126. Krivankova, L. and Bocek, P., Determination by capillary isotachophoresis of the iron(III)-EDTA complex and of free EDTA in scrubbing liquid, *Chem. Prum.,* 40, 134, 1990.
127. Swaile, D. F. and Sepaniak, M. J., Determination of metal ions by capillary zone electrophoresis with on-column chelation using 8-hydroxyquinoline-5-sulfonic acid, *Anal. Chem.,* 63, 179, 1991.
128. Karovicova, J., Polonsky, J., and Simko, P., Determination of preservatives in some food products by capillary isotachophoresis, *Nahrung,* 35, 543, 1991.
129. Foret, F., Sustacek, V., and Bocek, P., Separation of some triazine herbicides and their solvolytic products by capillary zone electrophoresis, *Electrophoresis,* 11, 95, 1990.
130. Jokl, V. and Petrzelkova, I., Capillary isotachophoresis of some beta-adrenolytic agents, *Cesk. Farm.,* 40, 65, 1991.
131. Karovicova, J., Polonsky, J., Drdak, M., and Pribela, A., Determination of nitrates in vegetables by capillary isotachophoresis, *Nahrung,* 34, 765, 1990.
132. Kenney, B. F., Determination of organic acids in food samples by capillary electrophoresis, *J. Chromatogr.,* 546, 423, 1991.
133. Kopacek, P., Kaniansky, D., and Hejzlar, J., Characterization of humic substances by capillary isotachophoresis, *J. Chromatogr.,* 545, 461, 1991.
134. Watarai, H., Microemulsion capillary electrophoresis, *Chem. Lett.,* 3, 391, 1991.
135. Chadwick, R. R. and Hsieh, J. C., Separation of cis and trans double-bond isomers using capillary zone electrophoresis, *Anal. Chem.,* 63, 2377, 1991.
136. Karovicova, J., Polonsky, J., and Pribela, A., Composition of organic acids of Sambucus nigra and Sambucus ebulus, *Nahrung,* 34, 665, 1990.
137. Karovicova, J., Drdak, M., and Polonsky, J., Utilization of capillary isotachophoresis in the determination of organic acids in food, *J. Chromatogr.,* 509, 283, 1990.
138. Krivankova, L. and Bocek, P., Determination of pyruvate, acetoacetate, lactate, and 3-hydroxybutyrate in plasma of patients with diabetes mellitus by capillary isotachophoresis, *J. Microcol. Sep.,* 2, 80, 1990.
139. Lutonska, P., Determination of the contents of carboxylic acids in silages by the capillary isotachophoresis method, *Agrochemia,* (Bratislava), 30, 218, 1990.
140. Vindevogel, J., Sandra, P., and Verhagen, L. C., Separation of hop bitter acids by capillary zone electrophoresis and micellar electrokinetic chromatography with UV-diode array detection, *J. High Resolut. Chromatogr.,* 13, 295, 1990.
141. Oefner, P., Haefele, R., Bartsch, G., and Bonn, G., Isotachophoretic separation of organic acids in biological fluids, *J. Chromatogr.,* 516, 251, 1990.
142. Green, J. S. and Jorgenson, J. W., High-speed zone electrophoresis in open-tubular fused silica capillaries, *J. High Resolut. Chromatogr. Chromatogr. Commun.,* 7, 529, 1984.
143. Green, J. S. and Jorgenson, J. W., Variable-wavelength on-column fluorescence detector for open-tubular zone electrophoresis, *J. Chromatogr.,* 352, 337, 1986.
144. Dovichi, N. J. and Cheng, Y. F., Ultramicro amino acid analysis: ranging below attomoles, *Am. Biotechnol. Lab.,* 7(2), 10, 12, 14, 1989.
145. Cheng, Y. F. and Dovichi, N. J., Subattomole amino acid analysis by capillary zone electrophoresis and laser-induced fluorescence, *Science,* 242, 562, 1989.

146. Kuhr, W. G. and Yeung, E. S., Indirect Fluorescence Detection of Native Amino Acids in Capillary Zone Electrophoresis, *Anal. Chem.,* 60, 1832, 1988.

147. Gozel, P., Gassmann, E., Michelsen, H., and Zare, R. N., Electrokinetic resolution of amino acid enantiomers with copper(II)-aspartame support electrolyte, *Anal. Chem.,* 59, 44, 1987.

148. Yu, M. and Dovichi, N. J., Attomole amino acid determination by capillary zone electrophoresis with thermooptical absorbance detection, *Anal. Chem.,* 61(1), 37, 1989.

149. Yu, M. and Dovichi, N. J., Sub-femtomole determination of DABSYL-amino acids with capillary zone electrophoretic separation and laser-induced thermo-optical absorbance detection, *Mikrochim. Acta,* III, 27, 1988.

150. Pawliszyn, J., Nanoliter volume sequential differential concentration detector, *Anal. Chem.,* 60, 2796-2801, 1988.

151. Pentoney, S. L., Huang, X., Burgi, D. S., and Zare, R. S., On-line connector for microcolumns: application to the on-column OPA derivatization of amino acids separated by capillary zone electrophoresis, *Anal. Chem.,* 60, 2625, 1988.

152. Wallingford, R. A. and Ewing, A. G., Capillary zone electrophoresis with electrochemical detection, *Anal. Chem.,* 59, 1762, 1987.

153. Heber, M., Liedtke, C., Korte, H., Hoffmannposorske, E., Donelladeana, A., Pinna, L. A., Perich, J., Kitas, E., Johns, R. B., and Meyer, H. E., Non-radioactive determination of pth and dabsyl phosphoamino acids by capillary electrophoresis, *Chromatographia,* 33, 347, 1992.

154. Wright, B. W., Ross, G. A., and Smith, R. D., Capillary zone electrophoresis with laser fluorescence detection of marine toxins, *J. Microcol. Sep.,* 1, 85, 1989.

155. Nickerson, B. and Jorgenson, J. W., High speed capillary zone electrophoresis with laser induced fluorescence detection, *J. High Res. Chromatogr. Chromatogr. Comm.,* 11, 533, 1988.

156. Wu, S. and Dovichi, N. J., High-sensitivity fluorescence detector for fluorescein isothiocyanate derivatives of amino acids separated by capillary zone electrophoresis, *J. Chromatogr.,* 480, 141, 1989.

157. Yu, M. and Dovichi, N. J., Attomole amino acid determination: capillary-zone electrophoresis with laser-based thermooptical detection, *Appl. Spectrosc.,* 43(2), 196, 1989.

158. Grossman, P. D., Wilson, K. J., Petrie, G., and Lauer, H. H., Effect of buffer pH and peptide composition on the selectivity of peptide separations by capillary zone electrophoresis, *Anal. Biochem.,* 173, 265, 1988.

159. Guttman, A., Paulus, A., Cohen, A. S., Grinberg, N., and Karger, B. L., Use of complexing agents for selective separation in high performance capillary electrophoresis. Chiral resolution via cyclodextrins incorporated within polyacrylamide gel columns, *J. Chromatogr.,* 448, 41, 1988.

160. Grossman, P. D., Colburn, J. C., and Lauer, H. H., A semiempirical model for the electrophoretic mobilities of peptides in free-solution capillary electrophoresis, *Anal. Biochem.,* 179, 28, 1989.

161. Liu, J., Cobb, K. A., and Novotny, M., Separation of precolumn ortho-phthalaldehyde-derivatized amino acids by capillary zone electrophoresis with normal and micellar solutions in the presence of organic modifiers, *J. Chromatogr.,* 468, 55, 1989.

162. Nickerson, B. and Jorgenson, J. W., Characterization of a post-column reaction-laser-induced fluorescence detector for capillary zone electrophoresis, *J. Chromatogr.,* 480, 157, 1989.

163. Terabe, S., Shibata, M., and Miyashita, Y., Chiral separation by electrokinetic chromatography with bile salt micelles, *J. Chromatogr.,* 480, 403, 1989.

164. Dobashi, A., Ono, T., Hara, S., and Yamaguchi, J., Enantioselective hydrophobic entanglement of enantiomeric solutes with chiral functionalized micelles by electrokinetic chromatography, *J. Chromatogr.,* 480, 413, 1989.

165. Fanali, S., Ossicini, L., Foret, F., and Bocek, P., Resolution of optical isomers by capillary zone electrophoresis: Study of enantiomeric and diastereoisomeric Co(III) complexes with ethylenediamine and amino acid ligands, *J. Microcol. Sep.,* 1, 190, 1989.

166. Gozel, P., and Zare, R. N., Resolution of DL-amino acids by capillary zone electrophoresis using chiral electrolytes, *ASTM Spec. Tech. Publ. 1009 (Prog. Anal. Lumin.),* 41, 1988.

167. Otsuka, K., Terabe, S., and Ando, T., Electrokinetic chromatography with micellar solutions. Separation of phenylthiohydantoin-amino acids, *J. Chromatogr.,* 332, 219, 1985.

168. Ueda, T., Mitchell, R., Kitamura, F., Metcalf, T., Kuwana, T., and Nakamoto, A., Separation of naphthalene-2,3-dicarboxaldehyde-labeled amino acids by high-performance capillary electrophoresis with laser-induced fluorescence detection, *J. Chromatogr.,* 593, 265, 1992.

169. Nickerson, B. and Jorgenson, J. W., High sensitivity laser induced fluorescence detection in capillary zone electrophoresis, *J. High Resolut. Chromatogr. Chromatogr. Comm.,* 11, 878, 1988.

170. Cheng, Y. F. and Dovichi, N. J., Subattomole amino acid analysis by laser-induced fluorescence: the sheath flow cuvette meets capillary zone electrophoresis, *ASTM Spec. Tech. Publ. 1066,* 151, 1990.

171. Waldron, K. C., Wu, S., Earle, C. W., Harke, H. R., and Dovichi, N. J., Capillary zone electrophoresis separation and laser-based detection of both fluorescein thiohydantoin and dimethylaminoazobenzene thiohydantoin derivatives of amino acids, *Electrophoresis,* 11, 777, 1990.

172. Sweedler, J. V., Shear, J. B., Fishman, H. A., Zare, R. N., and Scheller, R. H., Fluorescence detection in capillary zone electrophoresis using a charge-coupled device with time-delayed integration, *Anal. Chem.,* 63, 496, 1991.

173. Rose, D. R. J. and Jorgenson, J. W., Post-capillary fluorescence detection in capillary zone electrophoresis using o-pthaldialdehyde, *J. Chromatogr.,* 447, 117, 1988.

174. Ward, L. D., Reid, G. E., Moritz, R. L., and Simpson, R. J., Strategies for internal amino acid sequence analysis of proteins separated by polyacrylamide gel electrophoresis, *J. Chromatogr.,* 519, 199, 1990.

175. Otsuka, K. and Terabe, S., Enantiomeric resolution by micellar electrokinetic chromatography with chiral surfactants, *J. Chromatogr.,* 515, 221, 1990.

176. Foley, J. P., Critical compilation of solute-micelle binding constants and related parameters from micellar liquid chromatographic measurements, *Anal. Chim. Acta,* 231, 237, 1990.

177. Otsuka, K., Kawahara, J., Tatekawa, K., and Terabe, S., Chiral separations by micellar electrokinetic chromatography with sodium N-dodecanoyl-L-valinate, *J. Chromatogr.,* 559, 209, 1991.

178. Sepaniak, M. J., Swaile, D. F., and Powell, A. C., Instrumental developments in micellar electrokinetic capillary chromatography, *J. Chromatogr.,* 480, 185, 1989.

179. Balchunas, A. T. and Sepaniak, M. J., Extension of elution range in micellar electrokinetic capillary chromatography, *Anal. Chem.,* 59, 1466, 1988.

180. Tsuda, T., Kobayashi, Y., Hori, A., Matsumoto, T., and Suzuki, O., Postcolumn detection for capillary zone electrophoresis, *J. Chromatogr.,* 456, 375, 1988.

181. Nielen, M., W. and F., Impact of experimental parameters on the resolution of positional isomers of aminobenzoic acid in capillary zone electrophoresis, *J. Chromatogr.,* 542, 173, 1991.

182. Nishi, H., Fukuyama, T., and Matsuo, M., Resolution of Optical Isomers of 2,3,4,6-Tetra-O-acetyl-b-D-glucopyranosyl Isothiocyanate (GITC)-Derivatized DL-Amino Acids by Micellar Electrokinetic Chromatography, *J. Microcol. Sep.,* 2, 234, 1990.

183. Tanaka, M., Asano, S., Yoshinago, M., Kawaguchi, Y., Tetsumi, T., and Shono, T., Separation of racemates by capillary zone electrophoresis based on complexation with cyclodextrins, *Fresenius. J. Anal. Chem.,* 339, 63, 1991.

184. Ong, C. P., Ng, C. L., Lee, H. K., and Li, S. F. Y., Separation of Dns-amino acids and vitamins by micellar electrokinetic chromatography, *J. Chromatogr.,* 559, 537, 1991.

185. Otsuka, K. and Terabe, S., Effects of methanol and urea on optical resolution of phenylthiohydantoin-DL-amino acids by micellar electrokinetic chromatography with sodium N-dodecanoyl-L-valinate, *Electrophoresis,* 11, 982, 1990.

186. Wallingford, R. A. and Ewing, A. G., Amperometric detection of catechols in capillary zone electrophoresis with normal and micellar solutions, *Anal. Chem.,* 60, 258, 1988.

187. Wallingford, R. A. and Ewing, A. G., Retention of ionic and non-ionic catechols in capillary zone electrophoresis with micellar solutions, *J. Chromatogr.,* 441, 299, 1988.

188. Kaneta, T., Tanaka, S., and Yoshida, H., Improvement of resolution in the capillary electrophoretic separation of catecholamines by complex formation with boric acid and control of electroosmosis with a cationic surfactant, *J. Chromatogr.,* 538, 385, 1991.

189. Ong, C. P., Pang, S. F., Low, S. P., Lee, H. K., and Li, S. F. Y., Migration behavior of catechols and catecholamines in capillary electrophoresis, *J. Chromatogr.,* 559, 529, 1991.

190. Tanaka, S., Kaneta, T., and Yoshida, H., Separation of catecholamines by capillary zone electrophoresis using complexation with boric acid, *Anal. Sci.,* 6, 467, 1990.

191. Lee, K. J. and Heo, G. S., Application of home-made capillary zone electrophoresis system to the separation of organic molecules, *J. Korean Chem. Soc.,* 35, 219, 1991.

192. Tanaka, S., Kaneta, T., Taga, M., Yoshida, H., and Ohtaka, H., Capillary tube isotachophoretic separation of catecholamines using cyclodextrin in the leading electrolyte, *J. Chromatogr.*, 587, 364, 1991.

193. Olefirowicz, T. M. and Ewing, A. G., Dopamine concentration in the cytoplasmic compartment of single neurons determined by capillary electrophoresis, *J. Neurosci. Methods*, 34, 11, 1990.

194. Olefirowicz, T. M. and Ewing, A. G., Capillary electrophoresis in 2 and 5 .mu.m diameter capillaries: application to cytoplasmic analysis, *Anal. Chem.*, 62, 1872, 1990.

195. Chien, J. B., Wallingford, R. A., and Ewing, A. G., Estimation of free dopamine in the cytoplasm of the giant dopamine cell of Planorbis corneus by voltammetry and capillary electrophoresis, *J. Neurochem.*, 54, 633, 1990.

196. Olefirowicz, T. M. and Ewing, A. G., Capillary electrophoresis for sampling single nerve cells, *Chimia*, 45, 106, 1991.

197. Guzman, N. A., Trebilcock, M. A., and Advis, J. P., Increased sensitivity to analyze brain tissue constituents: use of capillary electrophoresis coupled to fluorescence microscopy, *Anal. Chim. Acta*, 249, 247, 1990.

198. Hernandez, L., Guzman, N., and Hoebel, B. G., Bidirectional microdialysis in vivo shows differential dopaminergic potency of cocaine, procaine and lidocaine in the nucleus accumbens using capillary electrophoresis for calibration of drug outward diffusion, *Psychopharmacology*, (Berlin), 105, 264, 1991.

199. Liu, Y. and Chan, K. F. J., High-performance capillary electrophoresis of gangliosides, *Electrophoresis*, 12, 402, 1991.

200. Nguyen, A. L., Luong, J. H. T., and Masson, C., Capillary electrophoresis for detection and quantitation of domoic acid in mussels, *Anal. Lett.*, 23, 1621, 1990.

201. Nguyen, A. L., Luong, J. H. T., and Masson, C., Determination of nucleotides in fish tissues using capillary electrophoresis, *Anal. Chem.*, 62, 2490, 1990.

202. Thibault, P., Pleasance, S., and Laycock, M. V., Analysis of paralytic shellfish poisons by capillary electrophoresis, *J. Chromatogr.*, 542, 483, 1991.

203. Mereish, K. A., Morris, S., McCullers, G., Taylor, T. J., and Bunner, D. L., Analysis of palytoxin by liquid chromatography and capillary electrophoresis, *J. Liq. Chromatogr.*, 14, 1025, 1991.

204. Yeo, S. K., Ong, C. P., and Li, S. F. Y., Optimization of high-performance capillary electrophoresis of plant growth regulators using the overlapping resolution mapping scheme, *Anal. Chem.*, 63, 2222, 1991.

205. Tsuda, T., Kobayashi, Y., Hori, A., Matsumoto, T., and Suzuki, O., Separation of polyamines in rat tissues by capillary electrophoresis, *J. Microcol. Sep.*, 2, 21, 1990.

206. Matsumoto, T., Tsuda, T., and Suzuki, O., Polyamine determination in clinical laboratories by high-performance liquid chromatography, *Trends Anal. Chem.*, 9, 292, 1990.

207. Huang, X., Shear, J. B., and Zare, R. N., Quantitation of ribonucleotides from base-hydrolyzed RNA using capillary zone electrophoresis, *Anal. Chem.*, 62, 2049, 1990.

208. Liu, J., Shirota, O., and Novotny, M., Capillary electrophoresis of amino sugars with laser-induced fluorescence detection, *Anal. Chem.*, 63, 413, 1991.

209. Liu, J., Shirota, O., Wiesler, D., and Novotny, M., Ultrasensitive fluorometric detection of carbohydrates as derivatives in mixtures separated by capillary electrophoresis, *Proc. Natl. Acad. Sci. U. S. A.*, 88, 2302, 1991.

210. Honda, S., Suzuki, S., Nose, A., Yamamoto, K., and Kakehi, K., Capillary zone electrophoresis of reducing mono- and oligo-saccharides as the borate complexes of their 3-methyl-1-phenyl-2-pyrazolin-5-one derivatives, *Carbohydr. Res.*, 215, 193, 1991.

211. Garner, T. W. and Yeung, E. S., Indirect fluorescence detection of sugars separated by capillary zone electrophoresis with visible laser excitation, *J. Chromatogr.*, 515, 639, 1990.

212. Deyl, Z., Miksik, I., and Struzinsky, R., Separation and partial characterization of Maillard reaction products by capillary zone electrophoresis, *J. Chromatogr.*, 516, 287, 1990.

213. Honda, S., Yamamoto, K., Suzuki, S., Ueda, M., and Kakehi, K., High-performance capillary zone electrophoresis of carbohydrates in the presence of alkaline earth metal ions, *J. Chromatogr.*, 588, 327, 1991.

214. Al Hakim, A., and Linhardt, R. J., Capillary electrophoresis for the analysis of chondroitin sulfate- and dermatan sulfate-derived disaccharides, *Anal. Biochem.*, 195, 68, 1991.

215. Ampofo, S. A., Wang, H. M., and Linhardt, R. J., Disaccharide compositional analysis of heparin and heparan sulfate using capillary zone electrophoresis, *Anal. Biochem.,* 199, 249, 1991.

216. Salomon, K., Burgi, D. S., and Helmer, J. C., Separation of seven tricyclic antidepressants using capillary electrophoresis, *J. Chromatogr.,* 549, 375, 1991.

217. Chmela, Z. and Stransky, Z., Analytical capillary isotachophoresis of some psychopharmaco-logical agents, *Cesk. Farm.,* 39, 172, 1990.

218. Snopek, J., Soini, H., Novotny, M., Smolkova, K. E., and Jelinek, I., Selected applications of cyclodextrin selectors in capillary electrophoresis, *J. Chromatogr.,* 559, 215, 1991.

219. Ong, C. P., Ng, C. L., Lee, H. K., and Li, S. F. Y., Determination of antihistamines in pharma-ceuticals by capillary electrophoresis, *J. Chromatogr.,* 588, 335, 1991.

220. Altria, K. D. and Smith, N. W., Pharmaceutical analysis by capillary zone electrophoresis and micellar electrokinetic capillary chromatography, *J. Chromatogr.,* 538, 506, 1991.

221. Arrowwood, S. and Hoyt, A. M. J., Determination of cimetidine in pharmaceutical preparations by capillary zone electrophoresis, *J. Chromatogr.,* 586, 177, 1991.

222. Kenndler, E., Schwer, C., Fritsche, B., and Poehm, M., Determination of arbutin in uvae-ursi folium (bearberry leaves) by capillary zone electrophoresis, *J. Chromatogr.,* 514, 383, 1990.

223. Pietta, P. G., Mauri, P. L., Rava, A., and Sabbatini, G., Application of micellar electrokinetic capillary chromatography to the determination of flavonoid drugs, *J. Chromatogr.,* 549, 367, 1991.

224. Swedberg, S. A., Use of nonionic and zwitterionic surfactants to enhance selectivity in high-performance capillary electrophoresis. An apparent micellar electrokinetic capillary chroma-tography mechanism, *J. Chromatogr.,* 503, 449, 1990.

225. Krivankova, L., Foret, F., and Bocek, P., Determination of halofuginone in feedstuffs by the combination of capillary isotachophoresis and capillary zone electrophoresis in a column-switching system, *J. Chromatogr.,* 545, 307, 1991.

226. Meier, P. and Thormann, W., Determination of thiopental in human serum and plasma by high-performance capillary electrophoresis-micellar electrokinetic chromatography, *J. Chromatogr.,* 559, 505, 1991.

227. Steuer, W., Grant, I., and Erni, F., Comparison of high-performance liquid chromatography, supercritical fluid chromatography and capillary zone electrophoresis in drug analysis, *J. Chro-matogr.,* 507, 125, 1990.

228. Swartz, M. E., Method development and selectivity control for small molecule pharmaceutical separations by capillary electrophoresis, *J. Liq. Chromatogr.,* 14, 923, 1991.

229. Lloyd, D. K., Cypess, A. M., and Wainer, I. W., Determination of cytosine-.beta.-D-arabinoside in plasma using capillary electrophoresis, *J. Chromatogr.,* 568, 117, 1991.

230. Guzman, N. A., Ali, H., Moschera, J., Iqbal, K., and Malick, W., Assessment of capillary electrophoresis in pharmaceutical applications. Analysis and quantification of a recombinant cytokine in an injectable dosage form, *J. Chromatogr.,* 559, 307, 1991.

231. Tsikas, D., Hofrichter, A., and Brunner, G., Capillary isotachophoretic analysis of .beta.-lactam antibiotics and their precursors, *Chromatographia,* 30, 657, 1990.

232. Yeo, S. K., Lee, H. K., and Li, S. F. Y., Separation of antibiotics by high-performance capillary electrophoresis with photodiode-array detection, *J. Chromatogr.,* 585, 133, 1991.

233. Kenndler, E., Schwer, C., and Kaniansky, D., Purity control of riboflavin-5′-phosphate (vitamin B2 phosphate) by capillary zone electrophoresis, *J. Chromatogr.,* 508, 203, 1990.

234. Tanaka, S., Kaneta, T., Yoshida, H., and Ohtaka, H., Capillary tube isotachophoretic separation of niacin derivatives, *J. Chromatogr.,* 521, 158, 1990.

235. Yik, Y. F., Lee, H. K., Li, S. F. Y., and Khoo, S. B., Micellar electrokinetic capillary chroma-tography of vitamin B6 with electrochemical detection, *J. Chromatogr.,* 585, 139, 1991.

236. Zweigenbaum, J., Separation of anionic surfactant components by capillary electrophoresis and micellar electrokinetic capillary chromatography using P/ACE System 2000, *Chromatogram,* 11, 9, 1990.

237. Stover, F. S., Applications of capillary electrophoresis for industrial analysis, *Electrophoresis,* 11, 750, 1990.

238. Wainright, A., Capillary electrophoresis applied to the analysis of pharmaceutical compounds, *J. Microcol. Sep.,* 2, 166, 1990.

239. Yik, Y. F., Ng, C. L., Ong, C. P., Khoo, S. B., Lee, H. K., and Li, S. F. Y., Analysis of environmental pollutants using high-performance capillary electrophoresis, *Bull. Singapore Natl. Inst. Chem.*, 18, 91, 1990.

240. Northrop, D. M., Martire, D. E., and MacCrehan, W. A., Separation and identification of organic gunshot and explosive constituents by micellar electrokinetic capillary electrophoresis, *Anal. Chem.*, 63, 1038, 1991.

241. Sepaniak, M. J., Swaile, D. F., Powell, A. C., and Cole, R. O., Capillary electrokinetic separations: influence of mobile phase composition on performance, *J. High Resolut. Chromatogr.*, 13, 679, 1990.

242. Ghowsi, K. and Gale, R. J., Application of field effect electro-osmosis to separation-based sensors, *Biosens. Technol., [Proc. Int. Symp.]*, 55, 1989.

243. Takigiku, R. and Schneider, R. E., Reproducibility and quantitation of separation for ribonucleoside triphosphates and deoxyribonucleoside triphosphates by capillary zone electrophoresis, *J. Chromatogr.*, 559, 247, 1991.

244. Karovicova, J. and Polonsky, J., Determination of synthetic colorants by capillary isotachophoresis, *Nahrung*, 35, 403, 1991.

245. Cole, R. O., Sepaniak, M. J., and Hinze, W. L., Optimization of binaphthyl enantiomer separations by capillary zone electrophoresis using mobile phases containing bile salts and organic solvent, *J. High Resolut. Chromatogr.*, 13, 579, 1990.

246. Fanali, S. and Bocek, P., Enantiomer resolution by using capillary zone electrophoresis: resolution of racemic tryptophan and determination of the enantiomer composition of commercial pharmaceutical epinephrine, *Electrophoresis*, (Weinheim, Fed. Repub. Ger.), 11, 757, 1990.

247. Leopold, E. and Gouesclou, L., Separation of chiral amino acids and peptides by capillary electrophoresis with P/ACE 2000, *Spectra*, 156, 27, 1991.

CHAPTER 4

Micellar Electrokinetic Chromatography

Kurt R. Nielsen and Joe P. Foley

CONTENTS

0-8493-9127-X/98/$0.00+$.50
© 1998 by CRC Press LLC

I. INTRODUCTION

Capillary zone electrophoresis (CZE) is a technique that was developed for very efficient separations of charged solutes. When a charged solute is exposed to an electric field, the solute will migrate with a characteristic velocity that is proportional to the electric field, as shown in Equation 1.

$$v = (\mu_{ep} + \mu_{eo})\ E \tag{1}$$

The proportionality constant, μ_{ep}, is termed the electrophoretic mobility and is the quotient of the net charge of the solute, q, and frictional drag coefficient, f. The second constant, μ_{eo}, is the coefficient of electroosmotic flow, a bulk displacement phenomenon explained in earlier chapters. It can be larger or smaller than μ_{ep} or even negligible (i.e., $\mu_{eo} \approx 0$), depending on the experimental conditions.

In CZE, molecules are separated based on the solutes' differences in electrophoretic mobility, resulting from differences in q, f, or both.

$$\mu_{ep} = \frac{q}{f}; \quad R_S \propto \frac{\mu_2 - \mu_1}{\mu_{avg} + \mu_{eo}} \tag{2}$$

Importantly, for neutral solutes q = 0 and hence $\mu_{ep} = 0$, and there is no selective transport. It was for this reason that Terabe and co-workers[1] introduced micellar electrokinetic chromatography (MEKC) in 1984. MEKC allows neutral and charged solutes to be separated simultaneously using CZE instrumentation.

When a micellar medium is used in CZE, it provides a pseudostationary phase with which solutes can interact. The most commonly used micellar phase is sodium dodecyl sulfate (SDS). The surfactant is negatively charged and, consequently, so are the micelles. This means that the micelles' electrophoretic velocity, $v_{mc(ep)}$, is toward the anode. Since under normal conditions with fused-silica capillaries the electroosmotic velocity, v_{eo}, is from anode to cathode, anionic micelles therefore oppose electroosmotic flow in the capillary. Under most conditions, however, anionic micelles still migrate along the separation axis toward the detector (near the cathode) because $v_{mc(ep)}$ is usually not large enough to overcome the electroosmotic flow, v_{eo}. More simply, the micelles form a phase that moves slower than the mobile phase, hence the name, *pseudostationary* phase. A similar situation can be obtained using a cationic surfactant and an electrode polarity opposite that of the previous case.

Micelles are aggregates of surfactant monomers. The surfactant monomers have two parts — a hydrophilic head group and a hydrophobic tail (Figure 4.1). There are many different types of surfactants and most of them can be grouped into classes based on the charge characteristics of the head group. In Figure 4.2 there are representative structures of some types of anionic, cationic, nonionic, and amphoteric surfactants. When the surfactant concentration exceeds a value known as the critical micelle concentration (cmc), the monomers in excess of the cmc begin to self-aggregate such that the hydrophilic head groups form an outer shell and the hydrophobic tails form a nonpolar core (Figure 4.1). This association of surfactant molecules is referred to as a micelle, and the average number of surfactant monomers per micelle is termed the aggregation number. Table 4.1 lists some common surfactants with values for the cmc and aggregation number. Note that the data are accurate for aqueous

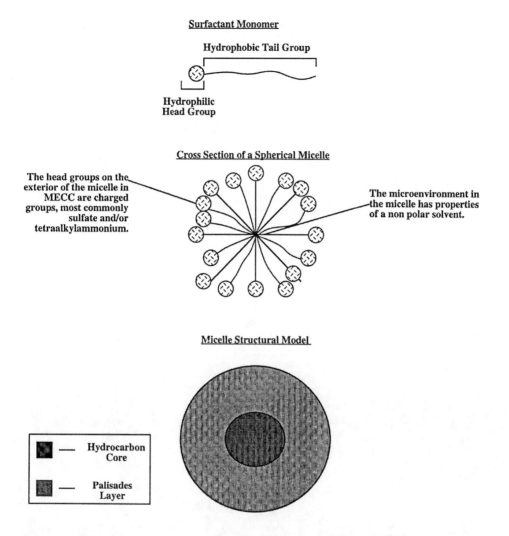

Figure 4.1 The top portion of the figure is an illustration of a surfactant monomer and a micelle. The bottom portion is a representation of a micelle that is used in modeling micellar solubilization equilibria.

solutions at ambient temperature and near-infinite dilution, and that ionic strength, organic solvents, temperature, etc., may have a significant effect on the cmc and aggregation number. It is also important to realize that micellization is a dynamic equilibria and that the micelles themselves have finite lifetimes.[2] Essentially the micelles form a second, less polar phase into which molecules can partition. This type of partitioning is analogous to that in solvent extraction or reverse-phase liquid chromatography (RPLC). The differential partitioning of neutral molecules between the buffered aqueous mobile phase and the micellar (pseudo)-stationary phase is the solve basis for their separation. Shown in Figure 4.3 is a hypothetical MEKC separation that could be obtained with an anionic surfactant like SDS and a fused-silica capillary at neutral pH.

As we shall discuss in more detail later, micelles and their properties can be manipulated to improve selectivity. In addition to the organic modifier approach, several other creative methods have been explored to alter selectivity including the addition of metal ions,[3] tetraalkylammonium salts,[4] or a second surfactant (i.e., mixed micelles).[5-12] These mixed micelles have been shown to be advantageous in MEKC.[11,13,14]

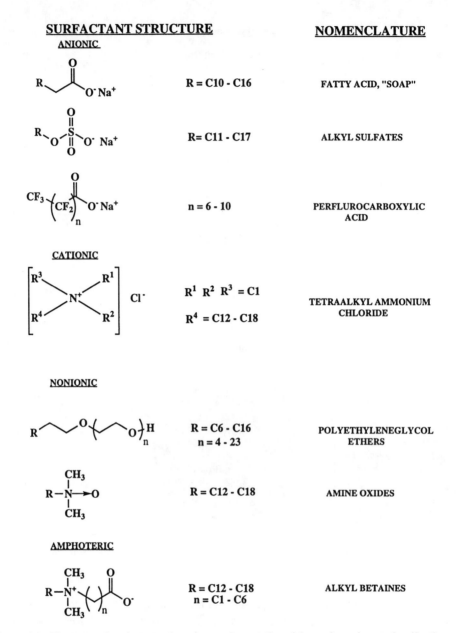

SURFACTANT STRUCTURE

NOMENCLATURE

ANIONIC

R = C10 - C16   FATTY ACID, "SOAP"

R= C11 - C17   ALKYL SULFATES

n = 6 - 10   PERFLUROCARBOXYLIC
                       ACID

CATIONIC

R^1 R^2 R^3 = C1
                       TETRAALKYL AMMONIUM
R^4 = C12 - C18   CHLORIDE

NONIONIC

R = C6 - C16   POLYETHYLENEGLYCOL
n = 4 - 23   ETHERS

R = C12 - C18   AMINE OXIDES

AMPHOTERIC

R = C12 - C18   ALKYL BETAINES
n = C1 - C6

Figure 4.2 The types of surfactants shown are representative of the various charge classifications.

MEKC has also inspired the use of other types of pseudostationary phases that aid in the separation of solutes, including those formed from bile salts,[15-17] cyclodextrins,[18-20] a variety of polymers,[21-25] microemulsions,[26] and suspensions (e.g., surfactant-coated chromatographic particles).[27] The variety of phases and the basic separation mechanisms utilized in CZE have counterparts in conventional liquid chromatography. To reflect this similarity, CZE techniques that employ a pseudostationary phase are more generally classified as electrokinetic chromatography (EKC).

EKC and MEKC are not restricted to the separation of neutral solutes. Charged species, which may differ in their electrophoretic mobilities in addition to their partitioning characteristics, can also be separated by MEKC; their transport and separation mechanisms are just

Table 4.1   Aggregation Number and CMC Values for Various Surfactants

Surfactant (charge type and name)	Acronym or trade name	Aggregation number	CMC moles L^{-1}
Anionic			
Sodium dodecylsulfate	SDS	62	8.1×10^{-3}
Sodium perfluorononanoic acid	NaPFN	NA	9.1×10^{-3} [a]
Potassium perfluorooctanoic acid	KPFO	NA	3×10^{-2}
Cationic			
Cetyltrimethylammonium bromide	CTAB	78	1.3×10^{-3}
Dodecyltrimethylammonium bromide	DTAB	55	1.5×10^{-2}
Nonionic			
Polyoxyethylene(6)dodecanol	NA	400	9×10^{-5}
Polyoxyethylene(23)dodecanol	Brij-35	40	1×10^{-4}
Zwitterionic			
N,N,-dimethyl-N-(carboxymethyl) dodecyl ammonium salt	Octyl betaine	NA	2.4×10^{-3}
N,N-dimethyl-N-(carboxyethyl) hexadecyl ammonium salt	NA	NA	2×10^{-5}

potentially more complex since both chromatographic and electrophoretic effects must be considered.

II. BASIC COLUMN CHROMATOGRAPHIC THEORY AND DEFINITIONS

A general knowledge of column chromatographic principles greatly facilitates one's understanding of MEKC. There are many sources that discuss these topics in greater detail.[28,29] The present discussion applies to column chromatographic processes with a true stationary phase. In MEKC the micelles are not stationary, and consequently column chromatographic theory needs to be adapted. Subsequent sections will address these modifications.

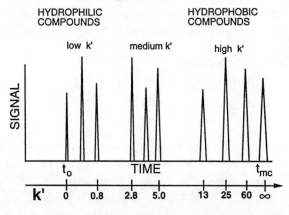

Figure 4.3   Hypothetical separation illustrating the finite elution window in MEKC and the relationship between retention factor (analyte polarity) and migration time.

The *degree of separation or resolution (R_s)* of two solutes is commonly defined as

$$R_s = \frac{\Delta t_R}{W_{avg}} = \left(\frac{t_{R1} - t_{R2}}{\dfrac{W_1 + W_2}{2}} \right) = \left(\frac{t_{R1} - t_{R2}}{2(\sigma_1 + \sigma_2)} \right) \tag{3}$$

where Δt_R is the difference between the two retention times of the solutes, and W_{avg} is the average peak width. More detailed discussions of resolution are available,[28,29] and a review on the uses of various R_s equations has been published.[30] A commonly used R_s equation based on chromatographic parameters is[30]:

$$R_s = \left(\frac{\sqrt{N}}{4} \right) \left(\frac{\alpha - 1}{\alpha} \right) \left(\frac{k'}{1 + k'} \right) \tag{4}$$

where *N is the column efficiency* (number of theoretical plates), α *is the selectivity*, and *k' is the retention factor.* The retention factor k' is a ratio of the number of moles of solute in the stationary phase divided by the number of moles in the mobile phase.

$$k' = \frac{n_{sp}}{n_{mp}} \tag{5}$$

Alternatively k' can be expressed in parameters that are readily obtained experimentally.

$$k' = \frac{t_R - t_o}{t_o} \tag{6}$$

where the terms t_o *and t_R are the retention (or migration) time of an unretained and retained solute, respectively.* As shown in Equation 7, t_R is linearly related to k'.

$$t_R = (1 + k') t_o \tag{7}$$

The *selectivity, α, of a separation* is a ratio of retention factors (k'_2/k'_1) for two adjacent compounds. Ideally the α term should be optimized to provide maximum resolution. Selectivity can be altered by changing the stationary phase or using mobile phase additives like an organic solvent.

The *column efficiency, N,* is given by,

$$N = \frac{L}{H} \tag{8}$$

L is the capillary length, and H is the height equivalent to a theoretical plate (HETP). Small H values indicate high efficiencies and better separating power. The number of theoretical plates for a peak can be calculated from readily available experimental parameters as follows:

$$N = 5.54 \left(\frac{t_R}{W_{0.5}} \right)^2 \tag{9}$$

where $W_{0.5}$ is the peak width at half height, t_R is the retention time, and a Gaussian peak profile is assumed. Other, similar Gaussian-based equations can also be employed,[31] but if peaks are tailed, and/or the peak shape varies widely over the results to be compared, use of such equations can result in errors in excess of 100% (due to underestimation of peak variance), and it is better to use

$$N = \frac{41.7(t_R / W_{0.1})^2}{b/a + 1.25} \tag{9a}$$

where b/a is an empirical asymmetry factor measured at 10% of peak height.[32,33] Although Equation 9a is based on the exponentially modified Gaussian, it is also fairly accurate for log-normal and other asymmetric profiles. And it is also more accurate for fronted peaks[34] than Equation 9, provided the reciprocal of the asymmetry factor (a/b) is used instead of b/a in Equation 9a.

The *peak capacity* n_c is the number of solutes that can be placed side-by-side along the separation axis within the constraints of the specific separation. As such, it represents an upper limit to the number of peaks that can be resolved; this limit is seldom realized in practice due to the random distribution of analytes along the separation axis. For chromatographic techniques, the peak capacity, n_c, is given by:

$$n_c = 1 + \frac{\sqrt{N}}{4} \ln \frac{t_\omega}{t_o} \tag{10}$$

where t_ω is the retention time of the last eluting peak. Capillary GC and gradient HPLC methods have very large peak capacities, sometimes into the hundreds of solutes. But MEKC as most often practiced does not have as large a peak capacity because of the finite migration time of the micelles (t_{mc}), which limits the value of t_ω. In this regard MEKC is much like size exclusion chromatography (SEC), in that there is a limited elution range (time window) over which solutes can be separated.

III. OVERVIEW OF SEPARATION MECHANISMS

A. Neutral Solutes

As we have already mentioned, the MEKC separation is based on a partitioning mechanism (Figure 4.4), which can be modeled as:

$$S_{aq} \rightleftharpoons S_{mc} \qquad P_{wm} = \frac{[S]_{mc}}{[S]_{aq}} \tag{11}$$

where $[S]_{aq}$ and $[S]_{mc}$ are the concentration of the solute, S, in the aqueous phase and micelle, respectively, and P_{wm} is the partition coefficient. If we now define a phase ratio β as the volume of micellar phase (V_{sp}) divided by the volume of the aqueous phase (V_{aq})

$$\beta = \frac{V_{sp}}{V_{aq}} \tag{12}$$

then a retention factor k' can be defined as:

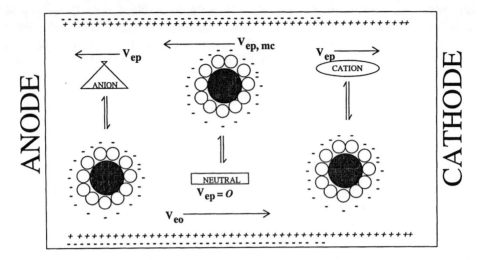

Figure 4.4   Longitudinal cross-section of a capillary in MEKC. The negative and positive charges at the wall of the capillary represent the ionized silanol groups and electrolyte ions, respectively. The single-headed arrows represent velocity vectors, while double arrows represent a potential partitioning equilibrium for each of the solute charge types shown.

$$k' = P_{wm}\beta = \frac{n_{mc}}{n_{aq}} \tag{13}$$

where n_{mc} and n_{aq} are the number of moles of solute in the micelle and aqueous phase, respectively.

An alternative model to describe a solute's interaction with a micelle is the binding model, where solute–micelle "binding" is defined to occur whenever the solute interacts with the micelle, whether in the interior of the micelle, in the palisades layer, or on the surface (Figure 4.5). An equilibrium binding constant, K_m, can be defined as follows:

$$S + M \underset{\leftarrow}{\rightarrow} SM \qquad K_m = \frac{[SM]}{[S_{aq}][M]} \tag{14}$$

where S is the unbound solute, M is the micelle, and SM is the solute–micelle complex. These binding constants can be measured in a variety of ways, i.e., from absorbance or fluorescence measurements, by micellar liquid chromatography (MLC), and by MEKC.[35] A compilation of K_m values determined from MLC data was reported by Foley for various combinations of solute and surfactant[36]; in a related study, Khaledi et al. compared the determination of K_m values by MLC and MEKC.[37]

B. Charged Solutes

In CZE, solutes are separated based on differences in μ_{ep} (Equation 2).[38,39] This is the predominant mechanism for separation in the absence of a stationary phase. If we now consider the retention mechanisms for charged solutes in MEKC, the picture is more complicated. In systems where the surfactant has a charge complementary to that of the solute, ion pairing may occur. When a solute and the surfactant have like charges, coulombic forces cause these molecules to repel each other. The molecule can traverse the column in several forms: (1) by

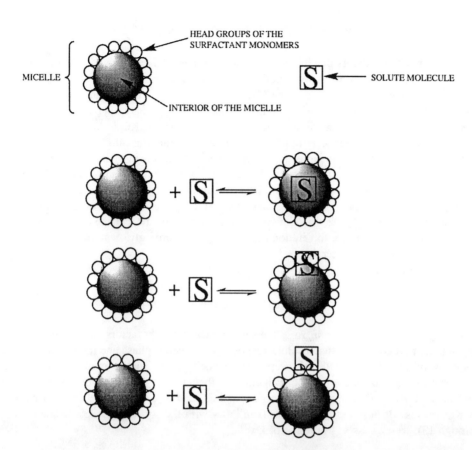

Figure 4.5 The binding of a solute to the micelle may be a result of any of the above interactions (top) partitioning into the interior of the micelle; (middle) partitioning into the micelle but remaining near the surface of the micelle; (bottom) surface interaction with the head groups of the surfactant monomers of the micelle.

itself in the aqueous phase; (2) ion-paired with the surfactant monomer; or (3) partitioned into the micellar phase. Depending on the experimental conditions, such as pH, type of surfactant, and organic modifier, all or some of the forms of the solute will be in equilibrium. The interactions for charged solutes that do *not* ion pair are illustrated in Figure 4.4.

It is important to appreciate that the separation of charged solutes by MEKC involves a combination of chromatographic and electrophoretic separation mechanisms in addition to voltage-driven transport. This results in a bias or difference in the rate of migration of cationic, neutral, and anionic analytes that does not normally exist in conventional chromatographic techniques such as HPLC or GC, i.e., the mobile phase velocities of cations, neutrals, and anions are the same in HPLC, whereas in MEKC they are different because of the additional electrophoretic velocity component of the charged species. To summarize, although our focus here is the chromatographic aspects of MEKC, one should not forget the role played by electrophoretic transport and separation mechanisms.

Some initial studies on the separation of charged solutes in MEKC examined the addition of tetraalkylammonium salts to the buffer.[4] The addition of this type of cation in general caused decreased retention for cationic solutes and increased retention for anionic solutes. Such an effect is also observed in ion-pair HPLC.[4] This was shown to be quite useful in cases were solutes could not be satisfactorily separated by CZE (i.e., solely on the basis of differences in μ_{ep}).

IV. THEORY FOR NEUTRAL SOLUTES

A majority of the theory that has been developed for MEKC has dealt with neutral solutes. Much of this preliminary work was published by Terabe et al.[40]

It is important to appreciate that the MEKC equations for neutral solutes are similar but not identical to Equations 4 through 7 and Equation 10 for conventional column chromatography. As we have already mentioned, the elution range in MEKC is finite, while in GC, HPLC, etc., the elution range is infinite. This is a very important difference and must be taken into account by the equations that describe solute behavior in MEKC. As illustrated in Figure 4.3, the time it takes an unretained solute to traverse the column is t_0, and the time it takes the micelles to traverse the column is t_{mc}. Neutral solutes will always have a retention time greater than or equal to t_0 and less than or equal to t_{mc}. An important quantity in MEKC is the elution range, t_{mc}/t_0. As the relative magnitudes of t_{mc} and t_0 change due to changes in the experimental conditions, the elution range changes accordingly. A more detailed discussion is presented in a later section.

The *average velocity*, v_{avg}, of a solute is given by:

$$v_{avg} = F_{aq}(v_{ep} + v_{eo}) + F_{mc}v_{mc,net} \tag{15}$$

where $F_{aq} = n_{aq}/n_{total}$ and $F_{mc} = n_{mc}/n_{total}$ are stoichiometric weight factors that express in mole fraction the portion of time that a solute spends in the aqueous phase and the micellar phase, respectively. The term v_{ep} is the electrophoretic velocity of the solute, and the terms v_{eo} and $v_{mc,net}$ are the electroosmotic and net velocities of the aqueous phase and the micelle. For a neutral solute, $v_{ep} = 0$. The weight factors F_{aq} and F_{mc} can be expressed either in terms of k' or K_m; because of most readers' greater familiarity with the partition models (Equations 11 through 13), we express v_{avg} in terms of k':

$$v_{avg} = \left(\frac{1}{1+k'}\right)v_{eo} + \left(\frac{k'}{1+k'}\right)v_{mc,net} \tag{16}$$

An equation for the *retention time, t_R*, can be derived using the simple relationship $v = L_{det}/t$, (L_{det} = length of capillary from inlet to detector), and is shown below.[1,40]

$$t_R = \left(\frac{1+k'}{1+\left(\dfrac{t_0}{t_{mc}}\right)k'}\right)t_0 \tag{17}$$

where $t_R = L_{det}/v_{avg}$ and $t_0 = L_{det}/v_{eo}$. The only difference between conventional column chromatography (Equation 7) and MEKC (Equation 17) is the extra term in the denominator of Equation 17. This is a consequence of the limited elution range in MEKC (nonzero value of t_0/t_{mc}). As t_{mc} approaches infinity (the micelles become stationary), the elution window gets larger, the extra term in the denominator approaches zero, and Equation 17 reduces to Equation 7.

Although *the retention factor, k'*, in MEKC was already defined in Equation 14, an alternative expression that is frequently more useful can be obtained by rearranging Equation 17:

$$k' = \left(\frac{t_R - t_o}{t_o \left(1 - \dfrac{t_R}{t_{mc}} \right)} \right) \tag{18}$$

The term $\left(1 - \dfrac{t_R}{t_{mc}} \right)$ is again a consequence of the limited elution range in MEKC; as t_{mc} approaches infinity, Equation 18 reduces to Equation 6.

The *resolution equation for neutral compounds* in MEKC is very similar to that for conventional column chromatography, but predictably there is an additional term that arises because of the limited elution range. As with most, if not all, resolution equations in chromatography, a Gaussian peak profile was assumed for its derivation.[40] The two terms on the right, in which k' appears, are collectively referred to as the retention term.

$$R_s = \frac{\sqrt{N}}{4} \left(\frac{\alpha - 1}{\alpha} \right) \left(\frac{k'_2}{1 + k'_{avg}} \right) \left(\frac{1 - \dfrac{t_o}{t_{mc}}}{1 + \dfrac{t_o}{t_{mc}} k'_{avg}} \right) \tag{19}$$

If two peaks are reasonably close together, i.e., if $k'_1 \approx k'_2 \approx k'_{avg} = k'$, then Equation 19 can be approximated by the following:

$$R_s = \frac{\sqrt{N}}{4} \left(\frac{\alpha - 1}{\alpha} \right) \left(\frac{k'}{1 + k'} \right) \left(\frac{1 - \dfrac{t_o}{t_{mc}}}{1 + \dfrac{t_o}{t_{mc}} k'} \right) \tag{20}$$

Note that as t_{mc} approaches infinity, the fourth term on the right-hand side of Equations 19 and 20 approaches unity, and the conventional R_s equation is the result (Equation 4).

An approximate expression for the *peak capacity* in MEKC can be obtained by substitution of t_{mc} for t_ω in Equation 10:

$$n_c = 1 + \frac{\sqrt{N}}{4} \ln \frac{t_{mc}}{t_o} \tag{21}$$

As the *elution range* (t_{mc}/t_o) increases, so does the peak capacity, albeit in a logarithmic fashion. In effect, the separation axis about which solutes can be distributed is lengthening. The peak capacity also depends on the square root of N.

Although the effect of the elution range on peak capacity is easy to appreciate both on an intuitive level and by inspection of Equation 21, its effect on resolution (Equation 20) is less obvious and is illustrated in Figure 4.6. An efficiency and selectivity of 75,000 and 1.03 were assumed arbitrarily for the calculation via Equation 20, along with an optimum value of k' calculated via Equation 24. The results are striking and show rather dramatically just how much the resolving power of MEKC can be increased merely by increasing the elution range. (Most MEKC practitioners achieve an elution range between 3 and 8.)

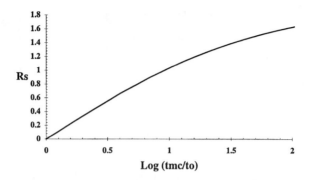

Figure 4.6 Maximum attainable resolution as a function of the elution range (t_{mc}/t_o), using Equation 20 with k′ = k'_{opt} (Equation 24) and arbitrarily assumed values of 75,000 and 1.03 for the efficiency (N) and selectivity (α), respectively.

Assuming $0 < t_o/t_{mc} < 1$ and identical values for N, α, and k, a direct comparison of Equations 4 and 19 shows that resolution will always be greater with conventional column chromatographic techniques (HPLC, GC, etc.) than with MEKC. Why then, should one bother to use MEKC?

The answer is that all factors other than t_o/t_{mc} are *not* equal. Many solutes, because of their limited volatility and/or thermal stability, are not amenable to GC. Moreover, selectivities in MEKC, although sometimes similar to reverse-phase liquid chromatography (RPLC), can also be quite unique. *Furthermore, much higher column efficiencies can be achieved in MEKC than in HPLC under average conditions*, i.e., 50,000 to 400,000 for MEKC vs. 5,000 to 20,000 for HPLC. Finally, as methods for increasing the elution range beyond the typical range of $3 < t_{mc}/t_o < 8$ (see also Table 4.2) gain acceptance,[25,41-47] the disadvantages of a finite elution range will virtually disappear.

Whereas Equations 17 through 21 have been written in terms of migration or retention times, they can just as easily be written (partially or almost completely) in terms of velocities[48] or mobilities,[49] e.g.,

$$R_s = \frac{\sqrt{N}}{4}\left(\frac{\alpha-1}{\alpha}\right)\left(\frac{k'}{1+k'}\right)\left(\frac{1-\dfrac{\mu_{mc,net}}{\mu_{eo}}}{1+\dfrac{\mu_{mc,net}}{\mu_{eo}}k'}\right) \tag{22}$$

With all of this background information, it is appropriate to summarize qualitatively how three parameters in Equations 19, 20, or 22 — N, α, and k′ — affect resolution (a detailed quantitative discussion follows immediately), assuming that any effects on each other can be neglected. For N and α, as in conventional GC or HPLC (Equation 4), an increase in either or both parameters results in a predictable increase in resolution. For k′, however, in distinct contrast to GC or HPLC, an increase does not always lead to a corresponding increase in resolution. Careful examination of Equations 19, 20, or 22 shows that an increase in the retention (k′) of a pair of solutes beyond an optimum value defined by Equation 24 below will reduce the resolution.

There have been many investigations of how resolution and/or one or more of its components (N, α, etc., in Equation 19) is affected by various experimental parameters in MEKC. A variety of conditions have been evaluated to assess their effect on N, including applied voltage,[50-52] electroosmotic flow velocity,[52] column dimensions,[51] buffer concentration,[51] temperature,[52-54] surfactant concentration,[51-53] retention factor R,[55] organic modifier content,[41,53,56]

Table 4.2 Elution Range (t_{mc}/t_o) of Various Surfactant Systems Under Different Experimental Conditions

Surfactant[a]	[Surf]/M	t_{mc}/t_o	Comment	Reference
SDS	0.075	1.5	0% 2-propanol	73
SDS	0.075	2.6	5% 2-propanol	73
SDS	0.075	7.1	10% 2-propanol	73
SDS	0.075	10.0	15% 2-propanol	73
SDS	0.075	16.7	20% 2-propanol	73
SDS	0.07	3.3[b]	0% modifier	67
SDS	0.07	3.8[b]	10% MeOH	67
SDS	0.07	6.6[b]	20% MeOH	67
SDS	0.07	4.5[b]	10% ACN	67
SDS	0.07	5.6[b]	20% ACN	67
SDS	0.025	2.5	untreated silica capillary	41
SDS	0.025	10.0	TCMS-reacted silica	41
SDS	0.03	3.6		40
SDS	0.15	5.6		40
SDS	0.05	3.5		236
SDS	0.10	4.6		236
SDS	0.10	4.7[b]		42
SDS	0.20	6.9[b]		42
SDS	0.30	11.0[b]		42
DTAB	0.03	2.4		236
DTAB	0.05	2.4		236
DTAB	0.10	2.6		236
SDS	0.05	3.1	15 kV	63
SDS	0.05	3.0	20 kV	63
SDS	0.05	2.8	30 kV	63
DTAC	0.05	2.4	≈15 kV	63
DTAC	0.05	2.4	20 kV	63
DTAC	0.05	2.3	30 kV	63
CTAC	0.05	2.8	15 kV	63
CTAC	0.05	2.6	20 kV	63
CTAC	0.05	2.4	30 kV	63
SDS/SB-12	0.02/none	2.86	20 kV	237
SDS/SB-12	0.02/0.005	2.69	20 kV	237
SDS/SB-12	0.02/0.01	2.65	20 kV	237
SDS/SB-12	0.02/0.02	2.64	20 kV	237
SDS/SB-12	0.02/0.04	2.45	20 kV	237
LiDS	0.05	2.72	15 kV	76
SDS	0.05	3.91	15 kV	76
LiDS	0.05	3.35	10% ACN, 15 kV	76
SDS	0.05	4.91	10% ACN, 15 kV	76
LiDS	0.05	4.38	20% ACN, 15 kV	76
SDS	0.05	6.81	20% ACN, 15 kV	76
KDS	0.05	10.5	20% ACN, 15 kV	76
SDS/Brij-35	0.02/0.012	>25	pH 7.0, 25 kV	46

[a] SDS = sodium dodecyl sulfate; DTAC = dodecyltrimethylammonium chloride; CTAC = cetyltrimethylammonium chloride; DTAB = dodecyltrimethylammonium bromide; Brij-35 = polyoxyethylene(23)dodecanol; SB-12 = N-dodecyl-N,N-dimethylammonium-3-propane-1-sulfonic acid.

[b] Data obtained from graphs; may be less precise than other data presented here.

and extracolumn effects.[52,57-62] Selectivity has also been examined extensively, and the effects of applied voltage,[63] pH,[64,65] organic modifier content,[66,67] surfactant identity,[9,63,68] and surfactant concentration[64,65,69] are well documented.

In both the above qualitative summary and later quantitative discussions of MEKC optimization theory, it was convenient to assume that α and N are independent of k′. These important assumptions warrant further explanation. The results of Row et al.[69] and Fujiwara et al.[64,65] clearly show that the selectivity for neutral solutes in a given surfactant system is virtually independent of k′. However, the assumption that N is independent of k′ is clearly a first-order approximation, since most researchers have predicted or reported a slight-to-moderate dependence.[51-53,55]

V. OPTIMIZATION OF SEPARATION PARAMETERS

A. Retention

1. Theory

A theory for the optimization of the retention of *neutral* solutes was reported in 1990[70] and is summarized below. (A similar theory for *charged* solutes is discussed near the end of Section VI [Equations 39–48].) The theory established the existence of an optimum k′ range within which good resolution would be obtained. It also defined quantitatively the upper and lower bounds for this range in terms of resolution and resolution per unit time, i.e., k'_{opt} for R_s and k'_{opt} for R_s/t_R. As shown in Equation 13, the optimization of retention (k′) in MEKC can then be accomplished (after measuring or estimating the elution range) by optimizing either or both of the following: (1) the phase ratio β by adjusting the surfactant concentration; (2) the solute partition coefficients (P_{wm}'s) by choosing a simple or mixed micellar system of appropriate hydrophobicity or modifying a readily available surfactant system like SDS with an organic solvent as necessary. With regard to option (2), although the main effect of an organic solvent is to lower the polarity of the aqueous phase and to decrease somewhat the electroosmotic flow, solvents can also modify the micelles, resulting in changes in selectivity as well as various micellar parameters (cmc, aggregation number, etc.). (This may be less of a concern for polymerized micelles and other pseudophases [cyclodextrins, starburst dendrimers] whose physical state is relatively insensitive to organic solvents.) When the initial selectivity and elution range are satisfactory, changing the surfactant concentration (option 1) is the preferred method for adjusting retention in MEKC.

Assuming that N and α are both independent of k′, the dependence of resolution on k′ is restricted to the last two terms of Equation 20

$$f(k') = \left(\frac{k'}{1+k'}\right)\left(\frac{1-\dfrac{t_o}{t_{mc}}}{1+\dfrac{t_o}{t_{mc}}k'}\right) \qquad (23)$$

The optimum k′ can be obtained explicitly by solving for the physically significant root of the first derivative of Equation 23[70]:

$$k'_{opt}\,(\text{maximum } R_s) = \sqrt{\frac{t_{mc}}{t_o}} \qquad (24)$$

In MEKC, as the retention factors (k′) of solutes begin to exceed k'_{opt} (Equation 24), the resolution will suffer. This possibility exists because compounds with $k' \gg k'_{opt}$ (Equation 24) may elute close to or even co-elute with the micelle ($t_R = t_{mc}$). Such a situation is not encountered in GC or HPLC where there is no upper limit in the separation window and, according to Equation 4, no k' is too large for good resolution, although in practice the peak width and analysis time become excessive and peak detection is difficult when $k' > 20$.

The optimization of resolution per unit time (R_s/t_R) is somewhat more complex than the optimization of resolution,[70] and only key equations are presented here. The derivation is generally quite similar to the approach used for other column chromatographic methods.[71] We omit the proof here and direct the interested reader to the original reference.[70]

The following physically significant root was obtained:

$$k'_{opt}(R_s/t_R) = \frac{-(1-t_o/t_{mc}) + \sqrt{(1-t_o/t_{mc})^2 + 16(t_o/t_{mc})}}{4(t_o/t_{mc})} \tag{25}$$

The k'_{opt} (R_s/t_R) as given by Equation 29 ranges between 1.19 and 2.00 for t_{mc}/t_o values from 2 to infinity. For conventional column chromatographic methods such as GC, HPLC, and SFC, the optimum k' for best R_s/t_R is 2.[71]

Using the above theory, it is now possible to define an optimum interval of retention factors, i.e., $k'_{opt\cdot interval} = k'_{low}$ to k'_{high}, where $k'_{low} = k'$ for best R_s/t_R (Equation 25) and $k'_{high} = k'$ for best R_s (Equation 24). This interval widens as the elution window (t_{mc}/t_o) becomes larger and is consistent with the effect of t_{mc}/t_o on the peak capacity (Equation 10). Alternatively, one can define an optimum interval of retention factors as that interval (k'_{low} to k'_{high}) over which the resolution is within an arbitrary percentage (e.g., 90% or 75%) of the maximum resolution obtainable at $k'_{opt} = \sqrt{t_{mc}/t_o}$.

Table 4.3 compares the retention factor interval (A) defined by k'_{opt} for best R_s/t_R (k'_{low}) and best R_s (k'_{high}) with two retention factor intervals (B, C) centered about the maximum resolution. At low-to-moderate elution ranges ($4 \leq t_{mc}/t_o \leq 30$), k'_{low} is very similar for intervals A and B; at higher elution ranges ($t_{mc}/t_o > 30$), k'_{low} is more similar for intervals A and C. In contrast to the similarities of k'_{low} among the three intervals in Table 4.3, k'_{high} is systematically higher for the resolution-centered retention factor intervals (B, C).

For low-to-moderate elution ranges, $k'_{opt} = \sqrt{t_{mc}/t_o}$ (Equation 24, k'_{high} in Interval A) can be significantly exceeded up to k'_{high} in intervals B and C with only a 10 or 25% loss in resolution, respectively, and only a small-to-moderate increase in analysis time due to the limited elution range (see Equation 17). For large to near-infinite elution ranges ($t_{mc}/t_o \geq 300$), however, k'_{high} for all three intervals is very large ($17 < k'_{high} < 144$); consequently, in practice the upper limit for k' may need to be lowered arbitrarily to keep analysis times reasonable.

2. Control

Surfactant. The control of R_s and R_s/t_R can be accomplished by proper manipulation of any variable that controls retention, but typically it will be the surfactant concentration,

Table 4.3 Comparison of Retention Factor Intervals for Neutral Solutes in MEKC

Elution range (t_{mc}/t_o)	Interval A: k' from best R_s/t_R (k'_{low}) to best R_s (k'_{high})		Interval B: R_s = 90% of maximum		Interval C: R_s = 75% of maximum	
	k'_{low}	k'_{high}	k'_{low}	k'_{high}	k'_{low}	k'_{high}
2	1.19	1.4	0.7	2.7	0.5	4.3
4	1.39	2.0	1.0	4.0	0.6	6.4
10	1.63	3.2	1.5	6.8	0.9	11.2
30	1.84	5.5	2.2	13.4	1.3	23.7
100	1.94	10.0	3.3	30.1	1.7	58.6
300	1.98	17.3	4.4	67.5	2.1	144

organic modifier, or possibly the temperature. As we show below, k' can be related directly to surfactant concentration, and an important consequence is that the surfactant concentration that provides the best R_s or R_s/t_R can be predicted from theory when a couple of key surfactant and solute parameters are available or can be measured. In short, one easy way to optimize R_s or R_s/t_R is to optimize the surfactant concentration.

Equation 13 is an expression for the retention factor. The phase ratio (Equation 12) can be written explicitly in terms of surfactant concentration [SURF], the critical micelle concentration (cmc), and the partial molar volume $(\overline{V})$ of the surfactant.

$$\beta = \frac{\overline{V}([SURF] - cmc)}{1 - \overline{V}([SURF] - cmc)} \tag{26}$$

Substitution of Equation 26 into Equation 13 and then solving for the surfactant concentration results in the following equation

$$[SURF] = \frac{k' + \overline{V}cmc(k' + P_{wm})}{\overline{V}(k' + P_{wm})} \tag{27}$$

Equation 27 is a very important result that shows *the one-to-one relationship between surfactant concentration and retention factor.* Since the parameters in Equation 27 are either experimentally measurable or physical constants, *a priori* prediction of an optimum surfactant concentration is possible. Assuming that $P_{wm} \gg k'$ (universally true except for very hydrophilic compounds), a convenient approximation to Equation 27 is

$$[SURF] = \frac{k'}{\overline{V}P_{wm}} + cmc \tag{28}$$

Now it is a rather simple matter to substitute either of the expressions for k'_{opt} (Equation 24 or 25) into either Equation 31 or 32 to obtain an expression for the optimum surfactant concentration. Since Equation 28 is simpler, we use it to obtain the following:

$$[SURF]_{opt} \text{ (for best } R_s) = \frac{\sqrt{\dfrac{t_{mc}}{t_o}}}{\overline{V}P_{wm}} + cmc \tag{29}$$

$$[SURF]_{opt} \text{ (for best } R_s / t_r) = \frac{k'_{opt(Eq.25)}}{\overline{V}P_{wm}} + cmc \tag{30}$$

The surfactant concentration cannot be varied over an infinitely wide range. The lowest usable surfactant concentration will be determined by the need for reproducible concentrations of micelles in order to obtain reproducible retention times. This lower limit will be dictated by the cmc and the precision with which it is known, but for our discussion we will assume the lower limit is 50% above the cmc, i.e., $[SURF]_{min} = 1.5$ cmc. The upper limit of the surfactant concentration is likely to be governed by such adverse solution properties as high viscosity or high conductance (Joule heating). From our own experience and that reported in the literature, the highest tolerable concentration for monovalent surfactants (SDS, CTAB, etc.) is probably about 0.15 M. Surfactant contributions to the Joule heating in MEKC can be minimized by partially substituting a zwitterionic surfactant for the original charged surfactant, particularly if the pH is close to the zwitterionic surfactant's pI (isoelectric point), assuming it is not a permanent zwitterion.

The optimization of resolution via surfactant concentration was demonstrated experimentally for three pairs of phenylthiohydantoin (PTH) derivatized amino acids, using two types of aqueous micellar systems: plain micelles of SDS and mixed SDS/Brij®-35 micelles. There were several advantages to using the mixed surfactant system: (1) lower operating currents for a given degree of solute retention; (2) smaller variation in t_{mc}/t_0 as the concentration of Brij-35 was varied; and (3) two to three times the efficiency of the SDS system. On the other hand, neither surfactant system without organic solvent was ideally suited for all the amino acids. The aqueous SDS system was better suited for moderately hydrophobic PTH-AAs, whereas the aqueous Brij-35/SDS system was more retentive and better suited for hydrophilic solutes.

Organic Modifier. Solutes of moderate-to-high hydrophobicity are difficult to analyze by MEKC using totally aqueous buffer systems. These types of solute elute near or with the micelle at t_{mc}. One way to increase the affinity of the solute for the mobile phase is to add an organic modifier to the buffer system. The use of an organic modifier will alter the retention, k', of the solutes. The mechanisms by which an organic modifier alters retention in MEKC are similar to those in reverse-phase liquid chromatography (RPLC). In RPLC the addition of an organic modifier alters the analyte's affinity for the stationary phase by changing: (1) the polarity of the mobile phase, and (2) the retention characteristics of the stationary phase, usually to a much lesser degree. There are four predominant interactions between molecules in the liquid phase: (1) dispersion or London forces, (2) induction forces, (3) hydrogen bonding, and (4) orientation forces.[28] The "polarity" of a molecule is its ability to interact using these four mechanisms. The total interaction of a solute molecule with a solvent molecule is described by all four interactions in combination. Solvent strength is directly related to polarity, thus "polar" solvents have an affinity for "polar" analytes. In RPLC the solvent strength increases with decreasing polarity, and this is also true with MEKC. Since an organic modifier is less polar than water, any combination of organic modifier and water will be less polar than pure water, hence the mobile phase will be stronger and retention will be reduced. The following relationship is often used in RPLC to describe k' as a function of organic modifier:

$$\log k' = \log k'_{aq} - S\phi \tag{31}$$

This equation shows how k' will be reduced as the percentage of organic modifier in the mobile phase is increased. And as k' is reduced, so is t_R. But one should be careful not to interpret Equation 31 in exactly the same way in RPLC and MEKC. In MEKC the addition of organic modifier alters not only the partitioning coefficient (P_{wm}) but also the phase ratio (β), the surfactant's cmc, the zeta potentials of the capillary wall and the micelles, and the solution viscosity.[67,72] Nevertheless, Equation 31 appears to be at least qualitatively valid for MEKC, including solvent gradient elution,[73] although the amount of organic solvent that can be employed in MEKC is typically limited to ≤35% for reasons explained below.

The effect of organic solvent on micelle formation and solute migration reproducibility govern the upper limit of organic solvent that can be used in MEKC. A fixed surfactant concentration (e.g., 100 m*M* SDS) can accommodate only a certain percentage of organic modifier before the micellization equilibria is eliminated (or adversely affected), resulting in the absence of micelles or poor reproducibility, respectively.

Two approaches to the challenge of lowering the retention of highly retained analytes sufficiently with organic solvent without adversely affecting the micellization equilibria have been reported in the literature: (1) the synthesis and subsequent use of polymerized micelles which are unaffected by large amounts of organic solvent[22,23,74,75]; and (2) the use of other

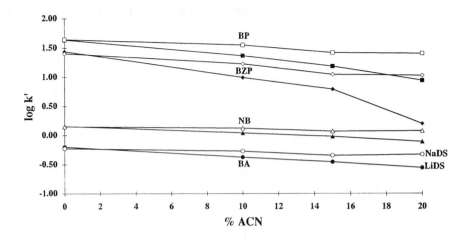

Figure 4.7 Effect of organic solvent on solute retention using 50 mM SDS (open symbols) and 50 mM LiDS (closed symbols). Solute identification: benzaldehyde (BA), nitrobenzene (NB), benzophenone (BZP), and biphenyl (BP). (Adapted from Ahuja, E. S. and Foley, J. P., *Anal. Chem.*, 67, 2315, 1995.)

commercially available conventional surfactants whose retention characteristics are more significantly altered by small amounts of organic solvent than the most popular ones (e.g., SDS, CTAB). Of the two approaches, the latter is probably more convenient. Along this line, Ahuja and Foley have shown that solute retention can be varied much more with dodecyl sulfate micelles in the presence of lithium counterions (i.e., LiDS) than when sodium is the predominant counterion.[76] Figure 4.7 compares the retention of four solutes in SDS and LiDS micellar solutions with phosphate buffer over the range of 0 to 20% acetonitrile added. In pure aqueous buffers, the retention with equimolar concentrations of SDS and LiDS is nearly identical, but as acetonitrile is added, the retention of all four solutes (and particularly the more highly retained ones) decreases more rapidly (note log scale on retention axis) in the LiDS system. For example, $k'_{benzophenone}$ decreases from 27.4 (no ACN) to 1.59 (20% ACN) in LiDS, whereas it decreases from 25.3 (no ACN) to 10.8 (20% ACN) in SDS.

The elimination of micelles and hence the MEKC separation mechanism by the addition of organic solvent to the buffer does not always mean that separations will no longer be possible. In two studies, sufficient organic modifier was added to suppress micellization and cause ionic surfactant monomers and neutral solutes to form charged solvophobic complexes that could be separated electrophoretically.[77,78] While according to our definition this is not an MEKC separation, it does fall under the more general category of electrokinetic chromatography (EKC).

B. Elution Range

1. Theory

A very significant limitation to the application of MEKC to complex samples is the low peak capacity due to the limited elution range (Equation 28). The problem of increasing the elution range can be solved in several ways.[46,79, 79a] One approach has been to use a different surfactant whose micelles can more effectively swim upstream. As $\mu_{ep,mc}$ approaches $-\mu_{eo}$, the micelles become "stationary" and the elution range becomes infinite. One such surfactant so employed was sodium decyl sulfate (STS), which forms small micelles with a more negative $\mu_{ep,mc}$. Unfortunately, the cmc increases as the chain length of the surfactant decreases, and higher surfactant concentrations are needed to generate STS micelles. Mobile phases with high surfactant concentrations generate large currents which enhance the Joule heating effect and result in poor reproducibility.[41]

2. Control

Kaneta et al. have reported that the addition of glucose to a buffer containing SDS results in an increased elution range because of an increase in the electrophoretic mobility of the micelle.[44] Cai and El Rassi have demonstrated that neutral octylglucoside micelles will complex alkaline borate and form a charged micelle that will function as a pseudostationary phase.[43] Among the interesting features of these micelles is that the elution range can be tuned by varying the borate concentration and/or the pH of the buffer. Several reports have also established that the elution range is a function of the surfactant counterion[45,66,76] and mixed micelle composition.[11,80,81]

Another way to increase t_{mc}/t_o is to decrease the electroosmotic flow. The electroosmotic flow can be reduced by decreasing the ζ (zeta) potential of the capillary wall. One parameter that can be used to manipulate the ζ potential is pH.[41,42] However, pH also affects the charge, mobility, retention, and selectivity of any acidic and/or basic analytes that may be present in a sample. Surface modification of the capillary is another method that can be used to reduce the ζ potential.[82] But this method is both tedious and time consuming; in addition, some solutes might then interact significantly with the capillary wall and be severely broadened, needing organic modifier to restore the efficiency.[41] Another way to control the ζ potential is to apply an additional electric field from outside the separation capillary; Tsai et al. have effectively utilized this technique in MEKC.[83,84]

Simultaneous manipulation of the micelle mobility and electroosmotic mobility is yet another way to control the elution range.[6] In particular, Ahuja et al. have demonstrated that a mixed micelle of Brij-35 and SDS can be used to achieve an infinite elution range.[46] The surfactant concentrations necessary for an infinite elution range are determined by selecting an SDS concentration, then varying the Brij-35 concentration and determining the electroosmotic and micellar velocity. Then the absolute values of the reciprocal velocities, electroosmotic and micellar, are plotted against the concentration of Brij-35 and least squares lines are fitted through the data. The point at which the two lines intersect is the concentration of Brij-35 necessary to achieve an infinite elution for the given SDS concentration. In this case, the peak capacity is increased significantly, albeit at the expense of analysis time and possibly efficiency.

One of the easiest ways to increase t_{mc}/t_o somewhat (by a factor of 1.5 to 2) is to add an organic modifier like methanol, acetonitrile, propanol, etc., to the buffer. It is a simple and effective way to significantly increase t_{mc}/t_o, although it usually results in reduced electroosmotic flow, thereby increasing the analysis time proportionally.[67] It may also result in a change in the selectivity, a disadvantage if the latter is already satisfactory.

A somewhat larger (2.5×) increase in the elution range was observed when up to 20 mM tetramethyl ammonium bromide (TMAB) was dissolved in a 50 mM SDS solution.[45] The presence of TMA reduced eof by 15% while reducing the micellar mobility by only 5% and changing the selectivity minimally for homologs and for most adjacent solute pairs; Joule heating increased by a factor of 1.8, however.

C. Selectivity

Selectivity in MEKC is defined as the ratio of analyte partition coefficients (P_{wm}) or retention factors (k'), the same as in GC and HPLC, and it is essentially the ability of a micellar system to separate two or more compounds that differ by one or more chemical groups. Although the *a priori prediction* of selectivity is not yet possible, it is easy to illustrate

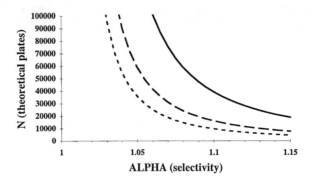

Figure 4.8 Number of theoretical plates (N) needed for baseline resolution ($R_s > 1.5$) for a given selectivity (α) and t_{mc}/t_o. The solid curve has $t_{mc}/t_o = 4$, the large dashed curve has $t_{mc}/t_o = 10$, and the small dashed curve has $t_{mc}/t_o = 25$.

the *importance* of selectivity. Figure 4.8 shows the number of theoretical plates (N) required to achieve baseline resolution ($R_s = 1.5$) for a given value of α, t_{mc}/t_o, and k' ($= k'_{opt}$ (R_s)). As α increases, there is a very dramatic drop in N needed for the separation. Also, as the elution range (t_{mc}/t_o) increases, fewer theoretical plates are needed.

It is often convenient to distinguish between two types of selectivity, hydrophobic and polar group selectivity, depending on whether the functional groups that differ among the compounds being separated are nonpolar or polar. For quantitative purposes, hydrophobic selectivity can be defined as the methylene selectivity or the selectivity between adjacent homologs. It is most easily measured from the ratio of retention factors of adjacent homologs or from the slope of a ln k vs. carbon number (homolog) plot. Polar group selectivity in MEKC is the ability of a micellar system to separate two or more compounds that differ by a polar functional group, and for convenience it is measured as the ratio of the retention factor of a derivative of benzene and benzene itself, i.e., $\alpha_{group} = k_{benzene\ derivative} \div k_{benzene}$. Alternatively, differences in polar group selectivity provided by different surfactants can be illustrated by graphing and comparing retention data for several classes of solutes (i.e., ln $k_{solutes\ w\ surf.\ \#2}$ vs. ln $k_{solutes\ w\ surf.\ \#1}$). For a given class of solutes, a systematic shift toward the y or x axis indicates a stronger interaction of the solute with surfactant #2 or #1, respectively.

Another way to characterize micellar selectivity is through linear solvation energy relationships.[85,86] By measuring the retention of a set of solutes with widely varying structures and well-characterized solvatochromic properties, the following relationship (or similar ones) between solute retention (log k') and solute structural characteristics (as represented by solvatochromic parameters and intrinsic molar volume) can be used to deduce chemical characteristics of the micelles themselves:

$$\log k' = a\alpha + b\beta + s\pi^* + \frac{mV_I}{100} + SP_o \tag{32}$$

The first two terms on the r.h.s. of the above equation reflect the contributions to retention of H-bonding, and the remaining terms are the contributions to retention from dipole–dipole interactions, cavity formation (nonspecific interactions), and the phase ratio of the system. The symbols a, b, s, and m are chemical properties of the micelle (H-bond acidity, H-bond basicity, dipolarity/polarizability, and cavity formation energy term), whereas the terms α, β, π^*, and V_I are solute properties (H-bond basicity, H-bond acidity, dipolarity/polarizability, and intrinsic molar volume). By performing multivariate linear regression on a relatively extensive retention data set, micelle characteristics a, b, s, and m can be measured and compared for various surfactant systems.

Table 4.4 Methylene Selectivity (Log$_e$ α_{-CH_2-}) of Alkylphenones in MEKC

Surfactant system	No organic modifier	15% MeOH	15% MeCN	15% 1-PrOH	Reference
25 mM SDS	0.845 (±0.009)	0.804 (±0.021)	0.714 (±0.012)	0.723 (±0.012)	238
40 mM SDS	0.861 (±0.025)	0.813 (±0.021)	0.762 (±0.018)	0.716 (±0.012)	238
55 mM SDS	0.870 (±0.028)	0.790 (±0.025)	0.776 (±0.021)	0.716 (±0.012)	238
70 mM SDS	0.880 (±0.032)	0.820 (±0.030)	0.785 (±0.021)	0.748 (±0.012)	238
	No modifier	10% MeCN	15% MeCN	20% MeCN	
50 mM SDS	0.857	0.843	0.838	0.836	76
50 mM LiDS	0.851	0.840	0.821	0.764	76
50 mM KDS	NA	NA	0.880	0.720	76
	No modifier	4% MeCN	10% MeCN	14% MeCN	
50 mM Mg(DS)$_2$	NA	0.885	0.791	0.752	66
	No additive	30 mM TPrAB	10 mM TPeAB	15 mM CTAB	
20 mM SDS	0.890	0.906	0.888	0.951	45
	No cosurfactant	10 mM SB-12	20 mM SB-12	40 mM SB-12	
20 mM SDS	0.916	0.986	0.936	0.900	11

Note: TPrAB = tetrapropylammonium bromide; TPeAB = tetrapentylammonium bromide; CTAB = cetyl-trimethylammonium bromide; SB-12 = N-dodecyl-N,N-dimethylammonium-3-propane-1-sulfonic acid.

Tables 4.4 and 4.5 summarize representative studies of methylene (hydrophobic) and polar group selectivity in MEKC. The results show that hydrophobic selectivity is similar to that achieved in reverse-phase liquid chromatography (RPLC), typically 0.45 to 1.2 in log$_e$ α_{-CH2-} units. Likewise, polar group selectivity varies considerably according to the buffer composition in MEKC, in much the same way that it varies according to the mobile phase composition in RPLC.

A more detailed discussion of α is beyond the scope of this chapter, but some general principles can be summarized. First, α is independent of applied voltage, assuming that Joule

Table 4.5 Comparison of Functional Group Selectivity for SDS and Brij-35/SDS Micellar Systems

Micellar medium	Benzene derivative					
	ϕ-COCH$_3$	ϕ-OH	ϕ-NO$_2$	ϕ-OCH$_3$	ϕ-CH$_3$	ϕ-Cl
	SDS					
40 mM	1.601*	0.501	1.274	1.601*	2.750	3.560
60 mM	1.593	0.500	1.271	1.593	2.694	3.458
80 mM	1.575	0.494	1.262	1.575	2.691	3.450
100 mM	1.567	0.495	1.249	1.567	2.673	3.433
	Brij-35/SDS[a]					
30 mM	0.598	0.773	1.122	1.366	2.725	4.519
40 mM	0.598	0.808	1.127	1.363	2.667	4.408
50 mM	0.595	0.827	1.125	1.360	2.653	4.331
60 mM	0.598	0.858	1.128	1.347	2.615	4.229

[a] SDS concentration in mixtures of Brij-35 was 20 mM.
*Due to coelution, the functional group selectivities of anisole and acetophenone in SDS are the same.
From Little, E. L., Ph.D. thesis, Louisiana State University (1992).

heating is not excessive. Second, α is also independent of surfactant concentration, assuming there are no solute-specific interactions with the buffered micellar media. (This assumption is valid for virtually all neutral analytes and for some charged analytes.) Finally, several chemical variables of the running buffer may be changed to alter selectivity (and sometimes the analytes' P_{wm}): (1) adding or changing organic modifiers (e.g., Table 4.4); (2) adding or changing surfactants and/or their counterions (e.g., Table 4.5); (3) changing the pH (for ionizable solutes and/or surfactants); and (4) the use of other mobile phase additives like cyclodextrins, crown ethers, metal ions, or co-surfactants.

D. Efficiency

1. Background

Many useful insights can be gained by viewing chromatographic processes from a molecular perspective. While a group of molecules is displaced along the separation axis, the individual molecules are essentially engaging in random, erratic motion. It is very difficult to predict the behavior of a single molecule, but it is possible to use a statistical approach to describe the behavior of a group of molecules (i.e., the solute zone). The use of a random walk model in chromatography has been used by researchers to describe the various mechanisms that contribute to zone spreading.[28,87,88]

2. Common Sources of Chromatographic Band Broadening

There are three basic sources of intracolumn band broadening (plate height) that are common to most forms of chromatography: (1) eddy dispersion, (2) longitudinal diffusion; and (3) resistance to mass transfer. The first, eddy dispersion, is characteristic of flow in a packed chromatographic bed. The irregularly shaped particles form irregularly shaped channels between particles which cause a nonuniform flow velocity. This source of plate height does not exist in unpacked open tubular columns and is not applicable to MEKC.

The second band-broadening phenomenon, longitudinal diffusion, results from the random Brownian motion (diffusion) that all solutes undergo in all directions as they are displaced along the column. Although diffusion in the radial direction is a desirable means of mass transport, solute diffusion along the separation axis, i.e., longitudinal diffusion, merely results in wider bands and hence poorer resolution.

The third source of band broadening, resistance to mass transfer, is itself made up of several components. The first is the finite rate of solute mass transfer from the mobile (aqueous) phase to the pseudostationary phase and vice versa. When a molecule partitions into/adsorbs onto the pseudostationary phase, its resulting velocity is different than its velocity in free solution. Since the adsorption–desorption process occurs randomly, the motion of solute molecules is quite erratic and can result in zone spreading if the kinetics are relatively slow.

A second component of resistance to mass transfer is the finite rate of solute mass transfer in the mobile phase. This is important in separation media in which the mobile phase velocity and/or analyte migration velocity is not uniform. Given the almost perfectly flat flow profile provided by electroosmotic transport in MEKC (except in the diffuse double layer near the capillary wall), this source of band broadening is always negligible for neutral analytes and usually negligible for charged analytes, except when excessive Joule heating results in radial temperature and viscosity gradients, the latter giving rise to radial electrophoretic velocity gradients for charged analytes.

3. Additional Band Broadening in MEKC

Explanations of band-broadening phenomena in MEKC have been and will continue to be an active area of research and debate.[51,52,55,89] Consequently, our discussion is based on the more popular proposed band-broadening mechanisms. We have purposely avoided excessive editorial comments in this section so as not to unduly influence the reader.

There are at least two other sources of band broadening unique to MEKC. First, micelles have a characteristic aggregation number indicative of the average number of surfactant monomers that make up a micelle. These aggregation numbers are characteristic not only of the identity of the surfactant but also the chemical environment that the micelle is in. For SDS under proper conditions the aggregation number is 62. It is important to keep in mind that there may be micelles that have more or fewer monomers than the aggregation number. This results in a size distribution of micelles and is termed the polydispersity. Since our systems utilize ionic buffers, the micelles are charged and any variation in the number of monomers in the micelle will give rise to a different charge for the micelle and hence a different electrophoretic mobility ($\mu_{ep,mc}$). In addition, the micelle shape may be affected, and this may also affect $\mu_{ep,mc}$. Terabe et al. have used random walk theory to relate polydispersity to zone spreading.[52] Davis has utilized an error propagation approach to estimate the magnitude of the affect of polydispersity on plate height and found only a second-order dependence of plate height on polydispersity.[55]

Second, because of the relatively high conductance of the micellar running buffers in MEKC, sufficient current is generated to cause heating of the capillary. And because the heat is primarily dissipated at the capillary walls, a radial temperature gradient exists across the capillary. In one approach to this problem of temperature-induced band broadening, the temperature dependence of the viscosity of the buffer in the capillary was considered to be the mechanism which determined the velocity profiles of the solute. It was subsequently shown that the temperature gradient, treated in this manner, was not a significant source of band broadening. In another approach, it was postulated that a radial gradient in k' exists because the P_{wm} is a function of temperature and will vary over the cross-section of the capillary. This causes solute molecules to spend a greater fraction of time in the mobile phase at the capillary center than near the walls of the capillary, and this means that a solute will have a greater v_{avg} at the center of the capillary, decreasing toward the capillary wall. Davis has used a random walk approach to characterize this potential band-broadening mechanism and termed it *transchannel mass transfer*.[55] This treatment of band broadening shows the proper dependence of H on column radius as previously observed.[51]

In one of the few experimental studies of band broadening in MEKC, Terabe et al. reported in 1989 a thorough investigation of numerous phenomena.[52] Essentially they found that at low electroosmotic velocities, the main source of band broadening was longitudinal diffusion, as in many forms of conventional chromatography. Yet as the v_{eo} was increased, other band-broadening mechanisms became apparent.

There has been considerable debate on the contribution of intermicellar mass transfer to plate height. Sepaniak et al. have reported that at surfactant concentrations close to the cmc, as the surfactant concentration is increased there is a dramatic increase in N.[51] They propose that, as the surfactant concentration increases, the number of micelles increases such that the distance between micelles decreases and improved mass transport results. But other researchers have presented evidence, both theoretical and experimental, that the observed effects under these conditions are due to overloading the micelles with solute. Another mass transfer mechanism is partitioning the solute into and out of the micelle, termed micellar mass transfer. But micellar mass transfer has also been determined to be virtually insignificant as a source of band broadening under the experimental conditions examined.[52,55]

Table 4.6 Suggestions for Obtaining the Highest Efficiency in MEKC

---

1. The use of organic solvents should be kept to a minimum.
2. Sample solvent and running buffer should be matched as closely as possible.
3. Inlet and outlet buffer reservoir levels should differ at most by 0.5 mm, and ideally should be exactly the same to avoid any hydrostatic flow effects.
4. To minimize H_{inj}, the length of the injection plug should not exceed 0.1% of the total capillary length, unless sample stacking is utilized. Injection times should also be relatively short.
5. When choosing a buffer and surfactant system, select those that minimize the overall conductance or ionic strength.
6. For a given buffer system, keep the applied voltage at the upper limit of the linear portion of the "Ohm's law plot" (a graph of current vs. voltage).
7. Capillary cuts should be as square as possible to avoid peak tailing, and samples should be filtered.
8. Interactive agents should be of the highest purity, in order to minimize H_{pd}.
9. When possible, use field amplified sample stacking.
10. If the length of the detector-illuminated volume of the capillary can be varied, it should be made as short as possible (e.g., ≤ 0.5 mm) to minimize H_{det}.
11. Judicious choice of the surfactant and/or its counterions can sometimes increase N significantly.

Although we have considered intracolumn sources of band broadening in some detail, there are of course extracolumn sources of band broadening due to injection and detection. Limited numbers of studies have been conducted on the effects of injection technique on plate height in MEKC.[58-62,90] Certainly, studies involving effects of injection technique on plate height in CZE are applicable to MEKC.[51,61,62,91-95] And one study done on electromigration injection technique illustrated that N can be maximized by using low voltages and short times of applied voltage. This results in a narrow sample plug being put at the beginning of the column and hence better efficiency.[58] It is a general principle that a narrow injection plug will result in better efficiency, and this means short injection times for other methods, like hydrostatic and vacuum injection, will reduce any contributions to σ^2_T from injection. There have also been a few recent reports on the application to MEKC of zone sharpening, i.e., "sample stacking" techniques.[59,60,90] The contribution to σ^2_T from the detector and related electronics has also been examined in detail.[52,61,62]

Although band broadening in MEKC is very interesting from a theoretical viewpoint, it is important to realize experimentally the anticipated high efficiencies. Based on a critical evaluation of the results of others as well as our own observations, we summarize in Table 4.6 some general guidelines for efficient separations.

E. Practical Strategies

There have been several reports on optimization of resolution by various methods. Unfortunately the term *optimization* means different things to different researchers. One should be aware that approaches to the optimization of separations based (1) on fundamental theories (e.g., k' optimization) or (2) on statistical designs of a small set of experiments (simplex, Plackett–Burman) provide complementary information on how to get the best separation. In order to truly optimize a separation, one approach should not be unduly favored over another.

The approach of McNair et al.[48] considered three ways to optimize R_s (Equation 20): (1) increase N; (2) increase α; (3) increase the retention term(s). Since α is very hard to predict and manipulate, their study focused on N and f(k'). At low values of v_{eo} and k' it was possible to get infinite R_s, but at the expense of longer analysis times. Their conclusion was that R_s could be maximized most effectively by lengthening the capillary rather than decreasing v_{eo}. Sandra and Vindevogel[96] utilized a Plackett–Burman statistical design to optimize R_s for a single set of four testosterone esters. This approach appears to be good for the simultaneous optimization of two or perhaps three parameters. Castagnola et al. have used a weighted variable-simplex algorithm to optimize the separation of PTH-amino acids.[97] Other systematic approaches such as the mixture design, exhaustive or interpretive response surface mapping (window diagrams), mixed lattice, adaptive searches, and step-search designs[98,99] may also be effective, but they have not yet been investigated. Some may be less suited for the discrete changes in solution variables (pH, surfactant concentration, etc.) that are more convenient with current commercial or in-house instrumentation (cf. mobile phase composition that can be varied continuously by most HPLCs).

A systematic evaluation of optimization of R_s by manipulating surfactant concentration has been reported.[6] Essentially this study is an experimental verification of a theory for neutral solutes proposed by Foley[70] and discussed in some detail in previous sections. The theory helps to predict the optimum surfactant concentration for a given pair of solutes. As Equation 24 shows, the optimum retention factor for a given separation is related to t_{mc}/t_0. One potential difficulty is that as the surfactant concentration is increased, the t_{mc}/t_0 value may vary considerably, depending on the surfactant system used. For the reported study, two surfactant systems were evaluated, one based on SDS and the other on a Brij-35/SDS mixed surfactant system. As noted earlier, there were several advantages to using the mixed surfactant system, but for the present discussion the interesting point is the difference in the behavior of t_{mc}/t_0 for the two systems. Over the concentration ranges studied, the t_{mc}/t_0 value for the Brij-35/SDS system *decreased* by only 15%, while the t_{mc}/t_0 value for the SDS system *increased* by nearly 300%. The optimization of R_s for systems that have variable t_{mc}/t_0 is less desirable because an iterative process is needed to converge to the optimum surfactant concentration. But only two iterations at most should be required for the proper convergence. A more detailed discussion of mixed micellar systems is presented in a later section.

VI. THEORY FOR CHARGED SOLUTES

While the studies cited earlier in our qualitative overview for charged solutes are informative, it is also important to understand migration behavior and resolution quantitatively in order to optimize separations in the best possible manner. Some issues pertaining to the mobility, migration, and/or retention of permanent ions and ionogenic (ionizable) compounds are addressed before we present a general theory for the resolution of charged analytes.

A. Mobility of Inorganic Ions Above and Below the CMC

Although the electrophoretic mobilities of many inorganic ions are sufficiently different to be separated by CZE, there are also occasions when MEKC is necessary. When MEKC is used to separate inorganic ions, the surfactant employed is often oppositely charged in order to avoid electrostatic repulsion which would tend to inhibit ion–micelle interactions. Electrostatic interactions, presumably at or near the surface of the micelle, are often dominant due to the polar character (hydrophilicity) of most inorganic ions.

In one study, cetyltrimethylammonium chloride was used above and below its cmc to separate the following monovalent anions: bromide, nitrate, bromate, iodide, and iodate.[100] The following relationship was derived to explain their effective mobility:

$$\frac{1}{\mu_{eff}} = \frac{K_{IA}}{\mu_{ep}}C_{sf} + \frac{1}{\mu_{ep}} \tag{33}$$

where μ_{eff} is the effective electrophoretic mobility, μ_{ep} is the electrophoretic mobility, K_{IA} is the ion association constant, and C_{sf} is the concentration of surfactant. Based on Equation 33, one might expect that a plot of $1/\mu_{eff}$ versus C_{sf} should yield a linear relationship with a slope of K_{IA}/μ_{ep} and intercept of $1/\mu_{ep}$. When the experimental data were analyzed, however, two distinct lines (with different slopes) were observed that intersected at the surfactant cmc, with the clearest distinction observed for iodide. The explanation is that the rate of change in μ_{eff} with increasing C_{sf} (a measure of the strength of the analyte–surfactant interaction) will often depend on the nature of the interaction (ion pairing [C_{sf} < cmc] vs. solute–micelle interactions [C_{sf} > cmc]). Also noteworthy was the change in selectivity (migration time) from below the cmc (bromide < nitrate < bromate = iodide < iodate) to above the cmc (iodate < bromate < bromide < nitrate < iodide).

Ion-exchange models have also been employed to describe MEKC separations of anions[101] and transition metal ions,[102] the latter using SDS and tartaric acid in the buffer.[102]

B. Migration and Retention of Ionizable Species

Khaledi et al. have very eloquently addressed the issues of retention mechanisms and optimization for both monoprotic acidic and basic solutes in MEKC using SDS surfactant systems.[103-105] The phenomenological approach utilized is very similar to the way such secondary chemical equilibria (SCE) have been treated in HPLC.[106,107] Despite the importance of the degree of analyte ionization on partitioning and mobility as described below, we remind the reader that selectivity does not depend solely on charge in MEKC and that manipulation of solute charge is only one aspect of optimizing the separation of ionizable species; all the previously discussed strategies for manipulating the selectivity of neutral solutes apply equally well to charged species.

The partitioning, P_{wm}, of a monovalent acidic solute will be a function of pH and can be expressed as follows[103,108]:

$$P_{wm\,acid} = \frac{P_{HA} + P_{A-}(K_a/[H^+])}{1+(K_a/[H^+])} \tag{34}$$

where P_{HA} is the partition coefficient for the neutral (protonated) form of the acid, P_{A-} is the partition coefficient for the anionic (deprotonated) form, K_a is the acid dissociation constant, and [H^+] is the hydrogen ion concentration. Equation 34 indicates that there is a sigmoidal relationship between $P_{wm\,acid}$ and pH and that $P_{wm\,acid}$ will decrease as pH is increased. The $P_{wm\,acid}$ decreases because the anionic form of the acid "likes" the aqueous environment and the like charges on the solute and micelle cause the species to be electrostatically repelled from each other.

An additional consideration needs to be addressed for acidic solutes: the dissociation constants (pK_a) of many acidic and basic analytes will change significantly from their values in aqueous solution when exposed to micellar media. The apparent K_a of an acidic substance in a micellar solution is given by[103,109]:

$$K_{a\,app} = K_a\left(\frac{K_{mA-}[M]+1}{K_{mHA}[M]+1}\right) \tag{35}$$

where K_{mA-} is the micellar binding constant for the ionized form of the acid, K_{mHA} is the micellar binding constant for the un-ionized form of the acid, and [M] is the concentration of the micelles. The $K_{a\,app}$ will be a function of micelle concentration and the physical characteristics of the solute that control binding to the micelle, K_m. Consider two solutes that have very similar aqueous pK_a values. When placed in a micellar solution, their pK_a values may change in different directions and/or by different amounts. This means that at a given pH their charge differences may be larger than in aqueous solution and consequently selectivity will be greater in MEKC than CZE for these solutes. This type of selectivity enhancement has been studied in MLC for a series of amino acids and peptides.[109] Conversely, two solutes with very different pK_a values in aqueous solution may have similar pK_a values in a micellar solution. The result in this situation would be a loss of selectivity.[103]

The retention factor for an acidic solute was also derived:

$$k'_{acid} = \frac{\mu - \mu_o}{\mu_{mc} - \mu} \tag{36}$$

where μ and μ_o are electrophoretic mobility of the solute in the presence and absence of micelles, and μ_{mc} is the electrophoretic mobility of the micelles. Note that in the interest of consistency with the original publications, all previous and subsequent equations in this section are written in terms of absolute rather than net electrophoretic mobilities since μ_{eo}, the other source of transport in the capillary, is a constant for all species.

Similar expressions for the partition coefficient and retention factor of monovalent basic solutes were also derived.[104] Because the ionized form of most bases is positively charged, it is necessary to take into account ion pairing with the negatively charged surfactant. The partition coefficient for a basic solute is:

$$P_{wm\,base} = \frac{P_{BH+}(K_b[H^+]/K_w) + P_B}{1 + (K_b[H^+]/K_w)(1 + K_I)} \tag{37}$$

where P_{BH+} is the partition coefficient of the ionized base, P_B is the partition coefficient of the un-ionized base, K_b is the aqueous base constant, K_w is the dissociation constant of water, and K_I is the ion pairing constant. Note that in the absence of ion pairing ($K_I = 0$), Equation 37 is exactly analogous to the equation for acidic solutes.

One should be cautious in making any conclusions on the dependence of $P_{wm\,base}$ with respect to pH. In systems where ion pairing can occur, the relative magnitude of P_{BH+} and P_B is unpredictable, thus making the dependence of $P_{wm\,base}$ with respect to pH very difficult to predict.

The retention factor for a basic solute is:

$$k'_{base} = \frac{\mu - \left(\dfrac{K_w + K_b[H^+]}{K_w + K_b[H^+](1 + K_I)}\right)\mu_o}{\mu_{mc} - \mu} \tag{38}$$

where all of the variables have been previously defined. In the absence of ion pairing ($K_I = 0$), Equation 38 is identical to Equation 36. Conversely, as the ion pairing of the protonated base and anionic surfactant becomes complete (K_I approaches infinity), the equation reverts to the one for k' of a neutral solute [where $\mu = k'\mu_{mc}/(1 + k')$]:

$$k'_{neutral} = \frac{\mu}{\mu_{mc} - \mu} \tag{39}$$

In the case of ionizable organic molecules, the pH and surfactant concentration can be varied to alter the solute charge and thereby manipulate the retention. This is in addition to other methods that may be used to alter retention like using organic modifiers.

The phenomenological approaches outlined above have been successfully utilized in the computer-assisted optimization for the separations of acidic and basic compounds.[105] In particular, this methodology was used to predict conditions for the separation of a group of

18 aromatic amines, which resulted in successful separation of these compounds with an analysis time of under 15 minutes.

C. General Theory for Charged Analytes

As we have mentioned, MEKC can also be used to effectively separate charged analytes in addition to neutral ones. All of the theories and equations that have been developed for neutral species can be expanded to encompass charged solutes. Consequently a more general theory for optimization of R_s that includes charged as well as neutral solutes can also be developed. Highlights of such a theory[110,111] are presented here. We have assumed that the influence of nonmicellar interactions (e.g., ion-pairing with surfactant monomers) on the mobility of charged analytes is negligible compared to influence of micellar interactions. Based on our experience and that of others,[112,113] this assumption is valid under most circumstances.

The retention time of a charged solute can be expressed as:

$$t_R = L / v_{(net)} = \left(\frac{1+k}{1+\mu_r + (t_o/t_{mc})k} \right) t_o \tag{40}$$

where $\mu_r = \mu_{ep}/\mu_{eo}$ is the relative electrophoretic mobility, i.e., the electrophoretic mobility of an analyte (positive, zero, or negative) relative to the coefficient of electroosmotic flow. For charged solutes that swim downstream ($\mu_{ep} > 0$), $\mu_r > 0$ and vice versa.

Technically, $\mu_r = \mu_{ep}/\mu_{eo}$ where both μ_{ep} and μ_{eo} are measured under MEKC conditions. However, it is not possible to measure μ_{ep} under MEKC conditions, because of the influence of the micelles on the analyte's migration. What is normally done is to measure μ_{ep} with the same buffer system but with no surfactant ($C_{surf} = 0$) or no micelles ($C_{surf} < cmc$). One then assumes that

$$\mu_{ep} \ (C_{surf} = 0 \quad \text{or} \quad C_{surf} < cmc) = \mu_{ep} \ (\text{MEKC conditions}) \tag{41}$$

The retention factor can be expressed in a manner similar to Equation 18 for neutral solutes.

$$k = \frac{t_R(1+\mu_r) - t_o}{t_o - (t_o/t_{mc})t_R} = \frac{t_R(1+\mu_r) - t_o}{t_o(1 - t_R/t_{mc})} \tag{42}$$

For neutral solutes, μ_{ep} is zero by definition and therefore so is μ_r. Thus Equations 40 and 42 reduce to Equations 17 and 18. As we will show below, the theory for neutral solutes is a special case of a more general theory that includes charged solutes.

Equation 40 for the migration time can be employed to derive a master resolution equation (MRE) that accounts for both chromatographic and electrophoretic separation mechanisms and is thus applicable to CZE, EKC, electrochromatography, and HPLC.[110,111] Using the commonly accepted definition of resolution[30] (Equation 3) and the number of theoretical plates, the following equation can be derived for the resolution of two solutes:

$$R_s = \frac{\sqrt{N}}{4} \left(\frac{(1 - t_o/t_{mc})(k_2' - k_1') + (\mu_{r,1} - \mu_{r,2}) + (\mu_{r,1}k_2' - \mu_{r,2}k_1')}{1 + t_o/t_{mc}k_1'k_2' + (1 + t_o/t_{mc})\frac{(k_1' + k_2')}{2} + \frac{(\mu_{r,1} + \mu_{r,2} + \mu_{r,1}k_2' + \mu_{r,2}k_1')}{2}} \right) \tag{43}$$

where for simplicity we have assumed Gaussian peak shapes and $N_1 = N_2 = N$.

To the best of our knowledge, the above equation is the first expression to appear that predicts the resolution resulting from simultaneous differences in the retention (k′) and relative electrophoretic mobility ($\mu_r = \mu_{ep}/\mu_{eo}$) of two solutes. It shows that resolution is proportional to (see numerator): (1) differences in k′; (2) differences in μ_r; and (3) a cross term that is positive, negative, or zero depending on whether the effects of k′ and μ_r on resolution are cooperative, neutral, or in opposition to one another.

The denominator of the above MRE can be simplified by defining the arithmetic means k'_{avg} and $\mu_{r,avg}$, and using the approximations (1) $k'_1 k'_2 \approx k'^2_{avg}$ and (2) $\dfrac{(\mu_{r,1} k'_2 + \mu_{r,2} k'_1)}{2} \approx$ $\mu_{r,avg} k'_{avg}$. This yields

$$R_s \approx \frac{\sqrt{N}}{4}\left(\frac{(1-t_o/t_{mc})(k'_2-k'_1)+(\mu_{r,1}-\mu_{r,2})+(\mu_{r,1}k'_2-\mu_{r,2}k'_1)}{(1+\mu_{r,avg}+t_o/t_{mc}k'_{avg})(1+k'_{avg})}\right) \tag{44}$$

Given the usual practice of expressing differences in chromatographic retention (k′₂ – k′₁) in terms of selectivity ($\alpha = k'_2/k'_1$), Equation 44 can be written as

$$R_s \approx \frac{\sqrt{N}}{4}\left(\frac{(1-t_o/t_{mc})(\alpha-1)k'_1+(\mu_{r,1}-\mu_{r,2})+(\mu_{r,1}k'_2-\mu_{r,2}k'_1)}{(1+\mu_{r,avg}+t_o/t_{mc}k'_{avg})(1+k'_{avg})}\right) \tag{45}$$

Although chromatographic and electrophoretic separation mechanisms are largely orthogonal, Equations 43 through 45 show that the simultaneous presence of both mechanisms is sometimes undesirable when only one of them contributes to the separation. When the separation is accomplished only by an electrophoretic mechanism ($\mu_{r,1} \neq \mu_{r,2}$ and $k'_1 = k'_2 = k'_{avg} > 0$), chromatographic retention serves only to lower resolution by increasing each of the terms in the denominator of Equations 44 through 45 via k'_{avg}. Likewise, when the separation is accomplished strictly by a chromatographic mechanism ($0 < k'_1 < k'_2$ and $\mu_{r,1} = \mu_{r,2} = \mu_{r,avg}$), electrophoretic transport decreases resolution for analytes that swim downstream ($\mu_{r,avg} > 0$) by increasing the first term in the denominator of Equation 44 or Equation 45. On the other hand, electrophoretic transport does increase resolution somewhat for analytes that swim upstream ($\mu_{r,avg} < 0$) by decreasing the first term in the denominator of Equation 44 or 45.

Equations 43 through 45 unify the fields of HPLC, CZE, MEKC, and electrochromatography because they are applicable to all of these separation modes. The more familiar resolution equations for HPLC and CZE can be obtained by use of the appropriate conditions and assumptions, although it would exceed space limitations to do so here.

When both chromatographic and electrophoretic separation mechanisms are in effect, any of the general MRE equations (Equations 43 through 45) may be employed to accurately predict resolution of charged solutes in MEKC. These equations are quite complex, however, and it is useful to consider two limiting cases: (1) resolution based solely on differences in electrophoretic mobility [$\mu_{r,1} \neq \mu_{r,2}$, $k'_1 = k'_2 = k'$]; and (2) resolution based solely on differences in retention [$k'_1 \neq k'_2$, $\mu_{r,1} = \mu_{r,2} = \mu_r$].

Applying the constraints of limiting case (1) to Equation 45 yields

$$R_s = \frac{\sqrt{N}}{4}\left(\frac{(1+k')(\mu_{r,1}-\mu_{r,2})}{(1+\sqrt{t_o/t_{mc}}k')^2+(1-\sqrt{t_o/t_{mc}})^2k'+(1+k')\mu_{r,avg}}\right) \tag{46}$$

This equation shows the effect of nonselective solute–micelle interactions ($k'_1 = k'_2 = k'$) and the associated finite elution range ($t_{mc}/t_o < \infty$) on resolution in a separation mode that is otherwise analogous to CZE.

Applying to Equation 45 the constraints of limiting case (2) yields

$$R_s = \frac{\sqrt{N}}{4}\left(\frac{\alpha-1}{\alpha}\right)\left(\frac{k'_2}{1+k'_{avg}}\right)\left(\frac{1+\mu_r-t_o/t_{mc}}{1+\mu_r+(t_o/t_{mc})k'_{avg}}\right) \tag{47}$$

While at first glance the number of compounds for which limiting case (2) is applicable might appear to be low, this category includes *chiral separations,* i.e., the separation of enantiomers by chiral second phases (micelles, bile salts, ionic cyclodextrins).

Equation 47 closely resembles the one developed by Terabe et al. for neutral solutes (Equation 19). The key difference between Equations 47 and 19 is that Equation 47, by virtue of the μ_r term that appears in both the numerator and denominator of the last term, is valid for both charged and neutral analytes, whereas Equation 19 (Terabe's expression) is applicable only to neutral solutes. Equation 19 is a limiting case of Equation 47, i.e., the separation of neutral solutes by MEKC (all $\mu_r = 0$). It will now be shown via Equation 47 that the resolving power of MEKC is superior for analytes that swim downstream, assuming their efficiencies are comparable.

The last two terms on the r.h.s. of the above equation (the retention "term") show a dependence on μ_r. This indicates that there may be some bias in the R_s based on μ_r. This point is best illustrated by Figure 4.9. The trend is very clear that as μ_r is varied from −1 to +1 the retention term increases and so does R_s.

An optimum k' with respect to R_s and R_s/t_R can also be derived based on the treatment used in Section V.

$$k'_{opt(Rs)} = \sqrt{\frac{1+\mu_r}{t_o/t_{mc}}} = \sqrt{t_{mc}/t_o(1+\mu_r)} \tag{48}$$

$$k'_{opt(Rs/tR)} = \frac{-(1+\mu_r-t_o/t_{mc})\pm\sqrt{(1+\mu_r-t_o/t_{mc})^2+16(t_o/t_{mc})(1+\mu_r)}}{4(t_o/t_{mc})} \tag{49}$$

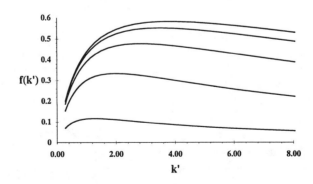

Figure 4.9 Bias in the retention term, f(k'), of the resolution equation for charged and neutral solutes (Equation 47): species that swim downstream ($\mu_r > 0$), don't swim (neutrals, $\mu_r = 0$), or swim upstream ($\mu_r < 0$). Curves of f(k') from top to bottom correspond to μ_r values of +0.8, +0.5, 0.0, −0.5, and −0.8.

The important conclusion from all of this is that a bias in the resolving power of MEKC potentially exists, unless the dependence of N on μ_r exactly cancels the dependence of the retention terms on μ_r (Equation 47). Assuming the retention dependence is dominant, it is better to analyze species that "go with the flow."[114] In other words, when electroosmotic flow is from anode to cathode (the usual case in MEKC with anionic surfactants such as SDS), cations will be better resolved than neutral species, and neutral species will in turn be better resolved than anionic compounds.

VII. PRACTICAL CONSIDERATIONS FOR PRESENT SYSTEMS

One very important requirement of a pseudostationary phase is that it migrates at a different velocity than the electroosmotic velocity. In short, the additives that form the pseudophase must be charged, and thus contribute to the total ionic strength of the solution. Consequently, the operating currents can be significantly higher than in CZE, a distinct disadvantage if adequate cooling of the capillary cannot be maintained. Pseudophases are a broader classification of mobile phase additives that can provide a variety of solute selectivities. The interested reader is directed to a more comprehensive review of pseudophases used in CE.[115]

A. Micelles

Micelles of ionic surfactants are by far the most widely utilized pseudophase in EKC. But the micelles are part of a dynamic equilibrium that can be affected by many experimental parameters. Temperature can affect the cmc, but many common surfactants exhibit minimum in cmc between 20°C and 30°C, and thus in those the cmc is not very temperature dependent. On the other hand, the cmc is often fairly sensitive to electrolyte concentration. An increase in electrolyte concentration causes an increase in the aggregation number and a decrease in the cmc.[116] In general a lower cmc is better because there will be a lower free monomer concentration, and this will reduce operating currents. From this point of view, SDS, the most popular surfactant at present for MEKC, is far from ideal, with a cmc of 8.1 mM in very dilute solutions.

Organic modifiers can affect the micelle in a variety of ways. As the percentage of a short-chain alcohol is increased, the cmc goes through a minimum and then increases. Other common organic solvents like acetonitrile, tetrahydrofuran, and acetone that can interact strongly with water's hydrogen bonding network inhibit micellization, thus raising the cmc.[117,117a] Long-chain alcohols and alkanes can raise or lower the cmc depending on the characteristics of the additive. In general, large amounts of organic solvent (>30% v/v) have detrimental effects on micelle formation, resulting in poor reproducibility and/or low efficiency.

The size and shape of the micelle is also influenced by the experimental parameters. As the surfactant concentration increases, micelles can undergo transitions from spherical to rodlike or cylindrical aggregates.[118] In addition, the counterion for a particular surfactant can influence the micellar properties.[119-129, 129a] The size and shape of the micelle not only affect the electrophoretic mobility (μ) of the micelle but also the solute partitioning mechanisms. In our laboratory, we have investigated the use of monovalent and divalent inorganic cations and organic cations as counterions for dodecylsulfate micelles. These studies include comparisons of (1) sodium and magnesium[66]; (2) lithium, sodium, and potassium[76]; and (3) different tetraalkylammonium counterions (tetramethylammonium, tetrapropylammonium, etc.).[45]

There are several theories on how to model micellar properties and the interested reader is directed to a review as it applies to EKC.[115] One common model depicts a micelle as composed of two regions: (1) a nonpolar core, with the characteristics of a liquid hydrocarbon solvent, and (2) a more polar palisades layer (Figure 4.1).[130-134]

The separation characteristics of a micellar phase can also be manipulated by changing the surfactant. A variety of different surfactants have been employed in MEKC, including anionic surfactants such as SDS,[135-140] cationic surfactants such as CTAB,[9,68,141-145] zwitterionic[11,146] and nonionic[6,46,147] surfactants in mixed micelles nonionic (e.g., SB-12, Brij-35, Tween), bile salts such as sodium cholate or sodium deoxycholate,[148-151] double chain anionic surfactants,[152] and chiral surfactants.[10,129a,153-160] Other commercially available surfactants have yet to be utilized but will undoubtedly be useful in MEKC.

The very complex nature of micellar systems puts constraints on the types and manner in which experimental parameters can be manipulated. Proper knowledge and understanding of these various processes are essential to method development and optimization.

B. Cyclodextrins

Cyclodextrins (CDs) have received much attention in the past few years as reagents in chromatographic processes. These compounds are cyclic oligosaccharides of D-glucose. The cyclodextrins come in three sizes, α, β, γ, that consist of 6, 7, and, 8 glucose residues, respectively. The cyclic arrangement of the sugar residues creates a cavity such that molecules of the proper size, shape, and polarity enter the cavity forming an inclusion complex.

CDs are interesting because they can provide a variety of separation selectivities. They can be used in chiral separations, as well as separations based on solute size and polarity. They were first employed in 1985, when Terabe et al. used a carboxymethyl derivative of β cyclodextrin as a "moving" stationary phase.[19] More recently, CDs have also been used simultaneously with SDS micelles (cyclodextrin-modified MEKC) to separate corticosteroids and aromatic hydrocarbons,[161] polyaromatic hydrocarbons (PAH),[162] and amino acids.[163] Cyclodextrin-modified MEKC has also been effectively utilized for chiral separations.[164,165]

Sepaniak et al. have described a form of electrokinetic chromatography using cyclodextrins (CD-EKC) that is very closely related to MEKC.[20] In this technique, there are two cyclodextrin phases. One traverses the capillary at the same velocity as the electroosmotic flow, while the second cyclodextrin phase traverses the capillary at a velocity different from the electroosmotic flow. Selectivity can be modulated by using a variety of CDs and organic solvents. In addition, separations are not adversely affected by high concentration of organic modifier (>40% v/v), as is the case in MEKC with a micellar solution of conventional surfactant. CD-EKC has been successfully applied to separations of polyaromatic hydrocarbons (PAH) and will undoubtedly be useful in the separation of positional isomers as well as enantiomers.

C. Other Additives

Crown ethers can also be used to provide a type of pseudophase. Crown ethers are large cyclic molecules that contain electron-donating heteroatoms like N, O, and S. Solutes typically form inclusion complexes with crown ethers such that the heteroatoms are the primary path of interaction. Crown ethers are probably most well known for their binding of metal ions, but nitrile, nitro, amine, and azo compounds can also be effectively complexed. Crown ethers are also capable of chiral recognition and have been used in CE to effect enantioselective separations.[166-173] Pseudophases like cyclodextrins and crown ethers are receiving an increasing amount of attention as their special characteristics (i.e., binding selectivities) can be more readily exploited.

Enantioselective separations can also be accomplished using other types of chiral pseudophases. Bile salt micelles have been used,[15,17,149,174-181] and the subject has been reviewed by Sepaniak and Cole.[182] Ternary metal complexes have been utilized effectively for enantioselective separations in CE.[183,184]

VIII. APPLICATIONS

Published applications of MEKC cover a wide range of disciplines, as noted in reviews pertaining to pharmaceutical and biomedical applications,[185] quantitation,[186] basic/fundamental concepts,[5] three comprehensive reviews,[187-190] and a book devoted entirely to MEKC.[191] An exhaustive treatment of MEKC is beyond the scope of this chapter; instead, we share with the reader some representative applications.

A. Biological

MEKC has many potential applications in the biological sciences. Among them are natural products, compounds of importance in today's world because of their medicinal properties. The analysis and identification of useful natural products from various sources is a rapidly expanding area of analytical chemistry. One such compound is taxol. Chan et al. used MEKC to separate taxol, cephalomannine, and baccatin II in crude extracts from both the bark and needle of the *Taxus* species.[192,193] The authors found that the MEKC method provided separation of taxol from taxinines, which was not possible using HPLC.[192] MEKC has also found applications in cancer research.[194] Another class of natural products are the flavonoids. MEKC has been used to separate a variety of flavonoids,[143,195] and flavonoid-O-glycosides,[68,196] including the separation of flavonol-2-glycosides and flavonol-3-glycosides in various plant extracts.[197] In one study, an MEKC method for flavonoids was developed that can be used for the characterization of the botanical and geographical origins of honey.[198] In another study, flavonol-2-glycosides were extracted from *Calendula officinalis* and *Sambucus nigra* and analyzed by both MEKC and HPLC.[199] Greater efficiencies and much shorter analysis times were observed with MEKC than HPLC.

Purines and substituted purines are of physiological importance. Perhaps the most widely recognized purines are caffeine and theophylline. The analysis of body fluids (saliva, blood, and urine) for these purines has been an area of considerable research and development. Thormann et al. have applied MEKC to the analysis of urine, blood, and saliva from infants and adults.[200] The solutes of interest were separated in a phosphate-borate buffer (pH ≈ 9.0) with 75 mM SDS. Saliva and serum samples could be analyzed by direct injection (i.e., no sample pretreatment), while urine samples required an extraction step before analysis. In other studies, acetylator phenotyping of urine was accomplished by MEKC using four caffeine metabolites, 5-acetylamino-6-formylamino-3-methyluracil (AFMU), 5-acetylamino-6-amino-3-methyluracil (AAMU), 1-methylxanthine (1X), and 1-methyluric acid (1U).[201] Linhares and Kissinger used MEKC to conduct *in vivo* pharmacokinetic studies on theophylline in ultrafiltrate samples from awake, freely moving rats.[178a] MEKC has also been used to analyze pharmaceutical[151,202,203] and food products[204] for caffeine and related compounds.

The analysis of DNA and DNA adducts is also possible with MEKC. Lecoq et al. reported the separation of the major constituents of nucleic acids and various derivatives of those nucleic acids.[205] Optimum conditions for the separation of some representative samples were reported. In one particular *in vitro* study, calf thymus DNA was exposed to genotoxic agents and then analyzed. Several modified bases were detected in the treated samples. The authors point out that one particular advantage that MEKC has over HPLC is that nucleosides and nucleotides elute separately. This allows the identification of solutes by the nuclease P1 enrichment procedure. In this procedure the normal nucleotides are digested to nucleosides. In addition, RNA contamination in digested DNA can be easily detected under the proper operating conditions. The ribonucleosides elute after the deoxyribonucleosides at a pH of 9.2 with 100 mM SDS, 10% acetonitrile, 20 mM phosphate-borate buffer, and an operating voltage of 12 kV, while at a pH of 7 the solutes coelute. In a study using similar compounds the retention of the solutes was manipulated by adding divalent inorganic ions to the buffered micellar solution.[3] One separation of 18 oligonucleotides, each with 8 bases, was achieved in less than 30 minutes using 20 mM Tris, 5 mM Na_2HPO_4, 50 mM SDS, 7 M urea, and 3 mM Zn (II), and an operating voltage of 22 kV. The Tris and Na_2PO_4 are the active buffering components, the urea helps to reduce solute–solute interactions, and the Zn (II) is added to alter selectivity (α) by chelating the solutes in a differential manner or by changing the characteristics of the micelle. Ramsey et al. investigated the use of MEKC for the separation of photoproducts of 2'-deoxyuridylyl-(3',5')-thymidine from normal nucleosides and nucleotides.[144] In addition, Lahey and St. Claire demonstrated that MEKC provided increased resolution for a mixture of 14 nucleotides and nucleosides in comparison to ion pair HPLC.[206]

There is an emerging role for MEKC in the clinical laboratory.[185,207-210] In particular, Nuñez et al. utilized MEKC to detect hypoglycemic (i.e., sulfonylurea) drugs in urine. Urine samples were prepared by a combination of liquid–liquid and using solid phase extraction (SPE). The samples were suspended in a phosphate/borate buffer solution containing the surfactant. Sodium cholate micelles provided the best resolution and resolution per unit time when compared to SDS micelles at pH = 8.5. Figure 4.10 is a comparison of electropherograms for a standard mixture of sulfonylurea drugs and a patient's urine sample (after extraction).

Alexander and Hughes utilized MEKC in conjunction with matrix-assisted laser desorption ionization (MALDI) mass spectrometry to characterize the stability of IgG antibodies.[211] When the antibody of interest, chimeric monoclonal antibody BR96, was exposed to a temperature of 60°C, it degraded considerably over a period of 166 hours, as illustrated in the series of electrokinetic chromatograms in Figure 4.11.

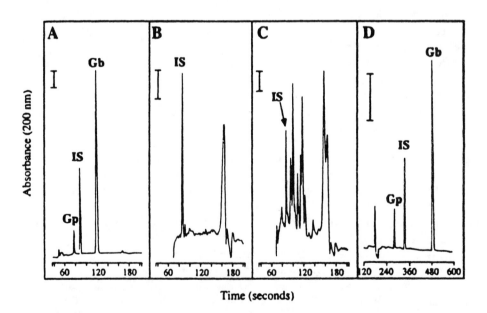

Figure 4.10 Separation of third-generation sulfonylurea drugs. (A) Separation of glipizide (Gp), glyburide (Gb), and internal standard (IS). (B) Separation of extract resulting from 1 ml of buffer spiked with internal standard. (C) Separation of extract resulting from 1 ml of normal urine with internal standard. (D) Separation of glipizide (Gp), glyburide (Gb), and internal standard (IS). A 40 cm × 50 µm i.d. bare fused-silica capillary was used with 20 mM phosphate/20 mM borate/50 mM SDS at pH = 8.5 and 30 kV applied voltage. Detection was at 200 nm. (From Nuñez, M., Ferguson, J. E., Machacek, D., Jacob, G., Oda, R. P., Lawson, G. M., and Landers, J. P., *Anal. Chem.*, 67, 3668, 1995. With permission.)

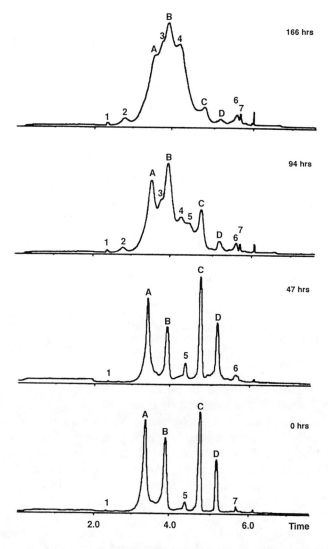

Figure 4.11 Electropherograms of heat-degraded chimeric BR96 monoclonal antibody after 0, 47, 94, and 166 hrs of thermal stress at 60°C. A 50 cm × 75 mm i.d. bare fused-silica capillary was used with 12 mM borate buffer at pH = 9.4 and 25 mM SDS and 30 kV applied voltage. Detection was at 214 nm. The labeled peaks were identified using MALDI-MS. (From Alexander, A. J. and Hughes, D. E., *Anal. Chem.*, 67, 3626, 1995. With permission.)

B. Environmental

Many chemicals exhibit some degree of toxicity, whether it be to animals, plants, or humans. Government agencies are continually issuing more stringent guidelines for the emission of potential environmentally hazardous chemicals, and consequently the need exists to determine such analytes rapidly and sensitively. Li et al. have analyzed one such set of nitroaromatic compounds.[212] The eight compounds were best separated with 30 mM SDS in a phosphate buffer (pH of 7.0). It was also demonstrated that structural isomers like 1-chloro-4-nitrobenzene and 2-chloro-4-nitrobenzene could be successfully separated. In another study, 11 priority phenols were analyzed by MEKC.[213] Takeda et al. used 2,4-diphenylhydrazine as a derivatization reagent for the analysis of short-chain aliphatic aldehydes in tap and river water by MEKC.[214] The detection limits for acetaldehyde and formaldehyde were 0.08 mg/mL and 0.05 mg/mL, respectively. Brumley et al. used LC-MS and MEKC for the analysis of synthetic dyes in water

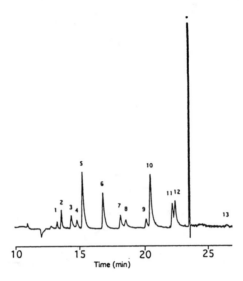

Figure 4.12 Separation of polycyclic aromatic hydrocarbons (PAHs). A 50 cm × 50 mm i.d. bare fused-silica capillary was used with 2% Elvacite 2669 (ionic block copolymer), 50% methanol, and applied voltage of 25 kV. The compounds are: (1) toluene (2) ethylbenzene (3) naphthalene (4) 1-methylnaphthylene (5) biphenyl (6) anthracene (7) pyrene (8) 1-nitropyrene (9) chrysene (10) 1,2-benzo[a]anthracene (11) perylene (12) benzo[a]pyrene (13) coronene. Detection at 254 nm for compounds 1–12 and 290 nm for compound 13. (From Yang, S., Bumgarner, J.G., and Khaledi, M.G., *J. High Resol. Chromatogr.*, 18, 445, 1995. With permission.)

and soil samples.[215] The authors found that MEKC functioned well as a screening tool, while the LC-MS was effective at identifying the compounds of interest; the two techniques gave complementary information about the samples being analyzed. Sepaniak et al. have also applied MEKC to a variety of environmental and biological samples.[216] Separation of hydrophobic polyaromatic hydrocarbons (PAH) with three or more rings is very difficult with MEKC. In order to circumvent the problems of separating PAHs using a micellar media, Yang et al. utilized a buffer containing an ionic polymer and 50% methanol (Figure 4.12).[217]

C. Forensic

A very unique application for MEKC is in forensic science,[218-225] which has been reviewed by Thormann et al.[226,227] In 1991 Northrop et al. reported the analysis of organic gunshot and explosive constituents.[218] The MEKC separation was very efficient, and the analysis time was under 10 minutes; a multiple wavelength detector was useful for the identification of the compounds. Some 26 different constituents were separated and identified, and a majority were nitrogen-containing aromatic compounds. MEKC has been demonstrated to have the potential for characterizing types of unfired explosives as well as the age and types of powders. Such analysis has been extended to the qualitative characterization of gunshot *residue* and the quantitation of its constituents.[220] Mussenbrock and Kleibohmer have also used MEKC to characterize explosive residues in soil.[224]

A number of classes of illicit drugs have been analyzed in a variety of matrices using MEKC.[228,231] Tagliaro et al. determined cocaine and morphine in hair samples,[232] Wernly and Thormann analyzed opioids, amphetamines, benzoylecgonine, and methaqualone in urine,[233-235] and Trenerry et al. used CD-MEKC to analyze methcathinone, cathine, propoxyphene, a variety of alpha-hydroxyphenethylamines and cocaine.[228] The CD-MEKC method could be used to analyze khat leaves (*catha edulis forsk*) for (–)-alpha-aminopropiophenone ((–)-cathinone), (+)-norpseudoephedrine (cathine), (–)-norephedrine, and trace levels of the phenylpentenyl-

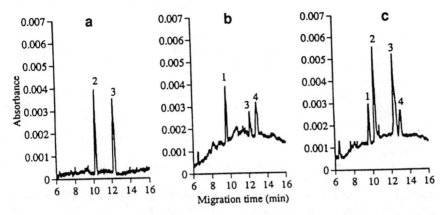

Figure 4.13 Electropherograms of (a) standard mixture of flumethasone and dexamethasone, (b) extract of dexamethasone horse urine, and (c) extract of dexamethasone horse urine spiked with standard mixture. A 50 cm × 75 mm i.d. bare fused-silica capillary was used with 40 m*M* borate, 10% methanol, 20 m*M* SDS at pH = 9.2 and an applied voltage of 30 kV. Detection was at 240 nm. (From Gu, X., Meleka-Boules, M., and Chen, C.L., *J. Cap. Electroph.*, 3, 43, 1996. With permission.)

amines (+)-merucathinone, (+)-merucathine, and possibly (–)-pseudomerucathine. Gu et al. have used MEKC for the determination of dexamethasone and flumethasone in horse urine (Figure 4.13).[239] Most states screen for the presence of these performance-enhancing gluco-corticoids before and/or after a race.

IX. CONCLUSIONS

Micellar electrokinetic capillary chromatography (MEKC) is a rapidly growing area within capillary electrophoresis due largely to its ability to simultaneously separate charged, neutral, and/or chiral compounds in a highly efficient manner. Although the thermodynamic factors that control resolution of neutral species are reasonably well understood, our grasp of kinetic factors (band broadening) is less complete. We can expect the applications of MEKC to remain somewhat ahead of our theoretical understanding summarized in this chapter, particularly for charged solutes. We look forward to additional theoretical contributions to MEKC from the separations community, and to the increase in our comprehension of MEKC that such research will provide.

REFERENCES

1. Terabe, S., Otsuka, K., Ichikawa, K., Tsuchiya, A., and Ando, T., Electrokinetic separations with micellar solutions and open tubular capillaries, *Anal. Chem.*, 56, 111, 1984.
2. Almgren, M., Grieser, F., and Thomas, J. K., Dynamic and static aspects of solubilization of neutral arenes in ionic micellar solutions, *J. Am. Chem. Soc.*, 101, 279, 1979.
3. Cohen, A. S., Terabe, S., Smith, J. A., and Karger, B. L., High-performance capillary electrophoretic separation of bases, nucleosides, and oligonucleotides: retention manipulation via micellar solutions and metal additives, *Anal. Chem.*, 59, 1021, 1987.
4. Nishi, H., Tsumagari, N., and Terabe, S., Effect of tetraalkylammonium salts on micellar electrokinetic chromatography, *Anal. Chem.*, 61, 2434, 1989.
5. Janini, G. M. and Issaq, H. J., Micellar electrokinetic capillary chromatography — basic considerations and current trends, *J. Liq. Chromatogr.*, 15, 927, 1992.

6. Little, E. L. and Foley, J. P., Optimization of the resolution of PTH-amino acids through control of surfactant concentration in micellar electrokinetic capillary chromatography — SDS vs. Brij 35/SDS micellar systems, *J. Microcol. Sep.,* 4, 145, 1992.

7. Okafo, G. N., Bintz, C., Clarke, S. E., and Camilleri, P., Micellar electrokinetic capillary chromatography in a mixture of taurodeoxycholic acid and beta-cyclodextrin, *J. Chem. Soc. Chem. Commun.,* 1189, 1992.

8. Saitoh, T., Ojima, N., Hoshino, H., and Yotsuyanagi, T., Solute partition study in aqueous sodium dodecyl sulfate micellar solutions for some organic reagents and metal chelates, *Mikrochim. Acta,* 106, 91, 1992.

9. Crosby, D. and El Rassi, Z., Micellar electrokinetic capillary chromatography with cationic surfactants, *J. Liq. Chromatogr.,* 16, 2161, 1993.

10. Terabe, S., Selectivity manipulation in micellar electrokinetic chromatography, *J. Pharmaceut. Biomed. Anal.,* 10, 705, 1992.

11. Ahuja, E. S., Preston, B. P., and Foley, J. P., Anionic/zwitterionic mixed micelles in micellar electrokinetic chromatography: sodium dodecyl sulfate/N-dodecyl-N,N-dimethylammonium-3-propane-1-sulfonic acid, *J. Chromatogr.,* 657, 271, 1994.

12. Smith, J. T. and El Rassi, Z., Micellar electrokinetic capillary chromatography with in situ charged micelles. 3. Evaluation of alkylglucoside surfactants as anionic butylboronate complexes, *Electrophoresis,* 15, 1248, 1994.

13. Rasmussen, H. T., Goebel, L. K., and McNair, H. M., Micellar electrokinetic chromatography employing sodium alkyl sulfates and Brij 35, *J. Chromatogr.,* 517, 549, 1990.

14. Wallingford, R. A., Curry, P. D., Jr., and Ewing, A. G., Retention of catechols in capillary electrophoresis with micellar and mixed micellar solutions, *J. Microcol. Sep.,* 1, 23, 1989.

15. Nishi, H., Fukuyama, T., Matsuo, M., and Terabe, S., Separation and determination of lipophilic corticosteroids and benzothiazepin analogs by micellar electrokinetic chromatography using bile salts, *J. Chromatogr.,* 513, 279, 1990.

16. Nishi, H., Fukuyama, T., Matsuo, M., and Terabe, S., Chiral separation of trimetoquinol hydrochloride and related compounds by micellar electrokinetic chromatography using sodium taurodeoxycholate solutions and application to optical purity determination, *Anal. Chim. Acta,* 236, 281, 1990.

17. Terabe, S., Shibata, O., and Miyashita, Y., Separation of enantiomers using bile salts in micellar electrokinetic chromatography, *J. Chromatogr.,* 480, 403, 1989.

18. Kuhn, R. and Hoffstetterkuhn, S., Chiral separation by capillary electrophoresis, *Chromatographia,* 34, 505, 1992.

19. Terabe, S., Ozaki, H., Otsuka, K., and Ando, T., Electrokinetic chromatography with 2-O-carboxymethyl-B-cyclodextrin as a moving "stationary" phase, *J. Chromatogr.,* 332, 211, 1985.

20. Sepaniak, M. J., Copper, C. L., Whitaker, K. W., and Anigbogu, V. C., Evaluation of a dual-cyclodextrin phase variant of capillary electrokinetic chromatography for separations of non-ionizable solutes, *Anal. Chem.,* 67, 2037, 1995.

21. Tanaka, N., Fukutome, T., Tanigawa, T., Hosoya, K., Kimata, K., Araki, T., and Unger, K. K., Structural selectivity provided by starburst dendrimers as pseudostationary phase in electrokinetic chromatography, *J. Chromatogr. A,* 699, 331, 1995.

22. Terabe, S. and Isemura, T., Effect of polymer ion concentrations on migration velocities in ion-exchange electrokinetic chromatography, *J. Chromatogr.,* 515, 667, 1990.

23. Terabe, S. and Isemura, T., Ion-exchange electrokinetic chromatography with polymer ions for the separation of isomeric ions having identical electrophoretic mobilities, *Anal. Chem.,* 62, 650, 1990.

24. Terabe, S., Ozaki, H., and Tanaka, Y., New pseudostationary phases for electrokinetic chromatography: a high-molecular surfactant and proteins, *J. Chin. Chem. Soc.,* 41(3), 251, 1994.

25. Bachmann, K., Bazzanella, A., Haag, I., Han, K. Y., Arnecke, R., Bohmer, V., and Vogt, W., Resorcarenes as pseudostationary phases with selectivity for electrokinetic chromatography, *Anal. Chem.,* 67, 1722, 1995.

26. Watarai, H., Microemulsion capillary electrophoresis, *Chem. Lett.,* 3, 391, 1991.

27. Bachmann, K., Gottlicher, B., Haag, I., Han, K. Y., Hensel, W., and Mainka, A., Capillary electrokinetic chromatography with a suspension of chromatographic particles, *J. Chromatogr. A,* 688, 283, 1994.

28. Giddings, J. C., *Unified Separation Science*, 1st ed., John Wiley & Sons, Inc., New York, 1991, 320.
29. Snyder, L. R. and Kirkland, J. J., *Introduction to Modern Liquid Chromatography*, Second Edition ed., John Wiley & Sons, Inc., New York, 1979,
30. Foley, J. P., Resolution equations for column chromatography, *Analyst*, 116, 1275, 1991.
31. Bidlingmeyer, B. A. and F.V. Warren, J., Column efficiency measurement, *Anal. Chem.*, 56, 1583A, 1984.
32. Jeansonne, M. S. and Foley, J. P., Review of exponentially modified Gaussian (EMG) function since 1983, *J. Chromatogr. Sci.*, 29, 258, 1991.
33. Foley, J. P. and Dorsey, J. G., Equations for calculation of chromatographic figures of merit for ideal and skewed peaks, *Anal. Chem.*, 55, 730, 1983.
34. Jeansonne, M. S. and Foley, J. P., Improved equations for calculation of chromatographic figures of merit for ideal and skewed peaks, *J. Chromatogr.*, 594, 1, 1992.
35. Delaguardia, M., Periscardells, E., Sancenon, J., Carrion, J. L., and Pramauro, E., Fluorimetric determination of binding constants between micelles and chemical systems, *Microchem. J.*, 44, 193, 1991.
36. Foley, J. P., Critical compilation of solute-micelle binding constants and related parameters from micellar liquid chromatographic measurements, *Anal. Chim. Acta*, 231, 237, 1990.
37. Kord, A. S., Strasters, J. K., and Khaledi, M. G., Comparative study of the determination of solute-micelle binding constants by micellar liquid chromatography and micellar electrokinetic capillary chromatography, *Anal. Chim. Acta*, 246, 131, 1991.
38. Jorgenson, J. W. and Lukacs, K. D., Zone electrophoresis in open tubular glass capillaries, *Anal. Chem.*, 53, 1298, 1981.
39. Jorgenson, J. W. and Lukacs, K. D., Zone electrophoresis in open-tubular glass capillaries: preliminary data on performance, *Hrc Cc, J. High Resolut. Chromatogr. Chromatogr. Commun.*, 4, 230, 1981.
40. Terabe, S., Otsuka, K., and Ando, T., Electrokinetic chromatography with micellar solution and open tubular capillary, *Anal. Chem.*, 57, 834, 1985.
41. Balchunas, A. T. and Sepaniak, M. J., Extension of elution range in micellar electrokinetic capillary chromatography, *Anal. Chem.*, 59, 1466, 1987.
42. Otsuka, K. and Terabe, S., Effect of pH on electrokinetic velocities in micellar electrokinetic chromatography, *J. Microcol. Sep.*, 1, 150, 1989.
43. Cai, J. and El Rassi, Z., Micellar electrokinetic capillary chromatography of neutral solutes with micelles of adjustable surface charge density, *J. Chromatogr.*, 608, 31, 1992.
44. Kaneta, T., Tanaka, S., Taga, M., and Yoshida, H., Effect of addition of glucose on micellar electrokinetic capillary chromatography with sodium dodecyl sulphate, *J. Chromatogr.*, 609, 369, 1992.
45. Nielsen, K. R. and Foley, J. P., Effect of the dodecyl counter ion on selectivity and resolution in micellar electrokinetic capillary chromatography: II. Organic counterions, *J. Microcol. Separ.*, 6, 139, 1994.
46. Ahuja, E. S., Little, E. L., Nielsen, K. R., and Foley, J. P., Infinite elution range in micellar electrokinetic capillary chromatography using a nonionic/anionic mixed micellar system, *Anal. Chem.*, 67, 26, 1995.
47. Wright, P. B. and Dorsey, J. G., Silver(I) mediated separations by capillary zone electrophoresis and micellar electrokinetic chromatography: Argentation electrophoresis, *Anal. Chem.*, 68, 415, 1996.
48. Rasmussen, H. T., Gobel, L. K., and McNair, H. M., Optimization of resolution in micellar electrokinetic capillary chromatography, *J. HRC&CC*, 14, 25, 1991.
49. Ghowsi, K., Foley, J. P., and Gale, R. J., Micellar electrokinetic capillary chromatography theory based on electrochemical parameters — optimization for 3 modes of operation, *Anal. Chem.*, 62, 2714, 1990.
50. Michaelsen, S., Moller, P., and Sorensen, H., Analysis of dansyl amino acids in feedstuffs and skin by micellar electrokinetic capillary chromatography, *J. Chromatogr., A*, 680, 299, 1994.
51. Sepaniak, M. J. and Cole, R. O., Column efficiency in micellar electrokinetic capillary chromatography, *Anal. Chem.*, 59, 472, 1987.

52. Terabe, S., Otsuka, K., and Ando, T., Band broadening in electrokinetic chromatography with micellar solutions and open-tubular capillaries, *Anal. Chem.,* 61, 251, 1989.

53. Balchunas, A. T., Swaile, D. F., Powell, A. C., and Sepaniak, M. J., Separations of compound of biological and environmental interest by micellar electrokinetic capillary chromatography, *Sep. Sci. Tech.,* 23, 1891, 1988.

54. Pyell, U. and Butehorn, U., Optimization strategies in micellar electrokinetic capillary chroma-tography. Optimization of the temperature of the separation capillary, *Chromatographia,* 40, 69, 1995.

55. Davis, J. M., Random walk theory of nonequilibrium plate height in micellar electrokinetic capillary chromatography, *Anal. Chem.,* 61, 2455, 1989.

56. Mclaughlin, G. M., Nolan, J. A., Lindahl, J. L., Palmieri, R. H., Anderson, K. W., Morris, S. C., Morrison, J. A., and Bronzert, T. J., Pharmaceutical drug separations by HPCE — practical guidelines, *J. Liq. Chromatogr.,* 15, 961, 1992.

57. Shihabi, Z. K. and Hinsdale, M. E., Sample matrix effects in micellar electrokinetic capillary electrophoresis, *J. Chromatogr. B-Bio. Med. Appl.,* 669, 75, 1995.

58. Burton, D. E., Sepaniak, M. J., and Maskarinec, M. P., The effect of injection procedures on efficiency in micellar electrokinetic capillary chromatography, *Chromatographia,* 21, 583, 1986.

59. Szucs, R., Vindevogel, J., Sandra, P., and Verhagen, L. C., Sample stacking effects and large injection volumes in micellar electrokinetic chromatography of ionic compounds — direct determination of iso-alpha-acids in beer, *Chromatographia,* 36, 323, 1993.

60. Nielsen, K. R. and Foley, J. P., Zone sharpening of neutral solutes in micellar electrokinetic chromatography with electrokinetic injection, *J. Chromatogr.,* 686, 283, 1994.

61. Otsuka, K. and Terabe, S., Extra-column effects in high-performance capillary electrophoresis, *J. Chromatogr.,* 480, 91, 1989.

62. Terabe, S., Shibata, O., and Isemura, T., Band broadening evaluation by back-and-forth capillary electrophoresis, *J. High Resolut. Chromatogr.,* 14, 52, 1991.

63. Burton, D. E., Sepaniak, M. J., and Maskarinec, M. P., Evaluation of the use of various surfactants in micellar electrokinetic capillary chromatography, *J. Chromatogr. Sci.,* 25, 514, 1987.

64. Fujiwara, S. and Honda, S., Determination of ingredients of antipyretic analgesic preparations by micellar electrokinetic capillary chromatography, *Anal. Chem.,* 59, 2773, 1987.

65. Fujiwara, S., Iwase, S., and Honda, S., Analysis of water-soluble vitamins by micellar electro-kinetic capillary chromatography, *J. Chromatogr.,* 447, 133, 1988.

66. Nielsen, K. R. and Foley, J. P., Effect of the dodecyl counter ion on selectivity and resolution in micellar electrokinetic capillary chromatography: SDS vs. Mg(DS)2, *J. Microcol. Sep.,* 5, 347, 1993.

67. Gorse, J., Balchunas, A. T., Swaile, D. F., and Sepaniak, M. J., Effects of organic mobile phase modifiers in micellar electrokinetic capillary chromatography, *J. HRC & CC,* 11, 554, 1988.

68. Morin, P. and Dreux, M., Factors influencing the separation of ionic and non-ionic chemical natural compounds in plant extracts by capillary electrophoresis, *J. Liq. Chromatogr.,* 16, 3735, 1993.

69. Row, K. H., Griest, W. H., and Maskarinec, M. P., Separation of modified nucleic acid constit-uents by micellar electrokinetic capillary chromatography, *J. Chromatogr.,* 409, 193, 1987.

70. Foley, J. P., Optimization of micellar electrokinetic chromatography, *Anal. Chem.,* 62, 1302, 1990.

71. Karger, B. L., Snyder, L. R., and Horvath, C., *Introduction to Separation Science,* John Wiley & Sons, Inc., New York, 1973, Chapter 5.

72. Sepaniak, M. J., Swaile, D. F., Powell, A. C., and Cole, R. O., Capillary electrokinetic separa-tions — influence of mobile phase composition on performance, *J. HRC & CC,* 13, 679, 1990.

73. Balchunas, A. T. and Sepaniak, M. J., Gradient elution for micellar electrokinetic capillary chromatography, *Anal. Chem.,* 60, 617, 1988.

74. Palmer, C. P., Khaled, M. Y., and McNair, H. M., Micellar electrokinetic chromatography with a polymerized micelle, *HRC-J. High Res. Chromatogr.,* 15, 756, 1992.

75. Wang, J. A. and Warner, I. M., Chiral separations using micellar electrokinetic capillary chro-matography and a polymerized chiral micelle, *Anal. Chem.,* 66, 3773, 1994.

76. Ahuja, E. S. and Foley, J. P., Influence of dodecyl sulfate counterion on efficiency, selectivity, retention, elution range, and resolution in micellar electrokinetic chromatography, *Anal. Chem.*, 67, 2315, 1995.

77. Walbroehl, Y. and Jorgenson, J. W., Capillary zone electrophoresis of neutral organic molecules by solvophobic association with tetraalkylammonium ion, *Anal. Chem.*, 58, 479, 1986.

78. Ahuja, E. S. and Foley, J. P., Separation of very hydrophobic compounds by hydrophobic interaction electrokinetic chromatography, *J. Chromatogr. A*, 680, 73, 1994.

79. Muijselaar, P. G. H. M., Claessens, H. A., and Cramers, C. A., Parameters controlling the elution window and retention factors in micellar electrokinetic capillary chromatography, *J. Chromatogr. A*, 696, 273, 1995.

79a. Janini, G.M., Muschik, G.M., and Issaq, H.J., Micellar electrokinetic chromatography in zero-electroosmotic flow environment, *J. Chromatogr. B*, 683, 29, 1996.

80. Ye, B., Hadjmohammadi, M., and Khaledi, M. G., Selectivity control in micellar electrokinetic chromatography of small peptides using mixed fluorocarbon hydrocarbon anionic surfactants, *J. Chromatogr. A*, 692, 291, 1995.

81. Bumgarner, J. G. and Khaledi, M. G., Mixed micellar electrokinetic chromatography of corticosteroids, *Electrophoresis*, 15, 1260, 1994.

82. Hjertèn, S. and Kubo, K., A new type of pH-stable and detergent-stable coating for elimination of electroendoosmosis and adsorption in (capillary) electrophoresis, *Electrophoresis*, 14, 390, 1993.

83. Tsai, P., Patel, B., and Lee, C. S., Micellar electrokinetic chromatography with electroosmotic step gradient elution, *Electrophoresis*, 15, 1229, 1994.

84. Tsai, P., Patel, B., and Lee, C. S., Direct control of electroosmosis and retention window in micellar electrokinetic capillary chromatography, *Anal. Chem.*, 65, 1439, 1993.

85. Yang, S. Y. and Khaledi, M. G., Linear solvation energy relationships in micellar liquid chromatography and micellar electrokinetic capillary chromatography, *J. Chromatogr. A*, 692, 301, 1995.

86. Yang, S. Y. and Khaledi, M. G., Chemical selectivity in micellar electrokinetic chromatography: Characterization of solute micelle interactions for classification of surfactants, *Anal. Chem.*, 67, 499, 1995.

87. Giddings, J. C., *Dynamics of Chromatography Part I Principles and Theory*, Marcel Dekker, New York, 1965.

88. Giddings, J. C., Generation of variance, "theoretical plates," resolution, and peak capacity in electrophoresis and sedimentation, *Sep. Sci.*, 4, 181, 1969.

89. Yu, L. X. and Davis, J. M., Study of high-field dispersion in micellar electrokinetic chromatography, *Electrophoresis*, 16, 2104, 1995.

90. Liu, Z. Y., Sam, P., Sirimanne, S. R., McClure, P. C., Grainger, J., and Patterson, D. G., Field-amplified sample stacking in micellar electrokinetic chromatography for on-column sample concentration of neutral molecules, *J. Chromatogr. A*, 673, 125, 1994.

91. Reijenga, J. C. and Kenndler, E., Computational simulation of migration and dispersion in free capillary zone electrophoresis. 2. Results of simulation and comparison with measurements, *J. Chromatogr. A*, 659, 417, 1994.

92. Vinther, A. and Soeeberg, H., Mathematical model describing dispersion in free solution capillary electrophoresis under stacking conditions, *J. Chromatogr.*, 559, 3, 1991.

93. Cohen, N. and Grushka, E., Influence of the capillary edge on the separation efficiency in capillary electrophoresis, *J. Chromatogr. A*, 684, 323, 1994.

94. Lin, B. C., Xu, X., and Luo, G. A., Zone broadening and simulation of migration process of peptides in capillary zone electrophoresis, *Sci. China Ser. B*, 37, 807, 1994.

95. Moore, A. W. and Jorgenson, J. W., Study of zone broadening in optically gated high-speed capillary electrophoresis, *Anal. Chem.*, 65, 3550, 1993.

96. Vindevogel, J. and Sandra, P., Resolution optimization in micellar electrokinetic chromatography — Use of Plackett-Burman statistical design for the analysis of testosterone esters, *Anal. Chem.*, 63, 1530, 1991.

97. Castagnola, M., Rossetti, D. V., Cassiano, L., Rabino, R., Nocca, G., and Giardina, B., Optimization of phenylthiohydantoinamino acid separation by micellar electrokinetic capillary chromatography, *J. Chromatogr.*, 638, 327, 1993.

98. Berridge, J. C., *Techniques for the Automated Optimization of HPLC Separations,* John Wiley & Sons, New York, 1985, 204.

99. Schoenmakers, P. J., *Optimization of Chromatographic Selectivity: Guide to Method Development,* Elsevier, Amsterdam, 1986, 345.

100. Kaneta, T., Tanaka, S., Taga, M., and Yoshida, H., Migration behavior of inorganic anions in micellar electrokinetic capillary chromatography using a cationic surfactant, *Anal. Chem.,* 64, 798, 1992.

101. Okada, T. and Shimizu, H., Retention mechanism of anions in micellar chromatography: Interpretation of retention data on the basis of an ion-exchange model, *J. Chromatogr. A,* 706, 37, 1995.

102. Okada, T., Interpretation of retention behaviors of transition-metal cations in micellar chromatography using an ion-exchange model, *Anal. Chem.,* 64, 589, 1992.

103. Khaledi, M. G., Smith, S. C., and Strasters, J. K., Micellar electrokinetic capillary chromatography of acidic solutes — Migration behavior and optimization strategies, *Anal. Chem.,* 63, 1820, 1991.

104. Strasters, J. K. and Khaledi, M. G., Migration behavior of cationic solutes in micellar electrokinetic capillary chromatography, *Anal. Chem.,* 63, 2503, 1991.

105. Quang, C. Y., Strasters, J. K., and Khaledi, M. G., Computer-assisted modeling, prediction, and multifactor optimization in micellar electrokinetic chromatography of ionizable compounds, *Anal. Chem.,* 66, 1646, 1994.

106. Foley, J. P. and May, W. E., Optimization of secondary chemical equilibria in liquid chromatography: theory and verification, *Anal. Chem.,* 59, 102, 1987.

107. Horvath, C., Melander, W., and Molnar, I., Liquid chromatography of iongenic substances with nonpolar stationary phases, *Anal. Chem.,* 49, 142, 1977.

108. Smith, S. C. and Khaledi, M. G., Prediction of the migration behavior of organic acids in micellar electrokinetic chromatography, *J. Chromatogr.,* 632, 177, 1993.

109. Khaledi, M. G. and Rodgers, A. H., Micellar mediated shifts of ionization constants of amino acids and peptides, *Anal. Chim. Acta,* 239, 121, 1990.

110. Foley, J. P., unpublished work.

111. Foley, J. P. and Ahuja, E. S., Electrokinetic chromatography, in *Pharmaceutical and Biomedical Applications of Capillary Electrophoresis,* 2, Lunte, S. M., Radzik, D. M., Eds., Pergamon Press, New York, 1996, 81.

112. Khaledi, M. G., personal communication, 1992.

113. Terabe, S., personal communication, 1992.

114. Foley, J. P., Resolving power of micellar electrokinetic chromatography and electrochromatography for charged solutes, manuscript in preparation.

115. Snopek, J., Jelinek, I., and Smolkova-Keulemansova, E., Micellar, inclusion, and metal-complex enantioselective pseudophases in high performance electromigration methods, *J. Chromatogr.,* 452, 571, 1988.

116. Hinze, W. L. and Armstrong, D. W., Eds., *Ordered Media in Chemical Separations,* vol. 342, American Chemical Society, Washington D.C., 1987.

117. Onori, G. and Santucci, A., Effect of temperature and solvent on the critical micelle concentration of sodium dodecylsulfate, *Chem. Phys. Lett.,* 189, 598, 1992.

117a. Seifar, R.M., Kraak, J.C., and Kok, W.T., Mechanism of electrokinetic separations of hydrophobic compounds with sodium docecyl sulfate in acetonitrile-water mixtures, *Anal. Chem.,* 69, 2772, 1997.

118. Menger, F. M., On the structure of micelles, *Acc. Chem. Res.,* 12, 111, 1979.

119. Jansson, M., Jonsson, A., Li, P. Y., and Stilbs, P., Aggregation in tetraalkylammonium dodecanoate systems, *Colloid Surface,* 59, 387, 1991.

120. Egorov, V. V. and Ksenofontova, O. B., Influence of various factors on size of micelles of cationic monomeric surfactants in water and toluene, *Colloid J. USSR-Engl. Tr.,* 53, 537, 1991.

121. Devijlder, M., On the specific effectiveness of added ions to induce changes in the structure of sodium dodecyl sulfate micelles, *Z. Phys. Chem.,* 174, 119, 1991.

122. Chung, J. J., Lee, S. W., and Choi, J. H., Salt effects on the critical micelle concentration and counterion binding of cetylpyridinium bromide micelles, *Bull. Kor. Chem. Soc.,* 12, 411, 1991.

123. Nusselder, J. J. H. and Engberts, J. B. F. N., Relation between surfactant structure and properties of spherical micelles — 1-alkyl-4-alkylpyridinium halide surfactants, *Langmuir,* 7, 2089, 1991.
124. Sugihara, G., Nagadome, S., Yamashita, T., Kawachi, N., Takagi, H., and Moroi, Y., Cationic surfactants with perfluorocarboxylates as counterion — Solubility and micelle formation, *Colloid Surface,* 61, 111, 1991.
125. Akisada, H., Kinoshita, S., and Wakita, H., The behavior of organic counterions in the micelle of alkylammonium decylsulfonates, *Colloid Surface,* 66, 121, 1992.
126. Berr, S., Jones, R. R. M., and Johnson, J. S., Effect of counterion on the size and charge of alkyltrimethylammonium halide micelles as a function of chain length and concentration as determined by small-angle neutron scattering, *J. Phys. Chem.,* 96, 5611, 1992.
127. Devijlder, M., On the specific influence of monovalent and divalent counterions on the micellar properties of an azo dye, *Z. Phys. Chem.,* 175, 109, 1992.
128. Gu, T. R. and Sjoblom, J., Surfactant structure and its relation to the Krafft point, cloud point and micellization — Some empirical relationships, *Colloid Surface,* 64, 39, 1992.
129. Sjoberg, M., Jansson, M., and Henriksson, U., A comparison of the counterion binding to ionic micelles in aqueous and nonaqueous systems, *Langmuir,* 8, 409, 1992.
129a. Peterson, A.G. and Foley, J.P., Influence of the inorganic counterion on the chiral micellar electrokinetic separation of basic drugs using the surfactant N-dodecoxycarbonylvaline, *J. Chromatogr. B.,* 695, 131, 1997.
130. Aamodt, M., Landgren, M., and Jonsson, B., Solubilization of uncharged molecules in ionic surfactant aggregates. 1. The micellar phase, *J. Phys. Chem.,* 96, 945, 1992.
131. Huang, J. F. and Bright, F. V., Microheterogeneity of sodium dodecylsufate micelles probed by frequency-domain fluorometry, *Appl. Spectrosc.,* 46, 329, 1992.
132. Weers, J. G., Solubilization in mixed micelles, *J. Am. Oil Chem. Soc.,* 67, 340, 1990.
133. Vethamuthu, M. S., Almgren, M., Mukhtar, E., and Bahadur, P., Fluorescence quenching studies of the aggregation behavior of the mixed micelles of bile salts and cetyltrimethylammonium halides, *Langmuir,* 8, 2396, 1992.
134. Qi, W. B. and Kang, J. T., The relationship between micellar sensitization and the structure of the organic chromogenic reagent. 2. Comparison of cadion with cadion 2B, *Acta Chim. Sin.,* 50, 32, 1992.
135. Pietta, P. G., Bruno, A., Mauri, P. L., and Gardana, C., Determination of sunscreen agents by micellar electrokinetic chromatography, *J. Pharmaceut. Biomed. Anal.,* 13, 229, 1995.
136. Nishi, H., Fukuyama, T., Matsuo, M., and Terabe, S., Effect of surfactant structures on the separation of cold medicine ingredients by micellar electrokinetic chromatography, *J. Pharm. Sci.,* 79, 519, 1990.
137. Ong, C. P., Ng, C. L., Lee, H. K., and Li, S. F. Y., Separation of Dns-amino acid and vitamins by micellar electrokinetic chromatography, *J. Chromatogr.,* 559, 537, 1991.
138. Takeda, S., Wakida, S., Yamane, M., Kawahara, A., and Higashi, K., Migration behavior of phthalate esters in micellar electrokinetic chromatography with or without added methanol, *Anal. Chem.,* 65, 2489, 1993.
139. Ong, C. P., Pang, S. F., Low, S. P., Lee, H. K., and Li, S. F. Y., Migration behavior of catechols and catecholamines in capillary electrophoresis, *J. Chromatogr.,* 559, 529, 1991.
140. Ahuja, E. S. and Foley, J. P., A retention index for micellar electrokinetic chromatography, *Analyst,* 119, 353, 1994.
141. Bjergegaard, C., Michaelsen, S., and Sorensen, H., Determination of phenolic carboxylic acids by micellar electrokinetic capillary chromatography and evaluation of factors affecting the method, *J. Chromatogr.,* 608, 403, 1992.
142. Lukkari, P., Siren, H., Pantsar, M., and Riekkola, M. L., Determination of 10 beta-blockers in urine by micellar electrokinetic capillary chromatography, *J. Chromatogr.,* 632, 143, 1993.
143. Bjergegaard, C., Michaelsen, S., Mortensen, K., and Sorensen, H., Determination of flavonoids by micellar electrokinetic capillary chromatography, *J. Chromatogr. A,* 652, 477, 1993.
144. Ramsey, R. S., Kerchner, G. A., and Cadet, J., Micellar electrokinetic capillary chromatography of nucleic acid constituents and dinucleoside monophosphate photoproducts, *HRC-J. High Res. Chromatogr.,* 17, 4, 1994.
145. Muijselaar, P. G. H. M., Claessens, H. A., and Cramers, C. A., Application of the retention index concept in micellar electrokinetic capillary chromatography, *Anal. Chem.,* 66, 635, 1994.

146. Hansen, S. H., Bjornsdottir, I., and Tjornelund, J., Separation of basic drug substances of very similar structure using micellar electrokinetic chromatography, *J. Pharmaceut. Biomed. Anal.,* 13, 489, 1995.

147. Croubels, S., Baeyens, W., Dewaele, C., and Vanpeteghem, C., Capillary electrophoresis of some tetracycline antibiotics, *J. Chromatogr. A,* 673, 267, 1994.

148. Qin, X. Z., Nguyen, D. S. T., and Ip, D. P., Separation of lisinopril and its RSS diastereoisomer by micellar electrokinetic chromatography, *J. Liq. Chromatogr.,* 16, 3713, 1993.

149. Cole, R. O., Sepaniak, M. J., Hinze, W. L., Gorse, J., and Oldiges, K., Bile salt surfactants in micellar electrokinetic capillary chromatography: application to hydrophobic molecule separations, *Report, Doe/er,* 7865, 1990.

150. Harman, A. D., Kibbey, R. G., Sablik, M. A., Fintschenko, Y., Kurtin, W. E., and Bushey, M. M., Micellar electrokinetic capillary chromatography analysis of the behavior of bilirubin in micellar solutions, *J. Chromatogr. A,* 652, 525, 1993.

151. Boonkerd, S., Lauwers, M., Detaevernier, M. R., and Michotte, Y., Separation and simultaneous determination of the components in an analgesic tablet formulation by micellar electrokinetic chromatography, *J. Chromatogr. A,* 695, 97, 1995.

152. Tanaka, M., Ishida, T., Araki, T., Masuyama, A., Nakatsuji, Y., Okahara, M., and Terabe, S., Double-chain surfactant as a new and useful micelle-forming reagent for micellar electrokinetic chromatography, *J. Chromatogr.,* 648, 469, 1993.

153. Otsuka, K. and Terabe, S., Enantiomeric resolution by micellar electrokinetic chromatography with chiral surfactants, *J. Chromatogr.,* 515, 221, 1990.

154. Ishihama, Y. and Terabe, S., Enantiomeric separation by micellar electrokinetic chromatography using saponins, *J. Liq. Chromatogr.,* 16, 933, 1993.

155. Nishi, H. and Terabe, S., Application of electrokinetic chromatography to pharmaceutical analysis, *J. Pharmaceut. Biomed. Anal.,* 11, 1277, 1993.

156. Mazzeo, J. R., Grover, E. R., Swartz, M. E., and Petersen, J. S., Novel chiral surfactant for the separation of enantiomers by micellar electrokinetic capillary chromatography, *J. Chromatogr., A,* 680, 125, 1994.

157. Swartz, M. E., Mazzeo, J. R., Grover, E. R., and Brown, P. R., Separation of amino acid enantiomers by micellar electrokinetic capillary chromatography using synthetic chiral surfactants, *Anal. Biochem.,* 231, 65, 1995.

158. Swartz, M. E., Mazzeo, J. R., Grover, E. R., Brown, P. R., and Aboulenein, H. Y., Separation of piperidine-2,6-dione drug enantiomers by micellar electrokinetic capillary chromatography using synthetic chiral surfactants, *J. Chromatogr. A,* 724, 307, 1996.

159. Peterson, A. G., Ahuja, E. S., and Foley, J. P., Enantiomeric separations of basic pharmaceutical drugs by micellar electrokinetic chromatography using a chiral surfactant, N-dodecoxycarbonylvaline, *J. Chromatogr. B,* 683, 15, 1996.

160. Peterson, A. G. and Foley, J. P., Determining thermodynamic quantities of micellar solubilization of chiral pharmaceutical compounds in aqueous solutions of N-dodecoxycarbonylvaline using micellar electrokinetic chromatography, *J. Microcol. Sep.,* 8, 427, 1996.

161. Nishi, H. and Matsuo, M., Separation of corticosteroids and aromatic hydrocarbons by cyclodextrin-modified micellar electrokinetic chromatography, *J. LC,* 14, 973, 1991.

162. Yik, Y. F., Ong, C. P., Khoo, S. B., Lee, H. K., and Li, S. F. Y., Separation of polycyclic aromatic hydrocarbons by micellar electrokinetic chromatography with cyclodextrins as modifiers, *J. Chromatogr.,* 589, 333, 1992.

163. Ueda, T., Mitchell, R., Kitamura, F., Metcalf, T., Kuwana, T., and Nakamoto, A., Separation of naphthalene-2,3-dicarboxaldehyde-labeled amino acids by high-performance capillary electrophoresis with laser-induced fluorescence detection, *J. Chromatogr.,* 593, 265, 1992.

164. Ueda, T., Kitamura, F., Mitchell, R., Metcalf, T., Kuwana, T., and Nakamoto, A., Chiral Separation of Naphthalene-2,3-Dicarboxaldehyde-Labeled Amino Acid Enantiomers by Cyclodextrin-Modified Micellar Electrokinetic Chromatography with Laser-Induced Fluorescence Detection, *Anal. Chem.,* 63, 2979, 1991.

165. Nishi, H., Fukuyama, T., and Terabe, S., Chiral separation by cyclodextrin-modified micellar electrokinetic chromatography, *J. Chromatogr.,* 553, 503, 1991.

166. Kuhn, R., Stoecklin, F., and Erni, F., Chiral separations by host-guest complexation with cyclodextrin and crown ether in capillary zone electrophoresis, *Chromatographia,* 33, 32, 1992.

167. Hohne, E., Krauss, G. J., and Gubitz, G., Capillary zone electrophoresis of the enantiomers of aminoalcohols based on host-guest complexation with a chiral crown ether, *HRC-J. High Res. Chromatogr.*, 15, 698, 1992.

168. Kuhn, R., Erni, F., Bereuter, T., and Hausler, J., Chiral recognition and enantiomeric resolution based on host guest complexation with crown ethers in capillary zone electrophoresis, *Anal. Chem.*, 64, 2815, 1992.

169. Kuhn, R., Steinmetz, C., Bereuter, T., Haas, P., and Erni, F., Enantiomeric separations in capillary zone electrophoresis using a chiral crown ether, *J. Chromatogr. A*, 666, 367, 1994.

170. Kuhn, R., Wagner, J., Walbroehl, Y., and Bereuter, T., Potential and limitations of an optically active crown ether for chiral separation in capillary zone electrophoresis, *Electrophoresis*, 15, 828, 1994.

171. Walbroehl, Y. and Wagner, J., Chiral separations of amino acids by capillary electrophoresis and high-performance liquid chromatography employing chiral crown ethers, *J. Chromatogr. A*, 685, 321, 1994.

172. Walbroehl, Y. and Wagner, J., Enantiomeric resolution of primary amines by capillary electrophoresis and high-performance liquid chromatography using chiral crown ethers, *J. Chromatogr., A*, 680, 253, 1994.

173. Castelnovo, P. and Albanesi, C., Determination of the enantiomeric purity of 5,6-dihydroxy-2-aminotetralin by high-performance capillary electrophoresis with crown ether as chiral selector, *J. Chromatogr. A*, 715, 143, 1995.

174. Terabe, S., Shibata, M., and Miyashita, Y., Chiral separation by electrokinetic chromatography with bile salt micelles, *J. Chromatogr.*, 480, 413, 1989.

175. Nishi, H., Fukuyama, T., Matsuo, M., and Terabe, S., Chiral separation of diltiazem, trimetoquinol and related compounds by micellar electrokinetic chromatography with bile salts, *J. Chromatogr.*, 515, 233, 1990.

176. Nishi, H., Fukuyama, T., Matsuo, M., and Terabe, S., Separation and determination of the ingredients of a cold medicine by micellar electrokinetic chromatography with bile salts, *J. Chromatogr.*, 498, 313, 1990.

177. Cole, R. O., Sepaniak, M. J., Hinze, W. L., Gorse, J., and Oldiges, K., Bile salt surfactants in micellar electrokinetic capillary chromatography — Application to hydrophobic molecule separations, *J. Chromatogr.*, 557, 113, 1991.

178. Lin, M., Wu, N. A., Barker, G. E., Sun, P., Huie, C. W., and Hartwick, R. A., Enantiomeric separation by cyclodextrin-modified micellar electrokinetic chromatography using bile salt, *J. Liq. Chromatogr.*, 16, 3667, 1993.

178a. Linhares, M.C. and Kissinger, P.T., Pharmacokinetic studies using micellar electrokinetic capillary chromatography with in vivo capillary ultrafiltration probes, *J. Chromatogr.*, 615, 327, 1993.

179. Boonkerd, S., Detaevernier, M. R., Michotte, Y., and Vindevogel, J., Suppression of chiral recognition of 3-hydroxy-1,4-benzodiazepines during micellar electrokinetic capillary chromatography with bile salts, *J. Chromatogr. A*, 704, 238, 1995.

180. Aumatell, A. and Wells, R. J., Determination of a cardiac antiarrhythmic, tricyclic antipsychotics and antidepressants in human and animal urine by micellar electrokinetic capillary chromatography using a bile salt, *J. Chromatogr. B-Bio. Med. Appl.*, 669, 331, 1995.

181. Clothier, J. G. and Tomellini, S. A., Chiral separation of verapamil and related compounds using micellar electrokinetic capillary chromatography with mixed micelles of bile salt and polyoxyethylene ethers, *J. Chromatogr. A*, 723, 179, 1996.

182. Cole, R. O. and Sepaniak, M. J., The use of bile salt surfactants in micellar electrokinetic capillary chromatography, *LC GC-Mag. Sep. Sci.*, 10, 380, 1992.

183. Gassmann, E., Kuo, J. E., and Zare, R. N., Electrokinetic separations of chiral compounds, *Science*, 230, 813, 1985.

184. Gozel, P., Gassman, E., Michelson, H., and Zare, R. N., Electrokinetic resolution of amino acid enantiomers with copper(II)-aspartame support electrolyte, *Anal. Chem.*, 59, 44, 1987.

185. Smith, N. W. and Evans, M. B., Capillary zone electrophoresis in pharmaceutical and biomedical analysis, *J. Pharmaceut. Biomed. Anal.*, 12, 579, 1994.

186. Altria, K. D., Quantitative aspects of the application of capillary electrophoresis to the analysis of pharmaceuticals and drug related impurities, *J. Chromatogr.*, 646, 245, 1993.

187. Kuhr, W. G., Capillary electrophoresis, *Anal. Chem.,* 62, 403R, 1990.
188. Kuhr, W. G. and Monnig, C. A., Capillary electrophoresis, *Anal. Chem.,* 64, 389R, 1992.
189. Monnig, C. A. and Kennedy, R. T., Capillary electrophoresis, *Anal. Chem.,* 66, 280R, 1994.
190. St. Claire, R. L., Capillary electrophoresis, *Anal. Chem.,* 68, 569R, 1996.
191. Sandra, P. and Vindevogel, J., *Introduction to Micellar Electrokinetic Chromatography,* Hüthig, Heidelberg, 1992.
192. Chan, K. C., Muschik, G. M., Issaq, H. J., and Snader, K. M., Separation of taxol and related compounds by micellar electrokinetic chromatography, *HRC-J. High Res. Chromatogr.,* 17, 51, 1994.
193. Chan, K. C., Alvarado, B., McGuire, M. T., Muschik, G. M., Issaq, H. J., and Snader, K. M., High-performance liquid chromatography and micellar electrokinetic chromatography of taxol and related taxanes from bark and needle extracts of Taxus species, *J. Chromatogr. B: Biomed. Appl.,* 657, 301, 1994.
194. Issaq, H. J., Chan, K. C., Muschik, G. M., and Janini, G. M., Applications of capillary zone electrophoresis and micellar electrokinetic chromatography in cancer research, *J. Liq. Chromatogr.,* 18, 1273, 1995.
195. Pietta, P., Mauri, P., Facino, R. M., and Carini, M., Analysis of flavonoids by MECC with ultraviolet diode array detection, *J. Pharmaceut. Biomed. Anal.,* 10, 1041, 1992.
196. Morin, P., Villard, F., and Dreux, M., Borate complexation of flavonoid-O-glycosides in capillary electrophoresis. 1. Separation of flavonoid-7-O-glycosides differing in their flavonoid aglycone, *J. Chromatogr.,* 628, 153, 1993.
197. Pietta, P., Gardana, C., and Mauri, P., Application of HPLC and MECC for the detection of flavonol aglycones in plant extracts, *HRC-J. High Res. Chromatogr.,* 15, 136, 1992.
198. Delgado, C., Tomasbarberan, F. A., Talou, T., and Gaset, A., Capillary electrophoresis as an alternative to HPLC for determination of honey flavonoids, *Chromatographia,* 38, 71, 1994.
199. Pietta, P., Bruno, A., Mauri, P., and Rava, A., Separation of flavonol-2-O-glycosides from Calendula-officinalis and Sambucus-nigra by high-performance liquid and micellar electrokinetic capillary chromatography, *J. Chromatogr.,* 593, 165, 1992.
200. Thormann, W., Minger, A., Molteni, S., Caslavska, J., and Gebauer, P., Determination of substituted purines in body fluids by micellar electrokinetic capillary chromatography with direct sample injection, *J. Chromatogr.,* 593, 275, 1992.
201. Guo, R. and Thormann, W., Acetylator Phenotyping via analysis of 4 caffeine metabolites in human urine by micellar electrokinetic capillary chromatography with multiwavelength detection, *Electrophoresis,* 14, 547, 1993.
202. Sun, P., Mariano, G. J., Barker, G., and Hartwick, R. A., Comparison of micellar electrokinetic capillary chromatography and high-performance liquid chromatography on the separation and determination of caffeine and its analogues in pharmaceutical tablets, *Anal. Lett.,* 27, 927, 1994.
203. Korman, M., Vindevogel, J., and Sandra, P., Application of micellar electrokinetic chromatography to the quality control of pharmaceutical formulations: The analysis of xanthine derivatives, *Electrophoresis,* 15, 1304, 1994.
204. Thompson, C. O., Trenerry, V. C., and Kemmery, B., Micellar electrokinetic capillary chromatographic determination of artificial sweeteners in low-Joule soft drinks and other foods, *J. Chromatogr. A,* 694, 507, 1995.
205. Lecoq, A. F., Leuratti, C., Marafante, E., and Dibiase, S., Analysis of nucleic acid derivatives by micellar electrokinetic capillary chromatography, *HRC-J. High Res. Chromatogr.,* 14, 667, 1991.
206. Lahey, A. and St. Claire, R. L., III, A comparison of ion-pairing LC and MECC in the separation of nucleosides and nucleotides, *Am. Lab.,* 22, 70, 1990.
207. Ji, A. J., Nunez, M. F., Machacek, D., Ferguson, J. E., Iossi, M. F., Kao, P. C., and Landers, J. P., Separation of urinary estrogens by micellar electrokinetic chromatography, *J. Chromatogr. B-Bio. Med. Appl.,* 669, 15, 1995.
208. Nuñez, M., Ferguson, J. E., Machacek, D., Jacob, G., Oda, R. P., Lawson, G. M., and Landers, J. P., Detection of hypoglycemic drugs in human urine using micellar electrokinetic chromatography, *Anal. Chem.,* 67, 3668, 1995.
209. Taylor, R. B. and Reid, R. G., Analysis of basic antimalarial drugs by CZE and MEKC. 1. Critical factors affecting separation, *J. Pharmaceut. Biomed. Anal.,* 11, 1289, 1993.

210. Noroski, J. E., Mayo, D. J., and Moran, M., Determination of the enantiomer of a cholesterol-lowering drug by cyclodextrin-modified micellar electrokinetic chromatography, *J. Pharmaceut. Biomed. Anal.*, 13, 45, 1995.
211. Alexander, A. J. and Hughes, D. E., Monitoring of IgG antibody thermal stability by micellar electrokinetic capillary chromatography and matrix-assisted laser desorption/ionization mass spectrometry, *Anal. Chem.*, 67, 3626, 1995.
212. Yik, Y. F., Lee, H. K., and Li, S. F. Y., Separation of nitroaromatics by micellar electrokinetic capillary chromatography, *HRC-J. High Res. Chromatogr.*, 15, 198, 1992.
213. Ong, C. P., Ng, C. L., Chong, N. C., Lee, H. K., and Li, S. F. Y., Retention of 11 Priority phenols using micellar electrokinetic chromatography, *J. Chromatogr.*, 516, 263, 1990.
214. Takeda, S., Wakida, S., Yamane, M., and Higashi, K., Analysis of lower aliphatic aldehydes in water by micellar electrokinetic chromatography with derivatization to 2,4-dinitrophenylhydra-zones, *Electrophoresis*, 15, 1332, 1994.
215. Brumley, W. C., Brownrigg, C. M., and Grange, A. H., Capillary liquid chromatography-mass spectrometry and micellar electrokinetic chromatography as complementary techniques in environmental analysis, *J. Chromatogr., A*, 680, 635, 1994.
216. Balchunas, A. T., Swaile, D. F., Powell, A. C., and Sepaniak, M. J., Separations of compounds of biological and environmental interest by micellar electrokinetic capillary chromatography, *Sep. Sci. Technol.*, 1988.
217. Yang, S., Bumgarner, J.G., and Khaledi, M.G., Separation of highly hydrophobic compounds in MEKC with an ionic polymer, *J. High Resol. Chromatogr.*, 18, 445, 1995.
218. Northrop, D. M., Martire, D. E., and MacCrehan, W. A., Separation and identification of organic gunshot and explosive constituents by micellar electrokinetic capillary electrophoresis, *Anal. Chem.*, 63, 1038, 1991.
219. Hargadon, K. A. and McCord, B. R., Explosive residue analysis by capillary electrophoresis and ion chromatography, *J. Chromatogr.*, 602, 241, 1992.
220. Northrop, D. M. and MacCrehan, W. A., Sample collection, preparation, and quantitation in the micellar electrokinetic capillary electrophoresis of gunshot residues, *J. Liq. Chromatogr.*, 15, 1041, 1992.
221. McCord, B. R., Jung, J. M., and Holleran, E. A., High resolution capillary electrophoresis of forensic DNA using a non-gel sieving buffer, *J. Liq. Chromatogr.*, 16, 1963, 1993.
222. Lurie, I. S., Klein, R. F. X., Dalcason, T. A., Lebelle, M. J., Brenneisen, R., and Weinberger, R. E., Chiral resolution of cationic drugs of forensic interest by capillary electrophoresis with mixtures of neutral and anionic cyclodextrins, *Anal. Chem.*, 66, 4019, 1994.
223. Miller, I., Gutleb, A. C., Kranz, A., and Gemeiner, M., Forensics on wild animals: Differentiation between otter and pheasant blood using electrophoretic methods, *Electrophoresis*, 16, 865, 1995.
224. Mussenbrock, E. and Kleibohmer, W., Separation strategies for the determination of residues of explosives in soils using micellar electrokinetic capillary chromatography, *J. Microcol. Sep.*, 7, 107, 1995.
225. Kennedy, S., Caddy, B., and Douse, J. M. F., Micellar electrokinetic capillary chromatography of high explosives utilising indirect fluorescence detection, *J. Chromatogr. A*, 726, 211, 1996.
226. Thormann, W., Molteni, S., Caslavska, J., and Schmutz, A., Clinical and Forensic Applications of Capillary Electrophoresis, *Electrophoresis*, 15, 3, 1994.
227. Thormann, W., Paper symposium — Clinical and forensic applications of capillary electro-phoresis, *Electrophoresis*, 15, 1, 1994.
228. Trenerry, V. C., Robertson, J., and Wells, R. J., The determination of cocaine and related substances by micellar electrokinetic capillary chromatography, *Electrophoresis*, 15, 103, 1994.
229. Weinberger, R. and Lurie, I. S., Micellar electrokinetic capillary chromatography of illicit drug substances, *Anal. Chem.*, 63, 823, 1991.
230. Krogh, M., Brekke, S., Tonnesen, F., and Rasmussen, K. E., Analysis of drug seizures of heroin and amphetamine by capillary electrophoresis, *J. Chromatogr. A*, 674, 235, 1994.
231. Lurie, I. S., Chan, K. C., Spratley, T. K., Casale, J. F., and Issaq, H. J., Separation and detection of acidic and neutral impurities in illicit heroin via capillary electrophoresis, *J. Chromatogr. B-Bio. Med. Appl.*, 669, 3, 1995.
232. Tagliaro, F., Poiesi, C., Aiello, R., Dorizzi, R., Ghielmi, S., and Marigo, M., Capillary electro-phoresis for the investigation of illicit drugs in hair — Determination of cocaine and morphine, *J. Chromatogr.*, 638, 303, 1993.

233. Wernly, P. and Thormann, W., Analysis of illicit drugs in human urine by micellar electrokinetic capillary chromatography with on-column fast scanning polychrome absorption detection, *Anal. Chem.*, 63, 2878, 1991.

234. Wernly, P. and Thormann, W., Drug of abuse confirmation in human urine using stepwise solid-phase extraction and micellar electrokinetic capillary chromatography, *Anal. Chem.*, 64, 2155, 1992.

235. Thormann, W., Lienhard, S., and Wernly, P., Strategies for the monitoring of drugs in body fluids by micellar electrokinetic capillary chromatography, *J. Chromatogr.*, 636, 137, 1993.

236. Otsuka, K., Terabe, S., and Ando, T., Electrokinetic chromatography with micellar solutions. Separation of phenylthiohydantoin-amino acids, *J. Chromatogr.*, 332, 219, 1985.

237. Ahuja, E. S., Preston, B. P., and Foley, J. P., Anionic-zwitterionic mixed micelles in micellar electrokinetic chromatography: Sodium dodecyl sulfate-N-dodecyl-N,N-dimethylammonium-3-propane-1-sulfonic acid, *J. Chromatogr. B-Bio. Med. Appl.*, 657, 271, 1994.

238. Little, E. L., Ph.D., Louisiana State University (1992).

239. Gu, X., Meleka-Boules, M., and Chen, C.L., Micellar electrokinetic capillary chromatography combined with immunoaffinity chromatography for identification and determination of dexa-methasone and flumethasone in equine urine, *J. Cap. Electroph.*, 3, 43, 1996.

CHAPTER **5**

Separation of Enantiomers

George N. Okafo

CONTENTS

0-8493-9127-X/98/$0.00+$.50
© 1998 by CRC Press LLC

I. INTRODUCTION

In the field of natural sciences, the stereochemical resolution of optically active molecules still remains an important and essential stage in the development of biologically active chemical entities as potential drugs. This is not unexpected as many biological interactions and reactions are subject to varying degrees of stereoselectivity. The interaction of individual stereoisomers (or enantiomers) of a chiral drug molecule with a chiral macromolecule (e.g., an enzyme) results in the formation of a pair of diastereoisomeric complexes, which differ energetically. Hence, the consequences of these stereoselective transformations will depend on the ability of the biological environment to discriminate between two different enantiomers.

Stereoselectivity in nature is evident in a number of biological processes, e.g., absorption, protein binding, selective tissue uptake, renal or biliary excretion in addition to metabolism. Metabolic transformations can be characterized by their stereochemical courses, e.g., prochiral to chiral transformations (e.g., aromatic oxidation of phenytoin to yield 4-hydroxyphenytoin[1]), chiral-to-chiral transformations (e.g., stereoselective oxidation of (S)-warfarin to yield (S)-7-hydroxywarfarin[2]), chiral to diastereoisomeric metabolites ((+)-hexobarbitol is converted to form β-3′-hydroxyhexobarbitol, whereas the (–)-enantiomer is preferentially metabolized to the α-3′-hydroxyhexobarbitol[3]), chiral-to-nonchiral transformations (e.g., deamination of amphetamine to yield phenylacetone[4]), and metabolic chiral inversions (e.g., chiral inversion of (R)-ibuprofen to (S)-ibuprofen in the mammalian gut[5]).

By definition, enantiomers are molecules that are identical in their atomic composition and bonding but different in the three-dimensional arrangement of their atoms. These compounds have essentially identical physical (except optical rotation) and chemical (except in a chiral environment) properties. Their structure usually consists of an atom (either carbon, nitrogen, or sulfur) that is surrounded by at least four different functional groups. Each optical isomer will rotate the plane of polarized light in the opposite direction giving rise to the (+)- and (–)- prefix nomenclature. For naturally occurring molecules like carbohydrates and amino acids, the absolute configuration is described by the terms D- (dextrorotatory) or L- (levorotatory). The assignment of the these prefixes to an unknown analyte can be empirically determined now using methods developed by Bijvoet by the chemical correlation of compounds of known configuration (e.g., (+)-tartaric acid salts) with those of unknown configurations.[6] For those compounds synthesized via a stereoselective route, the R- and S- notation developed from the Cahn, Ingold and Prelog Sequence Rules[7] is generally adopted.

The rapid development of analytical techniques for the chiral resolution of optically active molecules has in part been due to several factors. First, developing stereoselective synthetic routes to chirally pure drugs can sometimes be very expensive. Second, pure enantiomeric products are now required in diverse scientific areas such as mechanistic studies, catalysis, kinetics, and biochemical and geochemical studies. Third, the optical isomers from many chiral drugs have different pharmacokinetic and pharmacodynamic activities, and the presence of an unwanted enantiomer in a drug that has been administered as a racemate can be of much concern. In some cases, the presence of the unwanted isomer can either be harmless or provide other beneficial effects. In other cases, the contaminating enantiomer can disturb vital biological processes leading to toxicological effects.

A well-known and widely reported case of a chiral drug containing enantiomers which have been reported to have differing biological effects is thalidomide[8] (1). In the early sixties, this drug was administered to pregnant women as a racemic mixture to relieve the effects of morning sickness. Unfortunately, toxicological side effects such as peripheral neutritis and birth defects were observed. These harmful effects were later discovered by Blaschke et al.[9] to be due to the S-(–)-enantiomer of (1). In the case of terbutaline (2),[10] a drug that is a β2-selective adrenoceptor agonist, the (–)-enantiomer is a potent relaxant of tracheal smooth

(1)

(2)

(3)

muscle, whereas, (+)-terbutaline is several thousand times less effective against this biological target. Clenbuterol is a drug that belongs to the same class as (2), except that both enantiomers have opposite effects. The (+)-enantiomer is claimed to be a potent β2-selective antagonist, while the (−)-enantiomer retains the selective agonistic activity.[11] With compounds that have two chiral centers, four stereoisomers (i.e., two diastereoisomeric pairs) are possible which may have the same biological target but differ in their relative potencies. This is the case for formoterol (3), a *p*-trifluoromethyl anilide derivative with β-agonistic properties.[12] The order of potency for the four optical isomers with respect to smooth muscle relaxation in the bronchial airways is RR>>RS=SR>SS.

In view of the biological issues involved in the administration of drugs that are either racemates or a mixture of enantiomers and the need not to repeat terrible incidences such as those related to (1), regulatory authorities (e.g., the Food and Drug Administration (FDA) in the USA) are requiring much more detailed information on optically active drugs.[13] This includes knowledge of the stereochemical composition of a final drug compound and pharmacological and toxicological profiles of the individual enantiomers. To coincide with the current awareness of the importance of enantiomeric resolution of chiral drugs, rapid developments in new instrumentation and chemical methods for chiral analysis have occurred.

The current analytical methods used for the separation of optical isomers can either be chromatographic (e.g., gas chromatography [GC], high performance liquid chromatography [HPLC], thin-layer chromatography [TLC], and supercritical fluid chromatography [SFC]), or electrophoretic (paper [PE] and capillary electrophoresis [CE]). Both techniques rely upon the formation of transient (noncovalent) metastable diastereoisomers between each enantiomer of an optically active compound and a chiral component. Chiral resolution of the two enantiomers is possible when there are sufficient differences in the free energies of formation of the two diastereoisomeric complexes. In GC, HPLC, and SFC, the chiral component has been immobilized to an inert achiral backbone (usually silica-based) to form a chiral stationary phase (CSP). Examples of CSPs include Pirkle-type phases, derivatized celluloses, bonded protein phases, chiral ligand exchange systems, and cyclodextrin-bonded phases. The applications using these CSPs have been well reviewed.[14,15] The alternative approach involves the *in situ* generation of diastereoisomeric derivatives *via* the reaction of isomers with chiral reagents. The derivatization of a racemic mixture to form covalent diastereoisomers can be achieved using a range of chiral reagents that are selective toward functional groups like amines (e.g., (−)-α-methoxy-α-methyl-1-naphthylacetic acid[16]), carboxylic acids (e.g., S-(−)-α-methylbenzylamine[17]), and ketones and alcohols (e.g., (−)-methylchloroformate[18]). These resulting diastereoisomeric derivatives are then separated on an achiral phase as a result of differences in their physicochemical properties.

Paper electrophoresis[19,20] is not widely used for enantiomeric separations. This method has largely been replaced by capillary electrophoresis (CE). The application of CE to the separation of enantiomers is relatively new and offers many advantages over the more conventional analytical methods, such as speed, simplicity, high resolution, low cost, and small sample volume requirements. As a result, chiral method development in CE can be achieved more easily. As a qualitative and semiquantitative analytical tool for the separation of enantiomers, CE offers an approach that is complementary to many of the popular chromatographic methods. Chiral resolution in CE involves the addition of chiral compounds to the buffer system resulting in the formation of transient noncovalently bound diastereoisomers. Alternatively, covalent diastereoisomers can be generated *in situ via* reaction of the chiral analyte with optically active reagents.

II. CHIRAL RESOLUTION VIA THE FORMATION OF NONCOVALENTLY BOUND DIASTEREOISOMERS

The most commonly used chiral additives in CE that are able to form noncovalently bound diastereoisomers belong to the following groups: cyclodextrins (native and derivatized), acyclic carbohydrates (neutral oligosaccharides and charged polysaccharides), chiral crown ethers, chiral surfactants (naturally occurring and synthetic), protein additives, macrocyclic antibiotics (rifamycins and glycopeptide antibiotics), and enantioselective metal complexation.

A. Cyclodextrins

Cyclodextrins (CDs) are charged neutral cyclic oligosaccharides produced naturally from the degradation of starch and mediated by the enzyme cyclodextrin glycosyl transferase. The CDs are composed of between six and eight β-D-glucopyranosyl units linked *via* 1-4β glycosidic linkages to form a hollow truncated cone structure. The outer rims of the cone are lined by primary and secondary hydroxyl groups, whereas the interior cavity consists of the glycosidic bonds and carbon skeleton. Hence, the outer surfaces are hydrophilic, while the inner cavity is hydrophobic in nature. The common CDs contain either six (α-CD), seven (β-CD), or eight (γ-CD) glucose units and are available as white, inert, water soluble solids. The aqueous solubility of these oligosaccharides is due to the presence of the hydrophilic hydroxyl groups that line both openings of the CD. β-CD is the least soluble (1.8 g/100 mL at 25°C),[21] whereas γ-CD has the highest solubility in water (23 g/100 mL at 25°C).[21] Other properties of CDs that make them useful as CE additives include their low UV absorptivity at short wavelengths and their chemical stability over a large pH range.

Cyclodextrins are known to form inclusion complexes with a wide range of compounds. These interactions can result in several outcomes. First, in the case of nonpolar analytes, inclusion of the molecule into the cyclodextrin cavity can improve the aqueous solubility. Second, the chiral nature of the glucose units allows the possibility of stereoselective interactions which can lead to the chiral discrimination of optically active analytes. Both of these properties of CDs are important in capillary electrophoresis for the resolution of enantiomers. The chiral discrimination mechanism is still not clear, but it is generally accepted that there are two fundamental requirements for chiral recognition in aqueous media. The first is that the analyte must have the appropriate structural geometry to fit favorably into the CD cavity (this forms what is usually referred to as an inclusion complex) and, in general, the following simple rule applies: analytes with unsubstituted aromatic rings include favorably within the α-cavity (cavity diameter, 5 to 6 Å)[22]; for the cavity in β-CD (cavity diameter, 7 to 8 Å),[22] compounds with naphthalene groups or substituted phenyl rings are most suited; in the case of γ-CD, which has the largest cavity (cavity diameter, 9 to 10 Å),[22] most complex polycyclic structures are able to form inclusion complexes. The second structural requirement of the analyte is that polar groups should be on or in close proximity to the chiral carbon center for stereoselective bonding interactions to take place. Providing these interactions are stronger for one of a pair of enantiomers, chiral discrimination can occur. In CE, chiral separations will be dependent on either differences in the stability of the inclusion complexes formed between each enantiomer and the CD or in the mobility characteristics of the host–guest complexes formed.

The stereoselectivity and solubility properties of the CD molecule can be altered by derivatizing the rim hydroxyl groups to give either neutral (e.g., methyl, hydroxymethyl, or hydroxyethyl moieties) or charged (aminomethyl, carboxymethyl, succinyl, and sulphonyl groups) functional groups. By crosslinking CD monomers with reagents such as epichloro-hydrin, CD polymers can be formed. These polymeric derivatives are now commercially available and offer significant advantages over the monomeric form. As yet, there are no mathematical methods to allow the fast prediction of complexation mechanisms for chiral analytes in CE, hence, screening a range of different CDs will still be necessary to determine the chiral additive that yields the best selectivity and enantiomeric separation.

1. Native Cyclodextrins

Naturally occurring or native cyclodextrins are neutral in charge and hence migrate with the same velocity as the electroendosmotic flow (EOF). Complexation of the CD molecule with an ionic species results in the formation of a charged complex, which can then migrate under the influence of electrophoresis and electroendosmosis. The enantioresolution of DL-tryptophan and its N-methylated derivatives was performed under low pH (pH 2.5) conditions using 40 mM α-CD.[22] Chiral resolution was found to be dependent on the methyl substituents at the C-5 and C-6 positions and at the indole nitrogen of the tryptophan molecule which affects the inclusion and orientation of the analyte within the α-cavity.[23] Other amino acids such as tyrosine,[24] phenylalanine,[24] and DOPA,[24] have been resolved using an α-CD-based system. In all these examples, the later migrating peak corresponded to the enantiomer that formed the strongest complex with α-CD. Attempts to enhance enantioresolution using β-CD or γ-CD were unsuccessful because of poor inclusion of the phenyl ring within the larger cavities. The larger β- and γ-cavities are suitable for structures such as 1-N,N-dime-thylamino-naphthalene-5-sulfonate[25] (dansyl or DNS) and 1-cyano-2-substituted-benz[f]isoindole[26,27] (CBI) derivatives of DL-amino acids.

The size of the aromatic moiety and the nature of substituents attached play an important role in the chiral discrimination mechanism in CE. Both of these factors relate to the inclusion of the analyte within the CD cavity. For the antirheumatic immunomodulator thiazole derivative[28] (4) and some vinyl triazole drugs[29] (uniconazole, X,H, (5) and diniconazole X, Cl, (6)), optimum enantiomeric resolution was achieved using native β-CD and γ-CD in phosphate and borate buffer systems (Figures 5.1a and 5.1b). Chiral resolution of the enantiomers of uniconazole and diniconazole is particularly relevant because these drugs are marketed as racemic mixtures and each isomer exhibits different biological effects in plants.[28] The R-enantiomers are potent fungicides, whereas the S-enantiomers show high plant-growth regulating properties. Attempts to use other cyclodextrins proved unsuccessful primarily because of the poor fit of the molecule within the cavities of α-CD and derivatized CDs. The introduction of bulky or ionic sidechains and aromatic substituents either on the phenyl ring for the vinyl triazoles or on the biphenyl moiety in the thiazole derivative led to a loss of enantiomeric resolution.

Hempel and Blaschke[30] used fluorescence studies to monitor the inclusion of an analyte in a CD containing buffer. The nonbenzodiaxepine hypnotic drug, zopicolone ($R_1=R_2=CH_3$, (7)) exhibits fluorescence, and in the presence of β-CD, the emission intensities were significantly higher, indicating that inclusion had occurred.[30] Using a β-CD-based buffer system, the enantiomers of zopicolone and its desmethyl ($R_1=R_2$, H, (8)) and N-oxide (R_1), R_2, CH_3,

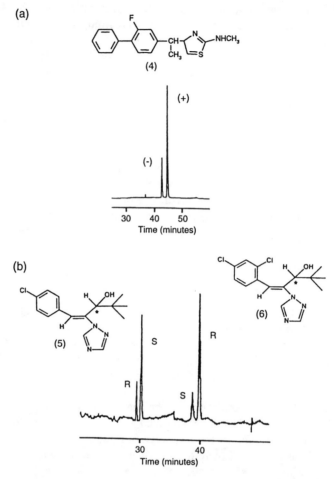

Figure 5.1 Electropherogram showing enantiomers of (a) thiazole derivative (**4**) and (b) diniconazole (**6**) and uniconazole (**5**). Separation conditions: buffer (a) 100 mM borate and 50 mM phosphate buffer, pH 9.0 with 100 mM SDS, 2 M urea and 50 mM β-CD, (b) 100 mM borate pH 9.0, 100 mM SDS, 50 mM γ-CD and 5% (v/v) 2-methyl-2-propanol; capillary, 570 mm × 0.075 mm i.d. (effective length, 500 mm); applied voltage, 15 kV; detection, 254 nm; temperature, 25°C. (From Furuta, R. and Doi, T., *J. Chromatogr.*, 708, 245, 1995; and Furuta, R. and Doi, T., *Electrophoresis*, 15, 1322, 1994. With permission.)

(**9**)) metabolites were resolved in urine samples from patients dosed with the drug (Figure 5.2). The achiral analog of (**7**), namely zolpiden, was used as a reference marker.

2. Derivatized Cyclodextrins

The physicochemical properties of cyclodextrins can be enhanced by chemically modifying the outer rim hydroxyl groups at the 2, 3, and/or 6 positions by the introduction of a range of charged and uncharged functional groups. These modifications influence the overall hydrophobic character of the CD, giving rise to changes in the shape and size of their cavities and their hydrogen-bonding ability. The dimensions of CD cavity, e.g., cavity depth and internal diameter, are significantly different compared to the native analog. The presence of these derivatized hydroxyl groups on the rim significantly increases the aqueous solubility of these cyclic oligosaccharides. For example, methylation of β-CD increases its aqueous solubility from 1.8 g/100 mL at 25°C to >33 g/100 mL at 25°C.[21] In addition, these chemical

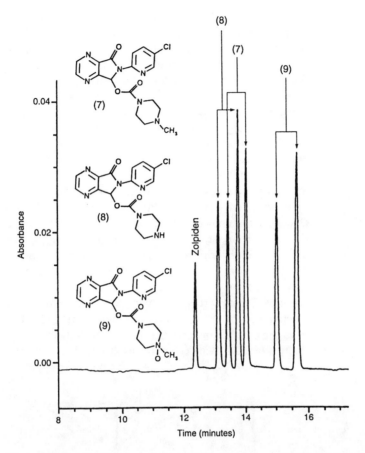

Figure 5.2 Electropherogram of the racemates of zopicolone (**7**), its metabolites N-desmethylzopicolone (**8**), zopicolone N-oxide (**9**), and zolpiden (an achiral reference marker). Separation conditions: buffer, 100 mM phosphate, pH 2.7 and 16.3 mM β-CD; capillary, 470 mm × 0.05 mm i.d. (effective length, 400 mm); applied voltage 18 kV; detection; LIF, λex 325 nm, λem, 450 nm; temperature, 20°C. (From Hempel, G. and Blaschke, G., *J. Chromatogr.*, 675, 139, 1996. With permission.)

modifications give rise to significant changes in the enantioselectivity of the CD, which are important for chiral separations in CE.

Alkylated cyclodextrins are commercially available, stable compounds that can be synthesized by the selective alkylation of hydroxyl groups at their 2-, 3-, 6-, 2,3- or 3,6- positions. Examples of these alkyl derivatives include methyl, ethyl, hydroxypropyl, naphthylcarbamoyl, and glycosylation. Characterization of these CD derivatives by H1 and C13-NMR[31] and mass spectrometric methods[31,32] (electrospray ionization, EI-MS) have shown that these compounds consist of mixtures of under- and over-alkylated CDs. In view of the difficulties in preparing pure derivatives, all commercially available derivatized CDs are supplied with specifications detailing the degrees of alkyl substitution (DS). A typical example is heptakis (2,6-di-O-methyl)-β-CD, which consists of 52.6% of the desired CD, 36.8% and 10.6% of the per- and dimethylated species.[33] For 2-monomethyl-β-CD, the desired CD derivative is present up to 76.9% with 15.4% of the undermethylated compound and 7.7% of the permethylated derivative.[33] In the case of trimethyl-β-CD, a much higher amount of the desired product is present (up to 92%).[33] When these CDs are dissolved in the BGE, the resulting mixture of chiral selectors can give rise to significant differences in chiral selectivity and enantiomeric resolution. Haskins et al.[32] recently showed, using EI-MS, that the composition of a commercially available sample of 2-hydroxypropyl-β-CD, when complexed with enantiomers of phenyl-alanine, propanolol, or tryptophan methyl ester, influenced its chiral selectivity toward these optically active analytes. Hence, it is important to consider the degree of substitution of these alkylated CDs when using these additives in chiral separations in CE. The significance of the DS of alkylated CD on enantioselectivity in chiral separations has been further demon-strated by Valko and co-workers.[34] They examined the effects of a range of 2-hydroxypropyl-β-CDs (HP-β-CD) with degrees of substitution ranging from 3.0 to 7.3 on the enantiomeric resolution of some optically active organic acids. Briefly, the results from their study showed that a small change in the DS value from, for example, 3.0 to 4.3 resulted in the complete loss of chiral resolution for certain analytes. Conversely, improvements in enantiomeric separation were also observed as the DS varied.

The contributions of the primary and secondary hydroxyl groups in these alkylated CDs in chiral discrimination has been demonstrated by Yoshinaga and Tanaka[33] for the enantio-meric resolution of twelve dansyl amino acid racemates. They compared native β-CD with 6-monomethyl-β-CD (which has the methyl group attached at the primary hydroxyl position) and found similar enantioselectivities for the two CDs. These results indicate that for these analytes, the secondary hydroxyl groups play a major role in the chiral recognition mecha-nism, whereas the contribution by the primary OH groups appears to be a minor one. Derivatization of the secondary hydroxyl groups at either the 2- and/or 3-positions profoundly influenced the enantioselectivity of the CD molecule. This is the case for 2- and 3-mono and dimethylated-β-CD used to separate the optical isomers of dansyl-DL-amino acids[33] (Figure 5.3). Interestingly, methylation of the 2-hydroxyl group resulted in the complete loss of chiral

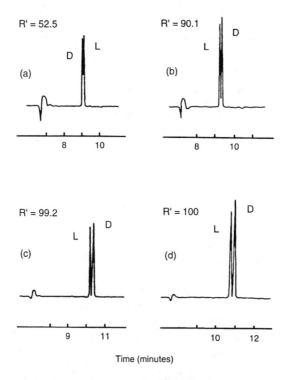

Figure 5.3   Chiral separations of dansyl-DL-leucine with (a) β-CD, (b) 3-MM-β-CD, (c) 2,3-DM-β-CD and (d) 2,3,6-TM-β-CD. Separation conditions: buffer 100 mM borate at pH 9.0 and 10 mM CD; capillary, 720 mm × 0.050 mm i.d. (effective length, 500 mm); applied voltage, 20 kV; detection 254 nm; temperature, 30°C. (From Yoshinaga, M. and Tanaka, M., *J. Chromatogr.*, 679, 359 1994. With permission.)

resolution for most of the analytes (except dansyl-DL-α-aminobutyric acid), whereas the 3-methyl CD derivatives were able to chirally resolve six dansyl derivatives. These data[33] suggested that after inclusion of the analyte *via* the dansyl moiety, interaction between polar groups (e.g., amide and/or carboxylates) on the analyte and the secondary hydroxyl groups on the rim of the CD were important for chiral discrimination.

A more recent example[35] of an alkylated CD derivative used in CE for chiral analyses is monosubstituted 1-(1-naphthyl)ethylcarbamoylated β-CD (NEC-β-CD). Although, this chiral selector showed enantioselectivity comparable to other CD derivatives toward analytes like phenylglycine, phenylalanine, and homophenylalanine derivatives, the presence of the UV absorbing naphthyl moiety can give rise to high background absorbances. Interestingly, the conjugated naphthyl ring system can give rise to the possibility of pi-pi interactions, which may contribute to the chiral discrimination mechanism. The glycosylation of α-CD (G1-α-CD) was shown[36] to stretch the cavity mouth of the CD and change the steric qualities of the host–guest interaction. This may explain the excellent separations achieved for racemates of binaphthyl derivatives using G1-α-CD.

A range of charged cyclodextrins is now commercially available which contains "rim" hydroxyl groups that have been either carboxyalkylated, succinylated, aminoalkylated, or converted to sulfoalkyl ether derivatives. For these ionic derivatives, the degree of substitution will vary and will influence the enantioselectivity of the CD. Typical values for carboxyalkyl derivatives are two to three carboxymethyl groups per CD ring (DS = 2 to 3). One major advantage of these charged CDs over their neutral relatives is that they can be used in different modes depending on the pH of the buffer system. At low pH values, they are useful for separating different ionic enantiomers, cations as well as anions, whereas, at higher pH values, the carboxylate and amine-based CD derivatives are largely present in their deprotonated forms, thus enabling the enantioresolution of neutral isomers. In this case, the CD acts as a carrier molecule and interacts with the positively charged analyte *via* ion pairing. One additional benefit afforded by charged CDs, particularly the sulfonated ether CD derivatives, over their neutral analogs is that the separation window is much larger. This is because for neutral CDs, the migration time window is limited to the difference between the electrophoretic mobilities of the analyte and that of a neutral species. In the case of separations involving charged CDs, this separation window is limited by the electrophoretic mobilities of the free analyte and that of the charged ionic CD derivative. This allows the resolution of analytes that only weakly interact and are poorly differentiated. One minor drawback in using these chiral selectors is that under buffer conditions where both the CD and the analyte carry the same charge, electrostatic repulsion can hinder chiral recognition.

Under acidic conditions (pH <4), carboxymethyl-β-CD (CM-β-CD) will be present as the protonated uncharged molecule. These conditions have been used to resolve the enantiomers of basic drugs[37] (e.g., ephedrine (**10**) and propanolol (**11**)). The migration of the analyte/CD complex is due to the electrophoretic mobility of the positively charged compound. The presence of the COOH groups at the CD rim may also play an important role in stereoselective hydrogen bonding and chiral discrimination. Similar observations were made for the chiral analysis of the amphetamine derivative, 3,4-methylene-dioxyethamphetamine[38] and some racemic binaphthyl derivatives.[39]

Under basic conditions, the carboxyl groups of CM-β-CDs and carboxyethyl-β-CD (CE-β-CD) are deprotonated and will be carried along by the EOF, thus making it possible to analyze neutral compounds. Chiral resolution in this case is dependent on differences in the inclusion mechanism of the two stereoisomers which will affect the overall mobility of the charged CD. For hexobarbitol[37] (**12**), binaphthol[37] (**13**), and an oxazolidinone derivative[36] (**14**), enantiomeric separation was achieved in a polyacrylamide-coated capillary using either 1.5% (w/v) CM-β-CD (Figure 5.4a) or 0.66%(w/v) CE-β-CD (Figure 5.4b). Interestingly, differences in the length of the alkyl spacer that links the COOH group to the CD gave rise

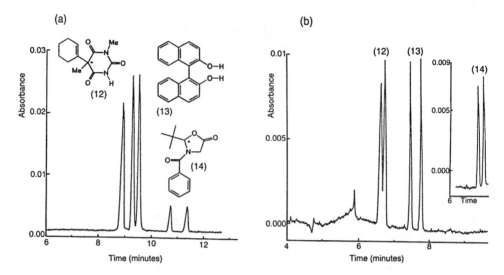

Figure 5.4 Separation of uncharged enantiomers of hexobarbitol (**12**), binaphthol (**13**), and an oxazolidinone derivative (**14**) using charged (a) CM-β-Cd or (b) CE-β-CD. Separation conditions:
buffer, 20 m*M* phosphate, 5.5 with either 1.5% (w/v) CM-β-CD or 0.66% (w/v) CE-β-CD;
capillary, 370 mm × 0.075 mm i.d. (effective length 300 mm) coated with 4% T polyacrylamide;
applied voltage, 15 kV; detection, 214 nm; temperature, ambient. (From Schmitt, T. and
Engelhardt, H., *Chromatographia*, 37, 475, 1993. With permission.)

to some variations in enantioselectivity depending on the analyte. The ion-pairing ability of
charged CD has also been demonstrated[37] at pH 5.8, where the CD is anionic in nature and
drugs such as propanolol are cationic. Using coated capillaries (with suppressed EOF), the
migration direction of the CD/analyte complex tends toward the anodic electrode, despite the
tendency of the basic analyte to move to the cathode. This ion-pairing behavior is dependent
on several factors such as CD concentration, the type of charged CD (small changes in the
concentration of succinylated CD can lead to changes in the migration direction of the cationic
analyte), and the analyte structure.

   Sulfonated and sulfated cyclodextrins have been used as chiral selectors for direct enantiomeric resolution. Generally, preparations of these CD derivatives consist of mixtures of
non-, mono-, di-, and higher substituted compounds. Characterization of these mixtures using
ion-spray MS[31] and C-13 NMR[31] reveals that CDs with one or two sulphoalkyl groups
predominate, and the average DS value for a derivatives, such as sulphopropyl ether, was
found to be 2.3. For other alkyl derivatives of these sulfonated CDs, the degree of substitution
can be as high as 4 for sulfobutyl ether β-CD (SBE-β-CD).[40] One useful property of these
CDs is that the sulfonic acid groups remain anionic over the entire pH range accessible to
CE experiments (i.e., pH 3 to 9), and the presence of up to four methylene units serves as a
spacer arm between the CD cavity and the sulfonate group. In addition, the solubility of this
CD derivative is greatly enhanced compared to the native oligosaccharide.

   Optimum enantiomeric resolution using sulphoalkyl ether CDs has been achieved using
low pH buffers because this allows the solubility of basic analytes, and the reduced EOF
provides an added benefit to the separation mechanism. Baseline resolution of the enantiomers
of adrenaline[40] (R, CH$_3$,(**15**)) and noradrenaline[40] (R,H, (**16**)) were achieved using 2.5 m*M*
SBE-β-CD dissolved in 200 m*M* phosphate buffer at pH 2.5 (Figure 5.5). In both examples,
the (−)-enantiomer migrated before the (+)-antipode, indicating a stronger interaction between
the latter isomer and the CD. For analytes such as tyrosine[40] and DOPA,[40] the observed
migration times were found to be longer than expected because the pH of the buffer system
was close to the pK_a values of the carboxyl groups of the two analytes, thus reducing the
overall electrophoretic mobility of the free analyte. It is well known that the groups that

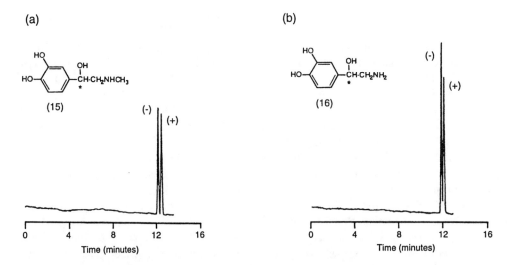

Figure 5.5 Electropherogram showing the separation of the enantiomers of (a) 1.4 mM racemic adrenaline (**15**) and (b) racemic noradrenaline (**16**). Separation conditions: 200 mM phosphate,
pH 2.5 and 2.5 mM SBE-β-CD; capillary, 600 mm × 0.025 mm i.d.; applied voltage 13 kV;
detection, 210 nm; temperature, ambient. (From Tait, R. J., Thompson, D. O., Stella, V. J.,
and Stobaugh, J. F., *Anal. Chem.*, 66, 4013, 1994. With permission.)

surround the chiral carbon center can directly influence chiral resolution; this is particularly
relevant from the standpoint of multipoint specific interactions that are required for chiral
discrimination. For racemic ephedrine[40] and pseudoephedrine,[40] both of these molecules have
chiral centers at the α-carbon and β-carbon with respect to the aromatic ring system. Chiral
separation using 1.5 mM SBE-β-CD gave four baseline resolved peaks for the two racemic
beta-blockers. In comparison with neutral CD derivatives, SBE-β-CD was found to be slightly
better, particularly for the enantiomers of ephedrine. Moreover, much lower concentrations
of the chiral additive were necessary. For sulphopropylated derivatives of CDs, slightly higher
concentrations of up to 10 mM at pH 7 were required to resolve the enantiomers of three
barbituric acid derivatives[31] and 4,5-dihydrodazepam.[31] An interesting application[41] of SBE-
β-CD is in the forensic analysis of illicit drugs such as amphetamine, methamphetamine,
methcathinone, and propoxyphene. To enhance enantioresolution, trace amounts of methanol
were added.

The cationic CD derivatives, 6A-methylamino-β-CD[42] (MENH-β-CD) and mono-(6-β-
aminoethylamino-6-deoxy-)-β-CD[43] (CDen) are positively charged at pH values less than 8.
The basic nature of these CDs allows them to be used for the analysis of anionic analytes
such as NSAIDS (suporfen, ibuprofen) and derivatized amino acids. Small amounts (between
2.5 and 5 mM) of MENH-β-CD[42] and CDen[43] were required for chiral separations, indicating
that they have higher stereoselectivities than their neutral counterparts. This is most likely
because the CD will tend to migrate in the direction opposite the negatively charged analyte,
hence enhancing complexation equilibria. In terms of structure, MENH-β-CD has a cavity
depth similar to that of β-CD and smaller than that of di-ME-β-CD due to the presence of
methoxy groups on the rim. These structural features may explain why the complexation
power of this CD derivative is lower than that of the methylated CDs.

3. Polymerized Cyclodextrins

Cyclodextrins can be polymerized by appropriate bi- or polyfunctional agents to form
oligomers, longer-chain polymers, or crosslinked networks. The chemical and physical properties of these polymeric species differ in many ways from their monomeric analogs, including

high water solubility (>20% w/v) for the lower molecular weight cyclodextrin oligomers and the formation of gel-like materials for the larger polymers (MW > 10,000 Da). In the context of chiral separations, these chiral polymeric derivatives show significant differences in enantioselectivity compared to monomeric CDs in CE. Commercially available polymerized cyclodextrins can be either neutral or charged in nature (carboxymethylated-β-cyclodextrin polymer). These polymers are prepared by linking monomeric cyclodextrin together using reagents like 1-chloro-2,3-epoxypropane. As in the case of their monomeric counterparts, enantiomeric resolution is influenced by experimental parameters such as pH, temperature, organic solvents, and concentration of polymeric additive.

Neutral β-cyclodextrin polymer has been used effectively to resolve the enantiomers of a range of cationic pharmaceutical racemates, e.g., selegine, clenbuterol, and beta-blockers.[44] Typical separation conditions were 50 mg/mL β-cyclodextrin polymer dissolved in 50 mM sodium phosphate at pH 2.5. The acidic conditions were required to allow protonation of these amine compounds in the presence of a substantially reduced electroendosmotic flow, hence the chiral additive acts as a pseudostationary phase. For some analytes, such as ephedrine,[44] it was necessary to add up to 30% (v/v) methanol and use elevated temperature (50°C) to achieve enantioseparation.

For some polycyclic compounds[45] (bupivacaine, terbutaline, and ketamine), the resolution was found to diminish under these conditions. Increasing the concentration of the additive to the background electrolyte (BGE) led to a decrease in the effective mobility for most analytes due to increased complexation with the chiral polymer. The resulting transient diastereoisomers migrate more slowly than the uncomplexed analyte. A consequence of the reduced electrophoretic mobility is increased resolution and selectivity as exemplified in the chiral separation of some racemic beta-blockers[45] (ephedrine, terbutaline, propanolol, and epinephrine).

The influence of analyte hydrophobicity is an important factor that determines the degree of interaction with the cyclodextrin polymer and enantiomeric resolution (Rs). For racemates of alkyl esters of tryptophan,[45] the value of Rs decreased as the steric bulk of the alkyl group decreased. One explanation for this may be due to hydrophobic interactions of the analyte inside and/or outside the cyclodextrin cavity. The best enantiomeric resolutions were obtained at pH values between 2 and 6.5, because under these conditions, the low EOF allows greater residence time for interactions between the analyte and the chiral selector. This is shown[46] for the enantiomers of trimetoquinol (**17**), its structural isomer (**18**), norlaudansoline (**19**), and laudanosine (**20**) in Figure 5.6. Enhanced enantioresolution has also been achieved by increasing the ionic strength of the buffer. The addition of surfactants (e.g., SDS or STDC) can be used to manipulate enantioselectivity, and this was particularly evident[46] for the enantiomers of laudanosine and norlaudanosine. Comparisons of monomeric cyclodextrins with their polymeric derivatives show greater enantioselectivity for the latter cyclodextrin compound. This has been attributed to greater molecular weight and the structure of the CD polymer. The decreased free rotation of each CD unit, a constant distribution of CD, hydrophobic interactions, and hydrogen bonding all contribute greatly to enhanced chiral discrimination.

The physicochemical properties of polymerized cyclodextrins can be altered by modifying the oligosaccharide by the introduction of charged functional groups. Polymerized carboxymethylated-β-cyclodextrin is a commercially available chiral selector that has a molecular weight range of between 6 and 8 kDa and with a carboxylate:cyclodextrin ratio of 2.[47] Its ionic nature dramatically increases the solubility of this polymer in water to >20% (w/v), and its properties as a chiral selector are strongly influenced by pH. Generally, increasing the pH value leads to a decrease in the Rs. For some basic racemates[47] like norephedrine, ketamine, norphenylephrine, and β-hydroxy-phenylethylamine, acidic buffer systems containing 2 to 4 mg/mL of the polymer resulted in diminished enantioresolution. This was explained by differences in the number of hydroxyl and nitrogen groups that are able to form hydrogen bonds with hydroxyl and carboxylic acid groups on the rim of the CD molecule.

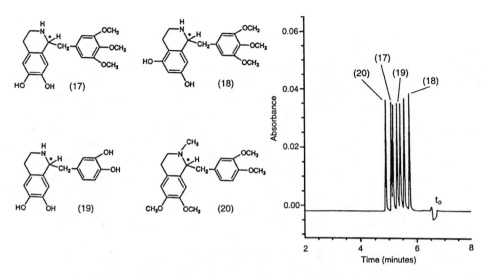

Figure 5.6 Separation of the enantiomers of trimetoquinol (**17**), its structural isomer (**18**), norlaudano-soline (**19**), and laudanosine (**20**). Separation conditions: buffer 25 mM phosphate, pH 6.5 with 2 M urea and 3% (w/v) β-CD polymer; capillary, 570 mm × 0.075 mm i.d. (effective length, 500 mm); applied voltage 20 kV; detection, 214 nm; temperature, 23°C. (From Nishi, H., Nakamura, K., Nakai, H., and Sato, T., *J. Chromatogr.*, 678, 333, 1994. With permission.)

However, enantioresolution for these basic racemates can be achieved by increasing the concentration of the polymeric additive to counteract the effects of low pH. For other analytes,[47] such as terbutaline and propanolol, an opposite effect was observed in which at pH 6.2 enantioresolution decreased because of the shorter times allowed for dynamic exchanges to occur. Another experimental parameter that can influence enantioresolution is the use of coated capillaries. Interestingly, a reversal of the voltage polarity under negligible electroendosmotic flow conditions in coated capillaries leads to a reversal in the migration order for the enantiomers of propanolol. This is useful if the analysis of low levels of a contaminating isomer are present in the desired enantiomer.

4. The Effects of Additives and pH on Chiral Resolution Involving Cyclodextrins

Varying the experimental conditions and the use of coadditives can have important effects on chiral resolution by improving peak shape and enhancing enantioselectivity in cyclodextrin-containing buffer systems in capillary electrophoresis (CE). Some of these parameters are cyclodextrin concentration, ionic buffer strength and buffer type, buffer pH, and the value of the applied voltage and the polarity. Examples of coadditives include surfactants, organic solvents, urea, other chiral compounds, alkylhydroxy-alkylcelluloses, coated capillaries, and polyacrylamide gels.

There are many reported applications involving chiral resolution that could only be achieved by the coaddition of surfactants to cyclodextrins in buffer systems. This mode of CE has been termed cyclodextrin-modified micellar electrokinetic chromatography (CD-MEKC). The detergent additives can themselves be optically active, like sodium taurodeoxy-cholate (STDC), or achiral, like sodium dodecyl sulfate (SDS), and are usually added to the CD-containing buffer systems at concentrations above their CMC. The enhancement in chiral resolution and enantioselectivity offered by CD-MEKC is due to the prolongation of the analyte migration times due to partitioning of the analyte between the aqueous and micellar phases. These equilibria processes occur in competition with differential inclusion of each enantiomer with the cyclodextrin. In applications involving the use of chiral detergent coadditives, additional stereoselective interactions with the chiral micellar surface further enhance the enantioselectivity of the system.

The enantiomers of compounds such as barbiturates, polycyclic aromatic compounds, glutethimide and its analogs (the potent nonsteroidal aromatase inhibitors), and NDA derivatized amino acids have been resolved[48] using phosphate/borate buffers containing SDS and CDs.

The use of bile salts like STDC in CD-containing buffers to enhance enantioresolution was first introduced by Okafo et al.[49] to resolve a range of ionic (DNS- and CBI-amino acids), cationic (RS-naphthylethylamine), and neutral (mephenytoin and its hydroxy metabolites) analytes. Since this first report,[49] the utility of the STDC/CD-MEKC system has been extended[50] to enantiomeric separations of β-agonists, β-antagonists, phenylethylamines, stimulants, and diclofensine (an antidepressant). In the majority of the reported examples, chiral resolution was possible only when the bile salt surfactant was added to the CD/buffer system. This is the case for the chiral analysis of S,R-diclofensine[50] (21) using HP-β-CD (average DS, 0.6) in the presence of 50 mM STDC (Figure 5.7). The improved enantioresolution was attributed to the partitioning of the analyte in the slower migrating pseudostationary micellar phase, resulting in specific optical interaction between the micelle polar interior and the CD buffer phase. Resolution diminished at high surfactant concentrations (up to 100 mM), probably due to competition between the surfactant monomers and the optically active analyte

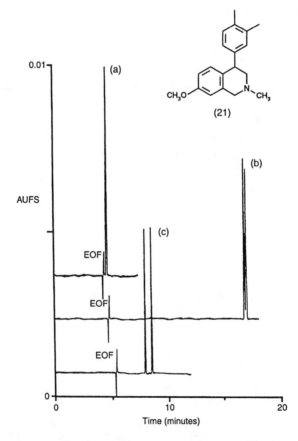

Figure 5.7 Electropherogram of a 300mg/mg S,R-diclofensine (21) injected into (a) 30 mm HP-β-CD, (b) 2 mM HP-β-CD and 50 mM STDC, and (c) 30 mM HP-β-CD and 50 mM STDC. Other separation conditions: buffer 50 mM borate; capillary, 1000 mm × 0.05 mm i.d. (effective length, 500 mm); applied voltage, 30 kV; detection, 200 nm. (From Aumatell, A. and Wells, R. J., *J. Chromatogra.*, 688, 329, 1994. With permission.)

for inclusion into the CD cavity. Generally, detergent concentrations between 10 and 60 mM were found to be optimum.

By varying the concentration of the CD additive in the BGE, optimum enantiomeric resolution can be obtained. Increasing the amount of CD in the buffer usually leads to a higher enantiomeric separation factor (the α value), resolution (Rs), and migration times. These observations can be explained in terms of the formation of stronger inclusion complexes and a reduced electroendosmotic flow as a result of the increased BGE viscosity. A model proposed by Wren[51] indicated that the optimum CD concentration is related to the affinity of the analyte interacting with the CD cavity. For the enantiomeric separation of R/S-N-t-BOC-DL-tryptophan (Figure 5.8), it was observed[52] that for strong analyte–CD interactions, a lower concentration of the additive was required, giving rise to faster and better separations. The anticoagulants, warfarin and 5-chlorowarfarin were resolved[53] using 6 mM and 3 mM Me-β-CD, respectively. Similar observations were obtained for the enantioseparation of racemic epinephrine.

The pH of the buffer can affect the ionic nature of the analyte, the magnitude of electro-endosmotic flow, and the charged nature of some derivatized CDs. In the case of optically active compounds with ionizable groups, such as amines and carboxylates, varying the acidity of the BGE can greatly influence stereoselective interactions (e.g., hydrogen bonding and electrostatic attraction) between the analyte and the hydroxyl groups on the CD rim. The pH of the BGE directly influences the magnitude of the electroendosmotic flow (EOF), and

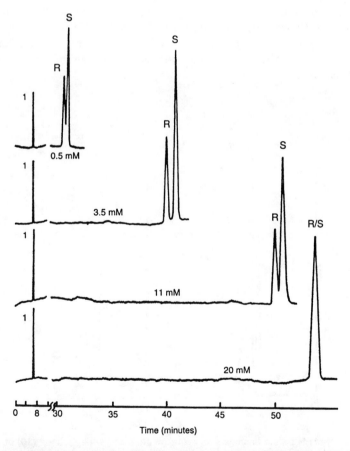

Figure 5.8 Separation of R- and S-N-t-BOC tryptophan by various concentrations of β-CD from 0.5 to 20 mM. Separation conditions, 20 mM HIBA, pH 4.5 and 0.5 to 20 mM β-CD; capillary, 600 mm × 0.100 mm i.d.; detection 254 nm; temperature, 25°C. (From Vespalec, R., Fanali, S., and Bocek, P., *Electrophoresis*, 15, 1523, 1994. With permission.)

generally, for amine-based drugs, enantiomeric resolution can be achieved at low pH values (pH 2 to 6). For anionic analytes,[54] more basic buffer conditions are recommended. This has been shown for some β-blocker drugs (e.g., racemic epinephrine) where enantiomeric reso-lution (Rs) increases dramatically with a lowering of the pH value. A combination of reduced EOF and differences in electrophoretic mobility between the free protonated amine and the CD/amine complex enhanced the Rs for these amine-based drugs. The enantiomeric resolution of barbiturates has been found to be particularly sensitive to changes in pH value.[55] An increase in pH from 7 to 9 results in the complete loss of resolution. These pH dependent observations were found to be related to the ionization (pK$_a$) of the phenolic hydroxyl moiety in these barbiturates. Under basic conditions, a polar phenolate anion is generated which will not favorably include within the CD cavity. Data from NMR and X-ray studies[56] confirmed that barbiturates include into the CD cavity *via* their nonpolar, aliphatic, or phenyl sidechains. In some cases, the stability of the analyte can be affected by pH. For hexobarbital,[56] enan-tioresolution was achieved using β-CD at pH >8 (where the compound is present in its deprotonated form). However, under these alkaline buffer conditions, urea-based drugs of this type were found to decompose due to hydrolysis.

Changes in the buffer pH can also have dramatic effects on the enantiomeric migration order. Schmitt and Engelhardt[37] showed that enantiomeric migration order of racemic (+/–)-ephedrine (**10**) could be changed depending on the pH of the BGE. Using polyacrylamide-coated capillaries (where EOF has been suppressed), it was shown that 1.5 to 2% (w/v) CM-β-CD dissolved in a citrate buffer at pH 2.7 resulted in (+)-ephedrine (**10**) migrating before the (–)-enantiomer. Conversely, at neutral pH (Tris at pH 7.2), the reverse migration order was observed (Figure 5.9). The reversal of migration order was due to differences in the charged state of the derivatized CD molecule, the analyte, and the analyte/CD complex under the buffer pH conditions.

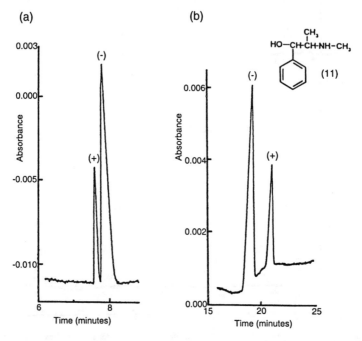

Figure 5.9 Effect of pH on the enantiomeric migration order of (+/–)-ephedrine (**10**) at (a) pH 2.7 and (b) pH 7.2. Separation conditions: buffer, (a) 20 m*M* citric acid pH 2.7 and 2% (w/v) CM-β-CD, (b) 100 m*M* Tris pH 7.2 and 1.5% (w/v) CM-β-CD; capillary, 370 mm × 0.075 mm i.d. (effective length, 300 mm); applied voltage, (a) 13 kV, (b) 10 kV; detection, 214 nm; temper-ature, 20°C. (From Schmitt, T. and Engelhardt, H., *Chromatographia*, 37, 475, 1993. With permission.)

Urea[48,56,57,58] is a useful additive in chiral separations in CE primarily because it allows higher concentrations of CD to be used in the BGE (in urea-containing buffer, the aqueous solubility of β-CD increases from approximately 10 mM to 16 mM). The direct consequence of increasing the concentration of CD amounts is the formation of stronger inclusion complexes. High concentrations of this additive (up to 7 M) have been used, which increases the viscosity of the BGE and leads to a lowering of the EOF, resulting in longer migration times. Unlike surfactant monomers, urea does not readily include in the CD cavity. The decreased EOF may also result in enhanced enantioselectivity. Another interesting property[58] of urea is the ability to decrease the current of BGEs *via* an unknown mechanism in buffers that contain this additive, thus allowing the use of higher separation voltages. The addition of 7 M urea to a buffer containing 10 mM DM-β-CD decreased the current by approximately half to 30 μA after an applied voltage of 20 kV when compared with the same buffer without the urea additive. Alkylated urea derivatives (e.g., methyl, ethyl, and 1,3-dimethylurea) have also been used, and these show similar enhancement in enantioselectivities for CD. However, they differ from unsubstituted urea in that a much greater suppression of EOF can be achieved. For example, the enantiomeric separation of DNS-DL-Leu and DNS-DL-Thr was significantly improved in the presence of 4 M 1,3-dimethylurea (1,3-DMU)[58] (Figure 5.10). The overall run times and Rs had increased significantly.

Water-soluble organic solvents such as methanol, acetonitrile, and propan-2-ol can be used effectively to enhance enantiomeric resolution in CE.[56] These solvents act by increasing the viscosity of the BGE and modifying the charge and hydrophobicity of the capillary inner wall leading to a reduced EOF and an extension of the migration time window. Other important effects include increasing the solubility of the analyte in the BGE and changing the solvation structure surrounding the analyte. The amount of organic solvent required is dependent on other factors, such as analyte type, buffer concentration, and the type of organic solvent. Generally, solvent concentrations ranging from 2 to 30% (v/v) have been used to enhance enantiomeric resolution of analytes such as warfarin (2% (v/v)) and propanolol (up to 30%

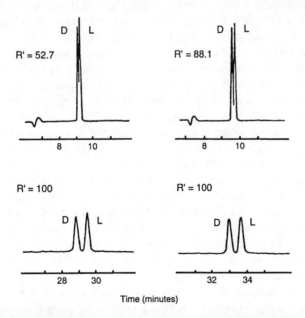

Figure 5.10 Chiral separation of (a) dansyl-DL-leucine and (b) dansyl-DL-threonine with 10 mM β-CD in the absence (upper traces) and presence (lower traces) of 4 M 1,3-dimethylurea. R values indicate the percentage of enantiomeric separation. Separation conditions: buffer, 100 mM phosphate–50 mM borate pH 9.0; capillary, 720 mm × 0.050 mm i.d. (effective length, 500 mm); applied voltage, 20 kV; detection, 220 nm; temperature, 30°C. (From Yoshinaga, M. and Tanaka, M., *J. Chromatogr.*, 710, 331, 1995. With permission.)

(v/v)). At high organic solvent concentrations, resolution decreases because this additive can interfere with the inclusion process and the current flow profile within the capillary.

The addition of chiral coadditives to BGE that already contain CDs has been shown to enhance enantioresolution and enantioselectivity. These compounds can either be simple organic molecules or more complex ones, such as crown ethers and other cyclodextrins. When simple compounds are used as coadditives, the mechanism is thought to involve the coinclusion of the chiral additive with the optically active analyte within the CD cavity. Stereoselective interaction between the two included molecules gives rise to the improved chiral discrimination. Two examples are D-camphor-10-sulfonate[48] and L-methoxyacetic acid,[48] which have been shown to markedly improve the enantioresolution of thiopental, pentobarbital, and some polycyclic aromatic compounds. The zwitterionic quaternary chiral amine L-carnitine has been shown[59] to improve the resolution of DNP-DL-Phe and AQC-DL-Leu mediated by CDs. Here, the mechanism may involve a complex interaction with the capillary silanol groups leading to a reversal of EOF. This effect is superimposed with interaction equilibria between the CD, the buffer system, and the analyte. The addition of chiral crown ethers to CD-based buffer systems has been shown to have a synergistic effect on enantiomeric separations. The isomers of DL-Trp can both form inclusion complexes with either α-CD[59] or the chiral crown ether 18-crown-6-tetracarboxylic acid[24] (18-C-6-TCA) giving rise to baseline separations. Buffers containing mixtures of both chiral selectors enhanced the Rs for D- and L-amino acids from 1.29 (for α-CD alone) and 5.67 (for 18-C-6-TCA alone) to 7.37. In a recent report, Anigbogu et al.[60] showed improved resolution of aminoglutethimide using a dual-CD phase system in CE consisting of 5 mM CM-β-CD and 1 mM β-CD. The improvement in resolution obtained using this mixed CD system relies upon the magnitude of the difference between the mobilities of the enantiomers upon interaction with each CD.

The ionic strength and nature of the buffer have been shown to play an important role in enantioseparations in CE. This has been effectively demonstrated for the separation of warfarin[53] and epinephrine[54] enantiomers. In the former example, the ionic strength of 20 mM TAPS was varied by the addition of sodium chloride, and it was noticed that at high salt concentrations (up to 125 mM) chiral resolution of warfarin enantiomers increased as a result of a decrease in the EOF and an increased migration time. In the case of racemic epinephrine,[54] using increasing amounts of Tris buffer (from 1 to 60 mM) caused the Rs to diminish, and optimum buffer ionic strengths were found to be between 10 and 40 mM. The choice of optimum buffer concentrations is also dependent on other experimental factors, such as the applied voltage. The application of high potential differences can increase Rs and decrease the migration time. However, at very high voltages (>25 kV), enantioresolution is lost because of thermal diffusion of the analytes and Joule heating within the capillary. The effect of increasing temperature on enantioresolution is not unexpected because, the host–guest complexation mechanism is a kinetically driven process and the stability of the resulting CD complex is temperature dependent. Lindner et al.[61] showed that lowering the capillary external temperature from 55 to 5°C dramatically improved the resolution of the enantiomers of DNB-DL-Phe using 30 mM HP-β-CD (Figure 5.11). Other temperature effects include changes in the viscosity of the buffer, which directly affects the EOF and overall run times.

The suppression of EOF has already been shown to enhance enantiomeric resolution, and this can be achieved in several ways, some of which have already been described. The addition of alkyhydroxycelluloses to the BGE reduces the EOF by irreversibly binding to the inner surface of the silica capillary. The resolution of some dansylated amino acids[36] and the antibiotic chloramphenicol[62] have been resolved in CD-containing buffer using 0.1% (w/v) hydroxypropylcellulose and methylhydroxyethylcellulose, respectively.

Immobilization of the CD molecule by incorporation in polyacrylamide (PAG) gel matrices can dramatically enhance chiral selectivity. Guttman et al.[57] described an experimental procedure that involved the addition of different CDs to the PAG prior to the free radical polymerization step. These PAG/CD systems dramatically improved the enantiomeric reso-

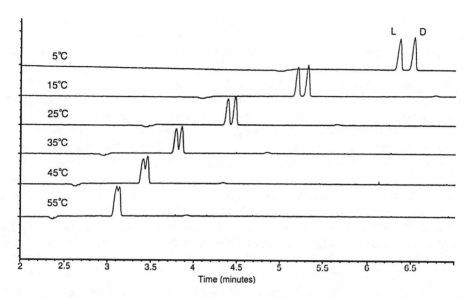

Figure 5.11 The effect of temperature on the enantiomeric separation of DNB-Phe enantiomers. Separation conditions: buffer, 20 mM sodium tetraborate and 30 mM HP-β-CD; capillary, 700 cm × 0.050 mm i.d.; applied voltage, 20 kV; detection, 214 nm; temperature, 5 to 55°C. (From Lindner, W., Bohs, B., and Seidel, V., *J. Chromatogr.*, 697, 549, 1995.

lution of three pairs of dansyl-DL-amino acids due to the reduced electrophoretic mobility of the inclusion complex.

B. Acyclic Carbohydrates

Noncyclic oligosaccharides and polysaccharides can form complexes with a wide variety of compounds. The formation of stereoselective host–guest complexation is probably the most common interaction of this type and is particularly important in enantiomeric separations in capillary electrophoresis. The earliest examples of the use of these chiral selectors was reported by D'Hulst and Verbeke[63] who used maltodextrins to resolve optically active acid racemates. These linear carbohydrates can be classified into two groups: neutral oligosaccharides and charged polysaccharides.

1. Neutral Oligosaccharides

Neutral oligosaccharides consist of maltooligosaccharides (also called maltodextrins or amyloses) generated from the enzymatic or acidic hydrolysis of starch. The D-(+)-glucose units are linked linearly *via* 1-4α glycosidic linkages to form these linear maltodextrins, which are characterized by the degree of polymerization.

Maltodextrins

The higher molecular weight oligomers have low solubilities in water. Maltoheptaose (where n, the number of glucose units, equals 5) is the highest commercially available maltodextrin (MD), whereas the other oligomers are usually sold as complex mixtures. The conformation adopted by these acyclic dextrins in aqueous solution is thought to be that of a flexible random coil consisting of extended helical segments connected by deformed nonhelical segments. In the presence of an interacting analyte and buffer salts, the flexible coil undergoes a conformational change to form a complete helix. The interior of this structure is hydrophobic (similar to the cavity in cyclodextrins), however, the inherent flexibility allows them to interact with a wider range of molecular structures than cyclodextrins. Examples of analytes include iodine,[64] fatty acids,[65] linear and branched alcohols,[66] DMSO,[67] and naphthyl derivatives.[68] Generally, the structural criteria required by an analyte for complexation with the chiral selector is a hydrophobic group which contains another nonpolar substituent in close proximity to a polar group.

Data from NMR studies[69] have provided much information in elucidating the mechanism of complexation between MDs and simple molecules. Many of the analytes studied were characterized by a downfield chemical shift which may reflect the formation of compact helices upon complexation. The exact mechanism of chiral discrimination involving MDs is unclear. However, it is thought that different mechanisms may be involved, e.g., electrostatic dipole–dipole, pi–pi interactions, and hydrogen bonding. The involvement of the helical cavity has also been well documented. Further evidence[69] has shown that the location of the chiral center between an aromatic moiety and a positive or negative charge in the chiral recognition process is important in the stabilization of the complex. For chiral separations involving maltodextrins (MD) in CE, the degree of enantiomeric resolution and selectivity is primarily dependent on the degree of polymerization (DP) and molecular weight of the oligomer. The dextrose equivalent (DE) value, usually quoted for most commercially available MDs, is a good indication of the distribution of dextrin oligomers present in a mixture. Commonly available MDs include, Glucidex2® (DE 2) and Dextrin® 10 (DE 10). These additives have been used[69] to resolve the enantiomers of a range of racemic coumarinic anticoagulant drugs (e.g., warfarin), acidic compounds (NSAIDS), and some basic drugs (e.g., verapamil and fluoxetine). Results from these studies[69] have shown that a decrease in the DE values enhanced enantiomeric resolution. A lower value for DE implies that a lower extent of the starting material (starch) is hydrolyzed, hence larger oligomeric fragments are produced.

In Figure 5.12, enantioseparation of the coumarinic anticoagulant drugs[63] (22) was achieved using a 3% (v/v) solution of Glucidex2 (DE 2) in a Tris-phosphate buffer. The improved enantiomeric resolution achieved by the use of higher molecular weight MD oligomers has been demonstrated by Soini et al.[69] A range of single maltodextrins ranging from maltotriose to maltoheptaose were used to resolve the enantiomers of warfarin, simendan, and ibuprofen. The best results were obtained using maltohexaose and maltoheptulaose, whereas the trimer, tetramer, and pentamer were unsuccessful in chiral separations. The improved enantioresolution using higher oligomers is probably due to the ability of oligomers (>5 glucose units in length) to overcome an induced-fit conformation change depending on the size of the complexing analyte. Spectroscopic titration studies[70] using two achiral UV absorbing probes, namely p-nitrophenyl and p-carboxyphenyl acetates were used to investigate the interaction of these two compounds to a range of MD from G1 to G7 (G = number of glucose residues). The kinetic data confirmed that stronger and more stable complexes were formed between the longer chain lengths. Moreover, hydrogen bonding makes an important contribution to the complexation mechanism.

MDs ranging from G2 to G7 have also been used[70] to separate the enantiomers of (S,R)-1,1-binaphthyl-2,2-dicarboxylic acid (S,R-BNC, (23)) (Figure 5.13). In this example it was observed that the migration order of S-BNC and R-BNC was reversed as the oligomer chain

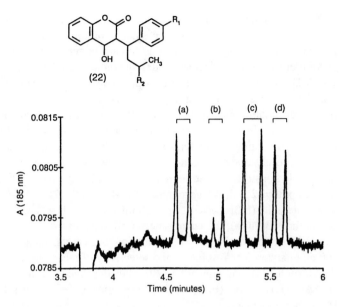

Figure 5.12 Enantiomeric separation of our racemic coumarinic anticoagulant drug derivatives (**22**), namely, (a), *p*-chlorophenocoumon (R_1,Cl; R_2,H), (b), *p*-chlorowarfarin (R_1,Cl;R_2,=O), (c), phenoprocoumon (R_1,R_2,H), and (d), warfarin (R_1,H,R_2,=O). Separation conditions: 10 mM Tris phosphate, pH 7 and 3% (w/v) Glucidex2®; capillary, 600 mm × 0.05 mm i.d.; applied voltage, 30 kV; detection, 185 nm; temperature; ambient.

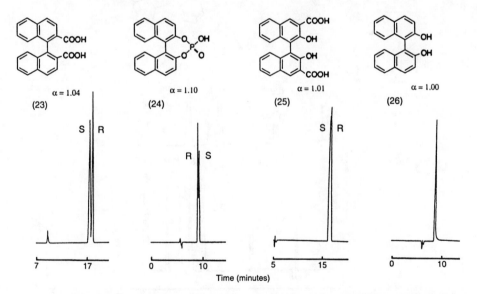

Figure 5.13 Enantiomeric separations of racemic binaphthyl derivatives BNC (**23**), BNP (**24**), HBNC (**25**), and BN (**13**). The α values indicated in the brackets indicate the degree of enantiomeric resolution. Separation conditions: buffer, 40 mM carbonate, pH 9.0 and 400 mM G_2; capillary, 300 mm × 0.05 mm i.d.; applied voltage, 15 kV; detection, 225 nm (BNC), 215 nm (BNP), 235 nm (HBNC), and 227 nm (BN); temperature, ambient. (From Kano, K., Minami, K., Horiguchi, K., Ishimura, T., and Kodera, M., *J. Chromatogr.*, 694, 307, 1995. With permission.)

length increased. Optimum enantiomeric resolution for racemic BNC was obtained using the maltotriose and maltoheptaose additives. For the smaller oligomers, the S-BNC forms the strongest complex with the carbohydrate, whereas for the larger fragments, R-BNC is the preferred complexing enantiomer. Molecular modeling studies[70] have indicated that the conformation of these MDs becomes more helical in structure as the number of glucose residues increases. This model suggests that the primary hydroxyl groups are located on the exterior of the helix with a hydrophobic interior. For maltoheptaose, a helical turn was found to be evident.[70] Close similarities in the conformations of acyclic and cyclic oligosaccharides indicate that chiral discrimination for analytes such as BNC may have similar mechanisms. The participation of stereoselective hydrogen bonding between the polar groups on the analyte and the MD has been demonstrated for some related binaphthyl derivatives;[70] only the polar compounds (1,1-binaphthyl-2,2-diyl hydrogenphosphate (BNP)(**24**), 2,2-dihydroxy-1,1-binaphthyl-dicarboxylic acid (HBNC)(**25**)) were chirally resolved, whereas for the neutral analog 1,1-bi-2-naphthol (BN)(**13**), the enantiomers were unresolved.

Increasing the MD concentration in the separation buffer can enhance enantiomeric resolution for certain analytes. The resolution and selectivity of racemates of ibuprofen (**26**), warfarin (**27**), ketoprofen (**28**), and simendan (**29**) were improved[69] using up to 15% (v/v) Dextrin 10 in a TAPS/Tris buffer (Figure 5.14). The opposite effect was observed[69] for

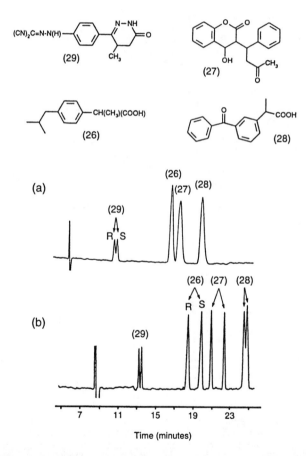

Figure 5.14 Enantiomeric separation of four NSAID racemates (ibuprofen (**26**), warfarin (**27**), ketoprofen (**28**), and simendan (**29**)) using Dextrin 10 concentrations of (a) 5% (w/v and (b) 15% (w/v). Peaks 1, 2, 3, and 4 relate to (**29**), (**26**), (**37**), and (**28**), respectively. Separation conditions: buffer, 20 m*M* TAPS/6.5 m*M* Tris and 4% (v/v) ethanol, pH 7.7, and 5 to 15% (w/v) Dextrin 10; capillary, 600 mm × 0.05 mm i.d. (effective length, 440 mm); applied voltage, 26 kV; detection, 220 nm; temperature, ambient. (From Soini, H. S., Stefansson, M., Riekkola, M.-L., and Novotny, M. V., *Anal. Chem.*, 66, 3477, 1994. With permission.)

simendan, where optimum separation conditions were achieved using lower concentrations (5% (v/v)) of the MD additive. This was thought to be due to stereoselective hydrogen bonding interactions of the polar nitrogen atom in simendan with hydroxyl groups on the oligosaccharide.

The nature of the buffer has a profound effect on the conformation of the oligosaccharide most probably due to solvation effects. This has been demonstrated[69] for the separation of (27) and (28) racemates using 20% (v/v) Dextrin 10 dissolved in different ionic strengths of a TAPS/Tris buffer system. Optimum enantioresolution and increased migration times were obtained[69] using low buffer concentrations of 20 mM TAPS/6.5 mM Tris at pH 7.7. This enhancement is caused by the limited availability of zwitterionic buffer species for both the capillary wall and dynamic complexation with Dextrin, and analytes may influence certain equilibria at the wall. Other observed effects include band broadening, particularly for the enantiomers of ibuprofen, indicating that multiple complexation of enantiomers with the Dextrin occur at different rates.

2. Charged Polysaccharides

Dextran sulfate, heparin, and chondroitin sulfate C are examples of charged polysaccharides that have been used in CE for chiral separations. Both chondroitin sulfate C and heparin are sulfated polysaccharides consisting of α1-4 linked glucosamine and uronic acid residues. These compounds are water soluble and are usually found in connective tissue or mast cell granules. Heparin is a naturally occurring glycosaminglycan derived from a variety of tissues including intestinal mucosa and lungs, whereas chondroitin sulfate C is derived from cartilage. For dextran sulfate, the monomeric units are composed of sulfated glucose residues linked *via* an α1-6 glycosidic link.

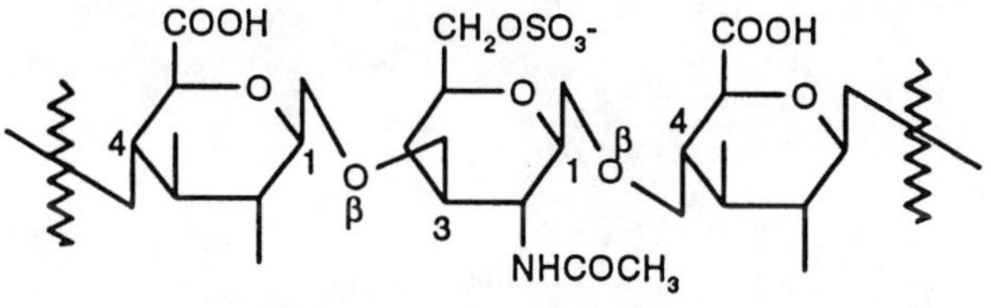

CHONDROITIN SULPHATE C

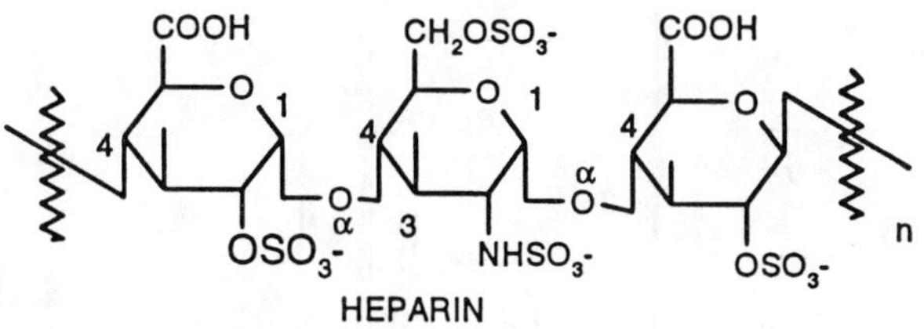

HEPARIN

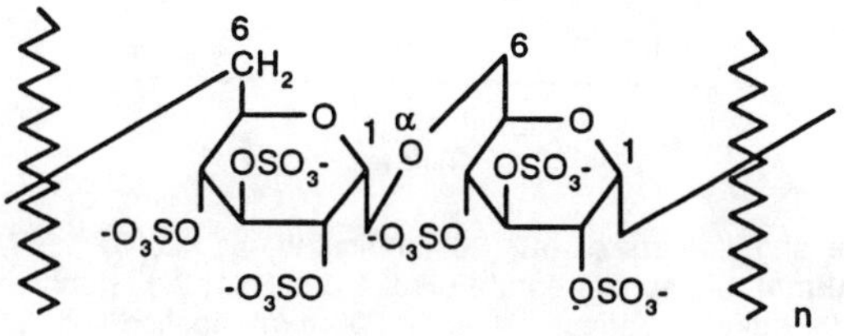

DEXTRAN SULPHATE

Molecular weights (MW) range from about 7300 (for dextran sulfate) up to 50,000 Da (for heparin and chondroitin sulfate C). Between the pH values 3 and 7, the three ionic polysaccharides are negatively charged, hence, their electrophoretic mobilities are dependent on the ionization of the carboxylic and sulfate groups. A comparison of the electrophoretic mobilities of the three charged polysaccharides indicates that chondroitin sulfate C has the lowest migration characteristics, which may play an important role in chiral resolution.

The three polysaccharides (chondroitin sulfate C, heparin, and dextran sulfate) exhibit wide enantioselectivity toward many chiral compounds. The racemates of compounds such as diltiazem derivatives, laudanosine analogs, trimetoquinol, some beta-blockers, and primaquine have all been resolved using these chiral selectors. Typically for chondroitin sulfate C, a low pH (pH 2.4) buffer system is required primarily because of its small ionic character under these acidic conditions. The enantioseparation of racemates of diltiazem (30) (Figure 5.15) and compounds related to trimetoquinol (17) have been resolved[71] using a phosphate–borate buffer system pH 2.4 in the presence of 3% (v/v) chondroitin sulfate C (Figure 5.15). In the case of heparin, more neutral conditions between pH 6.0 and 6.5 were required[72] for the enantiomeric separation of 8-chlorordiltiazem (31) and chlorpheniramine (32) (Figure 5.16). Similar conditions were necessary for resolving the enantiomers of trimetoquinol and diltiazem using dextran sulfate.[73]

Heparin has been used[72] to resolve the enantiomers of a range of antimalarial and antihistaminic drugs. These chiral compounds share the following common structural features which may be important in choosing this chiral additive. First, these analytes have a nitrogen-

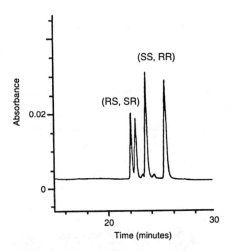

Figure 5.15 Separation of enantiomers of (a) (30) (indicated by SS, RR, SR, and RS) Separation conditions: 20 mM phosphate-borate pH 2.4 and 3% (v/v) chondroitin sulfate C; capillary, 570 mm × 0.075 mm i.d. (effective length, 500 mm); applied voltage, 20 kV; detection, (a) 235 nm and (b) 230 nm; temperature; 23°C. (From Nishi, H., Nakamura, K., Nakai, H., and Sato, T., *Anal. Chem.*, 67, 2334, 1995. With permission.)

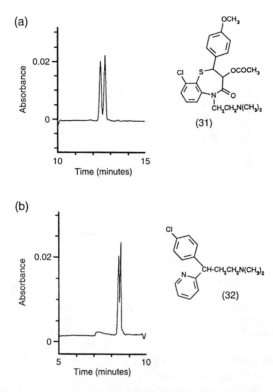

Figure 5.16 Separation of the enantiomers of (a) (**31**) and (b) (**32**) using heparin. Separation conditions: buffer, 20 m*M* phosphate-borate, (a) pH 6.0 and (b) 6.5; capillary, 570 mm × 0.075 mm i.d. (effective length, 500 mm); applied voltage, 20 kV; detection, 214 nm; temperature, 23°C. (From Stalcup, A. M. and Agyei, N. M., *Anal. Chem.*, 66, 3054, 1994. With permission.)

containing heterocyclic ring system. Second, many of the racemates that have been resolved contained at least one other amine group in their structure. And third, these analytes are mono-protonated in the pH range 5 to 6. For the antimalarial compounds, it was also shown[71] that ionization of the nitrogen atoms, relative ring size, and hydrophobicity play important roles in the chiral discrimination mechanism involving heparin. Similarly for the antihistamine drugs, the role of hydrophobicity was found to be relevant.[72] These observations are not surprising because data from X-ray crystallographic studies[72] have indicated that the width of the heparin molecule is similar to that of various cyclodextrins.

Varying the buffer pH value can have a profound effect on the resolution and migration times of chiral compounds analyzed using these charged polysaccharide additives. This is particularly evident for chondroitin sulfate C, where the ionic nature of this molecule and hence its contribution to solute migration and enantioseparation are influenced by the acidity of the buffer. This was examined[73] for the enantioseparation of (**30**) and (**17**) using 3% (v/v) chondroitin sulfate C in phosphate buffer in the pH range 3 to 7. Optimum enantioresolution for these racemates was obtained at low pH values between 3 and 5. Interestingly, for trimetoquinol, migration order for the R- and S-isomers changed between pH 2.4 and 2.8. This was due to differences in the ionic interaction between the cationic trimetoquinol and the charged additive under acidic conditions. Varying the pH value in buffer containing dextran sulfate has less dramatic effects.[71,73]

Increasing the concentration of the chiral additive enhances the enantiomeric resolution. Phosphate buffers containing up to 3% (v/v) chondroitin sulfate C showed[73] a marked improvement in enantioselectivity toward diltiazem (**30**) and 8-chlorodiltiazem (**31**) (Figure 5.17). This indicates that for effective interaction to occur between the additive and the solute,

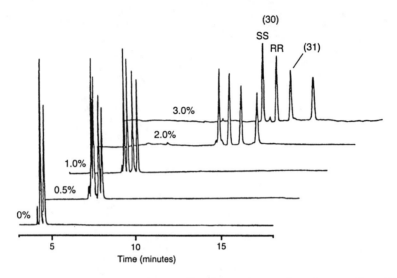

Figure 5.17 Effect of chondroitin sulfate C concentration on the enantiomeric separation of (**30**) and 8-chlorodiltiazem (**31**). Separation conditions: buffer, 20 mM phosphate-borate, pH 2.8 with 0 to 3% (w/v) chondroitin sulfate C; capillary, 570 mm × 0.075 mm i.d. (effective length, 500 mm); applied voltage, 20 kV; detection, (a) 235 nm; temperature, 23°C. (From Nishi, H., Nakamura, K., Nakai, H., and Sato, T., *Anal. Chem.*, 67, 2334, 1995. With permission.)

the concentration of the chiral additive should be at optimum levels of about 3% (v/v). Similar observations were made for dextran sulfate[71,73].

C. Chiral Crown Ethers

Chiral crown ethers are a class of chiral selectors that have been used successfully in chromatographic and capillary electrophoretic analysis to resolve the enantiomers of amino acid racemates and other primary amines. Crown ethers are large cyclic polyethers with an inclusion cavity formed by an electron-donating heteroatom (usually oxygen alone or oxygen and nitrogen) linked together by methylene units. By derivatizing the crown ether with chiral functional groups, chiral discrimination of optical isomers can occur either *via* a steric barrier mechanism or by stereoselective hydrogen bonding between the guest molecule and the carboxylic acids on the crown ether (to form host–guest inclusion complexes). The first useful chiral crown ether, namely (2R, 3R, 11R, 12R)-(+)-1,4,7,10,13,16-hexaoxacyclooctadecane-2,3,11,12-tetracarboxylic acid (abbreviated to 18-C-6-TCA) (**33**) was synthesized by Lehn and co-workers,[74] and it is now a commercially available product.

(**33**)

The size of the cavity is suitable for potassium ions (K$^+$), ammonium ions (NH$_4^+$) or protonated primary alkyl amines (RNH$_3^+$), which can hydrogen bond to form a tripod arrangement with electron-donating groups on the chiral crown ether.

The earliest reported applications of 18-C-6-TCA for enantiomeric separations in CE involving simple aromatic amino acids racemates were by Kuhn et al.[75,76] The utility of this chiral selector has now been extended to a range of more complex racemic analytes including allylic and alkenic GABA derivatives,[77] dipeptides,[78] dihydroxy-2-aminotetralins,[79,80] and aminomethyl-benzodioxanes.[79] All enantioseparations involving chiral crown ethers are carried out at high acidities (typically, pH 2) either with the additive alone or in the presence of a buffer system. Phosphate at low concentrations is usually the buffer of choice, but for certain applications,[78] e.g., separation of the dipeptide DL-leucyl-DL-leucine, a buffer consisting of Tris adjusted to pH 2 with citric acid was found to be most suitable. For the enantioseparation of glycyl-DL-dipeptides, no background buffer was required. The enantiomers of some DOPA[76] (34) and tyrosine analogs[76] (where X = F or H) (35) were resolved using 30 mM 18-C-6-TCA in a sodium phosphate buffer. Interestingly, DOPA derivatives where the group at the R_1 position were either alkyl, alkenic, or allylic were not chirally resolved. This lack of chiral resolution could be due to the low stability constant reported for complexes between the crown ether and primary amines of this type which have three substituents on the α-carbon. Kuhn et al.[75,76] showed that the enantiomers of the secondary amine ephedrine were not resolved, whereas enantioseparation occurred for the racemate of its primary analog norephedrine. These data support the chiral discrimination mechanism that involves complexation of the alkylammonium cation within the crown ether.

(34)   (35)

Increasing the concentration of the chiral crown ether in the buffer generally leads to an increase in the migration time and enantiomeric resolution. However, an optimum level was usually reached above which enantioresolution begins to diminish. This concentration is dependent on the analyte and the stability of the resulting inclusion complex. For some diastereoisomeric dipeptides[78] and 5,6-dihydroxy-2-aminotetralins,[79] this value is between 10 and 25 mM, whereas for GABA and DOPA analogs, a concentration of 30 mM was suitable. The increased migration times are due to a reduced electroendosmotic flow caused by removal of sodium ions that shield the capillary silanol groups by the crown ether. At concentrations of up to 100 mM, the Rs values improved for two of the four racemates. The later migrating peak corresponded to the isomer that forms a stronger complex with the crown ether. Unfortunately, at these high concentrations of this additive, precipitation of the crown ether can occur.

Other important experimental parameters that can affect enantiomeric resolution include temperature and applied voltage. As expected, migration times and hence resolution decrease with increasing temperature because of the lower viscosity of the buffer, which will lead to higher diffusion rates and sometimes band broadening. At high voltages, Joule heating becomes significant, and hence an optimum voltage is required.

As has already been mentioned, chiral discrimination by chiral crown ethers involves either a steric or hydrogen-bonding interaction between the analyte and the carboxylic acid moieties on the crown ether. In both of these mechanisms, the most important structural requirement for complexation is a primary amino functional group to allow hydrogen-bonding interactions at low pH values. Kuhn et al.[75,76] previously reported the enantioresolution of aromatic amino acid racemates where the chiral center was in close proximity to polar groups. In the case of glycyldipeptides,[78] where the asymmetric center is *delta* to the primary amine, best resolution was achieved for the bulkier branched isomers. Both diastereoisomeric peptides DL-leucyl-DL-leucine and DL-leucyl-DL-phenylanaline were baseline resolved into four stereoisomers using 25 mM 18-C-6-TCA. Similar observations were made for some analogs of DOPA[77] and tyrosine.[77] Here it was observed that compounds with three substituents on the carbon atom *alpha* to the amino group were not resolved. In addition, the introduction of a double bond in the position *beta* to the amino group improved steric differentiation. For the substituted aminotetralins,[80] the substitution position of the amino on the cyclohexane ring is important. The compound 2-aminotetralin was better resolved than the 1-amino analog.

For larger bicyclic and tricyclic aromatic amines such as the aminonaphthalenes[80] and aminophenanthrenes,[80] the effect of ring substitution and structure is even more evident. Dimethyl substitution at the 3-position for aminonaphthalene derivatives resulted in baseline resolution, whereas methyl substitution at the amino position prevents complexation. For the aminophenanthrenes, which have a higher hydrophobicity, ionization of the molecule was found to be as important as size and orientation for favorable host–guest complexation.

D. Chiral Surfactants

The use of chiral surfactants in micellar electrokinetic chromatography (MEKC) is now firmly established as an important separation mode in CE for optically active compounds. The separation system in chiral MEKC consists of two phases, the aqueous background electrolyte and a charged micellar phase which possesses chiral functional headgroups. The chiral discrimination mechanism is unclear. However, the first step may involve stereoselective interactions between functional groups on the chiral analyte and the optically active surface of the micelle. This may then be followed by inclusion of the preferred isomer into the hydrophobic interior of the micelle. The observed migration order relates to differential partitioning of both enantiomers between the two phases. The currently available chiral surfactants can be divided into two main groups: (1) detergents derived from naturally occurring molecules, e.g., bile salts, digitonin, saponins, and CHAPS derivatives, and (2) synthetic surfactants derived from simple chiral molecules such as amino acid enantiomers, derivatized glucose, and other charged chiral compounds.

1. Naturally Occurring Surfactants

There are many optically active molecules in nature that exhibit detergent-like properties and hence have found uses in chiral MEKC. Examples are bile salts, digitonin, saponins, and CHAPS derivatives. These compounds share the same steroidal molecular skeleton in their structure.

a. Bile Salts

The steroidal ring structure in bile salts consists of four ring systems with sidechains consisting of a carboxyl moiety bonded to either a taurine or a glycine residue. The conformation of the bile salt is such that the hydroxyl groups are oriented in the same position, perpendicular to the steroidal skeleton (**36**).

(36)

Bile Salt	R_1	R_2
(37)	OH	ONa
(38)	H	ONa
(39)	OH	$NHCH_2CH_2SO_3Na$
(40)	H	$NHCH_2CH_2SO_3Na$

Bile salts aggregate *via* the stepwise addition of detergent monomers held together by hydrophobic interactions between the nonpolar faces. The resulting primary bile salt micelles are generally smaller than the synthetic surfactant aggregates and are composed of up to 11 monomers. At high surfactant concentrations, larger cylindrical micelles are formed which are held together by intermolecular hydrogen bonding. Commonly used bile salts are sodium cholate (SC, (37)), sodium deoxycholate (SDC, (38)), sodium taurocholate (STC, (39)), and sodium taurodeoxycholate (STDC, (40)). The bile salts derivatized with taurine groups dissolve easily in water, whereas the others require buffers at pH >/ = 7. STDC (40) is the most widely used of all the bile salts because it has the broadest enantioselectivity and its effectiveness as a chiral selector may lie in its lack of a hydroxy group at the C-7 position and the ability of the sulfonate group on the taurine moiety to participate in electrostatic interactions. Stereoselective interactions, such as hydrophobic, steric, and repulsive, may also contribute to the utility of STDC as a chiral additive.

Examples of applications of STDC include resolution of dansyl-DL-amino acid racemates,[81] binaphthyl derivatives,[82] diltiazem (30) and related compounds,[83] (Figure 5.18) and some polycyclic compounds.[83] The chiral discrimination mechanism involving the use of bile salts is unclear, but many of the optically active analytes resolved using this detergent have rigid structures with restricted rotation often like polycyclic aromatic compounds. This reflects structural similarities between the bile salt and the types of analytes that are chirally resolved using this chiral selector.

b. Digitonin

Digitonin is a naturally occurring detergent extracted from the seeds *of Digitalis purpurea*. It has a steroid core structure that is glycosylated at the 3α-hydroxy and has two additional oxygen-containing rings (41).

Digitonin has no ionizable groups and hence is largely insoluble in water, but in the presence of a nonchiral surfactant such as SDS, a mixed micelle is formed which can then be utilized in aqueous buffer solutions. At pH 3.0, the digitonin/SDS mixed micelle system was able to resolve the enantiomers of some PTH-DL-amino acids.[84] One drawback of this method is the long migration times as a consequence of the reduced electroendosmotic flow at low pH.

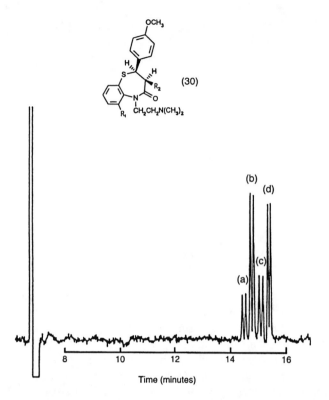

Figure 5.18 Separation of enantiomers of **(30)** and related compounds by MEKC with bile salts. Where, (a), R_1, Cl;R_2, OH, (b) R_1, Cl;R_2, OCOCH$_3$, (c) R_1, H; R_2, OH and (d), R_1, H; R_2, OCOCH$_3$. Separation conditions: buffer 20 mM phosphate-borate pH 7.0 and 50 mM STDC; capillary, 650 mm × 0.05 mm i.d. (effective length, 500 mm); applied voltage, 20 kV; detection, 210 nm; temperature, ambient. (From Nishi, H., Fukuyama, T., Matsuo, M., and Terabe, S., *J. Chromatogr.*, 515, 233, 1990. With permission.)

(41)

c. Saponins

Saponins are naturally occurring triterpene glucosides that possess detergent-like properties. Two commercially available examples are glycyrrhizic acid (**42**) and β-escin (**43**). The chemical structure of these detergents has the same steroidal skeleton with carboxylated carbohydrates attached to the α ring and alkyl functional groups attached at the ε-ring.

(42)   (43)

The triterpene moiety constitutes the hydrophobic part, whereas the sugar groups are hydrophilic. In the presence of nonchiral detergents, such as SDS, Tween-20, or octyl glucoside, the resulting mixture will form gels. However, enantiomeric separations were possible with the coaddition of octyl-β-D-glucoside and SDS to a phosphate/borate buffer solution containing glycyrrhizic acid. This mixed micelle system was used to resolve the enantiomers of some dansyl-DL amino acids.[85] Chiral mixed micelles consisting of β-escin and SDS were used to separate PTH-amino acid racemates successfully. A disadvantage of using these saponins in mixed micellar systems is their tendency to slowly form gels.

2. Synthetic Surfactants

Synthetic chiral surfactants have a chemical structure that consists of a charged polar headgroup linked to a hydrophobic alkyl chain usually a dodecyl unit. At concentrations above the critical micellar concentration (cmc), the detergent monomers aggregate to form micelles. These micelles are in dynamic association-dissociation equilibrium with the monomeric species in the bulk water phase. The structure of the outer surface of the micelle is defined by the nature of the charged headgroup. These can either be enantiomers of amino acid derivatives, other chiral acids, or sugar derivatives. These chiral detergents are soluble in aqueous solutions and can be added directly to the separation buffer. Six examples of synthetic chiral surfactants ((**44**) to (**49**)) have been reported recently.

The earliest examples[86] of chiral MEKC separations using this type of surfactant involved a chiral comicellar system consisting of equimolar concentrations of N-dodecanoyl-L-valinate (R = Valine, **44**) (SDVaL) and SDS to resolve the enantiomers of N-(3,5-dinitrobenzoyl)-O-isopropyl esters (DNTBP) of four amino acids racemates. The migration order indicated that the least hydrophobic amino acids migrated earlier than the nonpolar ones and that the later migrating DNTBP-L-isomers were interacting more strongly with the chiral component of the micelle. A similar mixed micelle system was used to resolve the enantiomers of some phenylhydantoin DL-amino acids (PTH). For PTH and DNTBP derivatives,[87] coadditives such

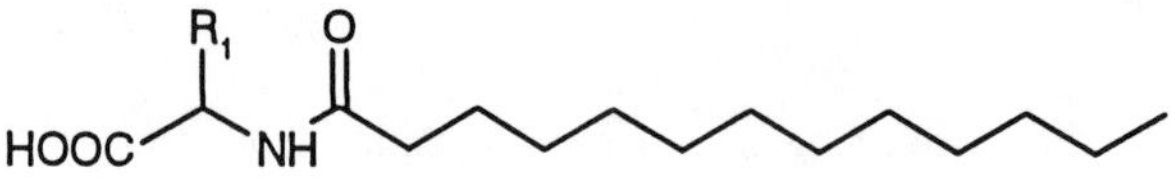

(44)

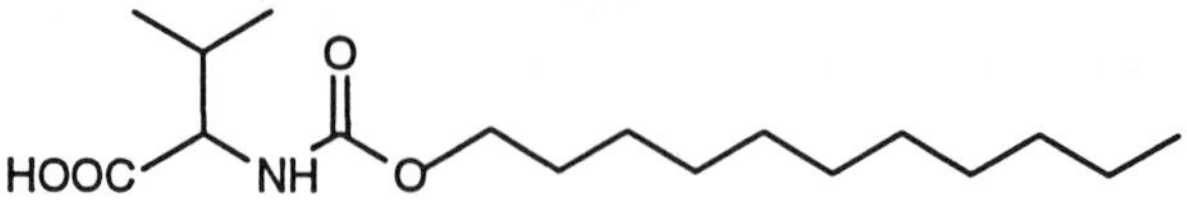

(45)

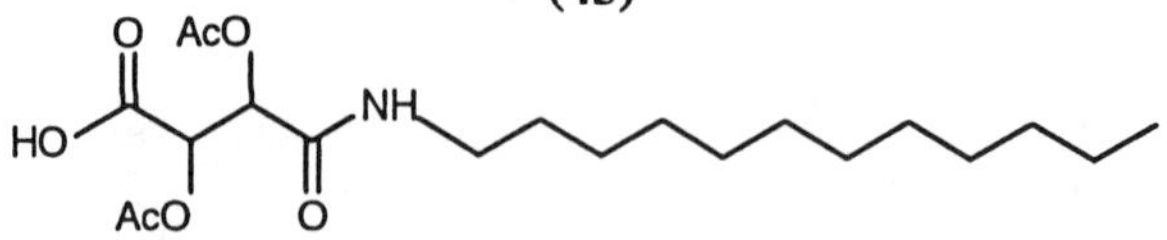

(46)

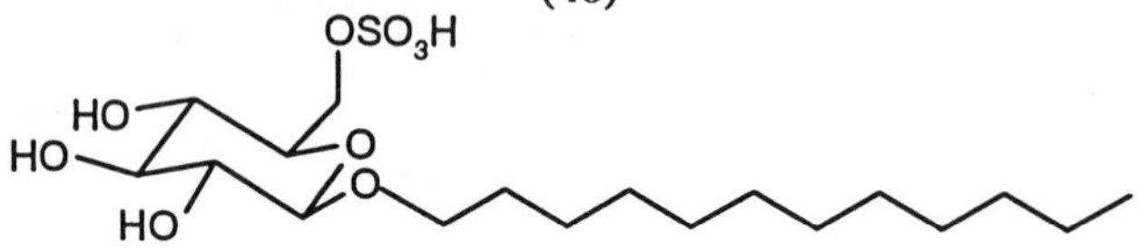

(47)

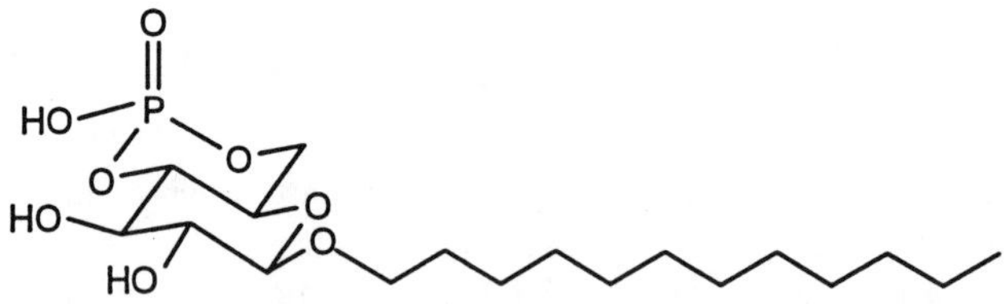

(48)

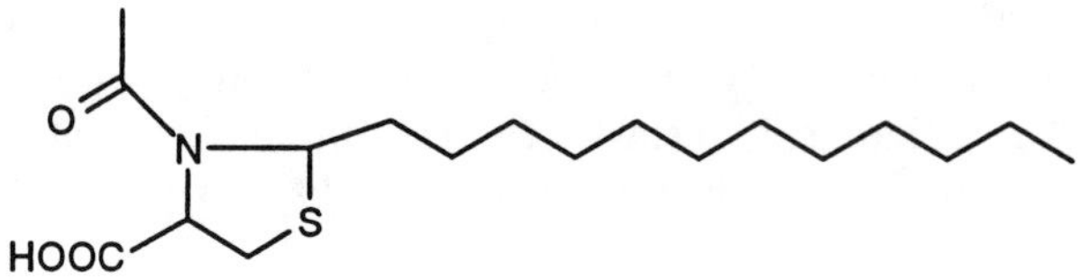

(49)

as methanol and urea were used primarily to increase the migration time window by reducing electroendosmotic flow and to improve peak shape. Other applications of this mixed micellar system include enantioseparations of warfarin and benzoin.[88] Micellar systems consisting solely of SDVal micelles are also capable of enantiomeric resolution. The structure of the amino acid will affect the resolution of the two isomers. Generally, increasing the steric bulk of the α-alkyl groups enhances chiral discrimination.[88] Moreover, the derivatizing group may also contribute to the enantioseparation mechanism. In the case of DNB derivatives, steric effects exerted by the benzoyl group and electrostatic binding interactions by nitro-groups were thought to be important.

A range of other amino acid ester derivatives has been used in the synthesis of chiral surfactants, e.g., L-threonine[88] (R = threonine, 44) and L-alanine[89] (R = alanine, 44). In comparison with SDVal, these detergents show lower α values for DNB-amino acid. The reasons for these differences are unclear. However, data from H1 NMR pyrene fluorescence studies[88] have indicated that differences in the shielding of the amide moieties from the bulk water phase and differences in the micellar packing arrangement were contributing factors. Moreover, increasing the steric bulkiness of the amino acid (e.g., SDVal) may lead to a less densely packed micelle which is much more susceptible to perturbation by the chiral analyte.

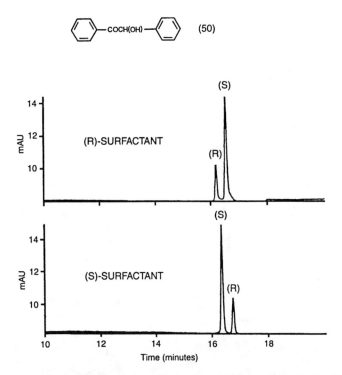

Figure 5.19 Migration order reversal of 3 (S): 1 (R) benzoin **(50)** using either (a) R-DDVal **(45)** or (b) S-DDVal **(45)**. Separation conditions: buffer 25 m*M* sodium phosphate and 25 m*M* sodium tetraborate, pH 8.8 and 25 m*M* DDVal; capillary, 350 mm × 0.05 mm i.d.; applied voltage, 30 kV; detection, 214 nm; temperature, ambient. (From Mazzeo, J. R., Grover, E. R., Swartz, M. E., and Peterson, J. S., *J. Chromatogr.*, 680, 125, 1994. With permission.)

Similar observations were made when surfactants synthesized from alanine and phenylalanine were compared.

Another recently introduced[90] class of synthetic chiral surfactants are (R)- and (S)-N-dodecoxycarbonyl-valines (DDVal) **(45)**. These detergents have structures similar to SDVal except that a carbamate bridge now links the valine headgroup with dodecanoyl alkyl chain. These compounds have been shown to resolve a range of twelve pharmaceutical drugs including beta-blockers, bupivacaine, and homatropine using a phosphate/borate buffer at pH 8.8 in the presence of 25 m*M* surfactant. The synthesis of both antipodes of DDVal allowed the reversal of the enantiomeric migration order of benzoin **(50)**[90] (Figure 5.19). This is particularly important in quantitation of low levels of chiral impurities in a sample consisting predominantly of one isomer. In comparison with SDVal, DDVal has a broader enantioselectivity and is useful at pH values >7.

Dodecyl-β-D-glucopyranosyl derivatives (**47** and **48**) represent a novel class of chiral surfactant recently introduced by Tickle et al.[91] The bulky glucose headgroup contains five chiral centers and up to three stereogenic hydroxyl groups which can be involved in hydrogen-bonding interactions. The charged moiety, either cyclic phosphate or sulfate groups attached at the C-6 position of the glucose ring, enables the surfactant to be highly soluble in aqueous buffer systems. Other advantages of this class of surfactant are the low UV absorptivity at low wavelengths and the low cmc values (between 0.5 and 1.0 m*M*). A variety of racemic mixtures were resolved[91] using this surfactant with structures ranging from dansyl derivatized amino acids to large binaphthyl derivatives. Figure 5.20a shows the separation of mephenytoin **(51)** and its 4-hydroxy derivatives **(52)** using a phosphate/borate buffer system at pH 8 in the presence of 45 m*M* of the phosphated surfactant **(48)**. For the enantioresolution of two barbiturates, namely hexobarbitol **(12)** and phenobarbitol **(53)**, using the sulfated detergent

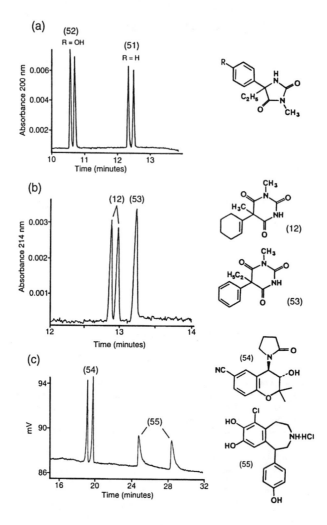

Figure 5.20 Electropherograms showing the enantiomeric resolution of (a) mephenytoin (**51**) and 4-hydroxy derivative (**52**) using the phosphated surfactant (**48**), (b) hexobarbitol (**14**) and phenobarbitol (**53**) using the sulfated surfactant (**47**) and (c) cromakalim (**54**) and fenoldapam (**55**) using surfactant (**49**). For (a) and (b) Separation conditions: buffer, 50 mM phosphate-borate, pH 8.0 and 45 mM surfactant; capillary, 570 mm × 0.05 mm i.d. (effective length 500 mm); applied voltage, 20 kV; detection, 214 nm; temperature, 25°C. For (c), Separation conditions: buffer, 50 mM borate/25 mM phosphate buffer pH 7.0 and 25 mM surfactant; capillary, 670 mm × 0.05 mm i.d.; applied voltage, 12.5 kV; detection, 254 nm; temperature, 20°C. (Panels a and b from Tickle, D. C., Okafo, G. N., Camilleri, P., Jones, R. F. D., and Kirby, A. J., An*al. Chem.*, 66, 4121, 1994. With permission.)

(**47**) (Figure 5.20b) only the hexobarbitol was chirally resolved, indicating the sensitivity of this separation system to changes in analyte structure, e.g., from a cyclohexane-containing ring system (i.e., hexobarbitol) to one containing a phenyl ring (i.e., phenobarbitol).

Other novel chiral surfactants include 2-undecyl-4-thiazolidine carboxylic acid (**49**), which can be synthesized from either L- or D-cysteine hydrochloride.[92] This additive was used to resolve the enantiomers of the racemic drugs cromakalim (**54**) and fenoldapam (**55**) (Figure 5.20c). As in the case of the R- and S-dodedoxycarbonylvalines, reversal of the migration order of related isomers was achieved simply by replacing an L-cysteine derived surfactant with the corresponding surfactant synthesized from D-cysteine. A surfactant synthesized from (R,R)-tartaric acid and long-chain alkyl amines has been used in the enantiomeric resolution of compounds with fused polyaromatic rings, e.g., binaphthol.[93] The dodecyl derivative of 6-

aminopenicillanic acid (6-APA) has been shown to be especially useful for the resolution of racemates related to warfarin and its hydroxymetabolites.[94]

The cholamide surfactants, N,N-bis(3-D-gluconamidopropyl)cholamide (Big CHAP) or deoxycholamide (deoxy Big CHAP), were recently used by El Rassi and Mechref[95] as chiral selectors in CE. These detergents are synthesized from cholic acid, and their structure consists of a steroidal portion with two polyhydroxylated chains. CHAPS and Big CHAPS have been used to resolve enantiomers of dansyl amino acids and some herbicides. The resolution efficiency of these chiral additives was found to be dependent on temperature and the use of organic solvents.

The stability and rigidity of synthetic detergent micelles can be enhanced by the formation of polymeric chiral micellar species. Poly(sodium N-undecylenyl)-1-valinate (pSUVal) consists of monomers of sodium undecenoyl-L-valinate covalently linked *via* UV irradiation of the terminal vinyl groups. Polymeric surfactant has no effective cmc in water (cf. SDVal) with a mean molecular weight of 10,000 Da. The lack of cmc value for this polymeric micelle has the advantage that chiral separations are not dependent on surfactant concentration, unlike SDVal, and typically 0.5% (w/v) of the detergent is used in most buffers. The rigidity of the pSUVal means that a stable pseudostationary phase can be formed in the buffer allowing interactions between solutes and the micelle to be simplified and enhanced. Analytes such as DNB amino acid isopropyl esters, (+/–)-1,1′-bi-2-naphthol and DL-laudanosine have also been chirally resolved using pSUVal.[87] The more compact structure of pSUVal means that solutes do not penetrate as deeply into the micelle compared to nonpolymerized micelles. Thus the observed broadened peaks which are normally associated with slower mass transfer rates are not generally observed for polymerized micelles. This is demonstrated[88] in Figure 5.21 for the enantioresolution of the binaphthyl derivative using either monomeric sodium N-undecylenyl-L-valinate or the polymeric analog, pSUVal. Varying experimental parameters such as pH and coadditives can influence enantiomeric resolution. Changing the buffer pH influences the conformation of the anionic polymeric micelle, making it less compact at higher pH values because of electrostatic repulsion. For the enantioresolution of DL-

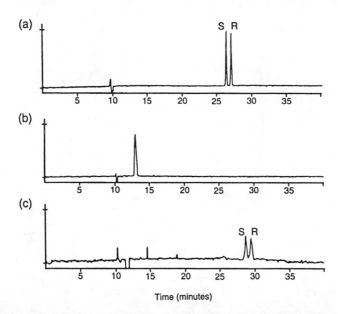

Figure 5.21 Comparison between polymerized micelle (pSUVal) and nonpolymerized micelle for the separation of (13). (a) 0.5% (w/v) pSUVal, (b) 0.5% (w/v) SDVal, and (c) 1% (w/v) SDVal. Separation conditions: buffer, 25 m*M* borate at pH 9.0; applied voltage, 12 kV; detection, 290 nm; temperature, ambient. (From Wang, J. and Warner, I. M., *Anal. Chem.,* 66, 3773, 1994. With permission.)

laudanosine,[96] increasing the pH from 9.0 to 10.0 improved resolution, probably because of more favorable interactions which can occur with the less compact polymeric micelle. The coaddition of SDS and urea to buffers containing pSUVal improved peak shape by reducing peak tailing and solute retention. Both of these coadditives can also have a disruptive effect on the water structure surrounding the micelles.

E. Protein Additives

Many proteins have been used successfully in high performance liquid chromatography (HPLC) as chiral selectors for the resolution of enantiomers. In HPLC, the proteins are usually permanently immobilized on a solid support forming a chiral stationary phase (CSP). Examples of protein stationary phases are bovine serum albumin (BSA), $\alpha 1$-acid glycoprotein (AGP, orosomucoid), human serum albumin (HSA), conalbumin (CON), fungal cellulases, ovomucoid (OVM), pepsin, and casein. The popularity of these types of CSPs is most likely due to their broad applicability to varied analytes and the fact that aqueous mobile phases (usually under reverse phase conditions) can be used, which in many cases is best suited for biological samples. Chiral discrimination in protein-based CSP involves many stereoselective bonding interactions (e.g., hydrogen bonding, pi–pi, and steric effects) which occur between groups or sites on the protein surface and the analyte. Despite their common use in HPLC, immobilized protein-based chiral columns is relatively expensive, and a range of different columns are sometimes required for successful separations. Other drawbacks include their relatively short lifetimes and low separation efficiencies. Figure 5.22 shows the increased

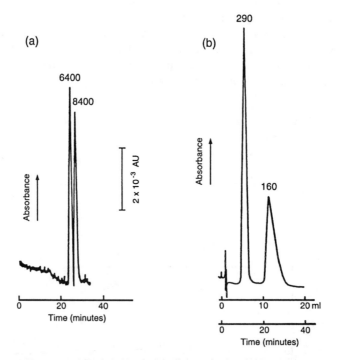

Figure 5.22 Comparison of separation efficiencies (given as plate numbers) achieved for the enantiomeric separation of N-2,4-DNP-DL-Glu in BSA-based systems in (a) CZE and (b) HPLC. Separation conditions: (a) polyacrylamide-coated capillary, 760 mm × 0.08 mm i.d. (effective length, 600 mm); buffer, 3 mg/mL albumin heated in 10 mM acetate-Tris solution for 30 min at pH 9.0 and 60°C and pH adjusted to 8.0; applied voltage, −9.5 kV; detection, 254 nm; temperature, ambient: (b) column, 150 mm × 3.3 mm i.d. Separon HEMABIO-BSA (Tessek, Prague) bonded albumin; mobile phase, 50 mM phosphate, pH 7.8 and 3% (v/v) 1-propanol; flow rate, 0.5 mL/min; temperature, ambient. (From Vespalec, R., Sustacek, V., and Bocek, P., *J. Chromatogr.*, 638, 255, 1993. With permission.)

separation efficiency achieved[97] for the chiral resolution of N-2,4-DNP-D,L-glutamic acid using a BSA-based system either with the protein bonded to an organic polymeric matrix in HPLC or as a buffer additive in capillary electrophoresis (CE).

In CE, the use of protein additives in the resolution of enantiomers has been conceived as an attractive and simple alternative to the HPLC approach. The term affinity electrokinetic chromatography (AEKC) has been applied to this mode of CE. In contrast to the LC approach, proteins are added to a buffered solution without the need for expensive permanent immobilization methods (cf. HPLC). Higher separation efficiencies, shorter run times, and smaller sample requirements are also important advantages of this method.

The mechanism of chiral discrimination involved in AEKC is very complex. It is clear that the conformation of the protein is vital for stereoselective recognition to occur. This has been demonstrated[98] for the enantiomeric resolution of labetolol using either cellobiohydrolase 1 (CBH1) immobilized to a continuous polymer bed in a chiral HPLC column or the free enzyme in AEKC. Baseline separation of the RR/SS and RS/SR isomers of labetolol could only be achieved[98] using the CE method. It appears that immobilizing the enzyme in this way may have induced some structural changes. Another example[99] is that for resolution of the isomers of R,S-ofloxacin (**56**) using HSA either coated onto a CE capillary or added to the running buffer. Chiral resolution was only achieved[100] using the system with HSA as an additive (Figure 5.23).

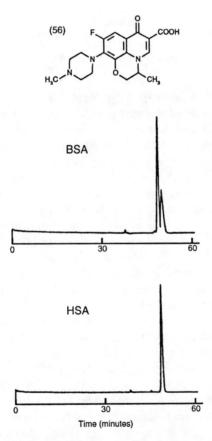

Figure 5.23 Comparison of electropherograms showing the enantiomeric separation of ofloxacin (**56**) using either BSA or HSA as the chiral selector. Separation conditions: buffer, 100 mM phosphate pH 7.0 and 0.3% (w/v) BSA or HSA; capillary, 500 mm × 0.05 mm i.d. (effective length, 450 mm); applied voltage, 10 kV; detection, 300 nm; temperature, ambient. (From Arai, T., Ichinose, M., Kuroda, H., Nimura, N., and Kinoshita, T., *Anal. Biochem.*, 217, 7, 1994. With permission.)

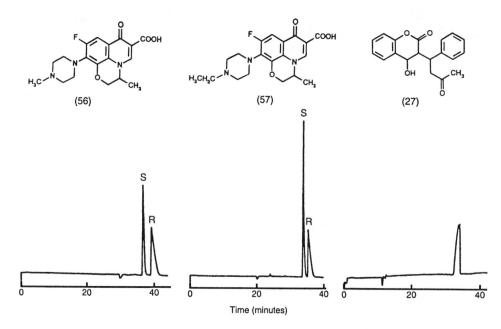

Figure 5.24 Electropherograms showing enantiomeric separations of (**56**), DR-3862 (**67**) and (**27**) using BSA. Separation conditions: buffer, 100 mM phosphate, pH 8.0 and 0.4% (w/v) BSA; capillary, 450 mm × 0.05 mm i.d. (effective length, 350 mm); applied voltage, 30 kV; detection, 300 nm; temperature, 35°C. (From Arai, T., Ichinose, M., Kuroda, H., Nimura, N., and Kinosihita, T., *Anal. Biochem.*, 217, 7, 1994. With permission.)

Some structural studies[100] of CBH1 have revealed a catalytic active center that can be inhibited by cellobiose and also diminishes enantioselectivity. Another factor that appears to be important in the separation mechanism is the relative hydrophobicity of the protein and the analyte. In the chiral separation of (**56**) and DR-3862 (**67**) (both novel quinolone antibacterial drugs) and warfarin (**27**) (an anticoagulant) using 0.4% (w/v) BSA, only the polar quinolone compounds were resolved[102] into their respective enantiomers (Figure 5.24). However, warfarin enantiomers were well separated on a BSA-conjugated HPLC column.[102] BSA is a hydrophobic protein and hence will interact strongly with the more nonpolar warfarin. It was concluded that too strong an affinity between the protein and the analyte prevents chiral discrimination. The primary structure of the protein has been shown to play an important role in chiral discrimination. Replacing BSA with HSA resulted in no enantiomeric separation of ofloxacin (OFX).[102] It appears that despite similarities in the physicochemical properties of both proteins, differences in their amino acid composition influences the chiral recognition process.

As additives in CE, proteins are not ideal principally because of their strong background absorbance at low UV wavelengths, typically between 190 and 220 nm, due to amide and aromatic groups. This results in low detection sensitivity and noisy baselines. In addition, the presence of many functional groups on the surface of the protein can lead to adhesion of the molecule to the inner surface of the capillary. The problem of background UV absorbance can be overcome in a number of ways. Valtcheve et al.[98] used cellulase (cellobiohydrolase I (CBH I) from the fungus *Trichoderma reesei*) added to a phosphate/buffer solution to resolve the enantiomers of a range of beta-blockers. To prevent detection of the protein in the detection cell, a plug of agarose was introduced to the injection side of the capillary close to the window. Another method[103] consisted of partially filling a polyacrylamide-coated capillary with the protein-containing separation buffer to within a few centimeters of the detection window. Under these conditions, the migration of the protein toward the window is negligible, and chiral discrimination of the basic racemates occurred as the compounds migrated through the separation buffer. This procedure termed, "partial zone separation technique," has been used to analyze a range of basic analytes such as epinastine, primaquine, chlorprenaline, and trimetoquinol using either BSA, OVM, AGP, or conalbumin. Another method reported by Lloyd et al.[104] involved the use of protein CSPs packed into a coated capillary. This form of electrochromatography using HSA and AGP immobilized onto packing particles (5 to 7 μm in size) was used to resolve the enantiomers of benzoin and temazepam. An alternative to the latter incorporates the biopolymer in a gel-like matrix. Birnbaum and Nilsson[105] used approximately 250 mM BSA crosslinked with gluteraldehyde in situ in a capillary to resolve the enantiomers of DL-tryptophan in high efficiency. Unfortunately, one disadvantage of this approach was the uncertain lifetime of the immobilized gel system.

Despite some disadvantages, the addition of proteins to the buffer system still offers the simplest approach to chiral resolution in AEKC. Prior to the start of any analysis using protein based buffers, it is important to first be aware of the isoelectric point (pI) of the biopolymer. One common feature shared by these proteins is their acidic pI values; in the pH range of 5 to 7, these biopolymers will have a net negative surface charge. These differences are primarily due to the number of exposed surface carboxyl functional groups and, in the case of the glycoproteins AGP, OVM, and cellulase, to variable sialic acid content. BSA has been used with great success for the enantiomeric separation of leucovorin, amino acids, mono- and dicarboxylic acids, and benzoin. Other proteins like AGP and ovomucoid have also been used.

1. The Effects of Additives on the Enantioselectivity of Proteins

The resolution of enantiomers in CE using protein-based buffer systems can be greatly influenced by varying the following experimental parameters, e.g., the protein concentration, the use of coadditives, pH variation, temperature, and the ionic strength and composition of the buffer. To optimize the separation of enantiomers in CE, the difference in their apparent mobilities should be maximized ($\Delta\mu_{app}$). The term $\Delta\mu_{app}$ is related to the protein concentration as shown by the following equations:

$$\mu_{app}, a = 1/(1 + K_1[P])\ \mu a \qquad \mu_{app}, p = 1/(1 + K_2[P])\ \mu p$$

$$\Delta\mu_{app} = \mu_{app,1} - \mu_{app,2}$$

where P is the protein concentration, K_1 and K_2 are the binding constants for two enantiomers, μa and μp are the electrophoretic mobilities of the analyte and the protein, respectively. The optimum protein concentration for maximum enantiomeric resolution ([P]opt) can be derived as:

$$[P]opt = (K_1\ K_2)^{-1/2}$$

Generally, values for K range between 1000 and 100,000 M^{-1}, [P]opt will range from 10 to 1000 μM. The value for [P]opt is also dependent on the type of analyte and protein used. Optimum protein concentrations for BSA range from 750 μM (for resolving the enantiomers of epinastine)[103] to between 50 to 300 μM (for native and derivatized DL-amino acids).[106] Figure 5.25 shows[100] the effects of increasing the BSA content from 0.1 to 0.5% in the buffer for the analysis of ofloxacin enantiomers (56). A similar observation was made[97] for the enantiomers of 2,3-dibenzoyl-D,L-tartaric acid. [P]opt values for other proteins ranges from 100 to 600 mM.

The effects of alcohols on the enantiomeric resolution has been demonstrated by Busch et al.[107] In this example, enhanced resolution was observed for R/S-benzoin and R/S warfarin when the concentration of 2-propanol in a phosphate buffer containing 50 μM BSA was increased from 0 to 6% (v/v). At solvent concentrations greater than 6% (v/v), the resolution

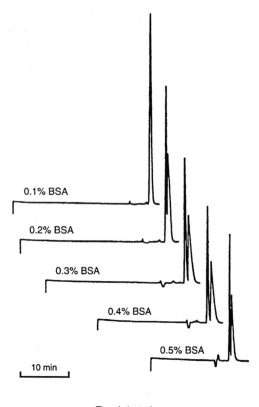

0.1% BSA

0.2% BSA

0.3% BSA

0.4% BSA

0.5% BSA

10 min

Time (minutes)

Figure 5.25 Electropherogram showing the effects of BSA concentration on the enantiomeric resolution of (**56**). Separation conditions: buffer, 100 mM phosphate, pH 7.0 and 0.1 to 0.5% (w/v) BSA; capillary, 450 mm × 0.05 mm i.d. (effective length, 350 mm); applied voltage, 10 kV; detection, 300 nm; temperature, 40°C. (From Arai, T., Ichinose, M., Kuroda, H., Nimura, N., and Kinosihita, T., *Anal. Biochem.*, 217, 7, 1994. With permission.)

began to diminish. In the case of D,L-tryptophan,[99] enantiomeric resolution was impaired with increasing solvent content. The presence of 1-propanol may adversely affect the interaction between the amino acid and the protein or induce a conformational change in the structure of BSA. Conversely, the addition of up to 25% (v/v) 2-propanol improved the resolution of the enantiomers of the beta-blocker R,S-alprenolol (**58**) using 40 mg/mL CBH I (Figure 5.26).[98] These observations[98,99,106] demonstrate the complex nature of the chiral recognition mechanism involving proteins.

The effect of pH on enantiomeric resolution in proteins is an important consideration. This is because the pH value can influence the ionization properties of functional groups on the protein surface and hence can influence the binding interactions between the chiral selector and the analyte. In addition, adhesion of the protein to the capillary inner wall can occur. For this reason, most experiments are carried out at pH values above the pI of the protein. A good example[107] is that of enantiomeric separation of eperisone using ovomucoid. This glycoprotein has a molecular weight of 28.8 KDa protein and a pI between 3.9 and 4.5. Analysis of the drug at a pH less than the pI resulted in no peak for the compound. Increasing the buffer pH to 8 resulted in improved enantioselectivity of ovomucoid and separation of the eperisone enantiomers. The significance of pH has also been shown[108] for the chiral analysis of leucovorin. This drug is a reduced folate that has been used with methotrexate in clinical therapy for cancer patients. The biologically active 6-S isomer has been resolved using BSA-containing buffer on a Peg-coated capillary under different pH values from 6.8

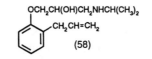

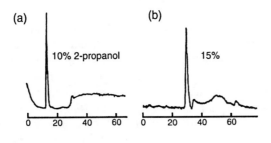

Time (minutes)

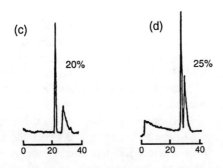

Time (minutes)

Figure 5.26 Electropherograms showing the effects of the concentration of 2-propanol on the enantiomeric separation of R,S-alprenolol (58). With (a) 10% (v/v), (b) 15% (v/v), (c) 20% (v/v),
and (d) 25% (v/v) 2-propanol. Other separation conditions: buffer, 400 mM phosphate pH
5.1 and 40 mg/mL CBH 1; capillary, 115 mm × 0.075 mm i.d. coated with non-crosslinked
polyacrylamide; applied voltage, 1 kV; detection, 220 nm; temperature, ambient. (From
Valtcheve, L., Mohammad, J., Petersson and Hjerten, S., *J. Chromatogr.*, 638, 263, 1993.
With permission.)

to 7.6. A general decrease in enantiomeric resolution was noticed under slightly basic conditions, and this was most likely due to reductions in separation efficiency and retention time
caused either by changes in the electroendosmotic flow or in the electrophoretic mobility of
the protein, or a combination of both effects.

Cationic coadditives have been used to great effect to enhance enantiomeric resolution
primarily by improving peak shape. The addition of 4.2 mM N,N-dimethyloctylamine
(DMOA) to a buffer containing 21 µM AGP dramatically improved the resolution of R/S-
promethazine (59).[106] This additive is thought to act by influencing the coulombic interaction
between the protein and the analyte. In addition, DMOA may alter the conformation of the
protein in order that groups on the protein surface might bind with the analyte more effectively.
The addition of O-phosphorylethanolamine (PEA) improved resolution of eperisone by reducing peak tailing and improving peak shape.[107] Unfortunately, a drawback in using PEA is
that instability of this compound can lead to poor day-to-day reproducibility.

For optimum enantiomeric resolution, differences in the electrophoretic mobilities of the
free and protein-bound analyte should be maximized. In circumstances where both the analyte
and the protein additive have the same negative charge, differences between the bound and

$$CH_2CH(CH_3)N(CH_3)_2$$

(59)

unbound ligands are minimal. In the example reported by Sun et al.[109] no resolution of the enantiomers of ibuprofen could be obtained using BSA alone. However, on addition of dextran to the protein solution both enantiomers were well separated. The polysaccharide may act by interacting strongly with the protein to greatly reduce its mobility, while smaller molecules are affected to a lesser extent. Similar observations were made for the enantiomeric separation of dansyl leucine, dansyl norvaline, and mandelic acid. For neutral racemates, Lloyd et al.[104] studied the effects of adding 0 to 10% dextran in HSA-containing buffers on the resolution of benzoin, promethazine, propiomazine, and thioridazine. Different effects were noticed for the four racemates, and it was concluded that these observations were probably due to a specific alteration or blocking of one binding site on HSA. Both promethazine and propiomazine have similar binding properties that are different from those of thioridazine and benzoin.[105] An extension of this method involved covalently linking BSA to a high molecular weight dextran polymer.[105] In this way, the protein was immobilized, and in the presence of reduced electroendosmotic flow, only the charged species migrated.

The ionic strength and type of buffer used in protein-based chiral separations can have a surprising effect on enantiomeric resolution in CE. Ishihama et al.[110] showed that increasing the phosphate buffer concentration from 10 to 50 μM in the presence of 60 μM OVM leads to a decrease in the resolution of eperisone enantiomers. The explanation proposed was that the increased ionic strength altered the conformation and hence the surface of the protein. High buffer concentrations will also affect ion–ion interactions. A range of buffer types containing OVM were also investigated consisting mainly of acetate, phosphate, and Tris at pH values 5 and 7. The best chiral separations were achieved using phosphate buffer alone. A similar observation was made by Arai et al.[102] for the enantiomeric separation of the quinolone antibacterial ofloxacin using 0.3% BSA. The biologically active S-(−)-enantiomers could only be resolved using 100 mM phosphate buffer at pH 7.0. Borate buffers were unsuitable because of their low buffering capacity at these pH values and the ability of these buffers to complex with the drug, which can inhibit analyte-protein interactions.

The voltage applied during chiral separations involving protein chiral selectors has been shown to have a significant effect on resolution. Generally, optimum enantiomeric separations were obtained at lower voltages as exemplified[99] for the separation of D- and L-tryptophan enantiomers using an HSA-containing buffer using applied voltages between 20 and 30 kV (Figure 5.27). In this example, acetone was used as an electroendosmotic flow marker. The detrimental effect to resolution observed at high field strengths was probably due to the generation of Joule heating and temperature gradient effects within the capillary which led to mixing and poor resolution.

F. Macrocyclic Antibiotics

The need for more diverse and potentially more powerful chiral additives in capillary electrophoresis is still a growing and active area of research. An important contribution to this are the macrocyclic antibiotics first introduced by Armstrong et al.[111,112] These compounds were first used as chiral selectors in TLC[113] and HPLC,[112] where they proved to be extremely efficient at chiral discrimination. This class of compound falls into two categories: ansamycin antibiotic (examples are rifamycin B and SV) and the amphoteric glycopeptide antibiotics (vancomycin, ristocetin A, and teicoplanin).

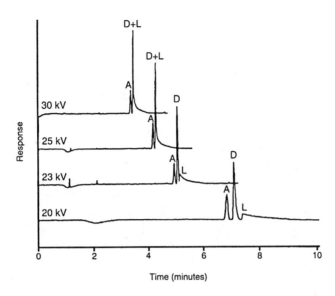

Figure 5.27 Electropherogram showing the effects of applied voltage on the enantiomeric resolution of DL-tryptophan (peak labeled D represents an acetone marker). Separation conditions: buffer, 25 mM phosphate buffer, pH 7.4 and 50 μM HSA; capillary, 700 mm × 0.05 mm i.d. (effective length, 450 mm); applied voltage, 20 to 30 kV; detection, 280 nm; temperature, ambient. (From Yang, J. and Hage, D. S., *Anal. Chem.*, 66, 2719, 1994. With permission.)

1. Rifamycins

Rifamycins belong to the ansamycin family. They are large cyclic structures produced by *Nocardia mediterranei*,[114] and their molecular weights range from approximately 600 to 2,200 Da. They are characterized by having a UV absorbing naphthohydroquinone ring that links the two ends of an aliphatic ring. The R group on the aromatic moiety can either be an oxyacetate (for rifamycin B (**60**, R, -OCH$_2$COOH)) or a hydroxyl group (for rifamycin SV (**61**, R, -OH)).

(60, 61)

Each rifamycin molecule has nine chiral carbon centers, four hydroxyl groups, one carbomethyl moiety, and one amide bond. The presence of multiple stereogenic sites and functional groups makes them ideal candidates for stereoselective interactions. The two carboxylic acid groups in rifamycin B have pK_a values of about 2.8 and 6.7. Rifamycin B is a yellow solid that is slightly soluble in water and more soluble in the lower alcohols. Aqueous solutions of this compound are orange in color depending on the experimental conditions (pH and cosolvents). Rifamycin B absorbs strongly in both the ultraviolet and visible spectral regions which appears to be independent of pH changes. Three maxima at 220, 350, and 425

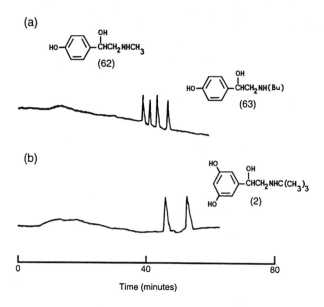

Figure 5.28 Electropherogram showing the enantiomeric separation of synephrine (**62**), bamethan (**63**), and terbutaline (**2**). Separation conditions: buffer, 100 m*M* phosphate, pH 7.0; capillary, 575 mm × 0.05 mm i.d. (effective length, 500 mm); applied voltage, 8 kV; detection, indirect at 254 nm; temperature, 22°C. (From Armstrong, D. W., Rundlett, K., and Reid, G. L., *Anal. Chem.*, 66, 1690, 1994. With permission.)

nm are observed, hence direct detection of analytes at near these wavelengths would be difficult. A low absorbance trough near the 350 nm band has been utilized for direct analysis of analytes.

Rifamycin B has an enantioselectivity toward positively charged chiral analytes, specifically amino alcohols[115,116] (e.g., alprenolol, metoprolol, and oxoprenolol) and beta-blockers (e.g., epinephrine). These analytes possess either primary or secondary amine groups located in close proximity to the chiral carbon center. In addition, compounds with up to two aromatic rings (e.g., propanolol) can be resolved. Interestingly, anionic analytes are not particularly well resolved using this macrocycle additive. As a result of the large background absorbance of buffers containing the rifamycin antibiotics, many of the applications reported for these chiral selectors utilize both direct and indirect detection methods during chiral analysis. Figure 5.28 shows the enantioseparation of three beta-blockers, namely synephrine (**62**), bamethan (**63**), and terbutaline (**2**) using 20 m*M* rifamycin B monitoring indirectly at 254 nm.[115] Data from the analysis of 17 biologically active amino alcohols have indicated that compounds with hydroxyl groups *alpha* to an aromatic ring were better resolved than those that were farther away. In addition, greater enantioselectivity was observed for secondary amines compared to their primary analogs. And thirdly, compounds with more than two aromatic rings were poorly resolved, which is indicative of size selectivity. The enantioselectivity of rifamycin SV differs from that of rifamycin B in that this chiral selector seems particularly suited for the enantioresolution of negatively charged analytes with at least two ring systems, e.g., hexobarbitol, glutethimide, and DNS-Asp.[116]

Varying the experimental parameters, such as the nature and amount of organic modifiers, buffer types and ionic strength, pH, and rifamycin concentration, affects enantiomeric resolution and chiral selectivity. With the exception of buffer pH, increasing any of the other parameters leads to an increase in migration run time and in some cases a corresponding reduction in enantiomeric resolution. Increasing the concentration of rifamycin B in the buffer increased overall run times and improved chiral resolution. Optimum concentration for rifamycins is about 25 m*M*. The effects observed for the addition of organic solvents is dependent on the concentration and type of solvent used. For example,[115,116] small amounts of 2-propanol were found to have a greater effect on Rs than larger volumes of a different organic modifier (e.g., methanol or ethanol). Many of these effects are not well understood.

The influence of pH is a complex one because of the many ionizable groups in these rifamycins. Therefore, pH changes not only affect the EOF, but also the charge and molecular recognition properties of both analyte and macrocycle. To achieve optimum enantioresolution, these pH effects should be balanced. For the chiral analysis of amino alcohols, pH values near 7 are recommended.[115,116]

2. Glycopeptide Antibiotics

Examples of glycopeptide antibiotics are vancomycin, ristocetin A, teicoplanin, and β-avoparcin. These compounds are a class of broad spectrum antibiotic active against Gram-positive bacteria, and the antibiotic activity results from selective binding to terminal D-Ala-D-Ala amino acid sequences in mucopeptides and inhibiting bacterial cell wall synthesis.

a. Vancomycin and Ristocetin A

Vancomycin (**64**) is produced from *Streptomyces orientalis*,[117] whereas ristocetin A is a fermentation product of *Nocardia luridia*.[118] Their structures consist of a cyclic glycopeptide (molecular weight, 1449 for vancomycin and 2066 for ristocetin A) in the shape of a basket and carbohydrate sidechains with an additional N-methyl leucine group for vancomycin.

(64)

There are a total of nine hydroxyl groups, two amines, seven amide bonds, five phenyl rings, and a total of eighteen chiral centers in vancomycin. Six of these groups are ionizable with the following approximate pK_a values, 2.9, 7.2, 8.6, 9.6, 10.5, and 11.7, and reports of the isoelectric point of vancomycin range from between 5 and 7.5. Hence, most chiral separations mediated by vancomycin have been performed within this pH range. All these groups are known to be useful for stereoselective molecular interactions with chiral analytes.

Ristocetin A (**64**), on the other hand, has 38 stereogenic centers, seven aromatic rings, six amide bonds, 21 hydroxyl groups, two primary amines, and one methyl ester. These varied functional groups form a four-macrocyclic-ringed compound that is soluble in aqueous acidic media and polar aprotic solvents. The pI for ristocetin A is slightly higher than that of

(65)

vancomycin at approximately 7.5. Both compounds are white solids soluble in water, methanol, and some aprotic solvents such as DMSO and DMF, but insoluble in the higher alcohols.

The spectral properties of both compound types are important considerations when used as chiral selectors in CE. Vancomycin and ristocetin A absorb strongly below 250 nm with weaker pH dependent bands at longer wavelengths. Between wavelengths 254 and 264 nm, a low absorbance trough is present, and at these wavelengths low concentrations of this additive can be used effectively with minimal background absorbance.

Armstrong et al.[119,120] have reviewed the application of these glycopeptide macrocycles for the enantioresolution of over 200 racemic compounds from different chemical classes. These include derivatized amino acids (AQC, DNS, DNP, N-DNP, PHTH, N-formyl, N-acetyl, N-benzoyl, CBZ, and DNB derivatives), NSAID drugs, and other carboxylic acid compounds. The enantioselectivities of ristocetin A and vancomycin are similar. However, some optically active carboxylates were resolved by the former antibiotic and not by the latter. Figure 5.29 shows a few examples of chiral acids[121] (3-hydroxy-4-methoxymandelic (**66**), 3-methoxy-mandelic (**67**), and mandelic acids (**68**)).

Other features that distinguish ristocetin A from vancomycin include shorter run times at similar concentrations (2 to 5 mM), enhanced enantioseparation in the presence of organic solvents, greater stability in aqueous solutions, and lower cost. Figure 5.30 shows the resolution of three AQC amino acid and DNS-Val racemates using 5 mM vancomycin in a phosphate buffer at pH 7.[120]

The effect of concentration of the macrocyclic antibiotic additive, pH, addition of organic solvent, and surfactants on enantiomeric resolution has been investigated in detail. In general, similar effects have been observed for both vancomycin[119] and ristocetin A.[120] Increasing the antibiotic concentration from 1 to 5 mM resulted in increased enantiomeric resolution, migration times, and effective electrophoretic mobilities of the analytes. These observations[119,120] are thought to be due to a decrease in the electroendosmotic flow which was caused either by reversible adsorption of the antibiotic (this was particularly prevalent for vancomycin) to the inner wall of the capillary or because of the resulting increase in viscosity on addition of the chiral selector. The interaction of vancomycin with the capillary wall can be suppressed by the addition of sodium dodecyl sulfate, which interacts with the

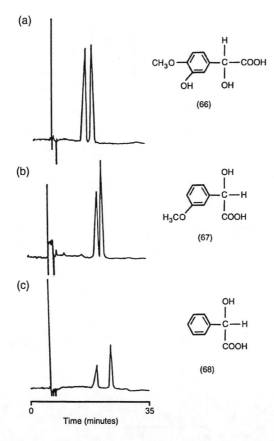

Figure 5.29 Electropherograms showing enantiomeric resolution of racemates of (a) 3-hydroxy-4-meth-oxymandelic acid **(66)**, (b) 3-methoxymandelic acid **(67)**, and (c) mandelic acid **(68)** using ristocetin A. Separation conditions: buffer, 100 mM phosphate, pH 6.0 and 2 mM ristocetin A; capillary, 325 mm × 0.05 mm i.d. (effective length, 250 mm) applied voltage, 5 kV; detection, 254 nm; temperature, ambient. (From Armstrong, D. W., Gasper, M. P., and Rundlett, K. L., *J. Chromatogr.*, 689, 285, 1995. With permission.)

macrocycle to prevent wall adhesion.[121] This has been shown to dramatically improve the enantiomeric resolution of a mixture of DNS-DL amino acids and, more interestingly, gives rise to a reversal of the migration order of the resolved enantiomers in the presence of up to 50 mM SDS. This change in migration order is due to changes in the mobility of the vancomycin upon binding to SDS. Optimum concentrations for vancomycin and ristocetin A were 5 and 2 mM, respectively.

Another important experimental parameter is pH because this governs the charge and mobility of both the analyte and the chiral additive. In separations involving vancomycin, an increase in pH decreased enantiomeric resolution probably because of changes in binding interactions, electrophoretic mobility differences between the bound and free analyte, or even the conformation of the chiral additive. Hence, enantioseparations were usually carried out at lower pH values, typically between 6 and 7. The effects of solvent coadditives such as methanol and 2-propanol were found to be complex and were dependent on the analyte.[119,120]

An interesting application of macrocylcic antibiotics in chiral discrimination was first reported by Carpenter et al.[122] and then by Chu et al.[123] They exploited the fact that vancomycin is known to stereoselectively bind to dipeptides of the D-configuration, and they used this property to determine binding constants for the mucopeptide n-acetyl-D-Ala-D-Ala. The results[122] showed clearly that vancomycin was highly selective toward the D-configured amino acid dipeptides. No binding was observed for the corresponding L-amino acid.

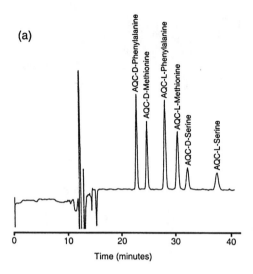

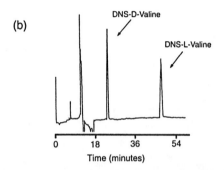

Figure 5.30 Electropherograms showing the enantiomeric resolution of (a) three racemic AQC amino acids and (b) racemic DNS-Val using vancomysin. Separation conditions: buffer, 100 mM phosphate with 5 mM vancomycin at (a) pH 7.0 and (b) pH 4.9; capillary, 325 mm × 0.05 mm i.d. (effective length, 250 mm); applied voltage, 5 kV; detection, 254 nm; temperature, 22°C.

b. Teicoplanin

Teicoplanin (**69**) is the third in the series of glycopeptide antibiotics that can be used as chiral selectors in capillary electrophoresis. It is produced *via* a fermentation process involving *Actinoplanes teichomyceticus,*[124] and like vancomycin and ristocetin A, it is highly active against pathogenic Gram-positive bacteria *in vitro*. Its structure is similar to that of vancomycin and ristocetin A in that it contains a semirigid aglycone structure based on a peptide unit that contains seven amino acids. However, it does differ from its relatives in that it contains D-configurations of mannose and glucosamine. In addition, the amine group of glucosamine has been substituted by a decyl or undecyl alkyl chain, thus making this macrocyclic antibiotic more hydrophobic than other common glycopeptide additives. The hydrophobic tail enables teicoplanin to aggregate in solution to form micellar structures with reported critical micelle concentrations between 0.21 and 0.84 mM depending on the pH value. Its structure consists of free amine and carboxylic acid groups giving rise to a pI value of 6.5. Moreover, the presence of 23 chiral carbon centers, and 7 aromatic groups contributes to the many stereoselective interactions that make teicoplanin a useful chiral additive in the separations of enantiomers in CE.

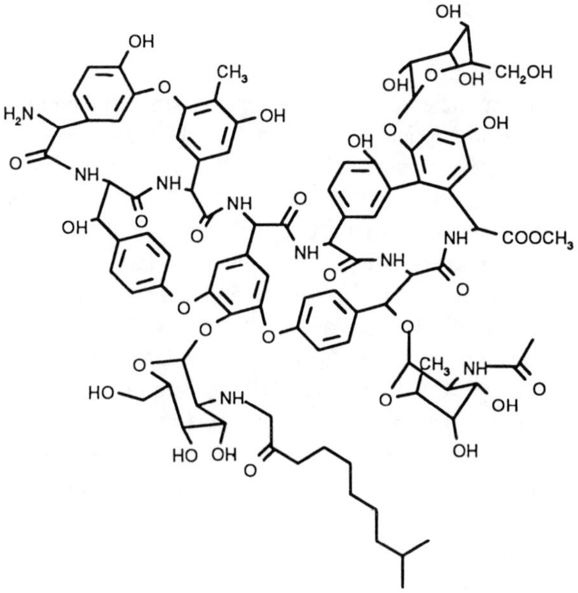

(69)

Like the other two glycopeptide macrocycles, the combination of the low extinction coefficient at 254 nm and the low concentrations used in chiral separation in CE minimizes background absorbance. The experimental conditions required for chiral separations using teicoplanin-containing buffer systems typically involve low additive concentrations (usually 2 mM) of the compound dissolved in high ionic strength phosphate buffers (100 mM) at pH 5.0. These conditions are required first to reduce background absorbance, second to reduce capillary wall interactions with the chiral additive, and third to stabilize the compound and allow ionization of functional groups (at higher pH values teicoplanin degrades rapidly). Low voltages are also recommended because Joule heat generated by the passage of current through the buffer can degrade the glycopeptide. The range of compounds separated by vancomycin and ristocetin A is also resolved by teicoplanin. In the case of the amino acids, the enantiomeric migration order is D-migrating before the L-antipode, indicating that the L-compound interacts strongly with the many stereoselective functional groups in teicoplanin.

The structure of the analyte was found to have significant effects on enantiomeric resolution.[125] Almost all the chiral analytes resolved by teicoplanin share the following structural features: they all possess either carboxylic or α-amino groups; an amide or carboxylic functional group is located in close proximity to the chiral carbon center. The location of aromatic rings in relation to the stereogenic center and the substitution pattern are also important structural considerations for chiral discrimination to occur.

The addition of low amounts of acetonitrile was found to enhance the enantioresolution by teicoplanin primarily because of a reduction of the electroendosmotic flow. This has been demonstrated[125] for the enantiomers of atrolactic acid (70), which was resolved using 2 mM teicoplanin in a 100 mM sodium phosphate buffer with 0 to 20% (v/v) acetonitrile added.

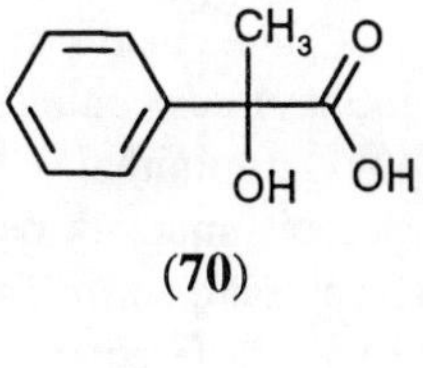

(70)

G. Enantioselective Metal Complexation

Enantioseparation by enantioselective metal complexation is based on the formation of mixed ternary diastereoisomeric chelate complexes consisting of a chiral bifunctional ligand, a transition metal ion, and the optically active analyte. This approach has been applied in a range of chromatographic (reverse-phase HPLC)[126,127] and electrophoretic[20,128] (CE and paper electrophoresis) methods. The structural requirement of the analyte is that it should have at least two polar groups within close proximity to each other and the asymmetric carbon center. Chiral ligands such as α-amino acids (e.g., N,N-didecyl-L-alanine, L-proline, and L-histidine) and dipeptides (e.g., L-aspartyl-L-phenylalanine methyl ester or aspartame) possess the structural components for enantioselective chelation with the metal ion. Transition metals such as Cu(II) and Zn(II) have been used successfully to resolve DNS-amino acid enantiomers. The use of other metal ions (e.g., Co(III) and Cr(III)), more commonly used in paper electrophoresis, are not suitable for use in CE because they form kinetically stable adducts. The structure of the chelate complex depends on whether the ligand is an amino acid or a dipeptide. Adducts formed from amino acids have a five-membered ring configuration, whereas those derived from aspartame are six-membered ring structures. Coordination occurs *via* interactions between the metal ion, the amine moiety, and the carboxyl group of the amino acids.

This chiral separation method has limited applications in CE with respect to the other chiral selectors (e.g., cyclodextrins and chiral surfactants) and has mainly been applied to the enantiomeric resolution of DNS-DL-amino acids, first reported by Zare et al.[128] Chiral resolution was found to be dependent on pH, ratio of metal ion to ligand, temperature, ligand type, and the addition of surfactants. Optimum conditions for the analysis of amino acid enantiomers were between pH 7 and 8. These buffer conditions facilitated a high enough EOF and allowed ionization of functional groups on both the ligand and the analyte. A lowering of the pH led to a loss of enantiomeric resolution because of competition for binding with the metal ion by protons, and the protonation of the ligand preventing the formation of the ternary complex. The desired metal:ligand ratio is 1:2. Above this value, the resolution diminishes because of increased competition between the amino acid ligand and the hydroxide ions for the metal. Generally, the choice of ligand determines the enantiomeric migration order. When the ligand is either aspartame or L-His, the migration order is D-amino acid followed by the L-isomer, whereas for amino acid additives with the opposite configuration, the migration order is reversed.

III. CHIRAL RESOLUTION VIA THE FORMATION OF COVALENT DIASTEREOISOMERS

The formation of covalent diastereoisomers is an alternative means of resolving enantiomers in capillary electrophoresis that does not necessarily require the use of chiral buffer additives. These derivatives are formed by the reaction of chiral reagents with optically active compounds that contain at least one derivatizable functional group, e.g., amino ($-NH_2$), carboxyl ($-CO_2H$) or a carbonyl group (R–C(O)–R). The formation of these derivatives can have additional benefits, such as improving the spectroscopic properties of compounds which would otherwise have low chromophoric or fluorophoric detection. Three criteria are required for the use of the chiral reagents in chiral purity analysis: (1) the reagents must be optically pure in order to avoid any ambiguous side reactions; (2) each enantiomer of the analyte should react with the reagent at similar rates; and (3) unwanted side reactions such as racemization and decomposition should be minimal.

The separation of covalent diastereoisomers is performed by CE in the presence of nonchiral surfactants such as SDS. The chiral discrimination mechanism relies on differences in the physicochemical properties of the diastereoisomeric derivatives giving rise to differential partitioning equilibria between the micellar and aqueous phases. Many of the reported[129-136] chiral derivatizing reagents have bulky alkyl or aromatic groups with one or more chiral centers. Upon reaction with the optically active analyte, differences between the two resulting diastereoisomers generated *in situ* from bulky reagents will be maximized and the resulting separation optimized. Many of these chiral reagents used in CE have originated from HPLC applications, and there are many detailed reviews describing these reactions.[137] In the following sections, some examples of chiral derivatization reagents used in CE will be discussed for the analysis of optically active analytes containing derivatizable amine, aldehyde, or carboxylic acid groups.[138] In addition, the use of CE to resolve the diastereoisomers of compounds that do not require derivatization and contain more than one chiral center will be mentioned.[139]

A. Chiral Amine Derivatization Reagents

The derivatization of amine groups appears to be by far the most common reaction type. One of the more popular chiral amine reagents used in MEKC is 2,3,4,6-tetra-O-acetyl-β-D-glucopyranosyl isothiocyanate[129,130] (GITC) (**71**), which reacts rapidly with primary and secondary amines forming diastereoisomeric thiourea derivatives:

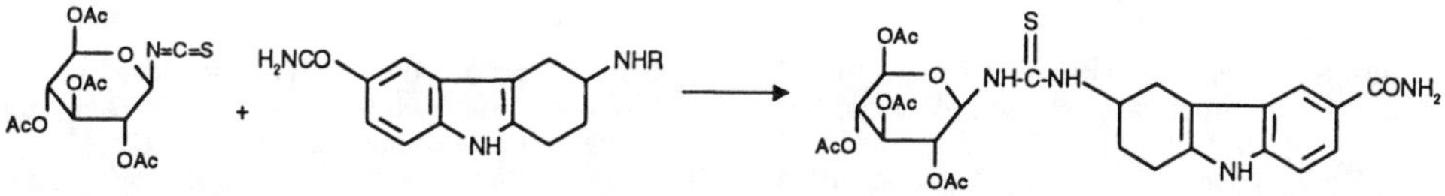

(71)   **(72)**

This derivatization has been applied to the analysis of racemates of amino acids[129] under neutral conditions using a phosphate/borate/SDS system. Interestingly, using the arabinopyranosyl analog of GITC, namely, 2,3,4-tri-O-acetyl-α-D-arabinosyl isothiocyanate (AITC) was found to reverse the enantiomeric migration order of AITC-derivatized amino acids.[131] In one report, GITC was used to resolve the enantiomers of some related antimigraine drugs, namely 3-aminoalkyl-6-carboxamido-1,2,3,4-tetrahydrocarbazoles (**72**), using a buffer containing β-cyclodextrin/STDC[130] (Figure 5.31). More important, this derivatization method was successfully used to monitor the enantiomeric enrichment of this drug by stereoselective crystallization of the preferred enantiomer as salts of (+)-camphorsulfonic acid.

Marfey's reagent[132,133] (or Na-(2,4-dinitro-5-fluorophenyl)-L-alaninamide) (**73**) is a useful compound that readily reacts with amines, particularly amino acids to form diastereoisomers that can easily be resolved using both free-zone and MEKC modes. An interesting aspect of these derivatives is that by varying the pH of the buffer and reversing the electrical polarity, the diastereoisomeric elution order can be altered. This is particularly useful in applications where the analysis of a diastereoisomer that elutes after the main compound is impaired by peak tailing.

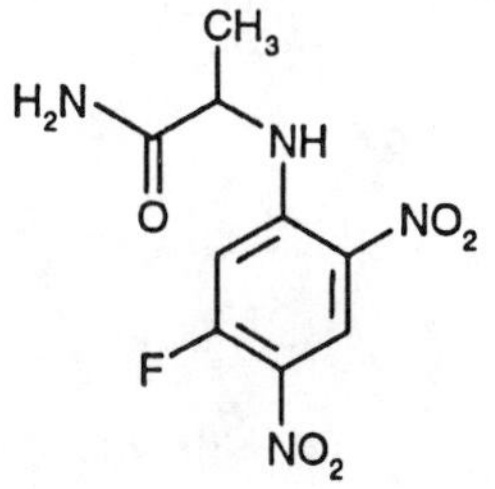

(73)

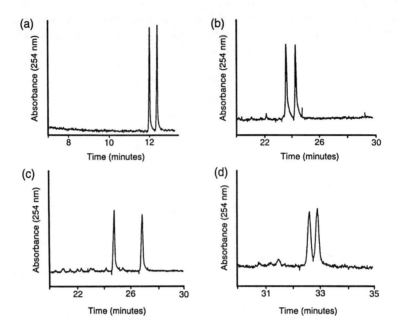

Figure 5.31 Electropherograms showing the diastereoisomeric separation of GITC derivatives of four structurally related antimigraine drugs (**72**). Where in (a) R,H, (b) R,CH$_3$, (c) R, C$_2$H$_5$, and (d) R, (CH$_2$)$_2$OCH$_3$. Separation conditions: buffer, 30 mM sodium phosphate, 10 mM boric acid, 50 mM STDC, and 20 mM β-CD; capillary, 570 mm × 0.05 mm i.d. (effective length, 500 mm); applied voltage, 25 kV; detection, 254 nm; temperature, 25°C. (From Okafo, G. N., Rana, K. K., and Camilleri, P., *Chromatographia*, 39, 627, 1994. With permission.)

In another report,[134] the reagent (+)-O,O′-dibenzoyl-L-tartaric acid anhydride (DBT anhydride) (**74**) was used to resolve the enantiomers of racemic amino acids. An interesting aspect of this study was that polymeric additives such as poly(vinylpyrrolidone) (PVP), poly(ethyne glycol) (PEG), and poly(acrylamide) (PAA) were used in the phosphate buffers systems to increase stereoselectivity and the range of analytes that could be separated. The enhanced diastereoisomeric resolution was thought to be due to retardation of the analyte *via* different interactions (either hydrophobic, dipole–dipole, pi–pi, or n–pi interactions) which significantly increase the migration time window. Optimum polymer concentrations were found to vary for each additive. The "ring" conformation adopted by the polymer may play a role in stereodiscrimination. Similar observations were made for the diastereoisomeric separation of some mono- and dibasic coumarinic drug derivatives.

(74)

Another chiral derivatizing reagent is [1-(9-fluorenyl)ethyl]chloroformate[135] (FLEC) (**75**). This compound is an optically active analog of the common derivatizing reagent 9-fluorenylmethylchloroformate (FMOC), which reacts with quaternary ammonium groups. As mentioned earlier, the fluorenyl skeletal structure enhances UV absorption and has strong fluorescence properties. FLEC has been applied to chiral analysis of D- and L-carnitine (3-hydroxy-4-N,N,N-trimethylammonium butyrate) (76) in microbiological and pharmaceutical formulations. In this application, the separation buffer consisted of a phosphate buffer at pH 2.6 with a tetrabutylammonium bromide (TBABr) additive. The addition of TBABr was necessary to reduce electroendosmotic flow leading to improved resolution of the derivatives. FLEC has also been used to resolve amino acid enantiomers using an SDS-containing buffer system. The chiral purity of FLEC has been determined by reacting the reagent with an achiral compound such as glycine. In this way approximately 0.1% of each enantiomer of FLEC could be detected in either (+)-FLEC or (–)-FLEC.[138] FMOC can also be used instead of FLEC, but a chiral selector in the buffer, such as HP-β-CD has to be used.

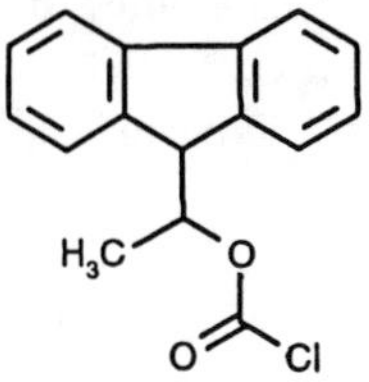

(75) 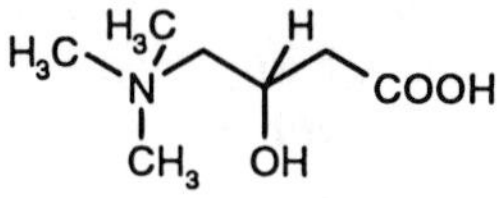   (76)

B. Chiral Aldehyde Derivatization Reagents

For the analysis of chiral analytes containing an aldehyde functional group, the reagent S-(–)-1-phenylethylamine (S-PEA) (77) is a useful,[136] commercially available reagent. This has been demonstrated in the diastereoisomeric separation of aldose enantiomers. Optimum reaction conditions for S-PEA derivatization involve heating the carbohydrate in the presence of the reducing agent sodium cyanoborohydride at 90°C for 1 h. The separation buffer conditions consisted of sodium borate, which complexes with geminal dihydroxyl functional groups forming negatively charged boronate esters. It was observed that optimization of pH, borate concentration, the addition of organic solvents, and temperature were necessary for maximum diastereoisomeric resolution.

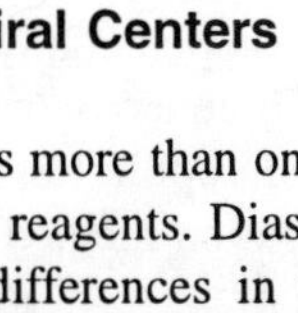

(77)

C. Compounds with Multiple Chiral Centers

For compounds that already possess more than one chiral center, there is usually no need to derivatize the compound with chiral reagents. Diastereoisomeric resolution can usually be achieved directly in CE because of differences in the physicochemical properties of the compounds. This is exemplified in the analysis of buthione sulfoximine[139] (BSO) (78), which, as an inhibitor of γ-glutamylcysteine synthase, blocks glutathione synthesis. BSO has two chiral centers at the α-carbon and sulfur atoms. Using an SDS/phosphate buffer system, a validated method was developed for monitoring levels of the two diastereoisomers (L,R- and L,S-BSO) in plasma samples from patients dosed with an unequal mixture of the two antipodes.

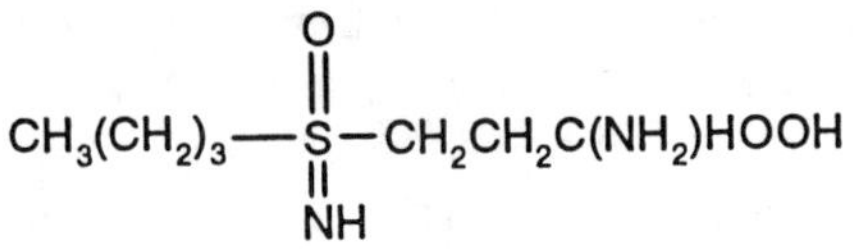

$$CH_3(CH_2)_3 - \overset{\overset{\displaystyle O}{\|}}{\underset{\underset{\displaystyle NH}{\|}}{S}} - CH_2CH_2C(NH_2)HOOH$$

(81)

IV. CHIRAL RESOLUTION BY CAPILLARY ELECTROCHROMATOGRAPHY

Electrochromatography is a new electrophoretic separation mode that combines the separation mechanisms in CE with those of liquid chromatographic systems. These separations are achieved either by using silica capillaries packed with micron-sized octadecyl silica particles or capillaries coated with a layer of material. For the resolution of enantiomers, the use of coated capillaries is the easiest approach. Coated capillaries are now commercially available for use in chromatographic techniques such as GLC,[140] SFC,[141] and more commonly GC.[142] In CE, Mayer and Schurig reported[143-145] the use of CD-coated capillary (Chirasil-Dex) to resolve the enantiomers of binaphthyl derivatives and 1-phenyethanol using a phosphate-borate buffer at pH 7.0. Other examples include NSAID drugs[143] and hexobarbitol.[144] Under these conditions, the stable immobilized phase consisted of permethylated β-cyclodextrin attached *via* an alkyl chain to an inert polysiloxane backbone in the silica capillary (**79**).

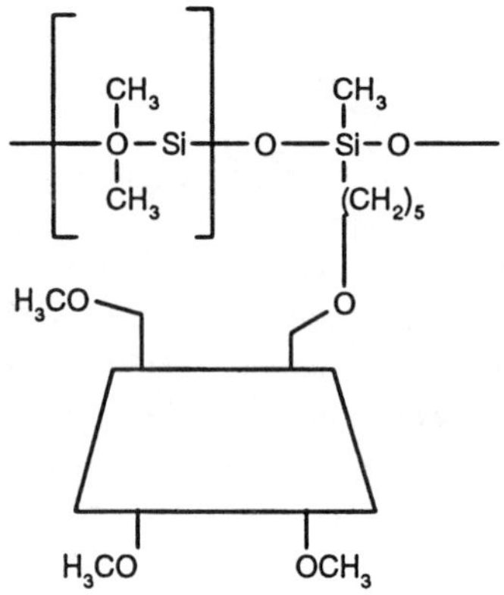

(79)

The separation mechanism involves enantioselective inclusion equilibria of each optical isomer with cyclodextrin-bonded coating on the capillary wall. A similar approach was reported by Szeman and Ganzler[146] for the enantioseparation of epinephrine using a CD-coated capillary. The alternative approach involving the use of capillaries packed with chiral HPLC stationary phase material has been reported by Li and Lloyd[147] using cyclobond I CSP packed into a silica capillary to separate the enantiomers of amino acid derivatives, benzoin, and hexobarbitol. Although this method is potentially very powerful, the instability of many of the HPLC chiral stationary phases in aqueous media, their relatively high cost, and the inherent difficulties in preparing packed capillaries may present possible drawbacks to the further development of this chiral separation method.

VI. CONCLUSIONS

Over the last few years, there has been a dramatic increase in applications of CE to the resolution of enantiomers of racemic molecules. This has been mainly due the development of a wider range of new chiral selectors, the growing awareness of the biological significance of chirality in drug molecules and the need for different analytical methodologies to separate the enantiomers of an even wider range of chiral molecules. New chiral selectors such as acyclic oligosaccharides and polysaccharides, chiral crown ethers, protein additives, macro-cyclic antibiotics, and chiral surfactants are now easily available. The enantioselectivity and enantiomeric resolution of these chiral additives has been exploited by varying the experi-mental conditions (e.g., pH, temperature, additive concentration, organic solvents, buffer ionic strength, urea additives, polymeric additives, and other chiral compounds) to achieve optimum separations. Chiral derivatization methods have also developed, mainly utilizing chiral reagents commonly found in liquid chromatographic applications. As our knowledge and understanding of the mechanisms involved in chiral recognition and discrimination increase, so will the development of chiral selectors with wider enantioselectivities.

ABBREVIATIONS

6-APA	6-Aminopenicillanic Acid
AEKC	Affinity Electrokinetic Chromatography
AGP	α1-Acid Glycoprotein
AQC	6-Aminoquinolyl-N-hydroxysuccinimidyl Carbamate
BGE	Background Electrolyte
BN	1,1-bi-2-naphthol
BNC	(S,R)-1,1-binaphthyl-,2-dicarboxylic Acid
BNP	1,1-binaphthyl-2,2-diyl hydrogenphosphate
BOC	Butyloxycarbonyl
BSA	Bovine Serine Albumin
BSO	Buthione Sulfoximine
18-C-6-TCA	18-Crown-6 Tetracarboxylic Acid
CBH I	Cellobiohydrolase I
CBI	1-cyano-2-substituted-benz[f]isoindole
CBZ	Carbobenzyloxy
CD	Cyclodextrin
Cden	mono-(6-β-aminoethylamino-6-deoxy)-β-Cyclodextrin
CD-MEKC	Cyclodextrin-modified Micellar Electrokinetic Chromatography
CE	Capillary Electrophoresis
CE-β-CD	Carboxyethyl-β-Cyclodextrin
CHAPS	N,N-bis(3-D-gluconamidopropyl)cholamide
CMC	Critical Micellar Concentration
CON	Conalbumin
CSP	Chiral Stationary Phase
CM-β-CD	Carboxymethyl-β-Cyclodextrin
DDVal	N-Dodecoxycarbonyl valine
DM-β-CD	Dimethyl-β-Cyclodextrin
DMF	N,N-dimethylformamide
DMOA	N,N-dimethyloctylamine
DMSO	Dimethyl Sulphoxide
1,3-DMU	1,3-Dimethyurea
DNB	Dinitrobenzoyl
DNP	Dinitrophenyl
DNS	Dansyl or 1-N,N-dimethylaminonaphthalene-5-sulfonate
DNTBP	N-(3,5-dinitrobenzoyl)-o-isopropyl esters
DOPA	3-(3,4-dihydroxyphenyl)alanine
DS	Degree of Substitution

DTB	(+)-O,O-dibenzoyl-L-tartaric acid
EI-MS	Electrospray Ionization Mass Spectrometry
EOF	Electroendosmotic Flow
FDA	Food and Drug Administration
FLEC	1-(9-fluorenyl)ethylchloroformate
FMOC	9-Fluorenylmethylchloroformate
GABA	γ-Aminobutyric Acid
GC	Gas Chromatography
HBNC	2,2-dihydroxy-1,1-binaphthyl dicarboxylic acid
HP-β-CD	2-Hydroxypropyl-β-Cyclodextrin
HPLC	High Performance Liquid Chromatography
HSA	Human Serum Albumin
MD	Maltodextrin
Me-β-CD	Methylated-β-Cyclodextrin
MEKC	Micellar Electrokinetic Chromatography
MENH-β-CD	6A-methylamino-β-Cyclodextrin
NDA	Naphthalene-2,3-Dicarboxaldehyde
NEC-β-CD	1-(1-naphthyl)ethylcarbamoylated-β-Cyclodextrin
NMR	Nuclear Magnetic Resonance
NSAID	Nonsteroidal Antiinflammatory Drug
OFX	Ofloxacin
OVM	Ovomucoid
PAG	Polyacrylamide Gel
PEA	O-Phosphorylethanolamine
PEG	Poly (ethyne glycol)
pSUVal	Poly(sodium N-undecylenyl-1-valinate
PTH	Phenylhydantoin
PVP	Poly(vinylpyrrolidine)
Rs	Resolution
SBE-β-CD	Sulphobutyl Ether-β-Cyclodextrin
SC	Sodium Cholate
SDC	Sodium Deoxycholate
SDVal	N-Dodecanoyl-L-valinate
SDS	Sodium Dodecyl Sulfate
SFC	Supercritical Fluid Chromatography
S-PEA	S-(–)-1-Phenylethylamine
STC	Sodium Taurocholate
STDC	Sodium Taurodeoxycholate
TAPS	(3-(Tris((hydroxymethyl)methyl)amino)-1-propane

RACEMIC ANALYTES AND NUMBERS

No.	Chemical Name
1	Thalidomide
2	Terbutaline
3	Formoterol
4	Thiazole derivative
5	Uniconazole
6	Dicoconazole
7	Zopicolone
8	Zopicolone desmethyl

RACEMIC ANALYTES AND NUMBERS *(continued)*

No.	Chemical Name
9	Zopicolone N-oxide
10	Ephedrine
11	Propanolol
12	Hexobarbitol
13	Binaphthol
14	Oxazolidone derivative
15	Adrenaline
16	Noradrenaline
17	Trimetoquinol
18	Trimetoquinol structural isomer
19	Norlaudanosoline
20	Laudanosine
21	Diclofensine
22	Couraminic anticoagulants
23	S,R-1,1-Binaphthyl-2,2-dicarboxylic acid
24	1,1-Binaphthyl-2,2-diyl hydrogenphosphate
25	2,2-Dihydroxy-1,1-binaphthyl dicarboxylic acid
26	Ibuprofen
27	Warfarin
28	Ketoprofen
29	Simendan
30	Diltiazem
31	8-Chlorodiltiazem
32	Chlorpheniramine
33	18-c-6-TCA
34	DOPA
35	Tyrosine analogs
36	Bile salt steroid skeleton
37	Sodium cholate
38	Sodium deoxycholate
39	Sodium taurocholate
40	Sodium taurodeoxycholate
41	Digitonin
42	Glycyrrhizic acid
43	β-Escin
44	N-Dodecanoyl-amino acid
45	N-Dodecoxycarbonylvaline
46	N-Dodecyltartarate derivatives
47	N-Dodecyl-β-D-glycopyranoside-6-hydrogen sulfate
48	N-Dodecyl-β-D-glycopyranoside-6-hydrogen phosphate
49	2-Undecyl-4-thiazole carboxylic acid
50	Benzoin
51	Mephenytoin
52	4-Hydroxymephenytoin
53	Phenobarbitol
54	Cromakalim
55	Fenoldapam
56	Ofloxacin
57	DR-3862

RACEMIC ANALYTES AND NUMBERS *(continued)*

No.	Chemical Name
58	Alprenolol
59	Promethazine
60	Rifamycin B
61	Rifamycin SV
62	Synephrine
63	Bamethan
64	Vancomycin
65	Ristocetin A
66	3-Hydroxy-4-methoxymandelic acid
67	3-Methoxymandelic acid
68	Mandelic acid
69	Teicoplanin
70	Atrolactic acid
71	GITC
72	Antimigraine drugs
73	Marfey's reagent
74	DBT Anhydride
75	FLEC
76	Carnitine
77	PEA
78	BSO
79	Chirasil-dex phase

REFERENCES

1. Jenner, P. and Testa, B., The influence of stereochemical factors on drug disposition, *Drug. Metab. Rev.*, 2, 117, 1973.
2. Testa, B. and Trager, W. F., in *Topics in Pharmaceutical Sciences*, Breimer, D. D. and Testa, B., Eds., Elsevier, Amsterdam, 1983, 99.
3. Miyano, K., Fujii, Y., and Toki, S., Stereoselective hydroxylation of hexobarbitol enantiomers by rat liver microsomes, *Drug. Metabol., Dispos.*, 8, 104, 1980.
4. Wright, J., Cho, A. K., and Gal, J., Metabolism of the enantiomers of amphetamines, *Xenobiotica*, 7, 257, 1977.
5. Kean, W. F., Lock, C. J. L., Rischke, J., Butt, R., Buchanan, W. W., and Howard-Lock, H., Effect of R and S enantiomers of naproxen on aggregation and thromboxane production in human platelets, *J. Pharm. Sci.*, 78, 324, 1989.
6. Bijvoet, J. M., Peerdeman, A. F., and Van Bommel, A. J., Determination of the absolute configuration of optically active compounds by means of x-rays, *Nature*, 168, 271, 1951.
7. Cahn, R. S., Ingold, C. K., and Prelog, V., Specification of molecular chirality, *Angew. Chem. Int. Ed.*, 5, 385, 1966.
8. Sunday Times (London) Insight Team, *Suffer the Children: The Story of Thalidomide*, Viking Press, New York, 1979.
9. Blaschke, G., Kraft, H. P., Fickentscher, K., and Kohler, F., Chromatographic racemic separation of thalidomide and teratogenic activity of its enantiomers, *Arzneim. Forsch.*, 29, 1690, 1979.
10. Jeppsson, A, B., Johansson, U., and Waldeck, B., Steric aspects of agonism and antagonism at β-adrenoreceptors: experimental with the enantiomers of terbutaline and pindolol, *Acta Pharmacol., Toxicol.*, 54, 285, 1984.
11. Keck, J., Kruger, G., and Noll, K., German Patent DT 2212600C3, 1976.

12. Trofast, J., Osterberg, K., Kallstrom, B. L., and Waldeck, B., Steric aspects of agonism and antagonism at β-adrenoreceptors: synthesis of and pharmacological experiments with the enantiomers of formeterol and their diastereomers., *Chirality*, 3, 443, 1991.
13. Announcement, FDA's policy statement for the development of new stereoisomeric drugs, *Chirality*, 4, 338, 1992.
14. Souter, R. W., *Chromatographic Separations of Stereoisomers*, Souter R. W., Ed., CRC Press, Boca Raton, FL, 1985.
15. Zief, M. and Crane, L. J., Eds., *Chromatographic Chiral Separations*, Marcel Dekker, New York, 1988.
16. Goto, J., Hasegawa, M., Nakamura, S., Shimeda, K., and Nambara, T., New derivatising reagents for the resolution of amino acids enantiomers by high performance liquid chromatography, *J. Chromatogr.*, 152, 143, 1978.
17. Stolenberg, J. K., Puglisi, C. V., Rubio, F., and Vane, F. M., High performance liquid chromatographic determination of stereoselective disposition of carprofens in humans, *J. Pharm. Sci.*, 70, 1207, 1981.
18. Annett, R. G. and Stumpy, P. K., L-(–)-menthyloxycarbonyl derivatisation of hydroxy acid methyl esters, *Anal. Biochem.*, 47, 638, 1972.
19. Fanali, S., Masia, P., and Ossicini, L., A paper electrophoretic study of ion-pair formation XVI. Resolution of optical isomers of some cobalt (III) complexes with amino acid ligands., *J. Chromatogr.*, 403, 388, 1987.
20. Ossicini, L. and Celli, C., A paper electrophoretic study of ion-pair formation XI. the behaviour of optical isomers in optically active electrolytes, *J. Chromatogr.*, 115, 655, 1975.
21. Szejtli, J., Cyclodextrins in drug formulations: part I, *Pharm. Technol. Int.*, 3, 15, 1991.
22. Szetli, J., Zsadon, B., and Cserhati, T., Cyclodextrin use in separation, in *Ordered Media in Chemical Separations*, Hinze, W. L. and Armstrong, D. W., Eds., American Chemical Society, Washington, DC, 1987, Chap. 11.
23. Nardi, A., Ossicini, L., and Fanali, S., Use of cyclodextrins in capillary zone electrophoresis for the separations of optical isomers: resolution of racemic tryptophan derivatives, *Chirality*, 4, 56, 1992.
24. Kuhn, R., Stoecklin, F., and Erni, F., Chiral separations by host-guest complexation with cyclodextrin and crown ether in capillary electrophoresis, *Chromatographia*, 33, 32, 1992.
25. Tanaka, M., Asano, S., Yoshinago, M., Kawaguchi, Y., Tetsumi, T., and Shono, T., Separation of racemates by capillary electrophoresis based on complexation with cyclodextrins, *Fresenius J. Chem.*, 339, 63, 1991.
26. Ueda, T., Mitchell, R., Kitamura, F., Metcalf, T., Kuwana, T., and Nakamoto, A., Separation of naphthalene-2,3-dicarboxaldehyde-labelled amino acids by high performance capillary electrophoresis with laser induced fluorescence detection., *J. Chromatogr.*, 593, 265, 1992.
27. Ueda, T., Kitamura, F., Mitchell, R., Metcalf, T., Kuwara, T., and Nakamoto, A., Chiral separation of naphthalene-2,3-dicarboxaldehyde-labelled amino acid enantiomers by cyclodextrin modified micellar electrokinetic chromatography with laser induced fluorescence detection, *Anal. Chem.*, 63, 2979, 1991.
28. Furuta, R. and Doi, T., Enantiomeric separation of a thiazole derivative by high performance liquid chromatography and micellar electrokinetic chromography, *J. Chromatogr.*, 708, 245, 1995.
29. Furuta, R. and Doi, T., Chiral separation of diniconazole, uniconazole and structurally related compounds by cyclodextrin-modified micellar electrokinetic chromatography, *Electrophoresis*, 15, 1322, 1994.
30. Hempel, G. and Blaschke, G. Enantioselective determination of zopicolone and its metabolites in urine by capillary electrophoresis, *J. Chromatogr.*, 675, 139, 1996.
31. Grosenick, H., Mayer, S., Juza, M., Jakubetz, H., and Schurig, V., Characterisation of β-cyclodextrin sulfopropyl ether and its application as a chiral additive in CE, poster presented at the 7th International Symposium on HPCE 1995.
32. Haskins, N. J., Saunders, M. R., and Camilleri, P., The complexation and chiral selectivity of 2-hydroxypropyl-β-CD with guest molecules as studied by electrospray mass spectrometry, *Rapid Commun. in Mass Spectrometry*, 8, 423, 1994.

33. Yoshinaga, M. and Tanaka, M., Use of selectively methylated b-cyclodextrin derivatives in chiral separation of dansylamino acids by capillary zone electrophoresis, *J. Chromatogr.*, 679, 359 1994.

34. Valko, I. E., Billiet, H. A. H., Frank., J., and Luyben, K. Ch. A. M., Effect of the degree of substitution of (2-hydroxy)propyl-β-cyclodextrin on the enantioseparation of organic acids by capillary electrophoresis, *J. Chromatogr.*, 678, 139, 1994.

35. Gahm, K-H. and Stalcup, A. M., Capillary electrophoresis study of naphthylethylcarbamoylated β-cyclodextrins, *Anal. Chem.*, 67, 19, 1995.

36. Sepaniak, M. J., Cole, R. O., and Clark, B. K., Use of native and chemically modified cyclodextrins for the capillary electrophoretic separation of enantiomers, *J. Liq. Chromatogr.*, 15, 1023, 1992.

37. Schmitt, T. and Engelhardt, H., Charged and uncharged cyclodextrins as chiral selectors in capillary electrophoresis, *Chromatographia*, 37, 475, 1993.

38. Cladrowa-Runge, S., Hirz, R., Kenndler, E., and Rizzi, A., Enantiomeric separation of amphetamine related drugs by capillary zone electrophoresis using native and derivatised b-cyclodextrins as chiral additives, *J. Chromatogr.*, 710, 339, 1995.

39. Chankvetadze, B., Endresz, G., and Blaschke, G., Enantiomeric resolution of anionic R/S-1,1-binaphthyl-2,2′-diyl hydrogen phosphate by capillary electrophoresis using anionic cyclodextrin derivatives as chiral selectors, *J. Chromatogr.*, 704, 234, 1995.

40. Tait, R. J., Thompson, D. O., Stella, V. J., and Stobaugh, J. F., Sulfobutyl ether β-cyclodextrin as a chiral discriminator for use with capillary electrophoresis, *Anal. Chem.*, 66, 4013, 1994.

41. Lurie, I. S., Klein, R. F. X., Cason, T. A. D., LeBelle, M. J. Brennersen, R., and Weinberger, R. E., Chiral resolution of cationic drugs of forensic interest by capillary electrophoresis with mixtures of neutral and anionic cyclodextrins, *Anal. Chem.,* 66, 4019, 1994.

42. Fanali, S. and Aturki, Z., Use of cyclodextrins in capillary electrophoresis for the chiral resolution of some 2-arylpropionic acids non-steroidal anti- inflammatory drugs, *J. Chromatogr.*, 694, 297, 1995.

43. Terabe, S., Electrokinetic chromatography: an interface between electrophoresis and chromatography, *Trends in Anal. Chem.*, 8, 129, 1989.

44. Ingelse, B. A., Everaerts, F. M., Sevick, J., Stransky, Z., and Fanali, S., A further study on the chiral separation power of a soluble neutral β-cyclodextrin polymer, *HRC J. High Resol. Chromatogr.*, 18, 348, 1995.

45. Ingelse, B. A., Everaerts, F. M., Desiderio, C., and Fanali, S., Enantiomeric separation by capillary electrophoresis using a soluble neutral β-cyclodextrin polymer, *J. Chromatogr.*, 709, 89, 1995.

46. Nishi, H., Nakamura, K., Nakai, H., and Sato, T., Chiral separation of drugs by capillary electrophoresis using b-cyclodextrin polymer, *J. Chromatogr.*, 678, 333, 1994.

47. Aturki, Z. and Fanali, S., Use of β-cyclodextrin polymer as a chiral selector in capillary electrophoresis, *J. Chromatogr.*, 680, 137, 1994.

48. Nishi, H., Fukuyama, T., and Terabe, S., Chiral separation by cyclodextrin modified micellar electrokinetic chromatography, *J. Chromatogr.*, 553, 503, 1991.

49. Okafo, G., Bintz, C., Clarke, S., and Camilleri, P., Micellar electrokinetic capillary chromatography in mixtures of taurodeoxycholic acid and β- cyclodextrin, *J. Chem. Soc., Chem. Commun.*, 1189, 1992.

50. Aumatell, A. and Wells, R. J., Enantiomeric differentiation of a wide range pharmacologically active substances by cyclodextrin-modified micellar electrokinetic capillary chromatography using a bile salt, *J. Chromatogr.*, 688, 329, 1994.

51. Wren, S. A. C., Theory of chiral separation in capillary electrophoresis, *J. Chromatogr.*, 636, 57, 1993.

52. Vespalec, R., Fanali, S., and Bocek, P., consequences of a maximum existing in the dependence of separation selectivity on concentration of cyclodextrin added as a chiral selector in capillary electrophoresis, *Electrophoresis*, 15, 1523, 1994.

53. Gareil, P., Gramond, J. P., and Guyon, F., Separation and determination of warfarin enantiomers in human plasma by capillary electrophoresis using a methylated β-cyclodextrin-containing electrolyte, *J. Chromatogr.*, 615, 317, 1993.

54. Peterson, T. E., Separation of drug stereoisomers by capillary electrophoresis with cyclodextrins., *J. Chromatogr.*, 630, 353, 1993.
55. Francotte, E., Cherkaoui, S., and Faupel, M., Separation of the enantiomers of some racemic nonsteroidal aromatase inhibitors and barbiturates by capillary electrophoresis, *Chirality*, 5, 516, 1993.
56. Fanali, S., Use of cyclodextrins in capillary zone electrophoresis resolution of terbutaline and propanolol enantiomers, *J. Chromatogr.*, 545, 434, 1991.
57. Guttman, A., Paulus, A., Cohen, A. S., Grinberg, N., and Karger, B. L., Use of complexing agents for selective separation in high performance capillary electrophoresis chiral resolution via cyclodextrins incorporated within polyacrylamide gel columns, *J. Chromatogr.*, 448, 41, 1988.
58. Yoshinaga, M. and Tanaka, M., Effect of urea addition on chiral separation of dansylamino acids by capillary electrophoresis with cyclodextrins, *J. Chromatogr.*, 710, 331, 1995.
59. Fanali, S. and Bocek, P., Enantiomer resolution by using capillary zone electrophoresis: resolution of racemic tryptophan and determination of the enantiomer composition of commercially available pharmaceutical epinephrine, *Electrophoresis*, 11, 757, 1990.
60. Anigbogu, V. C., Copper, C. L., and Sepaniak, M. J., Separation of stereoisomers of aminoglutethimide using three capillary electrophoretic techniques, *J. Chromatogr.*, 705, 343, 1995.
61. Lindner, W., Bohs, B., and Seidel, V., Enantioselective capillary electrophoresis of amino acids derivatives on cyclodextrin evaluation of structure-resolution relationships, *J. Chromatogr.*, 697, 549, 1995.
62. Snopek, J., Soini, H., Novotny, M., Smolkova-Keulemansova, E., and Jelinek, L., Selected applications of cyclodextrins selectors in capillary electrophoresis, *J. Chromatogr.*, 559, 215, 1991.
63. D'Hulst, A. and Verbeke, N. Separation of the enantiomers of coumarinic anticoagulant drugs by capillary electrophoresis using maltodextrins as chiral selectors, *Chirality*, 6, 225, 1994.
64. Mikus, F. F., Hixon, R. M., and Rundle, R. E., The complexes of fatty acids with amylose, *J. Am. Chem. Soc.*, 1115, 68, 1946.
65. Yamamoto, M., Sano, T., and Yasunaga, T., Interaction of amylose with iodine.II. Kinetic studies of the complex formation by temperature-jump method, *Bull. Chem. Soc. Jpn.*, 55, 1886, 1982.
66. Kowblansky, M., Calorimetric investigation of inclusion complexes of amylose with long-chain aliphatic compounds containing different functional groups, *Macromolecules*, 18, 1776, 1987.
67. Simpson, T. D., Dintzis, F. R., and Taylor, N. W., A V_7 conformation of dimethyl sulfoxide-amylose complex, *Biopolymers*, 11, 2591, 1972.
68. Aoyama, Y., Otsuki, J., Nagai, Y., Kobayashi, K., and Toi, H., Host-guest complexation of oligosaccharides: interaction of malto dextrins with hydrophobic fluorescence probes in water, *Tetrahedron Lett.*, 33, 3775, 1992.
69. Soini, H. S., Stefansson, M., Riekkola, M-L., and Novotny, M. V., Maltoligosaccharides as chiral selectors for the separation of pharmaceuticals by capillary electrophoresis, *Anal. Chem.*, 66, 3477, 1994.
70. Kano, K., Minami, K., Horiguchi, K., Ishimura, T., and Kodera, M., Ability of non-cyclic oligosaccharides to form molecular complexes and its use for chiral separation by capillary zone electrophoresis, *J. Chromatogr.*, 694, 307, 1995.
71. Nishi, H., Nakamura, K., Nakai, H., and Sato, T., Enantiomeric separation of drugs by mucopolysaccharide-mediated electrokinetic chromatography, *Anal. Chem.*, 67, 2334, 1995.
72. Stalcup, A. M. and Agyei, N. M., Heparin: A chiral mobile-phase additive for capillary electrophoresis, *Anal. Chem.*, 66, 3054, 1994.
73. Nishi, H., Nakamura, K., Nakai, H., Sato, T., and Terabe, S., Enantiomeric separation of drugs by affinity electrokinetic chromatography using dextran sulphate, *Electrophoresis*, 15, 1335, 1994.
74. Dietrich, J. M., Lehn, J. M., and Sauvage, J. P., Diaza-polyoxa-macrocycles et macrobicycles, *Tetrahedron Lett.*, 2885, 1969.
75. Kuhn, R., Stoecklin, F., and Erni, F., Chiral separations by host-guest complexation with cyclodextrin and crown ether in capillary electrophoresis, *Chromatographia*, 33, 32, 1992.

76. Kuhn, R., Erni, F., Bereuter, T., and Hausler, J., Chiral recognition and enantiomeric resolution based on host-guest complexation with crown ethers in capillary electrophoresis, *Anal. Chem.*, 64, 2815, 1991.

77. Walbroehl, Y. and Wagner, J., Chiral separations of amino acids by capillary electrophoresis and high performance liquid chromatography employing chiral crown ethers, *J. Chromatogr.*, 685, 321, 1994.

78. Schmid, M. G. and Gubitz, G., Capillary electrophoretic separation of the enantiomers of dipeptides based on host-guest complexation with a chiral crown ether, *J. Chromatogr.*, 709, 81, 1995.

79. Castelnovo, P. and Albanesi, C., Determination of the purity of 5,6-dihydroxy-2- aminotetralin by high performance capillary electrophoresis with crown ether as chiral selector, *J. Chromatogr.*, 715, 143, 1995.

80. Walbroehl, Y. and Wagner, J., Enantiomeric resolution of primary amines by capillary electrophoresis and high performance liquid chromatography using chiral crown ethers, *J. Chromatogr.*, 680, 253, 1994.

81. Terabe, S., Shibita, M, and Miyashita, Y., Chiral separation by electrokinetic chromatography with bile salts micelles, *J. Chromatogr.*, 480, 403, 1989.

82. Cole, R. O., Sepaniak, M. J., and Hinze, W. L., Optimisation of binaphthyl enantiomer separation by capillary zone electrophoresis using mobile phases containing bile salts and organic solvents, *J. High Resolut. Chromatogr.*, 13, 579, 1990.

83. Nishi, H., Fukuyama, T., Matsuo, M., and Terabe, S., Chiral separation of diltiazem, trimetoquinol and related compounds by micellar electrokinetic chromatography, *J. Chromatogr.*, 515, 233, 1990.

84. Otsuka, K. and Terabe, S., Enantiomeric resolution by micellar electrokinetic chromatography with chiral surfactants, *J. Chromatogr.*, 515, 221, 1990.

85. Ishihama, Y. and Terabe, S., Enantiomeric separation by micellar electrokinetic chromatography using saponins, *J. Liq. Chromatogr.*, 16, 933, 1993.

86. Dobashi, A, Ono, T., Hara, S., and Yamaguchi, J., Optical resolution of enantiomers with chiral micelles by electrokinetic chromatography, *Anal. Chem.*, 61, 1986, 1989.

87. Otsuka, K., Kawahara, J., Tatekawa, K., and Terabe, S., Chiral separations by micellar electrokinetic chromatography with sodium N-dodecanoyl-L-valinate, *J. Chromatogr.*, 559, 209, 1991.

88. Dobashi, A., Hamada, M., Dobashi, Y., and Yamaguchi, J., Enantiomeric separation with dodecanoyl-l-amino acidate micelles and poly (sodium (10-undecenoyl)-L-valinate) by electrokinetic chromatography, *Anal. Chem.*, 67, 3011, 1995.

89. Otsuka, K., Karuhaka, K., Higasimori, M., and Terabe, S., Optical resolution of amino acid derivatives by micellar electrokinetic chromatography with N- dodecanoyl-L-serine, *J. Chromatogr.*, 680, 317, 1994.

90. Mazzeo, j. R., Grover, E. R., Swartz, M. E., and Peterson, J. S., Novel chiral surfactants for the separation of enantiomers by micellar electrokinetic capillary chromatography, *J. Chromatogr.*, 680, 125, 1994.

91. Tickle, D. C., Okafo, G. N., Camilleri, P., Jones, R. F. D., and Kirby, A. J., Glucopyranoside-based surfactants as pseudostationary phases for chiral separations in capillary electrophoresis, *Anal. Chem.*, 66, 4121, 1994.

92. De Biasi, V, Senior, J, Zukowski, J. A., Haltiwanger, R. C., Eggleston, D. S., and Camilleri, P., Chiral discrimination in capillary electrophoresis using novel anionic surfactants related to cysteine, *J. Chem. Soc. Chem. Commun.*, 1575, 1995.

93. Dalton, D.D., Taylor, D. R., and Waters, D. G., Synthesis and use of novel chiral surfactants in micellar electrokinetic capillary chromatography, *J. Chromatogr.*, 712, 365, 1995.

94. Bouzige, M., Okafo, G. N., Dhanak, D., and Camilleri, P., The physico-chemical properties of novel surfactant derived from 6-amino-penicillanic acid and its use in capillary electrophoresis for chiral discrimination, *J. Chem. Soc. Chem. Commun.*, 671, 1996.

95. El Rassi, Z. and Mechref, Y., Seventeenth international Symposium on Capillary Chromatography and Electrophoresis, May 7–11, 1995, Wintergreen, Virginia, USA, 1995.

96. Wang, J. and Warner, I. M., Chiral separations using micellar electrokinetic chromatography and a polymerised chiral micelle, *Anal. Chem.*, 66, 3773, 1994.

97. Vespalec, R., Sustacek, V., and Bocek, P., Prospects of dissolved albumin as a chiral selector in capillary zone electrophoresis, *J. Chromatogr.*, 638, 255, 1993.

98. Valtcheve, L., Mohammad, J., Petersson, S., and Hjerten, S., Chiral separation of β-blockers by high performance capillary electrophoresis based on non- immobilised cellulase as enantioselective protein, *J. Chromatogr.*, 638, 263, 1993.

99. Yang, J. and Hage, D. S., Chiral separations in capillary electrophoresis using human serum albumin as a buffer additive, *Anal. Chem.*, 66, 2719, 1994.

100. Mohammad, J., Li, Y-M., El-Ahmad, M., Nakazato, K., Petersson, G., and Hjerten, S. *Chirality*, in press.

101. Petersson, G., Petersson, C., Stahlberg, J., Isaksson, R., and Jonsson, J., Department of Biochemistry and Pharmaceutical Analytical Chemistry, University of Uppsala, Personal Communication.

102. Arai, T., Ichinose, M., Kuroda, H., Nimura, N., and Kinosihita, T., Chiral separation by capillary affinity zone electrophoresis using an albumin-containing support electrolyte., *Anal. Biochem.*, 217, 7, 1994.

103. Tanaka, Y. and Terabe, S., Partial separation zone technique for the separation of enantiomers by affinity electrokinetic chromatography with proteins as chiral pseudo-stationary phases, *J. Chromatogr.*, 694, 277, 1995.

104. Lloyd, D. K., Li, S., and Ryan, P., Protein chiral selectors in free solution capillary electrophoresis and packed-capillary electrochromatography, *J. Chromatogr.*, 694, 285, 1995.

105. Birnbaum, S. and Nilsson, S., Protein-based capillary affinity gel electrophoresis for the separation of optical isomers, *Anal. Chem.*, 64, 2872, 1992.

106. Wistuba, D., Diebold, H., and Schurig, V., Enantiomer separation of DNP- amino acids by capillary electrophoresis using chiral buffer additives,. *J. Microcol. Sep.*, 7, 17, 1995.

107. Busch, S, Kraak, J. C., and Poppe, H., Chiral separations by complexation with proteins in capillary zone electrophoresis, *J. Chromatogr.*, 635, 119, 1993.

108. Barker, G. E., Russo, P., and Hartwick, R. A., Chiral separation of leucovorin with bovine serum albumin using affinity capillary electrophoresis, *Anal. Chem.*, 64, 3024, 1992.

109. Sun, P., Wu, N., Barker, G., and Hartwick, R. A., Chiral separations using dextran and bovine serum albumin as run buffer additives in affinity capillary electrophoresis *J. Chromatogr.*, 648, 475, 1993.

110. Ishihama, Y., Oda, Y., Asakawa, N., Yoshida, Y., and Sato, T., Optical resolution by electrokinetic chromatography using ovomucoid as a pseudostationary phase, *J. Chromatogr.*, 666, 193, 1994.

111. Armstrong, D. W., A new class of chiral selector for enantiomeric separations in LC, TLC, GC, CE and SFC, Pittsburgh Conference Abstracts, 572, 1994.

112. Armstrong, D. W., Tang, Y., Chen, S., Zhou, Y., Bagwill, C., and Chen, J-R., Macrocyclic antibiotics as a new class of chiral selectors for liquid chromatography, *Anal. Chem.*, 66, 1473, 1994.

113. Armstrong, D. W. and Zhou, Y., Use of a macrocyclic antibiotic as the chiral selector for enantiomeric separations by TLC., *J. Liq. Chromatogr.*, 17, 1695, 1994.

114. Sensi, P., Greco, A. M., and Ballotta, R., Rifamycins.I. Isolation and properties of rifamycin B and rifamycin complex, *Antibiot., Ann.*, 262, 1959-60.

115. Armstrong, D. W., Rundlett, K., and Reid, G. L., Use of a macrocyclic antibiotic, rifamycin B, and indirect detection for the resolution of racemic amino alcohols by CE., *Anal. Chem.*, 66, 1690, 1994.

116. Ward, T. J., Dann, C., and Blaylock, A., Enantiomeric resolution using the macrocyclic antibiotics rifamycin B and rifamycin SV as chiral selectors for capillary electrophoresis, *J. Chromatogr.*, 715, 337, 1995.

117. Higgins, H. M., Harrison, W. H., Wild, G. M., Bungay, H. R., and McCormick, M. H., Vancomycin: a new antibiotic VI. purification properties and properties of vancomycin, *Antibiotics Annual*, 906, 1957-1958.

118. Philip, J. E., Schrenk, and Hargie, M. P., Ristocetin A and B, two new antibiotics. Isolation and properties, *Antibiot. Ann.*, 669, 1956-57.

119. Armstrong, D. W., Rundlett, K. L., and Chen., J-R., Evaluation of the macrocyclic antibiotic vancomycin, as a chiral selctor for capillary electrophoresis, *Chirality*, 6, 496, 1994.

120. Armstrong, D.W., Gasper, M.P., and Rundlett, K. L., Highly enantioselective capillary electro-phoretic separations with dilute solutions of the macrocyclic antibiotic ristocetin A, *J. Chromatogr.*, 689, 285, 1995.

121. Rundlett, K. L. and Armstrong, D. W., Effect of micelles and mixed micelles on efficiency and selectivity of antibiotic-based capillary electrophoretic enantioseparations, *Anal. Chem.*, 67, 2088, 1995.

122. Carpenter, J. L., Camilleri, P., Dhanak, D., and Goodall, D., A study of the binding of vanco-mycin to dipeptides using capillary electrophoresis., *J. Chem. Soc., Chem., Commun.*, 804, 1992.

123. Chu, Y-H. and Whiteside, G. M., Affinity capillary electrophoresis can simultaneously measure binding constants of multiple peptides to vancomycin., *J. Org. Chem.*, 57, 3524, 1992.

124. Somma, S., Gastaldo, L., and Corti, A., Teicoplanin, a new antibiotic from Actinoplanes teichomyceticus nov. sp., *Antimicrob. Agents. Chemother.*, 26, 917, 1984.

125. Rundlett, K. L., Gasper, M. P., Zhou, E. Y., and Armstrong, D. W., Capillary electrophoretic enantiomeric separations using the glycopeptide antibiotic, teicoplanin, *Chirality*, 8, 88, 1996.

126. Lam, S., Stereoselective analysis of D-and L-dansyl amino acids as the mixed chelate copper(II) complexes by HPLC, *J. Chromatogr. Sci.*, 22, 416, 1984.

127. Lam, S., Chow, F., and Kamen, A., Reverse-phase high performance liquid chromatographic resolution of D- and L-amino acids by mixed chelate complexation, *J. Chromatogr.*, 199, 295, 1980.

128. Gozel, P., Gassmann, E., Michelsen, H., and Zare, R. N., Electrokinetic resolution of amino acid enantiomers with copper (II) aspartame support electrolyte, *Anal. Chem.*, 59, 44, 1987.

129. Nishi, H., Fukuyama, T., and Matsuo, M., Resolution of optical isomers of 2,3,4,6-tetra-O-acetyl-β-D-glucopyranosyl isothiocyanate (GITC) derivatised DL-amino acids by micellar electrokinetic chromatography, *J. Microcol. Sep.*, 2, 234, 1990.

130. Okafo, G. N., Rana, K. K., and Camilleri, P., Improved separation of diastereoisomers in capillary electrophoresis using a mixture of β-cyclodextrin and sodium taurodeoxycholate, *Chromatographia*, 39, 627, 1994.

131. Kinoshita, T., Kasahara, Y., and Nimura, N., Reversed phase high performance chromatographic resolution of non-esterified enantiomeric amino acids by derivatisation with 2,3,4,6-tetra-O-acetyl-β-D-glucopyranosyl isothiocyanate and 2,3,4-tri-O-acetyl-α-D-arabinopyranosyl isothio-cyanate, *J. Chromatogr.*, 210, 77, 1981.

132. Waetzig, H., Dette, C., Aigner, A., and Wilschowitz, L., Analysis of acetylcysteine by capillary electrophoresis (CE). Part 2: determination of side components., *Pharmazie*, 49, 249, 1994.

133. Dette, C., Waetzig, H., and Aigner, A., Analysis of acetylcysteine by capillary electrophoresis (CE), *Pharmazie*, 49, 245, 1994.

134. Schutzner, W., Fanali, S., Rizzi, A., and Kenndler, E., Separation of diastereomers by capillary zone electrophoresis in free solution with polymer additive and organic solvent component effect of pH and solvent composition, *J. Chromatogr.*, 719, 411, 1996.

135. Vogt, C., Georgi, A., and Werner, G., Enantiomeric separation of D/L-carnitine using HPLC and CZE after derivatisation, *Chromatographia*, 40, 287, 1995.

136. Noe, C. R. and Freissmuth, J., Capillary zone electrophoresis of aldose enantiomers:separation after derivatisation with S-(–)-phenylethylamine, *J. Chromatogr.*, 704, 503, 1995.

137. Blau, K. and Halket, J., Eds., *Handbook of Derivatives for Chromatography*, John Wiley & Sons, New York, 1993.

138. Engstrom, A., Wan, H., Andersson, P. E., and Josefsson, B., Determination of chiral reagent purity by capillary electrophoresis, *J. Chromatogr.*, 715, 151, 1995.

139. Sandor, V., Flarakos, T., Batist, G., Wainer, I. W., and Lloyd, D. K., Quantitation of the diastereoisomers of L-buthionine-(R,S)-sulfoximine in human plasma: a validated assay by capillary electrophoresis, *J. Chromatogr.*, 673, 123, 1995.

140. Schurig, V., Schmalzing, D., Muhleck, U., Jung, M., Schleimer, M., Mussche, P., Duvecot, C., and Buyten, J. C., Separation of enantiomers on immobilised polysiloxane-anchored permethylated-β-cyclodextrin (chirasil- dex) by supercritical fluid chromatography, *High Res. Chromatogr.*, 13, 713, 1990.

141. Schurig, V., Juvancz, Z., Nicholson, G. J., and Schmalzing, D., Separation of enantiomers on immobilised polysiloxane-anchored permethyl-β-cyclodextrin (CHIRASIL-DEX) by supercrit-ical fluid chromatography, *High Res. Chromatogr.*, 14, 58, 1991.

142. Schurig, V., Schmalzing, D., and Schleimer, M., Enantiomer separation on immobilised chirasil-metal and chirasil-dex by gas chromatography and supercritical fluid chromatography, *Angew. Chem., Int. Ed. Engl.*, 30, 987, 1991.

143. Mayer, S. and Schurig, V., Enantiomer separation by electrochromatography on capillaries coated with chirasil-dex, *J. High Res. Chromatogr.*, 15, 129, 1992.

144. Mayer, S., Schleimer, M., and Schurig, V., Dual chiral recognition system involving cyclodextrin derivatives in capillary electrophoresis, *J. Microcol. Sep.*, 6, 43, 1994.

145. Mayer, S. and Schurig, V, Enantiomer separation by electrochromatography in open tubular columns coated with chirasil-dex, *J. Liq. Chromatogr.*, 16, 915, 1994.

146. Szeman, J. and Ganzler, K., Use of cyclodextrins and cyclodextrin derivatives in high performance liquid chromatography and capillary electrophoresis, *J. Chromatogr.*, 668, 509, 1994.

147. Li, S. and Lloyd, D. K., Packed-capillary electrochromatography separation of the enantiomers of neutral and anionic compounds using β-cyclodextrin as a chiral selector: effect of operating parameters and comparison with free-solution capillary electrophoresis, *J. Chromatogr. A*, 666, 321, 1994.

CHAPTER 6

Determination of Inorganic Anions and Metal Cations

James S. Fritz

CONTENTS

I. INTRODUCTION

Since its introduction in 1975 and rapid development during the 1980s, ion chromatography has been the dominant method for separating and determining inorganic anions in analytical samples. Ion chromatography has also been widely used for the determination of metal ions, particularly for the alkali metals, magnesium, and the alkaline earths. Now, this preeminence is being challenged by capillary electrophoresis, which has an almost flat flow profile, whereas there is a parabolic flow through the ion chromatographic columns. This difference, coupled with the fact that there is no partition of analytes between liquid and solid phases in CE, makes capillary electrophoresis an attractive separation technique for inorganic ions.

In this chapter some fundamental principles of capillary electrophoresis will be reviewed. Then methods and techniques for determining inorganic anions and cations will be discussed.

II. PRINCIPLES

A. Introduction

The basic instrumentation, detection systems and general theory of capillary electrophoresis are discussed in earlier chapters of this book. However, several points that apply particularly to the CE behavior of inorganic ions should be mentioned.

Inorganic anions and cations are generally smaller and often have a higher charge than organic ions. Thus, inorganic ions tend to be more mobile than most organic ions. The electrophoretic mobilities of inorganic anions are an inverse function of their hydrated ionic radii.[1] Electrophoretic mobility is also affected by the charge on an ion and by the solvent medium. Tables of limiting ionic conductances are a convenient source for estimating electrophoretic mobilities of ions. Limiting equivalent conductance λ is converted to mobility with the help of the Faraday constant ($F = 9.6487 \times 10^4$ A sec equiv^{-1})

$$\mu_{ep} = \lambda_{equiv}/F \tag{1}$$

Kuhr has stated that the choice of separation buffer is one of the most critical parameters governing the efficiency of a separation.[2] He stressed the importance of keeping the sample concentration less than 1% that of the buffer concentration to avoid triangular peaks that result from electrophoretic "fronting."

It is also necessary to match the ionic mobilities of the electrolyte ions as closely as possible to the mobilities of the sample ions. Hjerten discussed peak asymmetry in CE with respect to the conductivity difference (Δk) observed at a boundary between a migrating analyte zone and a carrier electrolyte.[3] Equation 2 expresses the relationship between the sample ion concentration and the analyte and electrolyte ionic mobilities:

$$\Delta k = c_B |\mu_B(\mu_A - \mu_B)(\mu_R - \mu_B)| \tag{2}$$

where c_B is the sample ion concentration, and μ_B, μ_A, and μ_R are the respective ionic mobilities of the sample ion, the electrolyte co-ion, and the electrolyte counterion. According to Equation 2, peak shape in CE can be optimized not only by matching analyte and co-ion mobilities, but also by carefully selecting a carrier electrolyte counterion. This equation also indicates that, for any set of ionic mobilities, peak asymmetry increases with increasing concentration of the analyte ion in the migrating zone.

B. Buffers

A buffer must be present in the capillary electrolyte to ensure a constant pH throughout the capillary. The buffer also supplies ions needed to maintain a small electric current.

In addition to these factors, the buffer ion should have an electrophoretic mobility similar to that of the sample ions (see below). Finally, the buffer should not form complexes or precipitates with the sample ions. Some recommended buffers with their pK_a values, are as follows:

Basic solutions:	Borate	$pK_1 = 9.2$
	Bicarbonate	$pK_1 = 6.4$; $pK_2 = 10.4$
	Glycine	$pK_2 = 9.6$
	Taurine	$pK_2 = 8.7$
Acidic solutions:	Nicotinamide	$pK_a = 3.3$
	Glycine	$pK_1 = 2.6$
	β-Alanine	$pK_1 = 3.6$

C. Separation Modes

Separation of analyte ions is based on differences in electrophoretic mobility. However, electroosmotic flow is almost always present also and usually is greater than the electrophoretic flow. Two possibilities exist for separation:

1. Comigration. The vectors of μ_{ep} and μ_{eo} are in the same direction.
2. Counter migration. The vectors of μ_{ep} and μ_{eo} are in opposite directions. However, the magnitudes of the opposing vectors must be such that net flow will be through the detector.

These two modes of separation are illustrated in Figure 6.1 by vector diagrams representing electrophoretic and electroosmotic mobilities. In both instances the electroosmotic mobility is greater than the electrophoretic mobility of any of the sample anions. Note that the polarity of the applied potential is reversed in going from comigration to counter migration. In the comigration example, the direction of electroosmotic flow has been reversed by use of a chemical additive to coat the capillary surface with a positive charge.

It will be seen that the net mobilities are greater with comigration, thereby giving a faster separation than with counter migration. Electrophoretic efficiency is in fact better at higher velocities.[2] However, the effective column length is in a sense shortened in comigration because of the electroosmotic flow component which causes all ions to move at the same rate. This can limit the attainable resolution of peaks. In the counter migration mode the order of elution is opposite that in the comigration mode. Separations take longer in the counter migration mode but resolution of the sample ions should be better.

This may be illustrated by a numerical example. Suppose two sample compounds have electrophoretic velocities of 4.04 cm/min and 4.00 cm/min, respectively, and the electroosmotic velocity is 6.00 cm/min. For a capillary of 50 cm effective length, the migration times of the sample compounds with comigration are:

$$t_1 = \frac{50}{4.04 + 6.00} = 4.98 \, \text{min}$$

$$t_2 = \frac{50}{4.00 + 6.00} = 5.00 \, \text{min}$$

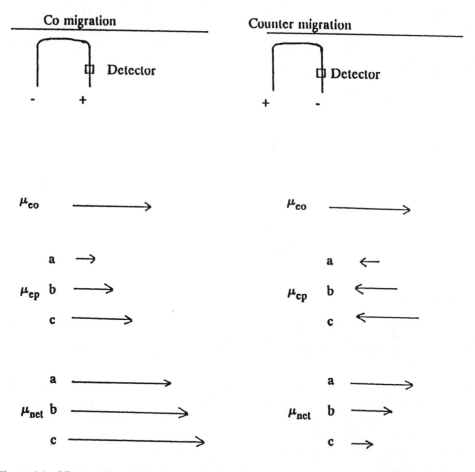

Figure 6.1 CE separation modes for sample anions.

$$\alpha = t_2 / t_1 = 1.004$$

With counter migration:

$$t_1 = \frac{50}{6.00 - 4.04} = 25.5 \, \text{min}$$

$$t_2 = \frac{50}{6.00 - 4.00} = 25.00 \, \text{min}$$

$$\alpha = t_1 / t_2 = 1.02$$

In most cases the larger separation factor (α) will more than compensate for the greater peak widths at longer migration times.

Indirect detection of inorganic sample ions is more prevalent than with organic ions because many inorganic ions lack the optical properties necessary for direct spectrophotometric detection. A "visualization" ion that absorbs in the UV or visible spectral region is added to the capillary electrolyte. When a sample ion zone passes through the detector the background absorbance of the visualization ion is reduced by an amount proportional to the concentration of the sample ion. The visualization ion must have a mobility close to that of the sample ions in order to give sharp, symmetric detection peaks.

III. SEPARATION OF ANIONS

A. Introduction

The first reports on applications of CE to the separation of inorganic anions appeared between 1979 and 1990. A brief summary of these earlier papers has been published.[4] The capillaries used in these early papers were made of PTFE, and isotachophoretic potential gradient detectors were adapted to detect migrating CE zones. Later, more efficient fused-silica capillaries became available. A major breakthrough occurred in 1990 when chromate was first added to the capillary electrolyte for indirect spectrophotometric detection of inorganic anions.[5] Conditions for separation of many inorganic and low-molecular-weight organic anions were reported. Since 1990, separation of inorganic and small organic anions has undergone very rapid development.

B. Electroosmotic Flow Modifiers

Most of the CE methods for anions use comigration with a negative power supply. For this to occur, the direction of electroosmotic flow must be reversed. This is usually accomplished by adding a reagent to the running electrolyte that will thinly coat the capillary surface, giving it a positive charge. Hydrated electrolyte anions flow through the detector toward the positive electrode, thus providing an electroosmotic flow in the desired direction.

Quaternary ammonium salts with three methyl or ethyl groups plus a long-chain alkyl group have been used as the electroosmotic flow modifiers. For convenience, such salts will be denoted as Q^+. Salts with chains of 10, 12, 14, 16, and 18 carbon atoms have been studied. However, C_{14} or TTAB (tetradecyl tetra ammonium bromide) and C_{16} or CTAB (cetyl tetra ammonium bromide) have been the most popular.

Coating the capillary surface with a Q^+ layer involves a dynamic equilibrium between the electrolyte solution and the capillary wall. The positive charge of the Q^+ is probably attracted to the negatively charged silanol groups with the long hydrocarbon chain sticking out from the wall. Additional Q^+ molecules are attracted hydrophobically to the hydrocarbon tails, with their N^+ end sticking out into the solution and away from the wall. This mechanism provides the net positive charge on the capillary surface needed to reverse the direction of electroosmotic flow.

The concentration of Q^+ added as a flow modifier typically ranges from 1 to 5 mM. In some cases increasing the concentration of Q^+ changes the migration order of inorganic anions.[6]

C. Reagents for Indirect Detection

A few inorganic anions absorb in the UV spectral region and can be detected by direct spectrophotometry. However, in most cases indirect detection must be employed. The principle of indirect photometric detection is straightforward. A low concentration of an anion that absorbs strongly in the visible or UV region is added to the running electrolyte. Such an ion is sometimes called a "visualization reagent." A background signal is established by the visualization reagent passing through the detector at a fixed rate. Within a sample ion zone the concentration of the visualization reagent is reduced by an amount proportional to the sample ion concentration, thus resulting in a detection peak of reduced absorbance. The reason that the visualization ion concentration is lower within a sample ion zone is that the total ionic current in the capillary must remain constant.

A second requirement of an ion used for indirect detection is that its mobility must match that of the sample ions as closely as possible. If the mobilities do not match reasonably well, peaks may be fronted or tailed. (See Section II.A). Since the mobilities of sample ions will differ, the mobility of the visualization reagent is matched as closely as possible to the mobility of the middle sample ions.

Chromate at a concentration of around 5 mM has been used very successfully for indirect detection of common inorganic anions at a wavelength of 254 nm. More recently, Shamsi and Danielson have proposed naphthalenedisfulonate (NDS) and naphthalene-trisulfonate (NTS) for indirect detection of inorganic and organic anions.[7] These additives have a high molar absorptivity and their solutions are said to be more stable when stored than chromate. As shown in Table 6.1, there is a good match in their migration times relative to those of common anions. Even though these are large anions, the migration times of NDS and NTS are short because of their 2- and 3-charges.

Sensitivity of detection is of course an important consideration. With indirect detection, sensitivity is governed by the molar absorptivity of the visualization ion (the probe ion), its charge and the transfer ratio, which is the number of moles of probe ion displaced by one mole of a sample ion. In many cases the transfer ratio is considerably less than theoretical,

Table 6.1 Relative Migration Time of Electrolyte and Analyte Anion Measured with Negative Polarity[a]

Electrolyte anion	Relative migration time[c,b]		
Naphthalenemonosulfonate (NMS)	11.37		
Naphthalenedisulfonate (NDS)	2.43		
Naphthalenetrisulfonate (NTS)	1.83		

	Rel. migration time[c,b]	Δt^d	
Analyte anion		NTS	NDS
Inorganic Anion			
Bromide	1.00	+0.83	+1.43
Chloride	1.05	+0.73	+1.38
Nitrite	1.14	+0.69	+1.29
Nitrate	1.19	+0.64	+1.24
Sulfate	1.33	+0.50	+1.10
Fluoride	1.84	−0.01	+0.39
Orthophosphate	2.52	−0.69	−0.09
Organic Acids			
Oxalate	1.39	+0.44	+1.04
Malonate	1.76	+0.07	+0.67
Formate	1.88	−0.05	+0.55
Fumurate	1.94	−0.11	+0.49
Maleate	1.99	−0.16	+0.44
Succinate	2.17	−0.34	+0.26
Citrate	2.24	−0.41	+0.19
Malate	2.28	−0.45	+0.15
Tartarate	2.53	−0.70	−0.10

[a] Measured with respect to bromide.
[b] Using 2 mM DETA, 100 mM H_3BO_3, 5 mM $Na_3B_4O_7 \cdot 10\ H_2O$, buffered at pH 8.
[c] Using 2 mM DETA, 4 mM NTS, 100 mM H_3BO_3, 5 mM $Na_3B_4O_7 \cdot 10\ H_2O$, buffered at pH 8.
[d] Relative migration time of electrolyte (NTS or NDS) minus relative migration time of analyte.

From Shamsi, S.A. and Danielson, N.D., *Anal. Chem.*, 66, 3757, 1994. With permission.

Table 6.2 Relative Migration Times of Several Anions and Visualization Reagents
(Probes) at pH 8 (Chromate = 1.00)

Anion	Relative migration	Probe	Relative migration
Bromide, iodide	1.01	Chromate	1.00
Chloride	1.02	Pyromellitate	1.13
Sulfate	1.03	Trimellitate	1.24
Nitrate	1.06	Phthalate	1.37
Citrate	1.17	Benzoate	1.80
Thiocyanate	1.21	p-Hydroxybenzoate	1.87
Hydrogen phosphate	1.24	p-Toluenesulfonate	2.00
Bicarbonate	1.39		
Acetate	1.45		
Propionate	1.56		
Butyrate	1.65		
Ethanesulfonate	1.66		
Propanesulfonate	1.79		
Butanesulfonate	1.92		
Pentanesulfonate	2.02		

Adapted from Buchberger, W., Cousins, S. M., and Haddad, P. R., *Trends in Anal. Chem.*,
13, 313, 1994.

leading to a lower than expected sensitivity. Equivalent-to-equivalent displacement occurs
only when the mobilities of the probe and sample ions are equal.

The use of transfer ratios and optional selection of a suitable probe ion have been discussed.[8] Migration times of a number of probes and analyte ions are given in Table 6.2. The
migration times of the probe and analyte ions must be similar in order to avoid peak distortion.

D. Examples of Anion Separation

During the early 1990s a number of papers were published on the separation of anions
utilizing chromate for indirect photometric detection.[6-11] Typical conditions for a separation
are as follows:

5.0 mM chromate adjusted to pH 8.0 with dilute sulfuric acid, 0.5 mM Waters electroosmotic
flow (EOF) modifier, 20 kV, and a 75 μm i.d. capillary. By going to a 50 μm i.d. capillary and
30 kV, the plate count for a separation increased by a factor of 2. Figure 6.2 shows the
electropherogram for a separation of 30 anions in a total elapsed time of only 3.1 min.

CE determination of anions can also be carried out on real samples. Figure 6.3 shows
peaks for several anions found in diluted urine. The analysis of inorganic and organic anions
in Kraft black liquor is shown in Figure 6.4. This is a viscous alkaline sample that would be
difficult to analyze by ion chromatography because of its wide range of organic acids, sulfur
species, and especially lignins that would foul an ion-exchange column.

1. Effect of 1-Butanol

In most cases, effective separation of anions requires reversal of the normal direction of
electroosmotic flow in fused-silica capillaries. This is usually accomplished by adding a
quaternary ammonium salt (Q$^+$) as an EOF modifier. Under typical conditions a concentration
of Q$^+$ >0.25 mM in the running electrolyte is needed to reverse the electroosmotic flow
direction. However, concentrations of this magnitude often cause a buildup of Q$^+$ on the
capillary surface over several runs that results in poor reproducibility.

Benz and Fritz[12] found that addition of a low percentage of n-butanol to the aqueous
electrolyte reduced the EOF. A combination of 4 to 5% n-butanol and a very low concentration
of Q$^+$ (typically 0.03 mM) was found to be particularly effective in reversing the EOF direction

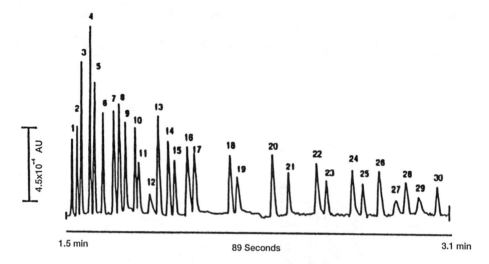

Figure 6.2 Peak identity and concentrations (ppm) for 30-anion electropherogram displayed in an 89-s electropherographic segment. Electromigration injection at 1 kV for 15 s. Peaks: 1 = thiosulfate (4); 2 = bromide (4); 3 = chloride (2); 4 = sulfate (4); 5 = nitrite (4); 6 = nitrate (4); 7 = molybdate (10); 8 = azide (4); 9 = tungstate (10); 10 = monofluorophosphate (4); 11 = chlorate (4); 12 = citrate (2); 13 = fluoride (1); 14 = formate (2); 15 = phosphate (4); 16 = phosphite (4); 17 = chlorite (4); 18 = galactarate (5); 19 = carbonate (4); 20 = acetate (4); 21 = ethanesulfonate (4); 22 = propionate (5); 23 = propane-sulfonate (4); 24 = butyrate (5); 25 = butanesulfonate (4); 26 = valerate (5); 27 = benzoate (4); 28 = L-glutamate (5); 29 = pentanesulfonate (4); 30 = D-gluconate (5). The electrolyte is a 5 mM chromate and 0.5 mM EOF modifier adjusted to pH 8.0. (From Jones, W. R. and Jandik, P., *J. Chromatogr.*, 546, 445, 1991. With permission.)

and in giving good separations of complex mixtures of anions. Figure 6.5 shows an excellent separation of 12 inorganic anions.

The most plausible mechanism[12] is one in which both butanol and Q^+ are adsorbed on the capillary surface by a dynamic equilibrium. The adsorbed butanol shifts the Q^+ solution–surface equilibrium so that the surface receives a net positive charge at a significantly lower Q^+ than it does in the absence of butanol.

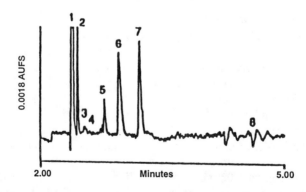

Figure 6.3 Pherogram of diluted urine. Sample preparation 50 × dilution in deionized water, solutes: 1 = chloride (67.7 ppm); 2 = sulfate (9.6 ppm); 3 = nitrate (1.0 ppm); 4 = oxalate (0.6 ppm); 5 = citrate (4.6 ppm); 6 = phosphate (11.5 ppm); 7 = carbonate (7.4 ppm); 8 = ascorbate (0.7 ppm). (From Wildman, B. J., Jackson, P. E., Jones, W. R., and Alden, P. G., *J. Chromatogr.*, 546, 459, 1991. With permission.)

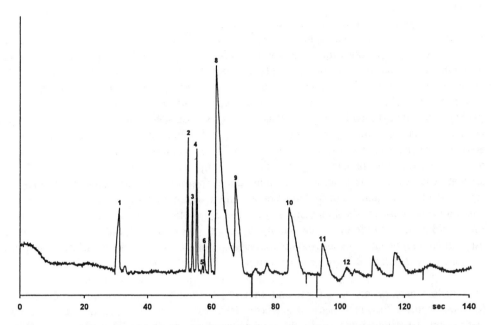

Figure 6.4 Fast separation of anions in Kraft black liquor. Conditions: 50 μm i.d. fused silica, L to detector 24.5 cm; electrolyte 5 mM chromate, 20% acetonitrile (v/v), pH 11.0, detection at 185 nm. Peak identification: 1 = hydroxide, 2 = thiosulfate, 3 = chloride, 4 = sulfate, 5 = sulfide, 6 = oxalate, 7 = sulfite, 8 = carbonate, 9 = formate, 10 = acetate, 11 = propionate, 12 = butyrate. (From Volgger, D., Zeman, A., Bonn, G., and Sinner, M., unpublished results, 1996.)

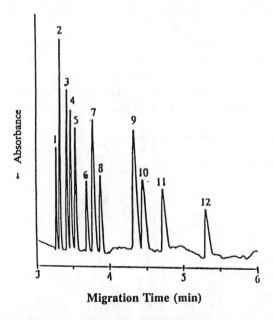

Figure 6.5 Separation of inorganic anions. Electrolyte: 5 mM chromate, 0.03 mM Q$^+$, 4% 1-butanol, pH 8.0; applied voltage: −30 kV, current: 23 μA, electromigration injection 5s/10 kV. Peaks: 1 = Br$^-$ (5 ppm), 2 = Cl$^-$ (5 ppm), 3 = SO$_4^{2-}$ (6 ppm), 4 = NO$_2^-$ (7 ppm), 5 = NO$_3^-$ (7 ppm), 6 = MoO$_4^{2-}$ (10 ppm), 7 = N$_3^-$ (10 ppm), 8 = ClO$_3^-$ (8 ppm), 9 = F$^-$ (8 ppm), 10 = HCOO$^-$ (8 ppm), 11 = ClO$_2^-$ (8 ppm), 12 = CO$_3^{2-}$ (7 ppm).

E. Suppressed Conductivity Systems

The success of suppressed conductivity detectors in ion chromatography has no doubt inspired researchers to develop a similar system for capillary electrophoresis. A suppressed conductivity capillary electrophoresis separation system has in fact been dubbed SUCCESS.[13] In this system a tubular cation-exchange membrane is placed at the end of a 60-cm-long fused-silica capillary. A static acid regenerant solution surrounds the membrane suppressor. The function of the suppressor is to convert an alkaline running electrolyte such as sodium borate into a slightly ionized species (boric acid) that has a much lower background conductivity. At the same time the counter ion sample anions of fairly strong acids are converted from Na^+ to the more highly conducting H^+.

Another system for suppressed conductivity detection has been devised by workers at the Dionex Corporation, Sunnyvale, CA.[14] In any system, the detector must be placed so that its functioning will not be disturbed by the electric field produced by the CE system itself. A circuit must be used that will allow one of the conductivity electrodes to be grounded. This detection method is also dependent on electroosmotic flow in the capillary to transport both analyte and buffer ions through the suppressor.

Direct conductivity detection of inorganic anions and cations is also possible using a novel sensor design developed at ThermoCapillary Electrophoresis (Franklin, MA).[15] A diagram of the system is given in Figure 6.6. The capillary starts at the injection side and ends at the detector cell assembly; the high voltage circuit is completed by plastic tubing running from the cell assembly to the receiving side receptacle. Special circuits are used to isolate the high voltage from the detector output.

The efficiency of combined CE-direct conductivity detection for ion analysis is illustrated by the separation of some 37 anions in Figure 6.7. The concentration of ions separated ranged from 1 to 7 ppm for the inorganic anions and 10 to 20 ppm for most of the organic anions.

F. Separation of Chloride Isotopes

Inorganic anions are almost always separated by comigration with the use of a negative power supply and addition of Q^+ to the running electrolyte to reverse the direction of electroosmotic flow. However, Avdalovic et al.[14] observed a partial separation of chloride ion isotopes. Lucy and McDonald[16] were able to achieve baseline resolution of $^{37}Cl^-$ and $^{35}Cl^-$ by using counter migration and making the electroosmotic and electrophoretic flow vectors

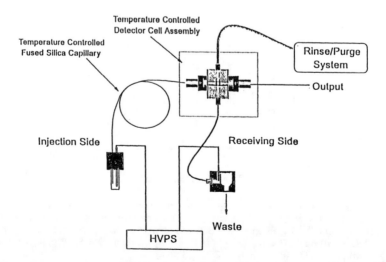

Figure 6.6 CE system with conductivity detection. (Courtesy ThermoCapillary Electrophoresis.)

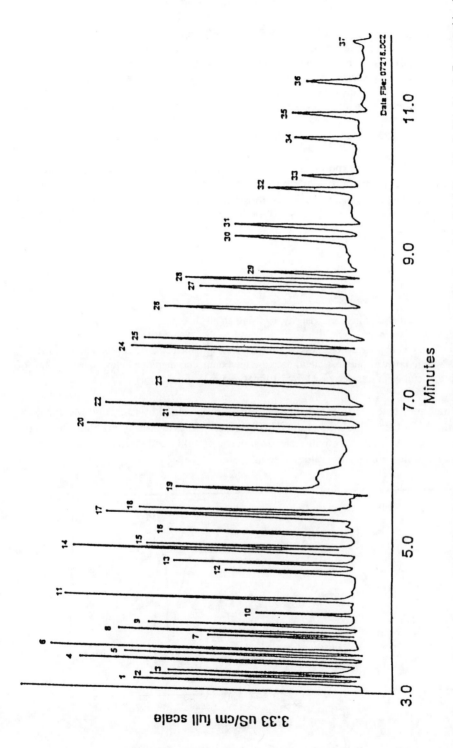

Figure 6.7 Separation of anions using direct conductivity detection. (Courtesy of ATI-Unicam Separations Group.) Peaks: 1. Bromide, 2. Chloride, 3. Ferrocyanide, 4. Nitrite, 5. Nitrate, 6. Sulfate, 7. Azide, 8. Oxalate, 9. Molybdate, 10. Tungstate, 11. 1,2,4,5 BTC, 12. Fluoride, 13. Tartrate, 14. Selenite, 15. Phosphate, 16. Citraconate, 17. Glutarate, 18. Phthalate, 19. Carbonate, 20. Acetate, 21. Chloroacetate, 22. Ethanesulfonate, 23. Dichloroacetate, 24. Propionate, 25. Propanesulfonate, 26. Crotonate, 27. Butanesulfonate, 28. Butyrate, 29. Toluenesulfonate, 30. Pentanesulfonate, 31. Valerate, 32. Hexanesulfonate, 33. Caproate, 34. Heptane- sulfonate, 35. MES, 36. Octanesulfonate, 37. d-Gluconate.

similar in magnitude (but opposite in direction). Net flow toward the detector and the positive electrode is possible in this case because of the high electrophoretic mobility of chloride.

The major conditions for the Lucy–McDonald separation were as follows:

1. A negative power supply was used (20 kV). Electrophoretic flow was toward the detector (+ electrode), and electroosmotic flow was in the opposite direction.
2. An alkaline pH was needed to keep the visualization reagent (chromate) ionized and to keep a fairly large electroosmotic vector counter to the electrophoretic vector. (See below.)
3. The concentration of borate (3 mM) was lower than generally used in CE. This was the minimum concentration that would maintain the desired pH. The optimum pH was found to be 9.2.
4. Chromate was added to the electrolyte for indirect detection of the chloride.

Careful control of these experimental parameters was found to be critical to the success of the separation. A fairly short migration time gave no separation of the 37, 35-chloride isotopes. It was necessary to increase the electroosmotic vector to obtain long migration times and obtain good resolution of the isotopes. Examination of the following equation revealed three possibilities:

$$\mu_{eo} = \frac{\sigma * \kappa^{-1}}{\eta} \tag{3}$$

1. Increase $\sigma*$, the surface charge density, by using an alkaline pH.
2. Decrease the viscosity, η.
3. Increase the double-layer thickness (κ^{-1}) by decreasing the electrolyte concentration in the bulk solution. It was not possible to decrease the borate concentration below about 3 mM, therefore chromate was decreased from 10 mM to 5 mM.

A separation of chloride isotopes similar to that first obtained by Lucy and McDonald[16] was reproduced in the author's laboratory[17] (Figure 6.8). It is worth noting that the larger ^{37}Cl$^-$ isotope apparently has the smaller electrophoretic mobility and therefore has the longer migration time in the counter migration system.

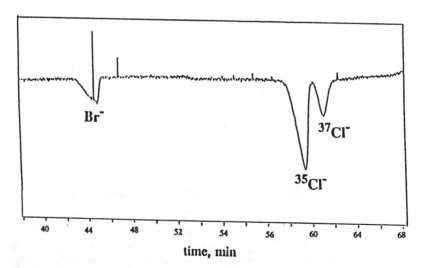

Figure 6.8 Separation of chloride isotopes by CE using counter migration. (Courtesy Youchun Shi.)

IV. SEPARATION OF METAL CATIONS

A. Principles

The simplest kind of CE separation of metal cations relies on differences in electrophoretic mobilities of the sample ions. A positive power supply is generally used so that both electrophoretic and electroosmotic migration vectors are in the direction of the fixed detector and the negative electrode. In this comigration mode, good separations can be obtained in a very short time — often <2 min. For more difficult separations, comigration effectively shortens the length of capillary available for actual separation and thus reduces the obtainable resolution. This is because the electroosmotic velocity is the same for all sample cations.

Indirect detection is commonly used for metal cations because most of these cations lack the high UV or visible absorptivity needed for direct photometric detection. Waters Associates introduced UV Cat 1 for indirect detection. This is an amine cation that absorbs strongly in the UV spectral region and has an electrophoretic mobility similar to 1+ and 2+ metal cations. Protonated phenylethylamine- or 4-methylbenzylamine are suitable visualization reagents for indirect detection of metal cations at moderately acidic pH values.

The relative electrophoretic migration of metal cations is readily predictable from a table of equivalent ionic conductivities.[18] Separation of alkali metal ions is quite easy, owing to large differences in mobility. The order of elution (K^+, Na^+, Li) is the opposite of that in ion chromatography. Magnesium(II) and the alkaline earth cations are also separated by CE. However, separation of the free metal cations of the divalent transition metals cannot be achieved by ordinary CE because the electrophoretic mobilities are too similar. The same is true for the trivalent lanthanide cations. Selective complexation must be used to separate either of the latter two groups.

Potassium(I) and ammonium cations cannot be separated in acidic solution because their mobilities are almost identical. However, as the pH of the electrolyte is increased, the NH_4^+ becomes progressively less protonated and its apparent mobility decreases. This occurs because some of the NH_4^+ is converted to nonionic NH_3. The K^+ is not affected by pH changes. At pH 8.5 the apparent mobilities of K^+ and NH_4^+ have become sufficiently different to permit a good separation. The effect of changing pH on CE separation of ammonium and alkali metal ions is shown in Figure 6.9.[18]

B. Separations Using Partial Complexation

1. Principles

The best way to separate groups of cations with very similar electrophoretic mobilities is to add a water-soluble complexing reagent to the capillary electrolyte. When done properly, much larger differences in effective mobility can be obtained by complexing the sample cations to different extents. The effective (observed) mobility of a metal is a combination of the mobilities of the free metal ion and the various complexes.

$$\mu_{eff} = a\mu_{M++} + b\mu_{ML+} + c\mu_{ML2} + \ldots + \mu_{eo} \qquad (4)$$

where a, b, and c are the mole fractions. Ions that are complexed to a greater extent move more slowly than those that have a lower mole fraction in a complexed form. Changes in pH and in the concentration of complexing reagent will affect the migration times of the metal ions to be separated.[19]

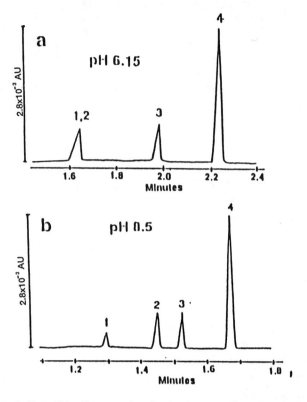

Figure 6.9   Effect of pH on NH_4^+–K^+ separation. Peaks: 1 = potassium, 2 = ammonium, 3 = sodium, 4 = lithium. (From Weston, A., Brown, P. R., Jandik, Jones, W. R., and Heckenberg, A. L., *J. Chromatogr.*, 593, 289, 1992. With permission.)

Jones et al.[20] obtained an excellent CE separation of 15 alkali, alkaline earth, and transition metal ions using 6.5 m*M* hydroxyisobutyric acid (HIBA) at pH 4.4 to partially complex some of the cations.

Phthalate, tartrate, and citrate have also been used in the separation of metal ions by CE.[21] Addition of methanol to the electrolyte lengthens the migration times and gives sharper separations. The electrophoretic mobilities were found to decrease almost linearly with methanol concentrations up to 25% by volume.[22] Figure 6.10 shows a separation of several metal ions with 2.5 m*M* tartrate as a complexing reagent in 10% methanol. Protonated phenylethylamine was used for indirect detection.

2. Separations Using HIBA

HIBA has been used extensively for CE separation of metal ions. Excellent separation of all the lanthanides has been obtained.[18,21,23] (See Figure 6.11).

Chen and Cassidy obtained better separations on silica C-18 bonded capillaries than on ordinary fused-silica capillaries.[24]

3. Separations Using Lactate

Lactate has the same α-hydroxy carboxylate complexing group as tartrate and HIBA, but it is a smaller molecule and forms somewhat weaker complexes than tartrate with most metal ions. Shi and Fritz found that a lactate system gave excellent separations for divalent metal ions and for trivalent lanthanides.[21] A brief optimization was first carried out to establish the

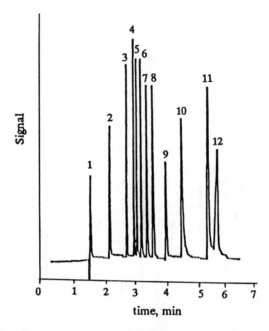

Figure 6.10 CE separation of metal ions. Electrolyte: 2.5 mM tartaric acid, 6 mM phenylethylamine, pH 5.5 10% MeOH. Capillary: 60 cm × 75 μm. Applied voltage: 30 kV. Peak identification: 1. K^+, 2. Na^+, 3. Li^+, 4. Mg^{2+}, 5. Ba^{2+}, 6. Sr^{2+}, 7. Mn^{2+}, 8. Ca^{2+}, 9. Cd^{2+}, 10. Co^{2+}, 11. Ni^{2+}, 12. Zn^{2+}.

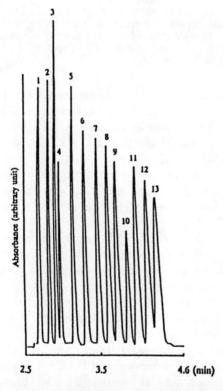

Figure 6.11 Separation of 13 lanthanides using HIBA. Electrolyte: 4 mM HIBA, 5 mM UV Cat 1, pH 4.3. Applied voltage: 30 kV. Peaks: 1 = La^{3+}; 2 = Ce^{3+}; 3 = Pr^{3+}; 4 = Nd^{3+}; 5 = Sm^{3+}; 6 = Gd^{3+}; 7 = Tb^{3+}; 8 = Dy^{3+}; 9 = Ho^{3+}; 10 = Er^{3+}; 11 = Tm^{3+}; 12 = Yb^{3+}; 13 = Lu^{3+}. (From Shi, Y. and Fritz, J. S., *J. Chromatogr. A*, 671, 429, 1994. With permission.)

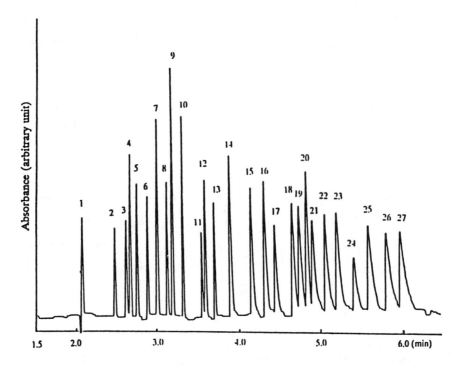

Figure 6.12 Separation of 27 alkali, alkaline earth, transition, and rare earth metal ions in a single run using lactate. Electrolyte: 15 mM lactic acid, 8 mM 4 methylbenzylamine, 5% methanol, pH 4.25. Applied voltage: 30 kV. Peaks: 1 = K$^+$; 2 = Ba^{2+}; 3 = Sr^{2+}; 4 = Na$^+$; 5 = Ca^{2+}; 6 = Mg^{2+}; 7 = Mn^{2+}; 8 = Cd^{2+}; 9 = Li$^+$; 10 = Co^{2+}; 11 = Pb^{2+}; 12 = Ni^{2+}; 13 = Zn^{2+}; 14 = La^{3+}; 15 = Ce^{3+}; 16 = Pr^{3+}; 17 = Nd^{3+}; 18 = Sm^{3+}; 19 = Gd^{3+}; 20 = Cu^{2+}; 21 = Tb^{3+}; 22 = Dy^{3+}; 23 = Ho^{3+}; 24 = Er^{3+}; 25 = Tm^{3+}; 26 = Yb^{3+}; 27 = Lu^{3+}. (From Shi, Y. and Fritz, J. S., *J. Chromatogr.*, 640, 473, 1993. With permission.)

best concentrations of lactate and UV visualization reagent and the best pH. Excellent separations were obtained for all thirteen lanthanides, alkali metal ions, magnesium and the alkaline earths, and several divalent transition metal ions. All of these except copper(II) eluted before the lanthanides. An excellent separation of 27 metal ions was obtained in a single run that required only 6 min (Figure 6.12).

The detection limit for these metal ions was in the range of 0.05 to 0.5 µg/mL. Light ions were in the low detection limit range, and heavy ions were in the high detection limit range. The average deviation in peak height from one run to another was <±5% at the concentrations used.

It was not possible to separate NH$_4^+$ and K$^+$ under the conditions used in Figure 6.12. These ions have almost identical mobilities and neither is complexed by lactate. Several investigators have found that ammonium and potassium cations can be separated by CE if a suitable crown ether is incorporated into the running electrolyte.[25-26] Figure 6.13 shows a separation of sixteen common metal ions, including ammonium and potassium, at pH 4.3 using an electrolyte containing both lactate and 18-crown-6.[22] The potassium(I) complex of 18-crown-6 has a lower mobility than the ammonium ion, making their separation possible. Addition of the crown ether to the lactate system also affected the migration times of several other metal ions. For example, 18-crown-6 resulted in a 9% increase in the migration time of potassium(I), a 15% increase in t_M for strontium(II), a 35% increase in t_M for barium(II), and an 18% increase in the migration time for lead(II).

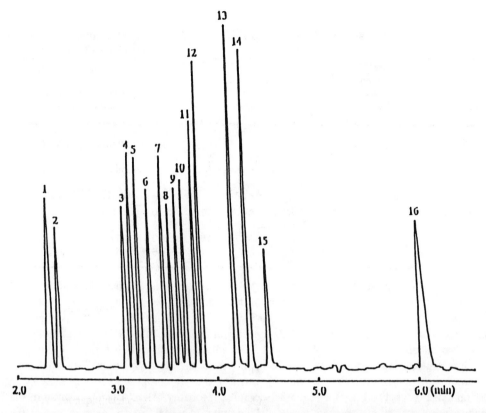

Figure 6.13 Separation of 16 common metal ions and ammonium. Electrolyte: 11 mM lactic acid, 2.6 mM 18-crown-6, 7.5 mM 4-methylbenzylamine, 8% methanol, pH 4.3. Capillary, 75 μm i.d. × 60 cm. Applied voltage: 30 kV. Peak identification: 1. NH_4^+, 2. K^+, 3. Na^+, 4. Ca^{2+}, 5. Sr^{2+}, 6. Mg^{2+}, 7. Mn^{2+}, 8. Ba^{2+}, 9. Cd^{2+}, 10. Fe^{2+}, 11. Li^+, 12. Co^{2+}, 13. Ni^{2+}, 14. Zn^{2+}, 15. Pb^{2+}, 16. Cu^{2+}.

4. The Separation Mechanism

In the separations just described, a reagent such as lactate or HIBA was added to partially complex metal ions and improve the quality of a separation by altering their net flow characteristics. For example, addition of lactate (L^-) to a solution containing metal ions (M_1^{++}, M_2^{++}, etc.) converts some fraction of the metal ions to lactate complexes (M_1L^-, M_2L^-, M_1L_2, M_2L_2). The free ions move along the capillary at rates proportional to their ionic mobilities, while the *complexed* metal ions move at a slower speed. Continuous equilibration between free and complexed metal ions causes each metal to move through the capillary as a tight zone. Separation occurs because of the different overall rates at which the zones move. The situation is somewhat analogous to HPLC except that in this case the "stationary phase" (metal lactate complexes) is not a real phase and is not stationary, but moves at a slower speed than the free metal cations.

The more completely a metal ion is complexed, the slower will be its rate of movement. If the fraction of metal in the free cation is too small, it may take a long time for the metal to emerge from the capillary and no good separation will be obtained. However, if the various metal ions are not sufficiently complexed (too large a fraction of free metal ion), ions of similar ionic mobility may be poorly separated. The question is how strongly metal ions should be complexed for optimal separation by CE.

Table 6.3 Fractions of Free (α_M) and Complexed (α_{ML3})
Rare Earth Metal Ions and Average Number of
Ligands ($\bar{n}$) in 4 mM HIBA Electrolyte Solution
at pH 4.3

Metal	α_M	α_{ML}	α_{ML2}	α_{ML3}	α_{ML4}	$\bar{n}$
La	0.578	0.360	0.612			0.482
Ce	0.448	0.496	0.052	0.004		0.612
Pr	0.407	0.496	0.093	0.005	0.000	0.697
Nd	0.333	0.572	0.085	0.010	0.000	0.772
Sm	0.296	0.520	0.170	0.013	0.001	0.903
Gd	0.250	0.481	0.244	0.024	0.001	1.045
Tb	0.181	0.470	0.307	0.040	0.002	1.212
Dy	0.141	0.384	0.387	0.084	0.004	1.426
Ho	0.122	0.365	0.413	0.093	0.006	1.494
Er	0.097	0.309	0.472	0.112	0.010	1.629
Tm	0.079	0.309	0.473	0.123	0.016	1.686
Yb	0.070	0.296	0.431	0.169	0.033	1.797
Lu	0.047	0.222	0.514	0.172	0.045	1.946

Consider the separation of rare earths using 4.0 mM hydroxyisobutyric acid (HIBA) at pH 4.3 as the complexing reagent (see Figure 6.11). Using published formation constants, the fraction of rare earths present in various chemical forms was calculated by a well-known method under the same conditions of pH and HIBA concentration used for the CE separation in Figure 6.11. The calculated distribution of chemical species for each rare earth is shown in Table 6.3.

Some interesting conclusions can be drawn from the information in Table 6.3. The predominating species are the free metal ion (M^{3+}), the 1:1 complex (probably ML_2^+), the 2:1 complex (probably ML^{2+}), and the 3:1 complex (probably ML_3). A small fraction of the higher rare earths is also present as the 4:1 complex. Another striking feature is that the average number of ligands associated with a rare earth ($\bar{n}$) increases rapidly with increasing atomic number. This occurs in a fairly regular manner as demonstrated by a plot of $\bar{n}$ against atomic number which has a linear regression correlation coefficient of 0.9958.

In CE the positively charged complexes, as well as the free metal cation, would be expected to move through the capillary by electrophoretic flow as well as by the electroosmotic flow that affects all species. However, the electrophoretic mobility should be slower for ML^{2+} than for M^{3+} and still slower for the larger ML_2^+. Even if the different species move at different rates, rapid equilibrium shifts should keep a tight zone for each of the rare earths. For example, as the faster moving M^{3+} starts to move ahead of the other species, it reequilibrates with the ligand (L) to form a larger fraction of the slower-moving species. At the back edge of the zone, the slower-moving complexes reequilibrate to give a greater fraction of M^{3+}. The average rate of movement should depend on the weighted average of the mobilities of the different species.

The proposed mechanism necessitates a very fast rate of equilibrium between the free metal ions and the various complexed species. This condition is fulfilled with lactate and HIBA for the metal ions studied. However, metal ions that have slow complexation kinetics cannot be determined by a partial complexation CE system. For example, aluminum(III) gave no peak in a lactate system.

A number of common metal cations can be separated by CE as the free ions.[22] For aluminum(III) this requires a very acidic pH in order to avoid hydrolysis.

5. Prediction and Separation of Separations

In a system where partial complexation of metal cations by an added ligand (L) occurs, the electrophoretic mobility will be the weighted sum of the free metal cation and the various complexed species.

$$\mu_{ep} = \alpha_M \mu_M + \alpha_{ML} \mu_{ML} + \alpha_{ML2} \mu_{ML2} + \ldots \tag{5}$$

where α is the molar fraction of each species. For any given ligand the various α values will depend on the metal ion, the concentration of ligand and the solution pH.

Quang and Khaledi[27] devised a general linear model to relate μ_{ep} to the concentration of added ligand and to the solution pH.

$$\mu_{ep} = k_o + k_1 C_{HL} + k_2 \cdot pH \cdot C_{HL} + k_3 C_{HL}^2 \tag{6}$$

where k_o, k_1, k_2, and k_3 are empirical parameters. Since Equation 6 contains only four parameters, five experimental points were used to build a model by using a least-squares approach.

The model developed was successful in predicting the electrophoretic mobilities, and thus the separation, of fourteen metal cations. By varying the pH and concentration of ligand (HIBA), separations could also be optimized. The pH range for the system used was 3.5 to 5.0. This range was determined by the lowest feasible pH for the HIBA-metal chelation system and at the upper end by the pH at which the visualization reagent (imidazole) would remain protonated.

C. Separation of Complexed Metal Ions

1. Principles

Another separation possibility is to quantitatively complex a group of sample metal ions by adding an excess of a complexing reagent and then separate the metal complexes by CE. The success of this approach is not dependent on a rapid rate of equilibration between the various metal species during the analytical migration. In fact, it is probably better to work with complexes that, once formed, will be stable to kinetic dissociation. If this is not the case, ligand exchange might occur during the electrophoretic separation.

The complexing reagent and separation conditions must be chosen carefully in order to prevent the various complexes from partially dissociating during the attempted separation. This is not always easy to accomplish because the sample solutions of metal ions are usually very dilute. Sometimes it is sufficient to simply add an excess of the complexing reagent to the sample and then adjust the pH to an appropriate value. In other cases, a certain concentration of the complexing reagent must also be added to the running electrolyte in order to keep the metal ions completely complexed during the CE run.

The successful separation of metal complexes requires some difference in mobility of complexes that may differ only by the metal ion that is surrounded by a bulky complexing agent. However, the efficiency of CE is so high that only small differences in analyte mobility are needed for a separation.

Complexation of a metal ion with an organic ligand usually produces a metal-organic complex with no charge. Except for separations by micelle electrokinetic chromatography (MEKC), electrophoretic separations require an ionic analyte. This is best achieved in the present situation by using a complexing reagent with a charged group, such as sulfonate or a quaternary ammonium group, in the chelating reagent. This charged group should be located in a somewhat remote position in the molecule where it will not interfere with the chelating groups. For example, the sulfonate group in 8-hydroxyquinoline-5-sulfonic acid imparts a negative charge to an otherwise neutral metal ion complex, but the sulfonate group is located away from the nitrogen and phenolic chelating groups.

Whenever possible the complexing reagent should form metal complexes that absorb in the visible or UV spectral region so that direct detection will be possible. The mere presence of a chromophore in the complexing reagent will not suffice; the metal complexes must absorb at a different wavelength than the reagent itself.

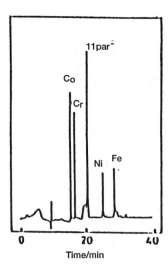

Figure 6.14 MEKC separation of metal-PAR chelates. (From Swaile, D. F. and Sepaniak, M. J., *Anal. Chem.*, 63, 179, 1991. With permission.)

2. Separation of Metal-Organic Complexes by MEKC

The reagent, 4-(2-pyridylazo) resorcinol, better known as PAR, forms neutral complexes with a rather large number of metal ions. One of the first CE separations of metal cations involved complexation with PAR and the separation of the metal complexes by MEKC.[28]

A low concentration of sodium dodecylsulfate (SDS) was added to the running electrolyte to provide the micelle needed for the separation. The micelle has an electrophoretic vector owing to its negative charge. Separation of the metal-PAR complexes is based on differences in partition coefficients between the free solution and the charged micelle. Figure 6.14 shows a separation of four metal-organic complexes.[29] The peaks are unusually sharp; a plate number of 1×10^5 per 60 cm is claimed.

Metal ion complexes of acetylacetone (acac) have also been separated by MEKC. The carrier solution contained 150 mM SDS, 100 mM Hacac, and 20 mM sodium tetraborate. Sharp peaks were obtained for VO $(acac)_2$, $Ni(acac)_3$, $Co(acac)_3$, and $Cu(acac)_2$.

3. 8-Hydroxyquinoline-5-Sulfonic Acid (HQS) Complexes

HQS forms stable complexes with a number of metal ions. The complexes have a negative charge by virtue of the sulfonate group. Swaile and Sepaniak[30] separated a few metal complexes of HQS by ordinary CE and used direct fluorescence detection. Timerbaev et al.[31] presented a detailed evaluation of HQS for CE separation of preformed chelates of transition and alkaline earth metal ions. They used counter migration with the electroosmotic flow toward the cathode and the electrophoretic migration in the opposite direction, toward the detector. It was necessary to add a tetra alkylammonium salt (Q^+) to reduce the electroosmotic flow to a degree where it was less than the electrophoretic flow.

Systematic investigation resulted in additional conclusions:

1. A reagent:metal ion ratio of approximately 3:1 in the sample is necessary for complete complexation. However, larger reagent excesses resulted in a higher signal:noise ratio (due to a higher background) and overlapping peaks.
2. The best pH range was 7.5 to 9.5. The complexes were less stable outside this range.

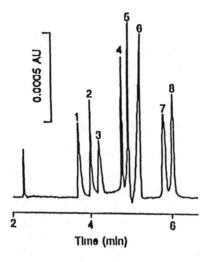

Figure 6.15 Reversed movement CZE separation of metal complexes of HQS. Capillary: fused silica, 35/42 cm × 75 μm i.d. Carrier electrolyte: 10 mmol/L borate buffer, 0.1 mmol/L HQS, pH 9.0. Voltage: + 15 kV. Detection: direct photometric at 254 nm. Injection: hydrostatic, 10 cm (1 sec). Metals: 1 = Mn; 2 = Cu; 3 = Al; 4 = Cd; 5 = Fe; 6 = Zn; 7 = Co; 8 = Ni. The first migrating peak (acetone) served as a marker of EOF velocity. (From Motomizu, S., Nishimura, S. I., Obata, Y., and Tanaka, H., *Anal. Sci.*, 7, 253, 1991. With permission.)

3. The concentration of HQS in the electrolyte must be kept sufficiently low (0.1 or 0.2 mM). Poor separations were obtained with no HQS in the electrolyte or with HQS concentration of 0.4 mM.

Another study showed that a different counter migration scheme gave still better separations.[32] In this method the net movement of the anionic complexes was toward the cathode under conditions where the electroosmotic mobility was greater than the electrophoretic. Eight metal ions were separated within 6 min. (Figure 6.15). Running the separation under MEKC conditions showed no apparent partitioning of the complexes into the micelle.

4. CDTA Complexes

Ethylenediaminetetraacetic acid (EDTA) forms very stable complexes with a very large number of metal cations. Furthermore, most complexes have a negative charge; divalent metal ions form complexes with a 2-charge, and trivalent metals form complexes with a 1- charge. These properties suggest the use of EDTA for CE separation of metal ions.

Several authors have found that cyclohexanediaminetetraacetic acid (CDTA) works considerably better than EDTA.[33-35] A complete separation of the lanthanides was obtained in only 12 min.[34]

Addition of ethylene glycol to the aqueous electrolyte greatly improves the resolution. An electrolyte containing 20 mM sodium borate buffer (pH 9.0), 5% ethylene glycol, and 1 mM CDTA was used to separate 23 metal-CDTA complexes within 10 min.[35] The metal ions separated included several that cannot be separated by partial-complexation techniques: aluminum(III), antimony(III), bismuth(III), chromium(III), mercury(II), palladium(II), silver(I), thallium(I), tin(IV), uranium(VI), vanadium (IV), (V), and zirconium(IV).

5. Positively Charged Metal Complexes

The reagent, 2,6-diacetylpyridine bis(N-methylene pyridineohydrazone) is easily prepared by a one-step reaction of 2,6-diacetylpyridine with Girard's reagent P. It forms very stable complexes with several metal ions containing four chelate rings. The two pyridinium groups give the metal complexes a positive charge. This reagent has been used for CE separation of 14 metal ion complexes in a single run.[36] The ions separated include iron(III), aluminum(III), uranium(VI), and others that cannot be separated by CE with partial complexing because of slow complexation equilibria.

V. CONCLUSIONS

Since the late 1970s ion chromatography (IC) has been the dominant analytical separation method for inorganic ions. Capillary electrophoresis is emerging as the major challenger of IC. So how do these two techniques compare with regard to practical, everyday chemical analysis?

The separation power of CE, for both anions and cations, is greater than that of IC. Separation of ~30 anions or cations in a single run of only a few minutes exceeds by a factor of two or three what can be done by ion chromatography. On the other hand, ion chromatography is a rugged technique that can be used to separate mixtures of several inorganic ions with excellent dependability. CE seems less rugged and requires considerable care to maintain the clean capillary surface necessary for reproducible separations. But in some respects CE is better suited than IC for analysis of dirty analytical samples of the real world because an unpacked capillary is less likely to permanently trap extraneous sample substances.

CE is a truly micro technique that can be used to separate much smaller samples than IC. The variety of detectors available for use in IC is greater than for CE, although this situation is gradually changing. The light path for spectrophotometric detection is much shorter in CE than in IC.

High concentrations of a sample matrix ion can be a problem in both techniques, but it tends to be much more serious in CE. In capillary electrophoresis, a high concentration of a sample ion may form a broad zone on the capillary or column that will overlap with those of much lower concentrations of other sample ions. This problem was illustrated in a recent paper in which the determination of 1 ppm each of Mg^{++}, Ca^{++}, Sr^{++}, and Ba^{++} was attempted by CE in the presence of 75 ppm of Na^+. The broad Na^+ peak covered up all of the divalent ion peaks except Mg^{++}.[23]

Pretreatment/preconcentration methods are needed that will separate ions of interest from much larger amounts of sample matrix ions. Such methods are needed particularly when CE is to be the final analytical method because a high concentration of a sample ion tends to form a broad zone within the capillary. To be practical, sample preconcentration methods need to be both convenient and fast.

At this writing, analytical laboratories have been slow to adopt CE for routine analysis. Perhaps scientists have been affected by the increasingly conservative environment in which we live. But it should be remembered that CE is still a young technique. One can safely predict that CE will have a bright future for practical chemical analysis.

ACKNOWLEDGMENTS

The author wishes to thank the following persons for their help in assembling material for this chapter: Youchun Shi, Andrei Timerbaev, Thomas Chambers, Xue Li, and Marilyn Kniss.

REFERENCES

1. Jandik, P. and Bonn, B., *Capillary Electrophoresis of Small Molecules and Ions,* VCH Publishers, New York, 1993, 20.
2. Kuhr, W. G., this book, Chap. 3.
3. Hjerten, S., Zone broadening in electrophoresis with special reference to high-performance electrophoresis in capillaries: An interplay between theory and practice, *Electrophoresis,* 11, 665, 1990.
4. Jandik, P. and Jones, W. R., Optimization of detection sensitivity in the capillary electrophoresis of inorganic anions, *J. Chromatogr.,* 546, 431, 1991.
5. Jones, W. R. and Jandik, P., New methods for chromatographic separations of anions, *Am. Lab.,* 22, 6, 51, 1990.
6. Jones, W. R. and Jandik, P., Controlled changes of selectivity in the separation of ions by capillary electrophoresis, *J. Chromatogr.,* 546, 445, 1991.
7. Shamsi, S. A. and Danielson, N. D., Naphthalenesulfonates as electrolytes for capillary electrophoresis of inorganic anions, organic acids, and surfactants with indirect photometric detection, *Anal. Chem.,* 66, 3757, 1994.
8. Buchberger, W., Cousins, S. M., and Haddad, P. R., Optimization of indirect UV detection in capillary zone electrophoresis of low-molecular-mass anions, *Trends in Anal. Chem.,* 13, 313, 1994.
9. Romaro, J., Jandik, P., Jones, W. R., and Jackson, P. E., Optimization of inorganic capillary electrophoresis for the analysis of anionic solutes in real samples, *J. Chromatogr.,* 546, 411, 1991.
10. Wildman, B. J., Jackson, P. E., Jones, W. R., and Alden, P. G., Analysis of anion constituents in urine by inorganic capillary electrophoresis, *J. Chromatogr.,* 546, 459, 1991.
11. Jones, W. R. and Jandik, P., Various approaches to analysis of difficult matrices of anions using capillary ion electrophoresis, *J. Chromatogr.,* 602, 385, 1992.
12. Benz, N. J. and Fritz, J. S., Studies on the determination of inorganic anions by capillary electrophoresis, *J. Chromatogr. A,* 671, 437, 1994.
13. Dasgupta, P. K. and Bao, L., Suppressed conductometric capillary electrophoresis separation systems, *Anal. Chem.,* 65, 1003, 1993.
14. Avdalovic, N., Pohl, C. A., Rocklin, R. D., and Stillian, J. R., Determination of cations and anions by capillary electrophoresis combined with suppressed conductivity detection, *Anal. Chem.,* 65, 1470, 1993.
15. Jones, W. R., Haber, C., Reineck, J., McGlynn, M., and Soglin, J., Small molecular weight analysis using capillary electrophoresis and an open-architecture conductivity detector, International Ion Chromatography Symposium, Podium presentation #56, Turin, Italy, 1994.
16. Lucy, C. A. and McDonald, T. L., Separation of chloride isotopes by capillary electrophoresis based on the isotope effect on ion mobility, *Anal. Chem.,* 67, 1074, 1995.
17. Shi, Y. and Fritz, J. S., Unpublished data, 1995.
18. Weston, A., Brown, P. R., Jandik, P., Jones, W. R., and Heckenberg, A. L., Factors affecting the separation of inorganic metal cations by capillary electrophoresis, *J. Chromatogr.,* 593, 289, 1992.
19. Weston, A., Brown, P. R., Heckenberg, A. L., Jandik, P., and Jones, W. R., Effect of electrolyte composition on the separation of inorganic metal cations by capillary ion electrophoresis, *J. Chromatogr.,* 602, 249, 1992.
20. Jones, W. R., Jandik, P., and Pfeifer, R., Capillary ion analysis, an innovative technology, *Am. Lab,* 5, 40, 1991.
21. Shi, Y. and Fritz, J. S., Separation of metal ions by capillary electrophoresis with a complexing electrolyte, *J. Chromatogr.,* 640, 473, 1993.
22. Shi, Y. and Fritz, J. S., New electrolyte systems for the determination of metal cations by capillary zone electrophoresis, *J. Chromatogr. A,* 671, 429, 1994.
23. Foret, F., Fanali, S., Nardi, A., and Bocek, P., Capillary zone electrophoresis of rare earth metals with indirect UV absorbance detection, *Electrophoresis,* 11, 780, 1990.
24. Chen, M. and Cassidy, R. M., Bonded phase capillaries and the separation of inorganic ions by high-voltage capillary electrophoresis, *J. Chromatogr.,* 602, 227, 1992.

25. Fukushi, K. and Hiro, K., Use of crown ethers in the isotachophoretic determination of metal ions, *J. Chromatogr.*, 523, 281, 1990.

26. Bächmann, K., Boden, J., and Haumann, I., Indirect fluorimetric detection of alkali and alkaline earth metal ions in capillary zone electrophoresis with cerium(III) as carrier electrolyte, *J. Chromatogr.*, 626, 259, 1992.

27. Quang, C. and Khaledi, M. G., Prediction and optimization of the separation of metal cations by capillary electrophoresis with indirect UV detection, *J. Chromatogr. A*, 659, 459, 1994.

28. Saitoh, T., Hoshino, H., and Yotsuyanagi, T., Separation of 4-(2-pyridylazo) resorcinolato metal chelates by micellar electrokinetic capillary chromatography, *J. Chromatogr.*, 469, 175, 1989.

29. Saitoh, K., Kiyohara, C., and Suzuki, N., Micellar electrokinetic capillary chromatography of metal complexes of acetylacetone, *Anal. Sci.*, 7 (Supplement), 269, 1991.

30. Swaile, D. F. and Sepaniak, M. J., Determination of metal ions by capillary zone electrophoresis with on-column chelation using 8-hydroxyquinoline-5-sulfonic acid, *Anal. Chem.*, 63, 179, 1991.

31. Tiberbaev, A. R., Buchberger, W., Semenova, O. P., and Bonn, G. K., Metal ion capillary zone electrophoresis with direct detection: determination of transition metals using an 8-hydroxyquinoline-5-sulfonic acid chelating system, *J. Chromatogr.*, 630, 379, 1993.

32. Timerbaev, A., Semenova, O., and Bonn, G., Metal ion capillary zone electrophoresis with direct UV detection: comparison of different migration modes for negatively charged chelates, *Chromatographia*, 37, 497, 1993.

33. Motomizu, S., Nishimura, S. I., Obata, Y., and Tanaka, H., Separation and determination of divalent metal ions with UV-absorbing chelating agents by capillary electrophoresis, *Anal. Sci.*, 7 (Supplement), 253, 1991.

34. Timerbaev, A. R., Semenova, O. P., and Bonn, G. K., Capillary zone electrophoresis of lanthanoid elements after complexation with aminocarboxylic carboxylic acids, *Analyst*, 119, 2795, 1994.

35. Timerbaev, A. R., Semenova, O. P., and Fritz, J. S., New possibilities on multi-element separation and detection of metal ions by CZE using pre-capillary complexation, *J. Chromatogr. A*, 756, 300, 1996.

36. Timerbaev, A. R., Semenova, O. P., Bonn, G. K., and Fritz, J. S., Determination of metal ions complexed with 2,6-diacetylpyridine bis(N-methylenepyridineohydrazone) by capillary electrophoresis, *Anal. Chim. Acta*, 296, 119, 1994.

CHAPTER 7

Analysis of Carbohydrates

Ziad El Rassi and Yehia S. Mechref

CONTENTS

0-8493-9127-X/98/$0.00+$.50

I. INTRODUCTION

High performance capillary electrophoresis (HPCE), in its various modes of operation including zone electrophoresis (CZE), isoelectric focusing (CIEF), and micellar electrokinetic chromatography (MEKC), is increasingly used in the separation and detection of a wide variety of carbohydrates. The broad acceptance of HPCE has been largely facilitated by the high intrinsic resolving power of electrophoresis and the suitability of the high separation efficiency of the capillary format for the separation of closely related carbohydrate structures.

There are additional merits and sound features for using HPCE in the analysis of carbohydrates as a complementary, or often as a stand-alone, analytical separation method. Similar to high performance liquid chromatography (HPLC) and polyacrylamide gel electrophoresis (PAGE), HPCE is also an aqueous-based separation method well suited for the analysis of inherently hydrophilic compounds such as carbohydrates. Furthermore, HPCE seems to possess several advantages over HPLC and PAGE by (1) providing higher separation efficiencies, (2) yielding shorter analysis time, (3) requiring small sample amounts, and (4) more important, consuming smaller amounts of expensive reagents and solvents.

However, to realize the full benefits of the many sound features of HPCE in the analysis of complex carbohydrate samples, two major difficulties have to be overcome. First, with the exception of a few carbohydrates including aldonic acids, uronic acids, sialic acids, amino sugars, and compositional sulfated sugars of glycosaminoglycans, which are naturally charged, most carbohydrate molecules lack readily ionizable charged functions, a condition that excludes their direct differential migration and eventual separation by electrophoresis. Second, most carbohydrate species neither absorb nor fluoresce, a property that hinders their sensitive detection by modern analytical separation techniques, including HPCE.

Various approaches have been introduced to render carbohydrates amenable to separation and detection by HPCE. These approaches have exploited many of the inherent properties of carbohydrates. The polyhydroxy nature of carbohydrates offers these species some unique alternatives for separation in electrophoresis including (1) *in situ* conversion of carbohydrates into charged species *via* complex formation with other ions such as borate and metal cations, and (2) the ionization of the hydroxyl groups at alkaline pH. Also, the reactivity of the reducing end and other functional groups of the sugar molecules (e.g., carboxylic acid groups and amino groups) that can be readily labeled with UV-absorbing or fluorescent tags has provided a means for tagging the carbohydrates with centers for sensitive detection. Furthermore, the electrochemical oxidation of carbohydrates at the surface of metallic electrodes provides another means by which underivatized carbohydrates can be sensitively detected.

The aim of this chapter is (1) to describe the basic aspects of the electrolyte systems used in CE of carbohydrates, (2) to discuss the advantages and disadvantages of the approaches and concepts that are most useful in the separation and detection of carbohydrates by CE, and (3) to review important applications.

For recent and detailed reviews on the various aspects of the capillary column technology, the interested reader may consult references 1 and 2. Furthermore, a special issue of the journal *Electrophoresis* on *Capillary Electrophoresis of Carbohydrate Species* has appeared recently.[3]

II. ELECTROLYTE SYSTEMS

Thus far, most CE analysis of "neutral" carbohydrates has been achieved by borate complexation and to a lesser extent by ionization at alkaline pH. Only one paper has appeared on the HPCE of sugars as alkaline earth metal ion complexes.[4]

A. Borate-Based Electrolytes

The anionic complex which results from the association between borate ions and carbohydrates has been known for quite a long time.[5-7] The Raman spectrum of the aqueous borate ion[8] and X-ray studies on boron minerals have shown that the borate ion has a tetrahedral symmetry existing as a tetrahydroxyborate ion, $B(OH)_4^-$.[5] This means that boric acid does not act as a proton donor but rather as a Lewis acid, accepting the electron pair of the base OH^- to form $B(OH)_4^-$ according to the following equilibrium:

$$B(OH)_3 \; + \; OH^- \; \rightleftharpoons \; B(OH)_4^-$$

$$(B) \hspace{6cm} (B^-)$$

I

This equilibrium is to a very large extent shifted to the right at an alkaline pH, where complexation is most effective, and consequently, it is the $B(OH)_4^-$ that undergoes complexation with carbohydrates. In general, the polyol-borate complex formation is represented by the following equilibria:

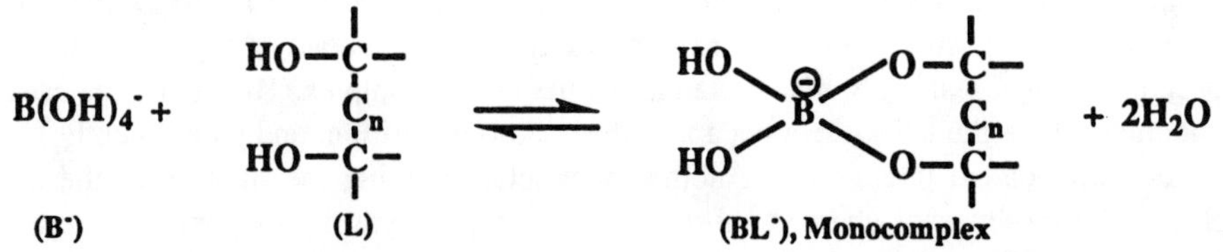

II

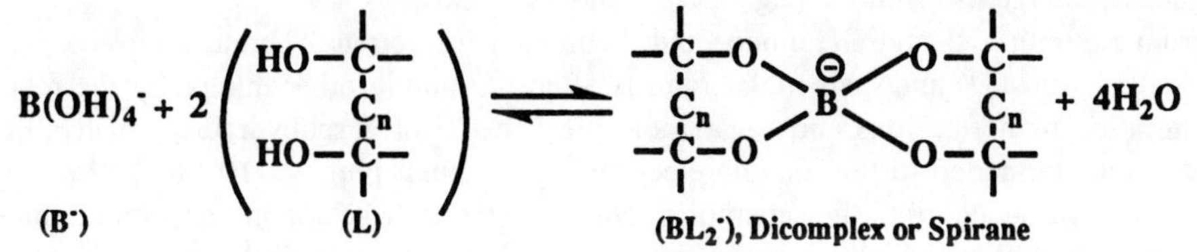

III

where BL^- and BL_2^- are the mono- and di-esters, respectively, L is the polyol, and n = 0 or 1. Equilibrium II is very much driven to the right, whereas III is dependent upon the position of the hydroxyl groups in the polyol.

As a result of the complexation between borate ions and carbohydrates, cyclic borate esters with either five or six atom rings are formed when n = 0 (for adjacent hydroxyl groups) or 1 (for alternate hydroxyl groups), respectively. In addition to these complexes, tridentate monocomplexes have been observed[9] and are dependent on the structure of the polyol.

The complex formation, as illustrated by Equilibria II and III, is possible only if two hydroxyl groups in the polyol molecule are favorably oriented and distanced from each other. In other words, the distance between the two hydroxyl groups in the polyol molecule should be of similar magnitude to the O–O distance in the borate ion, which is 2.36 and 2.39 Å for trigonal and tetrahedral boron, respectively.[10] This accounts for the higher stability of complexes formed with *cis*-oriented 1,2-diols of five-membered ring compounds (O–O = 2.49 Å) over the *trans* isomers (O–O = 3.40 Å).[11] Similarly, cyclic forms of sugars complex less strongly with borate ions than the open chain (acyclic) ones because of the O–O distance and conformational constraints. In other words, alditols form stronger complexes with borate than their counterpart aldoses under otherwise identical conditions.[12]

Generally, the strength of borate complexation increases with an increasing number of hydroxyl groups in the molecule. This is because increasing the number of hydroxyl groups will favorably adjust the orientation of the adjacent hydroxyl groups and consequently increase the stability of the borate–polyol complexes. In other words, for compounds with more than two adjacent hydroxyl groups the repulsion of these groups will prevent two adjacent groups from being 180° apart, and consequently shorten the O–O distance.

The distance between the two adjacent hydroxyl groups in the sugar molecule not only affects the stability of its borate complex but also the conformation of the adjacent groups. For acyclic polyol compounds, the stabilities of borate esters of 1,3-diols and 1,2-diols are similar; however, the *threo*-isomers of 1-2-diols form more stable complexes than the *erythro*-isomers, while the *syn*-isomer borate complexes are more stable than the *anti*-isomers for the 1,3-diols.[13,14]

The carbohydrate–borate complex formation is largely influenced by the presence of substituents in the polyol molecule as well as by their charges, locations, and anomeric linkages. Methylated sugars complex less with borate than the parent unsubstituted sugars, and consequently the electrophoretic mobilities of methylated sugars in zone electrophoresis are much lower.[15] The complexation is stronger with methyl-β- than methyl-α-D-glucopyranoside, and consequently the mobility of the α-anomer, compared to that of the β-anomer is lower in zone electrophoresis with alkaline borate. This may be due to the fact that in the α-anomer the glycosidic methoxyl group occupies an axial position and will interact strongly with the axial hydrogen atoms on C3 and C5, thus destabilizing the borate complex. This is not the case for the β-anomer where the substituent occupies an equatorial position, and consequently is free from strong nonbonded interactions. Another parameter affecting the stability of the borate complex is the presence of charged substituents in the polyhydroxy molecule. Generally, a decrease in the stability of a borate ester is observed as a result of coulombic repulsion between a negatively charged substituent (e.g., COO^-) and BO_4^- moieties.[14]

From Equilibria II and III, mono- and dicomplex (or spirane) borate esters coexist in aqueous solutions,[13,16] and their molar ratio is affected, among other things, by the relative concentration of borate ions and sugar molecules. In CE of carbohydrates, usually 0.1 to 0.2 M borate are added to the running electrolyte, and small plugs of 10^{-4} to 10^{-5} M sugar samples are introduced into the separation capillary. Under these conditions, anionic mono-complexes (i.e., BL^-), which are more favored at low sugar:borate ratio, are likely to predominate and thus migrate differentially under the influence of an applied electric field. However, it should be noted that whether the injected sugar samples form BL^- or BL_2^- or both while migrating in a borate medium, all sugar molecules will be associated with a negative charge. This is due to the fact that the mono- and dicomplex formations are dynamic. Nevertheless, the magnitude of the charge will be influenced by the position of the equilibrium and therefore by the stability of the complex. According to Equilibria I, II, and III, at constant sugar concentration, the amount of complex increases with borate concentration according to the law of mass action and also with pH due to the higher concentration of borate ions.

At a constant pH, resolution among various sugars analyzed by CE increases with increasing borate concentration. However, there is an optimum borate concentration at which maximum resolution is obtained for a multicomponent mixture[17] (see Figure 7.1). As can be seen in Figure 7.1b, 75 mM borate allowed the full resolution of seven sialooligosaccharides derived from gangliosides and derivatized with 7-aminonaphthalene-1,3-disulfonic acid. At a constant borate concentration, resolution among monosaccharide–borate complexes varies as a function of pH with an optimum in the pH range 10 to 11.[18]

In summary, under a given set of conditions, various sugars (charged or neutral) would undergo varying degrees of complexation with borate, leading to differences in the electrophoretic mobilities of the complexed solutes and hence separation. This has promoted the use of borate-based electrolytes as the separation electrolyte for the CE analysis of most saccharides.

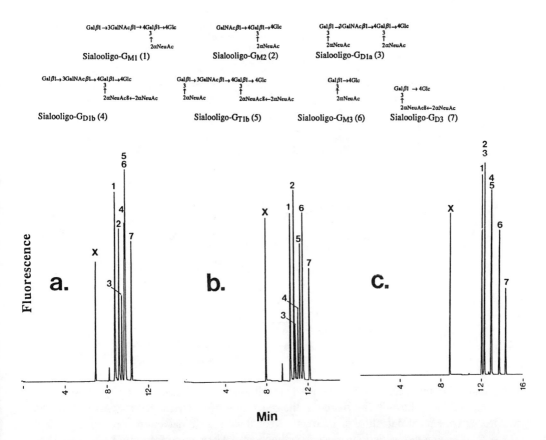

Figure 7.1 Electropherograms of ANDSA derivatives of sialooligosaccharides. Conditions, capillary, fused-silica, 50 cm (to detection point) 80 cm (total length) × 50 μm i.d.; running electrolytes, borate buffer of (a) 50, (b) 75, and (c) 125 mM, pH 10.0; voltage, 20 kV. Solutes, X = by-product, 1 = sialooligo-G_{M1}, 2 = sialooligo-G_{M2}, 3 = sialooligo-G_{D1a}, 4 = sialooligo-G_{D1b}, 5 = sialooligo-G_{T1b}, 6 = sialooligo-G_{M3}, 7 = sialooligo-G_{D3}. (From Mechref, Y., Ostrander, G. K., and El Rassi, Z., *Electrophoresis*, 16, 1499, 1995. With permission.)

B. Highly Alkaline pH Electrolytes

Another common approach to charging "neutral" carbohydrates is to perform the analysis at a highly alkaline pH using alkali-metal hydroxide solutions, such as lithium, sodium, or potassium hydroxide. Under these conditions, the differential electromigration of carbohydrates is presumably due to the ionization of the hydroxyl groups of saccharides, yielding negatively charged species called alcoholates.[19] The ionization constants for carbohydrates are in the range of 10^{-12} to 10^{-14}, i.e., pK_a = 12 to 14. The pK_a values of some typical sugars are listed in Table 7.1. Usually, reducing sugars (e.g., glucose, galactose, mannose, etc.) are the most easily ionized, while straight-chain alditols (e.g., glucitol, mannitol) have on the average about the same acidity as cyclitols (e.g., inositols) and glycosides (e.g., methylglucopyranosides) of similar molecular weight and hydroxyl content. The higher acidity of reducing sugars is caused by the higher lability of the hydrogen atom of the hemiacetal (anomeric) hydroxyl group, a condition that apparently originates from an electron-withdrawing polar effect (inductive effect) exerted upon this group by the ring oxygen.

One important feature of Table 7.1 is that the greater the number of hydroxyl groups, the greater the acidity. Glycerol has the same acidity as water, while lactose and maltose, which are reducing disaccharides, seem to be somewhat more acidic than aldopentoses (e.g., arabinose, ribose, lyxose, and xylose).

Table 7.1 Ionization Constants (Hydroxyl Group) of Carbohydrates in Water at 25°C

Compound	pK_a
D-Glucose	12.35
2-Deoxyglucose	12.52
D-Galactose	12.35
D-Mannose	12.08
D-Arabinose	12.43
D-Ribose	12.21
2-Deoxyribose	12.67
D-Lyxose	12.11
D-Xylose	12.29
Lactose	11.98
Maltose	11.94
Raffinose	12.74*
Sucrose	12.51
D-Fructose	12.03
D-Glucitol	13.57*
D-Mannitol	13.50*
Glycerol	14.40

* Measured at 18°C.

From Rendleman, J. A., *Adv. Chem. Ser.*, 117, 51, 1971. With permission.

Highly alkaline electrolyte solutions such as lithium, potassium, or sodium hydroxide at pH greater than 12 have been utilized in the separation of underivatized saccharides by CE with electrochemical detection or indirect UV or fluorescence detection only. As expected, the resolution among the various saccharides increased when going from pH 12.3 to pH 13.0 due to increasing ionization of the separated analytes.[20] Moreover, the nature of the alkali-metal ion influences the resolution of the sugar analytes[20] (see Figure 7.2). It should be noted that separations at extremely high pH can only be performed with naked fused-silica capillaries since most coated fused-silica capillaries will undergo hydrolytic degradation under such basic conditions.

C. Carbohydrate-Metal Cation Complexes

In the complex formation between metal cations and carbohydrates, the hydroxyl groups of sugars are thought to form coordinate bonds with the metal cations. Usually, strong complexation occurs between cations and a contiguous axial (a), equatorial (e), axial sequence of hydroxyl groups in carbohydrates as was ascertained from the electrophoretic movement of compounds containing this sequence and the immobility of many others lacking such an arrangement.[21] Other arrangements which favor coordination to cations include the rare 1,3,5-triaxial arrangement and the *cis,cis*-1,2,3-triol grouping on a five-membered ring. In the case of *cis,cis*-1,2,3-triol and in a twist configuration, the three oxygen atoms are in geometrical arrangement similar to the a,e,a arrangement.[21] The three oxygen atoms are then equidistant from the cations, thus favoring complex formation with large cations, while the smaller cations yield only weak complexes. On the other hand, some compounds may not have the a,e,a sequence in their most stable conformations (e.g., β-D-ribopyranose); therefore, they form weaker complexes than compounds having a,e,a sequence in their preponderant conformation (e.g., α-D-ribopyranose).[22,23] Moreover, if the difference in free energy between the noncomplexing and complexing conformation is large, complex formation will not take place, e.g., *myo*-inositol.

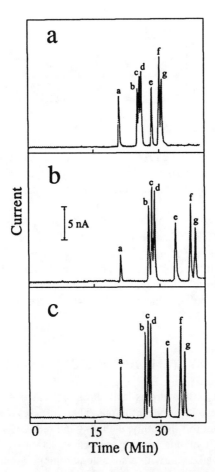

Figure 7.2 Electropherograms showing the effect of (a) 100 mM KOH, (b) 100 mM LiOH, and (c) 100 mM NaOH on the separation of six saccharides. The peaks correspond to (a) neutral marker (0.5% MeOH) (b) stachyose, (c) raffinose, (d) sucrose, (e) lactose, (f) galactose, and (g) glucose. The fused-silica capillary dimensions are 50 μm i.d. and 70 cm in length. The separation voltages were 8 kV, 10 kV, and 11 kV for KOH, NaOH, and LiOH, respectively. Injection is 10 s by gravity (10-cm height); the ADCP is performed at 0.6 V (vs. Ag/AgCl). (From Colón, L. A., Dadoo, R., and Zare, R. N., *Anal. Chem.*, 65, 476, 1993. With permission.)

For acyclic alditols, when three consecutive carbon atoms have the *threo-threo* configuration, the complex is most favored. An *erythro-threo* configuration is less favored for complex formation, and an *erythro-erythro* arrangement does not give rise to any noticeable complexation with cations. The more *threo* pairs of hydroxyl groups there are in the alditol, the stronger will be its complex.[22] Substitution of one of the hydroxyl groups by a methoxy group decreases the stability of the complex. Methylation of all three hydroxyl groups renders complex formation negligible.[22] This draws similarities to borate complexation with sugars.

The complex formation involving cyclic monosaccharides yields tridentate complexes, whereby no more than three oxygen atoms can coordinate to one cation as shown in Figure 7.3, where M[+] is the metal cation. Only in a few cases involving disaccharides has complexation been reported to occur at more than three oxygen atoms, and tetra- and even pentadentate complexes have been described.[22]

Discarding a few exceptional cases, the order of decreasing effectiveness of complexation of the carbohydrates is 1,3,5-triaxial triol > a,e,a triol on a six-membered ring > *cis-cis*-triol on a five-membered ring > acyclic *threo-threo*-triol > acyclic *threo* pair adjacent to a primary hydroxyl group > acyclic *erythro-threo*-triol > acyclic *erythro* pair adjacent to

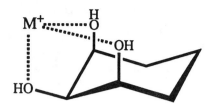

Figure 7.3 Metal–carbohydrate complex.

a primary hydroxyl group > acyclic *erythro-erythro*-triol > *cis*-diol on a five-membered ring > *cis*-diol on a six membered ring > *trans*-diol on a six-membered ring.[22] On the other hand, the order of decreasing complex formation of the cations is trivalent > divalent > univalent cations.[11]

The utility of electrolyte systems containing alkaline-earth metals for the separation of neutral carbohydrates by HPCE has been reported only in a single paper by Honda et al.[4] Separations in these media are mainly based on differences in the extent of complexation of the divalent metals with the carbohydrate solutes, and to a lesser extent on the hydrodynamic radius of the molecule. Although these systems provided a different selectivity than that achieved with borate buffers, the resolution was in general inferior to that of borate buffers.

III. DETECTION SYSTEMS AND PRECOLUMN DERIVATIZATION

As stated above, most carbohydrates lack chromophores in their structures, a condition that hinders their detection at low concentrations. Therefore, several detection strategies have been developed to surmount this difficulty, including indirect UV and fluorescence detection, electrochemical detection, dynamic labeling *via* complexation with absorbing or fluorescing ions, and precolumn derivatization with a suitable chromophore and/or fluorophore.

A. Detection of Underivatized Carbohydrates

1. *Direct UV Detection*

UV detectors have been used only occasionally in HPCE of underivatized carbohydrates. The major drawback of direct UV is the limited sensitivity associated with the inherent low molar absorptivities of most carbohydrates.

The molar absorptivities of mono- and oligosaccharides, at 195 nm, was shown to increase 2- to 50-fold (for instance from 2 to 25 for xylose and from 2 to 95 for myo-inositol) upon complexation with borate.[24] This increase was attributed to the fact that borate complexation shifts the equilibrium between carbonyl and cyclic sugar forms toward the carbonyl form. However, in that work the sugars were detected from relatively concentrated samples (0.75 mM to 7.5 mM), thus rendering the borate complexation not a very attractive detection approach for underivatized carbohydrates.

Only a few carbohydrates exhibit more or less significant absorbance in the low UV, including oligosaccharides containing *N*-acetylglucosamine, *N*-acetylgalactosamine, and sialic acid residues (i.e., glycans) which can be detected at 200 nm[25-28] or 185 nm.[29] Also, the glycosaminoglycan (GAG) hyaluronan could be detected at 200 nm at a rather modest sensitivity.[30] This is facilitated by the presence of a repeat unit of one D-glucuronic acid residue and one *N*-acetyl-D-glucosamine residue, linked by glycosidic bonds as shown here.

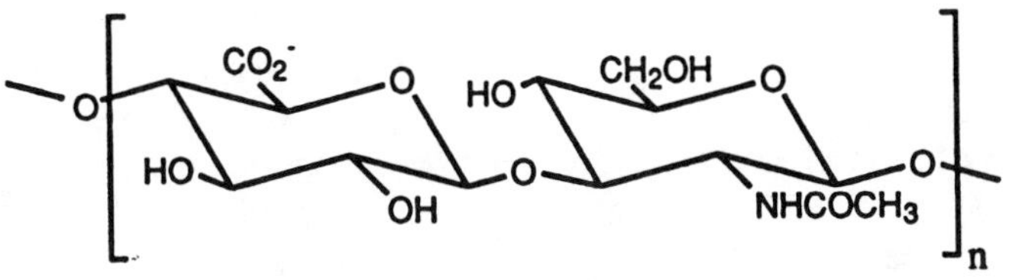

D-Glucuronic acid   N-Acetyl-D-glucosamine

Also, low-molecular-mass heparins and heparins with the repeating unit shown below could be detected at moderate sensitivity at 200 nm.[31]

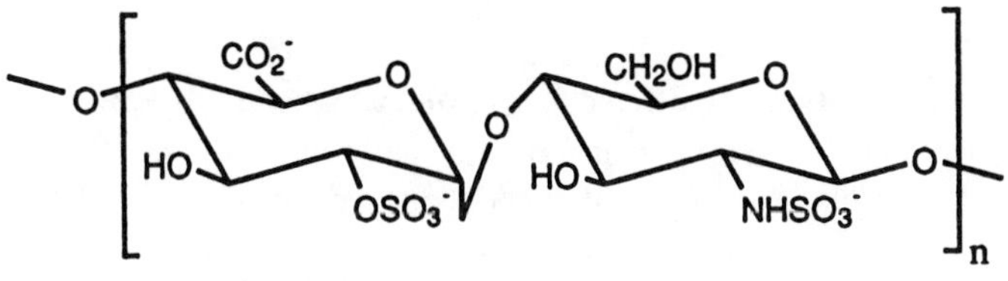

D-Glucuronate-   N-sulfo-D-glucosamine-
2-sulfate   6-sulfate

However, such low UV wavelengths impose serious restrictions on the choice of the composition of the running electrolytes by not allowing the use of many useful additives that may absorb extensively in the low UV.

Another class of oligosaccharides which can be detected directly in the UV consists of acidic di- and oligosaccharides derived from GAGs. These saccharides, which result from the enzymatic depolymerization of the large GAGs, possess unsaturated uronic acid residues at the nonreducing end that allows their direct UV detection at 232 nm (for structures, see Figures 7.17 and 7.19).[32-41] The molar absorptivity of this unsaturated uronic acid residue is approximately 5000 $Lmol^{-1}cm^{-1}$.[42]

Synthetic oligosaccharide fragments of heparin including di-, tetra-, penta-, and hexasaccharides, which lack the double bond in the uronic acid residue, were detected at 214 nm, but the limit of detection was one order of magnitude lower than that reached by indirect UV as explained below.[43]

2. Indirect UV and Indirect Fluorescence Detection of Underivatized Carbohydrates

Indirect detection schemes are inherently nondiscriminative (i.e., universal), thus permitting the detection of compounds that do not possess the necessary physical properties for direct detection, e.g., chromophores or fluorophores. A major advantage of indirect detection resides in the fact that it eliminates the need for pre- or post-column derivatization of the analytes, thus minimizing sample preparation.

In indirect detection, the analyte is thought to displace a component of the running electrolyte which is a chromophore or fluorophore. Usually, a detectable co-ion which possesses high molar absorptivity or high fluorescence quantum yield is added to the running electrolyte, thus providing a continuous detector response. Since charge neutrality must be maintained, an analyte of the same charge as the detectable co-ion will therefore displace this co-ion causing a decrease in the detector response.[44] Thus, the resulting peak is derived from the detectable background co-ion rather than from the analyte itself. On these bases, almost any detection scheme in CE can be made to function in the indirect mode by simply altering the composition of the running electrolyte and not the actual instrumentation usually used in direct detection.

The number of electrolyte co-ions displaced by one analyte molecule is defined as the transfer ratio (*TR*). Since a large background signal is required, the instability of the background signal has tremendous effects on the dynamic reserve (*DR*), which is defined as the ability to measure a small change on top of a large background signal. The *DR* is essentially the ratio of the background signal to the background noise. The concentration limit of detection (C_{lim}), expressed in concentration units, is given by[45]

$$C_{lim} = \frac{C_M}{DR * TR} \tag{1}$$

where C_M is the concentration of the detectable co-ion which generates the background signal. For a given system, the more stable the background signal (larger *DR*), the smaller the fractional change. Likewise, the more efficient the displacement process (large *TR*), the lower the C_{lim}. Moreover, the lower the C_M is, the greater the fractional change will be. It is desirable that the value of *TR* be close to unity.[45] When optimizing C_{lim}, it must be noted that the three parameters are not necessarily independent. For example, decreasing C_M will increase *TR*, but at the expense of decreasing *DR*. It should be noted that the *TR* is not necessarily the same for all analytes. In addition, it has been shown that the best detection sensitivity is achieved when the analyte ions have an effective mobility close to that of the detectable co-ion.[46] An in-depth discussion of this mode of detection is outside the scope of this chapter, and interested readers are advised to consult recent reviews[44,45,47] and Chapter 2.

Both indirect photometric and fluorometric detection modes have found use in CE of carbohydrates. In indirect detection, one of the critical factors is the selection of the detectable co-ion, which should meet several criteria including: (1) a high molar absorptivity at the detection wavelength and excitation wavelength used in UV and fluorescence, respectively, (2) a high quantum efficiency in fluorescence, preferably as close to unity as possible, (3) compatibility with the solvent system, i.e., it must be soluble and inert, (4) no interaction with the capillary wall, and (5) charge-bearing, preferably with a charge identical to that of the analyte being displaced. The last criterion will ensure a value close to unity for the transfer ratio, *TR*. To realize the full benefits of indirect LIF detection in terms of limits of detection, the stabilization of the laser power is of primary importance because it greatly improves the dynamic reserve, *DR*, by decreasing fluctuations in the background signal.

Garner and Yeung were the first to utilize indirect detection for CE analysis of carbohydrates[48] employing laser-induced fluorescence (LIF) detection. In that work, coumarin 343 was used as the background fluorescing co-ion which has a good solubility, high quantum efficiency, high molar absorptivity ($\varepsilon = 20,000$) at 442 nm, and is suitable to be used with a helium-cadmium (He-Cd) laser. Using this indirect LIF detection system, 640 femtomoles of three simple sugars could be detected and separated using 1 m*M* coumarin 343, pH 11.5. The high pH used ensured the partial ionization of the neutral carbohydrates, which is required for their electrophoretic separation as well as for their indirect detection. In indirect detection, the sensitivity is a function of the fraction of the analyte that is ionized. Therefore, the pH of the running electrolyte must be approaching 12 to have any substantial fraction, α, of the sugar solute in the ionized form. However, at this high pH the concentration of hydroxide ions is no longer negligible relative to the concentration of the detectable co-ion. The effect of the hydroxide ion can be approximated by[49]:

$$TR_{tot} = \frac{\alpha[\text{sugar}]}{[\text{FL}]+[\text{OH}^-]} \tag{2}$$

where TR_{tot} is the total transfer ratio, α[sugar] is the fraction of the sugar ionized, [FL] is the concentration of the detectable fluorophore, and [OH$^-$] is the hydroxide ion concentration. As can be seen in Equation 2, at constant fluorophore (or chromophore) and sugar concentrations, α in the numerator and [OH$^-$] in the denominator are competing functions of pH. The total transfer ratio goes through a maximum when plotted as a function of pH. This maximum is the most sensitive pH for the detection of a given sugar. The optimum detection pH for simple sugars was found to be 11.65, but 11.5 was used for detection because the rate of degradation of coumarin 343 was decreased without any appreciable decrease in detection efficiency. Using the detection system described above, the limit of detection was 2 femtomoles for fructose.[49]

Very recently, detection limits in the picogram range are possible for high-molecular-weight polysaccharides (e.g., dextran, amylose, amylopectin, etc.) when indirect LIF detection is employed.[50] An argon-ion laser source operating at 488 nm was used for excitation. Because of the highly alkaline electrolyte (pH 11.5) needed to ionize the polysaccharides, 1 mM fluorescein has to be added to the running electrolyte as the fluorophore in order to overcome the competition from the high concentration of hydroxide ions. In addition, only 9 out of 11 polysaccharides studied could be detected, which might be due to the weak ionization of polysaccharides at high pH.[50]

Although the above indirect detection schemes showed exceptional results in both sensitivity and efficiency, they involved the use of expensive laser equipment not available in most separation facilities. Indirect UV has been shown to be feasible for the detection of underivatized carbohydrates. Bonn and co-workers[51,52] have demonstrated the use of 6 mM sorbic acid at pH 12.1 as both the electrolyte and the detectable co-ion in the separation and indirect detection of several simple sugars. The alkaline pH ensured ionization of the sugars and, hence, their detection by means of charge displacement. The fact that sorbic acid has a high molar absorptivity (ε = 27,800 M^{-1} cm^{-1} at 256 nm) and carries a single charge ensured an enhanced detectability and a favorable transfer ratio, respectively. Under these conditions, a detection limit of 2 picomoles was obtained for glucose.

In a recent work, CE-indirect UV proved useful for the separation and detection of two aldonic acids.[53] Since these two analytes are readily ionized, 6 mM sorbic acid pH 5.0 was used as the co-ion and the running electrolyte. The detection limit of such a system was determined to be 18 femtomoles. This improvement in detectability can be attributed to the lower pH which allows for a larger TR, i.e., concentration of hydroxide ions is not a factor in such a system.

High hydroxide ion concentration in a highly alkaline buffer not only affects TR, but also results in a rapid increase of the low-frequency noise and baseline instability. This was found to be related to the Joule heat production and insufficient thermostating of the capillary tubing.[54] Detection could be greatly improved by using narrow capillaries (25 μm i.d.) and low voltages, and by thermostating the area surrounding the capillary column to allow a uniform heat dissipation along its length.[54] In addition, the noise was influenced by the composition of the background electrolyte. As shown in Figure 7.4, riboflavin and lithium were found to be the best chromophore and counterion, respectively, giving the best performance in terms of signal-to-noise ratio. This was because both compounds had the lowest mobility relative to the other chromophores and counterions tested. Optimization of all these factors allowed the pH range for the separation to be extended to 13 with a limit of detection of 50 μM,[54] which is one order of magnitude lower than previously reported values.[51] Moreover, the detection limit was improved by approximately 25 times at pH 12.3 by using a riboflavin-NaOH electrolyte system.

Recently, the results of CE experiments involving the determination of carbohydrates in fruit juices by CE-indirect UV mode of detection were compared with those obtained by high performance anion-exchange chromatography with pulsed amperometric detection (HPAEC-PAD).[55] In that work, potassium sorbate was chosen as the background electrolyte

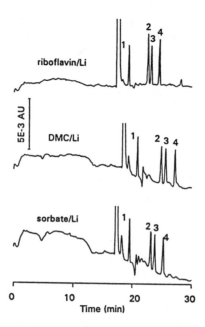

Figure 7.4 Influence of the monitoring ion on noise and baseline disturbances. Electrolytes: 12 m*M* absorbing co-ion, 63 m*M* LiOH; standard mixture of 1.0 m*M* of sugars. Peaks: 1 = sucrose, 2 = maltose, 3 = glucose, 4 = fructose. (From Zu, X., Kok, W. T., and Poppe, H., *J. Chromatogr. A*, 716, 231, 1995. With permission.)

(pH 12.2 to 12.3) and chromophore for indirect UV detection at 256 nm. HPAEC-PAD yielded concentration detection limits of 2 to 3 orders of magnitude lower than those with HPCE-indirect UV. However, the comparison was in favor of HPCE in terms of mass detection. The absolute amounts detectable for HPAEC-PAD was 25 to 50 pmol, while it was 0.9 to 1.1 pmol for HPCE-indirect UV. In the application of CE-indirect UV to the analysis of sugars in fruit juices, the detection sensitivity is not an issue since the sugar concentration is relatively high (100 gL^{-1}) and the sample must be diluted 1:50 prior to CE-indirect UV analysis.

Another chromophore that has been utilized very recently for indirect UV detection of carbohydrates is tryptophan.[56] This chromophore was shown to yield a twofold higher response of the sugars compared to sorbic acid when the background absorbances were kept at approximately the same level. Using an alkaline electrolyte system (pH > 12) containing tryptophan, nine mono- and oligosaccharides were separated and detected at 280 nm by CE-indirect UV in 20 min (see Figure 7.5). The linear dynamic range was reported to be 0.05 to 0.5 picomole. Figure 7.5 illustrates the effect of the size of the inner diameter of the capillary on the separation of the sugar mixture. The use of 50-μm i.d. capillary limited the concentration of tryptophan used due to the baseline noise that originated from the Joule heating effect (Figure 7.5A). A smaller i.d. capillary (25 μm) has more efficient heat dissipating properties, thus permitting the use of high tryptophan concentration without noticeable increase in baseline noise (Figure 7.5B). Consequently, the detection sensitivity was improved by using the 25-μm i.d. capillary.

Since the mechanism of indirect detection is based on displacement of the chromophore in the solute zone, the detectable co-ions are generally chosen to have the same charge as the analyte. However, an indirect response can also be observed using a chromophore with a charge opposite that of the analyte. The indirect response in such a system is due to a disturbance of the charge balance between the chromophore and buffer ions in the background electrolyte by the analyte. A positively charged chromophore used for indirect UV detection is *N*-benzylcinchonidium chloride (BCDC). However, and as expected, the detection sensi-

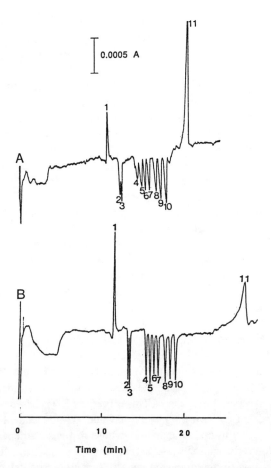

Figure 7.5 Electropherograms for separation of sugars. (A) on a 50-μm i.d. capillary ($L_{tot.}$ = 56 cm, $L_{det.}$ = 32 cm); running electrolyte, 1.0 mM tryptophan in 50 mM NaOH; solute amount, 4.65 pmole, each. (B) on a 25-μm i.d. capillary ($L_{tot.}$ = 57 cm, $L_{det.}$ = 34.5 cm). Running electrolyte, 5.0 mM tryptophan in 50 mM NaOH; solute amount 0.28 pmole; each; voltage, 7.0 kV; injection, 1 sec. Peak identification: (1) system peak (EOF), (2) stachyose, (3) raffinose, (4) melibiose, (5) cellobiose, (6) D-galactose, (7) D-glucose, (8) D-mannose, (9) D-xylose, (10) D-ribose, (11) marker. (From Lu, B. and Westerlund, D., *Electrophoresis*, 17, 325, 1996. With permission.)

tivity using BCDC was much lower than that attained using tryptophan because the displacement of BCDC is not as effective as that of tryptophan.[56]

Also, raffinose, melibiose, cellobiose, galactosamine, and glucosamine were CE analyzed and detected by indirect UV at 292 nm using 1.0 mM dimethylprotriptyline iodide.[57] The limits of detection were in the picomole range.

Other carbohydrates were detected by CE-indirect UV including eight heparin disaccharides and some synthetic sulfated disaccharide and oligosaccharide fragments of heparin.[43] CE-indirect UV was achieved by using either 5 mM 5-sulphosalicylic acid, pH 3.0, or 5 mM 1,2,4-tricarboxylbenzoic acid, pH 3.5, as the running electrolyte and chromophore. In contrast to direct UV detection, with indirect UV detection the signal obtained for the various synthetic pentasaccharides is nearly independent of their molecular structure, and the sensitivity is at least one order of magnitude higher than that of direct UV detection. Again, because of the low pH where the transfer ratio is at its optimum value, the limit of detection of synthetic pentasaccharide heparin fragments was below 5 fmol when performing the detection at 214 nm using 5 mM 5-sulphosalicylic acid, pH 2.5.[43] Even for heparin disaccharides possessing unsaturated uronic acid residues at the nonreducing end, CE-indirect UV at 214 nm in the presence of 5 mM 1,2,4,-tricarboxybenzoic acid at pH 3.5 yielded higher sensitivity than HPCE-direct UV at 230 nm when employing 200 mM sodium phosphate, pH 2.5, as the running electrolyte (see Figure 7.21).

Although the principle of indirect UV (or fluorescence) detection appears relatively simple and is significantly more sensitive than direct low wavelength UV of underivatized carbohydrates, several drawbacks can be pointed out. First, the instability of the detection system results in drift or disturbances of the baseline. Second, an indirect detection system requires working at a low concentration of background electrolyte in order to have efficient transfer ratios, a condition that results in lower efficiencies at higher sample concentration and the possibility of solute-wall interactions. A third disadvantage imposed by indirect detection in CE is the limitation in the selection of the composition and pH of the background electrolyte. In other words, there is not much room to manipulate selectivity and optimize separations. In fact, "neutral" carbohydrates are only partially ionized at the optimum pH normally used in indirect detection, i.e., pH ~ 12, a condition that does not favor their high resolution separation. Finally, another disadvantage is the limited linear dynamic range, typically under two orders of magnitude. Therefore, this mode of detection should be used only for fairly concentrated samples or whenever the analytes are not easily derivatized or cannot be detected otherwise.

3. Electrochemical Detection

Electrochemical techniques have been proven to be useful for the detection of underivatized carbohydrates. Moreover, electrochemical detection (ED) is an ideal mode of detection for microcolumn-based separation systems. This is because ED is based on an electrochemical reaction at the surface of the working electrode and is independent of cell volumes or pathlength. This is contrary to optical detectors whose response is dependent on pathlength. In particular, amperometric methods are among the most sensitive approaches currently available for the detection of underivatized sugars. Amperometric detection is based on the measurement of current resulting from the oxidation or reduction of analytes at the surface of an electrode in a flow-cell. (For detailed discussions see reference 58.)

Two approaches for ED have been shown useful for carbohydrate detection, namely amperometric detection at constant potential (ADCP)[20,59-61] and pulsed amperometric detection (PAD).[62-66] The principles of PAD have been described in detail by Johnson and LaCourse.[67] Platinum (Pt) and gold (Au) electrodes have been the most widely used metallic electrodes for the detection of carbohydrates and related species in HPLC.[67] The success of Pt and Au electrodes is due largely to the tendency of sugars to adsorb on their surfaces where sugars readily undergo electrochemical reactions at low potentials.[68] Unfortunately, this same adsorption phenomenon also constitutes one of the major disadvantages of Pt and Au in that accumulation of oxidation products generally leads to electrode poisoning and, unless overcome experimentally, a rapid decrease in analyte response occurs. As a result, anodic detection schemes using Pt and Au electrodes typically include routine desorption and conditioning steps in order to restore the electrodes to their full activity and provide a stable and reproducible response. Restoration of the electrode is generally achieved by serial application of brief positive and negative potential pulses to achieve oxidative desorption of adsorbed carbohydrate species with subsequent regeneration of clean and oxide-free surfaces. This technique is known as pulsed amperometric detection (PAD).

Recently, O'Shea et al.[62] utilized the PAD concept to detect carbohydrates after their CE separation in an off-column detection format with a gold wire microelectrode at the end of the capillary column. The PAD has been successful because the multistep waveform solves the problem of electrode poisoning typically found with the oxidation of carbohydrates at Au electrodes in the direct amperometric detection mode. This CE-PAD allowed a detection

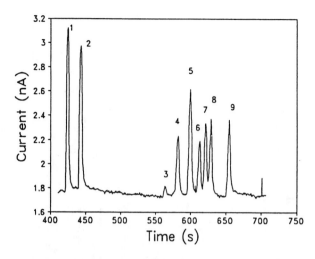

Figure 7.6 Electropherogram of carbohydrates with PAD. Experimental conditions: separation voltage, 30 kV over 10 μm × 60 cm capillary; electrode, 10 μm (in diameter) Au disk; electrode potential, 300 mV (vs. SCE) for 165 ms (sampling at 111 to 165 ms), 1200 mV for 55 ms, and −1000 mV for 165 ms; electrolyte, 0.1 M NaOH; electromigration injection, 30 kV for 3 s; sample concentrations, 1×10^{-4} M for inositol and 2×10^{-4} M for others. Peaks: 1 = inositol, 2 = sorbitol, 3 = unknown, 4 = maltose, 5 = glucose, 6 = rhamnose, 7 = arabinose, 8 = fructose, and 9 = xylose. (From Lu, W. and Cassidy, R. M., *Anal. Chem.*, 65, 2878, 1993. With permission.)

limit of 22.5 fmol. However, this detection limit was one to two orders of magnitude lower when a 10-μm disk gold electrode was utilized for the detection of carbohydrates separated in a 10-μm i.d. capillary.[65] Figure 7.6 shows the separation and detection of eight sugars using a gold working electrode. A linear working range was observed from 10^{-6} to 10^{-4} M for inositol.[65] Recently, Weber et al.[63] reported the applicability of PAD to the detection of complex carbohydrates such as glycopeptides with a detection limit of 2 μM (S/N = 3). Moreover, the same system was utilized for the analysis of carbohydrate-containing samples from a variety of biological sources.[64]

Very recently, La Course and Owens[66] reported an improvement in the design of PAD. They introduced the so-called integrated pulsed amperometric detection (IPAD) for the detection of carbohydrates after CE separation. In the case of regular PAD, the majority of the signal at high detection potentials originates from surface oxide formation, which is susceptible to solvent, ionic strength, temperature, and analyte effect, while the IPAD waveform utilizes a potential scan in the detection step which coulometrically rejects the substantial oxide formation signal. This led to overall improved baselines, eliminated oxide-induced artifacts, and enabled lower limits of detection.

Although the PAD approach has solved the problem of electrode poisoning usually encountered with the oxidation of carbohydrates at Pt or Au electrodes in the direct amperometric detection mode, the PAD detection system requires specialized pulse sequences, thus entailing expensive instrumentation. Also, other major drawbacks of PAD include problems involving charging currents and surface changes associated with potential pulsing, which do not allow the ultimate detectability to be achieved. An alternative to PAD is to use amperometric detection at a constant potential (ADCP). ADCP has long been proven a useful approach for the detection of electroactive species at the trace level using carbon electrodes. The principal difficulty encountered in this approach is that carbohydrates exhibit a large overpotential for oxidation at the carbon electrodes used in conventional liquid chromatography with electrochemical detection. This phenomenon drastically increases the potential required for the oxidation and thereby compromises both the selectivity and sensitivity of the detection. To overcome this problem, attention has been focused on the development of

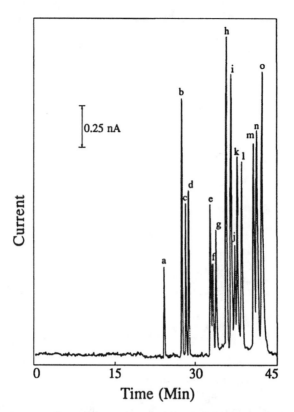

Figure 7.7 CZE/ADCP electropherogram of a mixture containing 15 different carbohydrates (80 to 150 μ*M*). Electrolyte, 100 m*M* NaOH; capillary, fused-silica, 73 cm (total length) × 50 μm i.d.; 10 s hydrodynamic injection, 10-cm height; voltage, 11 kV; analytes, a = trehalose, b = stachyose, c = raffinose, d = sucrose, e = lactose, f = lactulose, g = cellobiose, h = galactose, i = glucose, j = rhamnose, k = mannose, l = fructose, m = xylose, n = talose, o = ribose. (From Colón, L. A., Dadoo, R., and Zare, R. N., *Anal. Chem.*, 65, 476, 1993. With permission.)

new electrode materials[69] that permit the oxidation of carbohydrates at high pH and at relatively low potentials to provide optimum detector performance. In general, the Cu electrode was found to provide superior detection capabilities in terms of its range of response, detection limits, and especially stability, even in the 0.10 M NaOH at which the studies were performed. The Cu electrode had a detection limit of 3×10^{-8} M for glucose at +0.58 V (vs. Ag/AgCl) with a linear response range over four decades.[69]

ADCP with a Cu microelectrode was employed for the detection of carbohydrates after separation by CE.[20] The separation was performed in strongly alkaline solutions (i.e., pH 13) without prior derivatization or complexation with borate ions. The Cu microelectrode at +0.6 V (vs. Ag/AgCl) could be employed for hundreds of runs without deterioration. Because the pK's of most sugars are in the vicinity of 12 to 13, they are ionized at high pH and separated by CE under such conditions. Figure 7.7 illustrates the separation and detection of 15 different carbohydrates.

The limits of detection were calculated to be below 50 femtomoles for the 15 sugars studied with a linear dynamic range that extended over three orders of magnitude (e.g., μ*M* to m*M*). However, the reproducibility of this system is very low because of the difficulty associated with the electrode/capillary alignment during an electrophoresis run and from run-to-run. Recently, the design and characterization of a simple wall-jet electrochemical detector allowed the use of normal-size working electrodes, thus increasing the reproducibility of amperometric detectors without introducing significant peak broadening.[60] In this approach, a disk-shaped electrode consisting of metal wire with only its tip cross section exposed was positioned immediately in front of the capillary outlet. Detection was performed on the solution exiting the capillary and flowing radially across the face of the copper electrode. This design was shown to exhibit a 50-fold improvement in detection limit (ca. 1 fmol) over the conventional ED, and a five- to sixfold reproducibility improvement.[60]

So far, electrochemical detection of carbohydrates on a copper electrode has been focused almost exclusively on the detection of mono- and disaccharides. This was mainly because good CE separations have not been possible for underivatized polysaccharides in the simple aqueous metal-alkali hydroxide-based separation buffers employed with Cu electrodes. However, this problem was surmounted[59] by the inclusion of a cationic surfactant, such as cetyltrimethylammonium bromide (CTAB), in the running electrolyte. The surfactant CTAB provided an adequate differential migration of the analytes, and in addition reversed the direction of the EOF, thus permitting elution to be in order of increasing polysaccharide size. This method allowed the separation of samples containing 10 to as many as 1000 monosaccharide units. These samples included linear maltoses, enzymatically hydrolyzed starch, and commercially available dextrans of up to an average molecular weight of 18,300. Figure 7.8 illustrates the CE-ED analysis of maltooligosaccharide constituents of a commercial dextrin. The limit of detection for individual carbohydrates analyzed by this method was generally below the femtomole level.

A different approach for solving the problem associated with the alignment of the end of the separation capillary and the microelectrode is the use of an off-column mode of detection[70] instead of the end-column detection used by all the aforementioned approaches. A palladium-metal union was used as grounded cathode to decouple the high separation voltage before detection. With a second piece of capillary the field decoupler was connected to a homemade T-shaped detection cell. The solution is propelled by the EOF in the separation

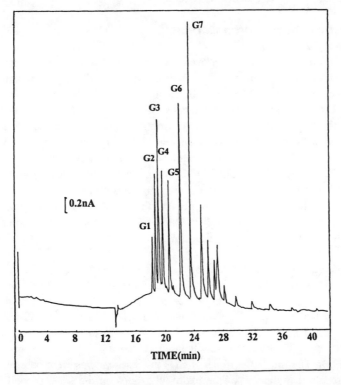

Figure 7.8 Electropherogram of Dextrin 15 (0.60 mg/mL). Running electrolyte, 100 mM NaOH with 10 mM CTAB; voltage, −10 kV; working electrode, 127 μm Cu disk at +0.60 V vs. Ag/AgCl. Numbers on the peaks reflect the number of glucose units. (From Zhou, W. and Baldwin, R. P., *Electrophoresis*, 17, 319, 1996. With permission.)

capillary *via* the small piece of coupling capillary toward the detector. Cuprous oxide-modified microelectrodes were utilized as the working electrodes at a constant potential of +0.6 V. One of the major drawbacks of this detection system is the zone broadening arising from the flow resistance of the coupling capillary. This flow resistance induces a back-pressure against the EOF in the separation capillary, thus distorting the uniform flow profile and consequently causing extra zone broadening. To preserve the flat EOF profile in the separation capillary, pressure compensation was applied. Under optimized conditions plate numbers of over 100,000 were obtained. Also, detection limits of 1 to 2 μM were found for various carbohydrates. An attractive feature of this detection mode is the ease of changing and installing capillaries and electrodes, thus improving reproducibility.

Regardless of whether the PAD or ADCP approach is used for the amperometric detection of carbohydrates, they both suffer from (1) the limitations imposed by the alkaline conditions needed for sensitive detection and differential electromigration which restrict the useful pH to a very narrow range, i.e., pH > 12, and (2) the nondiscriminative nature of amperometric detection which is known to yield a response not only for carbohydrates but also for other analytes including amino acids, peptides, organic acids, simple alcohols, and aliphatic amines,[67] a condition that may render peak assignment difficult and may lead to less accurate quantitative measurements. Nevertheless, HPCE-ED permitted the rapid determination of sugars in beverages,[20,61] the detection of glycopeptides,[63] the determination of glucose in human blood,[62] and the monitoring of the enzymatic oxidation of glucose with time.[61] Unfortunately, at the present time these electrochemical detectors are not available from commercial sources. This may become a reality as soon as further improvements are realized in their design.

B. Detection of Labeled Carbohydrates

1. Dynamically Labeled Carbohydrates

Recently, mixtures of α-, β-, and γ-cyclodextrins (CDs) were separated and detected by CE-LIF using 2-anilinonaphthalene-6-sulfonic acid (2,6-ANS) as the background electrolyte.[71] This detection scheme exploits the ability of cyclodextrins to form inclusion complexes with 2,6-ANS, thus allowing their simultaneous differential electromigration and LIF detection. A schematic diagram depicting this interaction is shown in Figure 7.9A. As the fluorophore complexed with the hydrophobic cavity of CDs, its fluorescence was enhanced, and consequently the CD-2,6-ANS adducts could be detected by the fluorescence increase as positive peaks (see Figure 7.9B). The LIF detection was performed with an argon ion laser operating at 363 nm for excitation, and the emission was collected at 424 nm. Under these conditions, the detection limits were determined to be 62, 2.4, and 24 μM for α-, β-, and γ-cyclodextrins, respectively. This may reflect that β-CD forms the strongest complex with 2,6-ANS. In fact, they eluted and separated in the order of increasing strength of complexation (see Figure 7.9B).

The same detection mechanism was also utilized for the UV detection of cyclodextrins and some of their derivatives, as well as the determination of formation constants of their inclusion complexes with UV absorbing labels.[72] Several background chromophores were utilized including benzylamine (λ_{max} = 210 nm), salicylic acid (λ_{max} = 230 nm), sorbic acid (λ_{max} = 254 m), and 1-naphthylacetic acid (NAA, λ_{max} = 222 nm). The CE analysis of α-, β-, and γ-cyclodextrins was attained in less than 18 min using 2 mM NAA or 5 mM sorbate solution (pH 12.2). The migration times of CDs in these systems were based on the charges acquired as a result of ionization of the hydroxyl groups on the CDs, as well as the charge acquired by the inclusion complex formed between the CDs and the background chromophore. Moreover, a mixture of α-, β-, dimethyl-β-, and trimethyl-β-CDs were also analyzed using

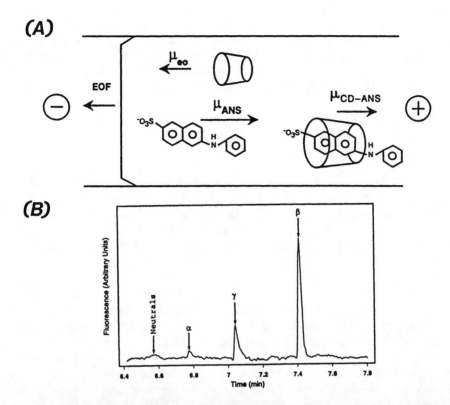

Figure 7.9 (A) Schematic diagram showing mobilities of 2,6-ANS, CD, and CD-2,6-ANS complex in a bare fused-silica capillary. (B) CE separation of α-, β- and γ-CD. Analysis buffer: 40 mM phosphate, pH 11.76, 1 mM 2,6-ANS. The analysis was carried out in a capillary of dimensions 50-μm i.d., 360-μm o.d., and 1 m in length in a field of 300 V cm^{-1}. The sample was introduced into the capillary by electrokinetic injection; 5 kV for 2 s from a sample containing 1.44 mg/mL α-CD, 0.017 mg/mL β-CD, and 0.24 mg/mL γ-CD. Detection was by fluorescence excited at 363 nm and monitored at 424 nm. (From Penn, S. G., Chiu, R. W., and Monnig, C. A., *J. Chromatogr. A,* 680, 233, 1994. With permission.)

50 mM salicylic acid or benzylamine solution (pH 6.0) as the background electrolyte. Again, the separations in these systems were based on the charges acquired by the CDs as a result of their inclusion complexation with the background electrolyte, which also facilitated their UV detection. The detection limits for the various CDs in the NAA and the salicylate systems were reported to be approximately 0.1 mM and 1 mM, respectively.

Very recently, large polysaccharides such as amylopectin and amylose were shown to complex with iodine, which provides both the charge needed for differential electromigration and the chromophore necessary for direct visible detection at 560 nm.[73] The primary basis for this process is the iodine binding affinity to carbohydrates, which can be manipulated through control of temperature and iodine concentration. (See Section IV.C for more details.)

The detection of GAGs at 240 nm as their copper (II) complexes is another example of dynamic labeling.[74] With the exception of keratan sulfate, which has no carboxylic groups that are the proposed sites for complexation with copper (II),[75] all other GAGs demonstrated the capability to bind to copper (II), thus forming a copper (II)-GAGs complex that has an absorbance maximum near 237 nm. Therefore, GAGs including heparin, heparan sulfate, chondroitin sulfate, dermatan sulfate, and hyaluronic acid were analyzed, as their copper (II) complexes, by reversed polarity CE at pH 4.7. Using this *in situ* labeling method, the highest detection sensitivity was achieved for GAGs having lower molecular weight as well as those having a high ratio of iduronic acid to glucuronic acid. This method has allowed the analysis of 10^{-9} g of copper (II)-heparin as well as the analysis of heparin oligosaccharides that lacked unsaturated uronic acid residues as a result of their preparation from heparin using controlled low pH nitrous acid depolymerization. For the carboxylated GAGs, the detection sensitivity was shown to be influenced by many factors including molecular weight, sulfation level, and type of uronic acid residues (glucuronic or iduronic acid).

2. Precolumn Derivatization

a. Important Criteria

Carbohydrates are generally tagged with a suitable chromophore or fluorophore to allow their detection at low levels. Since the tagging process brings about dramatic changes in the structure of the carbohydrate analytes, the selection of a suitable tag is crucial to the subsequent CE analysis of the analytes. This is due to the fact that derivatization not only allows sensitive detection but also imparts the derivatized solute with different structural properties. In this regard, it is preferred that the tag also supply the charge necessary for electrophoresis over a relatively wide range of pH or that the tag impart a hydrophobic character to the derivatives which allows the MEKC separation of derivatized carbohydrates. Other essential criteria for a successful precolumn derivatization include (1) high yields, (2) the formation of a single product for each species, (3) no detectable side products, (4) minimum sample work-out and cleanup, and (5) no cleavage of an essential sugar residue, e.g., sialic acid residue.

In precolumn derivatization reactions, it is generally preferred that the tagging occur at only one reactive functional group of the analyte, and it should be complete so that a single derivative is obtained in high yields. The polyhydroxy nature of sugars is attractive as far as the attachment of a tag to the molecule is concerned. This route to derivatization has been used extensively in gas-liquid chromatography in order to increase the volatility of carbohydrates and consequently facilitate their separation. However, derivatizing the hydroxyl groups would lead to multiple tagging of an analyte, and because hydroxyl groups vary in their relative reactivities, a distribution of derivatives rather than a single product would be obtained. In order to prevent multiple derivatization, other functional groups on the sugar molecule must be considered. The most popular sites for tagging include: (1) the carbonyl group in reducing sugars, (2) the amino group in amino sugars, and (3) the carboxylic moiety in acidic sugars. To produce a single product in a given precolumn derivatization, the tag must possess only one reactive site for attachment to the analyte. For UV detection, the tagging agent should exhibit a high molar absorptivity at a given wavelength with minimal interferences from the running electrolyte to ensure highly sensitive detection of the derivatized carbohydrates. Likewise, in fluorescence a tagging agent should exhibit high quantum efficiencies at a given excitation wavelength. These requirements become more crucial when dealing with the extremely small sample volumes encountered in nanoscale separation techniques such as CE.

b. Derivatization Schemes

Thus far, six different precolumn derivatization schemes have been introduced for the tagging of carbohydrates: (1) reductive amination (most widely used)[76,77] (Scheme 7.1); (2) condensation of carboxylated carbohydrates with aminated tags in the presence of carbodiimide[17,78-80] (Scheme 7.2); (3) base-catalyzed condensation between the carbonyl group of reducing carbohydrates and the active hydrogens of 1-phenyl-3-methyl-5-pyrazolone (PMP) or 1-(p-methoxy)phenyl-3-methyl-5-pyrazolone (PMPMP), forming bis-PMP and bis-

PMPMP derivatives, respectively[4,18,81,82] (Scheme 7.3); (4) reductive amination of reducing carbohydrates with amines to yield 1-amino-1-deoxyalditols followed by reaction with 3-(4-carboxybenzoyl)-2-quinolinecarboxaldehyde (CBQCA) in the presence of potassium cyanide[83,84] (Scheme 7.4); (5) reductive amination of reducing carbohydrates with amines to yield 1-amino-1-deoxyalditols followed by reaction with 5-carboxytetramethylrhodamine succinimidyl ester (TRSE)[85] (Scheme 7.5); and (6) condensation between the amine functional group in 5-dimethylaminonaphthalene-1-sulfonyl hydrazine (DHZ) with the aldehyde or ketone form of the sugar to form a stable imine or hydrazone product (Scheme 7.6).

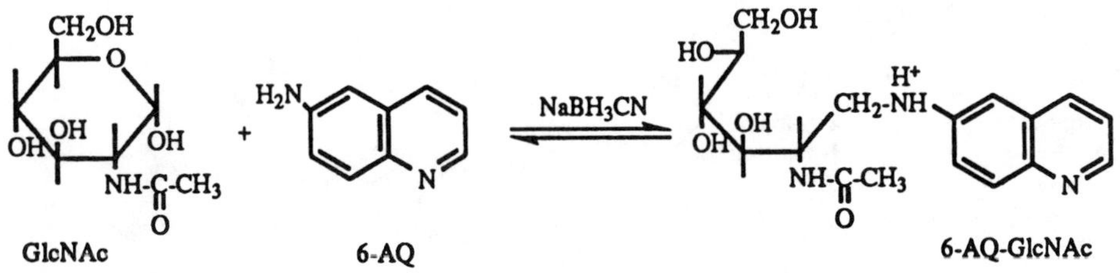

Scheme 7.1 Illustration of reductive amination where the reducing sugar is N-acetylglucosamine (GlcNAc) and the tag is 6-aminoquinoline (6-AQ).[86]

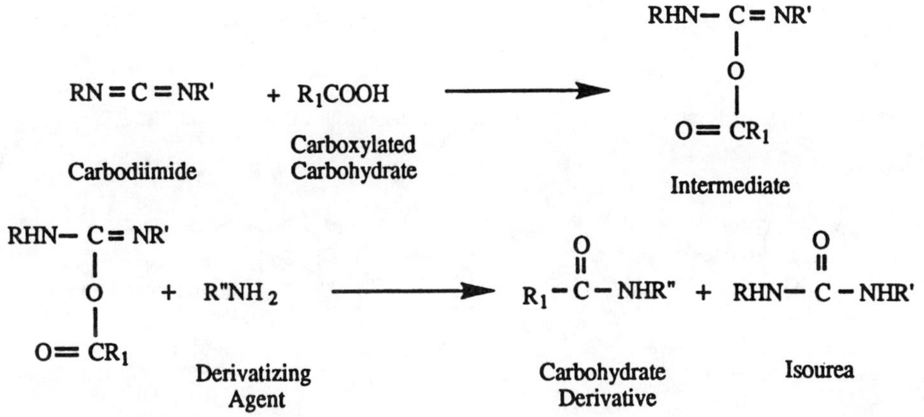

Scheme 7.2 Illustration of the selective precolumn derivatization of carboxylated carbohydrates via a condensation reaction between the carboxylic group of the saccharide and the amino group of the derivatizing agent in the presence of carbodiimide.[78]

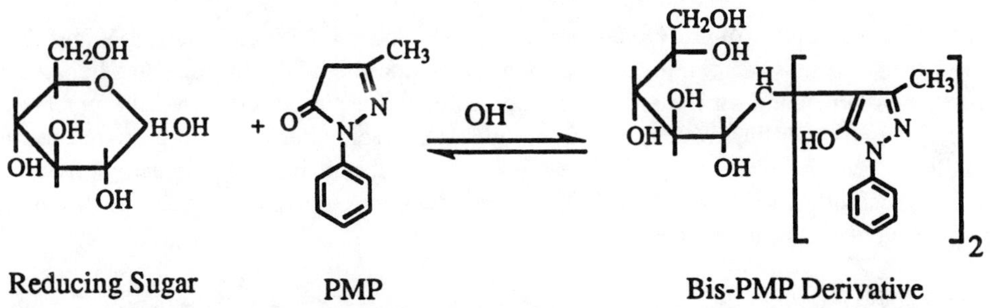

Scheme 7.3 Illustration of condensation reaction with PMP.[82]

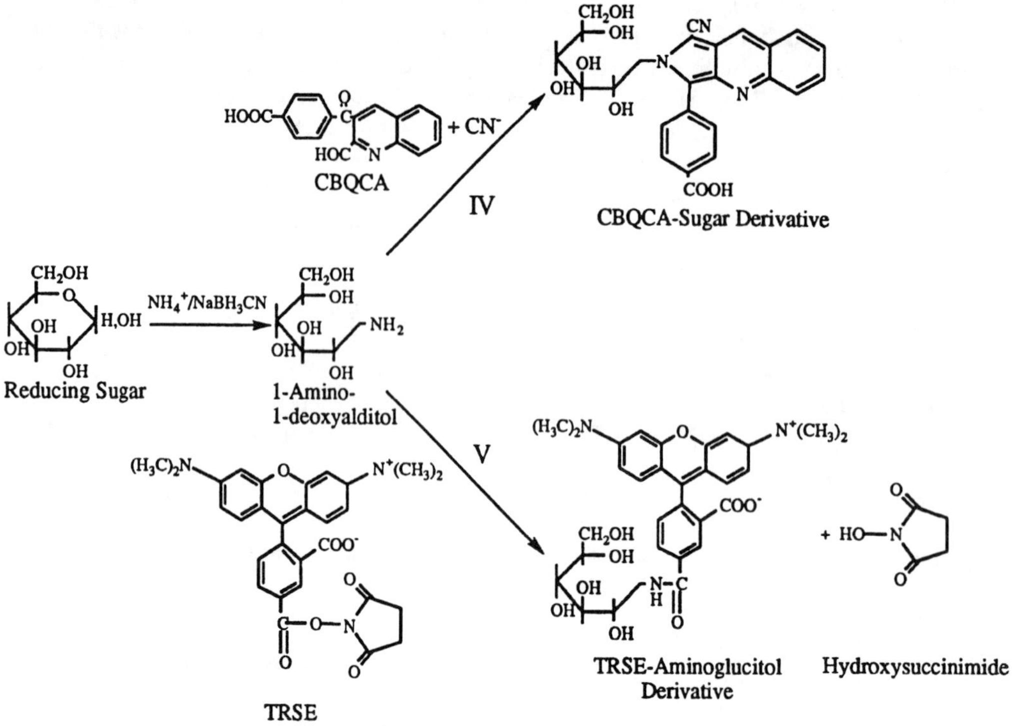

Schemes 7.4 and 7.5 Illustrations of precolumn derivatization with CBQCA[84] and TRSE,[85] respectively.

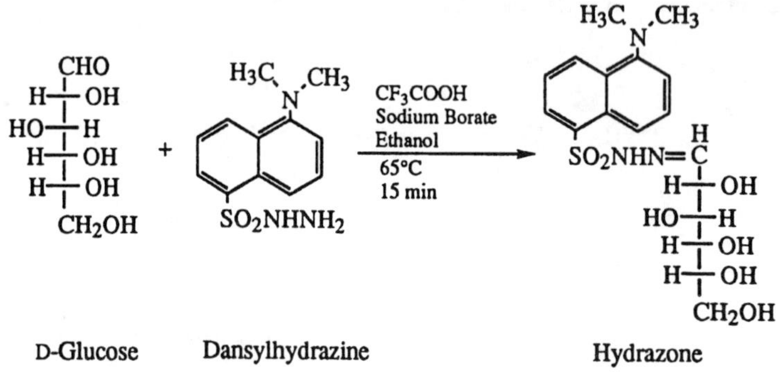

Scheme 7.6 Illustration of condensation reaction with dansylhydrazine.[87]

With the exception of precolumn derivatization in Scheme 7.3 which yields UV absorbing derivatives, all other precolumn derivatization can produce both UV absorbing and fluorescing derivatives depending on the spectral properties of the tag used. However, the precolumn derivatization according to Scheme 7.3 will yield fluorescing derivatives if a fluorescent tag similar to PMP in terms of chemical reactivity becomes available. While precolumn derivatization according to Schemes 7.4 and 7.5 involves two steps, the remaining are much simpler, requiring only one derivatization step.

c. Tagging Agents

Tables 7.2 and 7.3 list the tags that have been used so far for the precolumn derivatization of carbohydrates for their CE analysis. Table 7.2 includes tags which yield neutral or ionizable sugar derivatives, while Table 7.3 shows those yielding permanently charged sugar derivatives.

Table 7.2 Structures, Names, and Abbreviations of Tags Yielding Neutral and Ionizable
Sugar Derivatives

Name, abbreviation and number	Structure	Detection wavelengths and laser sources	Ref.
1-Phenyl-3-methyl-5-pyrazolone (PMP), **I**		$\lambda_{max} = 245$ nm	82
2-Aminoacridone (2-AA), **II**		$\lambda_{exc} = 425$ nm $\lambda_{em} = 520$ nm Argon ion laser 488 nm	157
p-Aminobenzonitrile, **III**		$\lambda_{max} = 285$ nm	90
2-Aminopyridine (2-AP), **IV**		$\lambda_{max} = 240$ nm $\lambda_{em} = 520$ nm He-Cd laser 325 nm	18,97
6-Aminoquinoline (6-AQ), **V**		$\lambda_{max} = 270$ nm $\lambda_{em} > 495$ nm He-Cd laser 325 nm	86,97
Ethyl-*p*-aminobenzoate, **VI**		$\lambda = 305$ nm	115

Table 7.2 Structures, Names, and Abbreviations of Tags Yielding Neutral and Ionizable Sugar Derivatives *(continued)*

Name, abbreviation and number	Structure	Detection wavelengths and laser sources	Ref.
S-(–)-1-Phenethyl-amine, **VII**	(structure)	$\lambda = 200$ nm	117
5-Dimethylamino-naphthalene-1-sulfonyl hydrazine (DHZ), **VIII**	(structure)	$\lambda_{em} = 500$ nm He-Cd laser 325 nm	87
3-(-4-Carboxybenzoyl)-2-quinolinecaboxyald-ehyde (CBQCA), **IX**	(structure)	$\lambda_{em} = 552$ nm Argon ion laser 457 or 488 nm He-Cd laser 442 nm	83,102 ,105
p-Aminobenzoic acid, **X**	(structure)	$\lambda_{max} = 285$ nm	116

Note: The spectral values listed in the table correspond to the sugar derivatives of the indicated tags and are extracted from the listed references. As can be seen in this table, most of the derivatives were excited at a wavelength dictated by the line of the laser source available, where in most cases the laser line yields only a fraction of the maximum absorption of the derivatives. The interested reader is advised to consult the listed references to find more details concerning UV and LIF detection of the various derivatives.

Besides their different spectral characteristics in terms of molar absorptivities and quantum efficiencies, and whether they are UV absorbing and/or fluorescing tags, the different tags listed in Tables 7.2 and 7.3 will yield sugar derivatives of varying electrophoretic mobility, separation efficiency, and selectivity. Limiting our discussion to labeling only "neutral" carbohydrates with the various tags, and distinguishing between the various schemes of precolumn derivatization listed above, we can state the following guidelines. PMP (tag I), which can only label carbohydrates according to Scheme 7.3, usually yields neutral sugar derivatives, which become negatively charged in aqueous basic solutions due to the partial dissociation of the enolic hydroxyl group of the PMP tag. As a result, the weakly ionized derivatives do not separate well, and borate-based electrolytes and/or micellar phases such as SDS are required to bring about differential electromigration, and ultimately, separation. Derivatization of neutral sugars with tags II and III according to Scheme 7.1 will also lead

Table 7.3 Structures, Names, and Abbreviations of Tags Yielding Permanently Charged
          Sugar Derivatives

Name, abbreviation, and number	Structure	Detection wavelengths and laser sources	Ref.
5-Carboxytetramethyl-rhodamine succinimidyl ester (TRSE), **XI**		$\lambda_{em} = 580$ nm He-Ne laser 543 nm	85,107
Sulfanilic acid (SA), **XII**		$\lambda_{max} = 247$ nm	78
2-Aminonaphthalene-1-sulfonic acid (2-ANSA), **XIII**		$\lambda = 235$ nm	138
5-Aminonaphthalene-2-sulfonic acid (5-ANSA), **XIV**		$\lambda = 235$ nm $\lambda_{em} = 475$ nm He-Cd laser 325 nm	97,138
7-Aminonaphthalene-1,3-disulfonic acid (ANDSA), **XV**		$\lambda_{max} = 247$ nm $\lambda_{exc} = 315$ nm $\lambda_{em} = 420$ nm Xenon-Mercury Lamp	78
4-amino-5-hydroxy-naphthalene-2,7-disulfonic acid (AHNS), **XVI**		$\lambda_{em} = 475$ nm He-Cd laser 325 nm	97

Table 7.3 Structures, Names, and Abbreviations of Tags Yielding Permanently Charged Sugar Derivatives *(continued)*

Name, abbreviation, and number	Structure	Detection wavelengths and laser sources	Ref.
3-Aminonaphthalene-2,7-disulfonic acid (3-ANDA), **XVII**		$\lambda = 235$ nm	138
8-Aminonaphthalene-1,3,6-trisulfonic acid (ANTS), **XVIII**		$\lambda_{exc} = 370$ nm $\lambda_{em} = 520$ nm He-Cd laser 325 nm	95,140
9-Aminopyrene-1,4,6-trisulfonic acid (APTS), **XIX**		$\lambda_{exc} = 455$ nm $\lambda_{em} = 512$ nm Argon ion laser 488 nm	98,99

Note: As in Table 7.2, the spectral values listed in this table correspond to the sugar derivatives of the indicated tags. Here also, most of the derivatives were excited at a wavelength dictated by the line of the laser source available, where in most cases the laser line yields only a fraction of the maximum absorption of the derivatives. For more details, see the listed references.

to neutral sugar derivatives, and again borate-based electrolytes and/or micellar phases are required for bringing about their separation. Sugars derivatized with tags IV, V, VI, and VII according to Scheme 7.1 can acquire a positive charge at acidic pH. In addition, tag VII is an enantiomeric reagent that allows the separation of sugar enantiomers. It should be mentioned that tag VI will yield derivatives that are not stable at high pH where the ester bond in the benzoate sugar derivative will rapidly undergo saponification to form the acid. Sugar derivatives of tags IV, V, VI, and VII obtained *via* Scheme 7.1 will electrophorese in borate buffers at alkaline pH. Since the sugar derivatives of tags IV, V, VI, and VII are neutral at basic pH, one can envision that they can be electrophoresed by MEKC at basic pH. Sugar derivatives of tag VIII (DHZ) obtained via Scheme 7.6, will carry a positive charge at acidic pH due to the protonation of the dimethylamino group of the tag. Again, since the sugar derivatives of DHZ are neutral at basic pH and carry a sizable hydrophobicity imparted by the tag, the DHZ-sugar derivatives can be electrophoresed in either a borate buffer or a micellar solution. CBQCA (tag IX), which is exclusively used in precolumn derivatization according to Scheme 7.4, will yield derivatives that are negatively charged at pH values above the pK_a of the carboxylic group. Tag X is an amphoteric solute, and therefore its sugar derivatives, labeled according to Scheme 7.1, will give derivatives that are positively charged below pH 3.0 and negatively charged above 3.8. TRSE (tag XI) sugar derivatives (Scheme 7.5) are positively charged at acidic pH and are zwitterions at neutral and basic pH. On the other hand, carbohydrates labeled with tags XII through XIX (Table 7.2) according to Scheme

7.1 are negatively charged over a wide range of pH due to their strong sulfonic acid groups and the very weak ability of their amino groups to become protonated (pK_a values ≤3). Thus, under a given set of conditions and for a given set of saccharides, different tagging agents will lead to sugar derivatives with different electrophoretic behavior and selectivity.

It should be emphasized that the sugar derivatives of all the tags will of course electrophorese at high pH in the presence of borate buffers. This will become clear as the discussion progresses, and more details will be provided in Section IV. Due to the presence of permanently ionized strong sulfonic acid groups in the tags listed in Table 7.3, their sugar derivatives will electrophorese over a wide range of pH. Although they do not require borate complexation to undergo differential electromigration, the sugar derivatives of the sulfonic acid-based tags (see Table 7.3) will certainly exhibit different electrophoretic behavior as borate complexes, and in turn different selectivity may be obtained.

d. Detection of the Derivatives and Some Features of the Various Precolumn Derivatizations

Due to the inherent high sensitivity of fluorescence, it is not surprising to see that laser-induced fluorescence (LIF) is the current trend for high-sensitivity detection of labeled carbohydrates after CE separation. Moreover, fluorescence is characterized by a good specificity and a relatively large linear dynamic range. However, one major drawback of fluorescence detection is the lack of satisfactory fluorophores in most analytes, a condition that dictates the chemical tagging of the analytes in order to achieve high sensitivity detection. This is especially true for carbohydrates. Also, another principal disadvantage of fluorescence as an analytical tool lies in its serious dependence on environmental factors, such as temperature, pH, ionic strength, and the presence of dissolved oxygen. Photochemical decomposition is rarely a problem when using fluorescence as a mode of detection because the analyte is only briefly exposed to the intense excitation radiation. Quenching, which is the reduction of fluorescence by a competing deactivating process resulting from the specific interaction between a fluorophore and another substance present in the system, can introduce significant errors when fluorescence is used as an analytical technique. The most noticeable forms of quenching in fluorescence detection are temperature, oxygen, and analyte concentration. As the temperature is increased, fluorescence decreases. This is due to the resulting increase in molecular motion and collisions which rob a molecule of energy through collisional deactivation. The change in fluorescence is typically 1% per 1°C. In CE, changes in the buffer composition, pH, ionic strength, and the operating voltage can all lead to large changes in temperature within the capillary unless the capillary temperature is controlled. For quantitative analysis using fluorescence as the mode of detection, care must be taken that the temperature at which the separation is performed is similar to that at which calibration was performed. Oxygen is another source of quenching. Oxygen, present in solutions at a concentration of 10^{-3} M, normally reduces fluorescence by 20%. In CE, oxygen quenching can be reduced or eliminated by degassing solvents and sample solutions before use. Concentration quenching causes many problems during a fluorometric determination. In order for fluorescence to be observed, absorption must occur first. When the absorption is too high, light cannot pass through the entire flow-cell to cause excitation. In order for a linear relationship to be observed between fluorescence and concentration, the absorbance must be kept below 0.05, otherwise a decrease in fluorescence can be observed. For a detailed description of the instrumentation used in fluorescence detection with HPCE, see Chapter 2 as well as recent reviews by Amankwa et al.[88] and Li.[89]

So far, precolumn derivatization by reductive amination (Scheme 7.1) has been the method of choice for labeling reducing saccharides. This has been facilitated by the availability of a large number of chromophores with amino functional groups as can be seen in Tables 7.2

and 7.3. Briefly, reductive amination of sugars involves the ring opening of the pyranose or furanose sugar to form the aldehyde or ketone, followed by a nucleophilic addition of the amino group of the labeling reagent to form a Schiff base (imine). Finally, reduction of this Schiff base by sodium cyanoborohydride results in the formation of a stable adduct as illustrated in Scheme 7.1. The yield of derivatization of carbohydrates by reductive amination as performed in solutions containing acetic acid was shown to depend on the acetic acid concentration.[76,90,91] This has prompted the investigation of the effect of the acid concentration as well as the pK_a of the acid used in the reductive amination of carbohydrates.[92] The organic acids were acetic, succinic, glycolic, L-malic, citric, malonic, and maleic acid, with pK_as in the range of 1.91 to 4.75. Generally, the efficiency of the acid catalyst increased with increasing acid strength. Consequently, derivatization of carbohydrates in the presence of organic acids of strength higher than acetic acid caused a higher yield (see Figure 7.10), an effect that is more pronounced for N-acetylamino sugars. This might be caused by the steric hindrance caused by the N-acetylamino group, thus affecting the ring opening or the Schiff base formation. Optimum yield was attained using citric acid as the catalyst and conversion of a few nanomoles of neutral saccharides to the APTS derivatives is achieved at 75°C in less than 60 min. The only drawback for using a strong acid at high concentration is the risk of removing sialic acid residues at the nonreducing ends of the oligosaccharides derived from glycoproteins.

The following discussion reviews some of the most important tags used in the labeling of carbohydrates by reductive amination and provides a brief description of the detectability of the sugar derivatives obtained by this labeling procedure. Typically, reductively aminated sugar derivatives of 2-aminopyridine (2-AP, tag IV) are usually detected in the UV at 240 nm at the 10 pmol level.[76] Also, these derivatives can be detected by fluorescence. 6-Amino-quinoline (6-AQ, tag V) is a useful tag for the detection and separation of carbohydrates by CE.[86] The 6-AQ derivatives showed a maximum absorbance at 270 nm, and the signal obtained was eight times higher compared to that obtained with 2-AP derivatives under otherwise identical conditions. The removal of excess 2-AP and 6-AQ is not required in CE since the derivatized carbohydrates migrate behind the excess tag, while in the case of acidic sugars, the derivatives and the derivatizing agent move in opposite directions.[93] Linhardt and co-workers[94] described the derivatization of N-acetylchitooligosaccharides with a negatively charged tag, 7-aminonaphthalene-1,3-disulfonic acid (ANDSA, tag XV) via reductive amination. The ANDSA-derivatives were excited at 250 nm using an arc lamp, and the fluorescence was collected at 420 nm with detection limits in the femtomole range. UV detection was also described for sugar derivatives of ANDSA.[94] ANDSA has a fairly strong absorbance at 247 nm ($\varepsilon = 3100$ M^{-1} cm^{-1}), but more important, it supplies the charge necessary for rapid electrophoretic analysis of sugars in CE. Chiesa and Horváth[95] utilized a similar tag, 8-aminonaphthalene-1,3,6-trisulfonic acid (ANTS, tag XVIII), for the derivatization and separation of maltooligosaccharides by CE which was first demonstrated in the derivatization of sugars separated in polyacrylamide slab gel electrophoresis.[96] ANTS not only supplies a strong chromophore, but also provides multiple negative charges even at low pH, which allows the rapid electrophoretic analysis. Using ANTS-derivatized glucose, as little as 15 femtomoles could be detected at 214 nm.[95] Using CE-LIF with a He-Cd laser at 325 nm, the limits of detection lie in the low attomole range, three orders of magnitude lower than in UV detection. Other aminonaphthalene sulfonic acid-based tags were introduced for the derivatization of carbohydrates by reductive amination[97] including the fluorescent tags 5-aminonaphthalene-2-sulfonic acid (5-ANSA, tag XIV) and 4-amino-5-hydroxynaphthalene-2,7-disulfonic acid (AHNS, tag XVI). A very recent and interesting development in fluorescent labeling of carbohydrates has been the introduction of APTS (tag XIX)[92,98-100] for the HPCE-LIF of mono- and oligosaccharides. (See Section IV for more discussions.)

The selective precolumn derivatization according to Scheme 7.2, which was introduced recently by El Rassi's research group for the derivatization of carboxylated carbohydrates,[17,78,80] offers several important features including (1) the formation of a stable amide

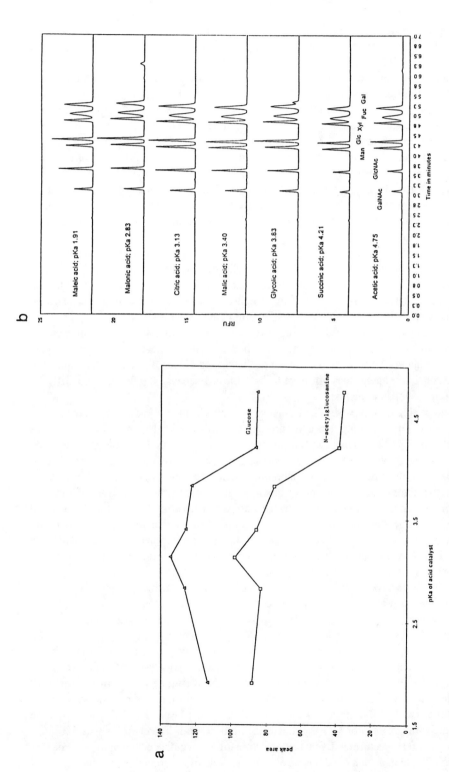

Figure 7.10. (a) The effect of pK$_a$ of organic acid on the adduct formation of APTS with GlcNAc and Glc at 75°C. (b) Electropherograms of seven APTS-derivatized monosaccharides in seven different organic acids. A mixture of 5 nmoles of each sugar was derivatized with 200 nmoles of APTS and 2 μmole NaBH$_3$CN in a 6 μL reaction mixture containing 0.6 M organic acid. The reaction mixtures were diluted with water to 200 μL and then a 25-fold dilution for CE-LIF analysis. Separation conditions: running electrolyte, 120 mM borate pH 10.2; voltage, 30 kV; current, 26 μA; capillary, 25 cm fused-silica × 19-μm i.d.; injection, 20 s pressure (0.5 psi); outlet cathode; excitation 488 nm; fluorescence emission collected at 520 nm. (From Evangelista, R. A., Guttman, A., and Chen, F.-T. A., *Electrophoresis*, 17, 347, 1996. With permission.)

bond between the amino group of the derivatizing agent and the carboxyl group of the carbohydrate molecule by acid-catalyzed removal of water in the presence of 1-ethyl-3-(3-dimethylaminopropyl) carbodiimide hydrochloride (EDAC), (2) the selective precolumn derivatization of sialylated saccharides at room temperature, thus avoiding the cleavage of the sialic acid residue from the carbohydrate molecule being derivatized, (3) quantitative yield as deduced from CE and mass spectrometry data,[78] and (4) the replacement of the weak carboxylic acid group of the saccharide by one or more strong sulfonic acid groups when tagging with sulfanilic acid (SA, tag XII) or 7-aminonaphthalene-1,3-disulfonic acid (ANDSA), a condition that allows the electrophoresis over a wide range of pH. Furthermore, using UV absorbance at 247 nm, the detection limits of acidic monosaccharides labeled with ANDSA or SA were 15 fmol or 30 fmol, respectively. For ANDSA derivatives, as low as 0.6 fmol could be detected with on-column fluorescence detection using a detector operated with a 200 W xenon-mercury lamp.[78]

The labeling procedure with CBQCA (Scheme 7.4, tag IX), which was first introduced by Novotny and co-workers,[83,101-103] allows the detection of submicromolar concentration of CBQCA-sugar derivatives by LIF. CBQCA-sugar derivatives have an excitation maximum at 456 nm, which conveniently matches the 442 nm line of an He-Cd laser, and an emission maximum near 552 nm. Using CBQCA, attomole levels of amino sugars have been analyzed by LIF-CE.[104] Very recently, Zhang et al.[105] utilized CBQCA for labeling aminated monosaccharides. The CBQCA-sugar derivatives were detected by LIF at the nanomolar concentration level. This limit of detection was achieved by utilizing a detection scheme that was based on a low scattering sheath flow cuvette as a postcolumn detector and two photomultiplier tubes that mutually exclude wavelength ranges to prevent the water Raman band from contributing to the background system. Such an arrangement had a limit of detection of 75 zeptomoles of fluorescently labeled 1-glucosamine.[106] One of the virtues of CBQCA derivatization is that the excess derivatizing agent does not have to be removed because the fluorogenic CBQCA does not interfere with the analysis.

The six most abundant hexoses (i.e., glucose, galactose, mannose, fucose, glucosamine, galactosamine) found in mammalian glycoproteins were derivatized with TRSE (tag XI) as shown in Scheme 7.5.[85] The detection limit of these derivatives was 100 molecules utilizing a post-column LIF detection in a sheath flow cuvette.[85] Although TRSE is a useful fluorescent label for aminated sugar monomers, the labeling reaction requires high sugar concentration to overcome the competition between the labeling reaction and the hydrolysis of the dye, thus rendering the analysis of some real sugar samples not feasible. In fact, the dye yielded two hydrolysis products as was reported by the authors.[85] In another report, six structurally similar oligosaccharides were labeled according to the aforementioned procedure and were baseline resolved by CE. Laser-induced fluorescence detection of these derivatized oligosaccharides allowed a detection limit of 50 molecules.[107]

The derivatization procedure illustrated in Scheme 7.6 was utilized to derivatize mono- and disaccharides directly with the fluorescent 5-dimethylaminonaphthalene-1-sulfonyl hydrazine (dansylhydrazine, DHZ, tag VIII), which reacts specifically with aldehydes and ketones. This derivatization procedure was originally performed by Avigad[108] for use in thin layer chromatography. Similar procedures were also utilized in the analysis of carbohydrates[109,110] and glycoproteins[111] by HPLC. The reaction was reported to be completed in only 15 min and allowed derivatization of sugars in the picomole range when performed in an organic medium or in a medium at high ratio in an organic solvent. The sugar-DHZ derivatives fluoresced at 530 nm when excited with a He-Cd laser at 325 nm. Under these conditions, the limit of detection of the derivatized mono- and disaccharides was 100 atto-moles. This method was utilized for the determination of glucose in small volumes of biological samples such as tear fluid. However, this derivatization procedure produces several by-products, which may interfere with the identification of the analyte peak.

IV. SEPARATION APPROACHES AND SELECTED APPLICATIONS

A. Monosaccharides

1. Underivatized Monosaccharides

Generally, CE analysis of intact monosaccharides has been performed mainly in the presence of either borate buffers or high pH electrolyte systems, which ensure the charging of the monosaccharides *via* borate complexation or ionization of the hydroxyl groups, respectively. The borate electrolyte systems have been shown compatible with direct UV detection at 195 nm[24] and RI detection.[112] The high pH electrolyte systems allowed indirect UV[51,52,54-56] or fluorescence[48] detection of monosaccharides as well as their electrochemical detection.[20,60-66,85,113]

As discussed above (see Section III), ED of monosaccharides requires highly alkaline pH electrolyte systems (e.g., alkali-metal hydroxide solutions, pH 12 to 13) for achieving high sensitivity detection.[20,60-66,85,113] The concentration and nature of the alkali-metal hydroxide largely affected the analysis time and the resolution of underivatized saccharides[20] (see Figure 7.2). When using ED for the sensing of underivatized saccharides, the addition of borate to bring about a better separation of closely related sugars was reported to decrease the anodic response which might be due to the reduced availability of oxidation sites present on carbohydrates.[20]

The utility of ED was also shown in the CE of charged monosaccharides including glucosaminic acid, glucosamine-6-sulfate, and glucosamine-6-phosphate.[62] Separation and detection were achieved by using 10 mM sodium hydroxide solution containing 8 mM sodium carbonate, pH 12. This electrolyte system allowed the determination of glucose in biological samples as complex as human blood after 1:50 dilution (85 μM); the only sample pretreatments were centrifugation and filtration.[62] Ye and Baldwin[61] reported the separation of mixtures containing glucose and galactose as well as their respective alditol and aldonic, uronic and aldaric acid derivatives using 50 mM sodium hydroxide and ED at a copper disk electrode. However, a higher sodium hydroxide concentration (250 mM) was used in order to achieve a baseline resolution of a set of eight alditols whose pK_a values are in the 13 to 14 range (see Figure 7.11). These separation and detection approaches have been applied to the analysis of the sugar contents of commercial apple juice as well as for monitoring the activity of glucose oxidase enzyme.[61]

Although they adversely affect the sensitivity of indirect UV or fluorescence detection, highly alkaline pH electrolytes are used in the detection schemes of underivatized saccharides to ensure optimum differential electromigration. The relatively low detection sensitivity of CE-indirect UV detection at highly alkaline pH is not an issue when the concentration of the saccharides to be determined is relatively high, such as the determination of the sugar content of fruit juices.[55]

While with borate buffers the separation of underivatized sugars is based on differences in the extent of complexation among the various solutes, the separation in the presence of high pH electrolyte systems can be explained by the lability of the proton of the monosaccharides, and by the charge-to-mass ratio. Using aqueous sodium hydroxide solution, pH 11.5, as the running electrolyte,[48] sucrose (pK_a = 12.51) was detected first, followed by glucose (pK_a = 12.35) and then fructose (pK_a = 12.03). This is the expected migration order when using positive polarity (cathodal EOF). The stronger acid fructose is moving at a higher velocity upstream against the EOF, thus eluting last. Glucose, a weaker acid, eluted before fructose. Sucrose, a disaccharide, moves upstream against the EOF at a lower velocity than glucose because of its higher molecular weight, thus eluting first. The influence of acidity and size of the molecule was also observed with a mixture of raffinose, deoxyribose, galactose,

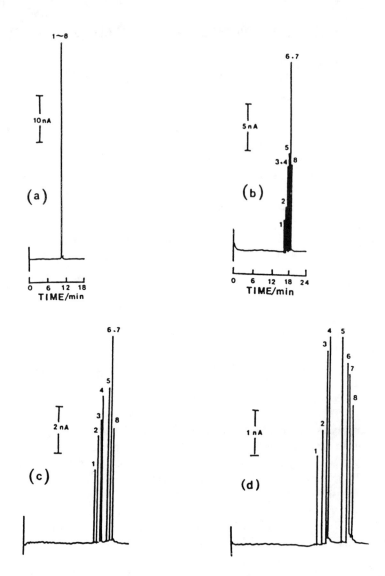

Figure 7.11 Electropherogram of alditol mixture. Conditions, running electrolyte, (a) 25 m*M* NaOH, (b) 100 m*M* NaOH, (c) 200 m*M* NaOH, and (d) 250 m*M* NaOH; working electrode, 100 µm Cu disk at + 0.60 V vs. Ag/AgCl; separation voltage, 15 kV; injection, by electromigration (15 kV for 3 s). Peaks identification: (1) ethylene glycol (1 m*M*), (2) glycerol (500 µ*M*), (3) inositol (160 m*M*), (4) erythritol (250 µ*M*), (5) galactitol (250 µ*M*), (6) adonitol (250 µ*M*), (7) glucitol (250 µ*M*), and (8) mannitol (250 µ*M*). (From Ye, J. and Baldwin, R. P., *J. Chromatogr. A,* 687, 141, 1994. With permission.)

glucose, and mannose.[51] They migrated in the order of increasing acidity, raffinose (a trisaccharide, pK_a = 12.74) < deoxyribose (pK_a = 12.52) < galactose (pK_a = 12.35) < glucose (pK_a = 12.35) < mannose (pK_a = 12.08).

Similarly, acidic monosaccharides can be separated from each other on the basis of their acidity differences using an electrolyte with a proper pH. In fact, two chiral 4-epimers, D-galactonic acid and D-gluconic acid, in an underivatized form could be separated with a resolution of 1.2 at pH values of 4.1 to 5.0. The addition of a chiral selector (β-cyclodextrin) did not improve the separation of the two aldonic acids.

2. Derivatized Monosaccharides

In most cases, the separation of closely related derivatized monosaccharides by CE may require the use of borate buffers even though the tag contains ionizable functional groups. For instance, twelve monosaccharides derivatized with 2-aminopyridine (2-AP) were separated as anionic borate complexes at pH 10.5 in about 25 min.[76] Obviously, the migration velocity of each derivative was primarily affected by the extent of complexation with borate. As typical examples, arabinose and ribose with *cis*-oriented hydroxyl groups at C3/C4 positions were more retarded than lyxose and xylose with *trans*-oriented hydroxyl groups. The same behavior was also observed with aldohexoses, e.g., galactose (*cis*) and glucose (*trans*), and hexuronic acids, e.g., galacturonic acid (*cis*) and glucuronic acid (*trans*). However, *N*-acetylhexosamines, e.g., *N*-acetylgalactosamine (*cis*) and *N*-acetylglucosamine (*trans*), showed the reverse effect concerning the 3,4-orientation of hydroxyl groups. This might be due to the contribution of the *N*-acetyl substituent at the C2 position.[76]

Monosaccharides derivatized with 2-AP can be migrated in electrophoresis at low pH in the presence of noncomplexing buffer (i.e., direct CZE) since the 2-AP tag imparts the analytes with a positive charge arising from the protonation of the amino group of the 2-AP-sugar derivative. However, due to equal charge-to-mass ratios, the 2-AP tagging of closely related monosaccharide isomers would only bring about group separation of the derivatives by direct CZE, i.e., in the absence of complex formation with a suitable complexing ion.[114] The positively charged 2-AP derivatives of galactose (Gal), mannose (Man), *N*-acetylglucosamine (GlcNAc), and *N*-acetylgalactosamine (GalNAc) were separated into groups using sodium phosphate electrolyte containing small amounts of tetrabutylammonium bromide, pH 5.0. In other words, 2-AP-Gal and 2-AP-Man differing only in the position of hydroxyl groups emerged together as one peak separated from the peak of 2-AP-GalNAc and 2-AP-GlcNAc, which have in their structure an additional acetyl group. The poor resolution of the 2-AP sugar derivatives at pH 5.0 may be due to the partial ionization of these rather weak bases. The 2-AP sugar mixture containing Man, Gal, GalNAc, and GlcNAc was resolved when a 200 mM borate buffer, pH 10.5, was used as the running electrolyte. Under these conditions, the 2-AP sugar derivatives are neutral, thus requiring a relatively high borate concentration in order to sufficiently complex them with borate and in turn allow their differential migration under the influence of an applied electric field.

Although monosaccharides derivatized with PMP are negatively charged in aqueous basic solutions, due to the dissociation of the enolic hydroxyl group, the PMP derivatives of isomeric aldopentoses or aldohexoses could not be separated because they have the same charge-to-mass ratio.[82] However, these PMP derivatives were readily converted to anionic borate complexes in a running electrolyte containing alkaline borate, and separated on the basis of the extent of complexation with borate. The mechanism of CE separation of the PMP-derivatives is the same as that of the 2-AP derivatives although the optimum pH for borate complexation was shifted to 9.5 as opposed to 10.5 in the case of 2-AP-derivatives, presumably because of the participation of the PMP substituent group in the complexation, perhaps through its hydroxyl group.[82] An electropherogram depicting a typical separation of a mixture of PMP-aldohexoses of the D-series is illustrated in Figure 7.12. When run alone, the aldopentoses were also well separated. However, the peaks of a few species of the pentose and hexose derivatives overlapped when a mixture of pentoses and hexoses was analyzed by CE.

Thirteen monosaccharides derivatized with CBQCA were separated by CE-LIF in less than 22 min with separation efficiencies that ranged from 100,000 to 400,000 per meter.[83] Besides tagging the sugar molecule with a fluorophore, the derivatization of monosaccharides with CBQCA provides each sugar derivative with an ionizable weak carboxylic acid group, thus allowing their electromigration. Therefore, the inclusion of only 10 mM borate in the

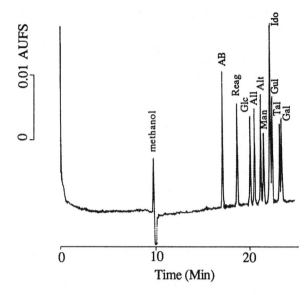

Figure 7.12 Separation of PMP-aldohexoses by CZE. Capillary, fused-silica tube, 63 cm (to the detection point), 78 cm total length × 50 μm i.d.; electrolyte, 0.2 *M* borate solution, pH 9.5; voltage, 15 kV; detection, UV at 245 nm. AB, amobarbital (internal standard); Reag, excess reagent (PMP); Glc, glucose; All, allose; Alt, altrose; Man, mannose; Ido, idose; Gul, gulose; Tal, talose; Gal, galactose. (From Honda, S., Suzuki, S., Nose, A., Yamamoto, K., and Kakehi, K., *Carbohydr. Res.*, 215, 193, 1991. With permission.)

running electrolyte (pH 9.4) was enough to magnify small steric differences between closely related isomers and bring about their separation by CE. Although the introduction of a negatively charged group would weaken the complex formation by virtue of coulombic repulsion, the extent of borate complexation with the hydroxyl groups of CBQCA sugar derivatives was still enough to produce differential migration among the various derivatized monosaccharides. Moreover, CBQCA was utilized for the derivatization of amino monosaccharides including D-glucosamine, D-galactosamine, 1-amino-1-deoxyglucosamine, 1-amino-1-galactosamine, 2-amino-2-deoxyglucosamine, 2-amino-2-galactosamine, 6-amino-6-deoxglucose, and D-galactosaminic acid.[104] The addition of borate to the running electrolyte was necessary to attain any separation between the CBQCA-D-glucosamine and -D-galactosamine which are structurally similar to each other, differing only in the orientation of a single hydroxyl group. Their baseline separation in borate buffer originates from their steric differences at the 4-hydroxyl group which permits their variable degree of complexation with borate.[104]

The concentration of borate in the running electrolyte required to bring about the separation of a given set of monosaccharide derivatives varies with the nature of the sugar derivatives being separated. For instance, the separation of six ANDSA-monosaccharides was best achieved in the presence of a 0.10 *M* borate buffer, pH 10 (see Figure 7.13).[79] The presence of two strong sulfonic acid groups in each derivative would weaken borate complexation to a much larger extent than in the preceding case (i.e., CBQCA-sugar derivatives) due to a higher coulombic repulsion. Under this condition, higher borate concentration is needed to overcome coulombic repulsive forces and bring about sufficient complexation with borate. The ANDSA-sugars shown in Figure 7.13 were obtained by the new and specific precolumn derivatization reaction for acidic monosaccharides which was recently introduced by Mechref and El Rassi (Scheme 7.2).[17,78,80] In addition to the improved detection sensitivity, the derivatization offered the advantage of replacing the weak carboxylic acid group of the sugar by the stronger sulfonic acid group of the tag, which is fully ionized at all pH. This allowed the electrophoresis of the sugar derivatives over a wide pH range and permitted the determination of acidic carbohydrates at the low femtomole levels by UV and fluorescence detection.[78]

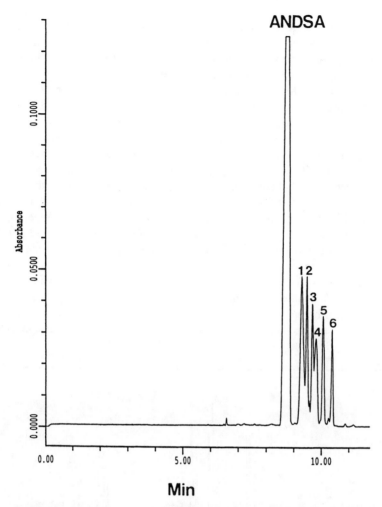

Figure 7.13 Electropherogram of ANDSA derivatives of acidic monosaccharides obtained on a dextran 150 kDa-coated capillary. Capillary, 47 cm total length (40 cm effective length) × 50 μm i.d.; electrolyte: 0.10 *M* borate, pH 10.0; pressure injection, 1 s; applied voltage, −15 kV; detection UV at 250 nm. Samples: 1 = D-glucuronic acid, 2 = D-glyceric acid, 3 = D-galactonic acid, 4 = D-galacturonic acid, 5 = D-gluconic acid, 6 = *N*-acetylneuraminic acid. (From Mechref, Y. and El Rassi, Z., *Electrophoresis,* 16, 617, 1995. With permission.)

A different situation was encountered in terms of the separation requirements when the derivatized monosaccharides had a sizable triply charged polyaromatic tag. This was the case of ten monosaccharides including *N*-acetylgalactosamine, *N*-acetylglucosamine, rhamnose, mannose, glucose, fructose, xylose, fucose, arabinose, and galactose which were derivatized with APTS (tag XIX) by the standard reductive amination procedure (Scheme 7.1) and subsequently separated by CE-LIF.[98,99] The various APTS-monosaccharide derivatives were more or less fully separated by four different buffer systems, but of course with different migration patterns and selectivities. The four buffer systems include: 135 m*M* borate, pH 10.2; 100 m*M* acetate buffer, pH 5.0; 120 m*M* Mops buffer, pH 7.0; and 50 m*M* sodium phosphate, pH 7.4.[99] Using 135 m*M* borate, pH 10.2, did not resolve APTS-arabinose and APTS-fucose. These two APTS derivatives were separated at much higher borate concentration but at the expense of much longer separation time. Again, the formation of anionic borate complexes with the APTS sugar derivatives may be hindered by the three negatively charged sulfonic acid groups of the APTS tag. With the exception of the borate buffer, where separation among the different APTS-monosaccharides is due to differences in borate complexation, the separations in the other buffers are based on the difference in the relative stereochemistry of the hydroxyl groups in aldoses which ultimately determines the hydrodynamic radius of the derivatized species, which is inversely proportional to the electrophoretic mobility. Increasing the Mops buffer concentration was shown to improve the resolution substantially; this might be due to the decrease in the EOF caused by increasing buffer concentration. As shown in Figure 7.14, the order of migration of the APTS derivatives with the Mops buffer (positive or normal polarity) was opposite that observed with the acetate buffer (negative or reverse polarity) and the analytes' migration toward the detector (at the cathode end) was due to the high cathodal EOF, which was strong enough to overcome the electrophoretic mobility of the analyte in the opposite direction. Phosphate buffer provided a good resolution and the migration order paralleled that obtained with the Mops buffer. In all cases, the analysis time in these buffer systems was shorter than that of the borate buffer system.[98] The tagging of sugars with APTS is an interesting development in the area of fluorescent labeling of carbohydrates for LIF detection.[98,99] APTS derivatization of monosaccharides not only provided a detection center for the monosaccharides with a substantially higher molar absorptivity and quantum efficiency than most of the commonly used fluorophore sugar derivatives, but also introduced three negatively charged groups which seem to ensure the separation of closely related structures such as those in Figure 7.14.

In the presence of borate in the running electrolyte, resolution and selectivity are largely affected by the nature of the tag. This may be due to a varying degree of destabilizing effect

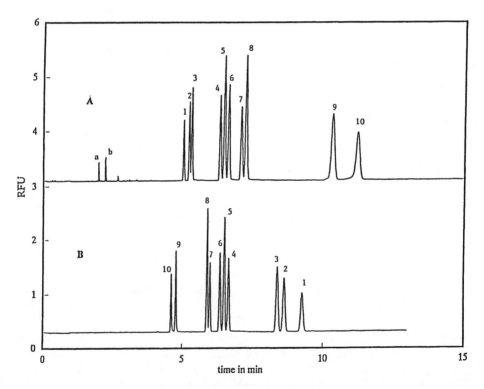

Figure 7.14 Electropherograms of 10 APTS-derivatized monosaccharides. Conditions: fused-silica capillary, 20 μm i.d. × 27 cm; laser source, 488 nm argon-ion, 2.5 mW; emission filter, 520 ± 20 nm and a notch filter at 488 nm; buffer, 100 mM sodium acetate, pH 5.0 in (A) and 120 Mops, pH 7.0 in (B); outlet, anode in (A) and cathode in (B); applied potential, 25 kV/14μA in (A) and 25 kV/19 μA in (B); peak identification, 1 = xylose, 2 = arabinose, 3 = ribose, 4 = fucose, 5 = rhamnose, 6 = glucose, 7 = galactose, 8 = mannose, 9 = N-acetylglucosamine, 10 = N-acetylglucosamine. (From Chen, F. and Evangelista, R. A., *Anal. Biochem.*, 230, 273, 1995. With permission.)

on the sugar-borate complex exerted by the tag. For example, glucose and mannose are well resolved when derivatized with ethyl 4-aminobenzoate[115] or 4-aminobenzonitrile.[90] However, their 4-aminobenzoic acid[52] or 2-aminobenzonitrile derivatives[90] are not resolved under the same separation conditions. Therefore, the selectivity and resolution of a certain system may be optimized not only by varying the pH or the borate concentration of the running electrolyte, but also by choosing the derivatizing agent most appropriate for the analysis of a given mixture of carbohydrates.

Precolumn derivatization of monosaccharides with p-aminobenzoic acid by reductive amination (Scheme 7.1) and their subsequent separation as borate complexes has been utilized for the quantitative determination of the monosaccharide composition of hemicellulose obtained from plant biomass by means of hydrothermal degradation.[116] The method was shown to be quantitatively accurate and reliable, and allowed the determination of the kind and quantity of glycosyl units present in hemicellulose.

Generally, and as discussed above, borate buffers find wide applicability in the separation of derivatized monosaccharides. The utility of borate complexation with derivatized monosaccharides has been recently applied to assist the enantiomeric separation of derivatized monosaccharides using borate buffer containing chiral selectors. D- and L-monosaccharides derivatized according to Scheme 7.1 with different fluorophores, namely 2-aminopyridine (2-AP, tag IV), 5-aminonaphthalene-2-sulfonic acid (5-ANSA, tag XIV), and 4-amino-5-hydroxynaphthalene-2,3-disulfonic acid (AHNS, tag XVI), were enantiomerically separated by CE as borate complexes in the presence of linear or cyclic dextrins.[97] 5-ANSA was shown to be the most suitable tag for the enantiomeric separation of the D- and L- forms of the carbohydrate derivatives when β-CD was used as the chiral selector.[97] Systematic studies on the effects of different CDs and modified β-CDs revealed the importance of the hydroxyl groups of the CD for chiral recognition since enantioselectivity was only observed with the underivatized CDs and hydroxypropyl-β-CD.

In another report, monosaccharides were also enantiomerically separated in a borate buffer after their derivatization with S-(–)-1-phenylethylamine[117] (tag VII). The derivatization followed the general reductive amination procedure (Scheme 7.1) and the derivatives were UV detected at 200 nm. The enantiomeric separation by this approach is based on the formation of diastereoisomers by the derivatization. Optimum enantiomeric separation of 16 aldohexoses was achieved using the running conditions described in the caption of Figure 7.15.[117]

Other approaches have been exploited in the CE separation of derivatized monosaccharides in order to provide an alternative to borate complexation. For instance, five PMP derivatives of monosaccharides (arabinose, ribose, galactose, glucose, and mannose) were analyzed by CE as complexes of divalent metal ions.[4] The five PMP-monosaccharides were resolved using an electrolyte system containing 20 mM calcium acetate. The separation in such a system is based on the relative ease of complexation of these derivatives with the metal ion, which depends on the orientation of their hydroxyl group (Section II.C). The migration order of the PMP-derivatives was PMP-ribose (erythro-erythro oriented hydroxyl groups), followed by PMP-lyxose (erythro-threo), then PMP-arabinose (threo-erythro), and finally PMP-xylose (erythro-threo). This complexation gives rise to a positive charge around the metal nucleus of the sugar–metal complexes, and consequently reduces the total negativity of the derivatives originating from the dissociation of the enolic hydroxyl group, thus migrating slower toward the anode than the unreacted reagent PMP. Due to electrostatic interaction between the divalent ions and the silanol groups on the walls of the capillary, a gradual inversion of the direction of the EOF from cathodal to anodal was observed. With anodal EOF (cathode to anode), the electrophoretic mobility of the derivatives is in the same direction as the EOF, a condition that leads to rapid separation. Other alkaline earth metal salts including barium, strontium, and magnesium acetate were also investigated. While the migration order of PMP-aldopentoses stayed the same regardless of the nature of the metal ion, as expected the electrophoretic mobility of the sugar–metal complexes was slightly higher with Ba^{2+} than

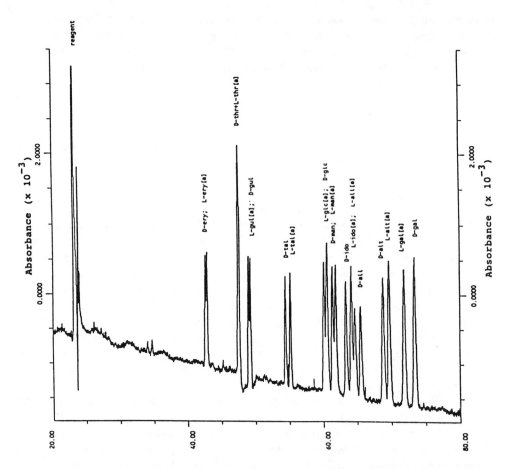

Figure 7.15   Separation of a mixture of 8 D-aldohexoses and 2 D-aldotetroses with rac-1-phenylethyl-
amine. Capillary: fused-silica, 107 cm (total length), 100 cm (to detector) × 50 μm i.d.;
electrolyte, 50 mM borate, pH 10.3 and 23% acetonitrile; voltage 30 kV; temperature, 25°C;
UV detection, 200 nm; injection, 3.0 s by pressure (3.45 kPa). In the electropherogram the
signals of derivatives of D-sugars with R-(+)-1-phenylethylamine are assigned as L-sugars
marked with [a]. They are enantiomers of the derivatives of L-sugars and S-(–)-1-phenyeth-
ylamine and have the same electrophoretic mobility in the separation system. (From Noe,
C. R. and Freissmuth, J., *J. Chromatogr. A*, 704, 503, 1995. With permission.)

with Sr^{2+} and Ca^{2+}, which is consistent with the fact that Ba^{2+} has a slightly larger ionic radius.
The separation efficiencies decreased in the following order: barium > calcium > strontium
>> magnesium.[4] Also, it seems that the binding of Ba^{2+} ions to the capillary surface is slightly
stronger than that of Ca^{2+}, while the binding of Sr^{2+} is the weakest. This was manifested by
a higher anodal EOF in the presence of barium acetate electrolyte, and consequently the
separation was faster.

   Although it is primarily performed to allow the high-sensitivity detection of the analytes,
a precolumn derivatization may also improve the compatibility of the solutes with certain
electrolyte systems. For instance, monosaccharides are generally too hydrophilic to be
solubilized in ionic surfactant-based micellar systems; however, their labeling with hydrophobic tags allows their separation by MEKC. In fact, eight different 2-aminoacridone (2-
AA, tag II) derivatized sugars (Scheme 7.1) namely, *N*-acetylglucosamine, galactose, mannose, fucose, glucose, *N*-acetylgalactosamine, ribose, and lyxose were separated by MEKC
in the presence of an electrolyte consisting of sodium taurodeoxycholate and sodium borate.[118]
The mechanism of migration of the 2-AA labeled monosaccharides in the MEKC system
was described as complex and may involve partition equilibria of the neutral species into

the micellar surface and electrophoretic movement of these molecules complexed with borate. This same system was also employed in the enantiomeric separations of 2-AA-galactose, -fucose, and -ribose enantiomers by the addition of β-CD to the sodium taurodeoxycholate and sodium borate buffer system. However, baseline resolution was only achieved for fucose enantiomers, and the analysis time was long (about 43 min). The replacement of taurodeoxycholate by a nonchiral surfactant such as sodium dodecyl sulfate (SDS) did not cause the enantiomeric separation.[118]

Another precolumn derivatization, which seems to impart hydrophilic saccharides with sufficient hydrophobicity to allow their separation by MEKC, is the PMP labeling (Scheme 7.3, tag I).[119] This precolumn labeling, which was first introduced by Honda et al.,[18] yields bis-PMP-adducts that strongly absorb in the UV. Sixteen PMP derivatives of monosaccharides and disaccharides were shown to separate quite nicely by MEKC (see Figure 7.16). The applicability of the SDS micellar system was extended to the identification and quantitation of monosaccharides obtained from carbohydrate hydrolyzates from glycoproteins.[119]

4-Aminobenzonitrile (a neutral derivatizing agent, tag III) is a third hydrophobic tagging agent that has been used to derivatize fourteen different saccharides including monosaccharides by reductive amination (Scheme 7.1), thus allowing their separation in about 5 min by

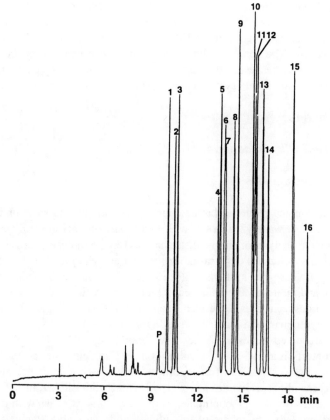

Figure 7.16 MEKC of standard mixture of 16 PMP-derivatized mono- and disaccharides. Capillary, 50 cm (to detection point)/72 cm (total length) × 50 μm i.d.; temperature, 30°C; running electrolyte, 30 mM Tris-phosphate, pH 7.5, containing 50 mM SDS; voltage, 20 kV; current, ~29 μA. Peak identification: P = PMP, 1 = mannose, 2 = rhamnose, 3 = lyxose, 4 = N-acetylmannosamine, 5 = lactose, 6 = maltose, 7 = cellobiose, 8 = glucose, 9 = glucosamine, 10 = mannosamine, 11 = xylose, 12 = altrose, 13 = galactose, 14 = fucose, 15 = galactosamine, 16 = 2-deoxy-D-ribose. (From Chiesa, C., Oefner, P. J., Zieske, L. R., and O'Neill, R. A., *J. Cap. Elec.*, 2, 175, 1995. With permission.)

MEKC using an SDS micellar phase.[90] The separation is based on the differential distribution of the neutral derivatives between the aqueous mobile phase and the micellar phase. While hydrophilic carbohydrate derivatives partitioned slightly inside the micelle and migrated first, hydrophobic derivatives such as deoxyaldohexoses partitioned strongly in the micelle and migrated very slowly toward the detector. Derivatives with intermediate hydrophobicity migrated within this migration time window. The final migration order reflected the strength of partition in the micellar phase which increased in the order ketohexoses < aldohexoses < aldopentoses < deoxyaldohexoses.[90] This migration order differs from that observed for PMP derivatized mono- and disaccharides using similar separation conditions.[119] While for 4-aminobenzonitrile derivatives the migration order was lactose, cellobiose, maltose, mannose, glucose, galactose, rhamnose, lyxose, fucose, and xylose, for PMP-derivatives the order was mannose, rhamnose, lyxose, lactose, maltose, cellobiose, glucose, xylose, galactose, and fucose. Therefore, in MEKC, and other electrolyte systems, the separation selectivity is strongly dependent on the nature of the derivatizing agent used.

In addition to the aforementioned MEKC systems, a cationic micellar phase composed of cetrimide hydroxide was also used for the CE analysis of 4-nitro-, 4-amino- and phenyl-substituted carbohydrates.[57] The selectivity of the cetrimide micellar system was compared to that of a highly alkaline buffer, borate buffer, and highly alkaline buffer containing a cationic polymer, e.g., polybrene. The MEKC system allowed the separation of anomers and element analogs. The migration order of the analytes in the highly alkaline buffer containing polybrene was different from that obtained in the MEKC system, which is due to the difference in the separation mechanism. The borate system provided the highest selectivity, but the peak efficiency was slightly lower because of the slow kinetics of the borate complexation. Nevertheless, the high selectivity of the borate system compensates for the low separation efficiency.

B. Oligosaccharides

1. Underivatized Oligosaccharides

a. Simple Oligosaccharides

As their underivatized monosaccharide counterparts, simple and short underivatized oligosaccharides have been analyzed in CE with highly alkaline pH electrolytes.[20,54,60] Typical examples include the analysis of simple disaccharides (e.g., trehalose, sucrose, lactose, lactulose, cellobiose), trisaccharides (e.g., raffinose), and tetrasaccharides (e.g., stachyose) by CE-ED using high pH electrolyte systems[20,60] (see Figure 7.7). In another approach, the disaccharides sucrose and maltose were analyzed at high pH by CE-indirect UV detection.[54] Very recently, other oligosaccharides have also been analyzed at high pH by CE-indirect UV detection using tryptophan as a marker [56] (see Figure 7.5).

Various underivatized oligosaccharides have been analyzed by CE using borate buffer and direct UV detection at 195 nm.[24] Arentoft et al.[120] also demonstrated the CE separation of underivatized oligosaccharides of the raffinose family as borate complexes using direct UV detection. These oligosaccharides are α-(1→6)-galactosides linked to C-6 of the glucose moiety of sucrose. Raffinose is the template of this homologous series with only one galactoside unit attached. Other members of this homologous series are stachyose, verbascose, and ajugose, which are formed by the successive binding of one, two, and three additional α-galactoside units, respectively, to C-6 of the terminating galactose of raffinose. The influence of various separation conditions including voltage, pH, temperature, and buffer composition on resolution, separation efficiency, migration time, and quantitative aspects were examined. Optimum separation in terms of resolution, separation efficiency, and acceptable analysis time was attained using 100 mM Na$_2$B$_4$O$_7$, pH 9.9, as the running electrolyte, 50°C and 10 kV. The importance of these oligosaccharides originates from the fact that their percentage in legume seeds reflects the quality or nutritive value of these seeds.[120]

b. Glycosaminoglycan-Derived Oligosaccharides

Glycosaminoglycans (GAGs), also called mucopolysaccharides, are unbranched polysaccharides of alternating uronic acid and hexosamine residues. They occur naturally in the cartilage and other connective tissues, which are collectively called the ground substance. GAGs exhibit a variety of biological functions and can be altered in disease states.[121,122] The most common GAGs include hyaluronic acid, chondroitin sulfates (chondroitin-4-sulfate and chondroitin-6-sulfate), dermatan sulfate, keratan sulfate, heparin, and heparan sulfate, which differ from each other in their template disaccharides. While the amino group of the hexosamine residue is either N-acetylated or N-sulfated, the uronic acid may be either D-glucuronic acid or L-iduronic acid. Moreover, the repeating disaccharide units (i.e., uronic acid-hexosamine disaccharide) for all GAGs except the nonsulfated hyaluronic acid are O-sulfated to varying degrees at C6 and/or C4 of the various glucosamine residues and at C2 of the uronic acid residues. This sulfation adds to the structural complexity of GAGs.

Because of their large size (average molecular weight 10,000 to 100,000) and structural complexity, GAGs are first broken down to oligosaccharides and then subjected to CE analysis. This involves the enzymatic degradation of GAGs with polysaccharide lyases, e.g., heparinases, chondroitinases, etc., thus yielding disaccharides bearing unsaturated uronic acids at the C4-C5, which allow their direct UV detection at 232 nm. Due to their inherent negative charge, GAG-derived disaccharides thus obtained can be readily analyzed by CE.

Capillary electrophoresis proved useful in the separation and quantitative determination of the disaccharides derived from chondroitin sulfate, dermatan sulfate, and hyaluronic acid.[35,34] Exhaustive treatment of these GAGs with polysaccharide lyases released nine different disaccharides bearing unsaturated uronic acids (Figure 7.17) with a molar absorptivity

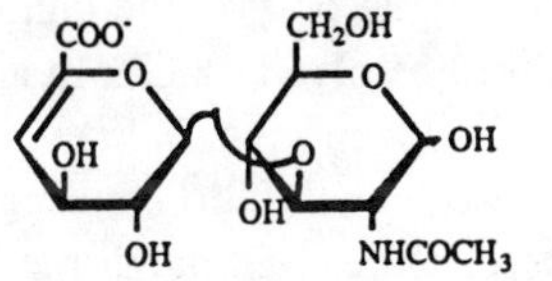

1 [Δdi-HA]

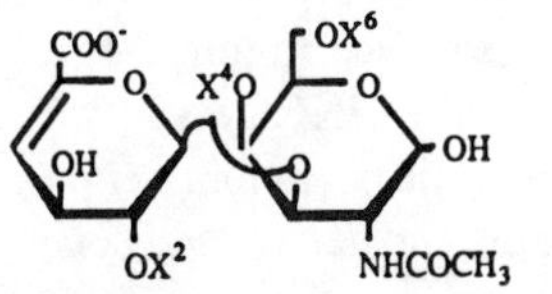

2 [Δdi-0S] where X^2, X^4, X^6 = H
3 [Δdi-6S] where X^2, X^4 = H, X^6 = SO_3^-
4 [Δdi-4S] where X^2, X^6 = H, X^4 = SO_3^-
5 [Δdi-UA2S] where X^4, X^6 = H, X^2 = SO_3^-
6 [Δdi-S_E] where X^4, X^6 = SO_3^-, X^2 = H
7 [Δdi-S_D] where X^2, X^6 = SO_3^-, X^4 = H
8 [Δdi-S_B] where X^2, X^4 = SO_3^-, X^6 = H
9 [Δdi-triS] where X^2, X^4, X^6 = SO_3^-

Figure 7.17 Disaccharides derived from chondroitin sulfates, dermatan sulfate, and hyaluronic acid by enzymatic depolymerization.

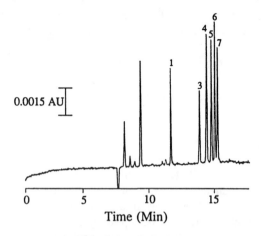

Figure 7.18 Electropherogram of oligosaccharides derived from hyaluronan by digestion with testicular hyaluronidase. Applied voltage, 15 kV; temp., 40°C; electrolyte, 40 m*M* phosphate containing 40 m*M* SDS and 10 m*M* borate, pH 9.0; capillary, 72 cm × 50 μm i.d.; the column was monitored at 200 nm. Peak 1 is the unsaturated disaccharide of hyaluronan (Δdi-HA), peaks 3, 4, 5, 6, and 7 are the saturated hexa-, octa-, deca-, dodeca-, and tetradecasaccharide of hyaluronan, respectively. (From Carney, S. L. and Osborne, D. J., *Anal. Biochem.*, 195, 132, 1991. With permission.)

of 5000 to 6000 M^{-1} cm^{-1} at 232 nm.[42] These disaccharides, having a net charge from –1 to –4, were well resolved by CE using a borate buffer, pH 8.8, primarily on the basis of their net charge and to a lesser extent on the basis of charge distribution. The use of phosphate buffer, in place of borate buffer, pH 8.8, yielded an increase in the separation time and a decrease in resolution. As expected, the nonsulfated disaccharides 1 and 2 eluted first from the capillary, followed by the monosulfated 3-5, the disulfated 6-8, and the trisulfated 9 disaccharides (for structures, see Figure 7.17). Despite the fact that the two nonsulfated disaccharides, structures 1 and 2, differed only in the chirality at the C4 in the hexosamine residue, these disaccharides were well resolved by CE. In another study,[34] the electrophoretic behavior of chondroitin disaccharides (see Figure 7.17) was investigated under various conditions including different pH, borate concentration, buffer ionic strength, and the inclusion of SDS micelles in the running electrolytes. Although the disaccharides are highly charged and too polar to partition in the SDS micelles, the presence of SDS in the electrolyte improved the resolution of the electrophoretic system. Also, the SDS-based electrolyte proved useful in the rapid separation (about 15 min) of six oligosaccharides derived from hyaluronan by digestion with testicular hyaluronidase (see Figure 7.18).

The utility of the borate buffer and MEKC system was exploited under various operating conditions in studying the electrophoretic behavior of eight commercial disaccharide standards derived from heparin, heparan sulfate, and derivatized heparins of the structure ΔUA2X(1→4)-D-GlcNY6X (where ΔUA is 4-deoxy-α-L-threo-hex-4-enopyransyluronic acid, GlcN is 2-deoxy-2-aminoglucopyranose, S is sulfate, Ac is acetate, X may be S, and Y is S or Ac).[36] Heparin and heparan sulfate are structurally similar, differing primarily in the relative content of *N*-acetylglucosamine, *O*-sulfation, and glucuronic acid. Using heparinases I, II, and III as the degrading enzymes, heparin and heparan sulfate can be depolymerized through an eliminative mechanism to yield eight different disaccharides shown in Figure 7.19.[36]

Using a borate buffer pH 8.8, two of the standard heparin/heparan sulfate disaccharides, having an identical charge of -2, ΔUA2S(1→4)-D-GlcNAc (structure 3) and ΔUA(1→4)-D-GlcNS (structure 4), were not fully resolved. The resolution of these two saccharides could be improved by preparing borate buffer in deuterated water or eliminating boric acid. Surprisingly, baseline resolution was achieved in SDS micellar solution in the absence of buffer. Since the two saccharides (structures 3 and 4, see Figure 7.19) are charged and polar, it is

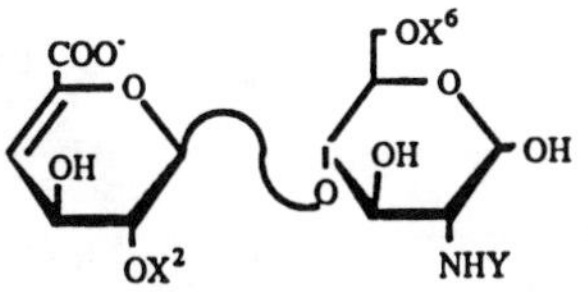

1. $X^2 = X^6 = H$, $Y = Ac$
2. $X^2 = H$, $X^6 = SO_3^-$, $Y = Ac$
3. $X^2 = SO_3^-$, $X^6 = H$, $Y = Ac$
4. $X^2 = X^6 = H$, $Y = SO_3^-$
5. $X^2 = X^6 = SO_3^-$, $Y = Ac$
6. $X^2 = H$, $X^6 = Y = SO_3^-$
7. $X^2 = SO_3^-$, $X^6 = H$, $Y = SO_3^-$
8. $X^2 = X^6 = Y = SO_3^-$

Figure 7.19 Disaccharide fragments obtained by enzymatic depolymerization of heparin and heparan sulfate using heparinases I, II, and III.

unlikely that the separation of these solutes was caused by differential partitioning in the SDS micelles. These electrophoretic systems were then applied to the determination of disaccharide composition of porcine mucosal heparin and that of bovine kidney heparan sulfate. Both GAGs were found to have an equimolar content of disaccharide $\Delta UA2S(1\rightarrow4)$-D-GlcNAc (structure 3) and $\Delta UA(1\rightarrow4)$-D-GlcNS (structure 4). These analyses required 15 ng of polysaccharides and a 20-min analysis time, and the results were comparable to those obtained by the more established strong anion exchange HPLC which requires 40 µg sample and a 90-min analysis time.[36]

Another important application of CE in the area of GAGs has been in the quality control of natural and synthetic heparin fragments.[32] Because of the anti-blood-clotting activity of heparin, the production of natural and synthetic heparin fragments for pharmaceutical use relies on the availability of analytical procedures for the efficient characterization of intermediates and final products. Using a low pH electrolyte system, namely 0.2 M phosphate, pH 4.0, and controlling the capillary column temperature at 40°C allowed the separation of the nine most common heparin disaccharides in a reversed polarity mode of separation (for structures, see Figure 7.19), the mapping of the oligosaccharides derived from heparin after heparinase treatment, and the assessment of the quality of synthetic heparin pentasaccharide preparations.[32] The separation in this system is based on differences in charge-to-mass ratios and in the hydrodynamic radius of the analytes.

The highly UV absorbing Δ-disaccharides derived from heparin and heparan sulfate by enzymic degradation with heparinase and both heparinases II and III were analyzed by CE using reversed polarity electric field and 15 mM phosphate buffer at pH 3.5.[123] Under this condition, the separation of all 12 known disaccharides (nonsulfated, monosulfated, disulfated, and trisulfated) carrying N-acetylated, N-sulfated, or unsubstituted glucosamine was achieved in a single run of 15 min (Figure 7.20).

Recently, CE has been shown useful in determining structural differences between heparin and heparan sulfate. This was achieved by first enzymatically degrading heparin and heparan sulfate to their corresponding disaccharides using heparinase I, II, and III, followed by the determination of the disaccharide composition of the digest by CE[124] using sodium phosphate buffer, pH 2.5, as the running electrolyte and a negative polarity applied electric field. The resulting electropherograms of the digested samples were interpreted with the aid of a reference mixture of disaccharide standards derived from heparin.[124] Also, Carney et al.[125] have used various electrophoretic techniques including CE to analyze and determine the fine structural functions of GAGs that are essential for antibody binding. CE was particularly

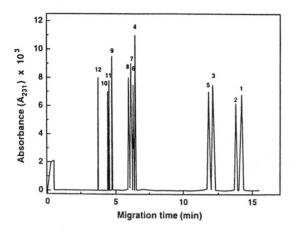

No	Disaccharide	R^2	R^6	Y
	Non-sulfated			
1.	aΔdi-nonS$_{HS}$	H	H	Ac
2.	Δdi-nonS$_{HS}$	H	H	H
	Monosulfated			
3.	aΔdi-mono6S$_{HS}$	H	SO$_3^-$	Ac
4.	Δdi-mono6S$_{HS}$	H	SO$_3^-$	H
5.	aΔdi-mono2S$_{HS}$	SO$_3^-$	H	Ac
6.	Δdi-mono2S$_{HS}$	SO$_3^-$	H	H
7.	Δdi-monoNS$_{HS}$	H	H	SO$_3^-$
	Disulfated			
8.	aΔdi-di(2,6)S$_{HS}$	SO$_3^-$	SO$_3^-$	Ac
9.	Δdi-di(2,6)S$_{HS}$	SO$_3^-$	SO$_3^-$	H
10.	Δdi-di(2,N)S$_{HS}$	SO$_3^-$	H	SO$_3^-$
11.	Δdi-di(6,N)S$_{HS}$	H	SO$_3^-$	SO$_3^-$
	Trisulfated			
12.	Δdi-tri(2,6,N)S$_{HS}$	SO$_3^-$	SO$_3^-$	SO$_3^-$

a = acetylated

Figure 7.20 (A) Typical electropherogram showing the separation of 12 heparin- and heparan sulfate-derived D-disaccharides obtained with a disaccharide standard mixture (1 ng/mL, injection 15 nL). Conditions: running electrolyte, 15 mM sodium dihydrogen orthophosphate buffered, pH 3.50; voltage, –20 kV; temperature, 25°C. Peaks identification, as in (B). (From Karamanos, N. K., Vanky, P., Tzanakakis, G. N., and Hjerpe, A., *Electrophoresis*, 17, 391, 1996. With permission.)

useful in the determination of the sulfation of GAGs and the sulfation of partly desulfated GAGs produced by methanolysis. This is particularly important since monoclonal antibodies developed against chondroitin sulfate have been an important tool in providing information relating to the variation of GAG structure during tissue development and in many diverse pathologies ranging from colon carcinoma to arthritis.[126,127]

In addition to the above applications, CE has been utilized for the elucidation of the structural differences between different low molecular weight (LMW) heparins (e.g., Fraxiparine, Fluxum, Fragmin, Sandoparin, and Enoxaparin) using oligosaccharide compositional analysis.[33] This was accomplished after complete depolymerization of heparin and LMW heparins with a mixture of heparin lyase I, II, and III followed by CE analysis using 10 mM sodium borate buffer, pH 8.81, containing 50 mM SDS. According to the authors,[33] the major mode of separation for such a system is zone electrophoresis, while the MEKC mode, resulting from the presence of SDS, is a minor contributor. As determined by CE, the oligosaccharide composition for the different LMW heparins varied, suggesting that LMW heparins have a significantly different proportion of antithrombin III binding sequence, which may explain their different biological activity. Although the identity of all cleaved components could not be elucidated because of the lack of standards, this method provides a reproducible fingerprint or map for a given heparin, and might be useful in resolving subtle composition differences between commercially available LMW heparin preparations. It can also be used in controlling the quality of these pharmaceutical compounds. It should also be noted that quantitative analysis was conveniently performed on 10 pmole of the oligosaccharide sample.

As discussed above, the addition of SDS to the running electrolyte at concentration above the cmc produced little change in the migration time of disaccharides from GAG and slightly improved the overall resolution. However, using cationic micelles (cetyltrimethylammonium bromide) seems to be a good choice for the separation of anionic GAG disaccharides.[37] Resolution improved with increasing CTAB concentration. Even though the GAG disaccharides eluted in the order of increasing number of charged groups of the disaccharides, the separation in such a system is believed to be based on the hydrophobic and ion-pairing interaction of the negatively charged disaccharides, the positively charged CTAB micelles, and the CTAB-covered capillary wall.[37] This system was used to determine the disaccharide composition of enzymatically degraded chondroitin sulfate A, B, and C as well as chondroitinase-treated mink skin samples.

Very recently, a comparative study on compositional analysis of two sets of eight unsaturated disaccharide standards derived from heparin/heparan sulfate (see Figure 7.19) and chondroitin/dermatan sulfate (see Figure 7.17) was carried out[38] using both normal and reverse polarity capillary electrophoresis. Reverse polarity CE completely resolved disaccharide mixtures into all components using a single buffer system composed of sodium phosphate, pH 3.48. At this pH, the EOF is negligible, and the solutes migrate by their own electrophoretic mobilities toward the grounded anode. In the same report, the separation of 13 heparin-derived oligosaccharides of sizes ranging from di- to tetrasaccharides using both normal and reverse polarities was reported. Mixtures containing oligosaccharides differing primarily in size (i.e., number of saccharide units) were better resolved by normal polarity[38] using 10 mM sodium borate buffer, pH 8.8, containing 50 mM SDS.

More recently, variously sulfated disaccharides derived from hyaluronan and chondroitin/dermatan sulfates were analyzed by CE using phosphate buffer at low pH and negative polarity applied electric field.[128] Under these conditions, baseline separation of the nine different sulfated disaccharides was obtained, while the two nonsulfated disaccharides exhibited peak splitting due to the anomeric forms of the hexosamines present in the reducing terminal of the nonsulfated disaccharides. Moreover, this system was shown useful in the determination of the various disaccharides derived from either glucuronic or iduronic clustered structures in dermatan sulfate by analyzing the dermatan sulfate digested by chondroitinase AC or B.[128]

Other applications of HPCE in the area of GAGs include: (1) the assay of sulfoesterase activity on sulfated disaccharides derived from chondroitin sulfate, dermatan sulfate, and heparin;[129] the high resolution of CE allowed the use of the assay on impure enzyme preparations containing high protein concentrations; (2) the use of CE as an analytical tool for monitoring chemical reactions of trisulfated disaccharides[130]; the reactions monitored were the acylation, pivaloylation, and benzylation of hydroxyl groups on heparin derived trisulfated disaccharides. The progress of these reactions was monitored using the borate/SDS buffer system, pH 8.8[33,38] and UV detection at 232 nm.

Additional applications of CE in the area of GAGs include the quantitative analysis of hyaluronan in human synovial fluid.[131] This involved the hydrolysis of the polymeric hyaluronan to the tetrasaccharide by the action of testicular hyaluronidase, followed by the separation of the product by CE using a borate/SDS based electrolyte system and UV detection at 200 nm.

While the above oligosaccharides could be readily detected at 232 nm *via* the unsaturated bond in the uronic acid residues, the CE analysis of sulfated synthetic low-molecular-weight heparin fragments necessitated the use of indirect UV.[43] This is because the sulfated synthetic oligosaccharides exhibit low molar absorptivities as a result of the absence of the double bond in their structures. The indirect UV detection involved the use of 5-sulfosalicylic acid or 1,2,4-tricarboxybenzoic acid as the running electrolyte and background chromophore. Sulfated disaccharides with unsaturated uronic acid residues derived from heparin were also analyzed by CE at low pH with indirect-UV mode of detection. The inherent charge possessed by most GAG disaccharides allowed their CE-indirect UV detection to be conducted using buffers at low pH, thus eliminating the hydroxide ion negative effect.[43] The sensitivity of indirect UV

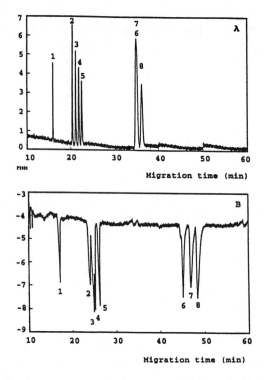

Figure 7.21   Capillary electrophoresis of eight heparin disaccharides using direct (A) or indirect (B) UV detection. Electrolytes: in (A), 200 mM sodium phosphate, pH 2.5; in (B), 5 mM 1,2,4-tricarboxybenzoic acid, pH 3.5. Injections: in (A) 9 nL from a solution containing 0.16 mg/mL of each saccharide; in (B) 1.8 nL from the same solution as in (A), except that the concentration was 0.1 mg/mL for each saccharide. Applied voltages: 131.5 Vcm^{-1} in (A) and 87.7 Vcm^{-1} in (B). Temperature, 25°C. Solutes: 1 = δUA2S→GlcNS6S, 2 = δUA2S→GlcNS, 3 = δUA1→GlcNS6S, 4 = δUA2S→GlcNAc6S, 5 = δUA2S→GlcNCOEt6S, 6 = δUA2S→GlcNAc, 7 = δUA→GlcNS, 8 = δUA→GlcNAc6S. (From Damm, J. B. L. and Overklift, G. T., *J. Chromatogr. A,* 678, 151, 1994. With permission.)

detection was reported to be at least one order of magnitude higher than that of direct UV detection.[43] Because the buffer systems used for CE-direct UV detection (phosphate buffer, pH 2.5) and CE-indirect UV detection (i.e., 5 mM 1,2,4-tricarboxybenzoic acid, pH 3.4) were different, the resolution power of the two systems was not the same.[124] Three of the disulfated disaccharides were not totally resolved in the indirect UV system, whereas they were in the direct UV system. However, two monosulfated disaccharides coeluted in the direct UV system, while they were baseline separated in the indirect UV system (see Figure 7.21).

c. Glycoprotein-Derived Oligosaccharides

Hermentin et al.[132] analyzed the oligosaccharides released from human α$_1$-acid glycoprotein (AGP) using both high-pH anion exchange chromatography with pulsed amperometric detection (HPAEC-PAD) and CE with UV detection at 190 nm. According to the authors, the CE analysis proved to be 4000 times more sensitive than HPAEC-PAD in terms of injected amounts, yet a much higher concentration of sample is required for analysis by CE. In fact, the carbonyl function of the N-acetyl and carboxyl groups present in the molecules enabled their direct UV detection at concentrations in the femtomole region. This approach has the advantage of avoiding derivatization and sample clean up. In that study, the authors compared the mapping profiles of AGP glycans released by conventional hydrazinolysis or by digestion with peptide-N-glycosidase F (PNGase F). Hydrazinolysis proved best, with practically no loss of N-acetylneuraminic acid using an automated apparatus, while the PNGase F digestion resulted in partial desialylation of the liberated N-glycans in the presence of SDS.[132]

In another separate report, Hermentin et al.[26] analyzed eighty underivatized sialooligosaccharides derived from glycoproteins by CE at 194 nm, and a carbohydrate-mapping database was established which would enable a carbohydrate structural analysis by simple comparison of migration times. Highly reproducible migration times could be achieved (RSD < 0.20%) by including mesityl oxide and sialic acid as two internal standards for the correction of migration time using a triple-correction method. The suitability and reliability of the database for the structural determination of sialylated N-glycans by comparison of corrected migration time was established by analyzing N-glycan pools of various glycoproteins, such as recombinant human urinary erythropoietin (baby hamster kidney), bovine serum fetuin, and human α_1-acid glycoprotein (AGP). The applicability of the database for structural determination of N-linked carbohydrates released from glycoproteins was further confirmed by CE measurements performed in a different laboratory and by a different analyst who used different CE instruments.[26]

MEKC has recently been investigated as an alternative mode of analysis of oligosaccharides released from glycoproteins. N-linked oligosaccharides released from recombinant tissue plasminogen activator (rt-PA) after N-glycanase digestion were separated by MEKC using SDS surfactant and direct UV detection at 200 nm.[27] The oligosaccharides consisted of neutral (high mannose) and mono- to tetra-negatively charged oligosaccharides. As one could expect, the neutral oligosaccharides separated on the basis of their differential partitioning into the micelles, whereas the separation of the sialylated oligosaccharides was mainly due to differences in their electrophoretic mobilities resulting from their different inherent negative charges. The addition of a divalent ion (Mg^{2+}) to the SDS electrolyte system enhanced the selectivity of the separation. This cation is known to form complexes with carbohydrates[4,133] and to be electrostatically attracted to the negative surface of micelles[134,135] and capillary walls. This attraction between the cation and the silanol groups of the capillary walls reduces EOF. On the other hand, divalent cations are expected to initiate a different micellar growth and to change the shape and size of the micelles by reducing the electrostatic repulsion between the polar heads of the surfactant. This change could be responsible in part for the enhanced selectivity.[27] Moreover, Mg^{2+}-carbohydrate complexation differs from one carbohydrate to another depending on the structure. Therefore, the enhanced selectivity of an Mg^{2+}/SDS electrolyte system is the result of those two effects. This electrolyte system was further utilized in the N-glycosylation mapping of rt-PA[28] to determine the difference between the oligosaccharide distribution of the two rt-PA variants which differ by the presence (type I) or the absence (type II) of oligosaccharides at the Asn-184 site.

N-Oligosaccharides from fetuin, rt-PA, and α_1-acid glycoprotein were separated by CE on the basis of their sialic acid content and their structures.[136] As the number of sialic acid residues increased in the oligosaccharides, the UV absorbance at 200 nm of the underivatized analytes was greatly enhanced. Within each group of sialylated N-glycans a significant separation was still attainable, indicating that the separation relies not only on a charge difference but also on a structural difference between sugar chains bearing the same number of sialic acid residues. The separation within each class of glycans may be attributed to variations in the type of linkage (α-2,3 or α-2,6) between the sialic acids and the galactose, to the oligosaccharide size, or to the peripheral fucose residue.

Since there is no enzymatic cleaving process available for the cleavage of O-linked oligosaccharides, the chemical process used for their cleavage (i.e., treatment with alkali in the presence of borohydride) results in reducing them to alditols. Thus, the released O-linked oligosaccharides lack a site for fluorescent labeling by reductive amination, and their detection is only possible by measuring UV absorbance at 185 nm.[29] Using this detection approach, several O-glycosidically linked monosialooligosaccharides were analyzed as their alditols by

HPCE.[29] Alkaline borate buffer yielded a migration profile for the oligosaccharides that was basically similar to that obtained in alkaline phosphate buffer, indicating no significant contribution of borate complex formation. However, neither electrolytes provided enough resolving power to separate N-acetyl and N-glycolylneuraminic acid containing oligosaccharide pairs. They were only resolved after the addition of 100 mM SDS to the borate buffer. The separation mechanism is based on changing the conformation of these oligosaccharides to different magnitude, thus resulting in variation of the molecular size.[29] This buffer system was utilized in a microscale analysis of sialooligosaccharides in bovine submaxillary mucin and swallow nest material.[29]

Moreover, neutral and sialylated O-linked oligosaccharides derived from human milk and urine were analyzed by CE using borate buffer and UV detection at 200 nm.[25] The separation of six standard O-linked oligosaccharides is depicted in Figure 7.22, using 35 mM sodium borate, 130 mM boric acid, pH 8.35. Apparently, the stability of the borate-oligosaccharides complexes of the O-linked oligosaccharide, whose structures are illustrated in Figure 7.22, differed in magnitude, thus providing enough selectivity to achieve baseline separation.

2. Derivatized Oligosaccharides

a. Simple Oligosaccharides

Several disaccharides, including gentibiose, maltose, lactose, cellobiose, and melibiose, were labeled with APTS by reductive amination and subsequently separated by CE using MOPS and borate buffers.[99] In MOPS buffer the disaccharide gentibiose migrated first, followed by maltose, lactose, cellobiose, and melibiose. Since the molecular weights of these disaccharides and their electrical charges are the same, the migration order is governed by the differences in the hydrodynamic volume arising from varying degrees of hydration due to varying positions of hydroxyl groups in the nonreducing end pyranose.[99] In borate buffer, the migration order was gentibiose, maltose, melibiose, cellobiose, and lactose. This migration order is dictated by the magnitude of the stability constant of the disaccharide–borate complexation. It is interesting to note that the migration order of these APTS-derivatized disaccharides is different from their underivatized counterparts in borate buffer, which was reported by Hoffstetter-Kuhn et al.[24] to be cellobiose, maltose, lactose, and finally gentibiose. The difference in migration order may be attributed to the effect of the tag on the stability of the borate complexes. This is further supported by the fact that the migration order of 4-aminobenzonitrile derivatives was reported by Schwaiger et al.[90] to be maltose, cellobiose, melibiose, and finally lactose under the same separation conditions. In the report by Chen and Evangelista,[99] two APTS-derivatized glucose tetrasaccharide isomers, [Glc-α(1-4)Glc-α(1-4)Glc-α(1-4)Glc-APTS] and [Glc-α(1-6)Glc-α(1-4)Glc-α(1-4)Glc], which differ only in one linkage at the nonreducing end were shown to resolve quite well in MOPS buffer but not in borate buffer. This means that borate forms a weak complex with both isomers and thus has no significant effect on the relative electrophoretic mobility of each species.

Recently, CE-LIF was applied to monitor enzyme products formed during the incubation of yeast cells with the trisaccharide α-D-Glc(1→2)α-D-Glc(1→3)α-D-Glc-O(CH$_2$)$_8$CONHCH$_2$CH$_2$NHCO-tetramethylrhodamine (-TMR). TMR is the fluorescent arm attached to the trisaccharide through the reductive amination of the analyte followed by the incubation with TMR.[107] After 5 h of incubating the yeast cells with the trisaccharide, the lysed yeast spheroplasts were injected and the components were separated and detected by CE-LIF. Most of this trisaccharide was converted to linker arm, intermediate disaccharide, and monosaccharide (see Figure 7.23). This resulted from the sequential activity of α-glucosidase I and II inside the yeast cell, which act specifically on α-D-Glc(1→2) and α-D-Glc(1→3) linkages, respectively.[107]

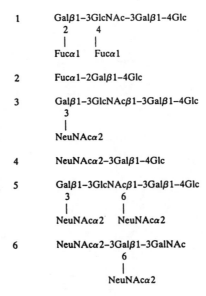

1 Galβ1–3GlcNAc–3Galβ1–4Glc
      2   4
      | |
   Fucα1   Fucα1

2 Fucα1–2Galβ1–4Glc

3 Galβ1–3GlcNAcβ1–3Galβ1–4Glc
      3
      |
   NeuNAcα2

4 NeuNAcα2–3Galβ1–4Glc

5 Galβ1–3GlcNAcβ1–3Galβ1–4Glc
      3   6
      | |
   NeuNAcα2   NeuNAcα2

6 NeuNAcα2–3Galβ1–3GalNAc
                    6
                    |
                NeuNAcα2

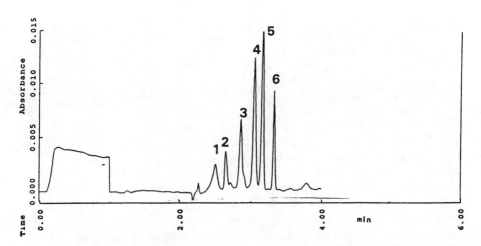

Figure 7.22 Separation of six O-linked oligosaccharides derived from human milk. Conditions: running electrolyte, 35 mM sodium borate, 130 mM boric acid, pH 8.35; voltage, 20 kV; sample degraded at 120°C for 30 min; capillary, pretreated capillary, 50 cm (to detection point)/57 cm (total length) × 75 μm i.d.; detection, UV 200 nm. (From Hughes, D. E., *J. Chromatogr. B*, 657, 315, 1994. With permission.)

Other simple oligosaccharides derivatized with 4-aminobenzonitrile,[90] 2-AP,[52,76] 2-AA,[118] CBQCA,[83] ethyl-*p*-aminobenzoate,[52] or *p*-aminobenzoic acid[52] were analyzed by CE using primarily borate buffer systems.

b. Homologous Oligosaccharides

The separation of derivatized homologous, ionic oligosaccharides or homologous oligosaccharides labeled with an ionic tag by CE is most often accomplished in the presence of regular noncomplexing electrolytes (e.g., phosphate, MES, Tris, etc.). This is because the members of these derivatized homologous oligosaccharides will exhibit significant differences

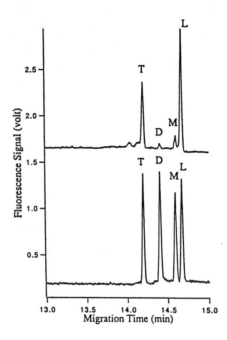

Figure 7.23 Electropherograms obtained from the analysis of lysed yeast spheroplasts (top) and a standard solution containing 10^{-9} M of each component (bottom). Electrolyte, 10 mM each of phosphate, borate, phenylboronic acid, and SDS, pH 9.3; capillary, 42 cm × 10 μm i.d.; voltage, 400 V/cm. Peaks T = α-D-Glc(1→2)α-D-Glc(1→3)α-D-Glc-O(CH$_2$)$_8$CONHCH$_2$CH$_2$NHCO-TMR; D = α-D-Glc(1→3)α-D-Glc-O(CH$_2$)$_8$CONHCH$_2$CH$_2$NHCO-TMR;M=α-D-Glc- O(CH$_2$)$_8$ CONHCH$_2$ CH$_2$ NHCO-TMR; L = H-O(CH$_2$)$_8$CONHCH$_2$CH$_2$NHCO-TMR. (From Le, X., Scaman, C., Zhang, Y., Zhang, J., Dovichi, N. J., Hindsgaul, O., and Palcic, M. M., *J. Chromatogr. A,* 716, 215, 1995. With permission.)

in the charge-to-mass ratios, thus ensuring sufficient differential migration and, consequently, separation. However, this is only true for up to a certain degree of polymerization.

Nashabeh and El Rassi[137] were the first to demonstrate the high resolving power of CE in the separation of pyridylamino derivatives of maltooligosaccharides having a degree of polymerization (d.p.) from 4 to 7. The positively charged sugar derivatives migrated ahead of the EOF marker and were separated according to their charge-to-mass ratio in the pH range 3.0 to 4.5. The inclusion of 50 mM tetrabutylammonium bromide in the electrolyte solution slightly decreased the EOF, and consequently allowed the separation of the maltooligosaccharides at pH 5.0. However, as the pH approached the pK$_a$ value of the derivatives (pK$_a$ = 6.71), the homologs practically coeluted and migrated virtually together with the EOF.

To examine the effect of the nature of the tagging agent on the spacing pattern between the migrating zones of homologs, a series of N-acetylchitooligosaccharides derivatized with either 2-AP or 6-AQ were analyzed by CE[86] using the electrolyte system introduced for the maltooligosaccharides[137] and a capillary having a polyether interlocked coating. As expected, since 2-AP and 6-AQ have similar characteristic charges, the resolution of homologs was virtually independent of the tagging agent (Figure 7.24).

On the contrary, for various tagging agents having different characteristic charges, the resolution between homologous oligosaccharides was shown to be dependent on the intrinsic charge of the derivatizing agent,[138,139] even in the presence of borate in the running electrolyte.[140] This effect is depicted in Figure 7.25, where 2-AP, 5-ANSA, and ANTS derivatives of dextran oligomers are analyzed by CE using 100 mM borate-Tris buffer, pH 8.65.[140] At this pH, the 2-AP, 5-ANSA, and ANTS derivatives possess the negative charges of zero, one, and three sulfonic groups, respectively. The average migration times for the individual ANTS oligomers was roughly one third of those observed with 2-AP derivatives, while 5-ANSA

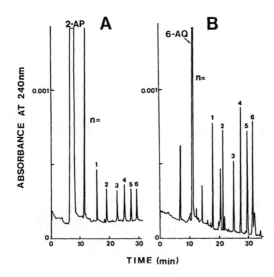

Figure 7.24 Electropherograms of pyridylamino (A) and quinolylamino (B) derivatives of N-acetylchito-oligosaccharides. Capillary, fused-silica tube with polyether interlocked coating on the inner walls, 50 cm (to the detection point), 80 cm total length × 50 μm i.d.; electrolyte, 0.1 M phosphate solution containing 50 mM tetrabutylammonium bromide, pH 5.0; voltage, 18 kV. 2-AP = 2-aminopyridine; 6-AQ = 6-aminoquinoline. (From Nashabeh, W. and El Rassi, Z., *J. Chromatogr.*, 600, 279, 1992. With permission.)

derivatives migration times were intermediate. In addition, the ANTS derivatives exhibited narrower peaks, greater resolution, shorter analysis time, and higher peak detection.

Another study examined the effects of the structure and charge of several naphthalene sulfonic acid-based derivatizing agents, such as ANTS, ANDSA, 3-aminonaphthalene-2,7-disulfonic acid (3-ANDA, tag XVI), 2-aminonaphthalene-1-sulfonic acid (2-ANSA, tag XII), and 5-ANSA, on the CE analysis of their maltooligosaccharide derivatives using a running electrolyte consisting of sodium phosphate, pH 2.5, in the presence of TEA.[138] The ANTS-derivatized maltooligosaccharides were separated for up to a d.p. of more than 30 glucose units in less than 30 min. 3-ANDA-derivatized maltooligosaccharides showed the same resolution, yet the separation was only achievable up to d.p. 30. On the other hand, for 2-ANSA and 5-ANSA-derivatized maltooligosaccharides only 20 components were resolved, and loss of efficiency was observed which might be attributed to the longer analysis time (about 40 min). These findings show the importance of having permanent multiple charges in the tag such as in ANTS and 3-ANDA. Although, 2-ANSA and 5-ANSA tags are structural isomers, the migration time of 2-ANSA is almost four times lower than that of 5-ANSA. The pK_a value for the primary amino group of 5-ANSA is higher than that of 2-ANSA because the sulfonic acid and amino groups are farther apart. This high pK_a value decreases the net negative charge on the molecule and, in turn, its mobility.

Also, the effect of the intrinsic charges of the fluorescent tags on the separation of negatively charged oligosaccharides, derived from partially hydrolyzed k-carrageenan, was also demonstrated by comparing the separation of ANTS- and 6-AQ-derivatized k-carrageenan oligosaccharides.[139] When the charge-to-mass ratio of oligosaccharides is increased by the end-label (i.e., ANTS), the migration order toward the anode is from smaller to larger oligomers, and the separation of larger oligosaccharides is further improved by using a sieving medium. The separation is based on the charge-to-mass ratio differences between oligomers. The derivatizing agent, ANTS, has three negative charges, and as a result it enhances the charge-to-mass ratio of the analytes; however; this enhancement is more significant in the case of small oligomers, and diminishes as the degree of polymerization of the oligomers

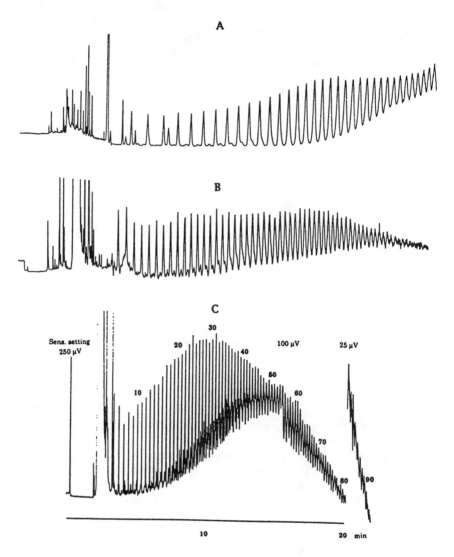

Figure 7.25 Influence of the fluorescent tag on the separation of a dextran standard with an average molecular weight of 18,000. The reagents used were (A) 2-AP, (B)ANA, and (C) ANTS. Conditions, running electrolyte, 100 mM borate, pH 8.65; voltage, −500 V/cm, capillary effective length 35 cm. (From Stefansson, M. and Novotny, M., *Anal. Chem.*, 66, 1134, 1994. With permission.)

increases. On the other hand, the migration order is entirely reversed when the charge-to-mass ratio is decreased by the end-label. This is the case with oligosaccharides tagged with 6-aminoquinoline, where the migration order starts with large oligomers and proceeds to smaller ones. This is due to the fact that the tag decreases the charge-to-mass ratio of the small oligomers more significantly than for the larger ones.[139]

An important operating parameter that largely affects the resolution of homologous oligomers is the magnitude of the EOF. In fact, the use of coated capillaries having very low or virtually no EOF improved the resolution of homologous oligosaccharides with a higher degree of polymerization. For instance, the separation of 2-AP derivatives of an oligogalacturonide homologous series with d.p. in the range 1 to 18 was best achieved on a coated capillary having a switchable (anodal/cathodal) EOF using 0.1 M phosphate solution, pH 6.5, as the running electrolyte, because at this pH the EOF is very low.[141] Similarly, 2-AP derivatives of isomaltooligosaccharides were completely separated from each other at least up to a d.p. of 20 using fused-silica capillaries in which the EOF was suppressed by chemically coating the capillary inner wall with linear polyacrylamide.[142]

Furthermore, Chiesa and Horváth[95] concluded that the use of polyacrylamide gel-filled capillaries does not contribute to enhancing the separation of derivatized homologous oligomers, which contradicts what was first shown by Novotny and co-workers.[101,143] In the work of Chiesa and Horváth,[95] ANTS-derivatized maltooligosaccharides were separated at pH 2.5 by open tubular CE, and the results were compared to those reported in the literature involving ANTS-derivatized maltooligosaccharides separated by polyacrylamide slab gel electrophoresis[96] and CBQCA-derivatized maltooligosaccharides separated in capillaries filled with highly concentrated polyacrylamide gel according to the procedure described by Novotny and co-workers.[101,143] The authors concluded that the presence of the gel shows no enhancement in the resolution of oligosaccharide derivatives containing at least up to 20 glucose units.[95] On the contrary, the mobility of the ANTS derivatives appears to be lower by a factor of 22 in the crosslinked gel than in free solution, without improving resolution. The separation of ANTS-maltooligosaccharides in open tubular CE was achieved with a background electrolyte consisting of 50 mM sodium/triethylammonium phosphate buffer, pH 2.50, containing 10.8 mM triethylamine (TEA).[95] Under these conditions, nearly 30 homologs were well resolved in less than 5 min. This separation electrolyte system was used elsewhere for the CE analysis of ANTS-derivatized dextran and galacturonic acid ladders.[91] The effect of the cationic additive (TEA) was mainly attributed to its interaction with the capillary wall (electrostatic binding). In fact, at sufficiently high concentrations of TEA (e.g., 50 mM), an inversion in the direction of EOF from cathodal to anodal was observed.[95] Because of the electrophoretic migration of the negatively charged ANTS derivatives, an anodal EOF has the beneficial effect of increasing the speed of separation. However, the resolution of the system decreased for d.p. greater than 12. This system allowed the ultrafast separation (in less than 10 s) of three short-chain maltooligosaccharides derivatized with ANTS.[95]

Very recently, Stefansson and Novotny[140] demonstrated the separation of large oligosaccharides of dextrans (although these oligosaccharides are branched, their electrophoretic behavior is first described here simply for the completeness of the discussion) in open tubular capillaries coated with linear polyacrylamide (i.e., zero EOF capillaries) using running electrolytes based on Tris-borate, pH 8.65. This work has clearly demonstrated the importance of capillaries with reduced EOF in achieving high resolution and high separation efficiencies (excess of 1 million theoretical plates/m) for large oligosaccharides (for further discussion, see the next section).

Despite the limited benefit in terms of resolution, generated from using sieving media in the separation of homologous oligosaccharides, ANTS-derivatized dextran oligosaccharides were separated by CE using polymer networks under various operating conditions. As a model system, ANTS-labeled wheat starch digest was analyzed by CE using 25 mM sodium acetate buffer, pH 4.75, containing 0.5% polyethylene oxide.[144] Although the polymer concentration in such a system was above the entanglement threshold value, separation was not based on the sieving mechanism as was evident from Ferguson plots. Therefore, the separation attained was caused by the charge-to-mass ratio differences of the oligosaccharides, and as mentioned above, the effect of the polymer network was only to slow down the velocity of the derivatives.

As in the case of monosaccharides, CE analysis of oligosaccharides derivatized with neutral or weakly ionizable tags requires the use of high pH electrolyte solutions or borate buffers. As a result, PMP-derivatives of homologous oligoglucans,[82] such as α-(1→3)-linked (laminara-) oligoglucans, α-(1→6)-(isomalto-) oligoglucans, and β-(1→4)- (cello-) oligoglucans were separated as borate complexes. All the homooligoglucans eluted in the order of decreasing size, and because of the unfavorable mass-to-charge ratio at high degrees of polymerization, resolution between the homologs decreased as the number of recurring units increased. As expected, the rate of migration varied among series because the extent of their complexation with borate is largely influenced by the orientation of hydroxyl groups, i.e., by the type of interglycosidic linkage of the various oligosaccharides. Alkali earth metal complexation with carbohydrates is another mechanism of separation that has been utilized for the CE analysis of the PMP derivatives of a series of isomaltooligosaccharides using an aqueous barium salt solution as the running electrolyte.[4] However, the selectivity and resolution of such a system was not comparable to that of the borate since separation of up to a d.p. of 9 was achieved with an alkali earth metal system, as opposed to a d.p. of 13 achieved in the presence of borate.[82]

As an application of the utility of CE in analyzing homologous oligosaccharides, Lee et al.[145] described a CE method that allows the characterization and determination of the action of chitinase (an endo-, exo-, and random glycosidase enzyme) toward a series of six homo-oligosaccharides of the structure $[N\text{-acetylglucosamine}\beta(1{\rightarrow}4)]_n$ (where n = 1 – 6) which were derivatized with ANDSA by reductive amination.

c. Applications Involving Linear and Branched Oligosaccharides Derived from Plants

The high resolution separation that HPCE provides for homologous linear oligosaccharides was also exploited in the separation of branched heterooligosaccharides derived from large xyloglucan polysaccharides (XGs) by enzymatic digestion.[86] The CE separation of 2-AP derivatives of xyloglucan oligosaccharides (2-AP-XG) obtained from cotton cell walls by cellulase digestion[146] is depicted in Figure 7.26. The peak numbers on the electropherogram (see Figure 7.26) correspond to the elution order in reverse-phase chromatography (RPC).[146] Unlike RPC where the elution order is mainly influenced by the size of the oligosaccharide and the hydrophobic character of the sugar residues, the migration order in CE is mainly governed by the number of sugar residues and the degree of branching. For instance, fragment 4, which is smaller in size with respect to fragment 3, was more retarded on the RPC column.[146] This may be attributed to the presence of fucosyl residue (i.e., 6-deoxygalactosyl) in structure 3, which is more hydrophobic than any other sugar residue in the molecule. The same reasoning

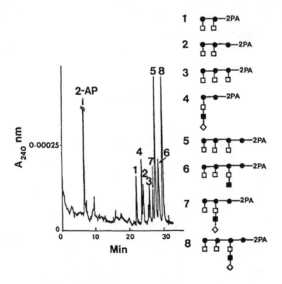

Figure 7.26 Capillary zone electrophoresis mapping of pyridylamino derivatives of xyloglucan oligosaccharides from cotton cell walls. Capillary, fused-silica tube with polyether interlocked coating on the inner walls, 50 cm (to the detection point), 80 cm total length × 50 µm i.d.; electrolyte, 0.1 *M* sodium phosphate solution containing 50 m*M* tetrabutylammonium bromide, pH 4.75; running voltage, 20 kV. Symbols: 2-AP, 2-aminopyridine; ●, glucose, □, xylose; ■, galactose; ◊, fucose. (From Nashabeh, W. and El Rassi, Z., *J. Chromatogr.*, 600, 279, 1992. With permission.)

can explain the elution order for fragments 6 and 7. Because all 2-AP-XG fragments possess the same charge, the migration profile of the 2-AP-XG derivatives in CE is in the order of increasing size, i.e., decreasing charge-to-mass ratio. However, for the same number of residues, but with slight differences in molecular weight, the less branched oligosaccharides migrated faster than the more branched ones. In fact, structure 4 (singly branched), which has a slightly higher molecular weight than the doubly branched structure 2, migrated first. This also explains the migration order of 7 (doubly branched) and 5 (triply branched). Generally, as the extent of oligosaccharide branching increases, the electrophoretic mobility decreases.

In order to interpret the electrophoretic behavior of the various 2-AP-XG and to quantitatively describe the effects of the various sugar residues on their electrophoretic mobility, Nashabeh and El Rassi[86] have introduced a mobility indexing system for the branched xyloglucan oligosaccharides with respect to the linear 2-AP derivatives of N-acetylchitooligosaccharides (2-AP-GlcNAc$_n$) homologous series, shown in Figure 7.24. The indexing system revealed that the addition of a glucosyl residue to the linear core chain of the oligosaccharide showed a change in the mobility index decrement similar to the addition of a xylosyl residue at the glucose loci and behaved as one half a GlcNAc residue in terms of its contribution to the electrophoretic mobility of the 2-AP-XG (see structures in Figure 7.26). However, the addition of a galactosyl residue to an already branched xylosyl residue exhibited less retardation than the addition of a glucosyl or xylosyl unit to the backbone of the xyloglucan oligosaccharide. The same observation was made when adding a fucosyl residue to a branched galactosyl residue. Thus, as the molecule becomes more branched, the addition of a sugar residue does impart a slightly smaller decrease in its mobility. This approach may prove valuable in correlating and predicting the structures of complex carbohydrates.

Very recently, the utility of CE in distinguishing complex oligosaccharides of very similar structures has been further exploited in the oligosaccharide mapping of laminarin[140] (a branched polysaccharide) after enzymatic cleavage with laminarinase, cellulase, and endoglucanases. The various ANTS-oligosaccharide maps obtained with the three different β-1,3-glucose-hydrolases show that cellulase and laminarinase seem to cause more complete hydrolyses than the endoglucanases (EG I and EG II). Also, EG I and EG II appeared to differ somewhat in their structural preferences. This shows again the importance of the high sensitivity and high resolution of CE in the characterization of structural preferences for different enzymes in the hydrolysis of a given polysaccharide.

Moreover, oligosaccharides of α-D-glucans (amylose, amylopectin, and pullulan) and β-D-glucans (exemplified by lichenan) were also derivatized with ANTS and analyzed by CE to evaluate their complexity using linear polyacrylamide-coated capillary.[147] The oligosaccharide maps were obtained using various borate-based electrolyte systems and after selective debranching using isoamylase, laminarinase, and cellulase enzymes. According to the authors, "complex glucan chains with numerous residual branches can potentially be assessed using this oligosaccharide mapping procedure." Since the capillary column was of the type with reduced EOF, a baseline separation of an intact amylose sample with d.p. close to 70 could be achieved.

Very recently, the utility of CE was exploited in the characterization of substrate specificity of endopolygalacturonase.[148] For this purpose, a CE system was developed for the separation of ANTS-derivatized neutral and acidic oligosaccharides according to their relative size using uncoated capillary and reversed polarity at pH 2.5. The low pH of the running electrolyte was needed to suppress both the EOF and the ionization of carboxylic acid groups for the acidic oligosaccharides. Under these conditions, oligomers of up to 20 galacturonic acid (GalA) residues in length were well resolved. Utility of the procedure was demonstrated by its application to characterization of the substrate specificity of endopolygalacturonases. The results have shown that both a fungal and a bacterial endopolygalacturonase need four adjacent nonesterified galacturonic acid residues in a pectin to be able to act.[148] In another report from the same laboratory,[149] a sensitive CE-based assay was described for the detection of pectate-depolymerizing enzymes using a fluorescent end-labeled pectate oligomer. The labeled oligomer was allowed to react with the enzyme either *in vitro* or *in vivo*, such as inside the intercellular spaces of a cotton cotyledon,[149] and after an appropriate incubation time, the products were CE analyzed. The site and mode of action of the pectate-depolymerizing activity could be inferred from the products. Both endo- and exopolygalacturonase activity and lyase activity were distinguished. Since only fluorescent oligomer and products from its labeled reducing end are detected, there is no interference from other compounds, and consequently only pectic enzyme activity is detected. On this basis, the assay demonstrated the presence of a considerable endo- and exopolygalacturonase activity in the intracellular spaces of cotton cotyledons.

d. Glycosaminoglycan-Derived Oligosaccharides

Although underivatized GAG-derived di- and oligosaccharides can be readily electrophoresed by CE and detected in the UV at 232 nm, precolumn derivatization of these species with a suitable chromophore or fluorophore will certainly improve their detectability. In addition, the tagging may provide additional properties to the GAG sugars which may enhance the separation potential.

Thus far, only a few research papers have appeared on the separation of derivatized GAG-derived oligosaccharides. Unsaturated disaccharides derived from GAGs using chondroitinase AC and ABC were labeled with PMP and subsequently separated by CE.[81] The conversion of these saccharides to their PMP derivatives improved the system sensitivity and selectivity (see Figure 7.27). The electrophoretic system also proved suitable for the quantitative estimation of human urinary chondroitin sulfates (see Figure 7.27b).

CBQCA is another derivatizing agent that has been used for the labeling of enzymatically degraded hyaluronic acid from human umbilical cords[101,143] and chondroitin sulfate A.[101] The CE analysis in both cases was performed in gel-filled capillaries with high gel concentration, and the resolution of such a system was shown to be superior to that of open tubular format of CZE.

e. Glycoprotein-Derived Oligosaccharides

Capillary electrophoresis is increasingly used in the separation and mapping of the oligosaccharide fragments of glycoproteins, i.e., glycans. Glycans are branched oligosaccha-

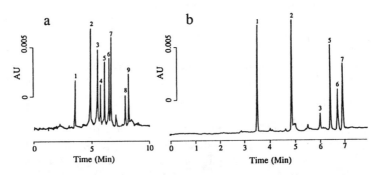

Figure 7.27 (a) Electropherogram of chondroitin ABC-digested mixture of chondroitin sulfates A-E, chondroitin and hyaluronic acid by CZE after derivatization with PMP. Capillary, fused-silica (51 cm × 50 µm i.d.); electrolyte, 100 mM borate buffer, pH 9.0; applied voltage 25 kV. Peaks: 1 came from the buffer for enzymatic digestion; 2, PMP (excess reagent); 3, PMP derivative of Δdi-0S; 4, PMP derivative of Δdi-HA; 5, sodium benzoate (internal standard); 6, PMP derivative of Δdi-4S; 7, PMP derivative of Δdi-6S; 8, PMP derivative of Δdi-SD; 9, PMP derivative of Δdi-SE. (b) Analysis of the PMP derivatives of unsaturated disaccharides derived from the GAG fraction of a urine sample digestion with chondroitinase ABC by CZE. Conditions and peak assignment are as in (a). (From Honda, S., Ueno, T., and Kakehi, K., *J. Chromatogr.*, 608, 289, 1992. With permission.)

rides implicated in many important biological processes such as cell-cell interaction, cell development and differentiation, and hormone-receptor and antigen-antibody interactions.[150,151] Moreover, many therapeutic proteins are often expressed with glycan moieties which affect their biological activity, lifetime, and specificity, thus making the correct glycosylation of these drugs an important issue.[152] Due to the increasing interest in protein glycosylation, various CE methods have been developed for the sensitive and efficient determination of the glycan moiety of glycoproteins.

Using ovalbumin as a model protein, Honda et al.[142] demonstrated the separation of glycans by CE with on-column fluorometric detection. The oligosaccharides of ovalbumin (hybrid and high-mannose type) were released with anhydrous hydrazine and tagged with 2-AP after re-N-acetylation. The CE analysis of the released oligosaccharides was performed using two different electrolytes, an acidic phosphate buffer whereby the derivatized glycans are positively charged due to the protonation of the amino group of the tag and an alkaline borate buffer which allows the *in situ* conversion of the derivatives to anionic borate complexes. Because of differences in their separation mechanisms, the two electrolytes yielded different selectivities. The phosphate system gave good resolution of the oligosaccharide derivatives that are different in their molecular size (i.e., oligosaccharides having a different number of monosaccharide units) but could not resolve solutes having the same degree of polymerization. On the other hand, CE as borate complexes separated the oligosaccharide derivatives based on structural differences of the outer monosaccharide residues. The greater the number of unsubstituted mannose units, the slower the migration of the derivatives.[142] However, the borate system failed to resolve high-mannose-type oligosaccharides having the same number of outer mannose residues. Similarly, hybrid-type oligosaccharides having the same number of peripheral mannose or galactose residues, but differing in the total number of monosaccharide units, were not resolved. In both cases, however, satisfactory separations were attained by direct CZE with low pH phosphate buffers.

Also, CE proved to be useful in elucidating the differences in glycan structures of a glycoprotein from different sources.[114] The oligosaccharides of both human and bovine AGP oligosaccharides were first cleaved from their corresponding glycopeptide fragments by treating the tryptic digest of the proteins with peptide-N-glycosidase F (PNGase F), and then were labeled with 2-AP by reductive amination. The CE mapping of the 2-AP derivatives of the liberated oligosaccharides from human and bovine AGP yielded two different electropherograms each containing well defined peaks and a few minor peaks, migrating after the excess 2-AP (see Figure 7.28). Both human and bovine AGP have been found to have the same sialic acid, galactose, and mannose content, but 50% of the sialic acid in bovine AGP is N-glycolylneuraminic acid and the fucose content is very low. These differences were unveiled by CE mapping of glycans derived from both types of AGP. Therefore, CE is a very powerful analytical tool in the field of glycan separation and characterization.

High mannose oligosaccharides released from RNase B by digesting the protein with PNGase F were analyzed by CE as their 2-AP derivatives.[86] One major peak in the CE map was identified as $(GlcNAc)_2$-Man_5 using a standard. The polypeptide chain of RNase B is known to have only one glycosylation site, which can accommodate five different high-mannose glycans. This explains the presence of several peaks in the CE map in addition to that of $(GlcNAc)_2$-Man_5 oligosaccharide.

Very recently, an electrophoretic system based on capillary gel electrophoresis (CGE) was described for the profiling of oligosaccharides enzymatically derived from ribonuclease B and labeled with APTS. The CGE system involving the use of an entangled polymer network exhibited a higher resolving power than open tubular CZE.[100] Optimum separation was attained using 25 mM acetate buffer, pH 4.75, containing 0.4% polyethylene oxide polymer and a neutrally coated capillary. This system yielded a high resolution of all the major components of the ribonuclease N-glycan pool and a baseline separation of the three positional isomers of mannose-7 and mannose-8 oligosaccharide (see Figure 7.29). According

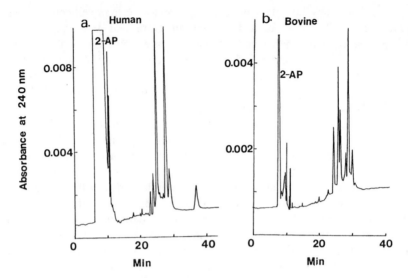

Figure 7.28 Capillary zone electrophoresis mapping of pyridylamino derivatives of human (a) and bovine (b) AGP oligosaccharides. Capillary, fused-silica tube with hydrophilic coating on the inner walls, 45 cm (to the detection point), 80 cm total length × 50 µm i.d.; electrolyte, 0.1 M phosphate solution, pH 5.0, containing 50 mM tetrabutylammonium bromide; running voltage, 18 kV; current, 80 µA; injection by electromigration for 2 s at 18 kV. (From Nashabeh, W. and El Rassi, Z., *J. Chromatogr.*, 563, 31, 1991. With permission.)

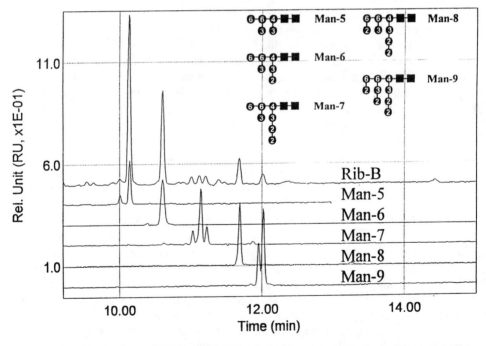

Figure 7.29 CGE separation of the APTS-labeled high-mannose type oligosaccharides released from bovine ribonuclease B (upper trace) and the individual standard structures (lower traces). Inset: structural representation of the high-mannose type N-linked oligosaccharides: ■ = GlcNAcβ1→4; ❹ = Manβ1→4; ❶ = Manα1→6; ❸ = Manα1→3; ❷ = Manα1→2. Conditions: 57 cm neutrally coated column (50 cm to detection point) × 50 µm i.d.; LIF detection: argon ion laser, excitation 488 nm, emission: 520 nm; separation buffer: 25 mM acetate buffer, pH 4.75, 0.4% polyethylene oxide; applied electric field = 500 V/cm.; i = 19 µA; 20°C. (From Guttman, A. and Pritchett, T., *Electrophoresis*, 16, 1906, 1995. With permission.)

to the authors, the separation is not related to a sieving effect as concluded from Ferguson plots, but rather to the change in the hydrodynamic volumes of the labeled glycans and viscosity of the separation medium.[100]

The capillary gel electrophoresis method was also effective in the separation and profiling of asparagine-linked glycans of bovine fetuin after enzymatic release with PNGase F and labeling with APTS by reductive amination.[153] The APTS-oligosaccharide derivatives were best separated at pH 4.75, yielding a map of four major peaks and several minor peaks. Two of the four major peaks were identified by spiking with standards and were found to be trisialylated triantennary structures with different sialylation linkage. The other two major peaks were thought to be tetrasialylated triantennary structures based on calculations considering their corresponding glucose unit values according to the method reported by Stack and Sullivan.[154]

In a recent report, the CE resolving power was contrasted to that of polyacrylamide slab gel electrophoresis (PAGE) in the separation and profiling of carbohydrates released from several glycoproteins including bovine fetuin, human α_1-acid glycoprotein, HIV envelope, and bovine ribonuclease B.[155] These carbohydrates were labeled with ANTS and analyzed by CE-LIF and PAGE. CE in the open tubular format (i.e., in the absence of an entangled polymer network) using acetate buffer, pH 4.75, yielded comparable results to those obtained with PAGE in terms of the number of migrating bands. In both cases, high resolution separations of the released and labeled carbohydrates were achieved. Therefore, these two methods are complementary, and each offers unique advantages. CE provides sample automation, fast analysis times, high resolution, and sensitivity with minimal sample consumption. On the other hand, PAGE provides the ability to perform comparative profiling of different oligosaccharide mixtures with a simple and versatile method.

Moreover, ANTS-derivatized complex oligosaccharides, both neutral and sialylated, were separated and detected by CE-LIF.[156] Concentration and mass detection limits of 5×10^{-8} M and 500 amol, respectively, could be achieved. However, the chemistry used in the precolumn labeling has a derivatization limit of 2.5×10^{-6} M. The linear relationship between the electrophoretic mobility and charge-to-mass ratios of the ANTS-derivatized oligosaccharides was used for peak assignment. Due to the relatively high separation efficiency, a migration time difference of 0.06 min was found to be sufficient for the baseline separation of the ANTS-derivatized complex oligosaccharides.[156]

The effect of the structure and charge of the derivatizing agent on the CE analysis of branched oligosaccharides was also investigated.[138] High-mannose-type oligosaccharides from bovine pancreatic ribonuclease B were labeled with aminonaphthalene mono-, di-, and trisulfonic acid (i.e., 2-ANSA, ANDSA, and ANTS, respectively) and separated by CE using phosphate buffer, pH 2.5, containing TEA. As can be seen in Figure 7.30, baseline resolution of the main components was obtained for all derivatives.

The resolution of two of the three structural isomers of Man7 indicates that the mechanism of separation is not strictly based on differences in charge-to-mass ratio but also on the three-dimensional structure.[138] Due to their higher charges, the analysis time for the ANTS derivatives is much faster (see Figure 7.30).

The nature of the derivatizing agent influences the choice of the CE mode to be used in the subsequent separation step. In fact, MEKC proved useful in the analysis of complex oligosaccharides labeled with a hydrophobic tag, 2-aminoacridone. The derivatized oligosaccharides derived from ribonuclease B, hen egg albumin, and fetuin were separated[157] using a borate buffer containing taurodeoxycholate surfactant that has been shown useful for the analysis of 2-AA-derivatized monosaccharides.[118] The separation mechanism of such a system is based more on the borate complex formation than on the partitioning of the derivatives in the micellar phase, as was determined by the absence of resolution when bicarbonate buffer was used instead of borate. However, the addition of the surfactant was shown to improve the separation efficiency. The pattern of separation of the major components of the 2-AA-

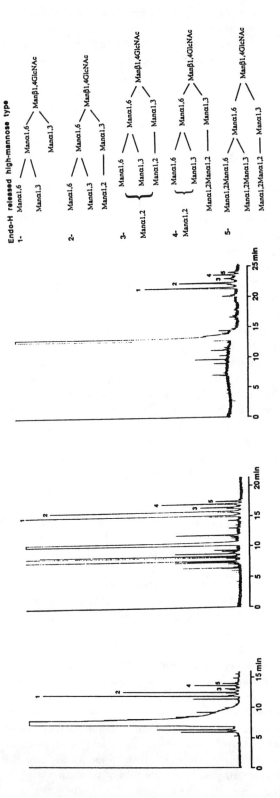

Figure 7.30 Electropherograms of high-mannose-type oligosaccharides from ribonuclease B derivatized with (a) ANTS, (b) 7-ANDA, and (c) 2-ANSA. Conditions: running electrolyte, 50 mM phosphate, ~36 mM TEA, pH 2.5; voltage: −20 kV; detection, UV, 235 nm; temperature: 25°C. (From Chiesa, C. and O'Neill, R. A., *Electrophoresis*, 15, 1132, 1994. With permission.)

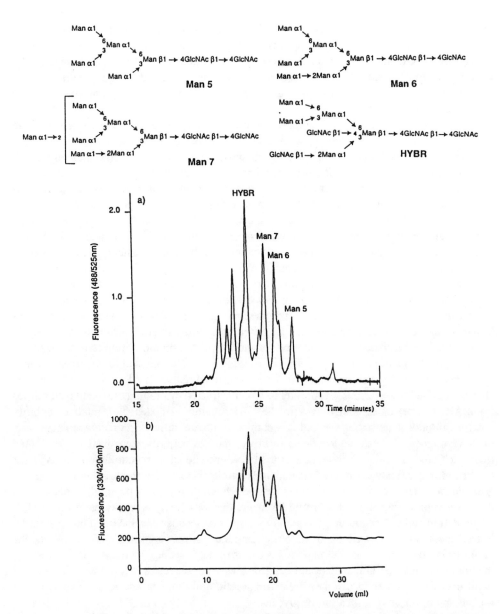

Figure 7.31 Analysis of derivatized N-linked oligosaccharides derived from hen egg albumin by (a) CE and (b) size exclusion chromatography using a GlycoMap column. CE conditions: running electrolyte, 500 m*M* sodium borate, pH 8.87, containing 80 m*M* taurodeoxycholate; capillary, 50 cm (to detection point)/57 cm (total length) × 50 μm i.d.; voltage, 20 kV; current, 50 μA, temperature, 22°C; injection, by hydrodynamic for 1 to 3 s. (From Camilleri, P., Harland, G. B., and Okafo, G., *Anal. Biochem.*, 230, 115, 1995. With permission.)

derivatized oligosaccharides derived from hen egg albumin is similar to that obtained by gel permeation (Figure 7.31).[157]

It should be noted that the time of analysis by the size exclusion chromatography (SEC) method is about 12 h compared to about 30 min by CE. However, the SEC method has the advantage of fraction collection for later enzymatic digestion and sequencing.[157] Four of the main peaks in Figure 7.31a were identified by spiking the 2-AA-oligosaccharides derived from the glycoprotein with authentic standards. The CE method was also utilized for fingerprinting the 2-AA derivatives of oligosaccharides derived from bovine fetuin and human IgG monoclonal antibodies,[158] and was shown to be useful in detecting differences in maps of the oligosaccharides released from glycoproteins from different culture systems.

Additional CE analysis of glycans was reported by Liu et al.[83] N-Linked oligosaccharides from bovine fetuin were released through hydrazinolysis and then derivatized with CBQCA. With on-column LIF detection, this tag permitted the CE of subpicogram amounts in a phosphate-borate buffer, pH 9.5.

Finally, to provide an HPCE methodology that makes it possible to draw inferences about structural characteristics of complex glycans and in turn expedite subsequent structural analyses, a novel method for identifying and quantifying desialylated N-glycosidically linked oligosaccharides in glycoproteins was introduced.[159-161] It is based on two-[160,161] or three-[159] dimensional mapping of different oligosaccharides by CE using different electrolyte systems in which the modes of separation differ from each other as much as possible. In the case of the two-dimensional mapping,[160,161] the oligosaccharides which were cleaved from various glycoproteins including fetuin, human transferrin, immunoglobulin G, human AGP, ribonuclease B (RNase B), and invertase can be classified into three major categories: the complex, high-mannose-, and hybrid-type glycans. The various oligosaccharides derivatized with 2-AP were separated using 100 mM phosphate buffer (pH 2.5) with polyacrylamide-coated fused-silica capillary (mode 1) or using 200 mM borate buffer (pH 10.5) with bare fused-silica capillary (mode 2). The mechanism of separation for mode 1 is based on differences in the charge-to-mass ratios as well as on differences in the hydrodynamic volumes of the derivatives. The mechanism of separation for mode 2 is based on differences in the stability constants of the *in situ* complexes formed between the borate ions and the hydroxyl groups of the derivatized oligosaccharides. The two-dimensional map of 2-AP-derivatized oligosaccharides derived from various glycoproteins was constructed by plotting the relative mobilities (to 2-AP-glucose) of the derivatives in the dual separation modes (i.e., relative mobility obtained in phosphate buffer versus that obtained in borate buffer). In this map, a few pairs of 2-AP-oligosaccharides could not be resolved, but the majority were well discriminated from each other. An important feature of the two-dimensional map is that the three types of oligosaccharides analyzed are distributed in separate regions. The high-mannose-type oligosaccharides are seen in the upper portion of the map, the hybrid-type are distributed from the lower right corner to the upper middle, while the complex-type are to the left-hand side of the hybrid-type. Furthermore, the complex-type oligosaccharides, having a larger number of outermost galactose residues, seem to occupy the upper portion of the map. Generally, the oligosaccharide position on the map moves toward the left as d.p. increases, and upward as the number of peripheral mannose and galactose residues increases. It was suggested that the two-dimensional mapping of relative electrophoretic mobilities might be a convenient tool for the approximate identification of glycans.[161]

Very recently, and along the same lines, a multidimensional mapping of the oligosaccharides derived from glycoproteins was proposed by Zieske et al.[159] The structures of 20 identified N-linked oligosaccharides derivatized with 2-AP have been assigned mapping positions from which comigrating unknown oligosaccharides can be characterized. This multidimensional mapping system had a different combination of buffer systems than those reported earlier,[160,161] and was shown to separate both charged and neutral oligosaccharides. One dimension involved electroosmotic flow-assisted CE in a sodium acetate buffer at pH 4.0. A second entailed separation based on borate complexation in a polyethylene glycol-containing electrolyte. A third dimension was based on sodium phosphate buffer, pH 2.5, which was able to resolve neutral oligosaccharides that were not separated by the other two dimensions. This multidimensional mapping might prove to be a valuable tool for the analysis and identification of glycoprotein-derived oligosaccharides.

f. Glycolipid-Derived Oligosaccharides

Very recently, Mechref et al.[80] reported the most suitable conditions for the selective precolumn derivatization of sialooligosaccharides derived from gangliosides, and the subsequent separation of the derivatives by CE. Seven sialooligosaccharides, whose structures and abbreviations are illustrated in Figure 7.1, were cleaved from gangliosides by ceramideglycanase and derivatized with ANDSA according to Scheme 7.2.[17,78] This precolumn derivatization, which involves the formation of a stable amide bond between the amino group of the ANDSA tag and the carboxylic acid group of the analyte, is very attractive for the labeling of sialooligosaccharides because it is readily achieved in an aqueous medium and at room temperature. Sialooligosaccharides are prone to desialylation at high temperature and in very acidic media. The ANDSA-sialooligosaccharide derivatives, which fluoresce at 420 nm when excited at 315 nm, were readily detected at the low femtomole level using an on-column

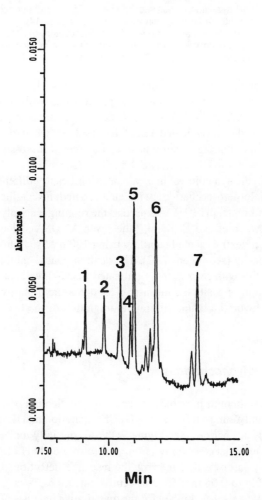

Figure 7.32 Electropherogram of ANDSA derivatives of sialooligosaccharides derived from gangliosides obtained on a dextran 150 kDa-coated capillary. Capillary, 47 cm total length (40 cm effective length) × 50 μm i.d.; running electrolyte: 0.10 M phosphate, pH 7.0; pressure injection, 1 s; applied voltage, −15 kV; detection, UV at 250 nm. Sample: 1 = Sialooligo-G_{D3}; 2 = Sialooligo-G_{M3}; 3 = Sialooligo-G_{T1b}; 4 = Sialooligo-G_{D1b}; 5 = Sialooligo-G_{D1a}; 6 = Sialooligo-G_{M2}; 7 = Sialooligo-G_{M1}. (From Mechref, Y. and El Rassi, Z., *Electrophoresis*, 16, 617, 1995. With permission.)

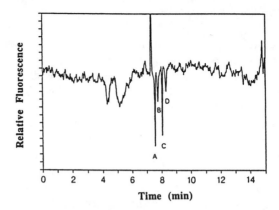

Figure 7.33 Separation and detection of polysaccharides in CE by indirect fluorescence. Capillary, 80 cm total length, 65 cm (to detection window) × 26 μm i.d., 140 μm o.d.; injection, electromigration 1 s at 15 kV; run at 15 kV with 1 mM fluorescein, pH 11.5. Solutes: A, dextran; B, comb-dextran; C, hydroxyethylamylose; D, amylose. (From Richmond, M. D. and Yeung, E. S., *Anal. Biochem.*, 210, 245, 1993. With permission.)

lamp-operated fluorescence detector. Since the precolumn derivatization with ANDSA replaces each weak carboxylic acid group of the parent sugar with two strong sulfonic acid groups, the various ANDSA-sialooligosaccharide derivatives are charged at all pH. Six of the seven sialooligosaccharides investigated could be resolved when using 100 mM sodium phosphate, pH 6.0, as the running electrolyte and an untreated fused-silica capillary. The two structural isomers sialooligo-G_{D1a} and sialooligo-G_{D1b} (see Figure 7.1) were not resolved, suggesting that their hydrodynamic volumes are not significantly different. The separation of the seven ANDSA-sialooligosaccharides with an uncoated fused-silica capillary was best achieved when 75 mM borate, pH 10.0, was used as the running electrolyte (see Figure 7.1b).[80] In another report by Mechref and El Rassi,[79] the seven ANDSA-sialooligosaccharides were perfectly separated in a dextran-coated capillary using 100 mM sodium phosphate buffer, pH 6.0 and a negative polarity (see Figure 7.32). The dextran-coated capillary used in that study exhibited a reduced EOF with respect to an untreated fused-silica capillary, and the reduced EOF was in the direction opposite the intrinsic electrophoretic mobility of the analytes. This condition could have favored a better differential migration of the two structural isomers.

C. Polysaccharides

1. Underivatized Polysaccharides

Thus far, only a few attempts have been made for the application of CE to polysaccharides. Recently, Richmond and Yeung[50] reported the HPCE separation and detection of high-molecular-weight native polysaccharides (see Figure 7.33). To partially ionize the various analytes and in turn achieve differential electromigration, an electrolyte of pH 11.5 was used. Detection was made possible by laser-excited indirect fluorescence detection. In general, migration times were reproducible to 0.05 min for consecutive injections.[50]

Capillary electrophoresis was utilized for the quantitative analysis of the glycosaminoglycan hyaluronan in human and bovine vitreous with UV detection at 200 nm.[30] A running electrolyte consisting of 50 mM disodium hydrogen phosphate, 10 mM sodium tetraborate, and 40 mM SDS, pH 9.0, was found to be optimum for assaying hyaluronan and its oligomers. This alkaline electrolyte ensures that hyaluronan migrates as a polyanion and minimizes the possibility of wall adsorption of both protein and polysaccharide components of the vitreous humor, while SDS binds to the proteins which then migrate away from hyaluronan. The signal corresponding to hyaluronan was confirmed by depolymerization of the native muco-

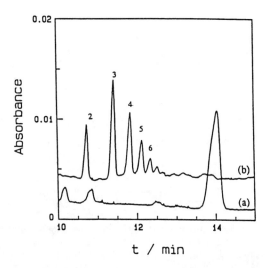

Figure 7.34 Electropherogram of (a) bovine vitreous, compared with (b) hyaluronidase digest of Hyaluronan. Peaks labeled according to the number of disaccharide units in the oligomer. Separation buffer: 50 mM disodium hydrogen phosphate, 40 mM SDS, 10 mM sodium tetraborate; capillary, 50 cm × 75 μm i.d.; voltage, 15 kV; detection at 200 nm. (From Grimshaw, J., Kane, À., Trocha-Grimshaw, J., Douglas, A., Chakravarthy, U., and Archer, D., *Electrophoresis*, 15, 936, 1994. With permission.)

polysaccharide by hyaluronidase. This resulted in the loss of the hyaluronan peak and the appearance of several new peaks corresponding to the oligomeric fragments which had shorter migration times (see Figure 7.34).

Also, complex polysaccharide mixtures were analyzed, for the first time, directly without derivatization by CE in strongly alkaline electrolyte solutions and electrochemically detected at a Cu microelectrode.[59] This was made possible by including the cationic surfactant cetyltrimethylammonium bromide which (1) resulted in reversing the direction of EOF, thus allowing the migration to be in order of increasing polysaccharide size, and (2) provided an ion-pairing mechanism between the negatively charged analytes and the positively charged surfactant which increased the differences in the charge-to-mass ratios among the different polysaccharides. The polysaccharides analyzed by this system included commercially available dextrans of up to an average molecular weight of 18,300.[59]

2. Dynamically Labeled and Derivatized Polysaccharides

Sudor and Novotny[103] reported the separation of neutral polysaccharides (e.g., chitosan, dextran, various water-soluble cellulose derivatives, etc.) labeled with CBQCA using capillaries coated with polyacrylamide and filled with an appropriate polymer solution such as linear polyacrylamide. In the presence of borate, the derivatized polysaccharides migrated readily in open tubular CE but showed little tendency to separate. Due to their large molecular size, the various polysaccharides investigated were not amenable to sieving using capillary gel electrophoresis because they did not penetrate the gel network. This hindrance was surmounted by causing the polysaccharides to migrate through solutions of entangled polymers, but the extent of separation in such sieving media was complicated by "biased reptation" behavior of the stretched-out macromolecules.[162] The biased reptation behavior was overcome by using pulsed field conditions where a potential gradient along the separation capillary was periodically inverted at a 180° angle, which brought about shape transitions and in turn favored the separation of polysaccharides according to molecular size. Figure 7.35 depicts the separation of six polydextran standards with molecular weights between 8,800 and 2,000,000 Da. This approach resembles what is known as pulsed-field electrophoresis, or

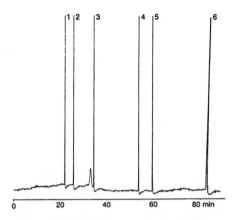

Figure 7.35 Separation of polydextran standards by pulsed-field capillary electrophoresis. Conditions: capillary, 20 cm (to detection point)/ 29 cm (total length) × 50 μm i.d., coated with linear polyacrylamide, and filled with linear polyacrylamide (5% T; 0% C); ratio of forward-to-reverse times, 2:1; applied voltage, ± 10 kV; and frequency, 3 Hz; running electrolyte, 50 mM sodium borate, 50 mM boric acid, 100 mM Tris, pH 8.81; electromigration injection, 5 s (1 kV). Peak identification: 1 = 8,800 Da, 2 = 39,100 Da, 3 = 70,000 Da, 4 = 503,000 Da, 5 = 667,800 Da, 6 = 2,000,000 Da. (From Sudor, J. and Novotny, M., *Proc. Natl. Acad. Sci. USA*, 90, 9451, 1993. With permission.)

field-inversion gel electrophoresis which was originally introduced for the separation of large DNA fragments in traditional gel slab electrophoresis.[163]

The electrophoretic migration of neutral and highly charged polysaccharides, such as chemically modified celluloses and heparins labeled with CBQCA, was regulated by secondary thermodynamic equilibria during capillary electrophoresis by using suitable buffer additives.[164] Electrophoretic migration of uncharged chemically modified cellulose was induced by the adsorption of hydrophobic and charged detergents onto the analyte components, and the process could be described by Langmuir adsorption isotherms. Different polymers were found to contain different adsorption sites, including multilayer formation. On the other hand, reduction of a high electrophoretic mobility of highly charged heparin was attained by the addition of ion-pairing reagents to the running electrolyte. Under these conditions, migration velocity and selectivity were influenced by the concentration and number of charges of the additive.[164]

Another approach for the electrophoresis and sensitive detection of polysaccharides has been the dynamic or *in situ* labeling of the analytes. Iodine dynamic labeling of polysaccharides has been shown recently to be an effective approach for CE separation and UV detection of amylopectin and amylose.[73] Iodine complexation with carbohydrates imparts charge and optical detection sensitivity at 560 nm. The starch–iodine complex consists of a helix of sugar residues surrounding a linear I_5^- core.[165,166] Unlike borate complexation, iodine binding is a cooperative interaction which exhibits strong chain-length dependence in both complexation and optical properties. The iodine binding constant increases nearly exponentially with glucan chain length, reaching a plateau at approximately 125 residues.[167] Moreover, the wavelength of maximum absorbance exhibits a red shift with increasing chain length.[168] These facts could be utilized to reveal information on the size and structure of the analytes independent of their electrophoretic behavior. Amylopectin and amylose were well resolved from each other in less than 10 min using uncoated capillaries and a separation electrolyte consisting of 20 mM citrate-phosphate, pH 6.0, containing 0.1 mg/mL potassium iodide and 0.1 mg/mL iodine. Amylopectin electrophoretic mobility is dependent on the iodine concentration as well as on the separation temperature, while that of amylose is not. This reflects the fact that the long chains of amylose are essentially saturated with iodine even at low iodine concentration. This system was also shown to be useful in the analysis of potato starch, amizo V, corn starch, and maltodextrin.[73]

Another approach to dynamic labeling of polysaccharides was reported very recently by Toida and Linhardt[74] and involved the separation of GAGs as their copper (II) complexes. This was based on the early report by Mukherjee et al.[75] that demonstrated the capability of all GAGs, except keratan sulfate, for binding to copper (II), thus forming a copper (II)-GAGs complex that has an absorbance maximum near 237 nm. The UV absorbance of these complexes was shown to be pH-dependent with the highest value observed at pH 4.5.[74] Therefore, GAGs including heparin, heparan sulfate, chondroitin sulfate, dermatan sulfate, and hyaluronic acid were analyzed as their copper (II) complexes by reversed polarity CE using 5 mM copper (II) sulfate, pH 4.7. Under these conditions, each GAG analyzed was migrated as a broad peak, which hindered the analysis of all GAGs in a single run even though the migration times for each GAG were different (except for heparin and LMW heparin). Several attempts were made to improve the resolution of the system including variation of copper (II) concentration, pH, and the separation voltage, as well as the addition of neutral, anionic, and cationic surfactants. Variation of copper concentration or pH of the separation electrolyte affected the sensitivity of the system but not the resolution. However, the addition of surfactants of any type decreased resolution. Increasing the separation voltage improved efficiency, but decreased resolution.

D. Glycopeptides and Glycoproteins

1. Glycopeptides

The high resolving power and unique selectivity of CE were also utilized for the separation and characterization of peptide and glycopeptide fragments of glycoproteins. Figure 7.36 illustrates the CZE mapping of the tryptic peptide fragments of human AGP as well as the submapping of its glycosylated and nonglycosylated fragments.[114] Prior to CE analysis, the whole tryptic digest of AGP was first fractionated into peptide and glycopeptide fragments by high performance lectin affinity chromatography on a Concanavalin A (Con A) column. Three pooled fractions were obtained; the first two being Con A nonreactive and Con A slightly reactive eluted with 20 mM phosphate pH 6.5 containing 0.1 M NaCl, while the third fraction interacted strongly with the Con A column and eluted with the heptanic sugar, i.e., methyl-α-D-mannopyranoside. The three fractions were then analyzed by CZE using a fused-silica capillary with fuzzy polyether coating and 0.1 M phosphate, pH 5.0, as the running electrolyte. As seen in Figure 7.36, the CZE mapping of the whole digest reveals the micro-heterogeneity of the glycoprotein as manifested by the excessive number of peaks for a protein of 181 amino acid residues with 20 trypsin cleavage sites (8 lysine and 12 arginine residues). In fact, by neglecting all sources of heterogeneity in AGP, the tryptic digest of such a glycoprotein should give 12 peptide and 5 glycopeptide fragments, and three single amino acids (two lysine and one arginine). However, more than any other serum glycoproteins, AGP is a highly heterogeneous protein with a peculiar structural polymorphism.[169] Substitutions were found at 21 of the 181 amino acids in the single polypeptide chain, which is responsible in part for multiple peptide and glycopeptide fragments in the tryptic digest. Another source of multiple fragments in the tryptic digest is the microheterogeneities of the oligosaccharide chains attached, as can be seen from the structure in Figure 7.37. In fact, the variation in the terminal sialic acid causes charge heterogeneity in the glycopeptide fragments cleaved at the same location by trypsin. The differences in the extent of glycosylation among a population of the protein molecules leads to fragments having the same peptide backbone but with or without carbohydrate chains, and the variation in the nature of the oligosaccharide chains at each glycosylation site yields several glycopeptides that have the same peptide backbone, but differ in their oligosaccharide structures.[114]

As can be seen in Figure 7.36, the CZE submapping of Con A reactive peptides produced peaks that are missing from the submaps of all other collected fractions, i.e., 0, 0′ and 1, but

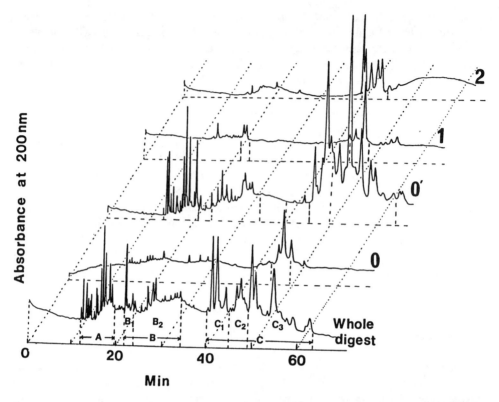

Figure 7.36 Capillary zone electrophoresis tryptic mapping and submapping of human AGP. Capillary, fused-silica tube with hydrophilic coating on the inner walls, 45 cm (to the detection point), 80 cm total length × 50 μm i.d.; electrolyte, 0.1 *M* phosphate solution, pH 5.0; running voltage, 22.5 kV; injection by electromigration for 4 s at 22.5 kV. Symbols: fraction 0 = Con A nonreactive (excluded from the column); fraction 0′ = Con A nonreactive (unretained by the column); fraction 1 = Con A slightly reactive (eluted with buffer); fraction 2 = Con A strongly reactive (eluted with the haptenic sugar). (From Nashabeh, W. and El Rassi, Z., *J. Chromatogr.*, 536, 31, 1991. With permission.)

whose components are found in the whole map (see area C1, Figure 7.36). This approach allows the monitoring of a group of peptides as well as the assessment of glycosylated fragments in the whole map. This methodology is also expected to work with other glycopro-

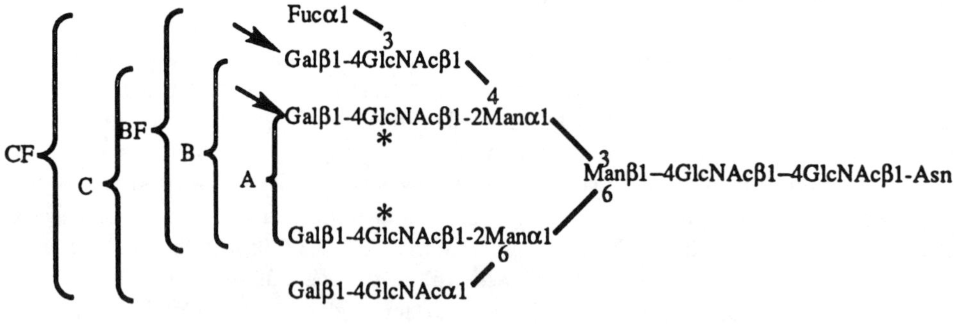

Figure 7.37 Primary structure of the carbohydrate classes of AGP. There are five carbohydrate classes attached to AGP having different degrees of branching and sialylation. Classes A, B, and C are the bi-, tri- and tetraantennary complex N-linked glycans, respectively, whereas BF and CF are the fucosylated B and C structures. Two additional glycans exist. One has two additional fucoses linked to the GlcNAc residues marked with asterisks, and one has an outer chain prolonged by Galβ1-4GlcNAc at either of the Gal residues marked with an arrow.

teins, and the CZE submapping of all the glycosylated tryptic fragments with different types of glycans may require the use of more than one lectin column in the prefractionation step.[114]

Also, CE has been shown useful in evaluating the glycopeptide microheterogeneity of recombinant human erythropoietin (r-HuEPO) expressed from Chinese hamster ovary (CHO) cells or *Escherichia coli*. This was achieved using 100 mM heptanesulfonic acid (ion-pairing agent) in 40 mM sodium phosphate buffer, pH 2.5.[170] The negatively charged heptanesulfonic acid forms ion pairs with basic amino acid residues, thus reducing analyte-wall interaction as well as altering analyte electrophoretic mobility. This led to improved peptide resolution, which allowed the evaluation of the heterogeneity of glycopeptides derived from r-HuEPO. The total tryptic map exhibited two regions, nonglycosylated and glycosylated peptides. Since r-HuEPO glycoprotein has three glycosylation sites, three glycopeptides are expected to result from tryptic digestion of this glycoprotein. However, the aforementioned electrophoretic system revealed the microheterogeneity of these glycopeptides by yielding at least 12 glycopeptide peaks in the peptide map.[170] Sialidase, which removes the sialic acid residue from both N- and O-linked oligosaccharides, and N-glycanase, which cleaves N-linked oligosaccharides, were used to simplify the level of complexity of the glycopeptide, thus permitting the characterization of the glycopeptides. According to the study, CE evaluation of glycopeptide microheterogeneity appears to be simpler, faster, and just as sensitive as other more frequently employed methods for glycopeptide analysis.

The utility of HPLC and CE using UV detection and PAD for the characterization of glycopeptides from recombinant coagulation factor VIIa (rFVIIa) has been demonstrated recently.[63] The combination of the more traditional methods of HPLC-UV and HPAEC-PAD with capillary electrophoresis methods based on CZE-UV and CZE-PAD allowed a better characterization of glycopeptide heterogeneity. In addition, this report demonstrated the potential of CZE-PAD in the analysis of PNGase F-treated glycopeptides where, in contrast to UV detection, both the peptides and the released carbohydrates can be detected simultaneously.[63] In another report, the same authors demonstrated the utility of CE-PAD in the analysis of a tryptic digest of recombinant tissue plasminogen activator (rt-PA).[64] CE-PAD provided structural information in terms of the glycosylation of rt-PA by simply using different detection potentials in sequential runs on a sample of the tryptic digest of rt-PA. This is based on the fact that at pH 9.3 both amino and hydroxyl groups give good responses at +0.39 V, while at +0.15 V hydroxyl groups give a significantly larger response than amino groups. Consequently, a change in the peak height for a certain sample, as the detection potential varies, furnishes structural information about the fragments of the tryptic digest, such as glycosylation.[64]

Finally, glycopeptides of the recombinant tissue plasminogen activator (rt-PA) carrying hybrid and complex type oligosaccharides were analyzed by CE using 100 mM tricine buffer, pH 8.2, containing 1.25 mM DAB.[136] The glycopeptides used were fraction collected from a tryptic digest of rt-PA by HPLC. The use of a borate buffer did not provide high resolution for the fractions analyzed, and the migration orders obtained with phosphate and borate buffers were the same, suggesting weak borate complexation with the carbohydrate moieties of the glycopeptides analyzed. The addition of a divalent cation such as DAB to a tricine buffer improved the resolution of the system by modifying the EOF as well as the electrophoretic mobility of the glycopeptides through electrostatic interaction.

2. Glycoprotein Glycoforms

Protein glycosylation can occur at two or more positions in the amino acid sequence, and the glycans even at a single position may be heterogeneous or may be missing from some molecules. Thus, glycoproteins exist as populations of glycosylated variants, referred to as glycoforms, whose relative proportions are found to be reproducible and not random. However, the glycoforms may be affected by several factors including the environment in which the protein is glycosylated, the manufacturing process, and the isolation procedures. This would affect the function of a glycoprotein, thus engendering the need for high resolution separation methods to allow the monitoring of glycoform populations especially for genetically engineered glycoprotein pharmaceuticals.

Several CE approaches have been described for profiling glycoprotein glycoforms. Iron-free transferrin glycoforms were separated by CZE and CIEF.[171] In both modes of CE, at least five components corresponding to the di-, tri-, tetra-, penta-, and hexasialo-transferrins differing from each other by one negative charge were resolved. The capillary columns used in this study were coated with linear polyacrylamide to suppress EOF and consequently provide better resolution and sharper focusing of the closely related glycoforms by CZE and CIF, respectively. To assess the presence of varying degrees of sialylation among the various glycoforms, the action of neuraminidase on the electrophoretic behavior of the various isoforms was monitored by CE. Neuraminidase is an exoglycosidase that specifically liberates the negatively charged sialic acids from the terminal nonreducing positions in glycans. As shown in Figure 7.38, the electrophoretic analysis of samples taken from the enzymatic

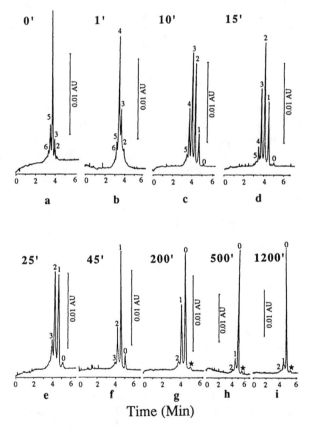

Figure 7.38 CZE of iron-free transferrin following incubation with neuraminidase. Capillary, 18.5 cm × 50 μm i.d.; electrolyte, 18 mM Tris-18 mM boric acid-0.3 mM EDTA, pH 8.4; running voltage, 8 kV. The samples for electrophoresis were taken after various incubation times: (a) 0; (b) 1; (c) 10; (d) 15; (e) 25; (f) 45; (g) 200; (h) 500; (i) 1200 min. The proportions of the transferrin isoforms (asialo, mono-, di-, trisialo, etc., marked 0, 1, 2, 3, etc.) changed with time. The sample taken after 20 h still contained transferrin molecules having one and two sialic acids. The small peak (labeled with an asterisk) appeared after 50 to 80 min, but did not increase in size on prolonged incubation time (g–i). (From Kilàr, F. and Hjerten, S., *J. Chromatogr.*, 480, 351, 1989. With permission.)

digestion at various time intervals demonstrated the gradual removal of sialic acid as manifested by the changes in the relative proportions of the different isoforms with time. The electrophoretic pattern of the final product was completely different from the starting material and showed one main component, the asialo-transferrin. Thus, in the case of transferrins, the major source of microheterogeneity seems to be the variation in the terminal sialic acid of the glycans. This microheterogeneity, which leads to broad or smeared bands in traditional electrophoresis, can be readily resolved by CE due to its high separation efficiencies. CIEF was also employed by the same authors for the analysis of iron-free transferrin and transferrin sample of known iron content.[172] This method gave a fast analysis and good resolution. Moreover, the molecular forms of transferrin (iron-free, monoferric, and diferric complexes) were easily identified by monitoring the focused protein zones at both 280 and 460 nm, since each form absorbs light differently.[172]

The microheterogeneity of glycoproteins results from sugar residues other than the sialic acid residues. This is the case of ribonuclease B (RNase B) whose glycan portions are of the high-mannose type. The separation of the five glycoforms of RNase B has been achieved through the formation of anionic borate complexes with the hydroxyl groups of the glycan moiety (see Figure 7.39a).[173] The relative proportions of the various glycoforms correlated with the relative proportions of the high-mannose glycan populations, i.e., Man_9-Man_5, determined by other more established analytical methods, e.g., mass spectrometry, high performance anion exchange chromatography (HPAEC), and size exclusion chromatography on Bio-Gel P4. To further substantiate the presence of the various glycoforms of RNase B, the time course for the digestion of the protein with *A. saitoi* $\alpha(1\text{-}2)$ mannosidase was monitored by CZE. Mannosidase is an exoglycosidase that specifically cleaves mannose from the nonreducing end of glycans. Figure 7.39b shows that after 25 h the glycoform populations carrying Man_9-Man_6 structures were all reduced to a single population carrying Man_5. Thus,

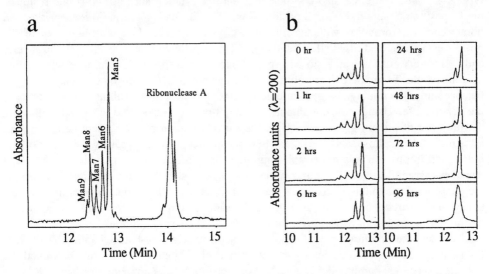

Figure 7.39   (a) CZE profile of ribonuclease showing the nonglycosylated form of the protein, ribonuclease A, and the glycoforms of the same protein, collectively known as ribonuclease B. RNase A is a contaminant of RNase B as supplied by Sigma. RNase B remained unaffected during the digestion of the oligosaccharide component of RNase B with *A. saitoi* $\alpha(1\text{-}2)$ mannosidase. (b) CZE profile of RNase B showing the time course for the digestion of the glycoprotein with the exoglycosidase, *A. saitoi* $\alpha(1\text{-}2)$ mannosidase. Capillary, fused-silica 72 cm × 75 μm i.d.; applied voltage, 1 kV for 1 min and 20 kV for 19 min, temp., 30°C; detection, UV at 200 nm; injection 1.5 s; electrolyte, 20 mM phosphate containing 50 mM sodium dodecylsulfate, 5 mM borate, pH 7.2. (From Rudd, P. M., Scragg, I. G., Coghil, E., and Dwek, R. A., *Glycoconjugate J.*, 9, 96, 1992. With permission.)

CE offers a direct method for analyzing glycoforms at the protein level with high resolution and precision. This method was recently utilized by Rudd et al.[174] to verify the purity of a single glycoform of RNase B resulting from treatment of the glycoprotein with mannosidase.

Another example of high-mannose-related microheterogeneity has been the various glycoforms of recombinant human bone morphogenetic protein 2 (rhBMP-2), which is a disulfide-linked homodimeric glycoprotein.[175] The separation of rhBMP-2 glycoforms by CE was achieved simply using phosphate buffer, pH 2.5, containing no additives. Under this condition, the rhBMP-2 sample yielded nine peaks which have been identified as glycoforms of rhBMP-2. The difference between any adjacent peaks is only one mannose residue ($M_r = 162$). The nine peaks obtained with intact rhBMP-2 reduced into one major peak when the endo-H digested rhBMP-2 was analyzed by CE. Endo-H is an endoglycosidase specific for the cleavage of high-mannose glycans from glycoproteins. This confirms that the microheterogeneity of the glycoprotein is due to the high-mannose-type carbohydrates. The migration order of the glycoforms was found to follow the increasing number of mannose residues in the analyte molecules. It was suggested that a mannose residue can effect the separation by providing higher friction coefficient rather than by a charge shielding effect.[175]

Phosphorylation of glycans seems to be another source of glycoprotein microheterogeneity. In fact, CE analysis of proteinase A glycoforms, both native and underglycosylated, revealed charge heterogeneities attributed to differences in the phosphorylation level of the carbohydrate moiety at Asn-68.[176] Both forms were separated into three distinct peaks that probably correspond to charge heterogeneities because of differences in carbohydrate phosphorylation.

The microheterogeneity of a glycoprotein is mostly the result of glycosylation and is largely unaffected by the presence of other functionalities in the protein. This is the case in ovalbumin, a phosphorylated glycoprotein. This protein has one asparagine residue that can accommodate at least nine different carbohydrate structures of the high-mannose and hybrid-type N-glycans. There are also two potential phosphorylation sites at two serine residues, one at position 68 and the other at position 344. The various glycoforms were separated *via* borate complex formation with the hydroxyl groups of the carbohydrate moieties of the protein using untreated fused-silica capillary.[177] This glycoprotein is a strongly acidic species and therefore would not undergo adsorption onto the naked capillary surface when using alkaline borate. To improve the resolution of the ovalbumin glycoforms, putrescine (i.e., 1,4-butanediamine), a doubly charged cationic species, was added in small amounts (1 mM) to the borate buffer. Using these conditions, five major protein peaks were separated, indicating the presence of protein glycoforms. Upon dephosphorylation of the glycoprotein with calf intestinal alkaline phosphatase or potato acid phosphatase, the five peaks were still resolved but shifted in position to a more rapid migration time, a behavior consistent with a loss of negative charge. Based on this observation, it was suggested that all ovalbumin glycoforms are phosphorylated to the same degree, and heterogeneity among ovalbumin isoforms resides solely in the carbohydrate structures. Also, the same electrophoretic system was shown to permit the separation of pepsin glycoforms, yet with a lower resolution.

Realizing the benefit of including an amine additive (e.g., 1,4-butanediamine, DAB) into the running electrolyte in achieving the separation of ovalbumin glycoforms prompted the investigation of other amine additives such as α,ω-bis-quaternary ammonium alkanes.[178] Three of the α,ω-bis-quaternary ammonium alkane additives, namely hexamethonium bromide (C_6MetBr), hexamethonium chloride (C_6MetCl), and decamethonium bromide (C_{10}MetBr), were examined, and the results were compared to those obtained using DAB as a buffer additive.[178] The alkyl chain length and the cation group of α,ω-bis-quaternary ammonium alkanes strongly influence the analysis time and resolution.[178] Under identical separation conditions, C_{10}MetBr was shown to yield a better resolution and shorter analysis time than C_6MetBr. Originally, with DAB it was thought that such an additive exerts its effect by mainly altering the EOF.[177] The fact that the additives repressed EOF similarly but the quaternary ammonium alkane additives allowed the resolution of ovalbumin glycoforms in half the time required with DAB suggests that the mechanism of action of these additives is not solely related to EOF repression. Other mechanisms may be involved, including protein-additive interactions, protein-wall interactions, additive-wall coating interactions, or any combination of these. In the same report, seven of the eight glycoforms of human chorionic gonadotropin (hCG) were resolved by CE using 1 mM C$_6$MetBr and 25 mM borate pH 8.4.[178] However, the eight glycoforms of the hCG were near-baseline resolved using 25 mM borate buffer, pH 8.8, containing 5 mM diaminopropane and separated in less than 50 min.[179]

In another very recent report, the effect of cationic buffer additives on the CE separation of serum transferrin (Tf) glycoforms from different species was examined.[180] For bovine Tf, the selectivity of borate buffer originating from carbohydrate-borate complexation allowed partial resolution of five peaks representing different sialylated forms of the glycoproteins. However, baseline resolution was attained only when DBA was included in the buffer. In both cases, the analysis was performed with an uncoated capillary. On the other hand, the selectivity of the system did not provide good resolution for the separation of Tf sialoforms from other species, not even with α,ω-quaternary ammonium alkane, which was shown earlier to provide comparable or better resolution than DBA.[178] Nevertheless, human Tf glycoforms both iron-saturated (huTf$_{sat}$) and iron-depleted (huTf$_{dep}$) were resolved in an uncoated capillary using a borate buffer containing millimolar concentrations of decamethonium bromide, an α,ω-bis-quaternary ammonium alkane buffer additive. Under these conditions, the various sialoforms of huTf$_{sat}$, which were presumably representing the hexa-, penta-, tetra-, and trisialic acid forms, were resolved, and the glycoforms were found to migrate differently than their iron-depleted counterparts. Although the resolution achieved under these conditions is high, the authors concluded that the lengthy analysis time is incompatible with the requirements of a clinical CE-based assay.

Along with the above separation strategies which include (1) the employment of either borate complexation or amine additives and (2) the use of specific enzymes directed either toward the glycan moieties or other protein functionalities, CE was applied successfully to the analysis of recombinant glycoprotein glycoforms. Yim[181] demonstrated the separation of the human recombinant tissue plasminogen activator (rt-PA) by CZE and capillary isoelectric focusing (CIEF). Tissue plasminogen activator is a fibrin-specific protein that has been approved for the treatment of myocardial infarctions. The CE analysis of two main glycosylation variants (type I and II) of the same glycoprotein showed different electrophoretic migration patterns. The study further elucidated the microheterogeneity of the glycoprotein as was manifested by the partial resolution of almost 15 glycoforms in a protein that has only four possible N-glycosylation sites. This report compared the CZE profile of an rt-PA sample to that of a desialylated rt-PA obtained through neuraminidase treatment, an approach similar to that introduced by Kilàr and Hjerten for human transferrin.[171] The desialylated rt-PA exhibited a much simpler CZE profile, indicating that the glycoprotein microheterogeneity is mostly the result of different levels of sialylation.

Very recently, the charged glycoforms of recombinant tissue-type plasminogen activator (rt-PA) were separated by capillary isoelectric focusing (CIEF) on the basis of their sialic acid content.[182] This method was validated as an alternative to slab gel isoelectric focusing (IEF). The pI of the rt-PA measured with the synthetic pI standards was in the pH range of 6.5 to 7.5, and the migration of the standards was affected by the presence of the protein. The method showed an acceptable recovery of >100%, and exhibited good sensitivity where 25 ng of protein could be resolved into constituent peaks. The method was linear over the 50 to 1000 µg/mL range. The age of the capillary was shown to affect peak migration time and, to a lesser extent, resolution. Overall, the data indicated that the methodology equals the performance of slab gel IEF, yet offers significant improvements in ease of operation and in time and amounts of reagents used.

Other studies on the potentials of CE in analyzing recombinant glycoprotein glycoforms include the report of Watson and Yao[183] on the use of CE in the separation of glycoforms of recombinant human granulocyte-colony-stimulating factor (rhGCSF) produced in Chinese hamster ovary cells. This glycoprotein contains only two O-linked carbohydrate moieties that differ only in having one or two sialic acid residues. Due to its relative simplicity compared to other more complex glycoproteins, the rhGCSF yielded two well-resolved and equal-sized peaks using phosphate-borate buffer, pH 8.0, and an untreated fused-silica capillary. Under these conditions, the acidic protein was repelled from the negatively charged capillary wall, and no apparent solute adsorption was observed. The resolution of the two equally present glycoforms was further enhanced upon adding 2.5 mM of 1,4-diaminobutane to the phosphate-borate buffer. The 1,4-diaminobutane apparently reduced the electroosmotic flow and further minimized solute–wall interaction. These effects allowed improved resolution but prolonged the separation time. As expected, the two glycoforms migrated in order of increasing numbers of sialic acids, since the separation was carried out in the positive polarity, i.e., anode to cathode. Of course, when the glycoforms were incubated with neuraminidase, a single peak was obtained eluting earlier than either of the original two sialylated glycoforms.

The high selectivity of CE was also demonstrated in the separation of the glycoforms of recombinant human erythropoietin (r-HuEPO),[184] a glycoprotein hormone produced in the kidney of adult mammals that acts on bone marrow erythroid progenitor cells to promote their development into mature blood cells. The glycoprotein has a polypeptide backbone of 165 amino acids and contains two types of carbohydrates: three N-linked complex oligosaccharides at the asparagine positions 24, 38, and 83, and one O-linked oligosaccharide chain at the serine position 126. The carbohydrate moieties make up 40% of the molecular weight of this 35 kDa glycoprotein. The effects of the pH, buffer type, organic additives, and capillary pre-equilibration time on the resolution of the microheterogeneity of r-HuEPO were investigated. The main factors for improving the resolution were the regulation of the EOF of the running buffer and the reduction of solute–wall interactions. For instance, in the pH range 6.0 to 9.0, pH 6.0 was found to give the best resolution. At this pH, a combination of two effects, reduction in the EOF and increase in the differences in charge between the glycoforms, resulted in enhanced separations. Optimum resolution of the four glycoforms was obtained by using a mixed acetate-phosphate buffer, pH 4.0.

In another study, optimum separation of the various glycoforms of r-HuEPO glycoprotein were obtained in the presence of an electrolyte consisting of 10 mM tricine-NaCl, 2.5 mM DAB, and 7 M urea, pH 6.2.[185] The migration order of the glycoforms was in the order of increasing number of sialic acid residues in the glycoforms. This was proved by comparison with the results obtained with gel isoelectric focusing and by spiking the r-HuEPO sample with individual glycoforms isolated by preparative isoelectric focusing.

James and co-workers[186] demonstrated the usefulness of MEKC in resolving recombinant interferon-γ (IFN-γ) glycoforms produced by Chinese hamster ovary cells. Separations were performed in uncoated fused-silica capillaries at alkaline pH in the presence of SDS micelles. Optimal separation was obtained with 400 mM borate buffer containing 100 mM SDS, pH 8.5. It was noted that optimum separation of IFN-γ glycoforms occurred at a pH close to the pI of the protein (8.5 to 9.0), because it is most susceptible to hydrophobic interaction at this pH. However, this electrolyte system did not resolve bovine serum fetuin nor α_1-acid glycoprotein glycoforms and showed partial resolution for ribonuclease B and horseradish peroxidase glycoforms.[186]

A combination of HPLC and CE was found useful in the analysis of the glycoforms of human recombinant factor VIIa (rFVIIa).[187] Again, the use of DAB was found to be essential for the separation of the various glycoforms. Under optimum conditions, rFVIIa migrated as a cluster of six peaks or more. The separation was reported to be primarily based on the different content of sialic acid of the oligosaccharide structures of rFVIIa, as was concluded from the CE separation of the neuraminidase-treated rFVIIa.

Very recently, the utility of CE in the area of glycoproteins was extended to include the profiling of oligosaccharide-mediated microheterogeneity of a monoclonal antibody using a borate buffer.[188] Under these conditions, complex formation between borate and the oligosac-charides of the monoclonal antibody furnishes information regarding the microheterogeneity of the analyte, since the analysis of the monoclonal antibody in phosphate buffer yielded one peak only. To further substantiate the glycosylation microheterogeneity of the antibody, the sugar moieties were enzymatically or chemically cleaved from the protein. The cleaved product (i.e., deconjugated protein) exhibited a different separation profile than the intact monoclonal antibody. This method was successfully employed for testing batch-to-batch consistency and for stability testing of the antibody. Additional information was obtained by matrix-assisted laser desorption ionization time-of-flight mass spectrometry which supple-mented CE for acquiring profound knowledge on the integrity of the glycoprotein.[188]

CE analysis of several glycoproteins including RNase B, ovalbumin, horse radish perox-idase, and lectin from *E. corallodendron* conducted under anodal EOF with polybrene-coated capillaries and acidic buffers was demonstrated by Kelly et al.[189] Baseline resolution of RNase B glycoforms was achieved using 2.0 M formic acid as the running electrolyte. These conditions were found entirely compatible for interfacing with an electrospray mass spec-trometry (ESMS). Mass spectral detection was useful in obtaining information on the structure of the analyte and in locating and identifying the carbohydrate attached to the protein. The CE-ESMS analysis of RNase B using 2.0 M formic acid is shown in Figure 7.40. The total ion current for the full mass scan acquisition (m/z 1300 to 2000) is presented in Figure 7.40A, while the extracted mass spectra for the peaks observed at 35.6, 35.2, and 40.8 min are portrayed in Figure 7.40B–D, respectively. The analyte ions are easily distinguished from the background signal, and a series of ions corresponding to multiply protonated forms of the molecule were observed in each mass spectrum.

The relative concentrations of RNase B glycoforms obtained using CE-ESMS is virtually indistinguishable from that obtained with CE-UV.[189] This method was also shown to be effective for the analysis of ovalbumin glycoforms which consist of two families of closely related *N*-linked carbohydrates, or for the analysis of other glycoproteins whose glycoforms contain complex-type carbohydrates.[189]

E. Glycolipids

The utility of HPCE in the separation of gangliosides, the sialic acid-containing glyco-sphingolipids, has been demonstrated by three different laboratories.[80,190,191] As shown in Figure 7.41, a ganglioside molecule has a hydrophilic sialooligosaccharide chain and a hydrophobic moiety, i.e., ceramide, that consists of a sphingosine and fatty acid. Due to this inherent structural feature, the gangliosides are amphiphilic solutes forming stable micelles in aqueous solutions with very low critical micellar concentration values (10^{-8} to 10^{-10} M).[192] Furthermore, gangliosides most often exist at low concentrations and their structures lack strong chromophores. Thus, two major obstacles must be overcome when developing a CE method for the analysis of gangliosides: (1) to be able to separate them as monomers and (2) to detect them at low levels. Thus far, CE separation of gangliosides as monomers has been addressed recently by Yoo et al.[191] and further elaborated by Mechref et al.,[80] while their detection at low levels was achieved by Mechref et al.[80]

On the other hand, Liu and Chan[190] applied CE to the study of the behavior of gangliosides in aqueous solutions. These researchers[190] demonstrated that CZE can separate some gangli-oside micelles, and consequently permitted the studies of the micellar properties of these amphiphilic species using untreated fused-silica capillaries and on-column direct UV detec-tion at 195 nm. The ganglioside micelles were successfully analyzed within 10 min with mass sensitivity in the order of 10^{-11} mol. Baseline resolution of a mixture of three ganglioside micelles, namely G_{M1}, G_{D1b}, and G_{T1b}, was achieved using 2.5 mM potassium phosphate, pH 7.40, as the running electrolyte (see Figure 7.42). The separation was mainly facilitated by the varying content of the sialic acid residues in the ganglioside micelles, which imparted

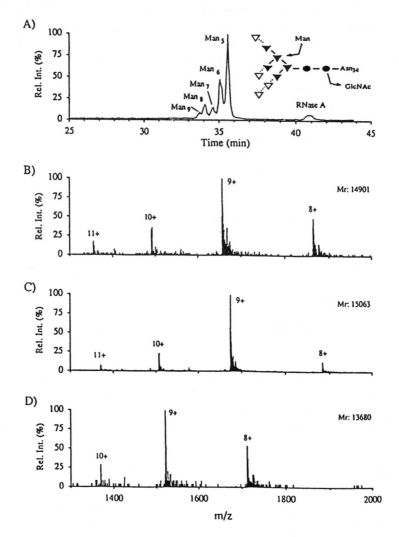

Figure 7.40 CE-ESMS analysis of RNase B. (A) total ion electropherogram for the full mass scan acquisition (m/z 1300 to 2000). Extracted mass spectra for peaks migrating at 35.2 (B), 35.6 (C), and 40.8 min (D). The calculated molecular weight is shown on the right corner of each spectrum. Conditions, capillary, fused-silica 110 cm × 50 μm i.d. coated with a solution of 5% polybrene and 2% ethylene glycol; running electrolyte, 2.0 *M* formic acid; injection, 6 pmol of RNase B. Sheath flow of 5 μL/min of an aqueous solution of 0.2% formic acid and 25% methanol. (From Kelly, J. F., Locke, S. J., Ramaley, L., and Thibault, P., *J. Chromatogr. A*, 720, 409, 1996. With permission.)

them with different electrophoretic mobilities. The observed migration velocities of the ganglioside micelles seemed to be largely unaffected in the pH range 7.0 to 11.0, and decreased monotonically when the pH was varied from 6.0 to 4.0. This is because of the diminishing dissociation of the carboxyl group of the sialic acid residues as the pH was decreased.[190] Increasing the ionic strength of the running electrolyte decreased the migration and broadened the peaks of the gangliosides. This may be due to the fact that increasing the ionic strength would decrease the electrostatic repulsion between the ganglioside monomers, thus increasing the number of monomers in the aggregates, and consequently the size of the ganglioside micelles. This increase in the size of the micelles would result in decreasing their migration velocity. The authors also studied the time course of mixed micelle formation of gangliosides. Both G_{D1b} and G_{T1b} were first separated into two distinct peaks after initial

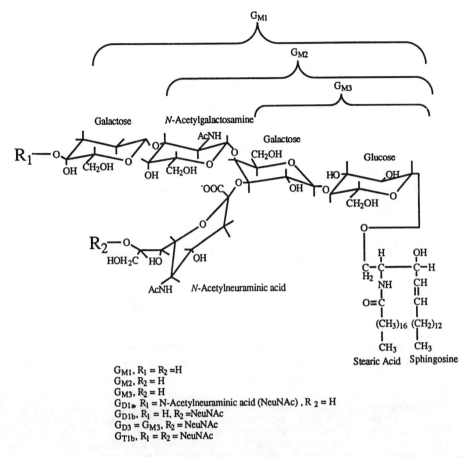

Galactose   N-Acetylgalactosamine

Galactose

Glucose

G_{M1}, $R_1 = R_2 = H$
G_{M2}, $R_2 = H$
G_{M3}, $R_2 = H$
G_{D1a}, $R_1 = $ N-Acetylneuraminic acid (NeuNAc) , $R_2 = H$
G_{D1b}, $R_1 = H$, $R_2 = $NeuNAc
$G_{D3} = G_{M3}$, $R_2 = $ NeuNAc
G_{T1b}, $R_1 = R_2 = $ NeuNAc

Figure 7.41 Structures of the gangliosides. (From Mechref, Y., Ostrander, G. K., and El Rassi, Z., *J. Chromatogr. A*, 695, 83, 1995. With permission.)

mixing. However, upon incubation at 37°C, complete fusion between both micellar peaks could be observed in less than 2.5 h (see Figure 7.42b). The fusion process was temperature-dependent. In fact, at 50°C the formation of mixed micelles between G_{D1b} and G_{T1b} was complete within 30 min. In contrast, no fusion of the ganglioside peaks was observed at 0°C even after 75 h. The mixed micelle formation seems to be dependent on the sialic acid content of the individual gangliosides. Whereas mixed micelle formation between G_{D1b} and G_{D1a} and that between G_{D1b} and G_{T1b} required 1.5 and 3.0 h, respectively, the aggregation of G_{M1} with polysialogangliosides (i.e., G_{D1b} and G_{T1b}) was 6- to 36-fold slower, under otherwise identical incubation conditions. In addition, no fusion was observed between G_{M1} and G_{M2} after 2 days of incubation. Based on these observations, it was suggested that polysialogangliosides (e.g., G_{D1a} and G_{T1b}) may have higher propensities than monosialoganglioside. Thus, the high resolution, high speed, and quantitative aspects of CZE were clearly demonstrated in monitoring processes that may have important implications in the distribution and function of gangliosides in biological membranes.

One of the many elegant features of CE is the ease with which the electrophoretic behavior of the analytes can be altered through simple addition of specific reagents to the running electrolyte. Among the many buffer additives described so far,[193] two kinds of additives are suitable for HPCE of gangliosides, namely cyclodextrins (CDs) and acetonitrile (ACN). While CDs can alter the electrophoretic behavior of a wide variety of compounds *via* inclusion complexes, the extent of which is determined by the solute hydrophobicity and size, ACN

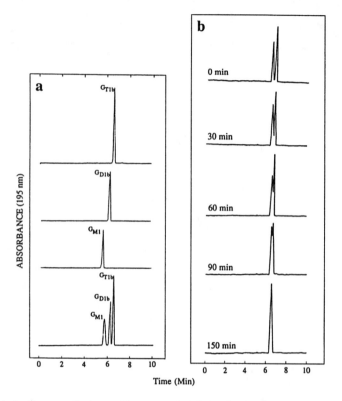

Figure 7.42 (a) Capillary electrophoresis of a mixture of G_{M1}, G_{D1b}, and G_{T1b}. Individual G_{M1}, G_{D1b}, and G_{T1b} micelles and a mixture of these three gangliosides shortly after mixing were analyzed by CE. The buffer was 2.5 mM potassium phosphate, pH 7.40. Detection was by UV at 195 nm. (b) Time course of mixed micelle formation between G_{D1b} and G_{T1b}. Equimolar concentrations of polysialogangliosides G_{D1b} and G_{T1b} (165 μM) in 2.5 mM potassium phosphate, pH 7.40, were mixed by vortexing and incubated in a water bath at 37°C. At time intervals, the electrophoretic patterns of the ganglioside mixtures were analyzed. Electrophoretic conditions are as in (a). (From Liu, Y. and Chan, K.-F. J., *Electrophoresis*, 12, 402, 1991. With permission.)

improves analyte solubility as well as selectivity, and controls electroosmotic flow.[193] On these bases, Yoo et al.[191] demonstrated the utility of CDs in the separation of native gangliosides at 185 nm. Among the various CDs, α-CD was the best buffer additive as far as separation is concerned. This may be due to the size of the cavity of α-CD that best fits the lipid moiety of the gangliosides.

Although the gangliosides could be detected at 185 nm,[191] the sensitivity is rather low to allow their detection in biological matrices where they are normally found in minute amounts. To overcome this difficulty, Mechref et al.[80] expanded the utility of the selective precolumn derivatization procedure which had been developed originally for the tagging of carboxylated monosaccharides[78] (see Scheme 7.2) to include the derivatization of gangliosides in order to improve their detectability. Moreover, novel electrolyte systems were also introduced for the analysis of derivatized gangliosides by CE. The derivatization involved the tagging of gangliosides with either SA (a UV-absorbing tag) or ANDSA (a UV-absorbing and also fluorescent tag) by a condensation reaction between the free amino group of the derivatizing agent and the carboxylic group of the sialic acid residue of the gangliosides to form a peptide link by acid-catalyzed removal of water in the presence of a carbodiimide, e.g., 1-ethyl-3-(3-dimethylaminopropyl) carbodiimide hydrochloride (EDAC). As with acidic monosaccharides, ganglioside derivatization with SA was shown to be complete in almost 1 h, while that with ANDSA required almost 2.5 h. The derivatization was also shown to be dependent on the concentration of EDAC and the tagging agent. Since gangliosides are available in minute quantities, the exact amounts of which were not known, there was always an excess of EDAC and tagging agent which resulted in the formation of a side product. The side product was an adduct formed between EDAC and derivatizing agent (i.e., ANDSA or SA) as concluded from NMR and pH studies by CE.[80] However, under controlled conditions and utilizing the ratio 1:1:2 of ganglioside:carbodiimide:derivatizing agent, the side product did not form. In fact, in the case of carboxylic monosaccharides, where the ratio of solute:carbodiimide:derivatizing agent was controlled in the 1:1:2 ratio, no side product was observed and quantitative derivatization was obtained.[78] The quantitative derivatization and absence of side product was again proved very recently in the case of the derivatization of phenoxy acid herbicides.[194] The derivatization was shown to occur uniformly on the sialic acid residues by monitoring the enzymatic digestion of SA-G_{T1b} with neuraminidase. The SA-G_{T1b} yielded three peaks upon digestion, which corresponded to SA-G_{M1}, SA-G_{D1b}, and SA-G_{T1b}, thus indicating that all three sialic acid residues of G_{T1b} were labeled.[80]

To separate the derivatized gangliosides in their monomeric forms, ACN and CD were added to the running electrolyte to break up ganglioside micelles. As can be seen in Figure 7.43, HPLC-grade acetonitrile at 50% (v/v) allowed the separation of three ANDSA-derivatized gangliosides in their monomeric forms, while in the absence of ACN the three ganglioside derivatives (i.e., ANDSA-G_{M1}, ANDSA-G_{D1a}, and ANDSA-G_{T1b}) migrated as a single broad peak.[80] The advantage of using ACN, whose UV cut-off is at 185 nm, was also illustrated in the ability to separate and detect underivatized gangliosides at 195 nm.[80] The use of ACN allowed the separation of the differently sialylated gangliosides shown in Figure 7.43; however, it did not provide enough selectivity to cause the separation of structural isomers ANDSA-G_{D1a} and ANDSA-G_{D1b}. Partial separation was attained using α-CD and 100 mM

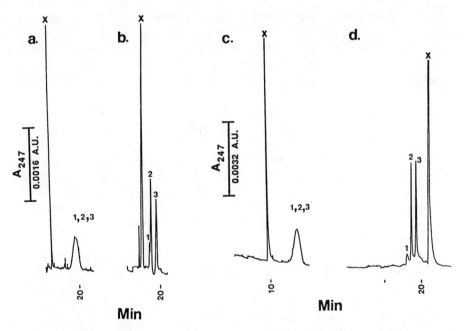

Figure 7.43 Electropherograms of standard gangliosides labeled with ANDSA at neutral (a and b) and high pH (c and d) in the presence (b and d) and absence (a and c) of acetonitrile in the running electrolyte. In (a) and (b): running electrolyte, 25 mM sodium phosphate, pH 7.0, at 0% (a) and 50% v/v (b) acetonitrile; voltage, 25.0 kV. In (c) and (d): running electrolyte, 10 mM sodium phosphate, pH 10.0 at 0% (c) and 50% v/v (d) acetonitrile; voltage, 20 kV; capillary, fused-silica, 50 cm (to detection point) 80 cm (total length) × 50 μm i.d. Solutes, (1) G_{M1}, (2) G_{D1a}, (3) G_{T1b}. For structures refer to Figure 7.41. (From Mechref, Y., Ostrander, G. K., and El Rassi, Z., *J. Chromatogr. A*, 695, 83, 1995. With permission.)

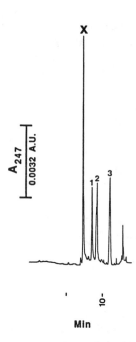

Figure 7.44 Electropherogram of standard ANDSA. Running electrolyte, 50 mM borate, pH 6.0 containing 5.0 mM MEGA surfactant and 15.0 mM α-CD; running voltage, 18.0 kV; capillary, fused-silica, 50 cm (to detection point) 80 cm (total length) × 50 μm i.d. Peaks: 1 = G_{D1a}; 2 = G_{D1b}; 3 = G_{D3}. (From Mechref, Y., Ostrander, G. K., and El Rassi, Z., *J. Chromatogr. A*, 695, 83, 1995. With permission.)

sodium borate, pH 10.0.[80] Other additives, such as polyvinyl alcohol and hydroxypropyl cellulose, improved the selectivity of the system, thus resolving the two isomers, i.e., ANDSA-G_{D1a} and ANDSA-G_{D1b}. Complete baseline separation of the two disialylated ganglioside isomers was attained using an *in situ* charged micellar system composed of decanoyl-*N*-methylglucamide (MEGA 10)/borate in the presence of α-CD,[80] as can be seen in Figure 7.44.

V. CONCLUSIONS

As this chapter reveals, the analysis of carbohydrates is one of the fascinating applications of CE. The major contributions to the advances made in CE of carbohydrates have been (1) the development of various detection systems and approaches, (2) the introduction of novel electrolyte systems, and (3) the design of improved capillary column technology and separation media. All of these achievements contributed to the widespread use of CE in carbohydrate analysis.

Several detection systems are being exploited, including (1) indirect UV and LIF detection, (2) electrochemical detection, and (3) precolumn labeling with suitable chromophores and fluorophores for the sensitive UV and fluorescence detection, respectively. However, in surveying the literature, precolumn derivatization seems to be the most elegant approach for the separation and detection of carbohydrates. In fact, six different reaction schemes have been introduced for labeling carbohydrates with various kinds of tags. All of these tagging processes have yielded the sensitivity required for the analysis of carbohydrates at moderate and low levels. Furthermore, the tagging of carbohydrates has imparted the derivatized carbohydrates with charges and/or hydrophobic functional groups that facilitated their efficient separation by various separation principles, thus leading to varying degrees of selectivity. It should be noted that multiply charged tags such as ANDSA, ANTS, and APTS are excellent labels for the CE of sugar derivatives not only because of their high detection sensitivity by either UV or LIF, but also because they yield derivatives that are readily separated by CE. Although major progress has been made in the area of precolumn labeling of carbohydrates, there is still room for improvement regarding the introduction of other tagging agents and optimizing the existing reaction schemes.

Although noncomplexing electrolyte systems have found some use in the electrophoresis of a wide range of carbohydrate species, the bulk of CE of carbohydrates is still accomplished primarily by borate complexation regardless of whether the carbohydrates are derivatized or underivatized. Borate complexation magnifies small structural differences among closely related carbohydrates, thus leading to a better resolution for multicomponent mixtures.

The ease with which the electrolyte systems can be modified and tailored to fit a given separation problem is another important feature of CE. In fact, glycolipids such as gangliosides which are not compatible with purely aqueous electrolyte solutions were readily separated in their monomeric forms in hydro-organic electrolyte systems.

Furthermore, the unique selectivity and high separation efficiencies of CE have proved extremely useful in the separation of all kinds of carbohydrate species. The technique provides an unsurpassed resolving power for profiling and mapping closely related oligosaccharides cleaved from glycoproteins, glycolipids, and glycosaminoglycans. This high resolving power has also allowed the efficient fingerprinting of complex glycoprotein glycoforms which in other separation techniques, such as ion-exchange HPLC and traditional gel electrophoresis, would yield smeared unresolved bands. In addition, with the introduction of precolumn labeling with suitable fluorescent tags for LIF detection, CE can reach nanomolar detection limits, thus making the technique extremely suitable for the analysis of minute amounts of carbohydrates. However, the major drawback of the technique is its limited preparative capability.

ACKNOWLEDGMENTS

The partial support by the Cooperative State Research, Education, and Extension Service, U.S. Department of Agriculture, under Agreement Nos. 94-37102-0989 and 96-35201-3342 is acknowledged.

REFERENCES

1. El Rassi, Z., Capillary electrophoresis of carbohydrates, *Adv. Chromatogr.*, 34, 177, 1994.
2. El Rassi, Z. and Nashabeh, W., High performance capillary electrophoresis of carbohydrates and glycoconjugates, in *Carbohydrate Analysis: High Performance Liquid Chromatography and Capillary Electrophoresis*, El Rassi, Z., Ed., Elsevier, Amsterdam, 1995, 267.
3. *Capillary Electrophoresis of Carbohydrate Species*, El Rassi, Z., Ed., VCH, Weinheim, 1996, Vol. 17.
4. Honda, S., Yamamoto, K., Suzuki, S. M., and U. Kakehi, K., High-performance capillary zone electrophoresis of carbohydrates in the presence of alkaline earth metal ions, *J. Chromatogr.*, 588, 327, 1991.
5. Weigel, H., Paper electrophoresis of carbohydrates, *Adv. Carbohydr. Chem.*, 18, 61, 1963.
6. Foster, A. B., Zone electrophoresis of carbohydrates, *Adv. Carbohydr. Chem.*, 12, 81, 1957.
7. Böeseken, J., The use of boric acid for the determination of the configuration of carbohydrates, *Adv. Carbohydr. Chem.*, 4, 189, 1949.
8. Edwards, J. O., Morrison, G. C., Ross, V. F., and Schultz, J. W., The structure of the aqueous borate ion, *J. Am. Chem. Soc.*, 77, 266, 1955.
9. Bell, C. F., Beauchamp, R. D. and Short, E. L., A study of the complexes of borate ions and some cyclitols using ^{11}B-N.M.R. Spectroscopy, *Carbohydr. Res.*, 147, 191, 1986.

10. Christ, C. C., Clark, J. R., and Evans, H. T., Studies of borate minerals (III): the crystal structure of colemanite, $CaB_3O_4(OH)_3 \cdot H_2O$, *Acta Cryst.*, 11, 761, 1958.

11. Frahn, J. L. and Mills, J. A., Paper ionophoresis of carbohydrates, *Aust. J. Chem.*, 12, 65, 1959.

12. Davis, H. B. and Mott, C. J. B., Interaction of boric acid and borate with carbohydrates and related structures, *J. Chem. Soc. Faraday*, 76, 1991, 1980.

13. Makkee, M., Kieboom, A. P. G., and Van Bekkum, H., Studies on borate esters. III. Borate esters of D-mannitol, D-glucitol, D-fructose and D-glucose in water, *Recl. Trav. Chim. Pays-Bas*, 104, 230, 1985.

14. van Duin, M., Peters, J. A., Kieboom, A. P. G., and van Bekkum, H., Studies on borate esters II, *Tetrahedron*, 41, 3411, 1985.

15. Foster, A. B., Separation of the dimethyl-l-rhamnopyranoses by ionophoresis, *Chemistry & Industry*, 828, 1952.

16. Kennedy, G. R. and How, M. J., The interaction of sugars with borate: an N.M.R. spectroscopic study, *Carbohydr. Res.*, 28, 13, 1973.

17. Mechref, Y., Ostrander, G. K., and El Rassi, Z., Capillary electrophoresis of carboxylated carbohydrates. Part 2. Selective precolumn derivatization of sialooligosaccharides derived from gangliosides with 7-aminonaphthalene-1,3-disulfonic acid fluorescing tag, *Electrophoresis*, 16, 1499, 1995.

18. Honda, S., Akao, E., Suzuki, S., Okuda, M., Kakehi, K., and Nakamura, J., High performance liquid chromatography of reducing carbohydrates as strongly ultraviolet-absorbing and electro-chemically sensitive 1-phenyl-3-methylpyrazolone derivatives, *Anal. Biochem.*, 180, 351, 1989.

19. Rendleman, J. A., Ionization of carbohydrates in the presence of metal hydroxides and oxides, *Adv. Chem. Ser.*, 117, 51, 1971.

20. Colón, L. A., Dadoo, R., and Zare, R. N., Determination of carbohydrates by capillary zone electrophoresis with amperometric detection at a copper microelectrode, *Anal. Chem.*, 65, 476, 1993.

21. Angyal, S. J. and Davies, K. P., Complexing of sugars with metal ions, *Chem. Commun.*, 10, 500, 1971.

22. Angyal, S. J., Complexes of metal cations with carbohydrates in solution, *Adv. Carbohydr. Chem. Biochem.*, 47, 1, 1989.

23. Leukinski, R. E. and Reuben, J., Studies of the binding of calcium and lanthanium ions to D-lyxose and D-ribose in aqueous solutions, *J. Am. Chem. Soc.*, 98, 3089, 1976.

24. Hoffstetter-Kuhn, S., Paulus, A., Gassmann, E. and Widmer, H. M., Influence of borate com-plexation on the electrophoretic behavior of carbohydrates in capillary electrophoresis, *Anal. Chem.*, 63, 1541, 1991.

25. Hughes, D. E., Capillary electrophoretic examination of underivatized O-linked and N-linked oligosaccharide mixtures and immunoglobulin G antibody-released oligosaccharide libraries, *J. Chromatogr. B*, 657, 315, 1994.

26. Hermentin, P., Doenges, R., Witzel, R., Hokke, C. H., Vliegenthart, J. F. G., Kamerling, J. P., Conradt, H. S., Nimtz, M., and Brazel, D., A strategy for the mapping of N-glycans by high-performance capillary electrophoresis, *Anal. Biochem.*, 221, 29, 1994.

27. Taverna, M., Baillet, A., and Baylocq-Ferrier, D., Analysis of neutral and sialylated N-linked oligosaccharides by micellar electrokinetic capillary chromatography with addition of a divalent cation, *Chromatographia*, 37, 415, 1993.

28. Taverna, M., Baillet, A., Schlüter, M. and Baylocq-Ferrier, D., N-glycosylation site mapping of recombinant tissue plasminogen activator by micellar electrokinetic capillary chromatogra-phy, *Biomed. Chromatogr.*, 9, 59, 1995.

29. Kakehi, K., Susami, A., Taga, A., Suzuki, S., and Honda, S., High-performance capillary electrophoresis of O-glycosidically linked sialic acid-containing oligosaccharides in glycopro-teins as their alditol derivatives with low-wavelength UV monitoring, *J. Chromatogr. A*, 680, 209, 1994.

30. Grimshaw, J., Kane, À., Trocha-Grimshaw, J., Douglas, A., Chakravarthy, U., and Archer, D., Quantitative analysis of hyaluronan in vitreous humor using capillary electrophoresis, *Electro-phoresis*, 15, 936, 1994.

31. Malsch, R., Harenberg, J., and Heene, D. L., High-resolution capillary electrophoresis and polyacrylamide gel electrophoresis of heparins, *J. Chromatogr. A*, 716, 259, 1995.

32. Damm, J. B. L., Overklift, G. T., Vermeulen, B. W. M., Fluitsma, C. F. and van Dedem, G. W. K., Separation of natural and synthetic heparin fragments by high performance capillary electrophoresis, *J. Chromatogr.*, 608, 297, 1992.

33. Desai, U. R., Wang, H., Ampofo, S. A., and Linhardt, R. J., Oligosaccharide composition of heparin and low-molecular-weight heparins by capillary electrophoresis, *Anal. Biochem.*, 213, 120, 1993.

34. Carney, S. L. and Osborne, D. J., The separation of chondroitin sulfate disaccharides and hyaluronan oligosaccharides by capillary zone electrophoresis, *Anal. Biochem.*, 195, 132, 1991.

35. Al-Hakim, A. and Linhardt, R. J., Capillary electrophoresis for the analysis of chondroitin sulfate- and dermatan sulfate-derived disaccharides, *Anal. Biochem.*, 195, 68, 1991.

36. Ampofo, S. A., Wang, H.-M., and Linhardt, R. J., Disaccharide compositional analysis of heparin and heparan sulfate using capillary zone electrophoresis, *Anal. Biochem.*, 199, 249, 1991.

37. Michaelsen, S., Schrøder, M.-B., and Sørensen, H., Separation and determination of glycosaminoglycan disaccharides by micellar electrokinetic capillary chromatography for studies of pelt glycosaminoglycans, *J. Chromatogr. A*, 652, 503, 1993.

38. Pervin, A., Al-Hakim, A., and Linhardt, R. J., Separation of glycosaminoglycan-derived oligosaccharides by capillary electrophoresis using reverse polarity, *Anal. Biochem.*, 221, 182, 1994.

39. Linhardt, R. J., Desai, U. R., Hoppensteadt, D., and Fareed, J., Low molecular weight dermatan sulfate as an antithrombotic agent; structure-activity relationship studies, *Biochem. Pharmacol.*, 47, 1241, 1994.

40. Linhardt, R. J., Capillary electrophoresis of oligosaccharides, *Methods Enzymol.*, 230, 265, 1994.

41. Linhardt, R. J., Liu, J., and Han, X.-J., Mapping and sequencing of oligosaccharides by electrophoresis, *Trends Glycosci. Glycotechnol.*, 5, 181, 1993.

42. Linhardt, R. J., Gu, K. N., Loganathan, D., and Carter, S. R., Analysis of glycosaminoglycan-derived oligosaccharides using reversed-phase ion-pairing and ion-exchange chromatography with suppressed conductivity detection, *Anal. Biochem.*, 181, 288, 1989.

43. Damm, J. B. L. and Overklift, G. T., Indirect UV detection as a non-selective detection method in the qualitative and quantitative analysis of heparin fragments by high-performance capillary electrophoresis, *J. Chromatogr. A*, 678, 151, 1994.

44. Yeung, E. S., Indirect detection methods: looking for what is not there, *Acc. Chem. Res.*, 22, 125, 1989.

45. Yeung, E. S. and Kuhr, W. G., Indirect detection methods for capillary separations, *Anal. Chem.*, 63, 275, 1991.

46. Foret, F., Fanali, S., L., O., and Bocek, P., Indirect photometric detection in capillary zone electrophoresis, *J. Chromatogr.*, 470, 299, 1989.

47. Yeung, E. S., Optical detectors for capillary electrophoresis, *Adv. Chromatogr.*, 35, 1, 1995.

48. Garner, T. W. and Yeung, E. S., Indirect fluorescence detection of sugars separated by capillary zone electrophoresis with visible laser excitation, *J. Chromatogr.*, 515, 639, 1990.

49. Kuhr, W. G. and Yeung, E. S., Optimization of sensitivity and separation in capillary zone electrophoresis with indirect fluorescence detection, *Anal. Chem.*, 60, 2642, 1988.

50. Richmond, M. D. and Yeung, E. S., Development of laser-excited indirect fluorescence detection for high-molecular-weight polysaccharides in capillary electrophoresis, *Anal. Biochem.*, 210, 245, 1993.

51. Vorndran, A. E., Oefner, P. J., Scherz, H., and Bonn, G. K., Indirect UV detection of carbohydrates in capillary zone electrophoresis, *Chromatographia*, 33, 163, 1992.

52. Oefner, P. J., Vorndran, A. E., Grill, E., Huber, C., and Bonn, G. K., Capillary zone electrophoretic analysis of carbohydrates by direct and indirect UV detection, *Chromatographia*, 34, 308, 1992.

53. Bergholdt, A., Overgaard, J., Colding, A., and Frederiksen, R. B., Separation of D-galactonic and D-gluconic acids by capillary electrophoresis, *J. Chromatogr.*, 644, 412, 1993.

54. Xu, X., Kok, W. T., and Poppe, H., Sensitive determination of sugars by capillary electrophoresis with indirect UV detection under highly alkaline conditions, *J. Chromatogr. A*, 716, 231, 1995.

55. Klockow, A., Paulus, A., Figueiredo, V., Amadò, R., and Widmer, H. M., Determination of carbohydrates in fruit juices by capillary electrophoresis and high-performance liquid chromatography, *J. Chromatogr. A*, 680, 187, 1994.

56. Lu, B. and Westerlund, D., Indirect UV detection of carbohydrates in capillary zone electrophoresis by using tryptophan as a marker, *Electrophoresis*, 17, 325, 1996.

57. Stefansson, M. and Westerlund, D., Capillary electrophoresis of glycoconjugates in alkaline media, *J. Chromatogr.*, 632, 195, 1993.

58. Rocklin, R., Electrochemical detection, in *A Practical Guide to HPLC Detection*, Parriott, D., Ed., Academic Press, New York, 1993, 145.

59. Zhou, W. and Baldwin, R. P., Capillary electrophoresis and electrochemical detection of underivatized oligo- and polysaccharides with surfactant-controlled electroosmotic flow, *Electrophoresis*, 17, 319, 1996.

60. Ye, J. and Baldwin, R. P., Amperometric detection in capillary electrophoresis with normal size electrodes, *Anal. Chem.*, 65, 3525, 1993.

61. Ye, J. and Baldwin, R. P., Determination of carbohydrates, sugar acids, alditols by capillary electrophoresis and electrochemical detection at a copper electrode, *J. Chromatogr. A*, 687, 141, 1994.

62. O'Shea, T. J., Lunte, S. M., and LaCourse, W. R., Detection of carbohydrates by capillary electrophoresis with pulsed amperometric detection, *Anal. Chem.*, 65, 948, 1993.

63. Weber, P., Kornfelt, T., Klausen, N. K., and Lunte, S. M., Characterization of glycopeptides from recombinant coagulation factor VIIa by high-performance liquid chromatography and capillary zone electrophoresis using ultraviolet and pulsed electrochemical detection, *Anal. Biochem.*, 225, 135, 1995.

64. Weber, P. L. and Lunte, S. M., Capillary electrophoresis with pulsed amperometric detection of carbohydrates and glycopeptides, *Electrophoresis*, 17, 302, 1996.

65. Lu, W. and Cassidy, R. M., Pulsed amperometric detection of carbohydrates separated by capillary electrophoresis, *Anal. Chem.*, 65, 2878, 1993.

66. LaCourse, W. R. and Owens, G. S., Pulsed electrochemical detection of nonchromophoric compounds following capillary electrophoresis, *Electrophoresis*, 17, 310, 1996.

67. Johnson, D. C. and LaCourse, W. R., Pulsed electrochemical detection of carbohydrates at gold electrodes following liquid chromatography separation, in *Carbohydrate Analysis: High Performance Liquid Chromatography and Capillary Electrophoresis*, El Rassi, Z., Ed., Elsevier, Amsterdam, 1995, 391.

68. Johnson, D. and LaCourse, W., Liquid chromatography with pulsed electrochemical detection at gold and platinum electrodes, *Anal. Chem.*, 62, 589, 1990.

69. Luo, P., Zhang, F., and Baldwin, R., Comparison of metallic electrodes for constant-potential amperometric detection of carbohydrates, amino acids and related compounds in flow systems, *Anal. Chim. Acta*, 244, 169, 1991.

70. Huang, X. and Kok, W. T., Determination of sugars by capillary electrophoresis with electrochemical detection using cuprous oxide modified electrodes, *J. Chromatogr. A*, 707, 335, 1995.

71. Penn, S. G., Chiu, R. W., and Monnig, C. A., Separation and analysis of cyclodextrins by capillary electrophoresis with dynamic fluorescence labeling and detection, *J. Chromatogr. A*, 680, 233, 1994.

72. Lee, Y. and Lin, T., Capillary electrophoresis analysis of cyclodextrins and determination of formation constants for inclusion complexes, *Electrophoresis*, 17, 333, 1996.

73. Brewster, J. D. and Fishman, M. L., Capillary electrophoresis of plant starches as the iodine complexes, *J. Chromatogr. A*, 693, 382, 1995.

74. Toida, T. and Linhardt, R. J., Detection of glycosaminoglycans as a copper (II) complex in capillary electrophoresis, *Electrophoresis*, 17, 341, 1996.

75. Mukherjee, D. C., Park, J. W., and Chakrabart, B., Optical properties of Cu (II) complexes with heparin and related glycosaminoglycans, *Arch. Biochem. Biophys.*, 191, 393, 1978.

76. Honda, S., Iwase, S., Makino, A., and Fujiwara, S., Simultaneous determination of reducing monosaccharides by capillary electrophoresis as the borate complexes of N-2-pyridylglycamines, *Anal. Biochem.*, 176, 72, 1989.

77. Hase, S., Pre- and post-column detection-oriented derivatization techniques in HPLC of carbo-hydrates, in *Carbohydrate Analysis: High Performance Liquid Chromatography and Capillary Electrophoresis*, El Rassi, Z., Ed., Elsevier, Amsterdam, 1995, 555.

78. Mechref, Y. and El Rassi, Z., Capillary electrophoresis of derivatized acidic monosaccharides, *Electrophoresis*, 15, 627, 1994.

79. Mechref, Y. and El Rassi, Z., Fused-silica capillaries with surface-bound dextran layer crosslinked with diepoxypolyethylene glycol for capillary electrophoresis of biological sub-stances at reduced electroosmotic flow, *Electrophoresis*, 16, 617, 1995.

80. Mechref, Y., Ostrander, G. K., and El Rassi, Z., Capillary electrophoresis of carboxylated carbohydrates. I. Selective precolumn derivatization of gangliosides with UV absorbing and fluorescent tags, *J. Chromatogr. A*, 695, 83, 1995.

81. Honda, S., Ueno, T., and Kakehi, K., High-performance capillary electrophoresis of unsaturated oligosaccharides derived from glycosaminoglycans by digestion with chondroitinase ABC as 1-phenyl-3-methyl-5-pyrazolone derivatives, *J. Chromatogr.*, 608, 289, 1992.

82. Honda, S., Suzuki, S., Nose, A., Yamamoto, K., and Kakehi, K., Capillary zone electrophoresis of reducing mono- and oligosaccharides as the borate complexes of their 3-methyl-1-phenyl-2-pyrazolin-5-one derivatives, *Carbohydr. Res.*, 215, 193, 1991.

83. Liu, J., Shirota, O., Wiesler, D., and Novotny, M., Ultrasensitive fluorometric detection of carbohydrates as derivatives in mixtures separated by capillary electrophoresis, *Proc. Natl. Acad. Sci. USA*, 88, 2302, 1991.

84. Novotny, M. V. and Sudor, J., High-performance capillary electrophoresis of glycoconjugates, *Electrophoresis*, 14, 373, 1993.

85. Zhao, J. Y., Diedrich, P., Zhang, Y., Hindsgaul, O., and Dovichi, N. J., Separation of aminated monosaccharides by capillary electrophoresis with laser-induced fluorescence detection, *J. Chromatogr. B*, 657, 307, 1994.

86. Nashabeh, W. and El Rassi, Z., Capillary zone electrophoresis of linear and branched oligosac-charides, *J. Chromatogr.*, 600, 279, 1992.

87. Perez, S. A. and Colón, L. A., Determination of carbohydrates as their dansylhydrazine deriv-atives by capillary electrophoresis with laser-induced fluorescence detection, *Electrophoresis*, 17, 352, 1996.

88. Amankwa, L., Albin, M. and Kuhr, W., Fluorescence detection in capillary electrophoresis, *Trends Anal. Chem.*, 11, 114, 1992.

89. Li, S. F. Y., Detection techniques, in *Capillary Electrophoresis. Principles, Practice and Appli-cations*, Elsevier, Amsterdam, 1992, Chap. 3.

90. Schwaiger, H., Oefner, P., Huber, C., Grill, E. and Bonn, G. K., Capillary electrophoresis and micellar electrokinetic chromatography of 4-aminobenzonitrile carbohydrate derivatives, *Elec-trophoresis*, 15, 941, 1994.

91. Klockow, A., Widmer, H. M., Amadò, R., and Paulus, A., Capillary electrophoresis of ANTS labeled oligosaccharides ladders and complex carbohydrates, *Fresenius J. Anal. Chem.*, 350, 415, 1994.

92. Evangelista, R. A., Guttman, A., and Chen, F.-T. A., Acid-catalyzed reductive amination of aldoses with 9-aminopyrene-1,4,6-trisulfonate, *Electrophoresis*, 17, 347, 1996.

93. Smith, J. T. and El Rassi, Z., Capillary zone electrophoresis of biological substance with fused silica capillaries having zero or constant electroosmotic flow, *Electrophoresis*, 14, 396, 1993.

94. Lee, K., Kim, Y., and Linhardt, R., Capillary electrophoresis for the quantitation of oligosac-charides formed through the action of chitinase, *Electrophoresis*, 12, 636, 1991.

95. Chiesa, C. and Horváth, C., Capillary zone electrophoresis of maltooligosaccharides derivatized with 8-aminonaphthalene-1,3,6-trisulfonic acid, *J. Chromatogr.*, 645, 337, 1993.

96. Jackson, P., Polyacrylamide gel electrophoresis of reducing saccharides labeled with the fluo-rophore 2-aminoacridone: subpicomolar detection using an imaging system based on a cooled charge-coupled device, *Anal. Biochem.*, 196, 238, 1991.

97. Stefansson, M. and Novotny, M., Electrophoretic resolution of monosaccharides enantiomers in borate-oligosaccharide complexation media, *J. Am. Chem. Soc.*, 115, 11573, 1993.

98. Evangelista, R. A., Liu, M., and Chen, F., Characterization of 9-aminopyrene-1,4,6-trisul-fonated-derivatized sugars by capillary electrophoresis with laser-induced fluorescence detection, *Anal. Chem.*, 67, 2239, 1995.

99. Chen, F. and Evangelista, R. A., Analysis of mono- and oligosaccharide isomers derivatized with 9-aminopyrene-1,4,6-trisulfonate by capillary electrophoresis with laser-induced fluorescence, *Anal. Biochem.*, 230, 273, 1995.

100. Guttman, A. and Pritchett, T., Capillary gel electrophoresis of high-mannose type oligosaccharides derivatized by 1-aminopyrene-3,6,8-trisulfonic acid, *Electrophoresis*, 16, 1906, 1995.

101. Liu, J., Shirota, O., and Novotny, M., Separation of fluorescent oligosaccharides derivatives by microcolumn techniques based on electrophoresis and liquid chromatography, *J. Chromatogr.*, 559, 223, 1991.

102. Liu, J., Shirota, O., and Novotny, M., Sensitive laser-assisted determination of complex oligosaccharides mixtures separated by capillary gel electrophoresis, *Anal. Chem.*, 64, 973, 1992.

103. Sudor, J. and Novotny, M., Electromigration behavior of polysaccharides in capillary electrophoresis under pulsed-field conditions, *Proc. Natl. Acad. Sci. USA*, 90, 9451, 1993.

104. Liu, J., Shirota, O., and Novotny, M., Capillary electrophoresis of amino sugars with laser-induced fluorescence detection, *Anal. Chem.*, 63, 413, 1991.

105. Zhang, Y., Arriaga, E., Diedrich, P., Hindsgaul, O., and Dovichi, N. J., Nanomolar determination of aminated sugars by capillary electrophoresis, *J. Chromatogr. A*, 716, 221, 1995.

106. Zhang, Y., Le, X., Dovichi, N. J., Compston, C. A., Palcic, M. M., Diedrich, P., and Hindsgaul, O., Monitoring biosynthetic transformations of N-acetyllactosamine using fluorescently labeled oligosaccharides and capillary electrophoretic separation, *Anal. Biochem.*, 227, 368, 1995.

107. Le, X., Scaman, C., Zhang, Y., Zhang, J., Dovichi, N. J., Hindsgaul, O., and Palcic, M. M., Analysis of capillary electrophoresis-laser-induced fluorescence detection of oligosaccharides produced from enzyme reaction, *J. Chromatogr. A*, 716, 215, 1995.

108. Avigad, G., Dansyl hydrazine as a fluorimetric reagent for thin-layer chromatographic analysis of reducing sugars, *J. Chromatogr.*, 139, 343, 1977.

109. Alpenfels, W. F., A rapid and sensitive method for the determination of monosaccharides as their dansyl hydrazones by high-performance liquid chromatography, *Anal. Biochem.*, 114, 153, 1981.

110. Mopper, K. and Johnson, L., Reversed-phase liquid chromatographic analysis of Dns-sugars. Optimization of derivatization and chromatographic procedures and applications to neutral samples, *J. Chromatogr.*, 256, 27, 1983.

111. Eggert, F. M. and Jones, M., Measurement of neutral sugars in glycoproteins as dansyl derivatives by automated high-performance liquid chromatography, *J. Chromatogr.*, 333, 123, 1985.

112. Bruno, A. E. and Krattiger, B., On-column refractive index detection of carbohydrates separated by HPLC and CE, in *Carbohydrate Analysis: High Performance Liquid Chromatography and Capillary Electrophoresis*, El Rassi, Z., Ed., Elsevier, Amsterdam, 1995, 431.

113. Huang, M., Plocek, J., and Novotny, M. V., Hydrolytically stable cellulose-derivative coatings for capillary electrophoresis of peptides, proteins and glycoconjugates, *Electrophoresis*, 16, 396, 1995.

114. Nashabeh, W. and El Rassi, Z., Capillary zone electrophoresis of α_1-acid glycoprotein fragments from trypsin and endoglycosidase digestions, *J. Chromatogr.*, 536, 31, 1991.

115. Vorndran, A. E., Grill, E., Huber, C., Ofner, P. J., and Bonn, G. K., Capillary zone electrophoresis of aldoses, ketoses and uronic acids derivatized with ethyl *p*-aminobenzoate, *Chromatographia*, 34, 109, 1992.

116. Huber, C., Grill, E., Oefner, P., and Bobleter, O., Capillary electrophoretic determination of the component monosaccharides in hemicelluloses, *Fresenius J. Anal. Chem.*, 348, 825, 1994.

117. Noe, C. R. and Freissmuth, J., Capillary zone electrophoresis of aldose enantiomers: separation after derivatization with S-(–)-1-phenylethylamine, *J. Chromatogr. A*, 704, 503, 1995.

118. Greenaway, M., Okafo, G. N., Camilleri, P., and Dhanak, D., A sensitive and selective method for the analysis of complex mixtures of sugars and linear oligosaccharides, *J. Chem. Soc., Chem. Commun.*, 1691, 1994.

119. Chiesa, C., Oefner, P. J., Zieske, L. R., and O'Neill, R. A., Micellar electrokinetic chromatography of monosaccharides derivatized with 1-phenyl-3-methyl-2-pyrazolin-5-one, *J. Cap. Elec.*, 2, 175, 1995.

120. Arentoft, A. M., Michaelsen, S., and Sørensen, H., Determination of oligosaccharides by capillary zone electrophoresis, *J. Chromatogr. A*, 652, 517, 1993.

121. Casu, B., Methods of structural analysis, in *Heparin: Chemical and Biological Properties, Clinical Applications*, Lane, D. A. and Lindahl, U., Eds.; Arnold, London, 1989, 25.

122. Kjellén, L. and Lindahl, U., Proteoglycans: structures and interactions, *Ann. Rev. Biochem.*, 60, 443, 1991.

123. Karamanos, N. K., Vanky, P., Tzanakakis, G. N., and Hjerpe, A., High-performance capillary electrophoresis method to characterize heparin and heparan sulfate disaccharides, *Electrophoresis*, 17, 391, 1996.

124. Damm, J. B. L., Overklift, G. T., and van Dedem, G. W. K., Determination of structural differences in the glycosaminoglycan chains of heparin and heparan sulfate by analysis of the constituting disaccharides with capillary electrophoresis, *Pharm. Pharmacol. Lett.*, 3, 156, 1993.

125. Carney, S. L., Caterson, B., and Penticost, H. R., The investigation of glycosaminoglycan mimotope structure using capillary electrophoresis and another complementary electrophoretic techniques, *Electrophoresis*, 17, 384, 1996.

126. Carney, S. L., Billingham, M. E. J., Caterson, B., Ratcliffe, A., Bayliss, M. T., Hardingham, T. E., and Muir, H., Changes in proteoglycan turnover in experimental canine osteoarthritic cartilage, *Matrix*, 12, 137, 1992.

127. Caterson, B., Mahmoodian, F., Sorrel, J. M., Hardingham, T. E., Bayliss, M. T., Carney, S. L., Ratcliffe, A., and Muir, H., Modulation of native chondroitin sulphate structure in tissue development and in disease, *J. Cell Sci.*, 97, 411, 1990.

128. Karamanos, N. K., Axelsson, S., Vanky, P., Tzanakakis, G. N., and Hjerpe, A., Determination of hyaluronan and galactosaminoglycan disaccharides by high-performance capillary electrophoresis at the attomole level. Applications to analyses of tissue and cell culture proteoglycans, *J. Chromatogr. A*, 696, 295, 1995.

129. Pervin, A., Kenan, G., and Linhardt, R. J., Capillary electrophoresis to measure sulfoesterase activity on chondroitin sulfate and heparin derived disaccharides, *Appl. Theor. Electrophoresis*, 3, 297, 1993.

130. Kerns, R. J., Vlahov, I. R., and Linhardt, R. J., Capillary electrophoresis for monitoring chemical reactions: sulfation and synthetic manipulation of sulfated carbohydrates, *Carbohydrate Res.*, 267, 143, 1995.

131. Grimshaw, J., Trocha-Grimshaw, J., Fisher, W., Rice, A., Smith, S., Spedding, P., Duffy, J., and Mollan, R., Quantitative analysis of hyaluronan in human synovial fluid using capillary electrophoresis, *Electrophoresis*, 17, 396, 1996.

132. Hermentin, P., Witzel, R., Doenges, R., Bauer, R., Haupt, H., Patel, T., Parekh, R. B., and Brazel, D., The mapping by high-pH anion-exchange chromatography with pulsed amperometric detection and capillary electrophoresis of the carbohydrate moieties of human plasma α_1-acid glycoprotein, *Anal. Biochem.*, 206, 419, 1992.

133. Goulding, R. W., Liquid chromatography of sugars and related polyhydric alcohols on cation exchangers, *J. Chromatogr.*, 103, 229, 1975.

134. Cohen, A. S., Terabe, S., Smith, J. A., and Karger, B. L., High-performance capillary electrophoretic separation of bases, nucleosides, and oligonucleotides: retention manipulation via micellar solutions and metal additives, *Anal. Chem.*, 59, 1021, 1987.

135. Oko, M. U., The effects of divalent metal ions on the micellar properties of sodium dodecyl sulfate, *J. Colloid Interface Sci.*, 35, 53, 1971.

136. Taverna, M., Baillet, A., Biou, D., Schlüter, M., Werner, R., and Ferrier, D., Analysis of carbohydrate-mediated heterogeneity and characterization of N-linked oligosaccharides of glycoproteins by high performance capillary electrophoresis, *Electrophoresis*, 13, 359, 1992.

137. Nashabeh, W. and El Rassi, Z., Capillary zone electrophoresis of pyridylamino derivatives of maltooligosaccharides, *J. Chromatogr.*, 514, 57, 1990.

138. Chiesa, C. and O'Neill, R. A., Capillary zone electrophoresis of oligosaccharides derivatized with various aminonaphthalene sulfonic acids, *Electrophoresis*, 15, 1132, 1994.

139. Sudor, J. and Novotny, M. V., End-label, free-solution capillary electrophoresis of highly charged oligosaccharides, *Anal. Chem.*, 67, 4205, 1995.

140. Stefansson, M. and Novotny, M., Separation of complex oligosaccharide mixtures by capillary electrophoresis in the open-tubular format, *Anal. Chem.*, 66, 1134, 1994.

141. Smith, J. T. and El Rassi, Z., Capillary zone electrophoresis of biological substances with surface-modified fused silica capillaries with switchable electroosmotic flow, *J. High Resolut. Chromatogr.*, 15, 573, 1992.

142. Honda, S., Makino, A., Suzuki, S., and Kaheki, K., Analysis of the oligosaccharides in oval-bumin by high-performance capillary electrophoresis, *Anal. Biochem.*, 191, 228, 1990.

143. Liu, J., Dolnik, V., Hsieh, Y.-Z., and Novotny, M., Experimental evaluation of the separation efficiency in capillary electrophoresis using open tubular and gel-filled columns, *Anal. Chem.*, 64, 1328, 1992.

144. Guttman, A., Cooke, N., and Starr, C. M., Capillary electrophoresis separation of oligosaccha-rides: I. Effect of operational variables, *Electrophoresis*, 15, 1518, 1994.

145. Lee, K.-B., Kim, Y.-S., and Linhardt, R. J., Capillary zone electrophoresis for quantitation of oligosaccharides formed through the action of chitinase, *Electrophoresis*, 12, 637, 1991.

146. El Rassi, Z., Tedford, D., An, J., and Mort, A., High-performance reversed-phase chromato-graphic mapping of 2-pyridylamino derivatives of xyloglucan oligosaccharides, *Carbohydr. Res.*, 215, 25, 1991.

147. Stefansson, M. and Novotny, M., Resolution of the branched forms of oligosaccharides by high-performance capillary electrophoresis, *Carbohydr. Res.*, 258, 1, 1994.

148. Mort, A. J. and Chen, E. M. W., Separation of ANTS-labeled oligomers containing galacturonic acid by capillary electrophoresis: application to determining the substrate specificity of endopo-lygalacturonases, *Electrophoresis*, 17, 379, 1996.

149. Zhang, Z., Pierce, M. L., and Mort, J. M., Detection and differentiation of pectic enzyme activity *in vitro* and *in vivo* by capillary electrophoresis, *Electrophoresis*, 17, 372, 1996.

150. Paulson, J. M., Glycoproteins: what are the sugar chains for?, *Trends Biochem. Sci.*, 14, 272, 1989.

151. Schachter, H., Branching of N- and O-glycans: biosynthetic control and functions, *Trends Glycosci. Glycotechnol.*, 4, 241, 1992.

152. Spellman, M. W., Carbohydrate characterization of recombinant glycoprotein of pharmaceutical interests, *Anal. Chem.*, 62, 1714, 1990.

153. Guttman, A., Chen, F., and Evangelista, R. A., Separation of APTS labeled asparagine-linked fetuin glycans by capillary electrophoresis, *Electrophoresis*, 17, 412, 1996.

154. Stack, R. S. and Sullivan, M. T., Electrophoretic resolution and fluorescence detection of N-linked glycoprotein oligosaccharides after reductive amination with 8-aminonaphthalene-1,3,6-trisulfonic acid, *Glycobiology*, 2, 85, 1992.

155. Guttman, A. and Starr, C., Capillary and slab gel electrophoresis profiling of oligosaccharides, *Electrophoresis*, 16, 993, 1995.

156. Klockow, A., Amadò, R., Widmer, H. M., and Paulus, A., Separation of 8-aminonaphthalene-1,3,6-trisulfonic acid-labeled neutral and sialylated N-linked complex oligosaccharides by cap-illary electrophoresis, *J. Chromatogr. A*, 716, 241, 1995.

157. Camilleri, P., Harland, G. B., and Okafo, G., High resolution and rapid analysis of branched oligosaccharides by capillary electrophoresis, *Anal. Biochem.*, 230, 115, 1995.

158. Harland, G. B., Okafo, G., Matejtschuk, P., Sellick, I. C., Chapman, G. E., and Camilleri, P., Fingerprinting of glycans as their 2-aminoacridone derivatives by capillary electrophoresis and laser induced fluorescence, *Electrophoresis*, 17, 406, 1996.

159. Zieske, L. R., Fu, D., Khan, S. H., and O'Neill, R. A., Multi-dimensional mapping of pyridy-lamine-labeled N-linked oligosaccharides by capillary electrophoresis, *J. Chromatogr. A*, 720, 395, 1996.

160. Suzuki, S., Kakehi, K., and Honda, S., Two-dimensional mapping of N-glycosidically linked asialo-oligosaccharides from glycoproteins as reductively pyridylaminated derivatives using dual separation modes of high-performance capillary electrophoresis, *Anal. Biochem.*, 205, 227, 1992.

161. Suzuki, S. and Honda, S., Two-dimensional mapping of N-glycosidically linked oligosaccha-rides in glycoproteins by high-performance capillary electrophoresis, *Trends Anal. Chem.*, 14, 279, 1995.

162. Slater, G. W., Rousseau, J., Noolandi, J., Turnel, C., and Lalande, M., Quantitative analysis of the three regimes of DNA electrophoresis in agarose gels, *Biopolymers*, 27, 509, 1988.

163. Carle, G. F., Frank, M., and Olson, M. V., Electrophoretic separation of large DNA molecules by periodic inversion of the electric field, *Science*, 232, 65, 1986.

164. Stefansson, M. and Novotny, M., Modification of the electrophoretic mobility of neutral and charged polysaccharides, *Anal. Chem.*, 66, 3466, 1994.
165. Teitelbaum, R. C., Ruby, S. L., and Marks, T. J., On the structure of starch-iodine, *J. Am. Chem. Soc.*, 100, 3215, 1961.
166. Rundle, R. E. and French, D., The complexation of starch and starch-iodine complex. II. Optical properties of crystalline starch fractions, *J. Am. Chem. Soc.*, 65, 558, 1943.
167. Thoma, A. and French, D., Starch-iodine-iodine interaction. I. Spectrophotometric investigations, *J. Am. Chem. Soc.*, 82, 4144, 1960.
168. Banks, W., Greenwood, C. T., and Khan, K. M., Interaction of linear, amylose oligomers with iodine, *Carbohydr. Res.*, 17, 25, 1971.
169. Schmid, K., α_1-*Acid Glycoprotein*, Vol. 1, Academic Press, New York, 1975, 183.
170. Rush, R. S., Derby, P. L., Strickland, T. W., and Rohde, M. F., Peptide mapping and evaluation of glycopeptide microheterogeneity derived from endoproteinase digestion of erythropoietin by affinity high-performance capillary electrophoresis, *Anal. Chem.*, 65, 1834, 1993.
171. Kilàr, F. and Hjerten, S., Separation of the human transferrin isoforms by carrier free high-performance zone electrophoresis and isoelectric focusing, *J. Chromatogr.*, 480, 351, 1989.
172. Kilàr, F. and Hjerten, S., Fast and high resolution analysis of human serum transferrin by high performance isoelectric focusing in capillaries, *Electrophoresis*, 10, 23, 1989.
173. Rudd, P. M., Scragg, I. G., Coghil, E., and Dwek, R. A., Separation and analysis of the glycoform populations of ribonuclease B using capillary electrophoresis, *Glycoconjugate J.*, 9, 86, 1992.
174. Rudd, P. M., Joao, H. C., Coghill, E., Fiten, P., Saunders, M. R., Opdenakker, G., and Dwek, R. A., Glycoforms modify the dynamic stability and functional activity of an enzyme, *Biochemistry*, 33, 17, 1994.
175. Yim, K., Abrams, J., and Hsu, A., Capillary zone electrophoretic resolution of recombinant human bone morphogenetic protein 2 glycoforms. An investigation into the separation mechanisms for an exquisite separation, *J. Chromatogr. A*, 716, 401, 1995.
176. Pedersen, J. and Biedermann, K., Characterization of proteinase A glycoforms from recombinant *Saccharomyces cerevisiae*, *Biotechnol. Appl. Biochem.*, 18, 377, 1993.
177. Landers, J. P., Oda, R. P., Madden, B. J., and Spelsberg, T. C., High performance capillary electrophoresis of glycoproteins: the use of modifiers of electroosmotic flow for analysis of microheterogeneity, *Anal. Biochem.*, 205, 115, 1992.
178. Oda, R. P., Madden, B. J., Spelsberg, T. C., and Landers, J. P., α,ω-Bis-quaternary ammonium alkanes as effective buffer additives for enhanced capillary electrophoretic separation of glycoprotein, *J. Chromatogr. A*, 680, 85, 1994.
179. Morbeck, D. E., Madden, B. J., and McCormick, D. J., Analysis of the microheterogeneity of the glycoprotein chorionic gonadotropin with high-performance capillary electrophoresis, *J. Chromatogr. A*, 680, 217, 1994.
180. Oda, R. P. and Landers, J. P., Effect of cationic buffer additives on capillary electrophoretic separation of serum transferrin from different species, *Electrophoresis*, 17, 431, 1996.
181. Yim, K. W., Fractionation of the human recombinant tissue plasminogen activator (rtPA) glycoforms by high-performance capillary zone electrophoresis and capillary isoelectric focusing, *J. Chromatogr.*, 559, 401, 1991.
182. Moorhouse, K. G., Rickel, C. A., and Chen, A. B., Electrophoretic separation of recombinant tissue-type plasminogen activator glycoforms: validation issues for capillary isoelectric focusing methods, *Electrophoresis*, 17, 423, 1996.
183. Watson, E. and Yao, F., Capillary electrophoretic separation of recombinant granulocyte-colony-stimulating factor glycoforms, *J. Chromatogr.*, 630, 442, 1993.
184. Tran, A. D., Park, S., Lisi, P. J., Huynh, O. T., Ryall, R. R., and Lane, P. A., Separation of carbohydrate-mediated microheterogeneity of recombinant human erythropoietin by free solution capillary electrophoresis, *J. Chromatogr.*, 542, 459, 1991.
185. Watson, E. and Yao, F., Capillary electrophoretic separation of human recombinant erythropoietin (r-HuEPO) glycoforms, *Anal. Biochem.*, 210, 389, 1993.
186. James, D. C., Freedman, R. B., Hoare, M., and Jenkins, N., High-resolution separation of recombinant human interferon-g glycoforms by micellar electrokinetic capillary chromatography, *Anal. Biochem.*, 222, 315, 1994.

187. Klausen, N. K. and Kornfelt, T., Analysis of the glycoforms of human recombinant factor VIIa by Capillary electrophoresis and high-performance liquid chromatography, *J. Chromatogr. A*, 718, 195, 1995.

188. Hoffstetter-Kuhn, S., Alt, G., and Kuhn, R., Profiling of oligosaccharides-mediated microheterogeneity of a monoclonal antibody by capillary electrophoresis, *Electrophoresis*, 17, 418, 1996.

189. Kelly, J. F., Locke, S. J., Ramaley, L., and Thibault, P., Development of electrophoretic conditions for the characterization of protein glycoforms by capillary electrophoresis-electrospray mass spectrometry, *J. Chromatogr. A*, 720, 409, 1996.

190. Liu, Y. and Chan, K.-F. J., High-performance capillary electrophoresis of gangliosides, *Electrophoresis*, 12, 402, 1991.

191. Yoo, Y. S., Kim, Y. S., Jhon, G.-J., and Park, J., Separation of gangliosides using cyclodextrin in capillary zone electrophoresis, *J. Chromatogr. A*, 652, 431, 1993.

192. Ulrich-Bott, B. and Wiegandt, H., Micellar properties of glycosphingolipids in aqueous media, *J. Lip. Res.*, 25, 1233, 1984.

193. Issaq, J. I., Janini, G. M., Chan, K. C., and El Rassi, Z., Approaches for the optimization of experimental parameters in capillary zone electrophoresis, *Adv. Chromatogr.*, 35, 101, 1995.

194. Mechref, Y. and El Rassi, Z., Capillary electrophoresis of herbicides. I. Precolumn derivatization of chiral and achiral phenoxy acid herbicides with a fluorescent tag for electrophoretic separation in the presence of cyclodextrins and micellar phases, *Anal Chem.*, 68, 1771, 1996.

CHAPTER 8

Separation of Peptides and Proteins

Herbert E. Schwartz, András Guttman, and Anders Vinther

CONTENTS

0-8493-9127-X/98/$0.00+$.50
© 1998 by CRC Press LLC

I. INTRODUCTION

Whereas CE is a useful separation technique for a wide variety of applications ranging from small inorganic ions to large biopolymers, a majority of the current installed base of CE instruments is dedicated to the analysis of proteins and peptides. Since the beginning of modern CE (early 1980s),[1,2] the protein application area of CE was also the first to be fully explored by researchers in the technique. The rationale to apply CE in free solution (i.e., FSCE, CE in the absence of a gel) for protein analysis was straightforward: proteins often yield inefficient, broad peaks with HPLC; CE, on the other hand, could potentially serve as a complementary technique to HPLC, with specific advantages of speed, resolution, and selectivity.[2] If Joule heating, diffusion, and adsorption could be adequately controlled, in principle very high separation efficiency (plate numbers of > 1 million) should be feasible with CE. The simplicity of a free solution CE format was also very appealing, especially for those who entered the field from a chromatography background.

The development of gel (polymer matrix)-filled capillaries for proteins (as well as DNA) began during the 1980s,[3,4] giving rise to a new mode of CE: capillary gel electrophoresis (CGE). Here the primary goal was to adapt widely used classical electrophoretic techniques such as SDS-PAGE to capillary-based instrumentation, thus opening the way to establish CE further in biochemistry, molecular biology, and pharmaceutical biotechnology laboratories. Now gel-filled and pre-coated capillaries are commercially available from various commercial sources.[5] In addition, CE instruments have become more sophisticated and flexible since their commercial introduction in 1988. These, and other instrument-related developments in CE over a 15-year period have led to many diverse applications of CE in the protein/peptide field.[6]

Today, CE is perhaps most often used for purity assays,[6] e.g., in the quality control of a drug manufacturing process where contaminants need to be screened or quantified. Other fields of research include structural studies (e.g., protein microheterogeneity), binding studies (e.g., drug-, DNA-, metal binding), process analysis, stability studies, immunoassays, and micropreparative work through automated fraction collection.[6] CE is also utilized in clinical research[7] and is being considered for implementation in the routine clinical laboratory,[8] e.g., for the analysis of serum proteins, hemoglobin variants, isoenzymes, and lipoproteins. A dedicated clinical analyzer based on multiple capillaries (Paragon™ CE) has recently been introduced by Beckman Instruments (Brea, CA).

It can be argued that of the different CE techniques applied to protein analysis, capillary isoelectric focusing (CIEF) has been the least developed. Indeed, the capillary format imposes some unique opportunities (and problems), and adaptation from a slab gel format to a (free solution) capillary configuration has not always been straightforward. In 1985, Hjerten and Zhu[9] were the first to explore the feasibility of IEF in capillaries. The advent of commercially available CE instrumentation, capillary coatings,[5] and, above all, a basic understanding of the CIEF separation mechanism (including the role of the EOF) has led to impressive results in recent years.

In this chapter, the three basic CE techniques for protein and peptides, i.e., FSCE, CGE, and CIEF, are discussed. While it is not the purpose of this chapter to provide an all-inclusive, comprehensive review of CE/protein/peptide-related work (for this, see for example, the "Application Reviews" and "Fundamental Reviews" in recent special issues of *Analytical Chemistry*), we will focus on a description of the main principles, methodologies, potential problems, and instrument requirements involved in the three CE separation modes applicable to proteins and peptides. In general, protein applications are the more difficult to develop, because, for example, proteins have a stronger tendency for adsorption to the capillary wall. Therefore this chapter focuses mainly on protein separations. However, most of the approaches described (including the influence of various additives on separation) can be used when developing peptide applications. Peptide analysis by CE is also dealt with in Chapters 1 through 3.

II. FREE SOLUTION CAPILLARY ELECTROPHORESIS (FSCE) OF PROTEINS AND PEPTIDES

FSCE denotes capillary electrophoresis (CE) performed in free solution — in contrast to CE performed in gels or polymer network solutions. A vast majority of the CE papers published so far concern analysis in free solution.

A. Capillary Zone Electrophoresis (CZE)

In the pioneering years of CE,[1-3] applications were performed almost exclusively in the zone electrophoresis mode (CZE). Although other CE modes have been introduced since then, CZE is still predominant — even for protein and peptide applications. This is partly because CZE is very simple to perform. The electrophoresis buffer (e.g., citrate, phosphate, borate, or zwitterionic buffers such as tricine, CAPS, or MES) is simply flushed through the narrow-bore capillary tube, and the sample is injected. When voltage is applied, electrophoretic separation takes place.

In CZE, analytes are separated based on differences in charge density, i.e., the electrophoretic mobility increases with decreasing analyte radii and increasing net charge. Because of the presence of an electroosmotic flow (EOF) — which generally is stronger than the electrophoretic flow of peptides and proteins in uncoated capillary tubes — analysis of anionic and cationic analytes in the same run is feasible. Neutral analytes migrate with the speed of the EOF. Therefore they are not separated by CZE, unless some other separation mechanism is also present, e.g., sieving or micellar interaction.

B. Micellar Electrokinetic Chromatography (MEKC)

Separations of uncharged molecules are good candidates for the MEKC mode of CE.[10] In MEKC, a surfactant is added to the separation buffer at a concentration above its critical micelle concentration. The micelle consists of surfactant molecules with hydrophobic tails pointing toward the center of the micelle and hydrophilic heads at the surface of the micelle in aqueous solutions. MEKC buffers are prepared by simply mixing the surfactant with the buffer solution. To use an HPLC analogy, the micelles act as a "moving stationary phase." They move through the capillary with a mobility determined by the net charge density of the micelle. A neutral molecule partitions between the hydrophobic interior of the micelle and the "free solution." When in free solution, the neutral molecule migrates with a mobility equal to the EOF. When in the micelle, the solute molecule migrates with the mobility of the micelle. This means that the solute elutes in the time window between the migration times corresponding to the EOF and the micelle. The anionic surfactant sodium dodecyl sulfate (SDS) has been most frequently used in MEKC. However, other surfactant molecules, including cationic surfactants such as cetyl trimethyl ammonium bromide (CTAB), are also used and may offer specific advantages for certain separation problems. Only free amino acids and small peptides are able to fully partition into the interior of most micelles. Larger peptides and proteins are too bulky. However, even for proteins, surfactant molecules might help to improve separation, possibly by an ion-pairing mechanism or hydrophobic interaction between the surfactant molecule and the protein.[11] For example, glycoforms of recombinant proteins or monoclonal antibodies have been successfully resolved using this approach.[6]

C. Modeling the Mobility of Peptides

pH is the most important parameter in electrophoretic separation of proteins and peptides. Changing the pH a tenth of a unit might improve resolution significantly. Instead of carrying out numerous experiments at different pH values in order to obtain a satisfactory resolution of the analytes, it would be desirable to be able to predict the optimum pH by theory. In 1966, Offord[12] proposed that the mobility of a peptide (μ) is proportional to its net charge (Z) and inversely proportional to the surface area of the molecule (assuming a spherical shape and a molecular weight (MW) proportional to the cubic root of the volume of the molecule):

$$\mu = k \frac{Z}{MW^{2/3}} \tag{1}$$

where k is a constant. The validity of this equation has been supported by CZE experiments, including data obtained when analyzing human growth hormone (hGH) peptide fragments.[13] Computer programs are available for calculating the charge of a peptide with a given amino acid sequence. For larger peptides and for proteins, a prediction of the electrophoretic behavior becomes more complex, which means that the Offord equation is not readily applicable to these types of molecules.

D. Factors to Consider in Method Development

Thus far, few routine protein applications in CE have found their way to quality control (QC) laboratories, despite publication of hundreds of protein and peptide applications in the separations literature. In principle, however, CE is highly suitable as a routine analytical tool in QC laboratories because of its high automation capability, low reagent costs, and short run times. Robust and efficient separations of proteins can be obtained by CE, but the development of such applications is not always as simple as it first may seem. There are several reasons for this, which we will discuss in the following sections.

1. Matrix Effects in CE and HPLC

First of all, it should be emphasized that CE and HPLC must be approached differently even though an electropherogram resembles a chromatogram. Among other things, CE and HPLC differ markedly with respect to separation principle (electrophoresis vs. chromatography), injection volume (low nanoliter vs. low microliter), and the importance of the sample matrix. Generally, matrix components do not affect separation in HPLC because nonadsorbing sample components are washed out in the void volume. In CE, matrix components are not washed out prior to analysis. Therefore, proteins as well as matrix components are electrophoretically separated when voltage is applied. As a matter of fact, differences in sample and electrophoresis buffer matrices are often important when optimizing peak shape, separation efficiency, and detection limits (see below).

2. Axial pH Gradients

Another point to consider when developing protein applications — in addition to the choice of electrophoresis buffer and pH adjusting reagent — is the buffer capacity. In CE, typically, field strengths of several hundreds V/cm are often applied. The electrode processes at the anode and cathode are

$$\text{Anode side: } 2H_2O \Leftrightarrow O_2 + 4H^+ + 4e^- \tag{2}$$

$$\text{Cathode side: } 2H_2O + 2e^- \Leftrightarrow H_2 + 2OH^- \tag{3}$$

Hence, H^+ and OH^- ions are formed, resulting in drifting pH values in the buffer reservoirs. In the "normal" CE mode (uncoated capillaries, EOF toward the cathode), electrophoresis buffers and samples are introduced at the anode side of the capillary. When the same buffer is used for multiple runs, the pH of the electrophoresis buffer will change from run to run, resulting in potential effects on resolution and analytical precision (migration times). The following example — which represents a typical CE run — serves to illustrate this point:

Current (I): 54 μA; voltage applied for (s): 60 min (3600 s);
volume in anode reservoir (V): 2 mL ($2 * 10^{-3}$ L),
Faraday's constant (F): 96,485 C/mol

Applying Faraday's law:

$$n = \frac{I \cdot t}{F} = \frac{54 \cdot 10^{-6} \cdot 3600}{96,485} = 2.0 \cdot 10^{-6} \text{ moles} \tag{4}$$

Within one hour, 2.0×10^{-6} mole H^+ (or 2.0 mM) is formed in the 2 mL-volume anode reservoir. Combining the above with the Acid-Base equation

$$pH = pK_a + \log \frac{[B]}{[A]} \tag{5}$$

where [A] and [B] are the concentration of the acid and its corresponding base, respectively (e.g., $H_2PO_4^-$ and HPO_4^{2-}), one can calculate the change in pH in the anode reservoir.

Figure 8.1 illustrates the axial pH gradient buildup. Three buffers, i.e., 10 mM phosphate, 100 mM phosphate, and 10 mM aspartic acid, were adjusted to pH 4.0. Each was flushed through the capillary and a voltage corresponding to a current of 10 μA (a fairly low current in CE) was applied for 150 min. During the voltage application, the pH was measured (as well as calculated) every 30 min. The 10 mM phosphate buffer caused the pH to decrease by 0.4 units at the anode reservoir. Increasing the buffer capacity — by increasing the phosphate concentration tenfold, or, even more effective, changing to a buffer with a higher capacity at the experimental pH (phosphate has pK_as at 2.15 and 7.21, whereas the side chain of aspartic acid has a pK_a of approximately 3.9) significantly reduced the decrease in pH.

The conclusion drawn from the above experiment is that an axial pH gradient buildup can result in poor run-to-run reproducibility of resolution and/or migration time when proteins are analyzed at pH values close to a pK_a value of the protein (pH ~ pK_a ± 1). The gradient is eliminated or minimized by using buffers with high buffer capacity, applying low currents (low applied voltages and/or low conductivity buffers), using relatively large buffer reservoirs, and by replenishing the buffer in the reservoirs between runs.

3. Radial pH Gradients

In addition to an axial pH gradient, a radial pH gradient exists in the capillary tube.[14] In aqueous solution, the capillary wall surface is negatively charged because of the presence of deprotonated silanol groups. H^+ counterions are arranged from the capillary surface toward the capillary axis in accordance with the Boltzmann equation: the highest H^+ concentration is found close to the capillary surface and the lowest concentration is found at the tube axis. The radial gradient is calculated[14] as

$$\Delta pH = \frac{230 \cdot \exp\left(\dfrac{1958}{T}\right) + 0.004605 \cdot T}{T} \cdot \mu_{EO} \tag{6}$$

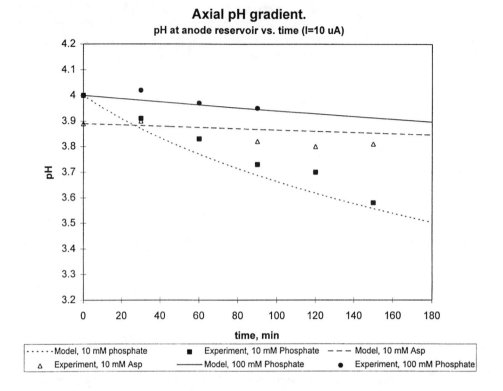

Figure 8.1   Axial pH gradient buildup. pH was measured (dots) in the anode reservoir and calculated (lines) for three buffers.

where T is the temperature in units K and μ_{EO} is the EOF mobility. Example: T = 30°C, μ_{EO} = 0.0005 cm^2/V/s results in a radial pH gradient of approximately one pH unit.

4. Factors Affecting Separation Performance

In their pioneering paper (1983), Jorgenson and Lukacs[2] suggested "that separation of proteins will be highly efficient." That statement is based on the assumption that only axial diffusion contributes to the zone dispersion process. In that case, the number of theoretical plates is proportional to the mobility of the protein and the applied voltage, and inversely proportional to the diffusion coefficient. Compared to small molecules, proteins have very small diffusion coefficients. Hence, in principle, very high plate counts should be obtainable. In fact, plate numbers exceeding 1 million have indeed been achieved in CZE; however, in general it is very difficult to obtain plate numbers in that range for proteins. The reason is that the axial diffusion is not the only contribution to the analyte peak width. Other factors — including radial diffusion, Joule heating of the buffer solution, the length of the sample plug, and adsorption of the proteins to the capillary surface — also play a role.

5. Strategies to Avoid Adsorption of Proteins

Proteins are polyelectrolytes, and adsorption usually occurs because of coulombic attractions between the negatively charged capillary surface and the positive charges on the protein molecule, as illustrated in Figure 8.2. The result in CE is either tailing peaks or even complete adsorption of the protein to the capillary surface, i.e., no peaks whatsoever.

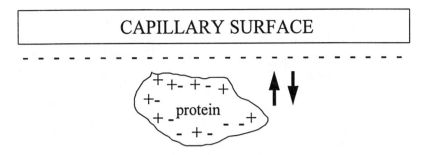

Figure 8.2 Proteins have a tendency to adsorb to the capillary surface because of electrostatic attraction.

One approach to avoiding proteins being adsorbed onto the capillary surface is to go to pH extremes in the electrophoresis buffer, i.e., operate either at high pH[15] or low pH[16] conditions. When analysis is performed above the isoelectric point (pI) of the protein, its net charge becomes negative even though the protein may contain both positive and negative charges. In this state, the protein molecule would tend to be repelled from the negatively charged capillary surface.

When pH is decreased to values below approximately 2, most of the negative charge on the inner capillary surface will be titrated off. This means that even though the protein is highly protonated at low pH, it will not be electrostatically attracted by the "neutral" capillary surface. An example of the "extreme pH" approach is shown in Figure 8.3. A human growth hormone (hGH, 191 amino acids, MW 22,125 Daltons, pI ~ 5) sample was allowed to degrade at elevated temperatures and was then analyzed by CE at pH 2.0 (below the pI of hGH) and pH 8.0 (above the pI of hGH). It can be seen that in both electropherograms a symmetric hGH peak is obtained, but only at pH 8.0 was resolution of hGH and a monodeamidated form (Asn→Asp) obtained. At pH 8.0 monodeamidated hGH contains one more negative charge than hGH because of the ionization of the Asp side chain carboxylate group.

A disadvantage of the "extreme pH approach" is that a wide range of usable pHs, i.e., from ~2 to a value above protein pI, is excluded. The more basic the protein, the wider that excluded range is. At pH extremes, proteins are either fully protonated or fully deprotonated.

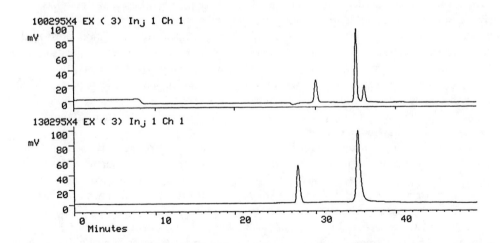

Figure 8.3 FSCE analysis of human growth hormone (hGH) in phosphate buffers at pH 2.0 (lower) and pH 8.0 (upper) in uncoated capillaries. (Courtesy of T. Kornfelt, Novo Nordisk A/S.)

Table 8.1 Capillary Coatings for Protein Separations

Coating type	Investigator	Reference
Methylcellulose	Hjerten	*J. Chromatogr.* 347, 191 (1985)
Polyacrylamide (through Si-O-Si-C bonds)	Hjerten	*J. Chromatogr.* 347, 191 (1985)
Polyacrylamide (through Si-C bonds)	Novotny	*Anal. Chem.* 62, 2478 (1990)
3-glycidoxypropyltrimethoxysilane	Jorgenson	*Science* 222, 266 (1983)
Epoxy-diol, maltose	Poppe	*J. Chromatogr.* 480, 339 (1989)
Polyethyleneglycol	Poppe	*J. Chromatogr.* 471, 429 (1988)
Polyvinylpyrrolidone	McCormick	*Anal. Chem.* 60, 2322 (1988)
Aryl-pentafluoro(aminopropyltrimethoxy)silane	Swerdberg	*Anal. Biochem.* 185, 51 (1990)
α-lactalbumin	Swerdberg	*J. High Res. Chromatogr.* 14, 65 (1991)
Polyether	El Rassi	*J. Chromatogr.* 559, 367 (1991)
Hydrophilic, C1, C8, C18	Dougherty	*Supelco Reporter,* Vol. X, No. 3 (1991)
Ion exchangers, polyacrylamide	Engelhardt	*J. Microcol. Sep. 3,* 491 (1991)
Neutral, hydrophilic	Karger	*J. Chromatogr.* 652, 149 (1993)
Hydrophilic & hydrophilic polymers	Lee	*Anal. Chem.* 65, 2747 (1993)

Hence, the separation of protein species varying only in subtle structural differences would be difficult, if not impossible, to achieve.

If the optimum for the separation is found at around neutral pH conditions, strategies other than the extreme pH approach should be considered. One of these is to coat the capillary surface either dynamically (using the rinse capability of the CE instrument) or covalently (by using chemical derivatization procedures prior to inserting the capillary in the instrument). The goal of these procedures is to minimize, eliminate, or reverse the surface charge on the capillary wall.

Covalent coating is generally performed by attaching hydrophilic polymers to the silanol groups on the surface. Robust and reproducible covalent coatings are often not easily made in the laboratory. However, nowadays several companies offer such coated capillaries.[5] They are typically developed for specific applications and are usable in a given pH range. Table 8.1 lists a selection of some of the coatings which have appeared in the CE literature.[6]

Instead of chemically modifying the capillary surface to avoid adsorption of the proteins, one can perform a "dynamic" coating by adding various substances to the electrophoresis buffer while using the rinse cycles of the CE instrument to equilibrate the columns. One simple way of doing this is to increase the electrophoresis buffer concentration or to add various salts to the buffer.[6,17] As the salt and protein compete for adsorption sites on the capillary wall, the higher the salt-to-protein ratio, the higher the probability of salt adsorption. A drawback of this approach is that the higher the salt concentration, the higher the conductivity of the buffer, and therefore the higher the power induction (at a constant applied voltage) and temperature increase in the capillary tube. This is especially pronounced in CE instruments lacking the ability to adequately remove Joule heating, i.e., with simple configurations based on ambient air-cooling of the capillaries.

The power induction can be greatly reduced by using zwitterionic salts such as tricine, CAPS, CHES, and betaine instead of ionic salts such as phosphate and borate.[18] The zwitterions yield very low conductivity, and therefore they may be used at rather high concentrations (up to more than 1 M, depending on their solubility) without causing high power inductions. As buffer components, zwitterions are frequently used in FSCE.

Cationic amines such as 1,4-diaminobutane (DAB) and 1,5-diaminopentane (DAP) have also been used to suppress protein-capillary surface interactions at concentrations up to 60 mM.[6] As mentioned above for the "high salt" approach, the idea is that the amine competes with the protein for adsorption sites on the capillary wall. The larger the protein, the higher the concentration of cationic amine must be in order to obtain the desired effect.

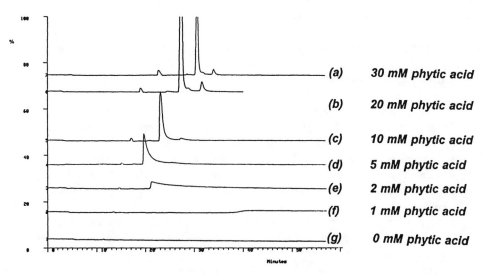

Figure 8.4   FSCE analysis of aprotinin by the use of phytic acid as ion-pairing reagent. Analysis was performed in 25 m*M* phosphate buffer at pH 7.0 with increasing amounts of phytic acid (0, 1, 2, 5, 10, 20, 30 m*M*, from bottom to top) added to the buffer. (From Okafo, G. N., et al., *Electrophoresis*, 16, 1917, 1995. With permission.)

Various cellulose derivatives (e.g., methylcellulose, hydroxyethylcellulose, and hydroxy-propylmethylcellulose — generally added at concentrations below 1%), polyethylene oxides, and ethyleneglycol (10 to 30%) decrease the zeta potential (i.e., a measure of the electric potential at the shear plane set up in the surface region of the electrical double layer) on the surface of the capillary wall. These additives also increase the viscosity of the buffer, thereby decreasing the tendency for protein adsorption.

Protein adsorption may also be reduced by addition of cationic surfactants such as cetyltrimethylammonium bromide (CTAB) to the buffer. Increasing the concentration first reduces, then eliminates, and finally reverses the charge on the capillary surface, thus creating a positively charged surface (the latter condition also causes a reversal of the EOF).

Finally, one can perform ion-pairing CE in order to suppress or eliminate protein adsorption. The anionic ion-pairing reagent, which is added to the buffer, reacts with cationic sites on the protein molecule, thus making it electrophoretically net negatively charged. An example of that approach is shown in Figure 8.4,[19] using phytic acid as the ion-pairing reagent. This molecule contains six phosphate groups with pK_a values in the range of 1.9 to 9.5, which allows one to manipulate its charge over a wide pH range. Aprotinin is a highly basic protein (pI ~ 10.5, MW 6511 Daltons, 58 amino acid residues). Initial FSCE analysis of aprotinin in a 25 m*M* phosphate buffer at pH 7.0 resulted in the protein being totally adsorbed onto the capillary surface. When phytic acid was added to the buffer, aprotinin eluted as a distinct peak — first severely tailed, but with increasing phytic acid concentrations the aprotinin peak became more symmetrical, even separating several aprotinin variants.

Ion pairing with phytic acid can also be used to optimize separation of peptides and predict their basic nature.[20] This is illustrated by the following example: Aprotinin was unfolded, reduced, pyridylethylated, and finally digested with endoproteinase Lys-c. In this way the expected five peptide fragments were formed: fragment 1, RNNFK (pI ~ 11.4); fragment 2, SAEDCMRTCGGA (pI ~ 4.4); fragment 3, AGLCQTVYGGCRAK (pI ~ 9.2); fragment 4, ARIIRYFYNAK (pI ~ 11.3); fragment 5, RPDFCLEPPYTGPCK (pI ~ 5.9). The digest was subjected to CE analysis in a 100 m*M* sodium phosphate buffer (pH 7.0) with varying concentrations of the dodecasodium salt of phytic acid. The mobilities of fragments 1, 4, and 3 (all with basic pI values) decreased with increasing concentration of phytic acid, whereas the mobilities of fragments 2 and 5 (acidic pI values) showed little or no change. This correlation of the ion-pairing tendency of phytic acid with the basic peptides was most useful for both the separation and preliminary identification of the peptide fragments.

6. Optimizing Resolution

Inherently, many of the strategies described above were developed not only to avoid adsorption of the protein to the capillary surface, but also to optimize resolution of the proteins in specific applications. In addition to the buffer additives already mentioned, several others have been published as being useful for optimization of specific applications. For example, Cu^{2+} and Zn^{2+} ions can interact with N, O, or S atoms in peptides/proteins and thereby affect separation as described for analysis of histidine containing peptides by addition of Zn^{2+} to the buffer.[21]

Addition of organic solvents such as methanol or acetonitrile changes the EOF mobility and the polarity of the buffer which may help to solubilize proteins and improve separation in certain applications.

The utility of cyclodextrins (CDs) in CE for chiral separations was first described by Guttman et al.[22] using gel-filled capillaries, while Terabe[23] reported their use for isomers in MEKC. CDs have the molecular shape of a truncated cone with a hydrophobic interior and a hydrophilic exterior. The CD and protein/peptide can interact in a host-guest manner. Figure 8.5 shows separation of stereoisomers and isoforms of the heptapeptide Leu-Glu-Asp-Gly-Ser-Pro-Arg (identical to one of the fragments obtained when digesting hGH with trypsin; amino acid residues 128 to 134). Separation of the stereoisomers was only achieved by addition of β-CD to the buffer.[24]

E. Improving Sample Detectability in FSCE

UV absorbance detectors are most often used in CE. Because only nanoliter volumes are introduced into the capillary, very sensitive detectors are required to achieve detectability of trace components. Typically the UV detectors in CE instrumentation have detection limits in the low femtomole range (1 femtomole corresponds to detection of a 1% impurity when injecting 10 nL of a 0.1 mg/mL protein solution having a MW of 10,000 Daltons). In HPLC, the detection sensitivity can be improved simply by adding a large volume of the sample solution onto the column. The protein is adsorbed onto the stationary phase until an appropriate elution strength is obtained during analysis. That approach is not readily applicable in CE because overload of the capillary and, consequently, low resolution would be obtained.

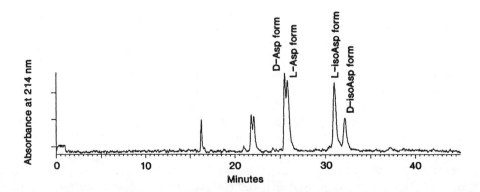

Figure 8.5 FSCE analysis of synthesized stereoisomers and isoforms of the heptapeptide Leu-Glu-Asp-Gly-Ser-Pro-Arg. The heptapeptides differed by the chemical and stereoisomers of the Asp residue as indicated in the figure. Analysis was performed in an uncoated capillary in a 25 mM aspartic acid, 10 mM β-CD buffer at pH 3.8.

However, as discussed below, there are some means to improve sensitivity when the protein cannot be concentrated prior to injection on the capillary.

Conceivably, one could replace the UV detector with a laser induced fluorescence (LIF) detector, which generally has a much better sensitivity than UV detectors.[25] However, this approach can only be used either when the protein has native fluorescence or when it has been derivatized with a suitable fluorophore. Because of the microheterogeneity of many proteins, derivatization is not trivial because multiple protein sites get tagged. Compared to DNA analysis, for example, the utility of LIF detectors for protein analysis is therefore rather limited.[25]

Another, perhaps more practical strategy for improving detectability is to use UV absorbance detection combined with various on-capillary protein stacking ("pre-concentration") techniques.

The first stacking method described in the CE literature was "conductivity stacking"[26] in which a sample plug with a lower specific conductivity than that of the electrophoresis buffer is introduced into the capillary tube. According to Ohm's Law, when voltage is applied, the field strength will be higher along the sample plug than along the electrophoresis buffer, because the current is constant throughout the capillary. Because the electrophoretic mobility is proportional to the field strength, the protein migrates faster in the sample plug than in the electrophoresis buffer. When reaching the interface between sample plug and electrophoresis buffer, the protein molecules are slowed down. Hence, the protein molecules are effectively concentrated in a lower volume than the sample volume in which they were initially applied. Using this simple on-column concentration method, typically a five- to tenfold increase in sensitivity is achievable. One must be aware that with this stacking method a higher power is induced in the sample zone than that in the electrophoresis buffer (i.e., the electric field strength is higher in the sample zone compared to the buffer). This is especially important when analyzing temperature-labile proteins. In such a case, on-capillary thermal degradation may occur during the stacking procedure.

Another way of stacking the protein molecules is to introduce them either at a very low or very high pH.[27] When voltage is applied they will migrate faster in the sample zone than in the electrophoresis buffer (due to a higher charge). The analytes are thus being stacked. The improvement in sensitivity by this method is more or less the same as with "conductivity stacking," i.e., ~ five- to tenfold. A drawback of pH stacking is that many proteins are labile at extreme pH values. Therefore, they might degrade in the sample solution.

One of the most effective ways of stacking the protein molecules is to combine zone electrophoresis with isotachophoresis (ITP). Using this technique, up to a 50- to 100-fold increase in detectability is feasible. Unlike zone electrophoresis, ITP is inherently a focusing (zone sharpening) technique. ITP can be used to concentrate the protein molecules in a shorter plug length than that originally introduced into the capillary. ITP stacking is followed by zone electrophoretic separation of the proteins into separate bands.

Typically, sample plug lengths in the order of 1 mm are introduced in CE during the injection procedure. Using ITP preconcentration for peptide samples, Schwer and Lottspeich[28] reported injection plug lengths in excess of 60 mm. The greater sample volume translates directly to much higher detectability of the peptides, while at the same time resolution is improved. Figure 8.6 illustrates how ITP-CZE can be carried out for positively charged (acidic) and also for negatively charged (basic) proteins.[29] For basic proteins, a leading electrolyte (containing ions with mobilities higher than the proteins) is flushed through the capillary. The protein sample is dissolved in leading electrolyte and subsequently injected. The injection end of the capillary (anode side) is immersed in terminating electrolyte (containing ions with a mobility lower than the proteins) and voltage is applied. After ITP stacking has taken place (concentrating the proteins at the border between the leading and terminating electrolytes), the injection end of the capillary is immersed in leading electrolyte, voltage is reapplied, and the proteins are separated based on zone electrophoresis. For acidic proteins the capillary is flushed with terminating electrolyte, the sample is dissolved in leading electrolyte, the injection

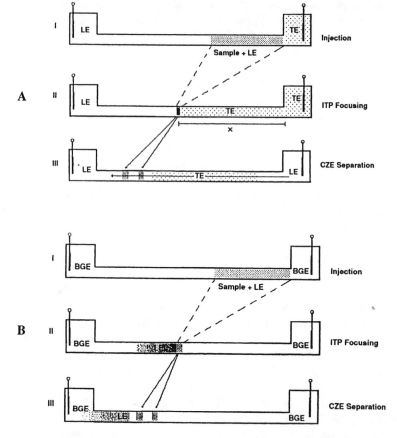

Figure 8.6 Schematic illustration of transient ITP sample preconcentration in CE. A) for basic proteins, B) for acidic proteins. LE = leading electrolyte, TE = terminating electrolyte, BGE = background electrolyte (in this case identical to terminating electrolyte).

end of the capillary is immersed in terminating electrolyte, and voltage is applied. After an initial ITP stacking period, the proteins will be separated into distinct bands based on zone electrophoresis. In this procedure, no change in buffer vials is required.

III. CAPILLARY GEL ELECTROPHORESIS (CGE)

A. Classical Electrophoresis: SDS-PAGE

From all the classical electrophoretic techniques, sodium dodecyl sulfate mediated polyacrylamide gel electrophoresis (SDS-PAGE) is the most used among researchers in the field. As early as 1967, Shapiro et al.[30] showed that molecular weights (MW) of proteins can be determined by measuring their electrophoretic mobility in sodium dodecyl sulfate (SDS) containing polyacrylamide gel. With some improvement in the methodology by Weber and Osborn[31] two years later, it became a standard technique for molecular weight determination of single protein chains. For most multichain proteins, boiling a slightly basic pH (~8) solution with 0.1% SDS in the presence of mercaptoethanol results in the breakage of disulfide linkages. Subsequently, the polypeptide chains dissociate while binding SDS in a random coil configuration.

Regular proteins bind SDS in a constant ratio of 1.4 g SDS/g protein. Therefore, SDS-complexed proteins have almost similar mass-to-charge ratios and can be readily separated according to their size on a slab gel during electrophoresis.[32] MWs of unknown proteins can then simply be estimated from a calibration curve based on protein MW standards. This procedure works well for proteins with MWs higher than 10,000 having no posttranslational modifications, such as glycosylation or phosphorylation. If the MW of the protein subunit is less than 10,000, the intrinsic charge of the amino acid backbone can play a significant role on the overall charge of the protein–SDS complex, thus disrupting the constant mass-to-charge ratio of the proteins — the requirement needed for a size-based electrophoretic separation.

Proteins with posttranslational modifications usually yield different charge-to-mass ratios than the expected 1.4 g SDS/g protein ratio.[33] With phosphorylation, the resulting charge-to-mass ratio is increased; with glycosylations the resulting mass-to-charge ratio decreases because the glycans do not bind SDS, but significantly contribute to the molecular mass of the protein. In these instances, MW determination by SDS gel electrophoresis is not straight-forward, i.e., a simple calibration curve of mobility data/MWs cannot be used), and special methods (i.e., the Ferguson method[34]) should be used to obtain satisfactory accuracy in MW estimation of unknown proteins.

B. Classical vs. Capillary Techniques

In conventional SDS-gel electrophoresis, the most commonly used separation medium is polyacrylamide gel, typically present in high concentration (>5%) with an appropriate crosslinker (0.1 to 0.3%).[32] Gradient gels with a gel concentration range of 5 to 25% are also popular.[33] The gel structure creates a molecular sieving effect, allowing the separation of different sized, charged particles in the gel matrix. In capillary gel electrophoresis (CGE) the same phenomenon is operable. Hence, SDS-PAGE is directly comparable to SDS-CGE. For reasons explained below, sieving polymers other than polyacrylamide are more useful in SDS-CGE.

In the early CGE work of Cohen and Karger,[4] crosslinked polyacrylamide gel was used in the capillary column — a separation of a protein mixture (MWs up to 35,000 Daltons) was obtained. One of the main drawbacks in the use of crosslinked polyacrylamide gels, however, was the lack of low-UV transparency of the gel in the on-column detection used in CGE. Thus, sample detectability was severely compromised. The lack of flexibility during and between CGE separations was also apparent. Only relatively clean samples could be injected because dirty samples easily plugged the gel-filled capillary. The plugging caused zone discontinuities, resulting in poor separation performance. Furthermore, crosslinked gels are heat sensitive. At slightly higher than room temperature, bubbles are typically formed in the capillary which result in a blockage of the current flow.

To avoid these problems, first linear polyacrylamide gels (with no crosslinker) were explored as a sieving matrix for CGE, and later on other hydrophilic linear polymer matrices, such as dextrans and polyethylene oxides (see Figure 8.7 for molecular structures) were employed in the appropriate concentrations required to yield molecular sieving of the SDS-protein complexes.[35-40] The latter polymer networks can be used in the low-UV region (down to 200 nm) needed for optimum sensitivity. In a recent paper by Righetti's group,[41] polyvinyl alcohol solutions were also found to be effective for SDS-CGE separations. Due to their relatively low viscosity, these linear, polymer-filled capillaries allow easy replacement of the sieving matrix in the column. Hence, when a "dirty" sample is injected and the column consequently becomes plugged, replacement of the gel (using the rinse feature on the CE instrument) should easily solve the problem.[42]

Figure 8.8 shows a typical SDS-CGE separation of a protein test mixture (MW: 14.4 to 94 kDaltons), employing a 3% polyethylene oxide sieving matrix. The inset shows the separation of the same test mixture using conventional SDS-PAGE slab gel electrophoresis.[40] It can be seen that the resolution of the two systems is roughly equivalent. However, the CGE system has many advantages over the slab gel approach. CE-based systems are fully auto-

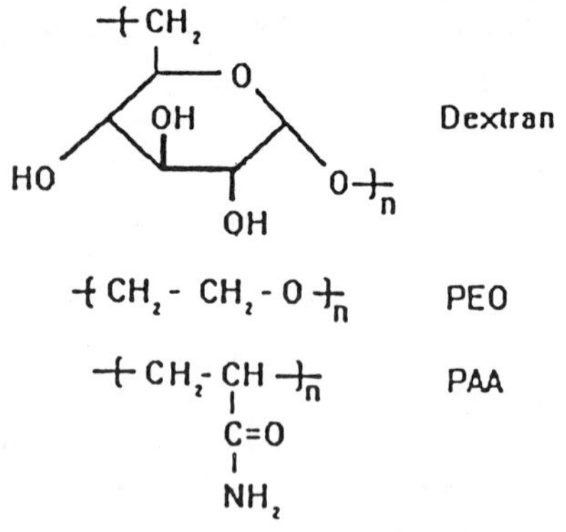

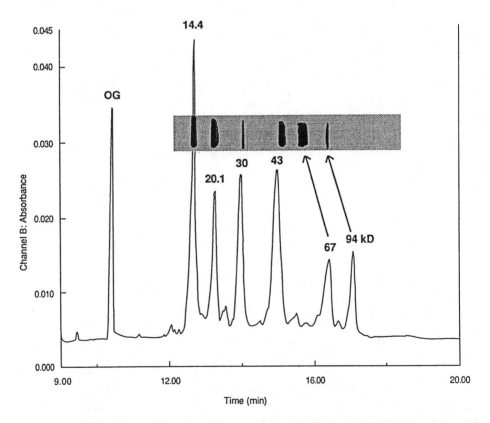

Dextran

PEO

PAA

Figure 8.7 Schematic structure of the monomer units of the polymers most frequently used in capillary SDS gel electrophoresis.

matable; HPLC-type electropherograms are obtained which in *real-time* are stored on the computer hard drive. The on-column UV detection in CE allows for direct, accurate quantitation of the protein peaks while no labor-intensive staining/destaining procedure (needed with slab gels) is necessary. Furthermore, the separation is fast (<20 min), extremely repro-

Figure 8.8 Capillary SDS gel electrophoresis of six protein standards on a 47-cm-long (40 cm to the detector) × 100 μm i.d. capillary. Numbers above the peaks correspond to their molecular weights in kDaltons (14.4: α-lactalbumin; 20.1: soybean trypsin inhibitor; 30: carbonic anhydrase; 43: ovalbumin; 67 bovine serum albumin; 94: phosphorylase B). OG: orange G, internal standard. Conditions: Running buffer: 100 mM Tris-CHES, pH 8.8, Sample buffer: 60 mM Tris-HCl, pH 6.6; Detection: UV 214 nm; E = 300 V/cm.

ducible, and the gel is easily replaced for the next CE run by simply instructing the instrument to implement a rinse step.

In order to determine the MW of the individual proteins, a calibration curve (e.g., MW vs. 1/migration time) can be constructed, similar to standard procedures based on SDS-PAGE. Then, from the migration time of the unknown protein peak, its MW can be easily assessed. Figure 8.9A shows a calibration curve (square symbols) based on a test mixture containing six proteins, i.e., α-lactalbumin (MW 14,400), carbonic anhydrase (MW 29,000), ovalbumin (MW 45,000), bovine serum albumin (MW 66,000), phosphorylase B (MW 97,400), β-galactosidase (MW 116,000), and myosin (MW 205,000). MWs of more than sixty other proteins (Table 8.2) were estimated with this CGE system, and the results are also plotted (as empty circles) in Figure 8.9A.

To compare the CGE system with the SDS-PAGE technique (coomassie blue staining[43]), a similar calibration curve and MW determination of the same proteins are shown in Figure 8.9B and Table 8.2, respectively. Table 8.2 shows that the estimated MWs obtained with the two methods are only slightly different for most proteins. Comparing Figures 8.9A and B, it can be concluded that with either the CGE or the slab gel method, similar separation performance and accuracy (MW estimation) can be obtained. Regression coefficients are similar in both cases, i.e., $R^2 = 0.995$ for CGE and 0.973 for SDS-PAGE. For both CGE and SDS-PAGE, the calibration curves can be used to estimate the MWs of most of the proteins with approximately 10% accuracy. Using the CGE method, however, the above-mentioned advantages (speed, automation, quantitation) come into play.[40] The implementation of the SDS-CGE technique in CE instruments has recently been facilitated by the introduction of commercially available kits containing coated capillaries, replaceable sieving buffers, protocols, etc.

1. Ferguson Method

Figure 8.9 clearly illustrates that, while in most cases the simple calibration curve method gives appropriate results, there are proteins which do not fit the line. Proteins with specific side groups, such as carbohydrates (glycoproteins), lipids (lipoproteins), or other prostetic groups, complex SDS differently than proteins lacking these groups.[33] This irregular binding of SDS results in altered charge-to-mass ratios for these molecules, causing inaccurate estimates in the apparent molecular mass, as is noticeable in Figure 8.9. As mentioned above, the Ferguson method[34] is usually used to correct for this discrepancy.

In SDS-CGE, the Ferguson method involves performing the electrophoretic separations at different gel concentrations, followed by plotting the logarithm of reciprocal migration times of the individual proteins as a function of the gel concentrations. Figure 8.10[44] depicts the Ferguson plot of the six standard proteins used in Figure 8.8. The slopes of the lines obtained from linear regression analysis of this plot yield the negative of the retardation coefficient ($-K_R$). A more universal calibration curve, the so-called K_R plot, is obtained by plotting log MW as a function of K_R (Figure 8.11).[44] K_R is assumed to be proportional to the effective molecular surface area (or to the radius of a spherical molecule with the same surface area) and not directly to the molecular mass.[33] It has been demonstrated that a linear relationship exists for the K_R vs. log MW data in SDS-PAGE[31] and also in SDS-CGE.[44]

In the case of SDS-PAGE, the application of the Ferguson method is extremely labor intensive and time consuming, primarily because of the requirement of making different gel concentrations in a slab format and the subsequent evaluation of the separated bands by staining/destaining procedures. However, with SDS-CGE, the Ferguson method becomes more practical: a set of migration time measurements with different gel concentrations can be made relatively quickly because the CE process is fully automatable, the runs are fast, and capillaries can be (re)filled using the rinsing capability of the instrument.[44] Table 8.3 compares MWs of several glycoproteins determined by SDS-CGE employing the simple calibration curve method and the Ferguson analysis. It is evident that the use of the Ferguson

(A)

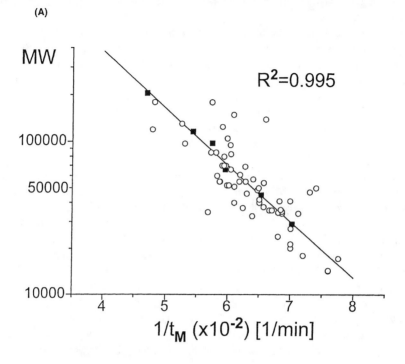

(B)

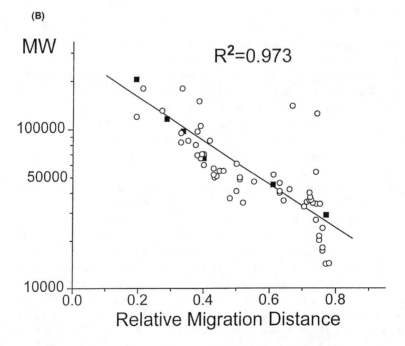

Figure 8.9 Analysis of 65 proteins (open circles) by capillary SDS gel electrophoresis (A) and SDS-PAGE (B). Regression plots were evaluated on the relative migration times and corrected migration distances of the proteins in the test mixture (squares). (From Guttman, A. and Nolan, J., *Anal. Biochem.*, 221, 285, 1994. With permission.)

Table 8.2 Comparison of the Relative Migration Distances and Reciprocal Corrected Migration Times of 65 Individual Proteins As Well As Their Estimated and Published Molecular Mass

	Protein	Literature MW	CGE 1/tM	CGE Calc MW	SLAB Rf	SLAB Calc MW
1	alpha-Lactalbumine	14200	7.6	14250	0.77	14200
2	Lysozime	14300	7.6	15900	0.78	<29000
3	Myoglobine	17200	7.76	17900	0.76	<29000
4	beta-Lactalbumine	18000	7.2	18500	0.76	18000
5	Soybean Trypsin Inhibitor	20100	7	21300	0.75	20000
6	Mycokinase	21400	7	28000	0.75	<29000
7	Trypsinogen	24000	6.8	33800	0.76	29500
8	Triosephosphate Isomerase	27000	7	27300	0.74	26600
9	Carbonic Anhydrase	29000	7.02	29700	0.77	28800
10	Tropomyosin	32700	6.39	47900	0.704	35000
11	Carboxypeptidase A	34000	7.12	32400	0.74	29000
12	Glycerophosphate Dehydrogenase	34000	6.86	32400	0.752	32500
13	Glucokinase	34500	6.8	36400	0.73	40000
14	Pepsin	34700	5.68	85000	0.52	45000
15	Malic Dehydrogenase	35000	6.86	34200	0.712	33800
16	Glyceraldehyde-3- phosphate Dehydr.	35700	6.66	32400	0.72	32500
17	Glyceraldehyde-3-phosphate Dehydr.	35700	6.7	36400	0.642	40500
18	Lactic Dehydrogenase	36000	6.84	32800	0.725	36500
19	Tropomyosin	37000	6.24	35800	0.48	48000
20	Asparagenase	37500	6.57	40500	0.722	37000
21	Creatinin Phosphokinase	40000	6.1	55700	0.63	40900
22	Aldolase	40000	6.5	36200	0.72	32500
23	Alkohol Dehydrogenase	41000	6.99	42300	0.63	40800
24	Enolase	41000	6.82	40500	0.5	45900
25	Luciferase	42000	6.5	38100	0.66	39600
26	Ovalbumin	45000	6.53	45500	0.61	42800
27	Carboxypeptidase	46000	6.3	44500	0.63	41000
28	Glyoxylate Reductase	47000	7.3	24400	0.553	43600
29	Fumarase	48500	6.47	44300	0.51	49000
30	alpha-Amylase	50000	6.49	59600	0.433	54000
31	Cytrate Synthetase	50000	7.4	22500	0.51	45600
32	Hexokinase	51000	6.1	55100	0.439	53000
33	Glutathione Reductase	52000	5.98	62200	0.43	54000
34	Anti Rabbit IgG gamma chain	52000	6.01	68800	0.611	41800
35	Concanavalin A	54000	6.57	40800	0.739	29400
36	Rabbit IgG	55000	6.18	69100	0.45	50900
37	Horse IgG	55000	6.28	63800	0.448	51700
38	Sheep IgG	55000	5.87	68900	0.46	50000
39	Goat IgG	55000	5.86	69600	0.46	50000
40	Pyruvate Kinase	57000	6.4	63800	0.43	54000
41	Monoamino Oxydase	60000	5.82	72400	0.4	59600
42	Phosphoglucose Isomerase	61000	6.19	67900	0.5	46000
43	Bovine Serum Albumin	66000	5.95	71100	0.393	62500
44	Albumin (crosslinked)	66000	6.04	63100	0.39	63400
45	Alkaline Phosphatase	69000	6.27	72900	0.38	65400
46	Amino Acid Oxydase	70000	5.9	71100	0.4	59600
47	Hemocyanin (crosslinked)	70000	5.96	65900	0.395	62000
48	Transkatalase	70000	5.93	74500	0.394	73400
49	Transferrin	80000	5.93	83600	0.375	67600
50	Urease	83000	6.04	82700	0.33	83000
51	Esterase	85000	5.72	88100	0.418	57000
52	Heptoglobin	85000	5.8	84200	0.352	86000
53	alpha-Actinin	95000	6.03	77100	0.33	81600
54	Amyloglucosidase	97000	5.3	118800	0.38	65000
55	Phosphorilase B	97400	5.74	89700	0.337	81600
56	Adenosine Deaminase	105000	5.98	70800	0.389	64000
57	beta Galactosidase	116000	5.43	116200	0.286	106700
58	Cytochrome Oxydase	120000	4.78	191900	0.194	200000
59	Uridine-5DPG4-Epimerase	125000	5.9	76800	0.741	29000
60	Pyruvate Carboxylase	131000	5.25	134900	0.271	130000
61	Glycocollate Oxydase	140000	6.59	39800	0.667	38800
62	Glucose Oxydase	150000	6.08	119700	0.385	64600
63	Aspartate Trasscarbamilase	180000	5.73	89100	0.333	83000
64	alpha-Macroglobulin	180000	4.81	188800	0.213	178000
65	Myosin	206000	4.69	213000	0.193	179800

From Guttman, A. and Nolan, J., *Anal. Biochem.*, 221, 285, 1994. With permission.

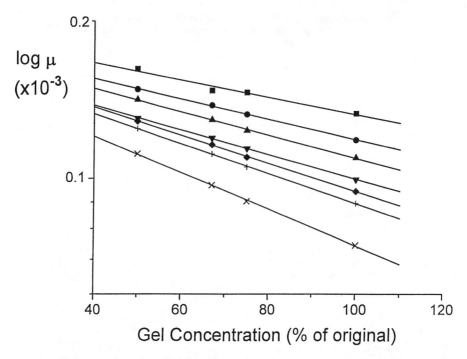

Figure 8.10 Ferguson plots of logarithm mobility vs. percent gel concentration for six proteins: α-lactalbumin (MW 14,400), carbonic anhydrase (MW 29,000), ovalbumin (MW 45,000), bovine serum albumin (MW 66,000), phosphorylase B (MW 97,400), β-galactosidase (MW 116,000), and myosin (MW 205,000).

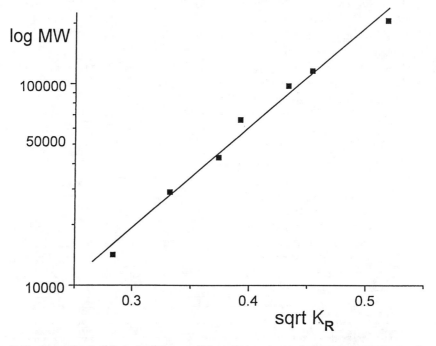

Figure 8.11 Square root of the retardation coefficient (K_R) vs. log molecular mass plot of the standard proteins of Figure 8.10. (From Guttman, A. et al., *J. Chromatogr.*, 676, 227, 1994. With permission.)

Table 8.3 Comparison of Estimated Molecular Weight Data of
Amylase and Light and Heavy Chain Subunits of
Human IgG Using Regular Standard Curve Method
and the Automated Ferguson Method

Protein	Molecular weight		
	Literature	SDS 14/200	Ferguson
Amylase	48,000	68,200 (42%)	54,900 (14%)*
IgG light chain	23,000	4,280 (87%)	27,200 (17%)
IgG heavy chain	55,000	68,500 (25%)	49,300 (10%)

* The values in parentheses after the SDS 14/200 and Ferguson
methods show the difference between the measured and literature values in percent.

From Guttman, A., Shieh, P., Lindahl, J., and Cooke, N., *J. Chromatogr.*, 676, 227, 1994. With permission.

analysis significantly increases the precision of the MW assessment of these test glycoproteins. Note that, in the instance of the IgG light chain, imprecision differences of 70% exist
between the two methodologies.[44]

C. Separation Mechanism

The role of the sieving polymer (chain length and concentration) in the separation mechanism of SDS-protein complexes was recently examined by Guttman.[45] Buffered solutions
of polyethylene oxides (MWs 100,000, 300,000, and 900,000) in various concentrations were
used as the separation media. These studies aimed to see if the separation can be described
by either the Ogston sieving theory, the reptation theory, or the reptation-with-stretching
theory. The separations were examined using the same standard protein test mixture as
mentioned above (Figure 8.9), and a possible separation mechanism of reptation with stretching was suggested.[45] Separation performance was improved when sieving polymer chain
lengths and/or concentration were increased.

D. Applications

By optimizing CGE conditions (capillary length, field strength), polyethylene oxide
mediated SDS-CGE electrophoresis can be used as a fast, high-resolution separation method
for purity control and MW assessment of protein molecules. Figure 8.12[46] shows the separation of a standard test mixture of six proteins ranging in molecular weight from 14,400 to
97,400 within just 3 min, without significantly compromising peak efficiency or resolution,
as can be seen by comparing the electropherogram with that of Figure 8.8. This makes the
SDS-CGE method an excellent technique for fast purity screening of proteins, e.g., in pharmaceutical biotechnology. An example is shown Figure 8.13. A potential blood substitute,
Hemolink™ (Hemosol, Inc., Etobicoke, Ontario, Canada), was checked for purity during its
manufacturing process. Hemolink is a proprietary product made by crosslinking highly
purified hemoglobin with a reagent prepared from raffinose. The hemoglobin is modified to
stabilize it against fragmentation into hemoglobin half-molecules, and to give it oxygen
binding and release properties similar to those of red blood cells. Various other applications
of SDS-CGE (as well as other CE techniques) in analytical biotechnology are described in
a recently published booklet.[7]

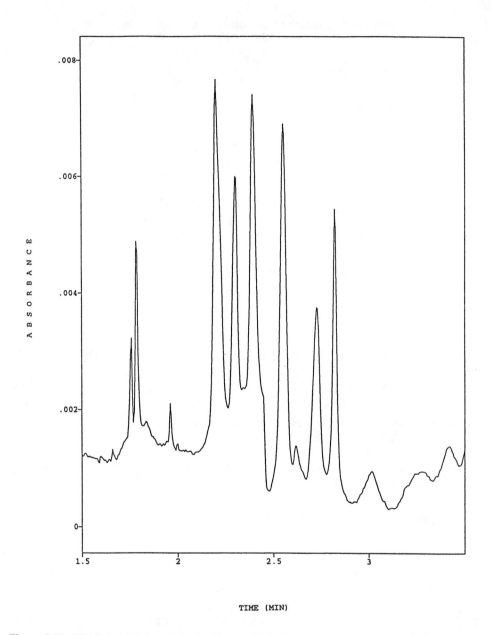

TIME (MIN)

Figure 8.12 Ultrafast separation of standard proteins of α-lactalbumin, soybean trypsin inhibitor, carbonic anhydrase, ovalbumin, bovine serum albumin, and phosphorylase B at high field strength. Conditions: 27 cm × 50 μm i.d., E = 888 V/cm. (From Benedek, K. and Guttman, A., *J. Chromatogr.*, 680, 375, 1994. With permission.)

IV. CAPILLARY ISOELECTRIC FOCUSING (CIEF)

A. Principle of IEF — CIEF vs. Slab Gel IEF

IEF is mostly applied to proteins (including enzyme isoforms, polyclonal and monoclonal antibodies, hemoglobin variants, and recombinant-DNA proteins), although peptides, whole cells and subcellular particles, viruses, and bacteria have also been analyzed by IEF. The technique is particularly useful to estimate the pI of an unknown protein through calibration

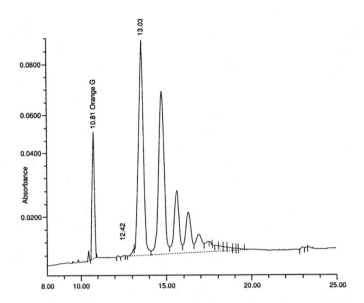

Figure 8.13 SDS-CGE of Hemolink, a crosslinked and polymerized purified hemoglobin-based oxygen carrier. Conditions as in eCAP SDS 200 kit (Beckman Instruments). Pressure injection, 60 s. Orange G was used as a reference standard. (Courtesy of Dr. David Wicks, Hemosol, Inc., Etobicoke, Ontario, Canada.)

with known protein standards. Many applications of IEF in the biomedical/clinical fields have been reported; the technique, in conventional as well as in capillary format, has recently been reviewed by Righetti,[47-49] and by Hjerten[50] and their co-workers. Hjerten and Zhu[9] were the first to show IEF separations using CE instrumentation.

In slab gel (or capillary) IEF, polyionic molecules such as proteins are separated by electrophoresis in a pH gradient. Basically, the technique exploits the difference in pI of different protein species. If the sample component has a net negative charge, migration is toward the anode. During the electrophoretic migration, it encounters progressively lower pH, thus picking up more positive charge. Eventually, a zone is reached where the net charge is zero. At this point (at the pI), migration stops and the sample component is focused in a narrow zone. Likewise, if a component has a positive charge, migration will be toward the cathode. Thus, each sample component migrates to its own isoelectric point.

In IEF, the pH gradient is provided by carrier ampholytes. Ampholytes consist of polyamino-polycarboxylic acids with slightly different pI values. In IEF on slab gels, poly-acrylamide or agarose gels are used as supporting and anticonvective media. It is also possible to *immobilize* pH gradients on a suitable matrix such as polyacrylamide. In this case, the buffering groups of the pH gradient are acrylamide derivatives which are copolymerized into the gel matrix.[48] IEF yields typically higher resolution than other modes of electrophoresis. With immobilized pH gradients, proteins differing by 0.001 pI unit have been separated.[48] Routinely, IEF provides resolution of 0.1 to 0.01 pI units. For even higher resolution, IEF can also be combined with other modes of electrophoresis, e.g., with SDS-PAGE or immu-noelectrophoresis ("two-dimensional electrophoresis").

As is isotachophoresis, IEF is a true electrophoretic *focusing* technique, i.e., the separated zones are *self-sharpened* during the electrophoretic separation. Protein molecules diffusing out of a focused zone will acquire a charge and be pulled back into the center of the zone where the net charge is zero. However, because of the local high concentration of protein in the focused zone, precipitation may be a problem. This effect can be minimized by reducing the protein concentration or the focusing time, or by adding surfactants.

In slab gel IEF, the separated zones are usually visualized by staining. CIEF, on the other hand, involves on-line detection (UV absorbance or fluorescence), without the need for a staining step.[50,51] Although the use of small i.d. capillaries impairs detection sensitivity by reducing the detector light path (Beer's Law), this is generally not a problem since the actual analyte concentrations in the zones are high. For example, focusing a protein into a 0.5 to 1.0 mm zone results in a 200- to 300-fold increase in protein concentration relative to the sample. As will be shown below, CIEF can be fully automated with the potential for substantial time savings for routine analyses. It is expected that in the near future, CE instrumentation with multiple capillaries (10 to 100) will become available. The development of capillary arrays and associated detection systems should further enhance the appeal of CIEF, especially for clinical/diagnostic applications where high sample throughput is desirable.

Except when EOF is used for zone mobilization (electroosmotic mobilization, Section IV.B.3), coated capillaries must be used in CIEF to eliminate the electroosmotic flow (EOF). Polyacrylamide-coated capillaries have commonly been used for this purpose[6,9]; commercially available polysiloxane capillaries (e.g., DB-1 or DB-17, J & W Scientific, Folsom, CA), dynamically modified with alkylcellulose derivatives, have also been shown to work well for many CIEF applications.[52-54] In addition to EOF suppression, these capillary coatings prevent undesirable adsorption of proteins to active sites on the capillary wall. The small residual EOF in coated tubes permits separations in very short capillaries. Ten- to twenty-cm capillaries with internal diameters of 25 μm have been used for many practical applications.[55]

As is the case in CZE separations, the ionic strength of the sample must be reasonably low in CIEF, e.g., < 50 mM for optimum performance. High ionic strength caused by buffer, salts, or ionic detergents often interferes with the focusing.[50,55] Long focusing times are then required which may result in peak broadening during mobilization. A high current generated due to the presence of salt also increases the risk of protein precipitation. Therefore, it is advisable to desalt samples with salt concentrations of 50 mM or greater by dialysis, gel filtration, or ultrafiltration. Recently, methods for desalting small sample volumes[56] and a special capillary cartridge with an on-line desalting device[57] were described for this purpose.

The protein concentration required in the sample depends upon sensitivity and the solubility of the protein components under focusing conditions. A concentration of 0.5 mg/mL per protein should provide adequate detectability and satisfactory focusing/mobilization performance. However, many proteins (e.g., immunoglobulins, membrane proteins) may precipitate during focusing at this starting concentration, since the final protein concentration in the focused zone may be as high as 200 mg/mL. In that case, the protein concentration in the sample must be more dilute.

The ampholyte composition is selected based upon the desired pH range in which the analytes are separated. For separating proteins with widely different isoelectric points, or for estimating the pI of an unknown protein, a wide-range ampholyte blend is advisable, e.g., pH 3 to 10. The final ampholyte concentration is typically 1 to 2%. For proteins with similar pI values, the use of narrow-range ampholyte mixtures should be considered to achieve optimum resolution. The strong absorbance of ampholytes at wavelengths below 240 nm makes detection of proteins in the low UV region impractical. Therefore 280 nm is generally used for absorbance detection in CIEF. This results in a loss in detector signal of as much as 50-fold relative to detection at 200 nm, but the high protein concentrations in the focused peaks more than compensate for the loss in sensitivity due to the selection of a 280 nm wavelength. A recent report from Shimura and Kasai[58] describes the utility of laser-induced fluorescence detection for the analysis of fluorescently labeled peptides which could be used as pI markers.

B. CIEF with Mobilization of Protein Zones

Since most CE instruments are designed with on-capillary detection at a fixed point along the capillary, CIEF must include a means of transporting the focused zones past that detection point in the capillary. Alternatively, it has been demonstrated that the *entire* capillary can be scanned with an appropriately designed detection system which would obviate the need for a separate mobilization step (see Section IV.C).

Three methods — all, in principle, adaptable to commercially available, automated instruments — have been used to mobilize focused zones. In chemical mobilization, changing the chemical composition of the anolyte or catholyte causes a shift in the pH gradient, resulting in electrophoretic migration of focused zones past the detection point. In hydrodynamic mobilization, focused zones are transported past the detection point by applying pressure or vacuum at one end of the capillary. In electroosmotic mobilization, focused zones are transported past the monitor point by exploiting the EOF, i.e., by electroosmotic pumping. Each of these approaches requires a different instrument configuration and a different strategy for optimizing the CIEF separation. Each of these methods will be discussed next.

1. Chemical Mobilization

In their first report on CIEF, Hjerten and Zhu[9] developed a two-step (i.e., focusing and mobilization) procedure. First, proteins were focused in the capillary. After completion of this process (monitored by a dropoff in current), displacement ("mobilization") of the zones out of the capillary was accomplished by means of changing the chemical composition of anolyte or catholyte solution, e.g., by adding acid, base, or salt to one of the vials in which the capillary is immersed. The change in anolyte or catholyte causes a shift in the pH gradient, resulting in migration of the zones past the detection point in the capillary. Typical anolyte solution and catholyte solutions are 10 to 20 mM H_3PO_4 and 20 to 40 mM NaOH, respectively.[55] For narrow-bore capillaries (e.g., 25 μm i.d.), field strengths of 300 to 1000 V/cm are generally used. During the focusing step, the initial current is relatively high; the current decreases as the number of charge carriers in the capillary decreases during focusing. Focusing is achieved rapidly (within a few minutes in short capillaries) and is accompanied by an exponential drop in current.

At the completion of the focusing step, the field is turned off and the anolyte or catholyte vials are replaced by a vial containing the mobilization reagent. At this point, high voltage is resumed. The choice of anodic *vs.* cathodic mobilization and the composition of the mobilization reagent depends upon the protein pI. Since the majority of proteins have isoelectric points between 5 and 9, cathodic mobilization is most often used.

A common mobilization reagent consists of a neutral salt solution, e.g., sodium chloride; sodium ion functions as the (nonproton) cation in the anodic mobilization, while chloride serves as the (nonhydroxyl) anion in the cathodic mobilization. The current initially remains low but gradually increases as chloride ions enter the capillary. When chloride is distributed throughout the capillary, the mobilization step has been completed as can be seen by the rapid rise in current.

In addition to using alkali salts, it was demonstrated by Zhu et al.[55] that zwitterions can also be used effectively for chemical mobilization. Cathodic mobilization requires a zwitterion with a pI between the pH of the anolyte and the pI of the most acid analyte protein. The use of zwitterions in chemical mobilization is advantageous because current levels are lower than those with salt mobilization (hence, less Joule heating, better performance).

In most commercial CE instruments, the detection window is positioned near the outlet of the capillary, thus creating a "short" and a "long" capillary segment. Because of their pI

386

CAPILLARY ELECTROPHORESIS

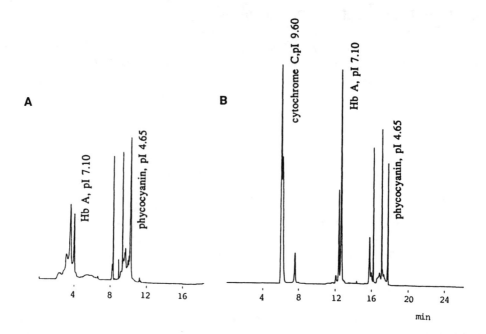

Figure 8.14 Use of TEMED to block the blind segment of the capillary. Chemical mobilization was used. A) CIEF separation of cytochrome c, hemoglobin A, and phycocyanin without TEMED. Cytochrome c is not detected. B) CIEF separation with 0.5% TEMED added to the sample. (From Zhu, M. et al., *J. Chromatogr.*, 559, 479, 1991. With permission.)

characteristics, certain proteins may focus past the detector window in the short segment of the capillary. These zones will not be detected in CIEF during the mobilization step. This is illustrated in Figure 8.14A where cytochrome c (pI 9.6) cannot be discerned in the electropherogram. With a 17-cm total capillary length, the "blind" section in the CE instrument corresponds to 4.6 cm /17 cm × 100% = 27% of the capillary. Hence, when pH 3 to 10 ampholytes are employed, proteins focusing in the pH 8 to 10 (blind) segment of the capillary will be undetected. In order to detect proteins at the basic end of the gradient during cathodic mobilization, the gradient must be altered so the effective length of the capillary (long segment — the distance from the capillary inlet to the detection point) covers the desired pH gradient. A basic compound, e.g., N,N,N',N'-tetramethylethylenediamine (TEMED), can be used to "block" the blind segment of the capillary. Yao-Jun and Bishop[59] described the use of this reagent in slab gel IEF. Using 0.5% TEMED, all proteins, including cytochrome c, are detected (Figure 8.14B). In the case of anodic mobilization of acidic proteins (pI less than 5), a similar strategy can be considered, e.g., the addition of an acidic compound such as an organic acid to the ampholyte mixture.

Since the early experiments by Zhu and Hjerten,[9] CIEF with chemical mobilization was further fine-tuned, theoretically examined,[60] and applied to a variety of substances including monoclonal antibodies,[61] hemoglobin variants,[49,62-64] human serum transferrin,[65] and glycoforms of recombinant proteins.[66] The CIEF methods discussed next (hydrodynamic and electroosmotic mobilization) were developed only a few years ago and appear well suited for implementation in automated CE instruments.

2. Hydrodynamic Mobilization

Instead of the above mobilization by chemical means, focused zones can also be moved past the detection point by physical (hydrodynamic) means. In their first CIEF experiments, Hjerten and Zhu[9] used a low-flow pump to displace the capillary contents while leaving the

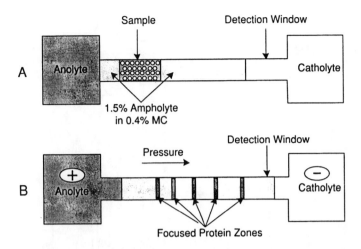

Figure 8.15   Principle of CIEF by simultaneous pressure/voltage mobilization. A) catholyte is backflushed past the detection point, and a sample plug is introduced into the coated capillary (no voltage). B) Focusing of sample is complete and the sample components are driven toward the detector by a low-pressure rinse. High voltage is applied during this step.

voltage on to maintain resolution of the focused protein zones. Recently, the principle of this method was demonstrated in modern CE instruments[52-54] using the vacuum or pressure-driven rinsing capabilities of the CE systems.

Figure 8.15 shows the principle of the pressure-driven method. A plug of sample is introduced into a short (typically 27 cm), 50 μm i.d., coated capillary which contains ampholytes and 0.4% methylcellulose (Figure 8.15A). The coated capillary and the use of methylcellulose as a buffer additive results in negligible EOF in the pH 2 to 10 range. The catholyte containing 20 mM sodium hydroxide is backflushed just past the detection point. As described above (Section IV.B.1) where TEMED was used, the base acts as a blocking agent so that focusing of basic proteins will not extend beyond the detection window, which would prevent their detection. After the focusing step (completed in about 3 min), a low pressure rinse is applied, moving the focused zones past the detection point (Figure 8.15B). During this step, the electric field is turned on to maintain the sharpness of the bands.

This CIEF system yields sharply focused protein peaks and appears to be very reproducible. RSDs for migration times < 0.6% were reported by Nolan.[53] Calculated pI values of proteins correlated well with reported pI's.[53,54] Chen and Wiktorowicz,[52] as well as Huang et al.,[54] demonstrated excellent linearity of a pI vs. mobility plot in the pH range of 2.75 to 9.5. An example of such a plot is shown in Figure 8.16, along with the corresponding electropherogram. The potential applicability of the method for certain peptides (typically difficult with conventional IEF) was also shown.

An example of a clinical application (provided by Dr. J. Hempe, Louisiana State University School of Medicine and Children's Hospital, New Orleans) of this approach is presented in Figure 8.17. Hemoglobins A_2, A, and A_{1c} are normally present in human blood. The patient's electropherogram shows evidence of sickle cell trait where hemoglobin A and S are both present. Hemoglobin F and S are elevated compared to normal levels. The short run times, excellent resolution, and automation capability promise that CIEF would be an excellent method for routine Hb screening.[67]

3. Electroosmotic Mobilization

The need for a simple, reproducible method which could be implemented in routine, QC-type applications have prompted some researchers[68-72] to explore one-step (i.e., no separate

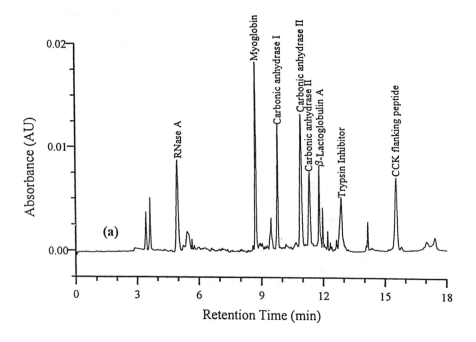

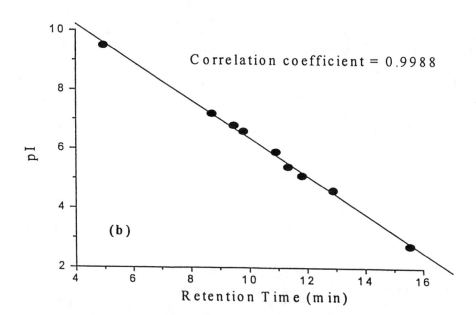

Figure 8.16   (Top) CIEF of protein standards with the pressure/voltage mobilization method. (Bottom) Linear plot of pI vs. migration time based on the above electropherogram. (From Huang, T.-L. et al., *Chromatographia*, 39, 543, 1994. With permission.)

focusing and mobilization steps) methods involving the EOF for mobilization of the focused zones. Both coated and untreated fused-silica capillaries have been used for this CIEF strategy. In one variant, Thormann et al.[68] introduced a large plug of sample (about 6% of the capillary volume) dissolved in an ampholyte solution into a 90-cm-long, untreated capillary which was filled with catholyte (20 mM sodium hydroxide, 0.3% hydroxypropylmethylcellulose). After

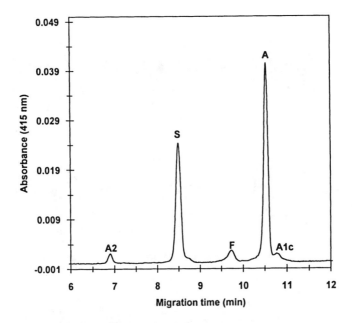

Figure 8.17   Electropherogram of a patient's sample showing an elevated hemoglobin F level. A 27 cm × 50 μm DB-1 coated capillary (J & W Scientific) was used. CIEF method as in Figure 8.15. (Courtesy of Dr. J. M. Hempe, Louisiana State University School of Medicine and Children's Hospital, New Orleans.[67])

placing the capillary between the catholyte and the anolyte (10 mM phosphoric acid), the electric field was turned on, causing the formation of a pH gradient and focusing the protein zones; simultaneously, the sample was swept toward the detection point by the EOF. In a later paper,[69] experimental conditions were further optimized, and the system was applied to the characterization of Hb patterns of normal adults, newborns, patients with diabetes, different hemoglobinopathies, and thalassemia syndromes. Qualitative comparisons of the CIEF method with HPLC and slab gel electrophoresis were made, and quantitative aspects of the method are under investigation by the authors.

In another variant of the EOF mobilization technique, Mazzeo et al.[70] described a one-step CIEF method in which moderate EOF was used to displace the protein zones out of the capillary. The method is different from the above-mentioned method of Thormann[68] in that, initially, the *entire* capillary is filled with sample and ampholytes. Whereas in their earlier papers uncoated capillaries were used, superior results (especially for the acidic proteins) were achieved with commercially available (Supelco), C-8 coated capillaries. The effect of surfactant-coated capillaries was also investigated by Yao and Regnier.[71]

Yowell et al.[72] used the electroosmotic mobilization method for the QC analysis of recombinant protein formulations. While in most CE work the short end of the capillary (i.e., between the detection point and the outlet) is considered ineffective in terms of separation performance, these authors demonstrated that this part of the capillary can also be used to initially focus the protein zones. The principle of their method is schematically shown in Figure 8.18. After filling the capillary with the sample mixture (which contained the basic blocking agent TEMED), the capillary was placed between the inlet (catholyte, 20 mM sodium hydroxide) and outlet (anolyte, 10 mM phosphoric acid). Subsequently, the voltage was turned on, focusing the protein zones in the short segment (7 cm) of the capillary. Here, the TEMED thus blocks the long segment (40 cm) of the capillary. Simultaneously, the EOF drives the zones past the detection point in the reverse direction of that used by the pressure-driven method (Figure 8.15).

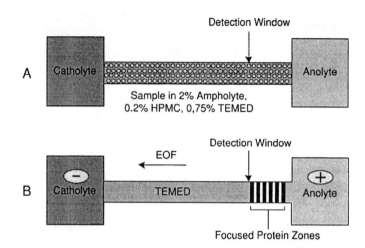

Figure 8.18 Principle of the CIEF method with electroosmotic mobilization. A) coated capillary is filled
with sample. B) Under high voltage, proteins are focused in the short end of the capillary
and mobilized past the detection point by the EOF.

As an illustration of this CIEF approach, Figure 8.19 from Pritchett[73] shows an electropherogram of a monoclonal antibody (MAb), anti-carcinoembryonic antigen, along with
protein standards. From the calibration curve, the pI's of the isoforms of the MAb were
estimated to be in the 6.73 to 7.46 range. A coated capillary (eCAP Neutral, Beckman
Instruments) was used to minimize MAb interaction with the capillary wall, while creating
moderate EOF conditions.[73]

C. CIEF Without Mobilization of Protein Zones: Scanning the Capillary

Several reports have recently appeared describing spectroscopic imaging of the focused
protein zones. It should be noted that, in contrast to the above-described CIEF approaches,
these CIEF methods cannot yet be performed with commercial CE instruments. While the
optical design of such instruments is unconventional and perhaps more complex than those
used in commercial instruments, the simplicity of monitoring only the focusing of the analytes
is an appealing concept. Chmelík and Thormann[74] devised an electrode array detector which

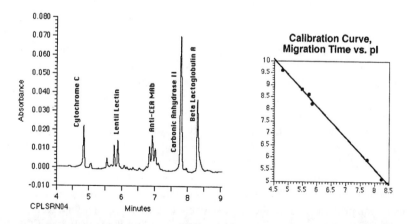

Figure 8.19 CIEF analysis of anti-CEA MAb. The method of Figure 8.18 was used with an eCAP neutral
coated capillary (Beckman Instruments). The calibration plot of the protein standards is also
shown.

could be used as an on-line imaging system for CIEF. However, the resolution obtained by this method was relatively low because of the limited number of electrodes.

Wu and Pawliszyn[75] designed a concentration gradient imaging detection system incorporating a low-power He-Ne laser and a CCD photodiode array. A segment of the capillary was illuminated by the defocused laser beam. The imaging system, based on Schlieren optics, was used to scan very short capillaries (4 cm) and separation/detection could be completed within a couple of minutes. This detection system was also used to study the dynamics of interaction between biological molecules.[76] As an example of the utility of this approach, the interaction between iron and bovine transferrin was studied. After focusing transferrin in the capillary, a plug of iron solution was introduced which resulted in extra peaks in the electropherogram due to the formation of iron–transferrin complexes. An advantage of using the concentration gradient imaging approach over adsorption detection is that a low-power, 633-nm He-Ne laser can be employed which eliminates the possibility of pumping (undesirable) energy into the reaction system.

In other papers from Wu and Pawliszyn, an absorption imaging detector[57,77] and a LIF imaging detector (for fluorescently labeled proteins[78]) were described. Recently, with the objective of increasing sample throughput, they also directed their efforts toward multicapillary systems. Fast (~3 min) separations of human hemoglobin, myoglobin, transferrin, carbonic anhydrase, and a MAb to fluorescein were shown in a capillary array system consisting of four short (4 cm) capillaries.[77] It was even possible to use 1-cm-long capillaries, which resulted in very fast (~30 s) separations of hemoglobin variants.[57] The instrumental design is shown in Figure 8.20. The outside coating of the capillaries was removed to allow detection by the CCD camera. Inside, the capillaries were coated with linear polyacrylamide. An on-line desalting and ampholyte mixing device was also described. The desalting device, consisting of dialysis hollow fibers, was incorporated into a cartridge which housed the capillaries and was compatible with the 1 to 10 µl sample volumes typically employed in CIEF. Desalting prevents the distortion of the pH gradient during the focusing process due to the presence of salt in the sample.

Wang and Hartwick[79] showed preliminary data involving a method in which the entire capillary was moved across the detection region, a technique they called "whole column detection." A similar approach was used by Beale and Sudmeier,[80] who built an LIF detection system based on epi-illumination and confocal optical detection. The capillary was mounted on a precision translational stage which moved the entire capillary through the probe beam.

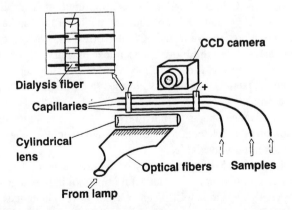

Figure 8.20 Design of CIEF instrument with multiple capillaries based on absorption imaging detection. The cartridge contains a desalting device incorporating dialysis hollow fibers. (From Wu, J. and Pawliszyn, J., *Anal. Chem.*, 67, 2010, 1995. With permission.)

A CIEF separation of FITC-labeled myoglobin was shown, as well as separations using other CE modes.

D. CIEF-Mass Spectrometry (CIEF-MS)

Mass detection of proteins in CZE-MS is typically in the high femtomole range. In practice, a protein concentration of 10^{-5} M is marginally detectable when dealing with biological samples. Because CIEF inherently is a sample concentration method, the possibility of combining CIEF with mass spectrometry is intriguing. Recently, Tang et al.[81] and Muller et al.[82] have explored CIEF-MS using an on-line and off-line approach, respectively. The system described by Tang et al.[80] involved an electrospray interface with a coaxial sheath flow configuration. Proteins were focused in a coated capillary, and the chemical mobilization technique was used to transfer the protein zone to the mass spectrometer. The method permitted MS analysis of very dilute protein samples with concentration detection limits in the 10^{-7} M range, a 100-fold increase over that possible with CZE-electrospray MS.

Foret et al.[82] developed an automated system for the off-line coupling of CIEF to a matrix-assisted laser desorption ionization time-of-flight (MALDI-TOF) mass spectrometer. After focusing and hydrodynamic mobilization, protein zones were directed to a specially constructed, computer-controlled fraction collector. This high-precision device consisted of a number of glass capillaries to collect the fractions mobilized (by pressure) in the CIEF step. Fractions eluting from the separation capillary were collected by separate glass tubes which were rotated into place at the appropriate time determined by the computer. The fraction collector incorporates a sheath liquid flow at the exit of the capillary, allowing continuous collection of multiple zones without voltage interruption. The fractions collected in the glass capillaries were subsequently analyzed by the MALDI-TOF mass spectrometer. The CE/mass spectrometric analysis of proteins is discussed in more detail in Chapter 2.

V. CONCLUSION

This chapter has reviewed the principal methods for analyzing proteins and peptides by CE. Clearly, in most cases the FSCE method is complementary to HPLC. In general, a comparable CE method is faster with better separation efficiency. HPLC continues to have the edge with regard to micropreparative work — only a handful of publications have appeared on fraction collection with CE. The capillary separation method is ideally suited for (very) small sample sizes (e.g., in QA and QC applications), while HPLC can be more easily scaled up from an analytical to a preparative format. CE also offers the possibility of solving unique contemporary problems, e.g., in the biopharmaceutical industry. A recent paper by Chu et al.,[83] for example, describes the use of affinity CE-MS for the separation and sequence determination of ligands from combinatorial libraries (e.g., peptides) that bind most tightly to a receptor (e.g., a drug).

CGE continues to advance thanks to much recent research in suitable polymer networks compatible with capillaries and detection systems. CE-equivalents to the slab gel SDS-PAGE method are now commercially available.

CIEF also has the potential to replace many of the established, classical IEF procedures and, in the near future, should become a practical routine tool in the bioanalytical and clinical laboratories. The recently developed CIEF approaches have advantages over slab gel IEF with regard to time savings, sample preparation, quantitation, and reproducibility. In addition, no staining/destaining of the bands needs to be performed because the detection takes place directly on the capillary, in *real time*, while the acquired data are directly transferred to a computer database.

Capillary array systems — allowing multiple samples to be run at the same time — would further increase sample throughput. This development is especially important for routine clinical work such as screening for hemoglobin variants or other blood proteins. CE-based immunoassays[84] should also be feasible in this format. Further miniaturization of CE instruments is also ongoing. For example, a microchip electrophoretic immunoassay, in which separations are performed in channels embedded in glass or fused silica, already has been reported.[85]

REFERENCES

1. Jorgenson, J.W. and Lukacs, K.D., Zone electrophoresis in open tubular glass capillaries, *Anal. Chem.*, 53, 1298, 1981.
2. Jorgenson, J.W. and Lukacs, K.D., Capillary zone electrophoresis, *Science*, 222, 266, 1983.
3. Hjerten, S., *Electrophoresis "83*, Hirai, H., Ed., Walter de Gruyter, Berlin, 1983, p. 71.
4. Cohen, A.S. and Karger, B.L., High performance capillary SDS polyacrylamide gel electrophoresis of peptides and proteins, *J. Chromatogr.*, 397, 409, 1987.
5. Heiger, D.N. and Majors, R.E., Capillaries and chemistries for capillary electrophoresis, *LC.GC*, 13, 12, 1995.
6. Schwartz, H.E. and Pritchett, T., *Separation of proteins and peptides by capillary electrophoresis: application to biotechnology*, Beckman Instruments, P/N 727484, Fullerton, CA, 1994.
7. Guzman, N.A., Gonzalez, C.L., Trebilcock, M.A., Hernandez, L., Berck, C.M., and Advis, J.P., The use of capillary electrophoresis in clinical diagnosis, in *Capillary Electrophoresis Technology*, Guzman, N.A., Ed., Marcel Dekker, New York, 1993, Chap. 22.
8. Klein, G.L. and Jolliff, C.R., Capillary electrophoresis for the routine clinical laboratory, in *Handbook of Capillary Electrophoresis*, Landers, J.P., Ed., CRC Press, Boca Raton, 1994, Chap. 16.
9. Hjerten, S. and Zhu, M., Adaption of the equipment for high-performance electrophoresis to isoelectric focusing. *J. Chromatogr.*, 346, 265, 1985.
10. Terabe, S., Otsuka, K., Ichikawa, K., Tsuchiya, A., and Ando, T., Electrokinetic separations with micellar solution and open-tubular capillaries, *Anal. Chem.*, 56, 111, 1984.
11. Vinther, A., Petersen, J., and Søeberg, H., Capillary electrophoretic determination of the protease Savinase in cultivation broth, *J. Chromatogr.*, 608, 205, 1992.
12. Offord, R. E., Electrophoretic mobilities of peptides on paper and their use in the determination of amide groups, *Nature*, 211, 591, 1966.
13. Rickard, E. C., Strohl, M. M., and Nielsen, R. G., Correlation of electrophoretic mobilities from capillary electrophoresis with physicochemical properties of proteins and peptides, *Anal. Biochem.*, 197, 197, 1991.
14. Vinther, A. and Søeberg, H., Radial pH distribution during capillary electrophoresis with electroosmotic flow. Analysis with high ionic strength buffers, *J. Chromatogr.*, 589, 315, 1992.
15. Lauer, H. H. and McManigil, D., Capillary zone electrophoresis of proteins in untreated fused silica tubing, *Anal. Chem.*, 58, 166, 1986.
16. McCormick, R., Capillary zone electrophoretic separation of peptides and proteins Using low pH buffers in modified silica capillaries, *Anal. Chem.*, 60, 2322, 1988.
17. Green, J. S. and Jorgenson, J. W., Minimizing adsorption of proteins on fused-silica in capillary zone electrophoresis by addition of alkali metal salts to the buffer, *J. Chromatogr.*, 478, 63, 1989.
18. Bushey, M. M. and Jorgenson, J. W., Capillary electrophoresis of proteins in buffers containing high concentrations of zwitterionic salts, *J. Chromatogr.*, 480, 301, 1989.
19. Okafo, G. N., Vinther, A., Kornfelt, T., Camilleri, P., Effective ion-pairing for the separation of basic proteins in capillary electrophoresis, *Electrophoresis*, 16, 1917, 1995.
20. Kornfelt, T., Vinther, A, Okafo, G.N., and Camilleri, P., Improved peptide mapping using phytic acid as ion-pairing buffer additive in capillary electrophoresis, *J. Chromatogr.*, 726, 223, 1996.
21. Stover, F. S., Haymore, B. L., and McBeath, R. J., Capillary zone electrophoresis of histidine containing compounds, *J. Chromatogr.*, 470, 241, 1989.

22. Guttman, A., Paulus, A., Cohen, A.S., Grinberg, N., and Karger, B.L., Use of complexing agent for selective separation in high performance capillary electrophoresis, *J. Chromatogr.*, 448, 41, 1988.

23. Terabe, S., Electrokinetic chromatography: an interface between electrophoresis and chromatography, *Trends Anal. Chem.*, 8, 129, 1989.

24. Vinther, A., Holm, A., Høeg-Jensem, T., Jespersen, A. M., Klausen, N. K., Christensen, T., and Sørensen, H. H., Synthesis of stereoisomers and isoforms of a tryptic heptapeptide fragment of human growth hormone and analysis by reverse-phase HPLC and capillary electrophoresis, *Eur. J. Biochem.*, 235, 304, 1996.

25. Schwartz, H.E., Ulfelder, K.J., Chen, F.-T.A., and Pentoney, S.L., Jr., The utility of laser-induced fluorescence detection in applications of capillary electrophoresis, *J. Cap. Elec.*, 1, 36, 1994.

26. Vinther, A. and Søeberg, H., A mathematical model describing dispersion in free solution capillary electrophoresis under stacking conditions, *J. Chromatogr.*, 559, 3, 1991.

27. Aebersold, R. and Morrison, H. D., Analysis of dilute peptide samples by capillary zone electrophoresis, *J. Chromatogr.*, 516, 79, 1990.

28. Schwer, C. and Lottspeich, F., Analytical and micropreparative separation of peptides by capillary zone electrophoresis using discontinuous buffer systems, *J. Chromatogr.*, 623, 345, 1992.

29. Foret, F., Szoko, E., and Karger, B. L., Analysis of proteins by capillary zone electrophoresis with on-column transient isotachophoretic preconcentration. *Beckman Technical Information Bulletin A-1740*, Fullerton, California, 1993.

30. Shapiro, A., Vinuela, E., and Maizel, J., Molecular weight estimation of polypeptide chains by electrophoresis in SDS polyacrylamide gels, *Biochem. Biophys. Res. Commun,* 28, 815, 1967.

31. Weber, K. and Osborn, M., Reliability of molecular weight determinations by dodecyl sulfate — polyacrylamide gel electrophoresis, *J. Biol. Chem.*, 244, 4406, 1969.

32. Chrambach, A., *The Practice of Quantitative Gel Electrophoresis*, VCH, Deerfield Beach, FL, 1985.

33. Andrews, A.T., *Electrophoresis*, 2nd ed., Clarendon Press, Oxford, 1986.

34. Ferguson, K.A., Starch-gel electrophoresis application to the classification of pituitary proteins and polypeptides, *Metab. Clin. Exp.*, 13, 985, 1964.

35. Tsuji, K., High performance capillary electrophoresis of proteins, *J. Chromatogr.*, 550, 823, 1991.

36. Widhalm, A., Schwer, C., Blass, D., and Kenndler, E., Capillary zone electrophoresis with a linear, non-crosslinked polyacrylamide gel: separation of proteins according to molecular mass, *J. Chromatogr.*, 546, 446, 1991.

37. Ganzler, K., Greve, K.S., Cohen, A.S., Karger, B.L., Guttman, A., and Cooke, N., High-performance capillary electrophoresis of SDS-protein complexes using UV-transparent polymer networks, *Anal. Chem.*, 64, 2665, 1992.

38. Werner, W., Demorest, D., Stevens, J., and Wiktorowicz, J.E., Size dependent separation of proteins denatured in SDS by capillary electrophoresis using replaceable sieving matrix, *Anal. Biochem.*, 212, 253, 1993.

39. Wu, D. and Regnier, F., SDS capillary gel electrophoresis of proteins using non-crosslinked gels, *J. Chromatogr.*, 608, 349, 1992.

40. Guttman, A., Nolan, J., and Cooke, N., Capillary sodium dodecyl sulfate gel electrophoresis of proteins, *J. Chromatogr.*, 632, 171, 1993.

41. Simo-Alfonso, E., Conti, M., Gelfi, C., and Righetti, P.G., Sodium dodecyl sulfate capillary electrophoresis of proteins in entangled solutions of poly(vinyl alcohol), *J. Chromatogr. A*, 689, 85, 1995.

42. Guttman, A., Horvath, J., and Cooke, N., Influence of temperature in the sieving effect of different polymer matrices in capillary SDS electrophoresis of proteins, *Anal. Chem.*, 65, 199, 1993.

43. Guttman, A. and Nolan, J., Comparison of the separation of proteins by sodium sodecyl sulfate-slab gel electrophoresis and capillary sodium dodecyl sulfate-gel electrophoresis, *Anal. Biochem.*, 221, 285, 1994.

44. Guttman, A., Shieh, P., Lindahl, J., and Cooke, N., Capillary sodium dodecyl sulfate gel electrophoresis of proteins II. On the Ferguson method in polyethylene oxide gels, *J. Chromatogr.*, 676, 227, 1994.

45. Guttman, A., On the separation of capillary sodium dodecyl sulfate-gel electrophoresis of proteins, *Electrophoresis*, 16, 611, 1995.

46. Benedek, K. and Guttman, A., Ultra-fast high-performance capillary sodium dodecyl sulfate gel electrophoresis of proteins, *J. Chromatogr.*, 680, 375, 1994.

47. Righetti, P.G., *Isoelectric Focusing: Theory, Methodology and Applications,* Elsevier Biomedical Press, Amsterdam, 1983.

48. Righetti, P.G. and Chiari, M., Conventional isoelectric focusing and immobilized pH gradients: an overview, in *Capillary Electrophoresis Technology*, Guzman, N.A., Ed., Marcel Dekker, New York, 1993, p. 89.

49. Righetti, P.G. and Gelfi, C., Isoelectric focusing in capillaries and slab gels: a comparison, *J. Cap. Elec.*, 1, 27, 1994.

50. Hjerten, S., Isoelectric focusing in capillaries, in *Capillary Electrophoresis: Theory and Practice,* Grossman, P.D. and Colburn, J.C., Eds., Academic Press, San Diego, 1992, Chap. 7.

51. Schwartz, H. E. and Pritchett, T., New approaches to capillary isoelectric focusing of proteins, *Biotechnology*, 12, 408, 1994.

52. Chen, S.M. and Wiktorowicz, J.E., Isoelectric focusing by free solution capillary electrophoresis, *Anal. Biochem.*, 206, 84, 1992.

53. Nolan, J., *Technical Information A-1750*, Beckman Instruments, Fullerton, 1993.

54. Huang, T.-L., Shieh, P.C.H., and Cooke, N., Isoelectric focusing of proteins in capillary electrophoresis with pressure-driven mobilization, *Chromatographia*, 39, 543, 1994.

55. Zhu, M., Rodriguez, R., and Wehr, T., Optimizing separation parameters in capillary isoelectric focusing, *J. Chromatogr.*, 559, 479, 1991.

56. Hjerten, S., Valtcheve, L., and Li, Y.-M., A simple method for desalting and concentration of microliter volumes of protein solutions with special reference to capillary electrophoresis, *J. Cap. Elec.*, 1, 83, 1994.

57. Wu, J. and Pawliszyn, J., A capillary cartridge with an on-line desalting device that allows fast sampling for capillary isoelectric focusing, *Anal. Chem.*, 67, 2010, 1995.

58. Shimura, K. and Kasai, K., Fluorescence-labeled peptides as pI markers in capillary isoelectric focusing with fluorescence detection, *Electrophoresis*, 16, 1479, 1995.

59. Yao-Jun, G. and Bishop, R., Extension of the alkaline end of a pH gradient in thin layer polyacrylamide electrofocusing gels by addition of N,N,N',N'-tetramethylethylenediamine, *J. Chromatogr.*, 234, 459, 1982.

60. Hjerten, S., Liao, J.-L., and Yao, K., Theoretical and experimental study of high performance electrophoretic mobilization of isoelectrically focused protein zones, *J. Chromatogr.*, 387, 127, 1987.

61. Wehr, T., Zhu, M., Rodriguez, R., Burke, D., and Duncan, K., High performance isoelectric focusing using capillary electrophoresis, *Am. Biotech. Lab.*, 8, 22, 1990.

62. Zhu, M., Rodriguez, R., Wehr, T., and Siebert, C., Capillary electrophoresis of hemoglobins and globin chains, *J. Chromatogr.*, 608, 225, 1992.

63. Zhu, M., Wehr, T., Levi, V., Rodriguez, R., Shiffer, K., and Cao, Z.A., Capillary electrophoresis of abnormal hemoglobins associated with α-thalassemias, *J. Chromatogr.*, 652, 119, 1993.

64. Conti, M., Gelfi, C., and Righetti, P.G., Screening of umbilical cord blood hemoglobins by isoelectric focusing in capillaries, *Electrophoresis*, 16, 1485, 1985.

65. Kilar, F. and Hjerten, S., Fast and high resolution analysis of human serum transferrin by high performance isoelectric focusing, *Electrophoresis*, 10, 23, 1989.

66. Yim, K.W., Fractionation of the human recombinant tissue plasminogen activator (rtPA) glycoforms by high performance capillary zone electrophoresis and capillary isoelectric focusing, *J. Chromatogr.*, 559, 401, 1991.

67. Hempe, J.M., *Technical Information A-1771-A*, Beckman Instruments, Fullerton, 1993.

68. Thormann, W., Caslavska, J., Molteni, S., and Chmelík, J., Capillary isoelectric focusing with electroosmotic zone displacement and on-column multichannel detection, *J. Chromatogr.*, 589, 321, 1992.

69. Molteni, S., Frischknecht, H., and Thormann, W., Application of dynamic isoelectric focusing to the analysis of human hemoglobin variants, *Electrophoresis,* 15, 22, 1994.

70. Mazzeo, J.R., Martineau, J.A., and Krull, I.S., Performance of isoelectric focusing in uncoated and commercially available coated capillaries, *Methods,* 4, 205, 1992.

71. Yao, X.-W. and Regnier, F.E., Polymer and surfactant-coated capillaries for isoelectric focusing, *J. Chromatogr.,* 632, 185, 1993.

72. Yowell, G.G., Fazio, S.D., and Vivilecchiia, R.V., Analysis of recombinant granulocyte macrophage colony stimulating factor dosage form by capillary electrophoresis, capillary isoelectric focusing and high performance liquid chromatography, *J. Chromatogr.,* 652, 215, 1993.

73. Pritchett, T., *Technical Information A-1769,* Beckman Instruments, Fullerton, California, 1993.

74. Chmelík, J. and Thormann, W., Isoelectric focusing field-flow fractionation and capillary isoelectric focusing with electroosmotic zone displacement, *J. Chromatogr.,* 632, 229, 1993.

75. Wu, J. and Pawliszyn, J., Capillary isoelectric focusing with a universal concentration gradient imaging system using a charge-coupled photodiode array, *Anal. Chem.,* 64, 2934, 1992.

76. Wu, J. and Pawliszyn, J., In vitro observation of interactions of iron and transferrin by capillary isoelectric focusing with a concentration gradient imaging detection system, *J. Chromatogr.,* 652, 295, 1993.

77. Wu, J. and Pawliszyn, J., Protein analysis by isoelectric focusing in a capillary array with an absorption imaging detector, *J. Chromatogr. B,* 669, 39, 1995.

78. Wu, X.-Z., Wu, J., Pawliszyn, J., Fluorescence imaging detection for capillary isoelectric focusing, *Electrophoresis,* 16, 1474, 1995.

79. Wang, T. and Hartwick, R.A., Whole column detection in capillary isoelectric focusing, *Anal. Chem.,* 64, 1745, 1992.

80. Beale, S.C., and Sudmeier, S.J., Spatial-scanning laser fluorescence detection for capillary electrophoresis, *Anal. Chem.,* 67, 3367, 1995.

81. Tang, Q., Harrata, A.K., and Lee, C.S., Capillary isoelectric focusing-electrospray mass spectrometry for protein analysis, *Anal. Chem.,* 67, 3515, 1995.

82. Foret, F., Muller, O., Thorne, J., Gotzinger, W., and Karger, B.L., Analysis of protein fractions by micropreparative capillary isoelectric focusing and matrix-assisted laser desorption time-of-flight mass spectrometry, *J. Chromatogr. A,* 716, 157, 1995.

83. Chu, Y.-H., Kirby, D.P., and Karger, B.L., Free solution identification of candidate peptides from combinatorial libraries by affinity capillary electrophoresis/mass spectrometry, *J. Am. Chem. Soc.,* 117, 5419, 1995.

84. Chen, F.-T.A. and Sternberg, J.C., Characterization of proteins by capillary electrophoresis in fused silica columns: review on serum protein analysis and application to immunoassays, *Electrophoresis,* 15, 13, 1994.

85. Koutny, L.B., Schmalzing, D., Taylor, T.A., and Fuchs, M., Microchip electrophoretic immunoassay for serum cortisol, *Anal. Chem.,* 68, 18, 1996.

CHAPTER 9

Separation of DNA

András Guttman and Herbert E. Schwartz

CONTENTS

0-8493-9127-X/98/$0.00+$.50

I. INTRODUCTION

Since the 1960s methods employing supporting media, such as polyacrylamide or agarose gels, have become standard in protein and nucleic acid analysis. The gel matrix acts as an anticonvective medium, reducing convective transport and diffusion so separated sample components remain positioned in sharp zones during the run. In addition, the gel acts as a molecular sieve, separating nucleic acids according to their size. Electrophoresis, as performed today in the majority of laboratories, is still typically a manual process although computerized scanning of gel traces is becoming more popular. Hence, electrophoresis is often time consuming and labor intensive. In addition, most electrophoretic techniques are qualitative, and accurate quantitation is often problematic. It can be expected that CE will replace classical electrophoretic techniques for DNA analysis, e.g., in various clinical, diagnostic, genetic, forensic, and biotechnology-related applications.

While the first papers on ds DNA analysis by CE only appeared in 1988,[1-3] their number since then has grown exponentially. The recent introduction of a commercially available highly sensitive laser induced fluorescence (LIF) detector has opened new perspectives for low-level DNA analysis, e.g., detectability into the zeptomole range.[4,5] This chapter describes the main principles of the various CE methods for DNA analysis, their instrumentation, and their applications for nucleosides, (oligo)nucleotides, and ds DNA fragments. While the emphasis is on DNA, some RNA separations are also discussed, as well as applications where RNA is reverse transcribed into DNA.

II. CE MODES FOR DNA ANALYSIS

A. Free Solution

Table 9.1 shows the modes of CE most useful for DNA separations, the principle of operation, and typical applications. CZE, MEKC, and ITP can be regarded as free solution techniques. For small DNA fragments, e.g., nucleosides, nucleotides, and small oligonucle-otides, free solution techniques (CZE, MEKC) can be applied — larger DNA molecules require gel or polymer network matrices for separation (i.e., the CGE mode). With CZE, only species with different mass-to-charge ratios can be separated unless advantage is taken of the EOF, e.g., to separate neutral species from charged ones.

Table 9.1 CE Modes for DNA Separations

Mode	Separation mechanism	Applications
Free solution		
Capillary Zone Electrophoresis (CZE)	Charge-to-mass ratio	Bases, nucleosides, nucleotides, (small) oligonucleotides
Micellar Electrokinetic Chromatography (MEKC)	Charge-to-mass ratio, partitioning into micelles	Same as CZE
Capillary Isotachophoresis (CITP)	Moving boundary, displacement	Preconcentration (for CZE, MEKC, CGE) nucleotides, nucleosides
Gel, polymer network		
Capillary Gel Electrophoresis (CGE)	Molecular sieving, reptation	Oligonucleotides, primers, probes, antisense DNA, PCR products, large ds DNA, point mutations, DNA sequencing

MEKC is particularly useful for the separation of relatively small, neutral molecules such as many pharmaceutical drugs. Bases, nucleosides, nucleotides, and small oligonucleotides (<10 bases) have also been separated by MEKC, as will be discussed in Section IV.A. The technique makes use of surfactants (e.g., 10 to 200 mM sodium dodecyl sulfate [SDS]) which are added to the run buffer in sufficiently high concentrations to form micelles in solution. During the high voltage CE run, the micelles usually migrate in a direction opposite the EOF. Analytes partition into the micelles differently according to their hydrophobicity, thus yielding different migration times. With MEKC, uncoated capillaries are most often used.

Isotachophoresis (ITP), a predecessor of modern CE, has also been applied to nucleotide and DNA separations.[5a,6] ITP is also an excellent preconcentration technique in conjunction with CZE, MEKC, or CGE. Using ITP preconcentration, 10- to 100-fold better sample detectability can be obtained.

B. Gels, Polymer Networks

1. Comparison with Slab Gels — Mechanism

In CE the separation and detection of DNA takes place simultaneously, i.e., in *real time*. Typically, this is not the case with slab gel techniques (some slab gel-based instrumentation also allows real-time detection of DNA, e.g., automated DNA sequencing). Slab gels do have the distinct advantage of running multiple samples simultaneously. This, of course, increases sample throughput compared to instrument designs where the samples have to be assayed one at a time, such as HPLC or CE. However, CE has, because of the higher fields applied, much faster (~10 to 100×) run times than slab gel electrophoresis. Instrumentation to run samples simultaneously on *multiple* CE columns has also recently been described and applied to PCR fragments and DNA sequencing (see Section III.B.1).

While the original purpose of using gels was to provide an anticonvective medium for the electrophoresis, the gels resolve DNA fragments (and also SDS-complexed proteins) by acting as molecular sieves. The amount of sieving can be controlled by adjusting the concentration of the gel. With polynucleotides, the phosphate group of each nucleotide unit carries a negative charge that is much stronger than any of the charges of the base above pH 7. Therefore, the mass-to-charge ratio of the polynucleotides is independent of the base composition, and, consequently nearly identical for closely related species. For that reason, free solution techniques (where the electrophoretic medium contains no gel or polymer network solution) have not proven successful in the electrophoresis of oligonucleotides or ds DNA.

The actual mechanism involved in DNA separations in electrophoresis has been, and is still, the topic of much discussion.[7-10] Fluorescent microscopy (video) allows individual DNA molecules to be monitored as they sieve through a gel under the influence of an electric field.[11] However, no single model can fully account for the dependence of DNA mobility on its molecular size and a number of experimental parameters (field strength, gel concentration, etc.). Two models which have been proposed are schematically shown in Figure 9.1. The first, commonly referred to as the *Ogston mechanism*, involves a coil of DNA percolating through a network of polymer fibers. It is thought that the coil moves through the gel as if it were a rigid particle. Its electrophoretic mobility is proportional to the volume fraction of the pores of a gel that the DNA can enter. Since the average pore size decreases with increasing gel concentration, mobility decreases with increasing gel concentration and increasing molecular weight. The Ogston theory does not account for the fact that a relatively large DNA molecule may stretch or deform so it can squeeze its way through the pores.

In the *reptation model*, the DNA is thought to squeeze through the mesh of the gel as if it were a snake going through an obstacle course. The "biased" reptation model (Slater and

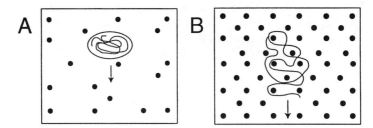

Figure 9.1 (A) Schematic diagram of a solute migrating through a polymer network by the Ogston mechanism. The DNA percolates through the mesh as if it were a rigid particle. (B) Schematic diagram of a solute migrating by the reptation mechanism. In this case, the DNA is forced to squeeze through the tubes formed by the polymer network. (From Grossman, P. D. and Soane, D. S., *Biopolymers*, 31, 1221, 1991. With permission.)

Noolandi[8]) can account for the experimentally observed fact that the mobility of DNA becomes independent of the molecular weight at high field strength (Figure 9.2). At high fields, the DNA stretches and the dependence of mobility on molecular size decreases. However, the reptation theory cannot explain certain effects observed in pulsed-field electrophoresis.[9] Different gel formulations have been used for separating DNA. With slab gels, the two most common ones are based on either polyacrylamide or agarose.

2. Capillary Gel Electrophoresis (CGE)

For most DNA fragments (e.g., restriction fragments, sequencing fragments, antisense DNA, chromosomal DNA), techniques based on molecular sieving and/or reptation are generally required. In CE, this can be achieved by using gel or polymer network-filled capillaries. These CGE applications typically are used in conjunction with special pretreated (coated) capillaries under moderate EOF conditions. However, the coating *per se* does not appear to be a prerequisite for good performance. Recently, excellent results were obtained with uncoated capillaries (used with dilute polymer solutions), for separations of intermediate-size (2 to 23 kbp) ds DNA.[12]

Gels may vary in consistency from viscous fluids to fairly rigid solids. In the electrophoresis literature terms such as polymer solutions, polymer networks, entangled polymer solutions, chemical gels, physical gels, and liquid gels have been used to describe gel media. In addition, the term "replaceable" gel has been introduced to describe relatively nonviscous gels which can be rinsed in and out of the capillary. In CE, two types of gel matrices can be distinguished: (1) a relatively high-viscosity, crosslinked gel which is chemically anchored to the capillary wall ("chemical" gel) and (2) a relatively low-viscosity, polymer network solution ("physical" gel). With both types, precoated capillaries are used to eliminate the EOF. Table 9.2 summarizes the main differences between these gels.

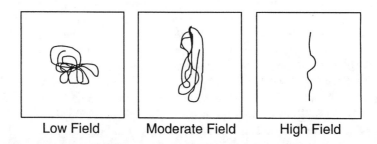

Low Field   Moderate Field   High Field

Figure 9.2 Schematic illustration showing the elongation of DNA under the influence of the electric field (reptation). (From Grossman, P. D. and Soane, D. S., *Biopolymers*, 31, 1221, 1991. With permission.)

Table 9.2 Characteristics of Gel Matrices Used in CGE

Chemical gels

- Crosslinked, chemically linked to the capillary wall
- Well-defined pore structure
- Pore size cannot be varied after polymerization
- Heat sensitive
- Particulates can damage gel matrix
- May provide extremely high resolution
- Not replaceable, generally high viscosity

Physical gels

- Not crosslinked, not attached to the capillary wall
- Entangled polymer networks of linear or branched hydrophilic polymers
- Dynamic pore structure
- Pore size can be varied
- Heat insensitive
- Particulates can be easily removed
- Gel is replaceable when a relatively low viscosity matrix is used

The chemical gels have a well-defined pore structure, they are rigid, and their pore size can be varied by adjusting the polymerization conditions (e.g., by varying the monomer and crosslinker concentration in the gels), as is well known from the practice of classical electrophoresis. As shown below, with these type of gels in CE, extremely high resolution separations of oligonucleotides can be achieved.[13,14]

The noncrosslinked, replaceable polymer networks have a dynamic pore structure and are more flexible. Polymer networks of variable viscosity can be made by carefully selecting the concentration and chain length of the linear polymers. Thus far, with CE, solutions consisting of linear PA, various alkylcelluloses, and agaroses have been mostly used. Other hydrophilic polymers used as sieving matrices for DNA separations include polyethylene oxide, polyvinyl alcohol, polyacryloylaminoethoxyethanol, Ficoll-400, polyethyleneglycol, and glucomannan.

Interestingly, while in most papers medium- to high-viscosity polymer solutions are used, very dilute polymer solutions (~0.010 to 0.001% hydroxyethylcellulose) are also effective in size-separating DNA.[10,12] A fully entangled polymer network does not appear to be a prerequisite for separation. A mechanism other than the Ogston and reptation models may be operative with DNA electrophoresis in these entangled polymer solutions. As envisioned by Soane and co-workers,[10,12,15,16] when DNA migrates through a polymer network, entanglement coupling between the DNA and the surrounding polymer chains occurs, as schematically shown in Figure 9.3. Ultimately, this coupled entanglement is thought to limit the resolution achievable with relatively large DNA fragments.[16]

An important advantage of the physical gels is that a low enough viscosity can be selected so that the contents of the capillary can be rinsed in and out of the column (by pressure or vacuum). As mentioned before, these gels are often referred to as replaceable matrices. When desired, a fresh gel can therefore be used for every sample injection. In addition, sample introduction is possible by either the pressure or the electrokinetic mode, in contrast to the chemical gels (and high-viscosity polymer networks) in which only the electrokinetic mode is possible (see Section III.A.2). This advantage of having the pressure injection mode available can be important in work dealing with quantitation. With polymer network solutions, CE can be performed at relatively high temperatures (50 to 70°C) and field strengths (1000 V/cm) without damaging the gel, as would be the case with chemical gels such as crosslinked PA gels. The physicochemical and mechanical aspects of gel-filled capillaries recently have been reviewed by various authors.[14,17-21]

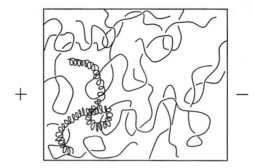

Figure 9.3 A schematic representation of the entanglement coupling interaction of DNA with the polymer chains of the sieving matrix. (From Barron, A. E. et al., *J. Chromatogr. A*, 652, 3, 1993. With permission.)

a. Polyacrylamide-Filled Capillaries

The polymerization of crosslinked and high-viscosity (i.e., >6%), linear PA is typically carried out *within* the fused-silica capillary tubing. For reasons of stability, the PA gel must be covalently bound to the wall of the column, e.g., by means of a bifunctional agent such as (3-methacryloxypropyl)-trimethoxysilane. After completion of the pretreatment procedure, the polymerization reaction can take place in the capillary.

Crosslinked PA gels are used mainly for oligonucleotide separations of up to 200 bases, usually under denaturing conditions of 7 to 9 M urea (Section IV.B). The typical gel concentration is 3 to 6% T with ~5% C. The resolving power depends on the length of the capillary (20 to 150 cm). Longer columns give higher resolution at the expense of longer separation time. Figure 9.4 from Guttman et al.[13] illustrates the ultra-high resolving power feasible with these types of gel capillaries. The 160-mer in the lower trace of Figure 9.4 (last peak) has a

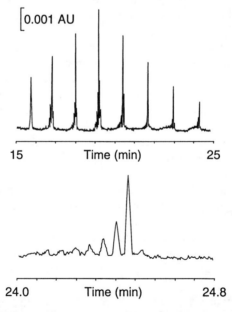

Figure 9.4 CGE of polydeoxythymidylic acid, p(dT)$_{20\text{-}160}$. The lower trace shows a blowup of the 24.0 to 24.8 min time interval, with the largest peak showing an efficiency of 30 million plates per m. (From Guttman, A. et al., *Anal. Chem.*, 62, 137, 1990. Copyright © 1990 American Chemical Society. With permission.)

plate count of 30 million plates per m, while the peak width is only a few seconds! This high resolving power of the crosslinked gels is the reason they often have been used in capillary-based DNA sequencing. However, medium viscosity, linear PA gels (~6% T) are easier to work with and can be replaced by pressure rinsing of the CE instrument.

Denaturing and the nondenaturing systems can both be used with PA gel capillaries. Denaturing PA gel-filled capillary columns are utilized mainly for size separation of short, ss DNA (up to several hundred bases, e.g., DNA primers and probes), and in DNA sequencing. The most commonly used denaturing agent is urea, while formamide is useful in some applications.[22] Some vendors (e.g., Beckman, Perkin-Elmer, BioRad) offer special kits for oligonucleotide analysis which contain prepacked capillaries, run buffers, standards, and urea as the denaturant. Nondenaturing PA gels may be useful when subtle differences based on the shape, size, and charge of the molecules are exploited.[23]

The low-viscosity, linear PA polymer networks are *not* covalently bound to the capillary wall. In this case, the capillary is precoated with a suitable polymer, e.g., with linear PA, which forms a monomeric layer on the inside surface of the capillary wall. With homemade replaceable capillaries, linear PA is generally used at low concentrations (1.5 to 6%). For the best polymerization reproducibility, it is recommended to prepare a high concentration gel (9 to 12%) that can be diluted to the appropriate concentration prior to use.[24,25] The performance of replaceable columns at various linear PA concentrations was examined by Pariat et al.[19] With a 6% linear PA capillary, the average peak efficiency was calculated as 4 million plates per m in the 51 to 267 bp region, making single bp resolution possible in this range. Comparisons of CGE with agarose slab gel electrophoresis for DNA digest analysis were made by Paulus and Husken.[26] CGE offered better resolution, especially in the <600 bp range.

Berka et al.[27] studied the sequence-dependent migration behavior of ds DNA with replaceable linear PA sieving matrices. DNA conformational effects were significant under high electric field conditions. Other CE parameters evaluated included polymer concentration, column temperature, and background electrolyte (denaturants, ions, intercalating dyes). Normal peak order (base number) could be restored by working with 3% PA, 200 V/cm, and a 50°C column temperature.

Nondenaturing, replaceable gel media are used (1.5 to 6% PA) for the separation of ds DNA molecules, such as restriction fragments or PCR products (see Section IV.E). Run-to-run reproducibility in these types of capillaries is excellent (typically ~0.2% RSD for migration times of DNA standards without an internal standard). Once conditioned, the replaceable capillaries can be used on a daily basis for months.

The size selectivity of sieving media (e.g., PA slab gels) is often characterized by Ferguson plots.[27a] Figure 9.5 from Heiger et al.[28] shows the PA gel concentration (expressed as % T) vs. the log mobility for linear PA in a CE system. The larger the DNA fragment, the steeper the slope, in accordance with the sieving theory. From the intercept of the plot, the free solution mobility (i.e., in the absence of a gel matrix) of a DNA fragment can be assessed. From Figure 9.5 it is also evident that the mobility of the DNA fragments in free solution, at 0% T, is virtually independent of the size of the DNA fragments examined. Schwartz et al.[29] showed similar Ferguson plots for sieving of DNA fragments in polymer networks of alkylcellulose.

b. Agarose, Alkylcelluloses, and Other Polymers

Agarose gels are characterized by large pore sizes, high mechanical strength, and biological inertness. As agarose is the medium of choice for the separation of relatively large DNA with slab gels, it seems logical that agarose would also be tried in capillaries. Compton and Brownlee[30] showed preliminary results of DNA separations with agarose. Since then, a few research groups (Schomburg[14]; Bocek and Chrambach[31-33]) have reported results with agarose-

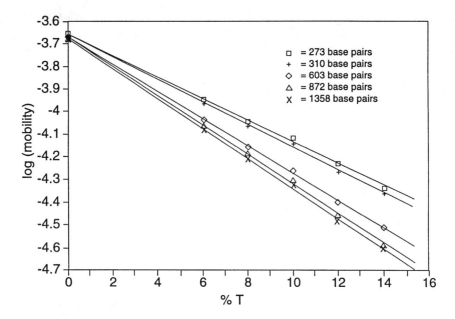

Figure 9.5 Ferguson plots for linear polyacrylamide. The lines represent the log mobility of various ϕX-174 Hae III fragments as a function of monomer composition. (From Heiger, et al., *J. Chromatogr.*, 516, 33, 1990. With permission.)

filled capillaries; they are not commercially available. Special purified grades of agarose must be used to avoid unwanted EOF in the capillary; in addition, agarose solutions must be optically clear for UV detection at 260 nm. Theoretical and practical aspects of DNA sieving in agarose have been published by Stellwagen[7] and Upcroft and Upcroft,[34] respectively. DNA entanglement in agarose gels was recently studied by Smisek and Hoagland.[9]

CGE of DNA restriction fragments with agarose concentrations between 0.3 and 2.6% at 40°C were described by Bocek and Chrambach.[31-33] The advantage of employing liquefied agarose above its "gelling" temperature is that the capillary is replaceable, i.e., it can be easily filled, rinsed, and refilled. Another advantage of this type of agarose is that its background absorbance at 254 to 260 nm is sufficiently low so that DNA detection at the ng level is possible. With an agarose sieving matrix, the inner surface of the capillary must be coated with a suitable polymer, e.g., linear polyacrylamide. Using this technique the effective size range for separation of ds DNA was limited to ~12 kbp.

Apart from polyacrylamide and agarose matrices, size separation of DNA in CE can be obtained by the use of various other entangled polymer solutions, e.g., various alkylcelluloses or polyethylene oxide.[34a] As with the PA-based sieving matrices, a precoated capillary is desirable with other sieving media. Commercially available polysiloxane-coated capillaries (e.g., DB-1 or DB-17 from J & W Scientific) can be used for this purpose, as well as others such as polyvinyl alcohol[14] or polyacrylamide.[35] Alkylcellulose additives form an additional "dynamic" coating on the inner surface of the capillary wall. Therefore, the EOF is significantly decreased or totally eliminated.[29]

The sieving of DNA through the medium can be manipulated by varying the chain length and the concentration of the polymer. This is illustrated in Figure 9.6 for a number of different polymers by plotting the mobility vs. the size (bp) of the DNA fragment on a semi-log scale. S-shaped curves with a linear middle section can be observed. The shallowness of the slope of the curve is a measure of the sieving power of the medium. The sieving depends on the viscosity of the medium and the polymer chain length. For example, it can be seen (panel A) that the sieving is better at higher concentration of the HPMC polymer. Comparison of panels A and B reveals that at the same concentration, the short-chain polymer (HPMC-100)

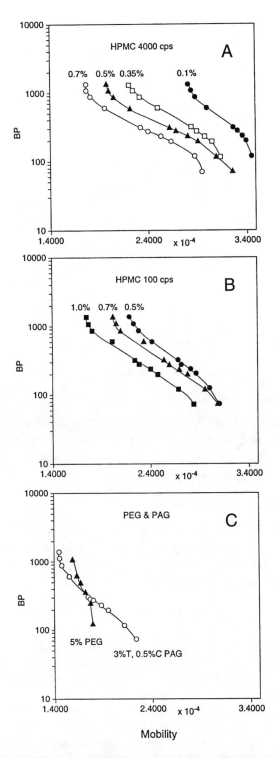

Figure 9.6 Effect of different polymer additives on molecular sieving. The plots (semilogarithmic scale) show the dependence of mobility on the base pair number. DNA fragments from the Hae III restriction digest of φX were used as size markers. Polymeric additives: (A) HPMC-4000 at 0.1, 0.35, 0.5, and 0.7%; (B) HPMC-100 at 0.5, 0.7, and 1.0%; (C) 5% PEG and polyacrylamide (3% T, 0.5% C). (From Schwartz, H. E. et al., *J. Chromatogr.*, 559, 267, 1991. With permission.)

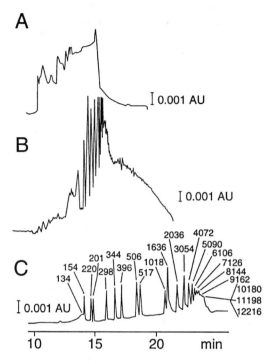

Figure 9.7   Effect of the concentration of HPMC-4000 on the CE separation of the 1-kbp DNA ladder. Concentration of HPMC-4000: (A) 0.1%, (B) 0.3%, and (C) 0.7%. (From Baba, Y. et al., *J. Chromatogr. A,* 653, 329, 1993. With permission.)

yields greater mobilities than the long-chain polymer (HPMC-4000), in agreement with earlier findings of Bode.[36] Panel C shows a steep curve for another sieving buffer consisting of 5% polyethylene glycol. This polymer solution would not be an effective sieving matrix for DNA separations. Other publications dealing with the effect of polymer chain length are from Barron et al.[10,15] and Baba et al.[37] Figure 9.7 shows the effect of polymer concentration on the CE separation of a 1- kbp DNA ladder.

III. INSTRUMENTATION AND METHODOLOGY

A. Conventional CE Instrumentation

1. *Capillaries: Coated vs. Untreated*

In CE, two types of forces drive the separation: (1) the electric field-driven force causing the electrophoretic migration and (2) the force exerted by the electroosmotic flow (EOF) through the capillary. The EOF bulk flow results from the charged inner capillary wall during application of an electric field. With untreated fused-silica tubing and an aqueous buffer, a negatively charged capillary surface results. The magnitude of the EOF is dependent on various experimental factors (e.g., pH of the buffer). Capillary coatings can reverse, reduce, or even totally eliminate the EOF. The latter is almost always the case in CGE. Coated capillaries now can be purchased from various commercial sources (see a recent survey in *LC.GC* magazine[38]). Special kits for DNA separations (including capillaries, reagents, and protocols) are also commercially available from manufacturers (e.g., Beckman, BioRad, Perkin-Elmer, Dionex).

a. Nucleotide Separations by CZE or MEKC

Separations can be performed with untreated as well as with coated capillaries. With untreated fused-silica capillaries and a typical neutral pH buffer (e.g., phosphate, pH 7) in nucleotide separations, the electrophoretic migration of the negatively charged nucleotides is toward the anode, opposite the direction of the EOF. Detection of the nucleotides will only take place if the existing EOF is larger than their electrophoretic mobility ($\mu_{EOF} > \mu_e$).

Using a solution of cationic surfactants (e.g., CTAB, DTAB), the surface of the fused-silica capillary can be modified *in situ* from a negative to a positive surface charge. The capillary wall is then said to be "dynamically" coated. When the concentration of the surfactant in the buffer is below its critical micelle concentration (cmc), CZE conditions exist. When the cmc is exceeded, a pseudostationary, micellar phase is created in the capillary, and MEKC conditions prevail; the charge on the capillary wall — and therefore the direction of the EOF — is reversed.

Generally, the strategy with permanently coated capillaries is to moderate or entirely eliminate the EOF. The result is often a more efficient and reproducible separation. Polyacrylamide-coated capillaries have been used for nucleotide separations in the CZE mode.[39] Similarly, commercially available capillaries[38] also should be useful for these types of separations.

b. ds DNA, Oligonucleotide Separations

With coated capillaries (and suppressed EOF), the order in which the DNA fragments elute from the gel-filled capillary column is according to their molecular size. Small, mobile DNA fragments have a short migration time (as in slab gel electrophoresis), whereas the larger DNA fragments are less mobile and have longer migration times. When untreated capillaries are used in conjunction with alkaline buffer conditions, the reverse order of elution has been observed because of the combined effects of the EOF and the molecular sieving/reptation.[12,40]

2. Sample Introduction: Stacking, Matrix Effects

a. Free Solution Methods

With slab gel methods, sample preconcentration ("enrichment") can be achieved with discontinuous buffer systems.[41] Similar tricks to obtain increased sample detectability can also be applied in CE.[42,43] This is especially useful when the analyte's concentration is very low (e.g., in biological media), thus UV detection is inadequate. (Another option is to use a different detection scheme, e.g., LIF.) Dissolving the sample components in a matrix which has a ~10× lower ionic strength than the run buffer may result in sharper peaks and 2 to 3× better detectability. Other similar stacking effects can be obtained by manipulating the pH of the sample vs. that of the run buffer[44] or by exploiting effects related to electrokinetic sample injection.[29]

More dramatic increases in sample detectability (~100×) can be achieved by applying the principles of isotachophoresis (ITP)[5a] to CE. ITP can be used as a preconcentration step prior to separation in the CZE mode. Using homemade systems, Foret and co-workers[45] have shown the utility of ITP-CE for the separation of nucleotides. ITP-CE can also be performed in commercial CE instruments.[46] The leading and terminating electrolytes must be carefully selected to achieve the desired zone focusing effect.

b. CGE Methods

Replaceable Gels. With replaceable gels, both pressure (typically, 2 to 20 sec, 0.5 psi) *and* electrokinetic injection are feasible. In CE, the pressure injection mode is generally

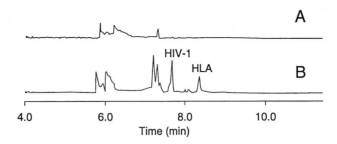

Figure 9.8 Effect of ultrafiltration on the PCR-amplified DNA peaks. (A) Untreated sample (no ultrafiltration), a coamplified HIV-1, HLA-positive (115 and 242 bp, respectively) control. (B) Desalted
sample. (From Schwartz, H. E. et al., *J. Chromatogr.*, 559, 267, 1991. With permission.)

recommended for quantitative work: the composition of the sample plug introduced into the
capillary is exactly that of the sample vial from which the injection took place. In addition,
sample preparation is simplified because no desalting needs to be performed. Butler et al.[47]
recently reported precision results with replaceable gels. Using pressure injection and an
internal standard, peak migration time precision was <0.1% RSD, whereas the area precision
was ~3% RSD. It should be noted, however, that electrokinetic injection often yields higher
peak efficiency than pressure injection.[29,47,48] When electrokinetically injected from a low
ionic strength sample solution, DNA fragments are effectively stacked against the relatively
viscous polymer network medium. In separations of small molecules, electrokinetic injection
may give rise to a sampling "bias" as sample components, because of their different mobilities,
move into the capillary at different speeds. However, since DNA fragments, at least in a
relatively narrow size range, essentially have the same mass-to-charge ratio in free solution,
no such sample bias occurs when these fragments are electrokinetically injected from an
aqueous solution.

In the practice of CE, the separation performance is often strongly dependent on the
composition of the sample solution. This is particularly important when samples are electrokinetically injected and variable amounts of salt are present. Desalting the sample by ultrafiltration — in conjunction with electrokinetic injection — may enhance sample detectability.[29,47,48] The ultrafiltration procedure removes low MW sample constituents, resulting in
efficient DNA peaks. Figure 9.8 shows the dramatic effect of desalting the sample for two
coamplified PCR products from an HIV-1 positive control cell line — no peaks are visible
in the trace corresponding to the untreated sample. However, it has been reported that
desalting, when used in conjunction with pressure injection, may also lead to loss of DNA
due to adsorption on the filter.[47] Thus, when possible, pressure injection, without desalting,
is preferable.

Recently, van der Schans et al.[49] studied sample matrix effects for analysis of PCR
products with replaceable gels and pressure injection. When the sample plug length was
increased, decreased efficiency — apparent as fronting peaks — was observed. Sharpening
of the peaks can be obtained by simply injecting a plug of low resistance, 0.1 *M* Tris-acetate
prior to the sample injection (Figure 9.9). The lower field conditions existing in the Tris-
acetate plug cause electrophoretic stacking of DNA fragments.

Coinjection of PCR products with a standard is a convenient method of verifying the
identity of the sample peaks. It was found that coinjection of a DNA standard with the PCR
sample can lead to sharpening of the sample peaks *or* the standard peaks, depending on the
order in which the plugs were loaded on the capillary. This effect is shown in Figure 9.10.
In the top trace, the 97-bp PCR sample is injected first, followed by the ϕX-174 Hae III DNA
standard. It can be seen that while the standard peaks are sharp, the PCR peak is broadened.
The opposite is seen in the lower trace where the injection order was reversed. The broadened
peaks are due to salt migrating from the PCR sample into the plug containing the DNA

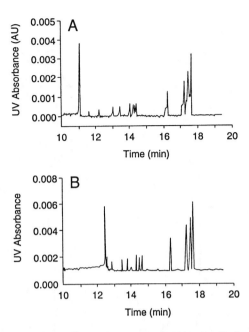

Figure 9.9 (A) Electropherogram of a 50 μg/mL φX-174 RF DNA Hae III dissolved in 20 mM NaCl. (B) Influence of presample injection of 0.1 M Tris-acetate, pH 8.3. Injection procedure: first injection: 10-s pressure injection of Tris-acetate; second injection: 20-s pressure injection of 50 μg/mL φX-174 Hae III sample in 20 mM NaCl. (From van der Schans, M. J. et al., *J. Chromatogr. A*, 680, 511, 1994. With permission.)

standards. During its migration through the capillary, the back end of the standard zone will migrate at a slower velocity relative to the front, because here the field strength is lower than at the front end of the plug.

Nonreplaceable gels. When working with capillaries containing high viscosity gels (e.g., >6% linear PA, viscous alkylcellulose), sample introduction is only practical by electrokinetic means (typically 0.015 to 0.15 Ws is applied). Pressure injection is limited with a highly viscous medium; the volume of the sample injected is inversely proportional to the viscosity of the buffer. Exceedingly long injection times would be required, making pressure injection impractical.

When using replaceable gels in conjunction with electrokinetic injection, effects due to the presence of salts or other substances in the sample matrix must be carefully considered. This is especially important in quantitative work. For example, when electrokinetic injections are made from samples which contain *different* salt concentrations, the amount of analyte introduced into the capillary will vary. This, in turn, has consequences for the accuracy of a drug assay. As noted by Srivatsa et al.,[50] with many pharmaceuticals for intravenous or ophthalmic use the products are formulated in isotonic salt solutions. In this case, it is important to use an external reference standard with the same sample matrix in order to accurately assay a drug product. While in CGE with electrokinetic injection the peak *migration time* precision generally is excellent (<0.2% RSD), *peak area* precision may exceed tolerable levels. Precision can be greatly improved, however, by the use of an internal reference standard.

Guttman and Schwartz[51] recently reported two injection-related artifacts with CGE. The first occurred with consecutive injections from the same, low-volume (10 to 200 μl) sample vial. Progressively smaller peak heights were obtained with each injection (Figure 9.11). Electrochemical processes occurring during the electrokinetic sample introduction process proved to be responsible for this effect. A simple solution was proposed: by performing an

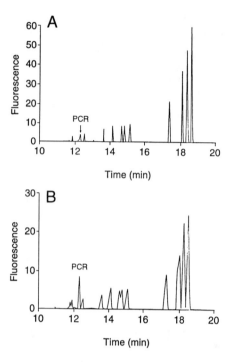

Figure 9.10 (A) Electropherogram of PCR sample and DNA standard. First injection: PCR sample (97 bp); second injection: φX-174 Hae III 10 μg/mL. (B) Electropherogram of DNA standard and PCR sample. First injection: φX-174 Hae III 10 μg/mL; second injection: PCR sample (97 bp). (From van der Schans, M. J. et al., *J. Chromatogr. A*, 680, 511, 1994. With permission.)

"intermediate" electrokinetic injection from a water vial prior to the actual sample injection, the trend toward decreasing peak heights with consecutive injections (as seen in Figure 9.11) was counteracted. The second artifact was related to the capillary edge (at the injection side). Oblique-edge capillaries resulted in poor peak performance compared to straight-edge capillaries.

3. Detection Options

In the majority of published DNA-related papers using CE, detection of DNA separations was achieved by UV absorbance. Other commercially available detection methods include UV-Vis (scanning) diode array detection, fluorescence, laser induced fluorescence (LIF), and on-line mass spectrometry.

a. UV Absorbance, Diode Array Detection

Optical detection techniques for CE recently have been reviewed by Pentoney and Sweedler.[52] In the vast majority of DNA as well as other applications of CE, UV-Vis absorbance detection has been used. Practically all commercial CE instrumentation is equipped with this detector, which is universal (i.e., suitable for many types of analytes) and also has adequate sensitivity for most applications. Diode array detection can offer further information about substance purity. However, in many bioscience applications, trace amounts of an analyte need to be determined in the presence of many other sample components, and detection may become problematic. The detection limit with UV detection is, among other factors, related to the small i.d. of the capillary; for example, a 200-bp DNA fragment typically has a minimal detectable concentration of ~0.5 μg/ml. Optimal stacking and/or ITP preconcentration meth-

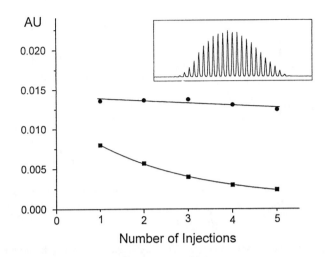

Figure 9.11 Peak height of the p(dA)$_{50}$ (absorbance units) vs. number of consecutive injections. Squares: 1.5 s, 7.5 kV injection from sample vial. Circles: 1.5 s, 7.5 kV injection from sample vial preceded by a 1.5 s, 7.5 kV injection from water. (From Guttman, A. and Schwartz, H. E., *Anal. Chem.*, 67, 2279, 1995. Copyright © 1995 American Chemical Society. With permission.)

ods (see Section III.A.2), as well as optimized optics, may improve the detection limits by a factor 2 to 10.

b. Laser Induced Fluorescence (LIF) Detection

Detection systems based on fluorescence, and especially laser induced fluorescence (LIF), may yield far lower detection limits than those based on UV absorbance. Fluorescence-based assays have the advantage of offering both excellent selectivity *and* very high sensitivity. Fluorescence-based detection in conjunction with fluorophore labeling systems have been developed for many biomedical applications, e.g., chromosome sorting, DNA sequencing, and DNA fingerprinting. Conventional light sources (e.g., lamps based on mercury-xenon, mercury, pulsed xenon, deuterium, and tungsten) have been used with CE.[4] However, the sensitivity is only marginally better than that of UV absorbance (about one order of magnitude). Lasers can focus the excitation light tightly on a small (~nL) detection volume. With laser sources and tailor-made instrumentation, it has become possible to examine the protein or nucleic acid contents of single human cells or even detect single molecules of stained DNA.

Recently, a LIF detector has become commercially available (Beckman Instruments). The P/ACE-LIF interface is supplied with a 488-nm argon ion laser but can be connected to various other laser sources.[4] The detector incorporates an ellipsoidal reflector to maximize the emission light collection efficiency. Thus far, most work in CE has been performed with the easy-to-use and relatively low-cost Ar-ion, He-Cd, and He-Ne lasers. These laser sources are also well suited for DNA and nucleotide work. The 488-nm emission of the Ar-ion laser matches the 490-nm peak of popular fluorescein-based labels (FITC, fluorescein succinimidyl ester). The 325-nm emission of the He-Cd laser matches OPA and dansyl labels. The compact and even less expensive "green or yellow" He-Ne lasers can be used with various rhodamine derivatives and are compatible with intercalating dyes such as ethidium bromide or propidium bromide.[53]

Low-cost, semiconductor ("diode") lasers can also be used with CE. Their wavelengths are in the 635 to 850 nm range; background fluorescence from biological sample matrices is strongly reduced at these long wavelengths. The analytically relevant blue light (i.e., wavelengths compatible with the popular fluorophore labels) can be obtained from these

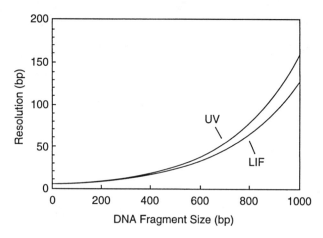

Figure 9.12 Comparison of the resolution capability for separation systems with (LIF trace)- and without (UV trace) intercalator. EnhanCE™ (Beckman Instruments) was used as the intercalating buffer additive. The resolution is expressed as base pairs resolved for a specific DNA fragment size.

lasers by "frequency doubling" techniques of (infra)red light. In order to extend the applicability of the red diode lasers to CE, new labeling reagents (based on thiazine-, oxazine-, and cyanine-type compounds) are being developed, as few analytes yield native fluorescence in the red or near-IR region.[54] Chen et al.[55] used a cyanine dye (Cy5) in DNA hybridization and sequencing (see also Figure 9.18, Section IV.C on DNA sequencing).

A number of detection schemes have been used for the detection of nucleic acids by fluorescence methods. One approach is based on the native DNA fluorescence in the low-UV region.[56] Native fluorescence allows analysis of the DNA molecule in its natural state, with excimer lasers (e.g., the pulsed KrF, 248-nm laser) as the excitation source. A later paper[57] describes some improvement in detectability (lower background signal) with this approach by using a sheath-flow arrangement while separations were performed at pH 2.8.

Another LIF detection scheme for DNA is based on indirect fluorescence. A fluorogenic CE buffer system such as salicylate can be used in conjunction with laser excitation (e.g., the 325-nm He-Cd laser). Examples of nucleotides[58] and ds DNA[59] have been demonstrated using this approach. However, the methods based on native and indirect fluorescence have not been widely used by other workers, and special, homemade instrumentation is required.

The most straightforward and currently popular LIF detection scheme for CE involves the use of fluorescent intercalators.[4,5,60] A nuclear stain is added to the CE buffer and/or sample and specifically interacts (intercalates) with sample DNA or RNA molecules. The dye inserts itself between the base pairs of DNA, thereby changing its persistence length, conformation, and charge, resulting in a reduction of electrophoretic mobility. As can be seen in Figure 9.12, intercalation may result in enhanced bp resolution. More important, the DNA-dye complex fluoresces strongly when excited by the appropriate laser wavelength, whereas the intercalator alone, as well as non-DNA sample components, generally do not. Hence, separation *selectivity* is often improved as visualized by a cleaner electropherogram than is possible with UV detection (see also Section IV.E.1, Figure 9.22).

The use of fluorescent intercalating dyes leads to two to three orders of magnitude enhanced sensitivity when compared to UV detection.[5,61-64] The gain in detectability is illustrated in Figure 9.13 for ϕX-174 Hae III RF DNA fragments — the UV trace shows an appreciable noise level, whereas in the LIF trace a "clean" baseline is obtained with no noise visible. In addition, the DNA concentration used for the UV detection was 20× higher than that of the LIF.

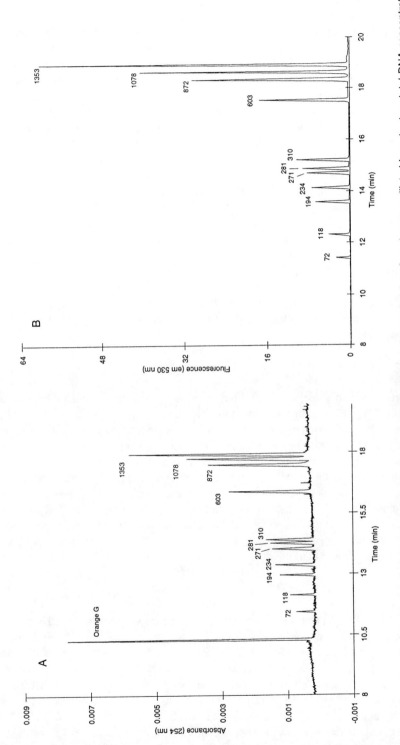

Figure 9.13 Separation of an Hae III restriction digest of φX-174 RF DNA using (A) UV and (B) LIF detection. Samples were diluted in water to a total DNA concentration of 200 μg/mL for UV detection; 10 μg/mL for LIF detection. Injection was by pressure for ten seconds. Buffer systems were the same, except for the addition of EnhanCE for LIF detection (B). (From Ulfelder, K. J., *Beckman Application Information Bulletin A-1748*, 1993. With permission.)

A family of monomeric as well as dimeric dyes have been utilized in recent CE-LIF applications.[53,63,65] The 488-nm Ar-ion laser is compatible with many of these DNA and RNA dyes. It has been reported[63] that the dimeric dyes (e.g., ethidium dimer, TOTO, YOYO) are best used in conjunction with monomeric intercalator additives, e.g., 9-aminoacridine; otherwise, broad peaks may result from the presence of multiple dye–DNA bonding.

Direct labeling of a DNA fragment with a suitable fluorophore is another suitable detection approach. Fluorescently labeled probes and primers are used in many molecular biology applications involving hybridization and PCR. DNA primers and probes are usually synthesized with a fluorescent label attached to the 5′ end of the molecule or with postsynthesis attachment of a dye using commercial DNA labeling kits. Unincorporated dye and/or failure sequences are generally removed by LC methods. For subsequent use in the PCR leading to fluorescent DNA products, conditions can be optimized such that primer purification is not necessary.

In DNA sequencing, fluorescent-labeled primers based on fluorescein, modified fluoresceins, Texas red, and tetramethylrhodamine are routinely used.[66] The fluorescein dyes match the 488-nm line of an Ar-ion laser, whereas the rhodamine dyes are compatible with the 543.5-nm line of an He-Ne laser or the 514.5-nm line of an Ar-ion laser. For the labeling of nucleotides, fluorophores such as dansyl[67] and fluorescein[68] have been used. Wang and Giese[69] recently described a phosphate-specific labeling of nucleotides with the fluorophore BODIPY Fl C$_3$ hydrazide (Molecular Probes, Eugene, OR). The labeled nucleotides were subsequently detected by CE-LIF. Adenine-containing nucleotides were analyzed with good selectivity by Tseng et al.[70] using a chloroacetaldehyde derivatization reaction to convert the analytes to fluorescent products.

c. Thermooptical Absorbance Detection

In an effort to obtain high detection sensitivity and accuracy (while avoiding derivatization) with either HPLC or CE, detection schemes based on thermooptical absorbance (TOA) have been developed.[66] This detection mode employs a laser pulse to repeatedly irradiate the sample. The absorbed light is converted by radiationless transitions into heat, i.e., the solvent temperature rises in the detection cell. These temperature changes translate into refractive index changes which are decoded by a lock-in amplifier. TOA detection has been used by Krattiger et al.[71] for the CE separation of various nucleosides and nucleotides absorbing at 257 nm. A frequency-doubled Ar-ion laser was used to optically pump the nucleic acids, while a diode laser or an He-Ne laser was used as a probe. A holographic optical element was used to guide the laser light to the detection site. Concentration detection limits (using 20 μm i.d. capillaries) with TOA were ~30× lower than possible with standard UV detection.

d. CE-Mass Spectrometry (CE-MS)

While CE-MS is not yet routinely used, new instrumentation and commercially available interfaces allow CE to be coupled to MS. Large biomolecules such as DNA, proteins, and carbohydrates are presently amenable by MS. CITP — which is an excellent preconcentration technique — can also be coupled to MS.[6] A recent review,[72] describes the current state of the art in this field. It has been demonstrated that free solution techniques such as CZE and MEKC can be coupled to MS by either fast atom bombardment (CE-FABMS) or electrospray (CE-ESMS) techniques. An example of the first technique was described by Wolf et al.[73] for the characterization of deoxynucleoside-polyaromatic hydrocarbon adducts. The analysis of ds antisense oligonucleotides was reported by Bayer et al.[74] using CE-ESMS. Since the exact mass of the double helix can be detected, the CE-ESMS method allows discrimination between specific and nonspecific interactions. The method offers potential in the study of hybridization reactions.

4. Fraction Collection from CE Capillaries

While CE is primarily an analytical technique, it can also be used for micropreparative purposes. For example, it is possible to collect fractions from protein digests and subsequently perform microsequencing to identify the peptides. In an early CE paper, Cohen et al.[3] collected a small quantity (less than a μg) of a 20-mer oligonucleotide primer from a capillary gel column (polyacrylamide, 8% T, 3.3% C); the collected fraction was subsequently used in a dot-blot assay. More recently, Kuypers et al.[75] used CGE (replaceable gel, 4% linear PA) to collect multiple peaks corresponding to denatured DNA from a mutated, 372-bp PCR product. The collected fractions were reamplified by PCR and subsequently analyzed again by CGE. It was found that the different peaks corresponded to different gene sequences.

In the majority of papers dealing with micropreparative CE, the electric field was kept constant during the collection of the fractions. However, with CGE of DNA, often very narrow peak widths — a few seconds wide — are obtained. This makes reproducible collection of peaks difficult. As shown by Guttman et al.,[13] by programming the electric field during the collection, the collection process can be simplified. This approach of field programming for fraction collection is demonstrated in Figure 9.14. A 47-mer oligonucleotide from a $p(dA)_{40-60}$ mixture was collected. During the micropreparative run (trace A) the field was maintained at 300 V/cm until just before the 47-mer reached the detector. At that point, the field was decreased tenfold, and the fraction was collected for 60 s. Trace B shows the reinjected collected fraction: only the 47-mer is visible together with the internal standard, a 20-mer. Typically, the micropreparative runs yield broader peaks than their corresponding analytical runs. The sample size injected for the micropreparative run resulting in the "overloaded" profile (trace A) was 6× higher than the analytical run (trace C).

B. Novel CE Instrumentation

1. Capillary Arrays

Sample throughput is a major concern in many analytical laboratories. In DNA analysis, this is particularly so for high-volume DNA sequencing (Section IV.C). Automation of sample preparation, sequence reactions (including electrophoresis), and data interpretation is necessary to achieve the ambitious goal of sequencing the entire human genome (~3 billion bp). With CE, the samples are loaded one at a time. Slab gels, on the other hand, can simultaneously be loaded with 24 to 36 samples. Instrumentation which would allow running several capillaries in parallel, together with robotics for sample handling, would dramatically increase the desired sample throughput with CE. DNA sequencing has been demonstrated in arrays of multiple (up to 100) capillaries.[76-80] Figure 9.15 shows a schematic diagram of a design based on a confocal fluorescence scanner. Capillary arrays have also been applied to DNA fragment sizing[63,79] and for the analysis of short tandem repeat alleles[81] suitable in forensic and genetic DNA typing (see also Section IV.E.1).

2. Pulsed-Field CE

Gel electrophoresis with dc fields can separate DNA fragments up to ~30 kb. Larger DNA fragments (from kb to ~6 Mb in size, chromosomal DNA) can be resolved by using ac fields, i.e., in pulsed-field gel electrophoresis (PFGE; see Gardiner[82] for a review). The problems

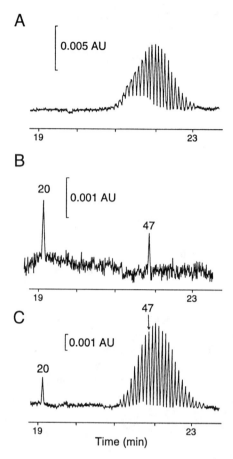

Figure 9.14   (A) Micropreparative CGE separation of a polydeoxyadenylic acid test mixture, p(dA)40–60;
(B) analytical run of the isolated p(dA)47 spiked with p(dA)20; and (C) analytical run of
p(dA)40–60 spiked with p(dA)20. (From Guttman, A. et al., *Anal. Chem.*, 62, 137, 1990.
Copyright © 1990 American Chemical Society. With permission.)

encountered with "normal" (dc) electrophoresis, when large DNA is separated have been
attributed to the stretching and reptation schematically shown in Figures 9.6 and 9.7. Under
these conditions the molecular sieving mechanism is no longer operative. However, by pulsing
the electric field, the reptation of DNA is counteracted. At the time of this writing, few papers
had appeared on the practical use of pulsed fields with DNA separations in CE.[83-86] Pulsed-
field CE separations with λDNA standards (8.3 to 48.5 kb) and λDNA concatamers (48.5 kb
to 1 Mb) were ~10 to 50× faster than typical slab gel separations of this kind. Heller et al.[86]
adapted a commercial CE instrument (Beckman P/ACE 2100) for field inversion. The pulses
were generated by using a programmable function generator and amplified by a high voltage
amplifier. Section IV.E.2 discusses the applications of pulsed-field CE.

3. CE on a Chip

A number of companies are involved in the research of biochips[87] which potentially could
be used in a wide variety of biomedical/diagnostic/forensic applications. CE technology could
be integrated in such devices, e.g., in simple, compact instruments designed to perform DNA
typing. Micromachined CE, in which the electrophoresis is carried out on a small glass plate
or chip, has already been applied to the analysis of antisense oligonucleotides, as reported
by Effenhauser et al.[88] CGE mode with LIF detection was employed for fast analysis (45 s)
of fluorescently labeled phosphorothioates.

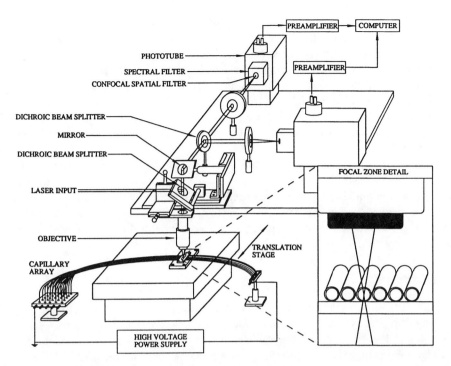

Figure 9.15 Diagram of a two-color, confocal fluorescence capillary array scanner. (From Huang, M. et al., *Anal. Chem.*, 64, 2149, 1992. Copyright © 1992 American Chemical Society. With permission.)

IV. APPLICATIONS

A. Mononucleotides and Nucleosides

Nucleotides and their nucleoside and base constituents play an important role in many vital biochemical processes. They are the activated precursors of DNA and RNA. Intracellular nucleoside metabolism is an important topic in AIDS research. A number of dideoxynucleoside analog drugs (azT, ddI, and ddC) are currently used in the treatment of HIV-1 positive individuals or are in human clinical trials (d4T and ddA). DNA "damage" at the nucleic acid base level is also actively studied.[89] Reliable, high-resolution analytical methods to quantitate nucleotides, nucleosides, and bases in biological samples (often at extremely low levels) are highly desirable in these areas. CE has demonstrated it can be a promising new tool in a number of recent studies. CE methods may complement existing HPLC methods; however, often superior resolution with shorter analysis time is possible with CE, while a minimal amount of sample is required for the sampling process.

1. DNA Adducts, DNA Damage

DNA damaged by covalent modifications or additions of xenobiotic components was studied by Norwood et al.[90] and Guarnieri et al.[91] The first group evaluated different CZE and MEKC conditions and demonstrated sample stacking techniques to increase detectability of benzo(a)pyrene-DNA adducts. Guarnieri et al.[91] measured 8-hydroxydeoxyguanosine by MEKC as a marker of DNA oxidation. Single-stranded DNA was incubated in the presence of an oxidizing agent and hydrolyzed by enzymatic digestion. MEKC did not, however, permit determination of extremely low levels of oxidized nucleosides generated by endogenous sources of free radicals.

There is increasing interest in measuring nucleotides at extremely low levels in biological matrices or even in single cells. DNA base damage studies (by either HPLC or CE) are often hampered by a lack of sensitivity, especially when only limited sample volume is available. A sensitive assay should, for example, be capable of detecting at least one DNA base modification in 10^4 to 10^6 normal bases within a few μg's of DNA.[89] CE with LIF detection is, in principle, sensitive enough to measure these type of modifications. Combined with suitable fluorophore chemistry, LIF detection can in principle provide orders of magnitude of improvement in sensitivity over UV. Preliminary reports[67,69] on nucleotide analysis by CE-LIF support this contention.

A few recent papers have described the use of CE coupled to mass spectrometry. CE was on-line coupled to FAB-MS by Wolf et al.[73] for the analysis of deoxynucleoside-polyaromatic hydrocarbon adducts. Zhao et al.[6] used both CZE and CITP for on-line electrospray ionization mass spectrometry of various nucleotides. CITP-tandem MS was utilized by these workers to provide both molecular weight and structural information of monophosphate dinucleosides.

2. Nucleoside Analog Drugs

Therapeutic drug monitoring of a nucleoside drug was described by Lloyd et al.[92] The antileukemic agent, cytosine-β-D-arabinoside (Ara-C), was determined in human serum. The authors found low-level (i.e., sub-μM) detection of Ara-C problematic. However, by using solid-phase extraction for concentration and sample cleanup, it was possible to determine Ara-C in the 1 to 10 μM range. This procedure removes most of the protein and allows doubling of the Ara-C concentration. The assay was validated over a concentration range of 1 to 10 μM. Response was linear in this range, with a correlation coefficient of 0.996 for the calibration plot. Compared to HPLC, the proposed assay has a rapid analysis time (no need to run a gradient to remove late eluting compounds) and is free from endogenous substances.

Rogan et al.[93] showed that CE is well suited to resolving enantiomers of nucleoside analog drugs. Determining the enantiomeric ratio of such drugs is important to regulatory agencies such as the U.S. Food and Drug Administration because one enantiomer may exhibit far greater efficacy (or toxicity) than the other. In the manufacturing process of an antiviral drug (Glaxo Wellcome), the racemic 2'-deoxy-3'-thiacytidine (BCH-189) undergoes an enantiospecific deamidation to yield the (–) enantiomer. The latter chiral drug is less toxic and therefore preferred for clinical use over the racemic drug. The time course of the enzymatic reaction could be followed by CE for more than two days. Compared to the HPLC method which uses expensive chiral stationary phases, the CE approach is simpler, whereas reliability and precision are reportedly excellent.

3. Nucleotides in Cell Extracts

Nguyen et al.[94] used a CZE method for the quantitation of nucleotide degradation in fish tissues. In most fish, ATP degrades rapidly to IMP which, in turn, degrades to inosine and the base hypoxanthine. IMP gives fish a pleasant fresh taste, while hypoxanthine accumulation results in an off-taste. Tissue extracts were assayed by both CZE and an enzymatic method. Good correlation between peak area and nucleotide concentration was found. The CZE method involved UV detection, untreated fused-silica capillaries, and a CAPS, pH 11 run buffer, generating a high EOF.

Nucleotide profiles in cell extracts were determined by Ng et al.[95] and Huang et al.[96] with CZE. Ng et al.[95] showed nucleotide profiles in human blood lymphocytes and leukemic cells. Reproducible area and migration time were obtained using a P/ACE instrument. A simple CZE system using an untreated fused-silica capillary with a borate, pH 9.4 run buffer was

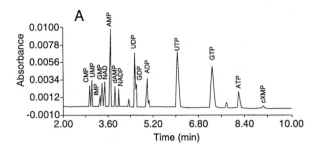

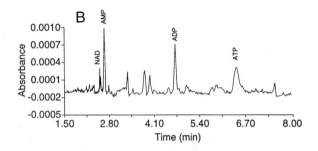

Figure 9.16 Separation of nucleotides by MEKC. (A) Separation of standard mixture of 15 nucleotides at 25 mM each (B) neutralized perchloric extract of rat tumor. (From Perrett, D., *Capillary Electrophoresis*, Camilleri, P. (Ed.), CRC Press, Boca Raton, Florida, 1993. With permission.)

employed. The negatively charged nucleotides are carried to the detector (cathode) by the EOF. Fourteen of the common ribonucleotides were determined in a CZE assay by Huang et al.[96] In this method, coated (polyacrylamide) capillaries resulted in negligible EOF using the mixed phosphate-Tris, pH 5.3 buffer. Electrophoretic flow carries the analytes to the detector end (anode, reversed polarity). Minimum detectable levels of the nucleotides were in the 1 to 10 µM range with UV detection at 254 nm. The method was applied to the quantitation of ribonucleotides in HeLa cells. In a later paper (Shao et al.[97]), coated capillaries were used for the determination of ribonucleotides in lymphoma cells. A similar method, involving (polyacrylamide) coated capillaries was described by Takigu and Schneider.[98] The authors discussed validation criteria, i.e., linearity and minimal detectable concentration (~1 µg/ml per nucleotide without stacking).

MEKC conditions have also been proposed for the analysis of nucleotides in cell extracts. It appears that cationic surfactants are more effective than their anionic counterparts such as SDS. Perrett and Ross[99] selected dodecyltrimethylammonium bromide (DTAB), resulting in a charge reversal on the capillary wall; 1 mM EDTA was added to the run buffer as a metal chelating agent to prevent metal–nucleotide interaction. Figure 9.16 shows that this method can be applied to acid extracts of cells. The upper trace shows the separation of a standard mixture of 15 nucleotides with a 50 mM phosphate (pH 7), 100 mM DTAB, 1 mM EDTA run buffer. A neutralized perchloric acid extract of rat tumor is shown in the bottom trace of Figure 9.16. The reported method gave a linear response up to 200 µM; migration time and peak area precision ranged from 2.2 to 5.5% and 3.3 to 6.1%, respectively. Similarly, Ramsey et al.[100] evaluated a number of cationic surfactants for the MEKC of nucleic acid constituents. Optimum resolution was here achieved by using tetradecyltrimethylammonium bromide (TTAB) as the micellar reagent. Loregian et al.[101] compared MEKC with HPLC for the quantitation of ribonucleotide triphosphates in four different cell lines. The MEKC method yielded ~1 million theoretical plates, with detectability down to 50 fmole.

B. Synthetic Oligonucleotides

1. Phosphodiester Oligonucleotides

CE is ideally suitable to the QC of oligonucleotides with 10 to 150 bases (primers or probes). Typically, crosslinked or relatively high-viscosity linear polyacrylamide gels (covalently bonded to coated fused-silica capillaries) have been used under denaturing conditions (urea, formamide, heat). The 7 M urea prevents secondary structure formation of oligonucleotides. The separation of a 119-mer oligonucleotide preparation on a poly-acrylamide column is shown in Figure 9.17. The upper trace shows the crude preparation, the lower trace the purified oligo. Separations of this type are difficult to perform with HPLC, which is generally limited to an upper range of about 70 bases.[102] In addition to the main component, the failure sequences are well resolved by the CGE capillary (N = 565,000 plates per m).

2. Antisense DNA

Antisense therapeutics are synthetic oligonucleotides whose base sequence is comple-mentary to a target sequence on a messenger RNA (mRNA) which encodes for disease-causing proteins, or to the double-stranded DNA from which the mRNA was transcribed. Another type of antisense, termed peptide nucleic acid (PNA), in which the deoxyri-bose–phosphate backbone is substituted for a peptide backbone composed of (2-amino-ethyl)glycine units, shows promise as a potent therapeutic agent and can also be analyzed by CGE.[103] Because of their potential use as drugs, stringent purity requirements are typically required for antisense DNA agents.

CGE results in excellent resolution of deletion sequences of a 20-mer phosphorothioate antisense product.[50] However, CGE cannot resolve phosphorothioates from their correspond-ing phosphodiesters (anion exchange chromatography is the method of choice here) because the separation mechanism in CGE is based on molecular sieving (size). Using a gel capillary (Beckman eCAP ss DNA 100), Srivatsa, et al.[50] demonstrated the utility of CGE for routine analysis of drug product formulations. In this report, the CGE method was tested for linearity, accuracy, selectivity, precision, and ruggedness.

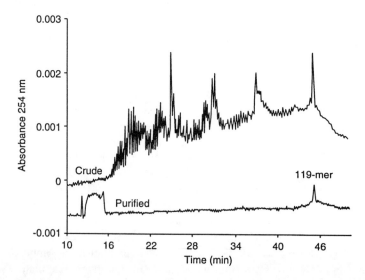

Figure 9.17 Analysis of a crude and purified synthetic 119-mer oligonucleotide using a nonreplaceable, polyacrylamide-based column (eCAP ssDNA 100, Beckman Instruments).

C. DNA Sequencing

Several research groups are currently exploring the use of CE as an alternative to slab-gel electrophoresis for automated DNA sequence determination.[66,104] The large surface-area-to-volume ratio of the capillary permits larger electric fields than are typically used with slab gels, resulting in very rapid and efficient separation of sequencing reaction products. Additionally, the capillary format is readily adaptable to automated sample loading and on-line data collection. With CE, detection of DNA sequencing fragments is performed by LIF. The sensitivity of the LIF detection allows sequencing reactions to be performed on the same template and reagent scale as that of manual DNA sequencing with autoradiographic detection. The identity of the terminal base of each DNA sequencing fragment is encoded in the wavelength and/or the intensity of the fluorescent emission. Instrumentation in which several capillaries are placed in parallel would dramatically increase the desired sample throughput with CE. DNA sequencing with arrays of multiple (20 to 100) capillaries has been reported by several groups.[77-80] A fast capillary scanning system — in which the separated bands are scanned longitudinally with a laser — was recently described by Kim et al.[105]

A large obstacle to the development of commercial CE-based DNA sequencers has been the stability of gel-filled capillaries. While they can provide extremely high resolving power, the crosslinked gels typically last only a few runs when sequencing reactions are loaded, after which time the entire column must be replaced. Recent developments in CGE column technology (in particular the replaceable gels) should eliminate the time-consuming and laborious procedures of the preparation and alignment of the capillaries.[22,106,107] With the replaceable matrix, it is possible to load a sequencing reaction, rapidly separate the DNA fragments at high field, and then reload the gel on the capillary prior to the next run. Figure 9.18 shows the CE separation of a single terminator, Sanger-Coulson reaction using a replaceable linear PA gel. A "red" diode-laser was used for excitation of the fluorophore (Cy5)-labeled DNA fragments.[55] The relatively low-viscosity (6% PA) gel matrix of these types of capillaries provide reproducible and fast separation of DNA fragments with sequence information extending to at least 400 bases. For DNA sequencing applications, typically a CE run buffer containing formamide and urea is used. A denaturing buffer of 30% formamide, 3.5 M urea has a lower viscosity than the typical 7 M urea buffer, and is therefore advantageous to use in a replaceable CGE formulation. In addition, increased decompression of sequences with secondary structures is obtained.

Polyethylene oxide (PEO)-based sieving matrices were used by Fung and Yeung[108] for high-speed DNA sequence analysis. The relatively low-viscosity matrix consisted of 1.4% PEO 600,000 M_n and 1.5% PEO 8,000,000 M_n, 1 X TBE (pH 8.2), and 3.5 M urea. The two size ranges of PEO were used to achieve good resolution of both small and large DNA fragments. Typical run times were 16 min (28 to 420 bases from the Sanger ladder).

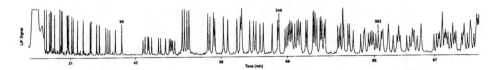

Figure 9.18 Electropherogram of fluorescently labeled (Cy5-20 primer) DNA fragments generated enzymatically (M13mp18 template) using ddG terminator. A LaserMax (Rochester, NY) red diode laser (639-nm excitation) was employed with an in-house-built CE-LIF instrument. The capillary was filled with a replaceable polyacrylamide gel. The pattern is recognizable beyond 400 bases. (Electropherogram courtesy of Dr. S. L. Pentoney, Jr., Beckman Instruments, 1994.)

D. Hybridization, Southern Blotting, DNA Binding

Southern blot hybridization, when used in conjunction with autoradiography, can be an extremely sensitive method for many DNA analyses, e.g., in clinical and forensic applications. It is also a very selective method, as the target DNA will only bind with a probe DNA whose sequence is complementary. However, the classical method is laborious and time consuming, often taking several days to complete. Brownlee, Sunzeri, and co-workers[109,110] were the first to show data demonstrating the feasibility of on-capillary hybridization with fluorescently labeled probes for the detection of target DNA (HIV-1 and HTLV-1) sequences. Using rhodamine and FITC-labeled probes, these sequences could be discriminated in a single run by fluorescence diode array detection at 563 nm and 519 nm, respectively. Also working with HIV-1 genome sequences, Bianchi et al.[111] analyzed PCR products by on-capillary hybridization. A 299-mer ss DNA fragment was hybridized with a complementary 28-mer, resulting in mobility shifts in the electropherogram.

Cohen and co-workers[112,113] have shown that it is possible to transfer Southern blotting from a slab gel to a CGE format, thereby greatly reducing the analysis time. In their first paper (Chen et al.[112]), preliminary results were reported with probes labeled with the fluorescent dye Joe. Analysis was made by LIF detection, using a 488-nm Ar-ion laser. In a later paper (Vilenchik et al.[113]), antisense DNA (phosphorothioate) was quantified down to 0.1 ng/ml using fluorescein-labeled probes. Although UV detection was also used, LIF detection (488-nm, Ar-ion laser) gave superior sensitivity. In addition, background DNA in the detection or sample cell, problematic with UV detection, does not affect LIF detection. The principle of the CE method used by Cohen and co-workers is illustrated in Figure 9.19 (UV traces). Sample preparations containing different amounts of target and probe (kept constant here) were injected onto two capillaries: one under nondenaturing CGE conditions (set A), the other under denaturing CGE conditions (set B). The electropherograms show three peaks: that of the target antisense DNA ("GEM"), that of the probe ("COM"), and that of the hybrid ("duplex"). As the GEM concentration increased, the duplex peak also increased in size, while the COM peak size decreased. In the set of denaturing electropherograms, the duplex peak, as expected, is not observed. The order of migration of the target and the probe is different than that under nondenaturing conditions, most likely because of secondary structure differences between the two. LIF detection of the antisense DNA proved quantitative and linear over three orders of magnitude. However, in the CE-LIF electropherograms, the fluorescein-labeled probe co-eluted with the duplex (as opposed to the UV traces where resolution is adequate, Figure 9.19). Using ethidium bromide (0.04 μM) as an intercalating dye, it was possible to increase the resolution sufficiently to enable quantitative analysis.

Identification of PCR products by precolumn and on-column hybridization was studied by Ulfelder et al.[114] A one-step hybridization and separation method would be highly desirable. Preliminary data showing the feasibility of this approach were demonstrated using sensitive LIF detection. While precolumn hybridization requires heat for ds DNA denaturation, the on-column method employed relatively low pH conditions (pH 3 to 4) — acidic enough to denature but not to depurinate the ds DNA target. The on-column method involved a sequential loading (injection) procedure of high ionic strength buffer, denatured target DNA, and probe in water.

CE can be used as a tool to study biomolecular, noncovalent reactions. The term affinity capillary electrophoresis (ACE) has been coined to describe the CE of receptor–ligand interactions, including those of antigen-antibody (immunoassays). Heegard and Robey[115] used ACE for the study of oligonucleotide–peptide interactions. Using dimeric peptides as probes, the binding was found to depend on both the size of the oligonucleotide and the specificity of the interaction.

Protein-DNA interactions have also been studied with CE. Protein-DNA interactions are involved in control of replication, recombination, modification, repair, and transcription processes. Methods for studying DNA-protein interactions include mobility shift assays,

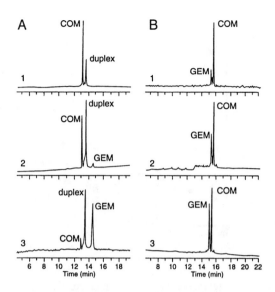

Figure 9.19 The separation of GEM, COM, and duplex by CGE. Electropherograms 1, 2, and 3 show different amounts of GEM with constant COM concentration. (A) Nondenaturing conditions. (B) Denaturing conditions. Conditions: (A) 9% T linear polyacrylamide column, effective length = 20 cm, applied electric field = 200 V/cm; (B) 13% T linear polyacrylamide, denaturing conditions, effective length = 15 cm, applied electric field = 400 V/cm. (Adapted from Vilenchik, et al., *J. Chromatogr. A*, 663, 105, 1994. With permission.)

where slab gel electrophoresis is used to detect a change in mobility of DNA when complexed to a protein. CE can be applied to these types of mobility shift assays, as shown by Maschke et al.[116] for the binding of an endonuclease, *Eco*RI with oligonucleotides. A free solution CE system with LIF detection (Ar-ion laser, 488-nm) was used. Joe-labeled oligonucleotides with the *Eco*RI recognition site, GAATTC, interact with the protein; the complex is detected as a faster-migrating fluorescent peak. Addition of excess unlabeled probe displaces the labeled probe in the complex, resulting in the disappearance of the fluorescent signal.

E. ds DNA Analysis

1. Relatively Small DNA, PCR Products

With the advent of PCR-related methods, the number of publications involving DNA fragment separations by CE is rapidly expanding. CE, and especially CE combined with LIF detection, will undoubtedly compete with or replace classical electrophoretic techniques in many PCR-related applications. From 1988 to 1991, the emphasis was on the development of suitable gels/polymer networks and capillary coatings, and fine-tuning of CE conditions to optimize separation performance.[28,29,117,118] At present, many DNA-related applications are being reported using these conditions and capillaries.

a. Clinical Diagnosis, DNA Polymorphisms, Point Mutations

Quantitation of viral load in patient specimens is important to assess the stage of disease progression or to monitor the effectiveness of drug therapy. Until recently, there were no reliable methods available to quantitate HIV-1 in the early onset phase of infection, i.e., at low copy numbers. PCR methods (e.g., competitive PCR) have recently allowed quantitation of proviral DNA or plasma viral RNA levels in patients with HIV-1 infection[119] using gel electrophoresis with video scanning of the ethidium bromide stained DNA. CE is ideally

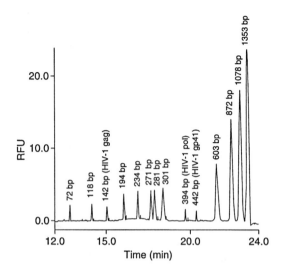

Figure 9.20 Detection of multitarget PCR-amplified HIV-1 gag, pol, env sequences by CE-LIF using the LIFluor dsDNA 1000 Kit. A φX-174 RF DNA standard was co-injected with the sample. (From Lu, W. et al., *Nature (London)*, 268, 269, 1994. With permission.)

suited to replacing slab gel techniques for these purposes. Competitive PCR is also adaptable to a CE format, as will be discussed in Section IV.E.1.b.

Brownlee, Sunzeri, and co-workers[109,110,120] developed PCR-based CE methods for the quantitation of multiple retroviral DNA sequences. UV and fluorescence (non-LIF) detection were employed using an instrument which permitted simultaneous diode array detection.[121] The sensitivity by UV or (non-LIF) fluorescence was not adequate, however, allowing detection of HIV-1 provirus at *very low* DNA copy numbers (e.g., for HIV proviral load in asymptomatic individuals). At that time (1991), LIF detection would have provided the extra sensitivity needed but was not yet commercially available. Recently, a French research team, Lu et al.,[122] used P/ACE with LIF detection for the quantitative analysis of PCR-amplified HIV-1 DNA or cDNA fragments. The LIFluor dsDNA 1000 kit (containing the EnhanCE™ intercalating dye) was used in the CE-LIF experiments. Quantitation of multitarget PCR fragments was demonstrated. Figure 9.20 shows the CE-LIF electropherogram of three HIV-1 sequences (142 bp, 394 bp, and 442 bp from the *gag, pol,* and *gp41* genes, respectively) together with a DNA standard. Figure 9.21 shows a dilution series of HIV-1 DNA templates ranging from 1 to 25,000 copies subjected to 40 PCR cycles. A linear range of three orders of magnitude was achieved using CE-LIF. The figure also shows a dilution series obtained from reverse-transcribed (RT) RNA from HIV-1 virions. Data of virion concentrations in sera of individuals infected with HIV-1 at different stages of infection were presented. The measurements by CE-LIF showed excellent correlation to the data acquired with the Southern blot hybridization method.

Separations of RT-PCR products from the polio virus RNA were shown by Rossomando et al.[62] Quantitation was achieved by comparing the corrected peak area for the RT-PCR product to a standard curve generated from known amounts of template RNA. The sieving buffer containing the intercalating dye, EnhanCE (Beckman Instruments), facilitated excellent separation of 53 bp, 71 bp, 97 bp, and 163 bp DNA fragments. The resolution by slab gel electrophoresis, on the other hand, while adequate for the 163 and 97 bp fragments, was inadequate for the other DNA fragments. Figure 9.22 compares UV (260-nm) vs. LIF detection of a 53-bp RT-PCR product derived from one strain of virus from the polio vaccine, Sabin-3. The migration times of the PCR products are longer for the LIF run, because in this case the DNA is intercalated with dye. In the LIF trace, interferences are far less prominent,

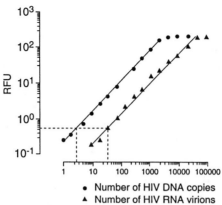

Figure 9.21 Linearity of LIF-CGE analysis of quantitative PCR or RT-PCR products. Serial dilutions of HIV-1 DNA templates ranging from 1 to 25,000 copies were subjected to a 40-cycle PCR with gag primers SK145/431. RNA extracted from serial dilutions of HIV-1 virions (ranging from 10 to 100,000 viral particles) was reverse transcribed with 20 pmol of 3′ primer SK431. (From Lu, W. et al., *Nature (London)*, 368, 269, 1994. With permission.)

and the pattern is unambiguous. PCR of target sequences often results in contaminating by-products which can interfere with UV detection.

Genetic linkage studies follow the inheritance of a particular trait (phenotype) in a family over several generations. The object is to correlate the trait with the presence of a specific DNA fragment (allele). By finding a marker that is close (linked) to the gene, persons predisposed to certain diseases can be identified. Ulfelder et al.[123] demonstrated the utility of CGE (with UV detection) for restriction fragment length polymorphism (RFLP) analysis of the ERBB2 oncogene (a candidate gene for a breast cancer gene at chromosome 17q). Polymorphic alleles can be identified by the presence or absence of a specific endonuclease recognition site. RFLPs typically have only two alleles at a given locus. Figure 9.23 shows the RFLP analysis of three individuals whose genomic, PCR-amplified DNA was digested with *Mbo*I. The top two traces represent two homozygous samples characterized by the presence of either the 500 bp or the 520 bp fragment. The heterozygous sample shows both fragments present, which were separated in a sieving buffer containing 0.5% HPMC.

In a similar study, Del Principe et al.[124] analyzed PCR-amplified products of the DXS 164 locus in the dystrophin gene. *Xmn*I digestion yielded polymorphic alleles of 740, 520, and 220 bp. The same sieving buffer as that used by Ulfelder et al.[123] was used. Another Italian research group, Gelfi et al.,[125-127] used CGE for a number of diagnostic PCR applications. In one paper,[125] an 8 bp deletion linked to congenital adrenal hyperplasia was investigated. CE separations were performed with a sieving buffer consisting of 6% linear polyacrylamide. The amplified PCR products were a normal, 135 bp fragment and a disease-linked, 127 bp fragment. In another paper,[126] the same sieving matrix was used for the detection of PCR fragments of basic fibroblast growth factor. Using a low-viscosity PA-based matrix, they also reported on mutations occurring in the dystrophin gene.[127] A CGE-based assay was developed which permitted the simultaneous analysis of Duchenne and Becker muscular dystrophies by resolving 18 DNA fragments ranging in size from ~100 to ~500 bp.

DNA microsatellites or short tandem repeats (STRs) are increasingly used as genetic marker systems in linkage studies. They are characterized by tandemly repeated, short (2 to 10 bases) sequences. A tetranucleotide repeat unit (GATT) linked to cystic fibrosis (CF) was studied by Gelfi et al.[128] The allelic forms, a hexamer (111 bp) and a heptamer (115 bp), were amplified by PCR and separated by polyacrylamide gradient slab gel electrophoresis and with CGE (6% linear PA). The hexamer was found linked to the CF-causing mutation in the gene. A sieving buffer consisting of polyacryloylaminoethoxyethanol was used to resolve PCR

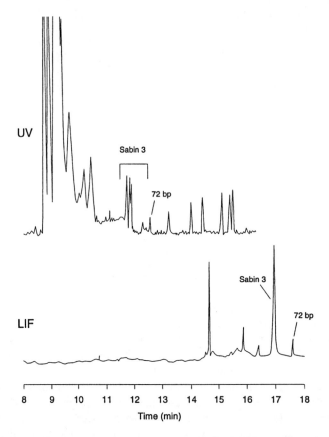

Figure 9.22 UV absorbance vs. LIF detection of the separation of a 53-bp RT-PCR product from the RNA of the Sabin-3 strain of the polio virus vaccine. A Hae III-digested φX-174 DNA marker was co-injected with the PCR product for size determination. The same Sabin-3 concentration was used for each analysis. The DNA marker concentration was 200 and 10 μg/mL for the UV and LIF analysis, respectively. (From Schwartz, H. E. et al., *J. Cap. Elec.*, 1, 36, 1994. Copyright ISC Technical Publications, Inc. With permission.)

fragments in the 450 to 550 bp range.[129] These fragments were derived from triplet (CAG) repeats in the Androgen receptor gene. An increase in the number of triplet repeats is linked to Kennedy's disease, a neurological disorder.

A number of techniques have been described to detect single point mutations in DNA.[130] Point mutation studies often require that the electrophoretic (or CE) conditions are chosen such that ds and ss DNA can be separated in one run. In denaturing gradient gel electrophoresis (DGGE), the mobility of a partially melted DNA on the slab gel is reduced compared to an unmelted molecule. A variant of DGGE, termed constant denaturant gel electrophoresis, was recently adapted in a CE format.[131] CGE was performed in capillaries containing a polymer network (6% linear PA) and a denaturant (3.3. *M* urea and 20% formamide). In a 10 cm portion of the capillary (the "denaturing zone"), the temperature was elevated; in the rest of the capillary ambient temperature conditions existed. Detection was by LIF. The critical role of temperature is illustrated in Figure 9.24. Homo- and heteroduplexes from two fluorescein-labeled DNA fragments (206 bp) originating from the human mitochondria were resolvable by tuning the temperature. The two homoduplexes differed by a single bp substitution (GC vs. AT). At 31°C, a single peak was obtained, indicating that all the duplexes were in the unmelted form. As the temperature was raised, the duplexes started to separate in the order of their melting stability. At 40°C, all the species migrated again in one single peak as their partially melted duplexes. The sensitivity of the CE method is such that 10^5 mutant species can be detected among 3×10^8 wild-type sequences.

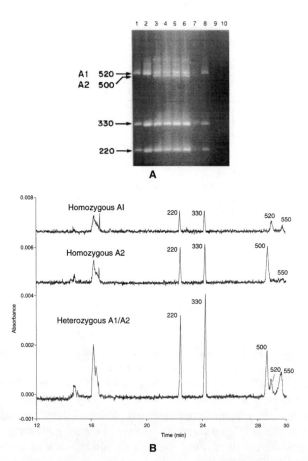

Figure 9.23 CE separation of PCR-amplified RFLP samples demonstrating MboI polymorphism. Top, homozygous for allele A1 (520-bp fragment); middle, homozygous for allele A2 (500-bp fragment); bottom, heterozygous for A1 and A2. Constant fragments of 220, 330, and 550 bp can be seen in all three runs as a result of incomplete Pvu II digestion of the 550-bp fragment. For comparison, an agarose gel is also shown. Lanes 1, 2, and 8: homozygous (A1). Lanes 3 through 6: heterozygous (A1 and A2). Lane 10: DNA size markers. (From Ulfelder, K. et al., *Anal. Biochem.*, 200, 260, 1992. With permission.)

A similar type of point mutation method, called heteroduplex polymorphism analysis (HPA), was reported by Cheng et al.[132] Duplexes and ss DNA were separated in a polymer network consisting of 0.5% HPMC and 4.8% glycerol. Ethidium bromide (3 μM) was added to increase resolution. The sensitivity of the CE method was not as high as the one discussed above, since UV detection instead of LIF was used.

Another technique for the screening of point mutations is single-strand conformation polymorphism analysis (SSCP). This technique, originally developed by Orita et al.[133] takes advantage of differences in mobilities between DNA fragments in nondenaturing gels. Point mutations in the DNA will cause conformational changes resulting in the mobility differences of ss DNA. Kuypers et al.[75] studied the p53 gene located on the short arm of chromosome 17. CGE was run on control and patient (multiple myeloma) samples using a 4% linear PA polymer network in the capillary. Denatured samples of normal DNA showed two peaks corresponding to the two ss species. The control cell line and patient samples revealed more complicated patterns of three to five peaks. Fraction collection by CE (Section III.A.4) was used to confirm the presence of different sequences in these peaks.

In the amplification refractory mutation system (ARMS) or allele-specific amplification (ASA), PCR is used to detect point mutations without requiring endonuclease digestion or Southern hybridization. The utility of CE-LIF to detect mutations in the phenylketonuria gene

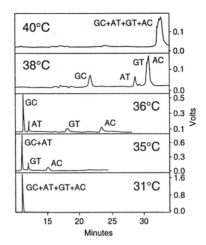

Figure 9.24 Constant denaturant CGE separation as a function of column temperature. The sample, an equimolar mixture of two homoduplexes (GC and AT) and two heteroduplexes (GT and AC), was prepared using fluorescein-labeled DNA fragments and run at the several temperatures indicated. (From Khrapko, K. et al., *Nucl. Acids Res.*, 22, 364, 1994. With permission.)

by the ARMS method was demonstrated by Arakawa et al.[134] Another paper by the same group[64] describes the utility of the CE-LIF method for diagnosis of medium-chain coenzyme A dehydrogenase (MCAD) deficiency, a disorder linked to sudden infant death and Rye-like syndrome. In most cases (90% of mutant alleles) the MCAD deficiency is caused by a single, A to G nucleotide change at position 985 in the gene. DNA fragments were amplified by two sets of allele-specific oligonucleotide primers. Mutant alleles yielded a single 175 bp fragment, normal alleles a 202 bp fragment, whereas heterozygous carriers produced both. The DNA fragments were well resolved within a 12-min run time on a capillary filled with lightly crosslinked PA. The CE-LIF method was linear over three orders of magnitude with a detection limit of ~10 ng/ml for a 603 bp DNA fragment. Compared to UV detection, LIF was 100× more sensitive.

b. Quantitation of Cellular mRNA, Competitive RNA-PCR

Accurate quantitation of PCR products — especially when dealing with low copy numbers — is often problematic. A group of methods, termed competitive PCR, have been described recently which effectively deal with the problem of accurate quantitation of PCR products.[119] In competitive PCR, a known amount of standard template DNA (the competitor) competes for the same primers with an unknown amount of target DNA. (In the case of RNA-PCR, the DNA is obtained by reverse transcription.) The competitor's sequence is chosen such that it is largely identical to the target sequence, except for the presence of a mutated restriction site or a small intron. During the amplification cycles, the target and competitor are exposed to the same PCR-related reaction variables; their product ratio should, therefore, remain constant, even after the products have reached a plateau. The amount of target DNA (or RNA) can be obtained through a simple interpolation procedure of an experimentally generated standard curve.

Fasco et al.[135] have recently demonstrated that CE-LIF is an attractive alternative to slab gel techniques: DNA fragments or PCR products, intercalated with the fluorescent dye YOYO-1 (an oxazole yellow dimer), can be detected at extremely low levels in *real time* with high efficiency and precision. CE-LIF allows accurate and precise quantitation of PCR products formed during competitive PCR reactions. With the CE method of Fasco et al.,[135] excellent peak efficiency was obtained (~10 bp resolution), and run times were less than 30 min. PCR product detectability with LIF is adequate for most clinical and diagnostic applications of competitive

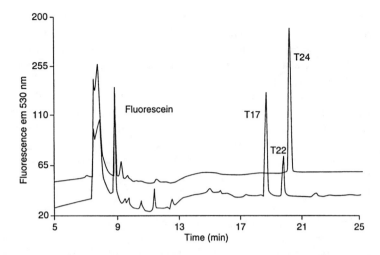

Figure 9.25 VNTR analysis of the amplified D1S80 locus for homozygous and heterozygous individuals using LIFluor dsDNA 1000 Kit and CE-LIF detection. (Courtesy of K. J. Ulfelder, Beckman Instruments.)

PCR. YOYO-1 was used in the CE run buffer for intercalating the DNA fragments. The CE-LIF method was applied to reverse-transcribed RNA from glyceraldehyde-3-phosphate dehydrogenase (GAPDH) and P4501A1 gene sequences.

Reyes-Engel and Dieguez-Lucena[136] developed procedures for quantifying mRNA. A specific biotinylated oligonucleotide was used to hybridize isolated RNA. The hybrid was captured on streptavidin magnetic beads and quantified by either CE (UV absorbance detection) or RT-PCR. The method correlated well with values obtained using a standard RT-PCR quantitative assay.

c. DNA Profiling in Forensic Science

CE-LIF has recently been applied in the analysis of genetic markers for human identification (see a review by Northrop et al.[137]). Because often extremely low levels of substances are investigated, LIF should be the detection method of choice. Therefore, several researchers have studied different polymer matrix–fluorescent dye systems to optimize separation efficiency and detectability. VNTR analysis of the amplified D1S80 locus (300 to 700 bp) is shown in Figure 9.25. D1S80 has a repeat unit of 16 bp. The EnhanCE dye was used in the polymer network-containing run buffer. The figure shows the alleles for homozygous and heterozygous individuals.

Srinivasan et al.[138] compared two asymmetric cyanine dyes, TOTO-1 and YOYO-1 (Molecular Probes, Eugene, OR) with the 488-nm Ar-ion laser. Three genetic marker systems (apolipoprotein B, 700 to 1000 bp with a 14 bp repeat; variable number tandem repeat (VNTR) locus D1S80, 300 to 700 bp with a 16 bp repeat; and mitochondrial DNA, 130 to 140 bp with a 2 bp repeat) were investigated for forensic applicability by PCR amplification. The PCR products were subsequently prestained with the fluorescent dye. (DNA was added to dye at a molar ratio of 5:1 DNA bp-to-dye and incubated for 20 min prior to analysis.) Capillaries containing easy-to-use, replaceable polymer network solutions (0.5% methylcellulose) were found superior to crosslinked, PA gels (3% C, 3% T).

McCord et al.[139] analyzed some other genetic marker systems of forensic interest, i.e., the human myelin basic protein gene, the Von Willenbrand Factor gene, and the HUMTHO1 gene located on chromosome 11. PCR-amplified alleles resulting from VNTRs with 4 bp repeat units in the 100 to 250 bp range were separated with replaceable, polymer network solutions. An asymmetrical dye, YO-PRO-1 (Molecular Probes, Eugene, OR) was added to the polymer network solution (1.0% hydroxyethylcellulose) and to each DNA sample for CE-LIF detection. A later study (Butler et al.[48]) focused on quantitative aspects of the CE-LIF method. Comparisons with other existing methods (slab gel, slot blot, fluorescence spectrophotometry) were also made. With an internal standard, peak migration time precision was <0.1% RSD. Peak area precision, using pressure injection, was ~3% RSD.

Genetic typing of STR polymorphism HUMTHO1 was reported by Wang et al.[81] CE separations were performed on arrays of five capillaries, and detection was by confocal LIF scanning (see also Section III.B.1). Target alleles were amplified using fluorescently labeled primers that fluoresce in the green region of the spectrum; unknown alleles, together with standard ladders, were amplified with different primers and designed to fluoresce in the red. This instrumental approach — which in principle could be extended to run 50 or more capillaries at the same time — greatly improves the sample throughput of the CE method.

d. DNA Profiling of Plants, Bacteria, and Fungi

Identification of bacteria in clinical samples using standard culturing techniques is both time consuming and cumbersome, because the bacteria must be grown in the laboratory and identified on the basis of nutritional development requirements. In addition, many bacteria are morphologically similar to one another. Avaniss-Aghajani et al.[140] have developed a method for identifying various bacterial species using PCR amplification of small subunit ribosomal RNA genes, which vary in sequence among bacterial species. PCR was accomplished using one set of primers, one of which was 5′-labeled with FITC. Subsequent fluorescently labeled PCR products were then subjected to digestion with restriction endonucleases, producing fragments of different length due to variations in sequence among the bacterial species. When analyzed by CE-LIF (Ar-ion laser at 488-nm), only the DNA fragments containing the terminal 5′-FITC label were detected. Length of the labeled restriction fragment was then used to identify a particular bacterial species. Figure 9.26 shows the CE analysis of PCR products for four bacterial species, after digestion with endonucleases *Msp*I and *Rsa*I. Use of this process will result in clear bacterial identification with significant time savings.

Size-selective DNA profiling and RFLP analysis of amplified polymorphic spacers originating from fungus rDNA was performed by Martin et al.[141] Inter- and intraspecific variation in the size and number of restriction sites of the amplified rDNA spacers from several fungi were examined, allowing the strains to be genotyped by CGE (UV detection). Marino et al.[142] showed results of soybean genotyping using CGE with crosslinked gels. Polymorphism was detected in dinucleotide, short tandem repeat sequences (STRs).

Kleparnik et al.[33] reported on the restriction analysis of bacteriophage DNA using CGE with agarose solutions. Seven representatives of the serogroup B *Staphylococcus aureus* bacteriophage were examined for genomic homology using DNA restriction fragment analysis. The high resolution power of CE provided an extension of the restriction patterns obtained by slab gel electrophoresis as small DNA fragments could be detected only by the CGE method. In addition, only 5 ng of DNA is required for a CE run, as opposed to 500 ng per lane with slab gels.

e. Plasmid Mapping

Restriction enzyme digestion of plasmids (plasmid mapping) is often used for confirmation of PCR products, in cloning experiments, and in biotechnology process control, e.g., to monitor genetic stability. Maschke et al. [143] employed CGE (UV detection) with replaceable, 6% linear PA gels for the mapping of four closely related plasmids. High-resolution plasmid maps were obtained with a number of different restriction enzymes. The smaller DNA fragments are better resolved by CGE than with agarose gel electrophoresis. Compared to the slab gels, CGE for plasmid mapping applications is quantitative, fast (run times <20 min), and consumes

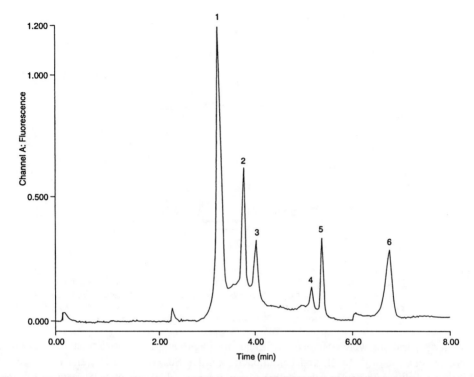

Figure 9.26 CE-LIF electropherogram of PCR-amplified SSU rRNA genes of *Flavobacterium okeanokoi-tes*, *Escherichia coli*, *Streptococcus faecalis*, and *Klebsiella pneumoneae* after digestion with MspI and RsaI. Numbered peaks correspond to the 5′ terminal restriction fragments of the digested PCR products. Peak 1: MspI digest from *F. okeanokoites*. Peak 2: RsaI digest from *E. coli*. Peak 3: MspI digest from *S. faecalis*. Peak 4: MspI digest from *K. pneumoneae*. (Electropherogram courtesy of E. Avaniss-Aghajani, UCLA.)

a minimal amount of sample for analysis. Comparisons between CGE (6% linear PA) and agarose slab gels for plasmid restriction digests were also published by Paulus and Husken.[26] In the analysis of plasmids, the polymer network concentration and/or chain length must be optimized to accommodate the larger sizes of the DNA fragments to be separated.

2. Relatively Large ds DNA (>2 kbp), Chromosomal DNA

In recombinant DNA technology, plasmid analysis is often used to control the genetic stability during fed-batch culture. Generally, the size range for plasmid analysis is larger than for PCR product analysis, i.e., 2000 to 22,000 bp. *E. coli* is the most frequently used host cell in the production of recombinant molecules. However, host cell–plasmid systems have limited genetic stability. During cell cultivation, therefore, the plasmid concentration (copy number) may decrease. Hebenbrock et al.[144] have shown that CGE is an excellent quantitative tool for monitoring the plasmid concentration during the cultivation of the *E. coli* strain containing the plasmid. The plasmid DNA concentration was estimated from the integrated peak areas of an internal standard (4363 bp) and the plasmid carrying the genetic information (a linearized, 13,000 bp restriction fragment). The CGE method used 4% linear PA polymer networks for separation of DNA fragments ranging from 3000 to 22,000 bp.

As in slab gel electrophoresis, the size range to be separated in CGE can be extended by diluting the polymer concentration in the sieving matrix. With linear PA-based gels, a 3% polymer network was found suitable to efficiently separate a 1 kbp DNA ladder (range: 72 bp to 12,216 bp), with some peaks exhibiting in excess of 4 million plates per m.[19] Similar

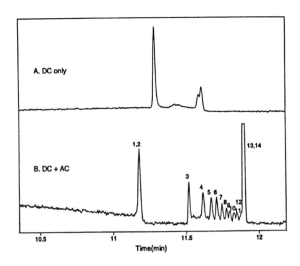

Figure 9.27 Separation of a megabase DNA standard in a dilute polymer solution (0.5× TBE containing 0.0033% hydroxyethylcellulose). (A) constant field, 100 V/cm; (B) pulsed field, 100V/cm + 18 Hz sine wave, 250% modulation. Peaks 1 through 14 represent DNA fragments in the 0.21 to 1.6 Mbp range. (From Kim, Y. and Morris, M. D., *Anal. Chem.*, 67, 784, 1995. Copyright © 1995 American Chemical Society. With permission.)

capillary separations were obtained by Strege and Lagu[35] and Baba et al.[37] on alkylcellulose polymer networks, by Bocek and Chrambach[32] on 2% SeaPrep agarose, and by Chiari et al.[145] on a polyacryloylaminoethoxyethanol replaceable gel. Pariat et al.[19] showed that low concentrations of linear PA (1.5%) can be used to further extend the size range, e.g., for the separation of λDNA Hind III restriction fragments. Other workers[10,12,84,85] have recently demonstrated that ultradilute solutions (~0.01%) of alkylcellulose can also be used for these types of separations.

Preliminary results of separations of very large DNA fragments were shown by Chrambach's group.[146] Linear PA-filled capillaries (0.1 to 0.9%) were used for three DNA-containing samples in the size range of 20 to 50 kbp, multiples of 50 kbp, and 3 to 6 Mb, respectively. However, as noted by the authors, various problems related to the sampling and the CE instrumentation renders the CGE method impractical for large DNA analyses. Pulsed-field CE in conjunction with dilute polymer solutions (see Section III.B.2)) should be a more promising approach as preliminary reports indicate.[83-86] Figure 9.27 illustrates the effect of field pulsing on the separation of large DNA molecules (up to 1.6 Mbp in size). A 0.5X TBE run buffer containing 0.0033% hydroxyethylcellulose was used to separate these large DNA molecules with good resolution in only 12 min.

REFERENCES

1. Kasper, T.J., Melera, M., Gozel, P., and Brownlee, R.G., *J. Chromatogr.*, 458, 303, 1988.
2. Cohen, A.S., Najarian, D., Smith, J.A., and Karger, B.L., Rapid separation of DNA restriction fragments using capillary electrophoresis, *J. Chromatogr.*, 458, 323, 1988.
3. Cohen, A. S., Najarian, D. R., Paulus, A., Guttman, A., Smith, J. A., and Karger, B. L., Rapid separation and purification of oligonucleotides by high-performance capillary gel electrophoresis, *Proc. Natl. Acad. Sci. USA*, 85, 9660, 1988.
4. Schwartz, H.E., Ulfelder, K.J., Chen, F.-T.A., and Pentoney, S.L., Jr., The utility of laser-induced fluorescence detection in applications of capillary electrophoresis, *J. Cap. Elec.*, 1, 36, 1994.
5. Schwartz, H.E. and Ulfelder, K., Capillary electrophoresis with laser-induced fluorescence detection of PCR fragments using thiazole orange, *Anal. Chem.*, 64, 1737, 1992.

5a. Everaerts, F. M., Beckers, J. L., and Verheggen, T. P. E. M. *Isotachophoresis: Theory, Instrumentation and Applications*, Elsevier, Amsterdam, 1976.

6. Zhao, Z., Wahl, J.H., Udseth, H.R., Hofstadler, S.A., Fuciarelli, A.F., and Smith, R.D., Online capillary electrophoresis-electrospray ionization mass spectrometry of nucleotides, *Electrophoresis,* 16, 389, 1995.

7. Stellwagen, N.C., in *Advances in Electrophoresis,* Vol. 1, Chrambach, A., Dunn, M.J., and Radola, B.J. (Eds.), VCH Publishers, New York, 1987.

8. Slater, G.W. and Noolandi, J., The biased reptation model of DNA gel electrophoresis: mobility vs. molecular size and gel concentration, *Biopolymers,* 28, 1781, 1989.

9. Smisek, D.L. and Hoagland, D.A., Electrophoresis of flexible macromolecules: evidence for a new mode of transport in gels, *Science,* 248, 1221, 1990.

10. Barron, A.E., Blanch, H.W., and Soane, D.S., A transient entanglement coupling mechanism for DNA separation by capillary electrophoresis in ultradilute polymer solution, *Electrophoresis,* 15, 597, 1994.

11. Smith, S.B., Aldridge, P.K., and Callis, J.B., Observation of individual DNA molecules undergoing gel electrophoresis, *Science,* 243, 203, 1989.

12. Barron, A.E., Sunada, W.M., and Blanch, H.W., The use of coated and uncoated capillaries for the electrophoretic separation of DNA in dilute polymer solutions, *Electrophoresis,* 16, 64, 1995.

13. Guttman, A., Cohen, A.S., Heiger, D.N., and Karger, B.L., Analytical and micropreparative ultrahigh resolution of oligonucleotides by polyacrylamide gel high-performance capillary electrophoresis, *Anal. Chem.,* 62, 137, 1990.

14. Schomburg, G., *Capillary Electrophoresis: Theory and Practice*, Camilleri, P. (Ed.), CRC Press, Boca Raton, FL, 1993, p. 255.

15. Barron, A.E., Soane, D.S., and Blanch, H.W., Capillary electrophoresis of DNA in uncrosslinked polymer solutions, *J. Chromatogr.,* 652, 3, 1993.

16. Bae, Y.C. and Soane, D., Polymeric separation media for electrophoresis: cross-linked systems or entangled solutions, *J. Chromatogr.,* 652, 17, 1993.

17. Baba, Y. and Tsuhako, M., Gel-filled capillaries for nucleic acid separations in capillary electrophoresis, *Trends Anal. Chem.,* 11, 280, 1992.

18. Guttman, A., *Handbook of Capillary Electrophoresis*, Landers, J.P. (Ed.), CRC Press, Boca Raton, FL, 1994, p. 129.

19. Pariat, Y.F., Berka, J., Heiger, D.N., Schmitt, T., Cohen, A.S., Foret, F., and Karger, B.L., Separation of DNA fragments by capillary electrophoresis using replaceable linear polyacrylamide matrixes, *J. Chromatogr.,* 652, 57, 1993.

20. Kenndler, E. and Poppe, H., Entangled, noncross-linked polymer solutions used as physical network in capillary zone electrophoresis, *J. Cap. Elec.,* 1, 144, 1994.

21. Mitnik, L., Laurence, S., Viovy, J. L., Heller, C., Systematic study of field and concentration effects in capillary electrophoresis of DNA in polymer solutions, *J. Chromatogr. A,* 710(2), 309, 1995.

22. Ruiz-Martinez, M. C., Berka, J., Belenki, A., Foret, F., Miller, A.W., and Karger, B.L., DNA sequencing by capillary electrophoresis with replaceable linear polyacrylamide and laser-induced fluorescence detection, *Anal. Chem.,* 65, 2851, 1993.

23. Guttman, A., Nelson, R.J., and Cooke, N., Prediction of migration behavior of oligonucleotides in capillary gel electrophoresis, *J. Chromatogr.,* 593, 297, 1992 A.

24. Kleemiss, M.H., Gilges, M., and Schomburg, G., Capillary electrophoresis of DNA restriction fragments with solutions of entangled polymers, *Electrophoresis,* 14, 515, 1993.

25. Guttman, A., Shieh, P., Hoang, D., Horvath, J., and Cooke, N., Effect of operational variables on the separation of proteins by capillary sodium dodecyl sulfate-gel electrophore, *Electrophoresis,* 15, 221, 1994.

26. Paulus, A. and Husken, D., DNA digest analysis with capillary electrophoresis, *Electrophoresis,* 14, 27, 1993.

27. Berka, J., Pariat, Y.F., Muller, O., Hebenbrock, K., Heiger, D.N., Foret, F., and Karger, B.L., Sequence dependent migration behavior of double-stranded DNA in capillary electrophoresis, *Electrophoresis,* 16, 377, 1995.

27a. Ferguson, K.A., *Metab. Clin. Exp.,* 13, 985, 1964.

28. Heiger, D.N., Cohen, A.S., and Karger, B.L., Separation of DNA restriction fragments by high performance capillary electrophoresis with low and zero crosslinked polyacrylamide using continuous and pulsed electric fields, *J. Chromatogr.*, 516, 33, 1990.

29. Schwartz, H.E., Ulfelder, K.J., Sunzeri, F.J., Busch, M.P., and Brownlee, R.G., Analysis of DNA restriction fragments and polymerase chain reaction products towards detection of the AIDS (HIV-1) virus in blood, *J. Chromatogr.*, 559, 267, 1991.

30. Compton, S.W. and Brownlee, R.G., Capillary electrophoresis, *Biotechniques*, 6, 432, 1988.

31. Bocek, P. and Chrambach, A., Capillary electrophoresis of DNA in agarose solutions at 40°C, *Electrophoresis*, 12, 1059, 1991.

32. Bocek, P. and Chrambach, A., Capillary electrophoresis in agarose solutions: extension of size separations to DNA of 12 kb in length, *Electrophoresis*, 13, 31, 1992.

33. Kleparnik, K., Mala, Z., Doskar, J., Rosypal, S., and Bocek, P., An improvement of restriction analysis of bacteriophage DNA using capillary electrophoresis in agarose solution, *Electrophoresis*, 16, 366, 1995.

34. Upcroft, P. and Upcroft, J.A., Comparison of properties of agarose for electrophoresis of DNA, *J. Chromatogr.*, 618, 79, 1993.

34a. Chang, H-I. and Yeung, E.S., Polyethylene oxide for high resolution and high speed separation of DNA by capillary electrophoresis, *J. Chromatogr. B*, 669, 113, 1995.

35. Strege, M. and Lagu, A., Separation of DNA restriction fragments by capillary electrophoresis using coated fused silica capillaries, *Anal. Chem.*, 63, 1233, 1991.

36. Bode, H.-J., The use of liquid polyacrylamide in electrophoresis. II. Relationship between gel viscosity and molecular sieving, *Anal. Biochem.*, 83, 364, 1977.

37. Baba, Y., Ishimaru, N., Samata, K., Tsuhako, M., High-resolution separation of DNA restriction fragments by capillary electrophoresis in cellulose derivative solutions, *J. Chromatogr. A*, 653, 329, 1993.

38. Heiger, D.N. and Majors, R.E., Capillaries and chemistries for capillary electrophoresis, *LC.GC Magazine*, 13, 12, 1995.

39. Huang, M., Liu, S., Murray, B.K., and Lee, M.L., High resolution separation and quantitation of ribonucleotides using capillary electrophoresis, *Anal. Biochem.*, 207, 231, 1992.

40. Grossmann, P.D. and Soane, D.S., Capillary electrophoresis of DNA in entangled polymer solutions, *J. Chromatogr.*, 559, 257, 1991.

41. Ornstein, L., Disc electrophoresis. I. Background and theory, *Ann. N.Y. Acad. Sci.*, 121, 321, 1964.

42. Chien, R.L. and Burgi, D.S., On-column sample concentration using field amplification in CZE, *Anal. Chem.*, 64, 489A, 1992.

43. Albin, M., Grossman, P.D., and Moring, S.E., Sensitivity enhancement for capillary electrophoresis, *Anal. Chem.*, 65, 489A, 1993.

44. Abersold, R. and Morrison, H.D., Analysis of dilute peptide samples by capillary zone electrophoresis, *J. Chromatogr.*, 516, 79, 1990.

45. Foret, F., Sustacek, V., and Bocek, P., On-line isotachophoretic sample preconcentration for enhancement of zone detectability in capillary zone electrophoresis, *J. Microcol. Sep. 2*, 299, 1990.

46. Foret, F., Szoko, E., and Karger, B.L., *Application Information A-1740*, Beckman Instruments, Inc., Fullerton, CA, 1993.

47. Butler, J.M., McCord, B.R., Jung, J.M., Wilson, M.R., Budowle, B., and Allen, R.O., Quantitation of polymerase chain reaction products by capillary electrophoresis using laser fluorescence, *J. Chromatogr. B*, 658, 271, 1994.

48. Butler, J.M., McCord, B.R., Jung, J.M., Wilson, M.R., Budowle, B., and Allen, R.O., Quantitation of polymerase chain reaction products by capillary electrophoresis using laser fluorescence, *J. Chromatogr. B*, 658, 271, 1994.

49. van der Schans, M.J., Allen, J.K., Wanders, B.J., and Guttman, A., Effects of sample matrix and injection plug on dsDNA migration in capillary gel electrophoresis, *J. Chromatogr. A*, 680, 511, 1994.

50. Srivatsa, G.S., Batt, M., Schuette, J., Carlson, R.H., Fitchett, J., Lee, C., and Cole, D.L., Quantitative capillary gel electrophoresis assay of phosphorothioate oligonucleotides in pharmaceutical formulations, *J. Chromatogr. A*, 680, 469, 1994.

51. Guttman, A. and Schwartz, H.E., Artifacts related to sample introduction in capillary gel electrophoresis affecting separation performance and quantitiation, *Anal. Chem.*, 67, 2279, 1995.

52. Pentoney, S.L., Jr. and Sweedler, J.V., *Handbook of Capillary Electrophoresis*, Landers, J.P. (Ed.), CRC Press, Boca Raton, FL, 1994, p. 147.

53. Kim, Y. and Morris, M.D., Separation of nucleic acids by capillary electrophoresis in cellulose solutions with mono- and bisintercalating dyes, *Anal. Chem.*, 66, 1168, 1994 B.

54. Jansson, M., Roeraarde, J., and Laurell, F., Laser-induced fluorescence detection in capillary electrophoresis with blue light from a frequency-doubled diode laser, *Anal. Chem.*, 65, 2766, 1993.

55. Chen, F.-T.A., Tusak, A., Pentoney, Jr., S., Konrad, K., Lew, C., Koh, E., and Sternberg, J., Semiconductor laser-induced fluorescence detection in capillary electrophoresis using a cyanine dye, *J. Chromatogr.*, 652, 355, 1993.

56. Milofsky, R.E. and Yeung, E.S., Native fluorescence detection of nucleic acids and DNA restriction fragments in capillary electrophoresis, *Anal. Chem.*, 65, 153, 1993.

57. McGregor, D.A. and Yeung, E.S., Detection of DNA fragments separated by capillary electrophoresis based on their native fluorescence inside a sheath flow, *J. Chromatogr. A*, 680, 491, 1994.

58. Kuhr, W.G. and Yeung, E.S., Optimization of sensitivity and separation in capillary zone electrophoresis with indirect fluorescence detection, *Anal. Chem.*, 60, 2642, 1988.

59. Chan, K.C., Whang, C.-W., and Yeung, E.S., Separation of DNA restriction fragments using capillary electrophoresis, *J. Liq. Chromatogr.*, 16, 1941, 1993.

60. Glazer, A.N. and Rye, H.S., Stable dye-DNA intercalation complexes as reagents for high-sensitivity fluorescence detection, *Nature*, 359, 859, 1992.

61. Ulfelder, K.J., *Application Information Bulletin A-1748A*, Beckman Instruments, Inc., Fullerton, CA, 1993.

62. Rossomando, E.F., White, L., and Ulfelder, K.J., Capillary electrophoresis: separation and quantitation of reverse transcriptase polymerase chain reaction products from polio virus, *J. Chromatogr. B*, 656, 159, 1994.

63. Zhu, H., Clark, S.M., Benson, S.C., Rye, H.S., Glazer, A.N., and Mathies, R.A., High-sensitivity capillary electrophoresis of double-stranded DNA fragments using monomeric and dimeric fluorescent intercalating dyes, *Anal. Chem.*, 66, 1941, 1994.

64. Arakawa, H., Uetanaka, K., Maeda, M., Tsuji, Matsubara, Y., Narisawa, K., Analysis of polymerase chain reaction-product by capillary electrophoresis with laser-induced fluorescence detection and its application to the diagnosis of medium-chain acyl-coenzyme A dehydrogenase deficiency, *J. Chromatogr. A*, 680, 517, 1994 B.

65. Figeys, D., Arriaga, E., Renborg, A., and Dovichi, N.J., Use of the fluorescent intercalating dyes POPO-3, YOYO-3, and YOYO-1 for ultrasensitive detection of double-stranded DNA separated by capillary electrophoresis with hydroxypropylmethyl cellulose and non-crosslinked polyacrylamide, *J. Chromatogr. A*, 669, 205, 1994.

66. Dovichi, N., *Handbook of Capillary Electrophoresis*, Landers, J.P. (Ed.), CRC Press, Boca Raton, FL, 1994, p. 369.

67. Lee, T., Yeung, E.S., and Sharma, M., Micellar electrokinetic capillary chromatographic separation and laser-induced fluorescence detection of 2′-deoxynucleoside 5′-monophosphate of normal and modified bases, *J. Chromatogr.*, 565, 197, 1991.

68. Li, W., Moussa, A., and Giese, R.W., Capillary electrophoresis with laser fluorescence detection for profiling damage to fluorescein-labeled deoxyadenylic acid by background, ionizing radiation and hydrogen peroxide, *J. Chromatogr.*, 633, 315, 1993.

69. Wang, P. and Giese, R.W., Phosphate-specific fluorescence labeling under aqueous conditions, *Anal. Chem.*, 65, 3518, 1993.

70. Tseng, H.C., Dadoo, R., and Zare, R.N., Selective determination of adenine-containing compounds by capillary electrophoresis with laser-induced flourescence detection, *Anal. Biochem.*, 222, 55, 1994.

71. Krattiger, B., Bruno, A.E., Widmer, H.M., and Dandliker, R., Hologram-based thermooptical absorbance detection in capillary electrophoresis: separation of nucleosides and nucleotides, *Anal. Chem.*, 67, 124, 1995.

72. Cai, J. and Henion, J., Capillary electrophoresis-mass spectrometry, *J. Chromatogr. A*, 703, 667, 1995.

73. Wolf, S.M., Vouros, P., Norwood, C., and Jackim, E., Identification of deoxynucleoside-polyaromatic hydrocarbon adducts by capillary zone electrophoresis-continuous flow-fast atom bombardment mass spectrometry, *J. Am. Soc. Mass. Spectrom.*, 3, 757, 1992.

74. Bayer, E., Bauer, T., Schmeer, K., Bleicher, K., Maier, M., and Gaus, H.-J., Analysis of double-stranded oligonucleotides by electrospray mass spectrometry, *Anal. Chem.*, 66, 3858, 1994.

75. Kuypers, A.W.H.M., Willems, P.M.W., van der Schans, M.J., Linssen, P.C.M., Wessels, H.M.C., de Bruijn, C.H.M.M., Everaerts, F.M., and Mensink, E.J.B.M., Detection of point mutations in DNA using capillary electrophoresis in a polymer network, *J. Chromatogr.*, 621, 149, 1993.

76. Zagursky, R.J. and McCormick, R.M., DNA sequencing separations in capillary gels on a modified commercial DNA sequencing instrument, *BioTechniques*, 9, 74, 1990.

77. Mathies, R.A. and Huang, X.C., Capillary array electrophoresis: an approach to high-speed, high-throughput DNA sequencing, *Nature*, 359, 167, 1992.

78. Ueno, K. and Yeung, E.S., Simultaneous monitoring of DNA fragments separated by electrophoresis in a multiplexed array of 100 capillaries, *Anal. Chem.*, 66, 1424, 1994.

79. Clark, S.M. and Mathies, R.A., High-speed parallel separation of DNA restriction fragments using capillary array electrophoresis, *Anal. Biochem.*, 215, 163, 1993.

80. Takahashi, S., Murakami, K., Anazawa, T., and Kambara, H., Multiple sheath-flow gel capillary-array electrophoresis for multicolor fluorescent DNA detection, *Anal. Chem.*, 66, 1021, 1994.

81. Wang, Y., Ju, J., Carpenter, B.A., Atherton, J.M., Sensabaugh, G.F., and Mathies, R.A., Rapid sizing of short tandem repeat alleles using capillary array electrophoresis and energy-transfer fluorescent primers, *Anal. Chem.*, 67, 1197, 1995.

82. Gardiner, K., Pulsed field gel electrophoresis, *Anal. Chem.*, 63, 658, 1991.

83. Sudor, J. and Novotny, M.V., Separation of large DNA fragments by capillary electrophoresis under pulsed-field conditions, *Anal. Chem.*, 66, 2446, 1994.

84. Kim, Y. and Morris, M.D., Pulsed field capillary electrophoresis of multikilobase length nucleic acids in dilute methyl cellulose solutions, *Anal. Chem.*, 66, 3081, 1994 A.

85. Kim, Y. and Morris, M.D., Rapid pulsed field capillary electrophoresis separation of megabase nucleic acids, *Anal. Chem.*, 67, 784, 1995.

86. Heller, C., Pakleza, C., and Viovy, J.L., *Poster P-427, Seventh International Symposium on High Performance Capillary Electrophoresis*, Wurzburg, Germany, 1995.

87. Wrotnowski, C., Biochip technology offers powerful tools for research and diagnostics, *Genetic Engineering News*, November 15, 1994, p. 8.

88. Effenhauser, C.S., Paulus, A., Manz, A., and Widmer, H.M., High-speed separation of antisense oligonucleotides on a micromachined capillary electrophoresis device, *Anal. Chem.*, 66, 2949, 1994.

89. Cadet, J. and Weinfeld, M., Detecting DNA damage, *Anal. Chem.*, 65, 675A, 1993.

90. Norwood, C.B., Jackim, E., and Cheer, S., DNA adduct research with capillary electrophoresis, *Anal. Biochem.*, 213, 194, 1993.

91. Guarnieri, C., Muscari, C., Stefanelli, C., Giaccari, A., Zini, M., and Di Biase, S., Micellar electrokinetic capillary chromatography of 8-hydroxydeoxyguanosine and other oxidized derivatives of DNA, *J. Chromatogr. B*, 656, 209, 1994.

92. Lloyd, D.K., Cypess, A.M., and Wainer, I.W., Determination of cytosine-β-D-arabinoside in plasma using capillary electrophoresis, *J. Chromatogr.*, 568, 117, 1991.

93. Rogan, M.M., Drake, D.M., Goodall, D.M., and Altria, K.D., Enantioselective enzymic biotransformation of 2'-deoxy-3'-thiacytidine (BCH 189) monitored by capillary electrophoresis, *Anal. Biochem.*, 208, 343, 1993.

94. Nguyen, A.-L., Luong, J.H.T., and Masson, C., Determination of nucleotides in fish tissues using capillary electrophoresis, *Anal. Chem.*, 62, 2490, 1990.

95. Ng, M., Blaschke, T.F., Arias, A.A., and Zare, R.N., Analysis of free intracellular nucleotides using high-performance capillary electrophoresis, *Anal. Chem.*, 64, 1682, 1992.

96. Huang, X., Shear, J.B., and Zare, R.N., Quantitation of ribonucleotides from base-hydrolyzed RNA using capillary zone electrophoresis, *Anal. Chem.*, 62, 2049, 1990.

97. Shao, X., O'Neill, K., Zhao, Z., Anderson, S., Malik, A., and Lee, M., Analysis of nucleotide pools in human lymphoma cells by capillary electrophoresis, *J. Chromatogr. A*, 680, 463, 1994.

98. Takigiku, R. and Schneider, R.E., Reproducibility and quantitation of separation for ribonucleoside triphosphates and deoxyribonucleoside triphosphates by capillary zone electrophoresis, *J. Chromatogr.*, 559, 247, 1991.

99. Perrett, D. and Ross, G.R., *Human Purine and Pyrimidine Metabolism in Man, Vol. 7, Part B*, Harkness, R.A. (Ed.), Plenum Press, New York, 1991, p. 1.

100. Ramsey, R.S., Kerchner, G.A., and Cadet, J., Micellar electrokinetic capillary chromatography of nucleic acid constituents and dinucleoside monophosphate photoproducts, *HRC*, 17, 4, 1994.

101. Loregian, A., Scremin, C., Schiavon, M., Marcello, A., and Palu, G., Quantitative analysis of ribonucleotide triphosphates in cell extracts by high-performance liquid chromatography and micellar electrokinetic capillary chromatography: a comparative study, *Anal. Chem.*, 66, 2981, 1994.

102. Warren, W.J. and Vella, G., Analysis of synthetic oligodeoxyribonucleotides by capillary gel electrophoresis and anion-exchange HPLC, *Biotechniques*, 14, 598, 1993.

103. Rose, D.J., Characterization of antisense binding properties of peptide nucleic acids by capillary gel electrophoresis, *Anal. Chem.*, 65, 3545, 1993.

104. Pentoney, S.L., Jr., Konrad, K.D., and Kaye, W., A single-fluor approach to DNA sequence determination using high performance capillary electrophoresis, *Electrophoresis*, 13, 467, 1992.

105. Kim, S., Yoo, H.J., and Hahn, J.H., Fast capillary-scanning system for detecting fluorescently labeled DNA sequencing fragments separated by capillary gel electrophoresis, *Chromatographia*, 40, 345, 1995.

106. Manabe, T., Chen, N., Terabe, S., Yohda, M., and Endo, I., Effects of linear polyacrylamide concentrations and applied voltages on the separation of oligonucleotides and DNA sequencing fragments by capillary electrophoresis, *Anal. Chem.*, 66, 4234, 1994.

107. Best, N., Arriaga, E., Chen, D.Y., and Dovichi, N.J., Separation of fragments up to 570 bases in length by use of 6% T non-cross-linked polyacrylamide for DNA sequencing in capillary electrophoresis, *Anal. Chem.*, 66, 4063, 1994.

108. Fung, E.N. and Yeung, E.S., High-speed DNA sequencing by using poly(ethylene oxide) solutions in uncoated capillary columns, *Anal. Chem.*, 67, 1913, 1995.

109. Brownlee, R.G., Sunzeri, F.J., and Busch, M.P., Application of capillary DNA chromatography to detect AIDS virus (HIV-1) DNA in blood, *J. Chromatogr.*, 533, 87, 1990.

110. Sunzeri, F.J., Lee, T.-H., Brownlee, R.G., and Busch, M.P., Rapid simultaneous detection of multiple retroviral DNA sequences using the polymerase chain reaction and capillary DNA chromatography, *Blood*, 77, 879, 1991.

111. Bianchi, N., Mischiati, C., Ferriotto, G., and Gambari, R., Polymerase-chain reaction: analysis of DNA/DNA hybridization by capillary electrophoresis, *Nucleic Acids Res.*, 21, 3595, 1993.

112. Chen, J.W., Cohen, A.S., and Karger, B.L., Identification of DNA molecules by pre-column hybridization using capillary electrophoresis, *J. Chromatogr.*, 559, 295, 1991.

113. Vilenchik, M., Belenky, A., and Cohen, A.S., Monitoring and analysis of antisense DNA by high-performance capillary gel electrophoresis, *J. Chromatogr. A*, 663, 105, 1994.

114. Ulfelder, K.J., Dobbs, M., and Liu, M.-S., *Poster P-224, Seventh International Symposium on High Performance Capillary Electrophoresis*, Wurzburg, Germany, 1995.

115. Heegaard, N.H.H. and Robey, F.A., Use of capillary zone electrophoresis for the analysis of DNA-binding to a peptide derived from amyloid P component, *J. Liq. Chromatogr.*, 16, 1923, 1993.

116. Maschke, H.E., Frenz, J., Williams, M., and Hancock, W.S., *Poster T121, Fifth International Symposium on High Performance Capillary Electrophoresis*, Orlando, FL, 1993.

117. Zhu, M.D., Hansen, D.L., Burd, S., and Gannon, F., Factors affecting free zone electrophoresis and isoelectric focusing in capillary electrophoresis, *J. Chromatogr.*, 480, 311, 1989.

118. Guttman, A. and Cooke, N., Capillary gel affinity electrophoresis of DNA fragments, *Anal. Chem.*, 63, 2038, 1991 A.

119. Piatak, M., Jr., Luk, K.-C., Williams, B., and Lifson, J.D., Quantitative competitive polymerase chain reaction for accurate quantitation of HIV-DNA and RNA species, *Biotechniques,* 14, 70, 1993.

120. Mayer, A., Sunzeri, F., Lee, T.-H., and Busch, M.P., Separation and detection of DNA polynucleotides using capillary electrophoresis: application to detection of polymerase chain reaction-amplified human immunodeficiency virus and HLA DNA, *Arch. Pathol. Lab. Med.,* 115, 1228, 1991.

121. Schwartz, H.E., Melera, M., and Brownlee, R.G., Performance of an automated injection and replenishment system for capillary electrophoresis, *J. Chromatogr.,* 480, 129, 1989.

122. Lu, W., Han, D.-S., Yuan, J., and Andrieu, J.-M., Multi-target PCR analysis by capillary electrophoresis and laser-induced fluorescence, *Nature,* 368, 269, 1994.

123. Ulfelder, K., Schwartz, H.E., Hall, J.M., and Sunzeri, F.J., Restriction fragment length polymorphism analysis of ERBB2 oncogene by capillary electrophoresis, *Anal. Biochem.,* 200, 260, 1992.

124. Del Principe, D., Iampieri, M.P., Germani, D., Menichelli, A., Novelli, G., and Dallapiccola, B., Detection by capillary electrophoresis of restriction fragment length polymorphism. Analysis of a polymerase chain reaction-amplified product of the DXS 164 locus in the dystrophin gene, *J. Chromatogr.,* 638, 277, 1993.

125. Gelfi, C., Orsi, A., Righetti, P.G., Zanussi, M., Carrera, P., and Ferrari, M., Capillary zone electrophoresis in polymer networks of polymerase chain reaction-amplified oligonucleotides: the case of congenital adrenal hyperplasia, *J. Chromatogr. B,* 657, 201, 1994 A.

126. Gelfi, C., Leoncini, F., Righetti, P.G., Cremonesi, L., di Blasio, A.M., Carniti, C., and Vignali, M., Separation and quantitation of reverse transcriptase-polymerase chain reaction fragments of basic fibroblast growth factor by capillary electrophoresis in polymer networks, *Electrophoresis,* 16, 780, 1995.

127. Gelfi, C., Orsi, A., Leoncini, F., Righetti, P.G., Spiga, I., Carrera, P., and Ferrari, M., Amplification of 18 dystrophin gene exons in DMD/BMD patients: simultaneous resolution by capillary electrophoresis in sieving liquid polymers, *BioTechniques,* 19, 254, 1995.

128. Gelfi, C., Orsi, A., Righetti, P.G., Brancolini, V., Cremonesi, L., and Ferrari, M., Capillary zone electrophoresis of polymerase chain reaction-amplified DNA fragments in polymer networks: the case of GATT microsatellites in cystic fibrosis, *Electrophoresis,* 15, 640, 1994 B.

129. Nesi, M., Righetti, P.G., Patrosso, M.C., Ferlini, A., and Chiari, M., Capillary electrophoresis of polymerase chain reaction-amplified products in polymer networks: the case of Kennedy's disease, *Electrophoresis,* 15, 644, 1994.

130. Perucho, M., *The Polymerase Chain Reaction,* Mullis, K.B., Ferre, F., and Gibbs, R.A. (Eds.), Birkhauser, Boston, 1994, p. 369.

131. Khrapko, K., Hanekamp, J.S., Thilly, W.G., Foret, F., and Karger, B.L., Constant denaturant capillary electrophoresis (CDCE): a high resolution approach to mutational analysis, *Nucleic Acids Res.,* 22, 364, 1994.

132. Cheng, J., Kasuga, T., Mitchelson, K.R., Lightly, E.R.T., Watson, N.D., Martin, W.J., and Atkinson, D., Polymerase chain reaction heteroduplex polymorphism analysis by entangled solution capillary electrophoresis, *J. Chromatogr. A,* 677(1) 169, 1994.

133. Orita, M., Iwahana, H., Kanazawa, H., Hayashi, K., and Sekiya, T., Detection of polymorphisms of human DNA by gel electrophoresis as single-strand conformation polymorphisms, *Proc. Natl. Acad. Sci. USA,* 86, 2766, 1989.

134. Arakawa, H., Uetanaka, K., Maeda, M., and Tsuji, A., Analysis of polymerase chain reaction product by capillary electrophoresis and its application to the detection of single base substitution in genes, *J. Chromatogr. A, 664,* 89, 1994 A.

135. Fasco, M.J., Treanor, C.P., Spivack, S., Figge, H.L., and Kaminsky, L.S., Quantitative RNA-polymerase chain reaction-DNA analysis by capillary electrophoresis and laser-induced fluorescence, *Anal. Biochem.,* 224(1), 140, 1995.

136. Reyes-Engel, A. and Dieguez-Lucena, J.L., Direct quantitation of specific mRNA using a selected biotinylated oligonucleotide by free solution capillary electrophoresis, *Nucleic Acids Res.,* 21, 759, 1993.

137. Northrop, D.M., McCord, B.R., and Butler, J.M., Forensic applications of capillary electrophoresis, *J. Cap. Elec.,* 1, 158, 1994.

138. Srinivasan, K., Girard, J.E., Williams, P., Roby, R.K., Weedn, V.W., Morris, S.C., Kline, M.C., and Reeder, D.J., Electrophoretic separations of polymerase chain reaction-amplified DNA fragments in DNA typing using a capillary electrophoresis-laser induced fluorescence system, *J. Chromatogr. A,* 652, 83, 1993.

139. McCord, B.R., McClure, D.L., and Jung, J.M., Capillary electrophoresis of polymerase chain reaction-amplified DNA using fluorescence detection with an intercalating dye, *J. Chromatogr A,* 652, 75, 1993.

140. Avaniss-Aghajani, E., Jones, K., Chapman, D., and Brunk, C., A molecular technique for identification of bacteria using small subunit ribosomal RNA sequences, *Biotechniques,* 17, 144, 1994.

141. Martin, F., Vairelles, D., and Henrion, B., Automated ribosomal DNA fingerprinting by capillary electrophoresis of PCR products, *Anal. Biochem.,* 214, 182, 1993.

142. Marino, M.A., Turni, L.A., Del Rio, S.A., and Williams, P.E., Molecular size determinations of DNA restriction fragments and polymerase chain reaction products using capillary gel electrophoresis, *J. Chromatogr. A,* 676, 185, 1994.

143. Maschke, H.E., Frenz, J., Belenkii, A., Karger, B.L., and Hancock, W., Ultrasensitive plasmid mapping by high performance capillary electrophoresis, *Electrophoresis,* 14, 509, 1993.

144. Hebenbrock, K., Schugerl, K., and Freitag, R., Analysis of plasmid-DNA and cell protein of recombinant *Escherichia coli* using capillary gel electrophoresis, *Electrophoresis,* 14, 753, 1993.

145. Chiari, M., Nesi, M., and Righetti, P.G., Capillary zone electrophoresis of DNA fragments in a novel polymer network: poly(N-acryloylaminoethoxyethanol), *Electrophoresis,* 15, 616, 1994.

146. Guszczynski, T., Pulyaeva, H., Tietz, D., Garner, M.M., and Chrambach, A., Capillary zone electrophoresis of large DNA, *Electrophoresis,* 14, 523, 1993.

CHAPTER 10

Capillary Electrophoresis in Biomedical and Pharmaceutical Research

David Perrett

CONTENTS

I. BACKGROUND

The demands placed by present day biomedical research on bioanalysts for rapid qualitative and quantitative analyses of increasingly smaller amounts of more complex samples is being met by a number of rapidly improving techniques. The late 1980s and early 1990s witnessed major advances not only in separation science, but also nuclear magnetic resonance spectroscopy and mass spectrometry as well as continuing improvements in areas such as immunoassay. Capillary electrophoresis and its derivative technologies, such as micellar electrokinetic chromatography (MEKC) and capillary electrochromatography (CEC), have or are becoming important tools to all workers performing analyses in biomedical and pharmaceutical environments.

The accurate and quantitative analysis of exogenous compounds, such as drugs, in biofluids is important to those studying drug metabolism or pharmacokinetics; clinical chemists often wish to quantify related groups of endogenous compounds such as amino acids, while others need to profile samples for many compounds, some of which may be unknown. Bioanalysts must balance many conflicting requirements when devising an assay for one or more analytes in biological samples. Goals common to most analytical methods are the optimization of sensitivity, sample throughput (speed), and obtaining the necessary selectivity and/or resolution for the samples and analytes of interest. Additionally, the analysis of biofluids for drugs, and more particularly for endogenous small molecules, is often complicated by the complex matrices in which the analytes may be dissolved, the often limited amounts of sample available, e.g., extracts of tissue biopsies and pediatric blood samples, the trace concentrations of some analytes in biological fluids, and the diversity of related groups of naturally occurring compounds. For example HPLC reveals at least 40 UV absorbing "purines" in human plasma with a many-fold variation in peak heights.

Although elegant chromatographic separations, e.g., amino acid chromatography, were developed during the 1950s, until the early 1970s chromatography and electrophoresis occupied equal standing in the biomedical laboratory, although neither technique was in routine use, except possibly the electrophoresis of serum proteins in clinical specimens and the amino acid analyzer for the determination of the amino acid composition of proteins. From the late 1970s biomedical separations, except for some macromolecules, increasingly relied upon HPLC using microparticulate silica. Using today's best technology, HPLC can achieve excellent separations since although efficiencies are only moderate (e.g., N < 10,000 plates per 10 cm column), high selectivity can be achieved because of the large number of combinations of stationary phases, eluents, and detectors now available. Using gradient elution, large numbers of related compounds can be resolved. On-line quantitation, as well as the possibilities of sample collection, are further benefits. The scale of use of HPLC hides the fact that it is not always chemically or physically the most appropriate method to employ, particularly for many biomolecules.

Excepting biopolymers that are relatively difficult to resolve chromatographically, "traditional electrophoresis," with its low efficiency toward small molecules, poor reproducibility, and poor quantitation, became a *Cinderella* technique. A few attempts were made to commercialize isotachophoresis, but they were in the main unsuccessful. Electrophoresis for protein separations, such as profiling serum proteins, and for oligonucleotides, such as DNA fragments, continued to be an important technique because quantitation was often not called for and many samples could be run in parallel. However the potential of electrophoresis for high resolution separations was always appreciated since the highest resolution achieved by any separation medium available was the two-dimensional separation of proteins and peptides using polyacrylamide gel electrophoresis plus isoelectric focusing.

As in the previous edition this chapter concentrates on the separation of small molecules, including drugs, in biofluids. After some general observations concerning the utility of CE for biological separations, it addresses how CE is being used in practice for a range of compound types relevant to bioanalysis. The separation of macromolecules will be mentioned only where relevant since detailed chapters on their separation are to be found elsewhere in this volume. The same proviso applies to other small molecules when assayed under controlled sample conditions, such as the separation of pharmaceuticals in quality control situations.

II. APPLICABILITY OF CAPILLARY ELECTROPHORESIS TO BIOLOGICAL SAMPLES

Most endogenous biomolecules are ionic or ionizable, whereas many drugs are relatively uncharged but frequently form ionic metabolites such as glucuronides. HPLC is not ideal for the separation of charged species. Ion-exchange chromatography (IEC) should be the appropriate separation mode, but traditional IEC on polymeric columns offers low efficiencies and ion-exchange modified silica tends to lose ion-exchange capacity. Reverse-phase high performance liquid chromatography (RPLC) is not appropriate for most ionic species unless they are either first derivatized or ion-pairing/ion-suppression is used. Before HPLC, electrophoresis textbooks[1-3] reported hundreds of separations that had been developed for charged biomolecules, some of which were used commonly in biochemical and clinical laboratories. However it is unlikely that the then current technologies, especially visualization, allowed them to be even semiquantitative.

The updating, using CE, of such electrophoretic assays for the analysis of ionic molecules would therefore appear totally appropriate. Among the first demonstrations of CE by Jorgenson and co-workers[4-6] were the separation of small biomolecules such as peptides. It is often overlooked that the potential of CE for clinical analysis was suggested from the publication of one of their three seminal papers in *Clinical Chemistry*.[6] Terabe's innovative modification of CE[7] to permit the separation of relatively hydrophobic small molecules by incorporating surfactant micelles into the electrolyte (MEKC) found immediate applications in drug analysis (see below) and, now, is increasingly employed in bioanalysis.

A. Resolution

The immediate attraction of CE for bioanalysis must be its high efficiency, giving the high resolution necessary to analyze complex biofluids and mixtures of biomolecules. CZE systems can readily achieve efficiencies orders of magnitude greater than those obtained by HPLC in similar or shorter periods of time. Developing 150,000 to 250,000 theoretical plates in 10 to 15 min is commonplace for many CZE separations, while up to 30 million plates have been reported for the capillary gel electrophoresis of oligonucleotides[8] and over 40 million for CEC. Such high efficiencies mean that CE using just one electrolyte is capable of bettering the resolution obtained by both isocratic and gradient HPLC. In addition, CE is simpler in terms of both operation and equipment. It requires significantly less sample and only small quantities of simple and inexpensive running buffers, as well as being potentially much faster and quantitative. Typically, we have observed a 30% increase in numbers of peaks resolved by CE compared to even gradient HPLC under comparable conditions of detection. This finding applies to a diverse selection of samples such as UV-absorbing and fluorescent compounds in urine and plasma and nucleotides in cell extracts. However, the majority of new peaks were minor, and it is unlikely that HPLC analyses were missing major components. Figure 10.1 compares the separation of human urine by gradient elution RPLC and CZE. Studies with a number of compound types have given similar findings.[9-11] Although faster and quantitative, CE does not yet have the resolution or "preparative" capabilities of traditional two-dimensional electrophoresis/chromatography systems.

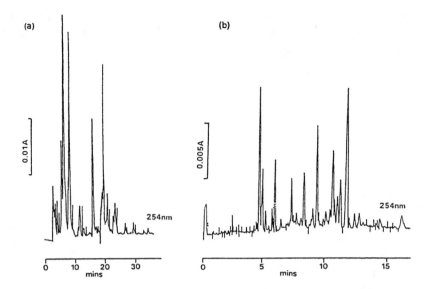

Figure 10.1 Comparison of separation of urinary components by (a) HPLC and (b) CE. HPLC conditions: Column: 250 × 4.6 mm 5 μm ODS Hypersil. Eluent: Linear gradient from 50 mM acetate pH 5 to 25% MeOH over 30 min. Flow rate: 1 mL/min detection dual wavelength 254 nm and 280 nm. Sample: 3 μl urine. CE conditions: 50 μm × 70 cm (50 cm to detector). Buffer: 50 mM phosphate pH 8. Load: 10 kV 5 sec (approximately 8 nL). Run: 20 kV. Detection: Isco CV4 at 254 nm.

B. Detectability

Under certain conditions, CE can out-perform HPLC with respect to detectability. For example, it is difficult, although not impossible, to perform HPLC separations at low UV wavelengths. Even with "low-UV grade" acetonitrile only moderate sensitivity toward peptides can be achieved without an excessive baseline rise using gradient elution with detection at 215 nm. Lower wavelengths (around 205 nm) can be used with isocratic HPLC, but sensitivity is then limited by flow induced noise. CE, on the other hand, operates without pulsatile flow, minimal refractive index changes, or temperature changes, and short pathlengths and can operate at 190 nm and even below. Detection at such wavelengths makes CE virtually universally applicable toward all molecules. Figure 10.2 compares the electropherograms for human urine obtained by monitoring at 195 nm, 250 nm, and 280 nm. These results were obtained by taking appropriate wavelength slices from the three-dimensional separation (Figure 10.2A). Clearly there is a large increase in the number of peaks detected at the lower wavelengths, but equally there is loss of selectivity. For some classes of biomolecules, such as certain sugars[12,13] and amino acids,[14] CE's ability to detect at such "nonspecific" wavelengths has allowed the development of simple assays, which usually require derivatization before HPLC.

C. Selectivity

Although the high resolution offered by CE is important for analytical specificity, selectivity can also be gained from suitable choice of other analytical conditions. Selectivity can be achieved by using an ever-increasing range of electrolyte modifiers ranging from surfactants[7] to deuterated water[15] and cyclodextrins.[16] (For full details of the various additives see other chapters in this book.)

Often forgotten is that analytical selectivity can be gained from employing more specific detectors such as electrochemical detection and fluorescence, which can offer both higher sensitivity and higher selectivity. Fluorescence is often more selective that single-electrode

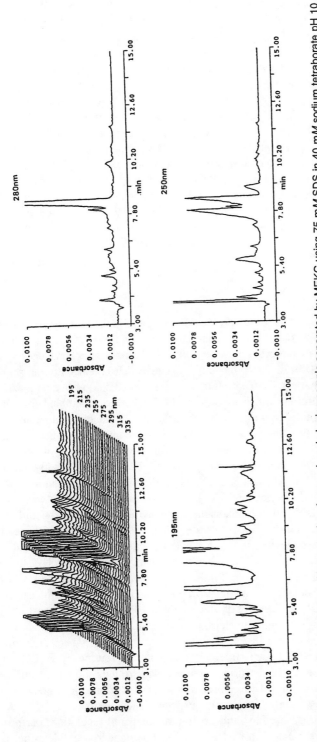

Figure 10.2 Scanning electropherogram of human urine. A normal pooled urine sample separated by MEKC using 75 m*M* SDS in 40 m*M* sodium tetraborate pH 10 at 20 kV in 50 μm × 50 cm (44 cm to detector). Load run: 20 kV. Detection: ThermoSeparations 2000 scanning from 195 nm to 335 nm. Other figures show selective cuts from the total electropherogram at 195 nm, 250 nm, and 280 nm to show sensitivity and selectivity changes with wavelength.

electrochemical detection, although redox mode electrochemistry using two electrodes is very selective. An increasing number of studies have used fluorescence detection with CE, mostly with laser-induced fluorescence (LIF), especially since a commercial LIF detector became available, although the highest sensitivities have usually been obtained on laboratory-assembled detectors. Unfortunately, the readily available lasers, e.g., He-Cd (442 nm or 325 nm), do not emit wavelengths particularly useful for exciting the native fluorescence of biomolecules that occurs typically between 270 and 290 nm. It is therefore necessary to resort to derivatization of the analyte, e.g., fluorescent intercalators such as YO-PRO for DNA analysis and amino reactive reagents for protein analysis, in order to match the fluorescence to the LIF detector. Various authors,[17,18] including ourselves, have modified HPLC xenon lamp detectors for use with CE. Although the sensitivity gain is below that found in HPLC,[19] the selectivity gains are equivalent. Improved cell designs including the use of rectangular capillaries and high intensity xenon-mercury lamps should offer significant gains in sensitivity, be cheaper than LIF, and offer a continuous spectrum much more usable in bioanalysis. However, little commercial interest is being shown in lamp-based fluorescence detection for CE. Recently Caslavska et al.[20] modified a tunable HPLC UV-detector fitted with a deuterium lamp to capture fluorescence and applied this dual detector to profiling body fluids. Although they found useful selectivity gains, they commented that the fluorescent signal was less than that obtained by absorbance detection.

D. Sensitivity

Without doubt the most important problem faced by CE in the field of biomedical analysis is still its relatively low concentration sensitivity. Initially, most attention in CE was given to the high efficiency of the process, which with on-column detection, leads to extremely high mass sensitivity. Unfortunately for most bioanalyses, it is concentration sensitivity not mass sensitivity that matters. Concentration sensitivity in CE is lower than mass sensitivity and is usually some tenfold below that found in HPLC, when considerable volumes of sample can be injected and detected in long pathlength flowcells. Unlike some other areas where CE is finding wide application, the bioanalyst cannot easily control either the concentration in the samples to be measured or the amount of sample available. We use tryptophan, a naturally occurring amino acid, to evaluate separation systems because it is readily separated by a variety of modes, including RPLC, ion-exchange, and electrophoresis, and it can be detected by UV, fluorescence, and electrochemistry. The separation of tryptophan was optimized for both CZE and RPLC. Table 10.1 compares their sensitivities using both UV and fluorescence detection. Clearly, CE has higher mass sensitivity than HPLC regardless of detector type, but concentration sensitivity is lower for CE with both detectors. Also notable is that the relative improvement in sensitivity achieved by using fluorescence detection with HPLC (155× better than UV) fell to only 6.5-fold with CE. Similar conclusions with respect to UV detectability can be derived from the detailed evaluation of commercial instruments performed by Watzig,[21] who showed that the UV sensitivity reported in Table 10.1 is typical of most CE systems.

At present this poor concentration sensitivity is sufficiently low to exclude CE from the analysis of many biomolecules in real samples such as plasma, where concentrations are often μmol/L and below, although such measurements are routine by HPLC. Many compounds can be detected by CE in urine, where concentrations are often mmol/L. Unless plasma analytes are extracted and concentrated in some way prior to injection, we are usually only able to determine strongly UV-absorbing species, such as urate.

Most studies, including those in Table 10.1, have compared separations performed by CE with the same sample separated on an HPLC system of standard configuration, i.e., 100 to 250 × 4.6 mm columns.[9-11,19,20] However, the more appropriate comparison is CE with microbore HPLC (μLC), but few such studies have been published. Cobb and Novotny[22] compared the separation of a tryptic digest of casein by CE and gradient elution μLC. μLC

Table 10.1   Comparison of Tryptophan Detection Limits Using UV and Fluorescence Detection for Both CE and HPLC

Parameter	UV detection[1] 280 nm		Fluorescence detection[2] ex 280 nm em 360 nm	
	CE[3]	HPLC[4]	CE[3]	RPLC[4]
Volume analyzed	8.3 nl	20 μl	8.3 nl	20 μl
Concentration sensitivity	9 μM	850 nM	1.4 μM	5.5 nM
Mass sensitivity*	75 fmol	17 pmol	10 fmol	110 fmol

[1] Isco CV4 Detector with capillary flowcell.
[2] Jasco 821-FP Detector with in-house designed capillary flowcell.
[3] **CE conditions.**
   Capillary: 50 μm × 70 (50) cm silica. Buffer: 50 mM sodium phosphate pH 8. Load: 10 kV 5 s. Run: 20 kV.
[4] **HPLC conditions.**
   100 × 4.6 mm 3 μm ODS-Hypersil Eluent: 50 mM ammonium acetate: MeOH (250:30) pH 6 1.2 mL/min.
   * Limits of detection expressed as S/N = 2.

and CE required 100 ng and 2 ng of tryptic digest, respectively, and their CE methodology was three times faster. Resolution was equivalent by both techniques. Ling et al.[23] compared a gradient μLC with fluorescence detection and a CE system with UV-detection for the separation of the ammonium 7-fluoro-2,1,3-benzoxdiazole-4-sulfonate (SBD-F) derivatives of thiols. Their CE system had a precision of approximately threefold that of the μLC system but was three times faster. There were useful differences in selectivity, and the resolution was better. Unfortunately, because of the different detection modes, absolute sensitivities could not be compared.

The question is therefore — how can concentration detection limits be increased in CE? A number of options are possible. Increased sensitivity can be obtained both instrumentally and by processing the sample before analysis.

First, a tenfold increase in sensitivity from detector improvements is considered realistic by many CE manufacturers. Improved lamp and detector electronics are possible, leading to less baseline noise although the limits of shot (electronic) noise may be approaching. Optical designs, particularly at the flowcell, should improve. The ubiquitous deuterium lamp has continued to be improved in order to operate better at lower wavelengths. It can be replaced by other light sources such as the zinc lamp for operation at even lower wavelengths (185 nm). However, the aim must be to reduce baseline noise relative to the absorption spectra of the analyte. Figure 10.3 shows the signal-to-noise ratio, determined in the author's laboratory, for sodium nitrite using a commercial CE system, and it shows the high energies available from a deuterium lamp in CE at low wavelengths. If all such improvements were incorporated, they may permit routine operation at less than 10 μAU f.s.d. with 1% baseline noise and put CE at least on a par with conventional HPLC. On the other hand, many such improvements would also benefit narrow bore HPLC and μLC systems.

Second, modifications to the capillary can be used. Compared to the 10 mm path found in most HPLC UV-detectors, the across capillary path (50 to 100 μm) is minute and causes much of the reduced sensitivity in CE. Given that the length occupied by a typical band in CE is 2 to 3 mm, using data from HPLC a path would not significantly reduce efficiency. Chervet et al.[24] used 3 mm Z-cells and increased sensitivity 14-fold compared with on-column detection, while the loss in resolution was <35%. This Z-cell has been linked to the standard optical configuration in CE, i.e., the ball lens, to give an even greater increase in performance.[25] Such a cell is available for some instruments, but for others its use requires a redesign of existing capillary cassettes or flowcell holders. Rectangular capillaries with inner dimensions from 16 × 195 μm to 50 × 1000 μm were studied by Tsuda et al.[26] Such capillaries have improved thermal characteristics, and the same efficiency but increased sensitivity up to 20-fold compared to normal circular capillaries. Tsuda et al.[26] employed rectangular glass

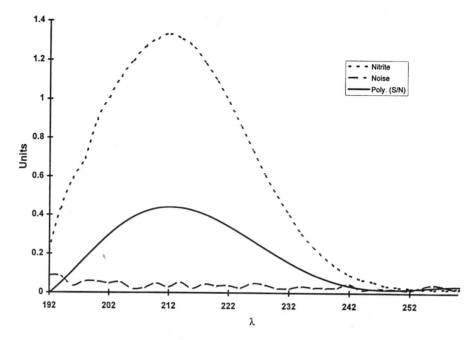

Figure 10.3 Variation in signal-to-noise ratio with wavelength for sodium nitrite using the Thermoseparations 2000 CE instrument equipped with a high-intensity UV lamp. Polynominal smoothing
was applied to the S/N curve.

capillaries, but deformed PTFE capillaries have been used to quantify immunoglobins where
the improved thermal characteristics were very important.[27] Rectangular silica capillaries are
now commercially available but have so far received little attention. The formation of a bubble
in the bore of the capillary to form the detector cell has been commercialized by Hewlett-
Packard.[28] Although the gain in sensitivity is only moderate (two- to fourfold) it is proving
popular and is available for most instruments. Other devices designed to increase pathlength
include direct[29] or indirect[30] axial illumination and multiple internal reflectance flowcells.[31]

The third option is to increase sample loading. Due to its low concentration sensitivity,
CE is usually operated at or near maximum loading. The need to maintain maximum resolution by keeping the concentration of sample at only 1% that of the buffer; increasing the
load further usually causes significant peak distortion. Even when not overloaded, efficiency
is dependent on the length of capillary originally occupied by sample.

Gains in loading can be achieved by sample stacking (i.e., loading from buffers at pH
values below that of the running electrolyte). It is commonplace to load the sample from a
water solution to achieve some pre-separation concentration. The various modes of stacking
have been reviewed in detail by Chien and Burgi.[32] These authors have shown that very large
sample volumes can be analyzed if, following loading into the capillary, the polarity is reversed
and one ion is backflushed from the capillary before normal CE is commenced.[33-34] Using
this method, up to an 85-fold increase in sensitivity with minimal resolution loss was possible.

For strongly ionic molecules, significant gains in sensitivity can be achieved by isotachophoretically focusing (and concentrating) relatively large volumes of sample into the
analytical capillary.[35-40] For nucleotides, Foret et al.[35] showed that more than 10 µl of a dilute
solution could be isotachophoretically loaded, giving an improvement in detection sensitivity
of 200-fold. At the same time, Dolnik et al.,[37] and later Foret et al.,[38] demonstrated the
applicability of cITP-CZE to the analysis of dilute protein mixtures, The group lead by Van-
der-Greef has amply demonstrated its utility with on-line MS.[39]

For hydrophobic compounds, isotachophoretic loading is not possible. So for many
workers, the simplest solution to increasing sample concentration, and often selectivity, is to

extract the sample in some way prior to CE analysis. It is often a simple matter to modify established liquid or solid phase extraction (SPE) procedures to give smaller, and therefore, more concentrated final volumes. Much attention is now being directed at performing SPE on-line with the capillary. The first attempt was to incorporate a small plug of reversed-phase HPLC packing silica into the load end of the capillary. Under appropriate conditions, this can lead to selective extraction and accumulation of analytes onto the ODS-silica. The retained analytes are then separated and quantified by CE using an eluent containing an appropriate amount of organic modifier.[40] Such capillaries are available from Waters, and increases in sensitivity of 250-fold are claimed. Simpler devices can be constructed by sandwiching near the injection end of the capillary a short plug (2 to 3 mm) of an appropriate solid phase extraction media[41] or a disc of extraction membrane.[42] During loading, up to a few µl of sample are passed through the SPE bed and, by careful choice of conditions, the analytes are trapped. The analytes are then released by a suitable electrolyte on starting the CE process. Using such devices, many-fold increases in sensitivity can be achieved. The downside to such devices are that it can take many minutes or even hours to draw sufficient sample into the SPE device; the devices must be reconditioned following use; they can be readily blocked; and they are not suitable for all samples. Solid phase extraction methods for CE have been comprehensively reviewed recently.[43]

III. SAMPLE PREPARATION

At best, CE requires little to no sample preparation. Optimum resolution is usually achieved when sample load volumes are small and sample mass concentrations are moderately high. Optimal peak shape results if the concentration of sample is less than 10% that of the background electrolyte.[44] Such criteria can often be met in pharmaceutical or biotechnology work but are less likely to be achieved in bioanalysis when sample make-up and analyte concentration are usually independent of the analyst. However minimal the sample preparation is, it is important to use solutions free of particulate matter that may easily block an analytical capillary. High-speed centrifugation (130,000 g for 1 min) or passage though a 0.45 µm filter is usually sufficient to remove particulate material.

Converting separations, particularly those developed with pure standards dissolved in water and resolved using free solution CE, to real biofluids often proves very difficult, with shifts in migration time ranging from a few seconds to almost total loss of sample. Reasons for this include:

1. The matrix affects on the loading of analytes especially when using electrokinetic injection
2. Matrix effects on electrophoretic mobilities
3. Effect of trace levels of organic extractant on MEKC separations
4. Changes to the capillary wall and therefore electroosmotic flow (EOF) due to other components in samples, particularly proteins
5. Adsorption of analytes onto the capillary, changing EOF
6. Capillary overload due to the total amount of material in the sample

A particular problem with all forms of electrophoresis is the effect of the ionic strength of the sample on electrophoretic mobilities. In the past, samples were routinely desalted before electrophoresis. Microdesalting techniques may be even more necessary with CE. The importance of desalting PCR products before CE analysis has been stressed,[45] and desalting can be achieved simply by placing the PCR product on a Millipore 0.025 µm VS membrane floating on a bath of water for a few minutes before injection. The concentration of salts in body fluids can be very high (e.g., 150 mmol/L in plasma) and very variable (50 to 1000 mM) in urine. Schoots et al.[9] studied the effect of NaCl concentrations over the range of 2.8

to 54.4 mmol/L in serum on migration times of common constituents in serum using CE. The greater the amount of NaCl, the later the migration of the UV-absorbing analytes in ultrafiltered serum, the electropherogram being affected by an increasing large negative peak associated with the migration of the Na^+ and Cl^- ions. Prior to analysis, they therefore diluted ultrafiltrates of clinical serum tenfold. It is, of course, much easier to control ionic strength of plasma than urine.

Some of the above effects can be avoided by sample preparation or appropriate choice of operating conditions or both. MEKC separations are less prone to the effects of sample salt concentration than CZE separations. This may be due to the role of the micelle in retaining more hydrophobic analytes, solubilizing traces or proteins while allowing the variable salts to rapidly migrate. If organic extraction is used with MEKC, it is necessary to remove all traces of extractant before CE analysis, because trace levels of organic extractants can change the micelle partition ratio.

Changes to the capillary walls are more difficult to deal with. Many workers advocate various capillary conditioning/washing routines to remove adsorbed proteins, etc. Such routines are not always necessary, but if used, they must be rigorously controlled and are best performed automatically, the problem being one of routinely returning the silica surface to the same degree of ionization each time.[46] These routines often take longer than the actual CE run and have the same sort of effect on sample throughput as gradient elution does in HPLC. Capillary conditioning is not necessary after every run but depends on the assay and analytes under investigation.[36] Not only do some proteins bind strongly, but some small molecules bind equally firmly in an ion-exchange manner. We have experienced variable losses of some amino compounds during analysis, presumably due to very strong binding onto surface silanols in a manner similar to protein binding. Many methods of reducing this binding by either coating the capillary walls and/or buffer additives have been suggested. A number of proprietary coated capillaries are available, but their chemical nature is usually unspecified. The use of these approaches with proteins was discussed in Chapter 8 and will not be repeated here.

The simplest way of reducing most of these variables is to employ some form of sample preparation involving selective isolation of the analytes of interest, which will usually be accompanied by desalting, deproteination, and possibly sample preconcentration. The methods available are acid precipitation of proteins, organic extraction, and solid phase extraction, etc. Table 10.2, which is derived both from the work of Blanchard[47] and the author's unpublished data, lists the efficiency of the most common extraction techniques at removing protein from typical blood fractions. The techniques are essentially the same for CE as for HPLC, except that it may be desirable to process smaller volumes. However, it may be equally important to avoid unnecessary dilution and addition of high concentrations of ionic species. A problem that is particular to CE is that many protein precipitating reagents leave residues in the sample that then give peaks, especially when monitoring at low wavelengths. So although Table 10.2 suggests that TCA is the most efficient precipitating reagent, it will leave large peaks that absorb below 225 nm on the electropherogram, and PCA, with a much cleaner blank, is a better choice with CE. When using CE, ultrafiltration is probably the simplest choice as a deproteinating method because modern membranes will yield 50 μl with just a few minutes' centrifugation. However, remember that it will not free protein-bound species.

Although protein removal is essential for some CE assays: CE can conversely also be used to reduce sample preparation. This is because differential loading due to mobility differences[48] can mean that only the compounds of interest are loaded using electrokinetic injection. The ability of MEKC to solubilize plasma proteins and impart a strong negative charge, release bound drugs, and resolve drug substances has been utilized to measure small molecules in unextracted biofluids containing large amounts of protein. Depending on detec-

Table 10.2   The Effectiveness of Common Protein Precipitating
             Procedures

Sample precipitant	Volume ratio of precipitant to sample required to remove 99% of protein
Human Plasma/Serum	
10% w/v Trichloroacetic acid	0.2
6% w/w Tungstate-sulfuric	0.7
5% w/v Phosphoric acid	3.0
Zinc sulfate-barium hydroxide	2.0
Acetonitrile	1.3
Acetone	1.4
Ethanol	3.0
Methanol	4.0
Saturated ammonium sulfate	ca. 3.0
Ultrafiltration	Not applicable[1]
Heating at 80°C	Not applicable[2]

	Volume ratio of precipitant to sample required to remove 97% of protein
Cell Extracts[3]	
5% w/v Trichloroacetic acid	4.5
10% w/v Trichloroacetic acid	3.8
15% w/v Trichloroacetic acid	3.3

[1] Provided there is no leakage, ultrafiltration will remove 98% of protein.
[2] Only poor removal of proteins ca. 95% in 10 min.
[3] The cells were human red cells of initial protein content 3.11 mg/mL.

Adapted from Blanchard, J., *J. Chromatogr.*, 226, 455, 1981, and unpublished data of the author.

tion wavelength the SDS-protein complexes, due to their size and overall negative charge, are seen to migrate significantly later than the small molecules of interest (Figure 10.4). Cefpiramide has been analyzed directly from plasma[49] following electrokinetic injection. Although electrokinetic injection can improve selectivity, hydrodynamic injection can be used as well. Many other drugs have now been measured in this way, e.g., the antibiotic aspoxicillin,[50] paracetamol[51] in plasma, and antipyrine in saliva.[52] For a more comprehensive list of un-extracted assays, see the review by Thormann.[35]

IV. QUANTITATIVE ASPECTS

Although there was initially some dispute over the precision of CE, this has now abated. The precision of CE using current instrumentation is now generally accepted as similar to that for HPLC with either manual or automatic valve injection of standard materials. An important point to realize is that with current integration practices peak areas in CE vary with migration velocity through the detector. Thus equimolecule amounts of related compounds such as adenosine nucleotides give increasing peak areas, in contrast to HPLC where the areas should be approximately the same. The use of spatial areas, i.e., the integrated peak area divided by the migration time,[54] corrects for this as well as other factors.[55] Most published validations have been for drug assays, and the extensive validation studies of Altria and co-workers should be read in this context.[56] Of particular importance in this respect are their intercompany studies which not only show good validation characteristics of a variety of assays but also the robustness of CE assays in transferring between laboratories and between

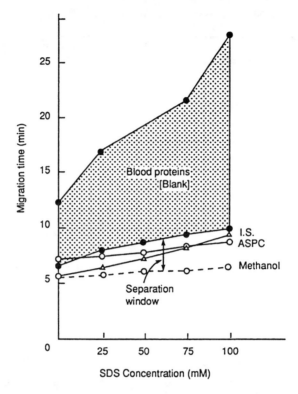

Figure 10.4 Diagram illustrating the effect of SDS concentration on the relative migration of the drug aspoxicillin and plasma proteins by MEKC. (From Nishi, H., Fukayama, T., and Matsuo, M., *J. Chromatogr.,* 515, 245, 1990. With permission.)

instruments.[57-58] Quantitative aspects of CE have been reviewed,[54] and factors to observe in improving the robustness of systems have been dealt with by a number of authors.[59-60]

So far, few studies have addressed quantitative precision with biological samples, but for CE, CVs are higher than for HPLC (e.g., 5 to 10%). However, for many studies on biological samples, such levels of precision are easily tolerated. Much of the imprecision appears to originate from drift of migration times leading to the need to correct to spatial areas. Matrix effects due to the sample, particularly urine, may be more difficult to deal with, and considerable thought may be needed to overcome them. Such effects usually manifest themselves as random changes or a continuous drift in migration times Possible approaches to limiting these effects are selective sample loading, e.g., electrokinetic injection, solid phase extraction prior to CE, migration time corrections, and the use of suitable internal standard procedures.[61,62]

V. SELECTED APPLICATIONS OF CAPILLARY ELECTROPHORESIS IN THE BIOSCIENCES

A growing number of applications of CE to biological matrices are being published. The topic has been reviewed in a number of recent publications,[63-65] and the theoretical background of this area of application was discussed some time ago.[66] Additionally, overviews of the application of CE to clinical analyses have appeared.[67-69] Because this field is growing so rapidly, the following is a *very* selective discussion of some of the main biomedical applications of CE.

Table 10.3   Assays of Selected Small Endogenous Molecules in Biological Fluids Using Capillary Electrophoresis

Analyte	Biofluid[1]	Sample prep[2]	CE mode	Detection	Ref(s)
Steroids	U		MEKC	UV	227–229
Nitrate and Nitrite	P, U	UF	CZE	UV	123–125, 230, 231
Inositol phosphates	P		CZE	Indirect UV	232
Tryptophan, Indoles	U	None	MEKC	LIF	233
Phenylalanine	S, U	L-L	CZE	UV	85, 234
Metals	S		CZE	Indirect UV	235, 236
Organic acids	S	UF	CZE	Indirect UV	237–239
Bilirubin	S	Neat	MEKC	LIF	240, 241
Vitamin C	S	Neat	CZE	UV	242
Vitamin A	S	Neat	CZE	LIF	243, 244
Vitamin B12	P	PCA	MEKC	LIF	245
Purine bases	P	PCA	CZE	UV	147, 246, 247
Uric acid	S	Neat	CZE, MEKC	UV	147, 248, 249
Hippuric acid	U	Neat	MEKC	Indirect UV	9, 250
Creatinine	S, U	Neat	CZE	UV	251–253
Porphyrins	U		CZE	LIF	113–115
Homovanillic acid	U	None	CZE	UV or FL	20
Methyl malonic acid	U			Indirect UV	254
Thiols	RBCs	PCA	CZE	FL, UV	108, 255, 256
Polyamines	Cells	L-L	CZE	Indirect UV	257

[1] S = serum, P = plasma, U = urine, RBC = red blood cells.
[2] SPE = solid phase extraction, TCA = trichloroacetic acid, PCA = perchloric acid, L-L = Liquid liquid extraction, IEC = ion exchange chromatography, HTAB = hexadecyltrimethylammonium bromide.

A. Small Endogenous Molecules

A number of classes of molecules of particular interest to the clinical chemist and the biomedical analyst are included in this section. Table 10.3 gives a selected list of analytes in this category with an outline of the CE methods used.

1. Nucleotides, Nucleosides, and Bases

This group of compounds were among the first to be separated by CE. Nucleotides are strong anions, while nucleosides and bases carry a positive charge over the pH range 2 to 6, are often neutral and hydrophobic in the region of pH 7, but can also be weak anions above pH 8. For nucleotides, the following approaches for their separation by CE can be employed.

1. Free solution CZE with normal polarity but high EOF
2. Free solution CZE with reversed polarity and possibly reduced EOF
3. MEKC with reversed polarity.

In a very early application of CZE, Tsuda et al.[70] used option (1) of high electroosmotic flow to overcome the natural migration of the negative nucleotides to the anode. They resolved a nucleotide mixture using a 50 mM borate pH 9 buffer in a glass capillary (75 µm i.d. × 70 cm) which needed to be coupled to a silica flowcell to detect the nucleotides at 254 nm. The separation was performed at +20 kV with electrokinetic injection. Under similar conditions, Perrett and Ross[10] achieved good resolution of standards but found that when cell extracts were injected, 30% variations in peak mobilities were obtained. This was presumed to be due to the high salt concentration of the extracts disturbing their separation compared to standards, although Tsuda has reported the separation of such extracts with no apparent problems.[71] A variety of coated capillaries have been used for the high-efficiency separation of nucleotides. A capillary coated with crosslinked polyacrylamide, giving minimal EOF, was reported to separate 14 nucleotides in 50 min with an LOD of 5.4 μM.[72] (This is significantly slower than many HPLC methods!) A Ucon-coated capillary was also prepared by the same group[73] and then applied to the determination of intracellular nucleotide pools in drug-treated cell lines.[74]

Hernandez et al.[75] separated some cyclic nucleotides, e.g., cAMP by CZE using 75 μm × 100 cm silica capillaries at 22 kV using 50 mM borate pH 8.3 buffer. Samples had to be monitored at 210 nm rather than 260 nm to gain sensitivity (maximum sensitivity = 40 μM). Even so, the levels in most biofluids or cell extracts are much lower, and it is difficult to envisage CE replacing binding or immunoassays for the measurement of cyclic nucleotides.

Liu et al.[76] studied the MEKC separation of nucleotides and compared dodecyltrimethyl-ammonium bromide (DTAB) and SDS systems with reversed polarity and obtained rapid and efficient separations with DTAB. To measure nucleotide pools in cells, Perrett and Ross[10] optimized the resolution of 15 natural nucleotides with respect to pH, DTAB concentration, applied potential, presence/absence of EDTA, and running temperature. Final conditions were 50 μm i.d. × 70 cm (63 cm to detector) capillaries, and the base electrolyte was 50 mM disodium phosphate pH 7, 100 mM DTAB, and 1 mM EDTA at 35°C. The addition of EDTA was important for good peak shape, presumably by inhibiting the formation of metal–nucleotide complexes. The response was linear to 200 μM, using electrokinetic loading. Area reproducibility ranged from 3.3 to 6.1%, while migration time reproducibility varied from 2.2% to 5.5%. Takigiku and Schneider[77] have also reported validation criteria for the separation of nucleotides using the SDS MEKC approach.

MEKC methods can be applied to acid-extracts of cells without the disturbances observed for the CZE method. They offer improved resolution and are operationally simpler than the anion exchange and ion-pair reverse-phase gradient HPLC methods usually employed. However, one problem remains in that DTAB, or an impurity in it, forms a yellow deposit on the anode, giving problematic currents.

Bases and nucleosides are neutral at pH 7, so separation by CE is enhanced using SDS. Oligonucleotides are negatively charged but separate with little selectivity. Incorporation of Mg or Cu ions in the buffer, as well as SDS, increased selectivity. Separation of 14 from 18 oligonucleotides was achieved in <30 min in the presence of 3 mM zinc and SDS. Nucleosides and bases are readily separated by CZE. At pH 9.4, at least 10 naturally occurring nucleosides and bases can be separated at 20 kV in a 44 cm × 75 μm capillary in as many minutes. Although usually detected at 260 nm, a tenfold increase in sensitivity is possible by operating at about 210 nm. At this wavelength, there is sufficient sensitivity to quantify several nucleosides and bases, including hypoxanthine and uric acid in biological samples including perchloric acid extracts of human plasma. This method represents the more general application of the many published methods for the assay of uric acid in plasma using CE.[78-80] Methyl xanthines such as caffeine are a subgroup of the purine bases, and a number of detailed studies have covered their resolution as drug substances[81] as well as the assay of caffeine in beverages.

Lecoq et al.[82] described the optimization of a MEKC separation of 10 deoxynucleosides in approximately 20 min. The final conditions were 20 mM lithium phosphate pH 9.6 with 100 mM SDS and 5% acetonitrile in a 57(50) cm × 50 μm capillary held at 35°C using a voltage of 15 kV with 254 nm detection. Free radical damaged nucleosides, such as 8-oxodeoxyguanosine, have been resolved by MEKC.[83]

In a paper that displayed one of CE's major advantages, i.e., the rapid and simultaneous separation of differently charged species, Grune and Perrett[84] resolved 19 nucleotides and related nucleosides and bases in a 44(40) cm × 75 μm fused-silica capillary in a 20 mM borate buffer pH 9.2 using 25 kV at 15°C with detection at 260 nm (Figure 10.5). Such a

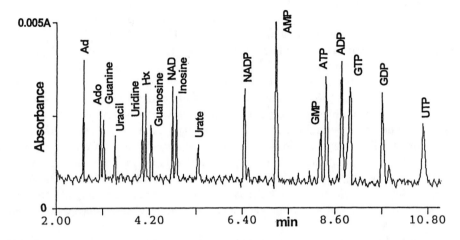

Figure 10.5 Separation of nucleotides, nucleosides, and bases by CZE. CE Conditions: capillary fused-silica 50 μm i.d. × 70 cm (63 cm to detector). Electrolyte 50 m*M* disodium phosphate pH 7 containing 1 m*M* EDTA and 100 m*M* DTAB. Temperature 35°C. Load hydrodynamic 5 s run 20 kV in A and 25 kV in B. Detection 255 nm (solid trace) 280 nm (dashed trace). A) Separation of standard mixture of 18 nucleotides at 25 μ*M* each.

rapid, efficient separation is not possible by HPLC because the only condition when the charges are even approximately similar is pH > 8, and silica-based HPLC columns are not normally usable in alkaline solution. In the separation, the charge on even the most anionic acids, such as urate, is minimized and falls below the EOF.

2. Amino Acids

The separation of the 18 amino acids found in typical protein hydrolysates and the even larger number found in physiological fluids has always been a major challenge in analytical biochemistry. Most amino acids do not absorb strongly in the mid-UV-spectrum, and a large number of derivatizing reagents have been developed. These range from the classic colorimetric reagent, ninhydrin, used in Moore and Stein type amino acid analyzers, to the many precolumn derivatization reagents developed for HPLC. In HPLC, precolumn derivatization not only improves detectability, but by reacting with the ionic amino group increases the amino acids' hydrophobicity, thus improving their separation by RPLC. Unfortunately, such reagents remove the main positively charged group by which amino acids can be readily separated electrophoretically.

Amino acids are cations at low pH and can therefore be readily separated by CZE, but only tryptophan, tyrosine, phenylalanine, and cyst(e)ine possess measurable absorbance in the range 230 to 300 nm. Some of this group can also be detected by fluorescence following CE as mentioned earlier. At wavelengths below 220 nm, other amino acids will absorb, and Bergman et al.[14] exploited this to separate and detect native amino acids. Six amino acids were separated by CZE at pH 2.5 and detected at 214 nm. The sensitivity toward neutral amino acids, such as glycine, serine, threonine, and proline, was approximately 1 pmol injected with femtomole sensitivity toward the aromatic amino acids and arginine. Tagliaro et al.[85] adapted this approach to rapidly screen for the inborn error of metabolism, phenylketonuria, by separating phenylalanine in ethanol-deproteinized serum using a pH 10 borate buffer. Tryptophan can be detected by electrochemical detection at a carbon fiber positioned at the end of the capillary. If a copper electrode is employed, most amino acids are electro-oxidizable.[86]

Another approach to the determination of native amino acids is indirect detection. In order to maintain ionic equilibrium in the electrolyte, a charged analyte must displace appropriate equivalents of electrolyte, and when the spectral properties of that analyte and the electrolyte are different, a negative peak results. The water peak seen near the front of some HPLC separations is an example of this effect. Kuhr and Yeung,[87] working with a background fluorophore, i.e., salicylate, in the electrolyte plus laser-induced fluorescence, separated and detected amino acids. Using the same principle but with a background electrophore of dihydroxybenzylamine and a carbon-fiber electrode, Olefirowicz and Ewing[88] also measured amino acids. The ultimate detector is of course the mass spectrometer, and demonstrations of its abilities to detect amino acids with various interfaces have appeared.[89]

Another approach, again mimicking column chromatography, is the separation of native amino acids by CE followed by post-column reaction, usually to form a fluorophore. The difficulties of post-column reactor design on the nanoscale compatible with CE have been tackled by a number of groups, including Rose et al.[90] and Pentoney et al.,[91] both of whom used amino acids as their model compounds. Others have combined post-column reaction with electrochemical detection,[92] while the use of post-column chemiluminescence detection has also been attempted.[93]

As with HPLC, pre-column derivatization combined with a CE separation has proved most popular. Most workers have used the same precolumn derivatization reagents, such as o-phthalaldehyde/thiol, fluorescamine, dansyl chloride, and phenylthiohydantoin (PTH), as used with HPLC, even though this means the use of MEKC. The use of fluorescent derivatizing agents has been reviewed by Novotny[94] and Albin et al.[95] The available reagents and their separation modes are summarized in Table 10.4. When fluorescent derivatization is combined with laser-induced fluorescence detection (LIF), some spectacular results have been reported. In particular, Dovichi and co-workers, using fluorescein-5-isothiocyanate derivatives and argon-ion laser excitation, reported detection limits for basic amino acids of less than 6000 molecules,[96] although they have achieved even lower limits of 10 yoctomoles (or six molecules) with other analytes, e.g., sulforhodamine 101.[97]

The separation of 22 PTH amino acids, using an electrolyte containing 50 mM SDS, was an early application of MEKC[98] which although successful had a limited migration window. Various approaches to increasing that window have appeared. Terabe et al.[99] added 4 M urea to the electrolyte to increase the window, while recently Little and Foley[100] found that mixed micelles containing both the anionic surfactant, SDS, and a neutral surfactant, Brij-35, improved the separation but showed that the mechanism was one of increasing separation efficiency while actually decreasing the separation window. Although considerable efforts have been expended in optimizing the separation of PTH amino acids, the routine use of such MEKC separations in the microsequencing of proteins has not yet been perfected. Only recently has a publication[101] actually detailing the microsequencing of peptides by CE been published. Given that PITC derivatives, which are UV-absorbing, have been routinely used for the HPLC analysis of amino acids in protein hydrolysates and biofluids for over a decade, it is surprising that only one early report of their partial separation by CE has appeared.[102]

A large number of methods for the resolution of amino-acid enantiomers have appeared, although in many cases amino acids were only the model or test compounds employed. However, few if any such methods have been applied to biological samples. For a detailed discussion of enantiomeric applications of CE, see Chapter 5.

All these approaches to amino acid analysis by CE might be inappropriate. Derivatization of the carboxylic terminal group might be more appropriate, leaving the amino acid a cation and therefore readily amenable to CZE. This approach does not appear to have been employed to date.

3. Carbohydrates

The role carbohydrates play in many biomolecules is increasingly being recognized, but due to the number of stereoisomers and the large number of combinations of the monomers,

Table 10.4 Pre-capillary Fluorescence Derivatizing Reagents for Amines

Derivatives	Buffer	pH	Micelle	Detector[1]	ex nm	em nm	Sensitivity[2]		Ref
							Conc.	Mass	
FITC	20 mM Borate	9.5	None	FL	490	519	12 nM	2.6 fmole	95
				LIF Ar laser	488	520	500 pM	100 amole	95
				FL	488	520	100 fM	200 zmole	258
Fluorescamine	20 mM Borate	9.5	100 mM SDS	FL	390	450		15.8 fmole	95
OPA/thiol	20 mM Borate	9.5	100 mM SDS	FL	350	400	210 nM	2.8 fmole	95
FMOC	20 mM Borate	9.5	25 mM SDS	FL	260	305	34 nM	0.5 fmole	95
NDA	10 mM Borate/20 mM KCi	9.5	None	LIF HeCd laser	442	490	40 nM	2.5 amole	259
Dabsyl	20 mM phosphate/ACN	7	5 mM SDS	Thermooptical			50 nM	40 amole	260
Dansyl	125 mM phosphate	6.86	None	FL	365	>470	180 nM	7.5 fmole	261
CBCQA	20 mM phosphate/borate	9.1	50 mM SDS	LIF HeCd	450	550	7 nM	10 amole	262
TBQCA	Phosphate/borate	7.2	50 mM taurocholate SDS	LIF Ar laser	465	550	20 pM		263
DTAF									264
TRTC	5 mM Borate	9.0	10 mM SDS	LIF He-Ne	544	590	1 pM	1 zmole	265

[1] FL = fluorescence detector usually with xenon lamp, LIF = laser induced fluorescence

[2] Sensitivity expressed in both sample concentration and molar mass of analyte determined for a typical amino acid at s/n = 2 or 3.

Note: FITC = Fluorescein isothiocyanate; OPA/thiol = o-phthalaldehyde + mercaptoethanol; TRTC = tetramethylrhodamine thiocarbamyl; NDA = naphthalene dicarboxaldehyde; CBCQA = 3-(4-carboxybenzoyl)-2-quinoline carboxaldehyde; TBQCA = 3-(4-tetraolebenzoyl)-2-quinolinecarboxaldehyde.

their analysis is difficult. Although carbohydrates being anions are readily separated by anion exchange HPLC, the utility of such methods has been limited by the number of similar sugars to be separated and their poor detectability using conventional detectors. An overview of the separation of sugars, etc., by CE has already been given in Chapter 7, and only some biomedical aspects will be covered here. Honda and colleagues pioneered the resolution and detection of sugars as their N-pyridylamine derivatives,[103] while others used direct detection.[104] For details of other approaches, such as indirect detection, see the introduction to reference 103. Many fluorescent derivatizing reagents for oligosaccharides are available, e.g., 2-aminopyridine, 2-aminoquinoline, while recently Camilleri et al.[105] introduced 2-aminoacridone as a fluorescent label for branched oligosaccharides.

Interesting applied studies are increasingly being reported. Carney and Osborne[12] separated chondroitin sulfate disaccharides and hyaluronan oligosaccharides using SDS micelles in pH 9 borate buffer with direct detection at 232 nm. Damm et al. compared CE and anion-exchange HPLC for the separation of natural and synthetic fragments of heparin and concluded that CE was the more attractive method.[106] They showed that no complexation was necessary and a good separation in 200 mM sodium phosphate buffer pH 3 and reversed polarity is possible with detection at 214 nm. A rapid assay with attomole sensitivity for hyaluronan and galactosaminoglycan disaccharides (GAGs) in tissue and cell proteoglycans was reported by Karamanos et al.[107] They first digested the GAGs with chrondroitinase followed by separation in 15 mM phosphate buffer pH 3 at –20 kV with detection at 232 nm. The LOD for a trisaccharide was 32 pmol/L. Rudd and colleagues resolved the complex number of glycoforms associated with ribonuclease-B.[13]

4. Thiols

A number of authors have derivatized thiols prior to CE. Jellum et al.[108] employed monobromobimane, which forms strong fluorophores with reduced thiols to monitor thiols such as cysteine and homocysteine in patients with inborn errors of metabolism. Disulfides were first reduced with dithiothreitol before reaction. Such amino acids derivatized on the –SH group remain strong cations and were readily separated by CZE in a 50 mM phosphate buffer (pH 7.5). Ling et al.[23] derivatized endogenous thiols and some thiol-containing drugs such as captopril with the fluorogenic reagent SBD-F, but due to the limitations of their CE apparatus, detected them by UV absorption at 220 nm. At pH 2.5, a separation of eight thiols was obtained in 7 min compared to 25 min using microbore HPLC. Recently a series of biological thiols and their disulfides were separated by CZE at pH 2.5 with detection at 200 nm,[109] while by monitoring at 320 nm, their unstable S-nitroso derivatives could also be detected — a claimed first for CE. However, extinction coefficients for thiols are relatively poor, and sensitivity was low compared to derivatization procedures.

Electrochemical detection of thiols is popular in HPLC, and recent attempts at transferring such methods to CE have been reported, the major challenge being to determine thiols and disulfides simultaneously following their separation by CE. This can be achieved by first reducing the disulfides at a negative electrode before determining the thiols formed, as well as any thiol already present at a positive amperometric electrode. The dual electrochemical detection of cysteine and cystine in CZE was reported by Lin et al.[110] and Zhou et al.[111] using microelectrodes. The determination of thiols alone using a palladium field-decoupler and chemically modified electrodes has also been reported.[112]

5. Porphyrins

Porphyrin levels are elevated in a number of clinical disorders, and their separation by CE has been demonstrated. Weinberger et al.[113] separated porphyrins in urine by MEKC using 100 mM SDS at pH 11 within 15 min, with detection either by absorbance at 400 nm or native fluorescence (ex 400 nm em > 550 nm). Fluorescence was about tenfold more sensitive than absorbance toward the test porphyrins, but even so, this was tenfold below a standard HPLC method. Grossly elevated levels of urinary porphyrins were found in a sample from a patient with *porphyria cutanea tarda* compared to normals. Vanberkel and colleagues separated the same compounds but employed electrospray ionization with mass spectrometry for the detector.[114] Interest in porphyrins continues, and recent advances in their detection by fluorescence[115] and separation by MEKC[116] have appeared.

6. Biologically Important Ions

Strong acid anions, such as chloride and sulfate, and weak acid anions, such as formate and oxalate, are ideal for electrophoretic separations because they possess high mobilities and can be rapidly separated. They have been regularly separated by isotachophoresis with conductivity detection, but sensitivity was low.

The analysis of both cations and anions in biological fluids is again receiving increased attention, although the problem is still one of detection. Ions are ideal candidates for CE. Metal ions have been determined by CZE with on-column chelation using 8-hydroxyquinoline-5-sulfonic acid.[117]

Researchers at Waters Inc. in a series of publications showed that CE with indirect detection offers extremely fast and sensitive resolution of complex mixtures of anions. Indirect detection uses chromate or phthalate or benzoate as the UV-chromophore dissolved in the electrolyte.[118] It is also usually necessary to reverse EOF by addition of flow modifiers such as alkyl ammonium salts, e.g., tetradecyltrimethylammonium bromide (TTAB) or hexamethonium hydroxyoxide (HMH).[119] This technique, named capillary ion analysis (CIA), is remarkably fast, allowing some 36 ions to be separated in under 3 min. Wildman et al.[120] applied this technique to urine and showed that chloride, sulfate, nitrate, citrate, phosphate, and carbonate could be determined in normal urine diluted 50-fold in less than 4 min. They suggest that the technique, which is a marked improvement over ion exchange chromatography, could be useful in aiding the diagnosis of renal stone disease and arsenic poisoning. With sample stacking under conditions of electrokinetic loading, sensitivities of less than 1ppb are possible and can be applied to biological samples. Reid[121] investigated the electrophoretic behavior of some organic anions of biochemical interest. Using CIA with indirect detection and TTAB as the EOF-reverser, the measurement of two ions that have traditionally proved difficult to quantify in urine, i.e., oxalate and citrate, was reported and validated by Holmes.[122]

With the discovery of the biological importance of nitric oxide in human physiology, much attention has focused on its measurement in biofluids. This is not readily possible and most researchers have therefore determined its principal metabolites, nitrate and nitrite. Many methods are available, but a particularly attractive one is to use a variant of CIA. Since both ions are UV-absorbing, there is no need for indirect detection. In the method of Leone,[123] plasma samples, diluted 1:10 with water, were electrokinetically injected and then separated in 72 cm × 75 μm silica capillary at −300 V/cm and detected at 225 nm in an HP bubble cell. Sensitivity extended down to a single μ*M*. Unfortunately, the method used a proprietary flow modifier OFM Anion-BT (Waters Co.) of undescribed composition. Other authors using the same additive have produced similar methods for these two anions in urine[124] and plasma.[125] It is possible to use submicellar concentrations of other EOF reversal additives, e.g., TTAB and TTAC to obtain equal results (Figure 10.6).

B. Separation of Small Exogenous Molecules

Many reports of the separation of drugs in standard solution by CE have been published both in journals and book chapters. (For some recent reviews, see references 64, 65, 69, and 126 through 129). No doubt an even larger number are unpublished within drug company

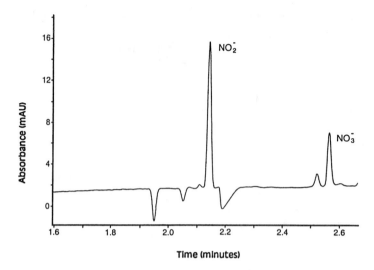

Figure 10.6 Separation of nitrate and nitrite in a normal urine sample. Separated by CZE using 1 m*M* TTAB in 150 m*M* NaCl at −15 kV in 75 μm × 50 cm (44 cm to detector). Load 5 sec HD Detection ThermoSeparations 2000 at 212 nm.

and equipment company files, the speed and resolution of CE assays being of prime importance in areas such as pharmaceutical quality control, monitoring chemical syntheses, formulation analysis, and the determination of isomeric purity. Assays for many different drug classes are possible and such assays are ideal for CE because, in the main, sample concentration and sample preparation are within the control of the analyst. The availability of CE instrumentation as well as trained personnel within laboratories is now leading to an increase in the numbers of "real-life" applications appearing in the literature, with the numbers of actual assays of drugs in biofluids growing apace. Table 10.5, derived from a database of over 3450 publications on CE maintained by the author,[130] lists those drug analytes where reports of actual assays in biofluids have been published, although it should be stated that not all these are fully validated assays.

A major factor leading to the relatively slow acceptance of CE for drug assays could be that the majority of present-day pharmaceuticals are relatively nonpolar, and reverse-phase HPLC has proved excellent for their analysis both in pharmaceutical formulations and in biofluids and now form standard operating procedures. Pharmaceutical analysts may therefore see no reason to change. However the resurgence of electrophoretic techniques for drug assays in biofluids is now real. The advantages of CE in this context are clear. CE is very useful for the separation of both charged drugs using CZE and uncharged drugs using MEKC. CE is proving very valuable for the study of charged metabolites such as glucuronide and sulfate conjugates.[131,132] At present, CE separations tend to be highly efficient but are relatively slow, and drug assays are no exception. However savings in time can be achieved by the reduction in sample preparation that can be afforded by the selectivity of CE both with regard to sample loading and system selectivity. Many drugs can be analyzed directly in plasma without any need to remove proteins, although other assays have used solid-phase extract techniques to pre-fractionate the samples before CE (see Table 10.5). Another advantage is that CE is capable of working at low UV-wavelengths, i.e., <215 nm, which enables the detection of molecules without obvious chromophores. Important advantages of CE are that not only does it use very little sample but it also consumes very little in the way of buffers and buffer additives, such as chiral reagents, compared to HPLC. Whereas as a gradient HPLC system might use 5 or more liters of organic solvent per week, a similar separation by CE could use as little as 100 mL. This saving may have cost, health and safety, and environmental advantages.

Table 10.5   Selected Assays of Drugs in Biological Fluids Using Capillary Electrophoresis

Drug(s)	Matrices	Sample prep.	CE mode **	Detection	Ref.
Lithium	Serum, plasma	UF	CZE	Conductivity	266
Methotrexate	Serum	SPE	CZE	LIF	267
Cefpiramide	Plasma	Neat	SDS	UV 280 nm	49
Aspoxicillin	Plasma	Neat	SDS	UV	50
S-carboxymethyl-L-cysteine	Urine	IEC	CZE	UV 206 nm	268
Barbiturates	Serum, plasma, urine	SPE	SDS	UV	269
Thiopental	Serum	Organic	SDS	UV 290 nm	270
Cimetidine	Serum, plasma	SPE	HTAB	UV 228 nm	271
Cytosine-β-D-Arabinoside	Serum	SPE	CZE	UV	272
Cefixime	Urine	Neat	CZE	UV	274
Illicit drugs	Urine	SPE	SDS	Diode array	136, 138, 274
Xanthines	Serum, urine, saliva	Neat	SDS	UV multiwave length	275, 276
Salicylate	Urine	Neat	SDS	FL	51
Paracetamol	Plasma, urine	Neat	SDS	UV 270 nm	51
Anthracyclines	Plasma	Organic	CZE	LIF	277
Antipyrine	Plasma, saliva	Neat	SDS	UV	52, 278, 279
Fosfomycin	Plasma	Neat	CZE	Indirect UV	280, 281
6-mercaptopurine	RBC		CZE	LIF	282
Codeine	Urine				283
Bupivacaine	Drain fluid	Organic	CZE	UV	279
LSD	Urine			CE-MS	284
Suramin	Serum	PCA	CZE	UV	285, 286
Sulphonamides	Urine		CZE		287

* Abbreviations as per Table 10.3.
** Mode or MEKC additive.

A few groups have developed assays for drugs in biofluids, but in general these have either been for drugs taken in relatively high doses, e.g., paracetamol, aspirin, or where sample extract clean-up has been employed to also increase the final analyte concentration. The work of Thormann and colleagues is noteworthy in this area. In a series of publications starting around 1991,[133] they have defined conditions for the assay of many therapeutic drugs in biofluids, particularly urine, and the need to confirm drug identities using spectral scanning methods. The majority of their publications have used the same simple MEKC separation, i.e., 9 mM borate, 15 mM sodium dihydrogen phosphate buffer with 50 mM SDS at 30 kV in 50 μm capillaries. The group at the Mayo Clinic was among the first to demonstrate that in order to get additional pharmacological information from drug assays in biofluids it was necessary to link CE with mass spectrometry. In a study on the neuroleptic drug haloperidol in man, they coupled CE to an electrospray ionization source.[134] Pharmacogenetics has become an important area of drug metabolism and therapy, and a number of authors have used CE to probe differences in the metabolism of certain drugs between individuals. Lloyd et al.[135] were the first to apply this approach when they determined the phenotype from the urinary ratio of two caffeine metabolites in a urine sample following ingestion of caffeine using MEKC in 70 mM SDS in phosphate-borate buffer pH 8.43. without needing sample preparation. Most of the drug assays reported in the literature are relatively slow, that is, taking the same time as similar HPLC assays, e.g., 10 to 15 min. With current commercial CE equipment, samples must be analyzed in series, so the throughput of samples is maybe 80 to 100 per day. This is well below the throughput achievable with ELISA assays for example. Much faster analyses of drugs by CE are possible by actively cooling short capillaries and removing all sample preparation. Perrett and Ross[52] were able to assay antipyrine at a rate of 60 samples per hour for liver function studies. The same rate could most probably be achieved for many other drug analytes.

These positive reports must be countered with a number of observations that suggest that much still needs to be done to make CE a robust technique when used with biofluids that will satisfy regulatory requirements. It took some 15 years for HPLC to be accepted by the Pharmacopoeia Commissions, but a number of CE methods are presently undergoing scrutiny by the FDA. The problems with CE have already been outlined at the start of this chapter, i.e., the need for improved sensitivity, particularly in the UV, the bias that can result from selective loading, and the variations in migration time that can result from matrix effects.

Application areas where CE could be valuable are in pediatric drug monitoring. CE is clearly compatible with *in vivo* microdialysis techniques. This technique, originally developed for neurochemical studies and now being investigated for the study of drug metabolism *in vivo*, yields only µL amounts of dialysate for analysis.

Because of its high resolution and small sample requirements, a number of police and government laboratories have started to use CE to profile seizures of illicit drugs. Others have used CE to screen for banned drugs in police suspects and athletes. Wernly and Thormann[136] adapted their system described above to detect illicit drugs in SPE-extracted urine using an MEKC separation in the presence of 60 mM SDS pH 9.1. Later they were able to demonstrate that MEKC could be used to confirm the presence of 11-nor-delta-9-tetrahydrocannabinol-9-carboxylic acid in urine following the ingestion of cannabis, since this is an important forensic test. Jumppanen et al.[138] employed two successive runs to screen for 15 illicit compounds, such as diuretics in samples from sportsmen. The samples were extracted by SPE followed by an MEKC separation using 60 mM CAPS pH 10.6 buffer. Tagliaro et al.[139] exploited the capabilities of CE to use small samples in an investigation into illicit drug usage, such as cocaine and morphine, by profiling human hair.

C. Chiral Separations

Because CE uses only relatively small quantities of electrolytes and other solvents, it is very amenable to being used with exotic electrolyte modifiers such as chiral discriminators. This combined with its inherent separating power is making CE the technique of choice for analytical isomeric separations, and a relatively large literature on chiral CE has grown up. The most popular chiral modifiers are the cyclodextrins,[140-141] but many other chiral selectors, such as bile salt micelles[142] and synthetic chiral surfactants,[143] have been studied. A detailed description of the mechanisms involved in chiral CE are given in Chapter 5.

Most of the applied chiral work in the literature has been concerned with the determination of isomeric purity under quality assurance conditions and not with biofluid assays. Prunonosa et al.[144] were the first to publish a chiral CE separation applicable to biofluids. Cicletanine, an antihypertensive agent, occurs in both S(+) and R(–) forms which were shown to give different pharmacodynamic patterns using CE. They resolved the isomers in organic extracts of human plasma by MEKC using 100 mM borate pH 8.6 plus 110 mM SDS and 25 mM γ-cyclodextrin. A number of other chiral assays in biofluids have since appeared and are summarized in Table 10.6. Nearly all have used cyclodextrins as the chiral additive.

D. Screening of Biofluids for Small Molecules

Column chromatography with long slow gradients can reveal over 100 compounds which absorb at around 260 nm in normal human urine, and a similar complexity can be found in plasma. The use of multiwavelength detection would increase these numbers dramatically. Clinical chemists use such information-rich separations to provide a fingerprint of the components of normal urine and plasma and look for variations in the pattern caused by disease or therapy. This type of screening is in the first instance usually only semiquantitative.

Table 10.6   Published Assays for the Resolution of Chiral Drugs in Biological Fluids Using Capillary Electrophoresis (up to March 1996)

Drug(s)	Matrix	Sample prep.	CE mode	Chiral additive	Detection	Ref.
Cicletanine	U, P	Organic	MEKC	25 mM β-CD	215 nm	144
Racemethorphan	Urine	SPE	MEKC	60 mM β-CD		288
Racemorphan	Urine	SPE	MEKC	60 mM β-CD		288
Bupivacaine	Serum		MEKC	60 mM β-CD		289
L-buthionine-(R,S)-sulfoximine	Serum		MEKC	60 mM β-CD		290
Dimethindene and metabolites	Urine		MEKC	60 mM β-CD		291
Mephenytoin and metabolites	Urine	Neat	MEKC	40 mM β-CD	192 nm	292
Zopiclone and metabolites	Urine					293
Warfarin	Plasma	L-L	CZE	8 mM methyl-CD	210 nm	294

* Abbreviations as per Table 10.3.

The high resolution and speed offered by CE would appear ideal for the analysis of such complex biofluids.

Hundreds of studies have been published on the identification of the UV-absorbing components of urine separated by HPLC, and the elution pattern of many components is well established. Yet it is correct to say that we are some distance from being able to claim the same about electropherograms of urine. Since physical identification is not readily possible with CE, at present the identification of CE peaks in complex electropherograms is difficult. Identification can be aided by (1) co-separation with known components or peaks isolated by HPLC, (2) the marked differences observed between CE and MEKC analyses, (3) the differences between electrokinetic and hydrodynamic loading in both these systems, (4) the use of spectral information gained from scanning detectors, (5) use of alternative detectors, and (6) directed enzymatic and chemical modification of the peak(s) of interest.

Nevertheless, CE has been employed in some disease states to measure accumulation of UV-absorbing species. Jellum et al.[108] determined the accumulation of adenylosuccinate and succinylamino-imidazole-carboxamid ribotide in the urine of a patient with the rare inherited disorder, adenylosuccinase deficiency, by CZE with detection at 214 nm. Although many other peaks were present, the grossly elevated levels of the two metabolites were easily discerned. Gross et al.[145] have recently employed this assay in a purine screening clinic. Confirmation of the deficiency requires the demonstration that the enzyme is lacking in patient cells, and CE has been employed in a microassay of adenylosuccinase.[146] The author has used MEKC to optimize the resolution of endogenous compounds in urine and used this system to demonstrate a variety of purine and pyrimidine disorders in urine from affected individuals. Figure 10.7 shows such a separation from a child with Lesch-Nyhan syndrome with excessive excretion of uric acid during treatment with allopurinol, hence the drug and its metabolite.

In uremia, many metabolites not normally measurable accumulate in the blood because of the underlying renal failure. Schoots et al.[9] separated some 10 peaks and identified uric acid, hippuric acid, and p-hydroxyhippuric acid in ultrafiltered serum from uremic patients, but only one peak, urate, was found in normal serum. They used CE at pH 6 in Teflon® capillaries which gave little or no electroosmotic flow. The system was monitored at 254 nm, a wavelength which is not ideal for these compounds. In 10 patients they observed reproducible migration times (<0.9%) and peak areas (<7%) for the major components plus excellent correlations with HPLC-derived data. The other major difference noted was that CE was three times faster than the reverse-phase gradient HPLC method previously used.

In most cases, levels of endogenous purines and pyrimidines circulating in blood are low and fall below the limit of detection of CE using UV detection, so screening for disease changes in blood is much more difficult. Grune et al.[147] optimized the separation of purine bases and nucleosides in human cord plasma using CZE in a pH 9 borate buffer and applied it to the

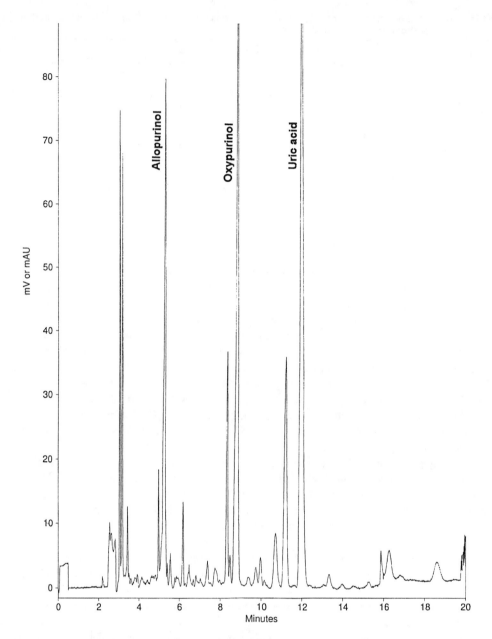

Figure 10.7 Separation of urine from a child with Lesch-Nyhan syndrome with excessive excretion of uric acid during treatment with allopurinol, hence the drug and its metabolite. CE conditions as in Figure 10.2.

accumulation of catabolites during ischemia. In order to achieve sufficient sensitivity, it was necessary to monitor at 210 nm rather than the more specific 260 nm and even then only a few peaks could be observed. In theory, CE should allow the determination of this group of compounds in a single pin-prick of blood, but this has so far not really been demonstrated.

E. Separation of Proteins and Peptides of Pharmaceutical Interest

Detailed descriptions of the techniques involved in the separation of macromolecules such as proteins and oligonucleotides have already been given elsewhere in this volume (see Chapters 8 and 9).

With the growth of biotechnology, it is increasingly important to control the quality of the proteins being produced with respect to product identification and purity as well as level and type of glycosylation. Proteins produced in this manner may be contaminated with trace amounts of other proteolytic or deamidation fragments. Among the earliest commercial applications of CE was the mapping of tryptic digests of recombinant human growth hormone to help achieve quality assurance goals.[148,149] Nielsen and Rickards, in particular, devised a number of procedures to help achieve such goals which are adaptable to many similar situations.[150,151] Other proteins assayed in this way include insulin[152] and calcitonin,[153] the tryptic peptides of which were separated in buffers dissolved in D_2O. Bihoreau et al.[154] recently used peptide mapping to control the purity of a human monoclonal anti-Rh(D) antibody produced for clinical studies. CE can also be used to monitor the stability assessment of peptide and protein drugs.[155] Even so, the sensitivity of CE with UV detection might be insufficient or require too much sample. Others[22,156] have developed on-line microreactors to digest proteins using immobilized trypsin. Yeung's group[157] employed native fluorescence to monitor tryptic digests of biopharmaceutical products. Since only tryptophan-containing peptides fluoresce, the resultant electropherograms are much simpler, often with less than half the number of peaks compared to UV detection.

Assaying the potency of a biopharmaceutical on a rapid but microscale is a major challenge to manufacturers, and CE is one analytical approach being employed. Nielsen et al. used CE to separate the antibody–antigen complexes formed against recombinant proteins[158] to help confirm product purity. Considerable attention has recently been paid to using CE to monitor the binding of molecules to macromolecules. When a ligand binds to a protein, the change in mobility leads to the separation of the bound complex from both the ligand and the free protein. The earliest studies[159] were applied to metal-binding proteins, such as calcium-binding proteins, but were soon followed by studies on drug binding, e.g., vancomycin.[160-162] Applications of affinity capillary electrophoresis have been reviewed.[163]

The role and position of attached sugars is very important for the biological functioning of biotechnologically derived proteins, and their positions need to be determined. Analysis of the microheterogeneity of a number of glycoproteins, e.g., chorionic gonadotropin,[164] has been performed by CE. The analysis of glycoproteins and other glycoforms has been reviewed by Oda et al.[165] The separation of sulfated and nonsulfated forms of peptides can also be readily achieved by CE and compares favorably with RPLC methods.[166] Similarly, phosphorylation sites can be monitored.[167] Trace contamination with other proteins has been monitored following derivatization with fluorescamine followed by a CE separation.[168]

Although not usually considered preparative, CE has been used to obtain pure fractions of some proteins.[153,169] Clearly, for many of the problems mentioned here the ultimate analytical solution is to link the separation to mass spectrometry, and that is now a main thrust of many biopharmaceutical researchers. Interfaces both on- and off-line to either electrospray[170,171] or matrix-assisted laser desorption time-of-flight (MALDI)[172] have appeared. See Chapter 2 for more details.

F. Separation of Proteins in Clinical Samples

Traditionally, electrophoresis has been used to qualitatively separate biopolymers such as proteins in clinical samples, and CE would appear to simplify and automate such separation with the added benefit of on-line quantitation. Although relatively little CE has been done so far by clinical research groups, what has been done has shown a number of problems. The principal ones are the relatively low efficiency of such separations, the binding of proteins to the capillary walls leading to changes in electroosmotic flow, and poor sample recovery. Nevertheless, sufficient work has been performed to suggest that if such difficulties can be overcome, this might be a natural area for CE. The following summarizes the present position with regard to the CE of clinically important proteins, but for a fuller discussion of the mechanisms of protein separation by CE see Chapter 8.

In 1937 Tiselius first applied the "new" technique of electrophoresis to plasma proteins, and plasma proteins continue to be routinely separated by cellulose acetate or slab gel, particularly agarose, electrophoresis, with appropriate staining to reveal the separated protein bands.[3] A number of groups have published studies on the separation of plasma and serum proteins by CE. Jorgenson and Lukacs[173] were probably the first to attempt to separate serum proteins in an open tubular glass capillary but found irreproducible migration times, which was somewhat improved when they modified the internal surface of the capillary with glycol. Later, both Chen et al.[174] and Gordon et al.[175] demonstrated that serum samples can give reproducible separations in unmodified fused-silica capillaries. Chen et al.[174] diluted their samples 1:10 with water before analysis, using borate buffer at pH 10, while Gordon et al.[175] diluted serum 40:1 with 1 mM boric acid pH 4.5 containing 20% ethylene glycol. Figure 10.8 compares the separation of plasma proteins by CE with a traditional gel separation. The separated proteins, in order of increasing migration time, were gamma-globulin, beta-globulin, alpha$_2$-globulin, alpha$_1$-globulin, and globulin. Both methods showed CE resolution to be only marginally better than that obtained by traditional agarose gel electrophoresis revealed by densitometry. However, CE offered a more rapid separation with immediate quantitation. Chen et al.[174] noted that using CE with detection at 214 nm, serum proteins have similar extinction coefficients, unlike their different staining of electrophoresis bands.

Various additives have been used to improve the separation with respect to time and resolution. Dolnik[176] investigated a number of novel buffers for the separation of serum proteins. The fastest separation of the five major protein bands took 3 min using 10 mM methylglucamine plus 5 mM lauric acid at 30 kV. In another buffer, 100 mM methylglucamine with 100 mM e-aminocaproic acid or 100 mM GABA, up to 10 protein zones could be resolved in 20 min. Jenkins et al.[177] compared the separation obtained using 50 mM borate pH 9.7 plus 1 mM calcium lactate with traditional methods for 1000 serum samples and favored the CE approach. Kim et al.[178] used a simple buffer system to compare protein patterns in disease. Reif et al.[179] have reviewed the separation of serum proteins by CE. Beckman has clearly identified the marketing of specific clinical/biochemical analyzers based on CE to be one of their priorities and has already introduced a dedicated clinical protein analyzer. In this new system, up to seven serums are analyzed simultaneously in parallel capillaries.

The separation of clinically important proteins need not be confined to blood plasma. Lal et al.[180] have resolved the proteins found in saliva by CE. Protein patterns in cerebrospinal fluid have been studied by Hiraoka et al.[181] and Cowdrey et al.[182] UV-absorbing compounds, including proteins in synovial fluid from patients with arthritic disease, have been separated by CE.[183] Recently, methods for the detection of plasma lipoproteins by CE have been published,[184] although most workers[185] have used CITP for these compounds. The protein content of human breast milk, determined using a high molarity buffer to prevent protein absorption, is shown in Figure 10.9.

The quantitation and identification of abnormal hemoglobins (Hb) is clinically important in a number of disease states. Genetically abnormal hemoglobins are found in the hemoglobinopathies and thalassemias, elevated levels of the trace glycosylated hemoglobin (HbA$_{1c}$) are used to monitor insulin therapy in diabetic patients, and carboxyhemoglobin levels are measured in toxicology. The hemoglobinopathies can be characterized at the level of the whole Hb, the intact globin, the peptide map formed from tryptic digestion of Hb, and finally the sequencing of abnormal tryptic peptides.

A number of workers have now reported separations of the intact Hb by CE. Chen et al.[175] separated the most common Hb variants at pH 8.6 in a fused-silica capillary within 10 min at only 7.5 kV with detection at 415 nm. Bolger et al.[186] achieved excellent resolution

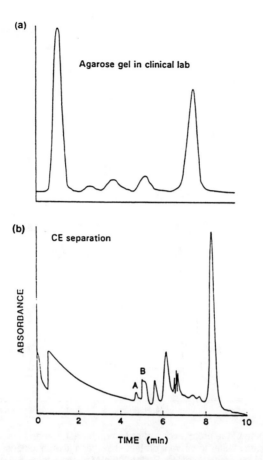

Figure 10.8 Analysis of plasma proteins by (a) traditional gel electrophoresis and (b) CE from a patient
with multiple myeloma. CE conditions: capillary fused-silica 75 μm i.d. × 37.5 cm (30.5 cm
to detector). Electrolyte 50 mM sodium tetraborate decahydrate pH 9.6. Temperature 20°C.
Sample serum diluted 40:1 with 1 mM boric acid pH 4.5 containing 20% ethylene glycol.
Load pressure 2 s Run 10 kV. Detection 200 nm. Note the reversal of peak order between
the two systems. (From Gordon, M. J., Lee, K. J., Arias, A. A., and Zare, R. N., *Anal. Chem.*,
63, 6, 1991. With permission.)

of hemoglobin variants using proprietary coated capillaries under isoelectric focusing conditions. Molteni et al.[187] used uncoated capillaries conditioned with methylcellulose to determine various hemoglobins including A1c, A, F, D, S, E, and A2. A mixture of ampholines pH 3.5 to 10 and Pharmalyte pH 6.7 to 7.7 (1:2 v/v) in a total concentration of 4.5% (w/v) gave separations which were complete in about 20 min at 20 kV. The system was simple, rapid, and compared well to gel IEF and HPLC methods.

Ferranti et al.[188] resolved the α, β, and γ globin chains from normal subjects and those with a variety of disorders within 5 minutes using a pH 2.5 buffer in a coated capillary. Law et al.[189] screened 227 samples using a high-pH system for the same separation.

Ferranti et al.[188] and ourselves[190] have produced tryptic maps of normal and abnormal Hb using CZE at pH 2.5, the analysis being complete in less than 20 min (Figure 10.10) compared to the 1 to 1.5 hours required for the commonly used reverse-phase gradient HPLC methods. The separation has been improved by the inclusion of phytic acid in the buffer.[191] As with other tryptic mapping techniques, identification of the CE peaks is difficult. Peaks can be identified by co-chromatography with known fractions collected from an HPLC separation, by spectral characterization,[190] but CE-MS is again proving to be the best approach.[192]

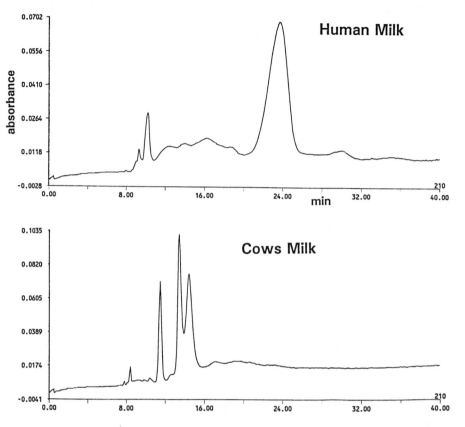

Figure 10.9 A comparison of the protein content of human breast milk and cow's milk determined using a high-molarity buffer to prevent protein absorption. CE conditions: fused-silica capillary 50 μm × 50 cm (44 cm to detector). Buffer 200 mM phosphate pH 6. Load 5 sec hydrodynamic. Sample milk diluted 1:5 with water. Run 10 kV. Detection 210 nm.

G. Single Cell and Single Molecule Analysis

The analysis of the intracellular chemical constituents of a single cell, along with the detection of single molecules, is a major challenge to analytical chemists. Single cell analyses by electrophoresis are not new. In 1965, measurements of the hemoglobins in a single erythrocyte were reported.[193] Most analyses of tissues and cells rely upon the physical extraction of molecules from at least milligram quantities prior to analysis. For many organs,

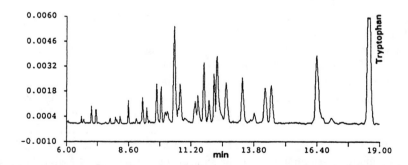

Figure 10.10 Separation of a tryptic digest of human hemoglobin A. CE conditions: fused-silica capillary 50 μm × 70 cm (50 cm to detector). Buffer 50 mM phosphate pH 2.5. Load 5 sec hydrodynamic. Sample approximately 1 mg/mL. Run 25 kV. Detection 200 nm. Tryptophan is an internal standard.

a milligram of tissue will comprise many hundreds, even thousands, of cells. However, a number of "giant" single cells do exist, and the neurons of certain species of snail, e.g., *Planorbis corneus* and *Helix aspersa,* are relatively large and therefore popular for this type of research. Using *Helix* cells Jorgenson's group[194] miniaturized the traditional approach of extracting whole cells and then subjecting the extract to microanalysis using either μLC or CE with high-sensitivity detection, particularly electrochemical. They have been able to profile over 28 amino acids and other amines in less than 20% of the extract obtained from a single cell following derivatization with NDA, followed by CZE-LIF analysis. Tseng et al. formed the fluorescent etheno derivatives of the adenine nucleotides in single oocytes and detected them by laser-induced fluorescence.[195]

In the cytoplasm of a cell, mass and concentration sensitivities are nearly equivalent. The concentration of many metabolites in the cytoplasm of neurons is of the order of 10 μM, so if the small pL volume of the cell can be sampled, the concentration will be high enough for CE. By etching small (<20 μm) capillaries to a point and then using micromanipulators, the cytoplasm of a single cell can be sampled and its contents analyzed directly. Ewing and co-workers developed this procedure and elegantly applied it to the measurement of the neurotransmitters 5-HT, dopamine, and related neurotransmitters in single neurons using electrokinetic injection and electrochemical detection at a carbon fiber electrode.[196-199] Figure 10.11 shows details of this procedure and a schematic of the etched capillary.

Another approach is to draw a few cells, or even a single cell, into the analytical capillary using a hydrodynamic load, lyse them *in situ* with a hypotonic solution which also forms the electrolyte, and then separate the components with appropriate high-sensitivity detection. This approach has been advanced principally by Yeung and co-workers. In their first study,[200] they determined intracellular proteins such as hemoglobin and carbonic anhydrase in individual erythrocytes using CE with either direct or indirect fluorescence detection. Later they applied the technique to determining the presence of hemoglobin A1$_c$ in single erythrocytes from diabetic patients.[201] Indirect fluorescence has allowed them to determine lactate and pyruvate in single erythrocytes by capillary electrophoresis.[202] They have subsequently adapted this single cell technique to other analytes in other cell types, e.g., catecholamines in adrenal chromaffin cells.[203] Gilman and Ewing[204] have also used this approach. Individual rat pheochromocytoma cells were lysed in the capillary reacted with NDA-CN reagent, and the resultant mixture of derivatized amino acids was separated in a 27 μm i.d. capillary with LIF detection. For aspartic acid, the content of an individual cell was found to be 180 amole. CZE has recently been used to monitor the transport of organic and inorganic solutes into single *Xenopus* oocytes, a technique that normally uses radioactive compounds.[205]

The drive, now, should be to make these techniques even more sensitive since snail neurons, although microscopic, are still 50 to 100 times larger than the equivalent mammalian cell. Single cell analysis has been reviewed by both Ewing[206] and Yeung.[207]

The problem with single cell analysis is one of heterogeneity among the cells. The questions raised are many. Is the cell selected typical of the whole? How many cells should be measured to gauge natural variation? Does homogenizing a number of cells give a truer answer? Whatever the result of such enquiries, the techniques being developed may well transfer to disciplines other than neurochemistry. For example the distribution of a pharmaceutical agent could be quantitatively mapped within a target organ.

Analytical chemists are now chasing what is possibly an even more difficult goal — single molecule analysis. The problems already outlined for single cells are still there, but there are also a number of new ones. The homogeneity of single enzyme molecules is now being questioned, since many forms, due to differing degrees of glycosylation and protein damage, are being shown to exist. Variability of intracellular lactate dehydrogenase isoenzymes in single human erythrocytes has already been shown by Yeung.[208] Measurement of individual molecules, whether small fluorescent molecules or large DNA, is also possible by CE. A fluorescent signal from single DNA has been shown.[209] Chen and Dovichi[210] have

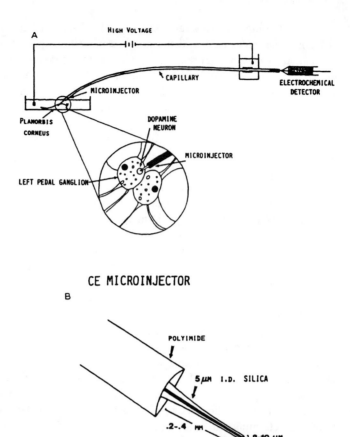

CE MICROINJECTOR

Figure 10.11 Schematic diagrams used for the injection, separation and detection of cytoplasmic samples from single cells. (a). The complete system with an exploded view of the ganglia of *Planorbis corneus* with the microinjector inserted directly into the dopamine neuron. (b) A diagram showing the end of an analytical capillary following etching in hydrofluoric acid. (From Chien, J. B., Wallingford, R., and Ewing, A. G., *J. Neurochem.*, 54, 633, 1990. With permission.)

shown that with LIF the counting of molecules is only 50% efficient. So whereas the injection of 10,000 molecules of B-phycoerythrin (17 zmol) leads to a peak area with 1% relative precision, the injection of 3000 molecules leads to irreproducible results, although single molecules can sometimes be observed. This molecular shot-noise is then a fundamental limit to chemical analysis!

H. Separation of Charged Particles and Whole Cells

Provided a particle is charged, it will exhibit an electrophoretic mobility when placed in a suitable field. The size of the particle is not particularly relevant. So far I have concentrated on the separation of molecules, but clearly the micelles used in MEKC also possess an electrophoretic mobility. Other solid particles such as polystyrene microspheres[211] and silica sols[212] have surface charges and have been resolved by CE. Viruses also possess charged protein coats and are amenable to separation by electrophoresis. Both Hjerten[212] and Grossman and Soane[214] demonstrated the ability of CE to separate intact tobacco mosaic virus particles. The potential of CE for the separation of other virus types is clearly present but as yet unexplored.

Intact biological organelles can be separated by CE. In an early study, Zhu and Chen[215] separated whole blood cells using 0.45 mm i.d. fluorinated ethylene-propylene copolymer (FEP) tubing and isotonic electrolytes such as 0.012% HEPES containing 5.1% glucose pH 7.2. At 20 kV human red cells gave a broad but distinct peak and could be resolved from those of other species which typically ran much later. Over the short term, reproducibility was good, but after a number of runs the FEP tubing was chemically degraded. Recently two further examples of whole cell separations have appeared; Ebersole and McCormick[216] separated and isolated viable bacteria by CZE, while Radko et al.[218] separated rat liver microsomes using polymer solutions. The use of CZE for the physical separation of other cell types, such as white blood cells, deserves further investigation in the light of current developments in coating technology.

I. Monitoring Enzyme Reactions and Immunoassays

Because of its low sample requirements, CE is ideal for monitoring many forms of chemical reaction, and enzyme assays are no exception. Such assays were first performed off-line with analysis of prepared samples[218] or with the development of temperature-controlled sample trays on-line with repeated monitoring of the same sample.[132] By incorporating the substrate into the electrolyte, it is possible to perform enzyme assays within the capillary.[219]

Recently a number of groups have attempted to monitor immunoassays by CE since the antibody-labeled antigen complex will have a different charge-to-mass ratio than either of its components and will therefore be separable electrophoretically. The aim of many of these groups is to automate clinical immunoassays either using multiple capillaries in parallel or microchip devices. Initial results on CE-immunoassays were presented by Schultz and Kennedy using insulin.[220] Subsequently CE-immunoassays for other biomolecules such as human growth hormone,[221] cortisol,[222] and drugs such as chloramphenicol,[222] theophylline,[224] and opiates,[225] have been published. Such assays required no cleanup or extraction and in the case of the cortisol assay were sensitive down to a few μg/L of analyte. This area of CE research has been reviewed.[174,226]

VI. CONCLUSIONS

CE is probably at about the same stage of development as HPLC was around 1980 with regard to separation technology but more advanced with respect to automation. However, many improvements both in its technology and its reliability as well as an increased understanding of the processes involved for CE are necessary to replace appropriate HPLC methods. The majority of the commercial CE systems are clearly derived from HPLC instrumentation, but it is hoped the second generation of instruments which are about to appear will be designed from scratch. They will, one hopes, be purpose-designed for the technique, offering, for example, detectors capable of working very close to the end of the capillary rather than 75% along its length, thus improving the available potential gradient, and sample vials capable of microsamples. The advancement of CE-MS would appear to be crucial for many of CE's roles in the future. LC-MS is now firmly established within research groups in the pharmaceutical industry, and it is to be expected that bench-top CE-MS systems will appear soon.

The importance of biomedical analysis to CE should be that it is the very area to which CE is best suited. Its potential can be judged from the those applications discussed in this chapter. The application of CE to bioanalysis is clearly at a crucial point. Its areas of application have now been outlined, a rapidly increasing number of applications are available, and provided the next few years see the commercialization of some of the research techniques described in this chapter, CE should become a powerful analytical tool for the types of compounds to which it is best suited.

REFERENCES

1. Durrum, E. L., Paper electrophoresis in *A Manual of Paper Chromatography and Paper Electrophoresis* (2nd Edition), Block, R. J., Durrum, E. L., and Zweig, G., Eds., Academic Press, New York, 1958, 580.
2. Smith, I., *Chromatography and Electrophoretic Techniques,* Vol II. *Zone Electrophoresis* (2nd Edition), Heinemann, London, 1968.
3. Andrews, A. T., *Electrophoresis* (2nd Edition), Oxford University Press, Oxford, 1986.
4. Jorgenson, J. W., and Lukacs, K. D., Zone electrophoresis in open-tubular glass capillaries, *Anal. Chem.,* 53, 1298, 1981.
5. Jorgenson, J. W. and Lukacs, K. D., High-resolution separations based on electrophoresis and electroosmosis, *J. Chromatogr.,* 218, 209, 1981.
6. Jorgenson, J. W. and Lukacs, K. D., Free-zone electrophoresis in glass capillaries, *Clin. Chem.,* 27, 1551, 1981.
7. Terabe, S., Otsuka, K., Ichikawa, K., Tsuchiya, A., and Ando, T., Electrokinetic separations with micellar solutions and open-tubular capillaries, *Anal. Chem.,* 56, 111, 1984.
8. Guttman, A., Cohen, A. S., Heiger, D. N., and Karger, B. L., Analytical and micropreparative ultrahigh resolution of oligonucleotides by polyacrylamide gel high-performance capillary electrophoresis, *Anal. Chem.,* 62, 137, 1990.
9. Schoots, A. C., Verheggen, T. P. E. M., DeVries, P. M. J. M., and Everaerts, F. M., Ultraviolet-absorbing organic anions in uremic serum separated by capillary zone electrophoresis, *Clin. Chem.,* 36, 435, 1990.
10. Perrett, D. and Ross, G. R., Capillary electrophoresis for the analysis of cellular nucleotides, in *Human Purine and Pyrimidine Metabolism VII Part B,* Harkness, R. A., Elion, G. B., and Zollner, N., Eds., Plenum Press, New York, 1991, 1.
11. Issaq, H. J., Janini, G. M., Atamna, I. Z., and Muschik, G. M., Separations by high performance liquid chromatography and capillary zone electrophoresis: A comparative study, *J. Liq. Chromatogr.,* 14, 817, 1991.
12. Carney, S. L. and Osborne, D. J., The separation of chondroitin sulfate disaccharides and hyaluronan oligosaccharides by capillary zone electrophoresis, *Anal. Biochem.,* 195, 132, 1991.
13. Rudd, P. M., Scragg, I. G., Coghill, E., and Dwek, R. A., Separation and analysis of the glycoform populations of ribonuclease-b using capillary electrophoresis, *Glycoconjugate,* 9, 86, 1992.
14. Bergman, T., Agerberth, B., and Jornvall, H., Direct analysis of peptides and amino acids from capillary electrophoresis, *FEBS Letters,* 283, 100, 1991.
15. Camilleri, P. and Okafo, G., Capillary electrophoresis in deuterated water solution, *J. Chromatogr.,* 541, 489, 1991.
16. Szeman J. and Ganzler K., Use of cyclodextrins and cyclodextrin derivatives in high-performance liquid chromatography and capillary electrophoresis, *J. Chromatogr.,* 668, 509, 1994.
17. Green, J. S. and Jorgenson, J. W., Variable-wavelength on-column fluorescence detector for open-tubular zone electrophoresis, *J. Chromatogr.,* 352, 337, 1986.
18. Kurosu, Y., Sasaki, T., and Saito, M., Fluorescence detection with an immersed flow cell in capillary electrophoresis, *J. High Res. Chromatogr.,* 14, 186, 1991.
19. Perrett, D., Faleye, J., and Ross, G., Comparison of capillary electrophoresis and HPLC with fluorescence detection for the determination of 5-HT, 5-HIAA and tryptophan in biological samples, in *Monitoring Molecules in Neuroscience,* Rollema, H., Westerink, B., and Drifhout, W. J., Eds., 152 Groningen, University Centre for Pharmacy, 1991.
20. Caslavska J., Gassmann, E., and Thormann, W., Modification of a tunable UV-visible capillary electrophoresis detector for simultaneous absorbance and fluorescence detection: Profiling of body fluids for drugs and endogenous compounds, *J. Chromatogr. A,* 709, 147, 1995.

21. Watzig, H. and Dette, C., Precise quantitative capillary electrophoresis: Methodological and instrumental aspects, *Fresenius J. Anal. Chem.*, 345, 403 1993 (n.b. updated in Poster at HPCE96, Orlando, 1996).
22. Cobb, K. A. and Novotny, M., High-sensitivity peptide mapping by capillary zone electrophoresis and microcolumn liquid chromatography, using immobilized trypsin for protein digestion, *Anal. Chem.*, 61, 2226, 1989.
23. Ling, B. L., Baeyens, W. R. G., and Dewaeles, C., Comparison of micro-LC and capillary zone electrophoresis for the analysis of thiols, *J. High Res. Chromatogr.*, 14, 169, 1991.
24. Chervet, J. P., Van-Soest, R. E. J., and Ursem, M., Z-shaped flow cell for UV detection in capillary electrophoresis, *J. Chromatogr.*, 543, 439, 1991.
25. Moring, S. E., Reel, R. T., and van-Soest, R. E. J., Optical improvements of a Z-shaped cell for high-sensitivity UV absorbance detection in capillary electrophoresis, *Anal. Chem.*, 65, 3454, 1993.
26. Tsuda, T., Sweedler, J. V., and Zare, R. N., Rectangular capillaries for capillary zone electrophoresis, *Anal. Chem.*, 62, 2149, 1990.
27. Izumi, T., Nagahori, T., and Okuyama, T., Capillary electrophoresis of immunoglobin using flat capillary tubing, *J. High Res. Chromatogr.*, 14, 351, 1991.
28. Heiger, D. N., Kaltenbach, P., and Sievert, H. J. P., Diode array detection in capillary electrophoresis, *Electrophoresis*, 15, 1234, 1994.
29. Taylor, J. A. and Yeung, E. S., Axial-beam absorbance detection for capillary electrophoresis, *J. Chromatogr.*, 550, 831, 1991.
30. Grant, I. H. and Steuer, W., Extended pathlength UV absorbance detector for capillary zone electrophoresis, *J. Microcol. Sep*, 2, 74, 1990.
31. Wang, T., Aiken, J. H., Huie, C. W., and Hartwick, R. A., Nanolitre-scale multireflection cell for absorption detection in capillary electrophoresis, *Anal. Chem.*, 63, 1372, 1991.
32. Chien, R.-L. and Burgi, D. S., On-column sample concentration using field amplification in CZE, *Anal. Chem.*, 64, 489A, 1992.
33. Burgi, D. S. and Chien, R.-L., Improvement in the method of sample stacking for gravity injection in capillary zone electrophoresis, *Anal. Biochem.*, 202, 306, 1992.
34. Chien, R.-L. and Burgi, D. S., Sample stacking of an extremely large injection volume in high performance capillary electrophoresis, *Anal. Chem.*, 64, 1046, 1992.
35. Foret, F., Sustacek, V., and Bocek, P., On-line isotachophoretic sample preconcentration for enhancement of zone detectability in capillary zone electrophoresis, *J. Microcol. Sep.*, 2, 229, 1990.
36. Stegehuis, D. S., Irth, H., Tjaden, U. R., and Van der Greef, J., Isotachophoresis as a on-line concentration pre-treatment technique in capillary electrophoresis, *J. Chromatogr.*, 538, 393, 1991.
37. Dolnik, V., Cobb, K. A., and Novotny, M., Capillary zone electrophoresis of dilute samples with isotachophoretic preconcentration, *J. Microcol. Sep.*, 2, 127, 1990.
38. Foret, F., Szoko, E., and Karger, B. L., On-column transient and coupled column isotachophoretic pre-concentration of protein samples in capillary zone electrophoresis, *J. Chromatogr.*, 608, 3, 1992.
39. Tinke, A. J., Reinhoud, N. J., Niessen, W. M. A., Tjaden, U. R., and Van-der-Greef, J., Online capillary isotachophoretic analyte focusing for improvement of detection limits in capillary electrophoresis/electrospray mass spectrometry, *Rapid Commun. Mass. Spectrom.*, 6, 560, 1992.
40. Cai, J. and El-Rassi, Z., On-line pre-concentration of triazine herbicides with tandem octadecyl capillaries — capillary zone electrophoresis, *J. Liquid Chromatogr.*, 15, 1179, 1992.
41. Beattie, J. H., Self, R., and Richards, M. P., The use of solid phase concentrators for on-line preconcentration of metallothionein prior to isoform separation by capillary zone electrophoresis, *Electrophoresis*, 16, 322, 1995.
42. Tomlinson, A. J. and Naylor, S., Systematic development of on-line membrane preconcentration-capillary electrophoresis-mass spectrometry for the analysis of peptide mixtures, *J. Cap. Elec.*, 2, 225, 1995.
43. Tomlinson, A. J., Guzman, N. A., and Naylor, S., Enhancement of concentration limits of detection in CE and CE-MS: A review of on-line sample extraction, cleanup, analyte preconcentration and microreactor technology, *J. Cap. Elec.*, 2, 247, 1995.

44. McLaughlin, G. M., Nolan, J. A., Lindahl, J. L., Palmieri, R. H., Anderson, K. W., Morris, S. C., Morrison, J. A., and Bronzert, T. J., Pharmaceutical drug separations by HPCE: Practical guidelines, *J. Liq. Chromatogr.*, 15, 961, 1992

45. Williams, P. E., Marino, M. A., Delrio, S. A., Turni, L. A., and Devaney, J. M., Analysis of DNA restriction fragments and polymerase chain reaction products by capillary electrophoresis, *J. Chromatogr. A*, 680, 525, 1994.

46. Lambert, W. J. and Middleton, D. L., pH hysteresis effect with silica capillaries in capillary zone electrophoresis, *Anal. Chem.*, 62, 1585, 1990.

47. Blanchard, J., Evaluation of the relative efficacy of various techniques for deproteinizing plasma samples prior to high-performance liquid chromatographic analysis, *J. Chromatogr.*, 226, 455, 1981.

48. Huang, X. H., Gordon, M. J., and Zare, R. N., Bias in quantitative capillary zone electrophoresis caused by electrokinetic injection, *Anal. Chem.*, 60, 375, 1989.

49. Nakagawa, T., Oda, Y., Shibukawa, A., and Tanaka, H., Separation and determination of cefpiramide in human plasma by electrokinetic chromatography with a micellar solution and an open tubular-fused silica capillary., *Chem. Pharm. Bull.*, 37, 707, 1989.

50. Nishi, H., Fukayama, T., and Matsuo, M., Separation and determination of aspoxicillin in human plasma by micellar electrokinetic chromatography with direct sample injection, *J. Chromatogr.*, 515, 245, 1990.

51. Perrett, D. and Ross, G., Capillary electrophoresis of drugs in biological fluids, in *BioAnalytical Approaches for Drugs Including Anti-Asthmatics and Metabolites,* Reid, E. and Wilson, I. D., Eds., Royal Society of Chemistry, Cambridge, U.K., 1992, p. 269.

52. Perrett, D. and Ross, G. A., Rapid determination of drugs in biofluids by capillary electrophoresis — Measurement of antipyrine in saliva for pharmacokinetic studies, *J. Chromatogr. A*, 700, 179, 1995.

53. Thormann, W., Molteni, S., Caslavska, J., and Schmutz, A., Clinical and forensic applications of capillary electrophoresis, *Electrophoresis*, 15, 3, 1994.

54. Goodall, D. M., Williams, S. J., and Lloyd, D. K., Quantitative aspects of capillary electrophoresis, *Trends in Analytical Chemistry*, 10, 272, 1991.

55. Lee, T. T. and Yeung, E. S., Compensating for instrumental and sampling biases occurring electrokinetic injection in capillary zone electrophoresis, *Anal. Chem.*, 64, 1226, 1992.

56. Altria, K. D. and Rudd, D. R., An overview of method validation and system suitability aspects in capillary electrophoresis, *Chromatographia*, 41, 325, 1995.

57. Altria, K. D., Clayton, N. G., Hart, M., Harden, R. C., Hevizi, J., Makwana, J. V., and Portsmouth, M. J., An inter-company cross-validation exercise on capillary electrophoresis testing of dose uniformity of paracetamol content in formulations, *Chromatographia*, 39, 180, 1994.

58. Altria, K. D., Clayton, N. G., Harden, R. C., Makwana, J. V., and Portsmouth, M. J., Inter-company cross validation exercise on capillary electrophoresis. Quantitative determination of drug counter-ion level, *Chromatographia*, 40, 47, 1995.

59. Altria, K. D. and Filbey, S. D., The application of experimental design to the robustness testing of a method for the determination of drug-related impurities by capillary electrophoresis, *Chromatographia*, 39, 306, 1994.

60. Hayashi, Y., Matsuda, R., and Terabe, S., Optimization of precision and throughput in micellar electrokinetic chromatography, *Chromatographia*, 37, 149, 1993.

61. Dose, E. V. and Guiochon, G., Internal standardisation technique for capillary zone electrophoresis, *Anal. Chem.*, 63, 1154, 1991.

62. Koh, E. V., Bissell, M. G., and Ito, R. K., Measurement of vitamin-C by capillary electrophoresis in biological fluids and fruit beverages using a stereoisomer as an internal standard, *J. Chromatogr.*, 633, 245, 1993.

63. Li, S. F. Y., Ng, C. L., and Ong, C. P., Pharmaceutical analysis by capillary electrophoresis, in *Advances in Chromatography* 35, Brown, P. R. and Grushka, E., Eds., Marcel Dekker, New York, 1995, p. 199.

64. Rabel, S. R. and Stobaugh, J. F., Applications of capillary electrophoresis in pharmaceutical analysis, *Pharm. Res.*, 10, 171, 1993.

65. Smith, N. W. and Evans, M. B., Capillary zone electrophoresis in pharmaceutical and biomedical analysis, *J. Pharm. Biomed. Anal.,* 12, 579, 1994.
66. Kleparnik, K. and Bocek, P., Theoretical background for clinical and biomedical applications of electromigration techniques, *J. Chromatogr. Biomed. Appl.,* 569, 3, 1991.
67. Landers, J. P., Clinical capillary electrophoresis, *Clin. Chem.,* 41, 495, 1995.
68. Xu, Y., Capillary electrophoresis, *Anal. Chem.,* 67, R463–R473, 1995.
69. Watzig, H. and Dette, C., Capillary electrophoresis (CE) — a review — Strategies for method development and applications related to pharmaceutical and biological sciences, *Pharmazie,* 49, 83, 1994.
70. Tsuda, T., Nakagawa, G., Sato, M., and Yag, K., Separation of nucleotides by high-voltage capillary electrophoresis, *J. Appl. Biochem.,* 5, 330, 1983.
71. Tsuda, T., Takagi, K., Watanabe, T., and Satake, T., Separation of nucleotides in organs of guinea pig by capillary zone electrophoresis, *HRC&CC J. High Resolut. Chromatogr. Chromatogr. Comm.,* 11, 721–723, 1988.
72. Huang, M. X., Liu, S. F., Murray, B. K., and Lee, M. L., High resolution separation and quantitation of ribonucleotides using capillary electrophoresis, *Anal. Biochem.,* 207, 231, 1992.
73. O'Neill, K., Shao, X., Zhao, Z., Malik, A., and Lee, M. L., Capillary electrophoresis of nucleotides on Ucon-coated fused silica columns, *Anal. Biochem.,* 222, 185, 1994.
74. Liu, S., Huang, M., Lee, M. L., North, J. A., and Murray, B. K., Analysis of intracellular ribonucleotide pools in a drug-treated human cancer cell line using capillary zone electrophoresis, *J. Microcol. Sep.,* 6, 49, 1994.
75. Hernandez, L., Hoebel, B. G., and Guzman, N. A., Analysis of cyclic nucleotides by capillary electrophoresis using ultraviolet detection, in *Analytical Biotechnology. Capillary Electrophoresis and Chromatography* (ACS Sym Series No 434), Horvath, C. and Nikelly, J., Eds., American Chemical Society, Washington, DC, 1990, p. 50.
76. Liu, J., Banks, F., and Novotny, M., High-speed micellar electrokinetic capillary chromatography of common phosphorylated nucleosides, *J. Microcol. Sep.,* 1, 136, 1989.
77. Takigiku, R. and Schneider, R. E., Reproducibility and quantitation of separation for ribonucleoside triphosphates and deoxyribonucleoside triphosphates by capillary zone electrophoresis, *J. Chromatogr.,* 559, 247, 1991.
78. Masson, C., Luong, J. H. T., and Nguyen, A. L., Analyses of uric acid in urine and reconstituted serum by capillary electrophoresis, *Anal. Lett.,* 24, 377, 1991.
79. Miyake, M., Shiukawa, A., and Nakagawa, T., Simultaneous determination of creatinine and uric acid in human plasma and urine by micellar electrokinetic chromatography, *J. High Res. Chromatogr.,* 14, 180, 1991.
80. Guzman, N. A., Berck, C. M., Hernandez, L., and Advis, J. P., Capillary electrophoresis as a diagnostic tool: Determination of biological constituents present in urine of normal and pathological individuals, *J. Liq. Chromatogr.,* 13, 3843, 1990.
81. Thormann, W., Minger, A., Molteni, S., Caslavska, J., and Gebauer, P., Determination of substituted purines in body fluids by micellar electrokinetic capillary chromatography with direct sample injection, *J. Chromatogr.,* 593, 275, 1992.
82. Lecoq, A. F., Montanarella, L., and Di-Biase, S., Separation of nucleosides and nucleotide-3'-monophosphates by micellar electrokinetic capillary chromatography, *J. Microcol. Sep.,* 5, 105, 1993.
83. Guarnieri, C., Muscari, C., Stefanelli, C., Giaccari, A., Zini, M., and Di-Biase, S., Micellar electrokinetic capillary chromatography of 8-hydroxydeoxyguanosine acid and other oxidized derivatives of DNA, *J. Chromatogr. B — Biomed. Appl.,* 656, 209, 1994.
84. Grune, T. and Perrett, D., Rapid simultaneous measurement of nucleotides, nucleosides and bases in tissues by capillary electrophoresis, in *Purine and Pyrimidine Metabolism in Man* VIII, Sahota, A. and Taylor, M. W., Eds., Plenum Press, New York, 1995, p. 805.
85. Tagliaro, F., Moretto, S., Valentini, R., Gambaro, G., Antonioli, C., Moffa, M., and Tato, L., Capillary zone electrophoresis determination of phenylalanine in serum — A rapid, inexpensive and simple method for the diagnosis of phenylketonuria, *Electrophoresis,* 15, 94, 1994.
86. Engstrom-Silverman, C. E. and Ewing, A. G., Copper wire amperometric detector for capillary electrophoresis, *J. Microcol. Sep.,* 3, 141, 1991.

87. Kuhr, W. G. and Yeung, E. S., Indirect fluorescence detection of native amino acids in capillary zone electrophoresis, *Anal. Chem.,* 60, 1832, 1988.

88. Olefirowicz, T. M. and Ewing, A. G., Capillary electrophoresis with indirect amperometric detection, *J. Chromatogr.,* 499, 713, 1990.

89. Lu, W., Yang, G., and Cole, R. B., Determination of amino acids by on-line capillary electrophoresis-electrospray ionization mass spectrometry, *Electrophoresis*, 16, 487, 1995.

90. Rose, D. J. and Jorgenson, J. W., Post-capillary fluorescence detection in capillary zone electrophoresis using o-phthaldialdehyde, *J. Chromatogr.,* 447, 117, 1988.

91. Pentoney, S. L., Huang, X. H., Burgi, D. S., and Zare, R. N., On-line connector for microcolumns: application to the on-column o-phthaldialdehyde derivatization of amino acids separated by capillary zone electrophoresis, *Anal. Chem.,* 60, 2625, 1988.

92. Zhu, R. and Kok, W. T., Post-column reaction system for fluorescence detection in capillary electrophoresis, *J. Chromatogr. A*, 716, 123, 1995.

93. Baeyens, W. R. G., Ling, B. L., Imai, K., Calokerinos, A. C., and Schulman, S. G., Chemiluminescence detection in capillary electrophoresis, *J. Microcol. Sep.*, 6, 195, 1994.

94. Novotny, M., Recent advances in the isolation and structural studies of biomacromolecules using microcolumn techniques, *J. Microcol. Sep.*, 2, 7, 1990.

95. Albin, M., Weinberger, R., Sapp, E., and Moring, S., Fluorescence detection in capillary electrophoresis — evaluation of derivatising reagents and techniques, *Anal. Chem.,* 63, 417, 1991.

96. Chen, D. Y., Adelhelm, K., Cheng, X. L., and Dovichi, N. J., Simple laser-induced fluorescence detector for sulforhodamine 101 in a capillary electrophoresis system — Detection limits of 10 yoctomoles or six molecules, *Analyst*, 119, 349, 1994.

97. Wu, S. and Dovichi, N. J., High-sensitivity fluorescence detector for fluorescein isothiocyanate derivatives of amino acids separated by capillary zone electrophoresis, *J. Chromatogr.*, 480, 141, 1989.

98. Otsuka, K., Terabe, S., and Ando, T., Electrokinetic chromatography with micellar solutions: Separation of phenylthiohydantoin-amino acids, *J. Chromatogr.,* 332, 219, 1985.

99. Terabe, S., Ishihama, Y., Nishi, H., Fukuyama, T., and Otsuka, K., Effect of urea addition on micellar electrokinetic chromatography, *J. Chromatogr.,* 545, 359, 1991.

100. Little, E. L. and Foley, J. P., Optimization of the resolution of PTH-amino acids through control of surfactant concentration in micellar electrokinetic chromatography: SDS vs. BRIJ-35/SDS micellar systems, *J. Microcol. Sep.,* 4, 145, 1992.

101. Kim, N. J., Kim, J. H., and Lee, K. J., Application of capillary electrophoresis to amino acid sequencing of peptides, *Electrophoresis*, 16, 510, 1995.

102. Rohlicek, V. and Deyl, Z., Simple apparatus for capillary zone electrophoresis and its application to protein analysis, *J. Chromatogr.*, 494, 87, 1989.

103. Honda, S., Iwase, S., Makino, A., and Fujiwara, S., Simultaneous determination of reducing monosaccharides by capillary zone electrophoresis as the borate complexes of N-2-pyridylglycamines, *Anal. Biochem.*, 176, 72, 1989.

104. Hoffstetter-Kuhn, S., Paulus, A., Gassmann, E., and Widmer, E. M., Influence of borate complexation on the electrophoretic behavior of carbohydrates in capillary electrophoresis, *Anal. Chem.,* 63, 1541, 1991.

105. Camilleri, P., Harland, G. B., and Okafo, G., High resolution and rapid analysis of branched oligosaccharides by capillary electrophoresis, *Anal. Biochem.*, 230, 115, 1995.

106. Damm, J. B. L., Overklift, G. T., Vermeulen, B. W. M., Fluitsma, C. F., and Van-de-Dem, G. W. K., Separation of natural and synthetic heparin fragments by high-performance capillary electrophoresis, *J. Chromatogr.*, 608, 297, 1992.

107. Karamanos, N. K., Axelsson, S., Vanky, P., Tzanakakis, G. N., and Hjerpe, A., Determination of hyaluronan and galactosaminoglycan disaccharides by high-performance capillary electrophoresis at the attomole level. Applications to analyses of tissue and cell culture proteoglycans, *J. Chromatogr. A*, 696, 295, 1995.

108. Jellum, E., Thorsrud, A. K., and Time, E., Capillary electrophoresis for diagnosis and studies of human disease, particularly metabolic disorders, *J. Chromatogr.*, 559, 455, 1991.

109. Stamler, J. S. and Loscalzo, J., Capillary zone electrophoretic detection of biological thiols and their s-nitrosated derivatives, *Anal. Chem.,* 64, 779, 1992.

110. Lin, B. L., Colon, L. A., and Zare, R. N., Dual electrochemical detection of cysteine and cystine in capillary zone electrophoresis, *J. Chromatogr. A*, 680, 263, 1994.

111. Zhou, J. X., O'Shea, T. J., and Lunte, S. M., Simultaneous detection of thiols and disulfides by capillary electrophoresis electrochemical detection using a mixed-valence ruthenium cyanide-modified microelectrode, *J. Chromatogr. A*, 680, 271, 1994.

112. Huang, X. J. and Kok, W. T., Determination of thiols by capillary electrophoresis with electrochemical detection using a palladium field-decoupler and chemically modified electrodes, *J. Chromatogr. A*, 716, 347, 1995.

113. Weinberger, R., Sapp, E., and Moring, S., Capillary electrophoresis of urinary porphyrins with absorbance and fluorescence detection, *J. Chromatogr.,* 516, 271, 1990.

114. Vanberkel, G. J., McLuckey, S. A., and Glish, G. L., Electrospray ionization of porphyrins using a quadrupole ion trap for mass analysis, *Anal. Chem.,* 63, 1098, 1991.

115. Wu, N., Li, B. H., and Sweedler, J. V., Recent developments in porphyrin separations using capillary electrophoresis with native fluorescence detection, *J. Liq. Chromatogr.*, 17, 1917, 1994.

116. Yao, Y. J., Lee, H. K., and Li, S. F. Y., Optimization of separation of porphyrins by micellar electrokinetic chromatography using the overlapping resolution mapping scheme, *J. Chromatogr.*, 637, 195, 1993.

117. Swaile, D. F. and Sepaniak, M. J., Determination of metal ions by capillary zone electrophoresis with on-column chelation using 8-hydroxyquinoline-5-sulphonic acid, *Anal. Chem.*, 63, 179, 1991.

118. Jandik, P. and Jones, W. R., Optimisation of detection sensitivity in the capillary electrophoresis of inorganic anions, *J. Chromatogr.*, 546, 431, 1991.

119. Harrold, M. P., Wojtusik, M. J., Riviello, J., and Henson, P., Parameters influencing separation and detection of anions by capillary electrophoresis, *J. Chromatogr.*, 640, 463, 1993.

120. Wildman, B. J., Jackson, P. E., Jones, W. R., and Alden, P. G., Analysis of anion constituents of urine by inorganic capillary electrophoresis, *J. Chromatogr.,* 546, 459, 1991.

121. Reid, R. H. P., Electrophoretic behaviour of a group of organic anions of biochemical interest in a functionally coherent series of buffers, *J. Chromatogr. A*, 669, 151, 1994.

122. Holmes, R. P., Measurement of urinary oxalate and citrate by capillary electrophoresis and indirect ultraviolet absorbance, *Clin. Chem.*, 41, 1297, 1995.

123. Leone, A. M., Francis, P. L., Rhodes, P., and Moncada, S., A rapid and simple method for the measurement of nitrite and nitrate in plasma by high performance capillary electrophoresis. *Biochem Biophys Res Comm*, 200, 951, 1994.

124. Janini, G. M., Chan, K. C., Muschik, G. M., and Issaq, H. J., Analysis of nitrate and nitrite in water and urine by capillary zone electrophoresis, *J. Chromatogr. B — Biomed. Appl.*, 657, 419, 1994.

125. Ueda, T., Maekawa, T., Sadamitsu, D., Oshita, S., Ogino, K., and Nakamura, K., The determination of nitrite and nitrate in human blood plasma by capillary zone electrophoresis, *Electrophoresis*, 16, 1002, 1995.

126. Nishi, H. and Terabe, S., Application of electrokinetic chromatography to pharmaceutical analysis, *J. Pharm. Biomed. Anal.*, 11, 1277, 1993.

127. Lloyd, D. K., Pharmaceutical and biomedical application of capillary electrophoresis, in *Pharmaceutical and Biomedical Applications of Liquid Chromatography*, Riley, C. M., Lough, W. J., and Wainer, I. W., Eds., Pergamon Press, Oxford, 1994, p. 3.

128. Altria, K. D., Application of capillary electrophoresis to pharmaceutical analysis, in *Capillary Electrophoresis Guidebook* (Series: *Methods in Molecular Biology* Vol. 52), Altria, K. D., Ed., Humana Press, Totowa, New Jersey, 1996, p. 265.

129. Li, S. F. Y., Ng, C. L., and Ong, C., Pharmaceutical analysis by capillary electrophoresis, in *Advances in Chromatography* 35, Brown, P. R. and Grushka, E., Marcel Dekker, New York, 1995, p. 199.

130. Perrett, D., *A Database of the Capillary Electrophoresis Literature,* The Chromatographic Society, Nottingham, U.K., 1996.

131. Wernly, P., Thormann, W., Bourquin, D., and Brenneisen, R., Determination of morphine-3-glucuronide in human urine by capillary zone electrophoresis and micellar electrokinetic capillary chromatography, *J. Chromatogr. — Biomed. Appl.*, 616, 305, 1993.

132. Taylor, M. R., Westwood, S. A., and Perrett, D., Phase II drug metabolites, *J. Chromatogr.*, in press, 1996.

133. Thormann, W., Meier, P., Marcolli, C., and Binder, F., Analysis of barbiturates in human serum and urine by high-performance capillary electrophoresis micellar electrokinetic capillary chromatography with on-column multi-wavelength detection, *J. Chromatogr.*, 545, 445, 1991.

134. Tomlinson, A. J., Benson, L. M., Johnson, K. L., and Naylor, S., Investigation of the metabolic fate of the neuroleptic drug haloperidol by capillary electrophoresis electrospray ionization mass spectrometry, *J. Chromatogr. — Biomed. Appl.*, 621, 239, 1993.

135. Lloyd, D. K., Fried, K., and Wainer, I. W., Determination of n-acetylator phenotype using caffeine as a probe compound — a comparison of high-performance liquid chromatography and capillary electrophoresis methods, *J. Chromatogr. — Biomed. Appl.*, 578, 283, 1992.

136. Wernly, P. and Thormann, W., Analysis of illicit drugs in human urine by micellar electrokinetic capillary chromatography with on-column fast scanning polychrome absorption detection, *Anal. Chem.*, 63, 2878, 1991.

137. Wernly, P. and Thormann, W., Confirmation testing of 11-nor-delta-9-tetrahydrocannabinol-9-carboxylic acid in urine with micellar electrokinetic capillary chromatography, *J. Chromatogr.*, 608, 251, 1992.

138. Jumppanen, J., Siren, H., and Riekkola, M. L., Screening for diuretics in urine and blood serum by capillary zone electrophoresis, *J. Chromatogr.*, 652, 441, 1993.

139. Tagliaro, F., Poiesi, C., Aiello, R., Dorizzi, R., Ghielmi, S., and Marigo, M., Capillary electrophoresis for the investigation of illicit drugs in hair — Determination of cocaine and morphine, *J. Chromatogr.*, 638, 303, 1993.

140. Sepaniak, M. J., Cole, R. O., and Clark, B. K., Use of native and chemically modified cyclodextrins for the capillary electrophoretic separation of enantiomers, *J. Liq. Chromatogr.*, 15, 1023, 1992.

141. Snopek, J., Soini, H., Novotny, M., Smolkovakeulemansova, E., and Jelinek, I., Selected applications of cyclodextrin selectors in capillary electrophoresis, *J. Chromatogr.*, 559, 215, 1991.

142. Terabe, S., Shibata, M., and Miyashita, Y., Chiral separation by electrokinetic chromatography with bile salt micelles, *J. Chromatogr.*, 480, 403, 1989.

143. Tickle, D. C., Okafo, G. N., Camilleri, P., Jones, R. F. D., and Kirby, A. J., Glucopyranoside-based surfactants as pseudostationary phases for chiral separations in capillary electrophoresis, *Anal. Chem.*, 66, 4121, 1994.

144. Prunonosa, J., Obach, R., Diezcascon, A., and Gouesclou, L., Determination of cicletanine enantiomers in plasma by high-performance capillary electrophoresis, *J. Chromatogr. Biomed. Appl.*, 574, 127, 1992.

145. Gross, M., Gathof, B. S., Kolle, P., and Gresser, U., Capillary electrophoresis for screening of adenylosuccinate lyase deficiency, *Electrophoresis*, 16, 1927, 1995.

146. Salerno, C. and Crifo, C., Microassay of adenylosuccinase by capillary electrophoresis, *Anal. Biochem.*, 226, 377, 1995.

147. Grune, T., Ross, G. A., Schmidt, H., Siems, W., and Perrett, D., Optimized separation of purine bases and nucleosides in human cord plasma by capillary zone electrophoresis, *J. Chromatogr.*, 636, 105, 1993.

148. Frenz, J., Wu, S. L., and Hancock, W. S., Characterization of human growth hormone by capillary electrophoresis, *J. Chromatogr.*, 480, 379, 1989.

149. Nielsen, R. G., Riggin, M., and Rickard, E. C., Capillary zone electrophoresis of peptide fragments from trypsin digestion of iosynthetic human growth hormone, *J. Chromatogr.*, 480, 393, 1989.

150. Grossman, P. D., Colburn, J. C., Lauer, H. H., Nielsen, R. G., Riggin, R. M., Sittampalam, G. S., and Rickard, E. C., Application f free-solution capillary electrophoresis to the analytical scale separation of proteins and peptides, *Anal. Chem.*, 61, 1186, 1989.

151. Nielsen, R. G. and Rickard, E. C., Method optimisation in capillary zone electrophoretic analysis of hGH tryptic digest fragments, *J. Chromatogr.*, 516, 99, 1990.

152. Nielsen, R. G., Sittampalam, G. S., and Rickard, E. C., Capillary zone electrophoresis of insulin and growth hormone, *Anal. Biochem.*, 177, 20, 1989.

153. Camilleri, P., Okafo, G. N., Southan, C., and Brown, R., Analytical and micropreparative capillary electrophoresis of the peptides from calcitonin, *Anal. Biochem.*, 198, 36, 1991.

154. Bihoreau, N., Ramon, C., Vincentelli, R., Levillain, J. P., and Troalen, F., Peptide mapping characterization by capillary electrophoresis of a human monoclonal anti-Rh(D) antibody produced for clinical study, *J. Cap. Elec.*, 2, 197, 1995.

155. Van-de-Noetelaar, P. J. M., Jansen, P. S. L., Melgers, P. A. T. A., Wagenaars, G. N., and Tenkortenaar, P. W. B., Stability assessment of peptide and protein drugs, *J. Controlled Release*, 21, 11, 1992.

156. Amankwa, L. N. and Kuhr, W. G., On-line peptide mapping by capillary zone electrophoresis, *Anal. Chem.*, 65, 2693, 1993.

157. Lee, T. T., Lillard, S. J., and Yeung, E. S., Screening and characterization of biopharmaceuticals by high-performance capillary electrophoresis with laser-induced native fluorescence detection, *Electrophoresis*, 14, 429, 1993.

158. Nielsen, R. G., Rickard, E. C., Santa, P. F., Sharknas, D. A., and Sittampalam, G. S., Separation of antibody-antigen complexes by capillary zone electrophoresis, isoelectric focusing and high performance size exclusion chromatography, *J. Chromatogr.*, 539, 177, 1991.

159. Kajiwara, H., Application of high-performance capillary electrophoresis to the analysis of conformation and interaction of metal-binding proteins, *J. Chromatogr.*, 559, 345, 1991.

160. Carpenter, J., Camilleri, P., Dhanak, D., and Goodall, D., A study of the binding of vancomycin to dipeptides using capillary electrophoresis, *J. Chem. Soc. Chem. Comm.*, 4, 804, 1992.

161. Chu, Y. H., Avila, L. Z., Biebuyck, H. A., and Whitesides, G. M., Use of affinity capillary electrophoresis to measure binding constants of ligands to proteins, *J. Med. Chem.*, 35, 2915, 1992.

162. Kraak, J. C., Busch, S., and Poppe, H., Study of protein drug binding using capillary zone electrophoresis, *J. Chromatogr.*, 608, 257, 1992.

163. Takeo, K., Advances in affinity electrophoresis, *J. Chromatogr. A*, 698, 89, 1995.

164. Morbeck, D. E., Madden, B. J., and McCormick, M. J., Analysis of the microheterogeneity of the glycoprotein chorionic gonadotropin with high-performance capillary electrophoresis, *J. Chromatogr. A*, 680, 217, 1994.

165. Oda, R. P., Smisek, D. L., and Landers, J. P., Capillary electrophoretic analysis of glycoproteins and glycoprotein-derived oligosaccharides, *Advances in Chromatography*, 36, xx, 1995.

166. Hortin, G. L., Griest, T., and Benutto, B. M., Separation of sulfated and nonsulfated forms of peptides by capillary electrophoresis: Comparison with reversed-phase HPLC, *BioChromatography*, 5, 118, 1990.

167. Yannoukakos, D., Meyer, H. E., Vasseur, C., Driancourt, C., Wajcman, H., and Bursaux, E., Three regions of erythrocyte band protein are phosphorylated on tyrosines: characterization of the phosphorylation sites by solid phase sequencing combined with capillary electrophoresis., *Biochim. Biophys. Acta*, 1066, 70, 1991.

168. Guzman, N. A., Moschera, J., Bailey, C. A., Iqbal, K., and Malick, A. W., Assay of protein drug substances present in solution mixtures by fluorescamine derivatization and capillary electrophoresis, *J. Chromatogr.*, 598, 123, 1992.

169. Hecht, R. I., Coleman, J. F., Morris, J. C., Stover, F. S., and Demorest, D., Micropreparative capillary zone electrophoresis of recombinant human interleukin-3, *Prep. Biochem.*, 19, 363, 1989.

170. Kelly, J. F., Locke, S. J., Ramaley, L., and Thibault, P., Development of electrophoretic conditions for the characterization of protein glycoforms by capillary electrophoresis electrospray mass spectrometry, *J. Chromatogr. A*, 720, 409, 1996.

171. Herold, M. and Wu, S.-L., Automated peptide fraction collection in CE, *LC-GC*, 12, 531, 1994.

172. Foret, F., Muller, O., Thorne, J., Gotzinger, W., and Karger, B. L., Analysis of protein fractions by micropreparative capillary isoelectric focusing and matrix-assisted laser desorption time-of-flight mass spectrometry, *J. Chromatogr. A*, 716, 157, 1995.

173. Jorgenson, J. W. and Lukacs, K. D., Capillary zone electrophoresis, *Science*, 222, 266, 1983.

174. Chen, F. T. A. and Sternberg, J. C., Characterization of proteins by capillary electrophoresis in fused-silica columns — Review on serum protein analysis and application to immunoassays, *Electrophoresis*, 15, 13, 1994.

175. Gordon, M. J., Lee, K. J., Arias, A. A., and Zare, R. N., Protocol for resolving protein mixtures in capillary zone electrophoresis, *Anal. Chem.*, 63, 6, 1991.

176. Dolnik, V., Capillary zone electrophoresis of serum proteins: Study of separation variables, *J. Chromatogr. A*, 709, 99, 1995.

177. Jenkins, M. A., Kulinskaya, E., Martin, H. D., and Guerin, M. D., Evaluation of serum protein separation by capillary electrophoresis: Prospective analysis of 1,000 specimens, *J. Chromatogr. B — Biomed. Appl.*, 672, 241, 1995.

178. Kim, J. W., Park, J. H., Park, J. W., Doh, H. J., Heo, G. S., and Lee, K. J., Quantitative analysis of serum proteins separated by capillary electrophoresis, *Clin. Chem.*, 39, 689, 1993.

179. Reif, O. W., Lausch, R., and Freitag, R., High performance capillary electrophoresis of human serum and plasma proteins, in *Advances in Chromatography* 34, Brown, P. R. and Grushka, E., Eds., Marcel Dekker, New York, 1994, p. 1.

180. Lal, K., Xu, L., Colburn, J., Hong, A. L., and Pollock, J. J., The use of capillary electrophoresis to identify cationic proteins in human parotid saliva, *Arch. Oral Biol.*, 37, 7, 1992.

181. Hiraoka, A., Miura, I., Hattori, M., Tominaga, I., and Machida, S., Capillary-zone electrophoretic analyses of the proteins and amino acid components in cerebrospinal fluid of central nervous system diseases, *Biol. Pharm. Bull.*, 16, 949, 1993.

182. Cowdrey, G., Firth, M., and Firth, G., Separation of cerebrospinal fluid proteins using capillary electrophoresis: A potential method for the diagnosis of neurological disorders, *Electrophoresis*, 16, 1922, 1995.

183. Grimshaw, J., Trocha-Grimshaw, J., Fisher, W., Kane, A., McCarron, P., Rice, A., Spedding, P. L., Duffy, J., and Mollan, R. A. B., An investigation of human synovial fluid by capillary electrophoresis, *J. Cap. Elec.*, 2, 40, 1995.

184. Lehmann, R., Liebich, H., Grubler, G., and Voelter, W., Capillary electrophoresis of human serum proteins and apolipoproteins, *Electrophoresis*, 16, 998, 1995.

185. Schmitz, G. and Mollers, C., Analysis of lipoproteins with analytical capillary isotachophoresis, *Electrophoresis*, 15, 3, 1994.

186. Bolger, C. A., Zhu, M., Rodriquez, R., and Wehr, T., Performance of uncoated and coated capillaries in free zone electrophoresis and isoelectric focusing of proteins, *J. Liq. Chromatogr.*, 14, 895, 1991.

187. Molteni, S., Frischknecht, H., and Thormann, W., Application of dynamic capillary isoelectric focusing to the analysis of human hemoglobin variants, *Electrophoresis*, 15, 22, 1994.

188. Ferranti, P., Malorni, A., Pucci, P., Fanali, S., Nardi, A., and Ossicini, L., Capillary zone electrophoresis and mass spectrometry for the characterization of genetic variants of human hemoglobin, *Anal. Biochem.*, 194, 1, 1991.

189. Law, H. Y., Ng, I., Ong, J., and Ong, C. N., Use of capillary electrophoresis for thalassemia screening, *Biomedical. Chromatogr.*, 8, 175, 1994.

190. Ross, G. A., Lorkin, P., and Perrett, D., Separation and tryptic digest mapping of normal and variant haemoglobins by capillary electrophoresis, *J. Chromatogr.*, 636, 69, 1993.

191. Okafo, G., Perrett, D., and Camilleri, P., Improved resolution of tryptic digest fragments from haemoglobin variants using phytic acid in free zone capillary electrophoresis, *Biomed. Chromatogr.*, 8, 202, 1994.

192. Ferranti, P., Malorni, A., and Pucci, P., Structural characterization of hemoglobin variants using capillary electrophoresis and fast atom bombardment mass spectrometry, in *Hemoglobins*, Pt B (Series: *Methods in Enzymology*) 231, Everse, J., Van-de-Griff, J. D., and Winslow, R. M., Eds., 45, 1994.

193. Matioli, G. T. and Niewisch, H. B., Electrophoresis of hemoglobin in single erythrocytes, *Science*, 150, 1824, 1965.

194. Kennedy, R. T. and Jorgenson, J. W., Quantitative analysis of individual neurons by open tubular liquid chromatography with voltammetric detection, *Anal. Chem.*, 61, 436, 1989.

195. Tseng, H. C., Dadoo, R., and Zare, R. N., Selective determination of adenine-containing compounds by capillary electrophoresis with laser-induced fluorescence detection, *Anal. Biochem.*, 222, 55, 1994.

196. Chien, J. B., Wallingford, R., and Ewing, A. G., Estimation of free dopamine in the cytoplasm of the giant dopamine cell of Planorbis by voltammetry and capillary electrophoresis, *J. Neurochem.*, 54, 633, 1990.

197. Olefirowicz, T. M. and Ewing, A. G., Dopamine concentration in the cytoplasmic compartment of single neurons determined by capillary electrophoresis, *J. Neuroscience Methods,* 34, 11, 1990.

198. Olefirowicz, T. M. and Ewing, A. G., Capillary electrophoresis in 2 and 5μm diameter capillaries: application to cytoplasmic analysis, *Anal. Chem.,* 62, 1872, 1990.

199. Chen, G. Y., Gavin, P. F., Luo, G. A., and Ewing, A. G., Observation and quantitation of exocytosis from the cell body of a fully developed neuron in planorbis corneus, *J. Neuroscience,* 15, 7747, 1995.

200. Hogan, B. L. and Yeung, E. S., Determination of intracellular species at the level of a single erythrocyte via capillary electrophoresis with direct and indirect fluorescence detection, *Anal. Chem.,* 64, 2841, 1992.

201. Lillard, S. J., Yeung, E. S., Lautamo, R. M. A., and Mao, D. T., Separation of hemoglobin variants in single human erythrocytes by capillary electrophoresis with laser-induced native fluorescence detection, *J. Chromatogr. A,* 718, 397, 1995.

202. Xue, Q. F. and Yeung, E. S., Indirect fluorescence determination of lactate and pyruvate in single erythrocytes by capillary electrophoresis, *J. Chromatogr. A,* 661, 287, 1994.

203. Chang, H. T. and Yeung, E. S., Determination of catecholamines in single adrenal medullary cells by capillary electrophoresis and laser-induced native fluorescence, *Anal. Chem.,* 67, 1079, 1995.

204. Gilman, S. D. and Ewing, A. G., Analysis of single cells by capillary electrophoresis with on column derivatization and laser-induced fluorescence detection, *Anal. Chem.,* 67, 58, 1995.

205. Nussberger, S., Foret, F., Hebert, S. C., Karger, B. L., and Hediger, M. A., Nonradioactive monitoring of organic and inorganic solute transport into single Xenopus oocytes by capillary zone electrophoresis, *Biophys. J.,* 70, 998, 1996.

206. Gilman, S. D. and Ewing, A. G., Recent advances in the application of capillary electrophoresis to neuroscience, *J. Cap. Elec.,* 2, 1, 1995.

207. Yeung, E. S., Chemical analysis of single human erythrocytes, *Acc. Chem. Res.,* 27, 409, 1994.

208. Xue, Q. F. and Yeung, E. S., Variability of intracellular lactate dehydrogenase isoenzymes in single human erythrocytes, *Anal. Chem.,* 66, 1175, 1994.

209. Haab. B. B. and Mathies, R. A., Single molecule fluorescence burst detection of DNA fragments separated capillary electrophoresis, *Anal. Chem.,* 67, 3253, 1995.

210. Chen, D. Y. and Dovichi, N. J., Single molecule detection in capillary electrophoresis: Molecular shot noise as a fundamental limit to chemical analysis, *Anal. Chem.,* 68, 690, 1996.

211. VanOrman, B. B. and McIntire, G. L., Analytical separation of polystyrene nanospheres by capillary electrophoresis, *J. Microcol. Sep.,* 1, 289, 1989.

212. McCormick, R. M., Characterization of silica sols by capillary zone electrophoresis, *J. Liq. Chromatogr.,* 14, 939, 1991,

213. Hjerten, S., Elenbring, K., Kilar, F., Liao, J. L., Chen, A. J. C., Siebert, C. J., and Zhu, M. D., Carrier-free zone electrophoresis, displacement electrophoresis and isoelectric focusing in a high-performance electrophoresis apparatus, *J. Chromatogr.,* 403, 47, 1987.

214. Grossman, P. D. and Soane, D. S., Orientation effects on the electrophoretic mobility of rod-shaped molecules in free solution, *Anal. Chem.,* 62, 1592, 1990.

215. Zhu, A. and Chen, Y., High-voltage capillary zone electrophoresis of red blood cells, *J. Chromatogr.,* 470, 251, 1989.

216. Ebersole, R. C. and McCormick, R. M., Separation and isolation of viable bacteria by capillary zone electrophoresis, *Bio-Technology,* 11, 1278, 1993.

217. Radko, S. P., Sokoloff, A. V., Garner, M. M., and Chrambach, A., Capillary electrophoresis of rat liver microsomes in polymer solutions, *Electrophoresis,* 16, 981, 1995.

218. Banke, N., Hansen, K., and Diers, I., Detection of enzyme activity in fractions collected from free solution capillary electrophoresis of complex samples, *J. Chromatogr.,* 559, 325–335, 1991.

219. Wu, D. and Regnier, F. E., Native protein separations and enzyme microassays by capillary zone and gel electrophoresis, *Anal. Chem.,* 65, 2029–2035, 1993.

220. Schultz, N. M. and Kennedy, R. T., Rapid immunoassays using capillary electrophoresis with fluorescence detection, *Anal. Chem.,* 65, 3161, 1993.

221. Shimura, K. and Karger, B. L., Affinity probe capillary electrophoresis — Analysis of recombinant human growth hormone with a fluorescent labelled antibody fragment, *Anal. Chem.*, 66, 9, 1994.

222. Schmalzing, D., Nashabeh, W., and Fuchs, M., Solution-phase immunoassay for determination of cortisol in serum by capillary electrophoresis, *Clin. Chem.*, 41, 1403, 1995.

223. Blais, B. W., Cunningham, A., and Yamazaki, H., A novel immunofluorescence capillary electrophoresis assay system for the determination of chloramphenicol in milk, *Food Agric. Immunol.*, 6, 409, 1994.

224. Chen, F. T. A. and Evangelista, R. A., Feasibility studies for simultaneous immunochemical multianalyte drug assay by capillary electrophoresis with laser-induced fluorescence, *Clin. Chem.*, 40, 1819, 1994.

225. Steinman, L., Caslavska, J., and Thormann, W., Feasibility study of a drug immunoassay based on micellar electrokinetic capillary chromatography with laser induced fluorescence detection: Determination of theophylline in serum, *Electrophoresis*, 16, 1912, 1995.

226. Pritchett, T., Evangelista, R. A., and Chen, F. T. A., Capillary electrophoresis-based immunoassays — a speedy alternative to solid-phase immunoassays, *Bio-Technology*, 13, 1449, 1995.

227. Ji, A. J., Nunez, M. F., Machacek, D., Ferguson, J. E., Iossi, M. F., Kao, P. C., and Landers, J. P., Separation of urinary estrogens by micellar electrokinetic chromatography, *J. Chromatogr. B — Biomed. Appl.*, 669, 15, 1995.

228. Abubaker, M. A., Petersen, J. R., and Bissell, M. G., Micellar electrokinetic capillary chromatographic separation of steroids in urine by trioctylphosphine oxide and cationic surfactant, *J. Chromatogr. — Biomed. Appl.*, 674, 31, 1995.

229. Chan, K. C., Muschik, G. M., Issaq, H. J., and Siiteri, P. K., Separation of estrogens by micellar electrokinetic chromatography, *J. Chromatogr. A*, 690, 149, 1995.

230. Janini, G. M., Chan, K. C., Muschik, G. M., and Issaq, H. J., Analysis of nitrate and nitrite in water and urine by capillary zone electrophoresis, *J. Chromatogr. B — Biomed. Appl.*, 657, 419, 1994.

231. Bjergegaard, C., Moller, P., and Sorensen, H., Determination of thiocyanate, iodide, nitrate and nitrite in biological samples by micellar electrokinetic capillary chromatography, *J. Chromatogr. A*, 717, 409, 1995.

232. Buscher, B. A. P., Tjaden, U. R., Irth, H., Andersson, E. M., and Van-der-Greef, J., Determination of 1,2,6-inositol trisphosphate (derivatives) in plasma using iron(III)-loaded adsorbents and capillary zone electrophoresis-(indirect) UV detection, *J. Chromatogr. A*, 718, 413–419, 1995.

233. Chan, K. C., Muschik, G. M., and Issaq, H. J., Separation of tryptophan and related indoles by micellar electrokinetic chromatography with KrF laser-induced fluorescence detection, *J. Chromatogr. A*, 718, 203, 1995.

234. Dolnik, V., Capillary zone electrophoresis of pathological metabolites in phenylketonuria, *J. Microcol. Sep.*, 6, 63, 1994.

235. Che, P., Xu, J., Shi, H., and Ma, Y. F., Quantitative determination of serum iron in human blood by high-performance capillary electrophoresis, *J. Chromatogr. B — Biomed. Appl.*, 669, 45–51, 1995.

236. Takatsu, A., Eyama, S., and Uchiumi, A., Determination of aluminum in serum by capillary zone electrophoresis with laser-induced fluorescence detection, *Chromatographia*, 40, 125–128, 1995.

237. Dolnik, V. and Dolnikova, J., Capillary zone electrophoresis of organic acids in serum of critically ill children, *J. Chromatogr. A*, 716, 269, 1995.

238. Hiraoka, A., Akai, J., Tominaga, I., Hattori, M., Sasaki, H., and Arato, T., Capillary zone electrophoretic determination of organic acids in cerebrospinal fluid from patients with central nervous system diseases, *J. Chromatogr. A*, 680, 243, 1994.

239. Shirao, M., Furuta, R., Suzuki, S., Nakazawa, H., Fujita, S., and Maruyama, T., Determination of organic acids in urine by capillary zone electrophoresis, *J. Chromatogr. A*, 680, 247, 1994.

240. Wu, N., Sweedler, J. V., and Lin, M., Enhanced separation and detection of serum bilirubin species by capillary electrophoresis using a mixed anionic surfactant-protein buffer system with laser-induced fluorescence detection, *J. Chromatogr. B — Biomed. Appl.*, 654, 185, 1994.

241. Wu, N., Wang, T. S., Hartwick, R. A., and Huie, C. W., Separation of serum bilirubin species by micellar electrokinetic chromatography with direct sample injection, *J. Chromatogr. — Biomed. Appl.*, 582, 77, 1992.

242. Koh, E. V., Bissell, M. G., and Ito, R. K., Measurement of vitamin-C by capillary electrophoresis in biological fluids and fruit beverages using a stereoisomer as an internal standard, *J. Chromatogr.*, 633, 245, 1993.

243. Ma, Y. F., Wu, Z. K., Furr, H. C., Lammikeefe, C., and Craft, N. E., Fast minimicroassay of serum retinol (vitamin-A) by capillary zone electrophoresis with laser-excited fluorescence detection, *J. Chromatogr. — Biomed. Appl.*, 616, 31, 1993.

244. Shi, H. L., Ma, Y. F., Humphrey, J. H., and Craft, N. E., Determination of vitamin A in dried human blood spots by high-performance capillary electrophoresis with laser-excited fluorescence detection, *J. Chromatogr. B — Biomed. Appl.*, 665, 89, 1995.

245. Lambert, D., Adjalla, C., Felden, F., Benhayoun, S., Nicolas, J. P., and Gueant, J. L., Identification of vitamin-B12 and analogues by high-performance capillary electrophoresis and comparison with high-performance liquid chromatography, *J. Chromatogr.*, 608, 311, 1992.

246. Wang, C. C., McCann, W. P., and Beale, S. C., Measurement of adenosine by capillary zone electrophoresis with on-column isotachophoretic preconcentration, *J. Chromatogr. B — Biomed. Appl.*, 676, 19, 1996.

247. Shihabi, Z. K., Hinsdale, M. E., and Bleyer, A. J., Xanthine analysis in biological fluids by capillary electrophoresis, *J. Chromatogr. B — Biomed. Appl.*, 669, 163, 1995.

248. Masson, C., Luong, J. H. T., and Nguyen, A. L., Analyses of uric acid in urine and reconstituted serum by capillary electrophoresis, *Anal. Lett.*, 24, 377, 1991.

249. Miyake, M., Shiukawa, A., and Nakagawa, T., Simultaneous determination of creatinine and uric acid in human plasma and urine by micellar electrokinetic chromatography, *J. High Res. Chromatogr.*, 14, 180, 1991.

250. Lee, K. J., Lee, J. J., and Moon, D. C., Application of micellar electrokinetic capillary chromatography for monitoring of hippuric and methylhippuric acid in human urine, *Electrophoresis*, 15, 98, 1994.

251. Xu, X., Kok, W. T., Kraak, J. C., and Poppe, H., Simultaneous determination of urinary creatinine, calcium and other inorganic cations by capillary zone electrophoresis with indirect ultraviolet detection, *J. Chromatogr. B*, 661, 35–45, 1994.

252. Petucci, C. J., Kantes, H. L., Strein, T. G., and Veening, H., Capillary electrophoresis as a clinical tool — Determination of organic anions in normal and uremic serum using photodiode-array detection, *J. Chromatogr. B — Biomed. Appl.*, 668, 241, 1995.

253. Lee, K. J., Heo, G. S., and Doh, H. J., Determination of creatinine in serum by capillary zone electrophoresis, *Clin. Chem.*, 38, 2322, 1992.

254. Marsh, D. B. and Nuttall, K. L., Methylmalonic acid in clinical urine specimens by capillary zone electrophoresis using indirect photometric detection, *J. Cap. Elec.*, 2, 63, 1995.

255. Huang, X. J. and Kok, W. T., Determination of thiols by capillary electrophoresis with electrochemical detection using a palladium field-decoupler and chemically modified electrodes, *J. Chromatogr. A*, 716, 347, 1995.

256. Piccoli, G., Fiorani, M., Biagiarelli, B., Palma, F., Potenza, L., Amicucci, A., and Stocchi, V., Simultaneous high-performance capillary electrophoretic determination of reduced and oxidized glutathione in red blood cells in the femtomole range, *J. Chromatogr. A*, 676, 239, 1994.

257. Zhou, G., Yu, Q. N., Ma, Y. F., Xue, J., Zhang, Y., and Lin, G. C., Determination of polyamines in serum by high-performance capillary zone electrophoresis with indirect ultraviolet detection, *J. Chromatogr. A*, 717, 345–349, 1995.

258. Arriaga, E., Chen, D. Y., Cheng, X. L., and Dovichi, N. J., High-efficiency filter fluorimeter for capillary electrophoresis and its application to fluorescein thiocarbamyl amino acid derivatives, *J. Chromatogr. A*, 652, 347, 1993.

259. Nickerson, B. and Jorgenson, J. W., High speed capillary zone electrophoresis with laser induced fluorescence detection, *J. High Res. Chromatogr.*, 11, 533, 1988.

260. Yu, M. and Dovichi, N. J., Attomole amino acid determination by capillary zone electrophoresis with thermooptical absorbance detection, *Anal. Chem.*, 61, 37, 1989.

261. Green, J. S. and Jorgenson, J. W., Variable-wavelength on-column fluorescence detector for open-tubular zone electrophoresis, *J. Chromatogr.,* 352, 337, 1986.

262. Liu, J., Hsieh, Y.-Z., Wiesler, D., and Novotny, M., Design of 3-(4-carboxybenzoyl)-2 quinolinecarboxaldehyde as a reagent for ultrasensitive determination of primary amines by capillary electrophoresis using laser fluorescence absorbance detection, *Anal. Chem.,* 63, 408, 1991.

263. Camilleri, P., Dhanak, D., Druges, M., and Okafo, G., High sensitivity detection of amino acids using a new fluorogenic probe resolution of peptides and proteins, *Anal. Proc.,* 31, 99, 1994.

264. Lalljie, S. P. D. and Sandra, P., MEKC analysis of FITC and DTAF amino acid derivatives with LIF detection, *Chromatographia,* 40, 513, 1995.

265. Zhao, J. Y., Chen, D. Y., and Dovichi, N. J., Low-cost laser-induced fluorescence detector for micellar capillary zone electrophoresis: Detection at the zeptomol level of tetramethylrhodamine thiocarbamyl amino acid derivatives, *J. Chromatogr.,* 608, 117, 1992.

266. Huang, X. H., Gordon, M. J., and Zare, R. N., Quantitation of lithium in serum by capillary zone electrophoresis with an on-column conductivity detector, *J. Chromatogr.,* 425, 385, 1988.

267. Roach, M. C., Gozel, P., and Zare, R. N., Determination of methotrexate and its major metabolite, 7-hydroxymethotrexate, using capillary zone electrophoresis and laser-induced fluorescence detection, *J. Chromatogr.,* 426, 129, 1988.

268. Tanaka, Y. and Thormann, W., Capillary electrophoretic determination of S-carboxymethyl-L-cysteine and its major metabolites in human urine: Feasibility investigation using on-column detection of non-derivatized solutes in capillaries with minimal electroosmosis, *Electrophoresis,* 11, 760, 1990.

269. Thormann, W., Meier, P., Marcolli, C., and Binder, F., Analysis of barbiturates in human serum and urine by high-performance capillary electrophoresis micellar electrokinetic capillary chromatography with on-column multi-wavelength detection, *J. Chromatogr.,* 545, 445, 1991.

270. Meier, P. and Thormann, W., Determination of thiopental in human serum and plasma by high-performance capillary electrophoresis — micellar electrokinetic chromatography, *J. Chromatogr,* 559, 505, 1991.

271. Soini, H., Tsuda, T., and Novotny, M. V., Electrochromatographic solid-phase extraction for determination of cimetidine in serum by micellar electrokinetic capillary chromatography, *J. Chromatogr.,* 559, 547, 1991.

272. Lloyd, D. K., Cypess, A. M., and Wainer, I. W., Determination of cytosine-β-d-arabinoside in plasma using capillary electrophoresis, *J. Chromatogr.,* 568, 117, 1991.

273. Honda, S., Taga, A., Kakehi, K., Koda, S., and Okamoto, Y., Determination of cefixime and its metabolites by high-performance capillary electrophoresis, *J. Chromatogr.,* 590, 364, 1992.

274. Reinhoud, N. J., Tjaden, U. R., Irth, H., and Van-der-Greef, J., Bioanalysis of some anthracyclines in human plasma by capillary electrophoresis with laser-induced fluorescence detection, *J. Chromatogr. — Biomed. Appl.,* 574, 327, 1992.

275. Thormann, W., Minger, A., Molteni, S., Caslavska, J., and Gebauer, P., Determination of substituted purines in body fluids by micellar electrokinetic capillary chromatography with direct sample injection, *J. Chromatogr.,* 593, 275, 1992.

276. Zhang, Z. Y., Fasco, M. J., and Kaminsky, L. S., Determination of theophylline and its metabolites in rat liver microsomes and human urine by capillary electrophoresis, *J. Chromatogr. B — Biomed. Appl.,* 665, 201, 1995.

277. Reinhoud, N. J., Tjaden, U. R., Irth, H., and Van-der-Greef, J., Bioanalysis of some anthracyclines in human plasma by capillary electrophoresis with laser-induced fluorescence detection, *J. Chromatogr. — Biomed. Appl.,* 574, 327, 1992.

278. Brunner, L. J., DiPiro, J. T., and Feldman, S., Serum antipyrine concentrations determined by micellar electrokinetic capillary chromatography, *J. Chromatogr.,* 622, 98, 1993.

279. Wolfisberg, H., Schmutz, A., Stotzer, R., and Thormann, W., Assessment of automated capillary electrophoresis for therapeutic and diagnostic drug monitoring: determination of bupivacaine in drain fluid and antipyrine in plasma, *J. Chromatogr.,* 652, 407, 1993.

280. Baillet, A., Pianetti, G. A., Taverna, M., Mahuzier, G., and Baylocq-Ferrier, D., Fosfomycin determination in serum by capillary zone electrophoresis with indirect ultraviolet detection, *J. Chromatogr. — Biomed. Appl.,* 616, 311, 1993.

281. Leveque, D., Gallion, C., Tarral, E., Monteil, H., and Jehl, F., Determination of fosfomycin in biological fluids by capillary electrophoresis, *J. Chromatogr. B — Biomed. Appl.,* 655, 320, 1994.

282. Rabel, S. R., Stobaugh, J. F., and Trueworthy, R., Determination of intracellular levels of 6-mercaptopurine metabolites in erythrocytes utilizing capillary electrophoresis with laser-induced fluorescence detection, *Anal. Biochem.,* 224, 315, 1995.

283. Hufschmid, E., Theurillat, R., Martin, U., and Thormann, W., Exploration of the metabolism of dihydrocodeine via determination of its metabolites in human urine using micellar electrokinetic capillary chromatography, *J. Chromatogr. B — Biomed. Appl.,* 668, 159, 1995.

284. Cai, J. Y. and Henion, J., Elucidation of LSD in vitro metabolism by liquid chromatography and capillary electrophoresis coupled with tandem mass spectrometry, *J. Anal. Toxicol.,* 20, 27, 1996.

285. Garcia, L. L. and Shihabi, Z. K., Suramin determination by capillary electrophoresis, *J. Liq. Chromatogr.,* 16, 2049, 1993.

286. Hettiarachchi, K. and Cheung, A. P., Precision in capillary electrophoresis with respect to quantitative analysis of suramin, *J. Chromatogr. A,* 717, 191, 1995.

287. Simoalfonso, E. F., Ramisramos, G., Garciaalvarezcoque, M. C., and Esteveromero, J. S., Determination of sulphonamides in human urine by azo dye precolumn derivatization and micellar liquid chromatography, *J. Chromatogr. B — Biomed. Appl.,* 670, 183, 1995.

288. Aumatell, A. and Wells, R. J., Chiral differentiation of the optical isomers of racemethorphan and racemorphan in urine by capillary zone electrophoresis, *J. Chromatogr. Sci.,* 31, 502, 1993.

289. Soini H., Riekkola, M. L., and Novotny, M.V., Chiral separations of basic drugs and quantitation of bupivacaine enantiomers in serum by capillary electrophoresis with modified cyclodextrin buffers, *J. Chromatogr.,* 608, 265, 1992.

290. Lloyd, D. K., Quantitation of the stereoisomers of L-buthionine-(R,S)-sulfoximine in biological matrixes by capillary electrophoresis, *Anal. Proc.,* 29, 169–170, 1992.

291. Heuermann, M. and Blaschke, G., Simultaneous enantioselective determination and quantification of dimethindene and its metabolite n-demethyl-dimethindene in human urine using cyclodextrins as chiral additives in capillary electrophoresis, *J. Pharm. Biomed. Anal.,* 12, 753, 1994.

292. Desiderio, C., Fanali, S., Kupfer, A., and Thormann, W., Analysis of mephenytoin, 4-hydroxymephenytoin and 4-hydroxyphenytoin enantiomers in human urine by cyclodextrin micellar electrokinetic capillary chromatography — Simple determination of a hydroxylation polymorphism in man, *Electrophoresis,* 15, 87, 1994.

293. Hempel, G. and Blaschke, G. Enantioselective determination of zopiclone and its metabolites in urine by capillary electrophoresis, *J. Chromatogr. B — Biomed. Appl.,* 675, 139, 1996.

294. Gareil, P., Gramond, J. P., and Guyon, F., Separation and determination of warfarin enantiomers in human plasma samples by capillary zone electrophoresis using a methylated β-cyclodextrin-containing electrolyte, *J. Chromatogr. — Biomed. Appl.,* 615, 317, 1993.

APPENDIX I

Manufacturers and Suppliers of CE Equipment and Components

Since the first edition of this book, there have been many changes among the suppliers of CE equipment. The supply of instruments has become increasingly focused in the hands of a few big firms, and many smaller or more peripheral suppliers have left the field. 1996 also saw the first of the next generation of instruments that are both CE and electrochromatography instruments. The number of specialist suppliers of parts, though, has slightly increased.

1. Fully Automatic CE Systems

Advanced Molecular Systems
188 Woodvale Court
Vallejo, CA 945591
USA

Beckman
1050 Page Mill Road
Foster City, CA 94404
USA

BioRad
3300 Regatta Boulevard
Richmond, CA 94804
USA

Dionex Corporation
P.O. Box 3603
Sunnyvale, CA 94088
USA

Hewlett Packard
Analytical Division
Hewlett-Packard Str 8
76337 Waldbronn 2
Germany
(with CEC capability)

Perkin Elmer
(Applied BioSystems Division)
850 Lincoln Centre Drive
Foster City, CA 94404
USA

ThermoSeparations
3333 North First Street
San Jose, CA 95134
USA
(also CEC system)

Waters
34 Maple Street
Milford, MA 01757
USA

0-8493-9127-X/98/$0.00+$.50

2. Manual and/or Modular CE Systems

ATI Instruments
1001 Fourier Drive
Madison, WI 53717
USA

ThermoSeparations
3333 North First Street
San Jose, CA 95134
USA

BioRad
3300 Regatta Boulevard
Richmond, CA 94804
USA

3. Instrumental Components for CE

Glassman High Voltage Inc.
Route 22 (East) Salem Park
P.O. Box 551
Whitehouse Station, NJ 08889
USA

4. Polyimide-Coated Fused-Silica Capillary Tubing for CE

Source Manufacturers

Chrompack
Frankfurt, Germany

Microquarz (Siemens)
Munich, Germany

Polymicro Technologies
3033 N 33rd Drive
Phoenix, AZ 85017
USA

Scientific Glass Engineering
7 Argent Place
Ringwood, Victoria 3134
Australia

SRI
P.O. Box 1290
Eatontown, NJ 07724-1290
USA

Modified Capillaries

For some applications, e.g., DNA separations, capillary manufactured for GC purposes can be used. Specialist suppliers for CE are:

BioRad
3300 Regatta Boulevard
Richmond, CA 94804
USA
(Surface-modified capillaries)

Scientific Glass Engineering
7 Argent Place
Ringwood, Victoria 3134
Australia
(Surface-modified capillaries)

J & W Scientific
91 Blue Ravine Road
Folsom, CA 95630
USA
(Gel-filled capillaries)

SRI
P.O. Box 1290
Eatontown, NJ 07724-1290
USA

Waters
34 Maple Street
Milford, MA 01757
USA
(Trace-enrichment capillary)

Supelco
Supelco Park
Bellefonte, PA 16823
USA
(Surface-modified, UV-transparent)

5. CE Grade Chemicals and Buffer Additives

CycloLab
1525 P.O. Box 435
Budapest, Hungary
(Cyclodextrins)

SRI
P.O. Box 1290
Eatontown, NJ 07724-1290
USA
(CE-quality buffers)

Fluka Chemie
Industriestrasse 25
CH-9470 Buchs
Switzerland
(CE-quality buffers)

Waters
34 Maple Street
Milford, MA 01757
USA
(Buffer additives)

Merck
Germany
(CE-quality buffers)

6. Isotachophoresis Systems

Villa Labeco
Spisska Nova Ues
Slovakiafor
(Coupled column systems, ITP-CZE)

Shimadzu
7102 Riverwood Drive
Columbia, MD 21046
USA

7. Software for CE

Chromapon Inc.
P.O. Box 4131
Whittier, CA 90607-4131
USA
(Simulator software, data systems)

LCResources
2930 Camino Diablo
Suite 110
Walnut Creek, CA 94596
USA
(Training software)

Detectors

ISCO
P.O. Box 5347
Lincoln, NE 68505
USA
(UV detector)

Kontron Instruments
Via G Fantoli 16/15
20138 Milano
Italy
(UV detector)

JASCO
2967-5 Ishikawa-cho
Hachioji, Tokyo 192
Japan
(UV detector)

LCResources
2930 Camino Diablo
Suite 110
Walnut Creek, CA 94596
(Longpath UV cells)

Power Suppliers

There are many manufacturers of high-voltage power supplies that can be used with various degrees of adaptation for CE work. The following are some of those most commonly used by researchers building their own CE systems. Some supply 0–30 kV units specially for CE.

Bertan
121 New South Road
Hicksville, NY 11801
USA

Brandenburg
Astec House
Genesis Business Park
Albert Drive
Woking GU21 5RW
United Kingdom

SRI
P.O. Box 1290
Eatontown, NJ 07724-1290
USA
(Modeling software)

Buffer Solutions for Capillary Electrophoresis

Buffers suitable for use in capillary electrophoresis are now commercially available. Merck KGaA (Darmstadt, Germany) and Fluka Chemical Corporation (Ronkonkoma, New York) provide a wide range of buffers. Of course, most of these are easily prepared in a laboratory. For laboratory preparations, the buffer constituents must be of the highest chemical purity. Moreover, because particles provide an excellent nucleation medium for bubble formation, it is essential that all particulate material be removed, in particular dust and bacterial growth, by filtering all buffers through a 0.45 μm or smaller filter prior to use in capillary electrophoresis. It is always advisable to prepare fresh buffer solutions. If buffers are to be stored, plastic rather than glass containers should be used: leaching of ionic material from glass can cause changes of pH with time. It is also preferable not to dilute buffer preparations, as this tends to shift the pH toward 7. Thus, on dilution of buffer solutions, pH has to be remeasured and adjustments made by the addition of the smallest quantity of acid or base. Of course, if an accurate measurement of a range of concentrations of the buffer constituents is required, then the appropriate buffer constituents are weighed, and solutions in water are freshly prepared in all cases.

For successful measurements, it is also advisable to dissolve analytes at a buffer concentration about one tenth lower than that of the running buffer. It an analyte dissolves in water, this is the best medium in which it can by injected into the capillary. Increasing the ionic strength of the buffer leads to sharper peaks, reduces protein–wall interactions, and keeps peptides and proteins in solution. However, an excessive increase in the ionic strength of buffers unavoidably leads to Joule heating so that cooling of the capillary is critical in obtaining good CE results. Reconditioning of capillaries between runs with 0.1 M sodium hydroxide and/or 0.033 M phosphoric acid solutions is recommended for reproducible results.

The following is a list of buffers, covering a wide range of pH (2.5 to 11), that are suitable for capillary electrophoresis. The list is by no means comprehensive. For a larger variety of buffers and additives suitable for the separation of particular classes of compounds, the reader is advised to refer to the various chapters and the references cited therein.

pH	Buffer	Application
2.5 to 6.0	20 or 50 mM sodium citrate	Free zone
2.5 to 7.0	20 or 50 mM sodium phosphate	Free zone
7.0	100 mM boric acid/50 mM sodium phosphate/50 mM SDS	MEKC
7.0 to 9.0	50 mM sodium phosphate – sodium tetraborate/50 mM sodium taurodeoxycholate	MEKC

0-8493-9127-X/98/$0.00+$.50
© 1998 by CRC Press LLC

pH	Buffer	Application
8.0 to 9.3	20 mM sodium tetraborate	Free zone
8.5 to 9.5	50 mM sodium tetraborate/50 to 100 mM SDS	MEKC
10.0 to 11.0	20 mM CAPS	Free zone
11	20 mM glycine/NaOH	Free zone
8.0	50 mM Tris – borate/2.5 mM EDTA/0.5% methylcellulose	DNA restriction fragments

Capillary Electrophoresis Experiments

This appendix contains 14 experiments set up by some of the authors (and their colleagues) of the preceding chapters. The style chosen by these authors has been largely unchanged by the editor. Experiments have been designed to provide practical experience in the various modes of operation in capillary electrophoresis (CE), in particular free zone, MEKC, and gel electrophoresis. The range of experiments is, of course, not exhaustive but is adequate to illustrate the usefulness of CE as an analytical tool.

The aims of each experiment are clearly stated. In every case, it is essential to refer to the appropriate chapter in order to supplement the information given and to assess the value of the experiment undertaken. Appropriate questions are asked at the end of each exercise to ensure a comprehensive understanding of the basis and application of CE.

It is hoped that, with the help of these experiments, CE can be introduced to students at an early stage of their college or university course.

Contributions prepared by authors are as follows:

A. Vinther and T. Kornfelt	Experiments 1, 2, 9, and 10
G.N. Okafo	Experiments 3, 6, and 7
Joe Foley	Experiments 4 and 5
Z. El Rassi and Y. Mechref	Experiment 8
P. Thibault, J.F. Kelly, and S.J. Locke	Experiment 11
H.E. Schwartz	Experiment 12
D. Perrett and J. Gibbons	Experiments 13 and 14

Experiment 1:
Electroosmotic Mobility, Electrophoretic Mobility, and Absolute vs. Relative Peak Area

1. AIMS

- Calculation of electroosmotic mobility.
- Calculation of electrophoretic mobilities.
- Calculation of relative areas.
- Understanding the information given in electropherograms with respect to polarity and charge density of the analytes.
- Understanding the importance of using relative peak areas instead of absolute peak areas when calculating the purity of a compound.

2. EXPERIMENTAL

Analysis was performed in the zone electrophoresis mode. The capillary was uncoated fused silica, the buffer pH was 7.0, and voltage was applied in "normal" mode, i.e., cathodic side at the detector end of the capillary and anodic side at the introduction end of the capillary.

- Total capillary length, L_t: 50 cm
- Effective capillary length (to detector), L_d: 40 cm
- Applied voltage, U: 20 kV
- Migration time of a neutral analyte, t_{EO}:   2.0 min
- Migration time of impurity A, t_A: 2.5 min
- Migration time of product analyte of interest, t_P: 3.0 min
- Migration time of impurity B, t_B: 3.1 min
- Migration time of impurity C, t_C: 5.0 min
- Peak area of impurity A, A_A: 250
- Peak area of principal peak, A_P: 2100
- Peak area of impurity B, A_B: 310
- Peak area of impurity C, A_C: 500

3. CALCULATIONS AND DISCUSSIONS

3.1. Mobilities

The apparent velocity, v_{APP}, of an analyte is the sum of the electroosmotic velocity, v_{EO}, and the electrophoretic velocity, v_{EP}:

$$v_{APP} = v_{EO} + v_{EP} = \mu_{EO}*E + \mu_{EP}*E$$

where μ is the mobility, and E is the electric field strength (U/L$_t$).

Calculate the electroosmotic mobility (the mobility of neutral analytes):

$$\mu_{EO} = \text{_____}$$

Calculate the electrophoretic mobility of the product and impurities A, B, and C.

Analyte	Electrophoretic mobility
Product	
Impurity A	
Impurity B	
Impurity C	

Comment on the following points:

a) Were the analyte and its impurities positively or negatively charged? Which polarity would an analyte migrating prior to the electroosmotic flow have?
b) Which one of the peaks had the highest appearent mobility?
c) Which one of the peaks had the highest numeric electrophoretic mobility?
d) All the analytes had equal mass (size). Which one of the compounds had the highest numeric charge?

3.2. Absolute vs. Relative Peak Areas

In contrast to HPLC, in CE relative rather than absolute peak areas are used for quantitative determinations. In HPLC, all analytes are propelled past the detector at an identical velocity equal to the pump flow rate. In CE, the analytes migrate past the detector based on their appearant velocity. This means that an analyte which migrates at a velocity half of that for another analyte will spend twice as much time in the detector. In this way, the slowest migrating analyte will be detected as having an area twice that of the fastest analyte in case equal amounts were injected. One can account for this simply by dividing the peak area by the migration time of the analyte. This is particularly important when one calculates the purity of a specific analyte. That is illustrated by the following example.

Calculate the amount of product and each of the three impurities in percentage of the total amount (sum of impurities A, B, C, and product) when absolute respectively relative areas are used.

Analyte	Area	Area,%	Area/migration time	
			Relative area	Relative area,%
Product				
Impurity A				
Impurity B				
Impurity C				

Comment on the following points:

e) In which case (absolute or relative areas) was the highest purity of the product obtained? Explain the reason for this.
f) If one does not account for different velocities of the analytes, will the error be greatest when the product and its impurities migrate close to each other or far away from each other?

4. CONCLUSION

The experiment should have demonstrated the following points:

1) Calculation of electroosmotic and electrophoretic mobilities
2) Interpretation of electropherograms to assign polarities to analytes
3) The use of relative peak areas (peak area divided by migration time) instead of absolute peak areas when quantitative calculations are performed

Experiment 2:
Axial pH Gradient Buildup

1. AIMS

- Understanding the electrode processes in capillary electrophoresis.
- Understanding the importance of using buffers with high capacity.
- Use of Faraday's Law in CE.

2. EXPERIMENTAL

2.1. CE Conditions

Capillary (uncoated fused-silica):

- Internal diameter: 50 μm
- Total length: 45 cm

Buffers:

- pH 7.2, 10 mM phosphate: 10 mM phosphoric acid, pH adjusted with NaOH.
- pH 7.2, 100 mM phosphate: 100 mM phosphoric acid, pH adjusted with NaOH.
- pH 9.5, 10 mM phosphate: 10 mM phosphoric acid, pH adjusted with NaOH.
- pH 9.5, 100 mM phosphate: 100 mM phosphoric acid, pH adjusted with NaOH.

Temperature: ambient

2.2. Run Conditions

- Step 1: Condition a capillary by rinsing 10 min with 0.1 M NaOH followed by 10 min with deionized water and 10 min with the pH 7.2, 10 mM phosphate buffer.
- Step 2: Fill the buffer reservoirs with the pH 7.2, 10 mM phosphate buffer. Fill to a volume typically used in the specific apparatus (typically a few mLs in each reservoir). Measure the pH in the anode reservoir. Then apply a voltage corresponding to a current of approximately 30 μA. The voltage must be applied for 150 min. However, each 30 min the voltage is turned off, pH measured in the anode reservoir, and voltage reapplied.
- Step 3: repeat step 2, but change the volume in the reservoirs. In this step only half the normal volume should be used.
- Step 4: repeat step 2, except that the pH 7.2, 100 mM phosphate buffer is used.
- Step 5: repeat step 2, except that the pH 9.5, 10 mM phosphate buffer is used.
- Step 6: repeat step 2, except that the pH 9.5, 100 mM phosphate buffer is used.

0-8493-9127-X/98/$0.00+$.50
© 1998 by CRC Press LLC

3. RESULTS, CALCULATION, AND DISCUSSION

3.1. Theory

In CE, field strengths of several thousands of volts per meter are often applied. The electrode processes at the anode and cathode are:

$$\text{Anode side: } 2\ H_2O \leftrightarrow O_2 + 4\ H^+ + 4e^-$$

$$\text{Cathode side: } 2\ H_2O + 2\ e^- \leftrightarrow H_2 + 2\ OH^-$$

H^+ and OH^- ions are formed at the anode and cathode, respectively. One can calculate the number of moles formed (n) at a certain current (I) and time (t) by the use of Faraday's law:

$$n = \frac{I \cdot t}{F}$$

F is the Faraday constant 96485 C/mole. By dividing the number of moles by the volume in the reservoir, one can calculate the amount of mM formed. When Faraday's law is combined with the general Acid-Base equation for the buffer in the buffer reservoirs, one can calculate the change in pH at the anode and cathode.

3.2. Results

Depict the measured pH values vs. time for all five experiments.
Calculate the change in pH from start to end for each of the five experiments.

Buffer	Volume	pH start	pH end	ΔpH
pH 7.2, 10 mM phosphate	Full			
pH 7.2, 10 mM phosphate	Half			
pH 7.2, 100 mM phosphate	Full			
pH 9.5, 10 mM phosphate	Full			
pH 9.5, 100 mM phosphate	Full			

Calculate the amount and mMs of H^+ formed in the anode reservoir from start to end of the experiment.

Buffer	Volume	I	t	n	mM
pH 7.2, 10 mM phosphate	Full				
pH 7.2, 10 mM phosphate	Half				
pH 7.2, 100 mM phosphate	Full				
pH 9.5, 10 mM phosphate	Full				
pH 9.5, 100 mM phosphate	Full				

3.3. Comment on the Following Points

a) In which experiment was the highest change in pH measured?
b) In which experiment was the lowest change in pH measured?
c) Was the change in pH highest when using full or half volume in the reservoir?
d) Explain the reason for the different changes in pH measured from experiment to experiment.
e) Mark the correct answers by a circle below:

The axial pH gradient is minimized by

- Using a **high/low** buffer concentration.
- Using a buffer with a pK_a value **close to/far away** from the buffer pH.
- Using **high/low** currents.
- **Replenishing/using the same buffer volume** from run to run.
- Using **large/small** buffer volumes.

4. CONCLUSION

This experiment should have demonstrated:

1) That the choice of buffer in CE is extremely important;
2) That pH continuously changes in the buffer reservoirs when voltage is applied due to electrode processes;
3) That by proper selection of buffer and other run conditions one can minimize/eliminate the changes in pH in the electrode reservoirs;
4) That Faraday's law can be applied to calculate the number of H^+ and OH^- moles formed;
5) That when low reproducibilities are observed in a specific application, one explanation could be use of low-capacity buffers.

Experiment 3:
Determination of pK$_a$ of Substituted Benzoic Acids by Capillary Electrophoresis (CE)

1. AIM

To measure the dissociation constant (pK$_a$) of some substituted benzoic acids (benzoic acid, *p*-methoxybenzoic acid, and *p*-chlorobenzoic acid) using capillary electrophoresis (CE).

2. EXPERIMENTAL

2.1. CE Conditions

Capillary (uncoated fused-silica):

- Internal diameter: 50 μm
- Total length   57 cm
- Length from injection to detection   50 cm

The capillary is conditioned by rinsing with 0.1 *M* NaOH for 2 min, followed by 3 min with the appropriate running buffer.

Detection: UV absorbance at 230 nm.

Applied Voltage: A voltage of 10 kV is applied during the analysis generating currents ranging from 35 to 70 μA.

Samples: A 5-sec high-pressure injection of 0.001 *M* solution of benzoic acid is prepared by dissolving the solid in an appropriate volume of a 50/50 mix of water and methanol.

Buffers: To maintain the same buffer ionic strengths, 75 m*M* NaCl is added to all buffers. pH values are adjusted using either HCl or NaOH. Typical buffers that can be used are listed below. Typically, 20 ml of buffer contained 87.8 mg NaCl (1.5 mmoles) and either 62.4 mg NaH$_2$PO$_4$ (0.4 mmoles) or 24.0 mg glacial acetic acid (0.4 mmoles).

Buffer components	Buffer pH
20 m*M* sodium phosphate	6.0 to 7.0
20 m*M* sodium acetate	4.0 to 5.5
20 m*M* sodium phosphate	1.5 to 3.5

Temperature: 25°C.

2.2. CE of Substituted Benzoic Acid Derivatives

- Rinse the capillary with 0.1 *M* NaOH followed by the appropriate buffer;

0-8493-9127-X/98/$0.00+$.50
© 1998 by CRC Press LLC

- Inject methanolic solution of benzoic acid derivative and apply a separation voltage of 10 kV; run time = 20 min;
- Record the migration times for benzoic acid derivative (t_{ba}) and methanol (t_m);
- Repeat above procedure for the other buffers.

3. CALCULATION OF ELECTROPHORETIC MOBILITY AND PK$_a$

3.1. Calculation of Electrophoretic Mobility (μ_e) of Benzoic Acid Derivative

The electrophoretic mobility (μ_e) of an analyte can be calculated from the migration time (t).

$$t_{obs} = t_{eof} + t_e$$

$$t_e = t_{obs} - t_{eof}$$

$$\mu_e = L^2/V \cdot t_e$$

where t_{obs} = the observed migration time for benzoic acid derivative (in min)
t_{eof} = the electroendosmotic flow time (i.e., migration time for methanol) (in min)
t_e = the effective migration for benzoic acid derivative (in mins)
μ_e = the effective mobility for benzoic acid derivative (in mins)
L   = the total capillary length (in cm)
V   = the applied voltage (in volts)

Record the buffer pH value, μ_e and [H⁺] in the table below:

Buffer pH	μ_e	[H⁺]

3.2. Determination of Dissociation Constant (K$_a$)

The electrophoretic mobility (μ_e) of an analyte is related experimentally to the hydrogen ion concentration by the following equation:

$$1/\mu_e = 1/(K_a\mu_{oa})[H^+] + 1/\mu_{oa}$$

where K_a = dissociation constant for benzoic acid
μ_{oa} = electrophoretic mobility of protonated benzoic acid
[H⁺] = the hydrogen ion concentration determined from pH

4. RESULTS

- Plot a graph of $1/\mu_e$ against [H⁺]
  - μ_{oa} is determined from the intercept
  - K_a (and hence pK_a) is calculated from the slope
- Plot μ_e against pH. This should give a characteristic sigmoidal plot.

Ka is estimated by reading the pH value at the halfway point between the two flat parts of the sigmoidal curve.

- Compare the pK_a values determined using the CE method with literature values[1] for benzoic acid (4.19), *p*-methoxybenzoic acid (4.47), and *p*-chlorobenzoic acid (3.98).

4. REFERENCES

1. *Handbook of Chemistry and Physics,* 1980, Weast, R. C. and Astle, M. J., Eds., CRC Press, Boca Raton, FL, Vol. 64, 1992, 775.

Experiment 4:
Retention Index in MEKC

1. AIM

To calculate retention index (RI) values for different neutral and charged analytes and evaluate different homologous series as RI standards.

Suggested reading:

1) E. S. Ahuja and J. P. Foley, *Analyst*, 119 (1994) 353–360.
2) Pim G. H. M. Muijselaar, H. A. Claessens, and C. A. Cramers, *Anal. Chem.*, 66 (1994) 635–644.
3) L. S. Ettre, *Anal. Chem.*, 36 (1964) 31.

Recommended reading:

Pacakova, V. and Ladislav. F., *Chromatographic Retention Indices: An Aid to Identification of Organic Compounds*, 1st Ed., Ellis Horwood, New York, 1992, pp. 1–285.

2. EXPERIMENTAL

2.1. Materials

Fused-silica capillary tubing of 75 μm i.d. × 363 μm o.d. and cut to a length of 40 to 45 cm, with an optical detection window installed if necessary.

2.2. Chemicals

1. Homologous series (RI standards)
   A. Alkylphenones (**C8-C14**), (obtained from Aldrich Chemical Company)
   B. For the optional experiment nitroalkanes (**Cl-C6**) (also from Aldrich); decanophenone (**C16**) is used as the tmc marker for both series of analytes.

2. Surfactant: sodium dodecyl sulfate (SDS)

3. Analytes
   A. Neutrals
      1. m-nitrotoluene
      2. benzaldehyde
      3. benzophenone

   B. Charged
      1. ascorbic acid
      2. folic acid

 4. Buffer(s)
    A. 100 mM Sodium phosphate (pH 7) (stock solution)

2.3. Procedure

The capillary should be activated by successive 15-min rinses with 1.0 M NaOH, 0.1 M NaOH, and running buffer. The applied voltage will be 12.0 kV in all experiments (in terms of Joule heating — no greater than 1.5 W/m).

The running buffer will be made up of 50 mM SDS (MW 288.4), 10 mM sodium phosphate. Dilutions will be required from the stock solutions of phosphate buffer and SDS. The running buffer components should be placed in a 100 mL volumetric and sonicated. (Note: the solutions are already prepared for you. See below.)

Stock solutions of analytes are prepared to a concentration of about 10 mg/mL in acetonitrile. These are then diluted with the running buffer to a final concentration of about 0.2 to 0.4 mg/mL. The final solution should also contain the t_{mc} marker, decanophenone. The t_o marker will be acetonitrile. Analyte dilutions are to be done as follows (assuming a 4 mL sample vial):

1) Obtain one 4-mL vial
2) Take 4 drops of the stock analyte (homologs) solution and put it in the 4-mL vial
3) Take 3 drops of the test analyte and put it in the same 4-mL vial
4) Fill the 4-mL vial with the 50 mM SDS/10 mM phosphate buffer solution
5) Put this solution in the syringe filter and filter with an 0.20 μm filter into a new 4-mL vial

Part 1

In the first part of this experiment, retention indices will be obtained for the neutral analytes. Alkylphenones will serve as the RI standards (detection will be at 254 nm). Filter all solutions using the 0.20 μm filters. This includes the 50 mM SDS/10 mM phosphate running buffer as well as the analyte solution, which should contain the alkylphenones and one of the three neutral solutes. Solutions should be filtered into appropriate sample vials.

Load the solutions in the instrument as shown by your instructor. You will do one run with the following operational parameters:

1) Repetition Number — 1
2) Injection time — 2 sec (hydrostatic)
3) Analysis time — 15 min
4) Applied voltage — the highest voltage that gives a power per unit length of less than 1.5 W/m

Repeat with another neutral analyte.

Using the same running buffer, repeat using the charged analytes one at a time. In order to calculate retention factors for the charged analytes, both μ_{eo} and μ_{ep} must be measured. μ_{ep} (electrophoretic mobility with no micelles present) must be obtained. This can be accomplished running each charged analyte using a 2 mM SDS/10 mM sodium phosphate running buffer at the operating conditions listed above.

Once you have calculated the retention factors for the analytes and the retention index markers, the retention index (RI) is calculated using the following equation:

$$RI = 100z + 100 \frac{\ln\ k'(x) - \ln\ k'(z)}{\ln\ k'(z+1) - \ln\ k'(z)}$$

where k' is the capacity factor, x refers to an analyte of interest, z refers to the alkylphenone with z carbon atoms that emerge before x, and $z + 1$ refers to an alkylphenone with $z + 1$ carbon atoms that emerges after x.

Part 2

Repeat the above experiment using 25 mM SDS instead of 50 mM SDS. Note, you do not have to repeat the μ_{ep} portion of the experiment using charged analytes.

Make sure you record the migration times of all the analytes, including t_o and t_{mc}. Also record the operating currents for all the different running buffer systems. You will need this information in order to do the calculations.

Optional — Retention Indices Using 1-Nitroalkanes as RI Markers

If time permits, instead of using the alkylphenones as retention index standards, use the 1-nitroalkane homologous series (C1-C6).

Changes in operating conditions:

1) UV detection will be at 214 nm
2) Neutral test analytes: anisole, benzophenone
3) t_{mc} marker will be decanophenone
4) Charged test analytes: 4-ethylbenzene sulfonic acid, toluic acid

All other conditions will be the same as those run for the alkylphenone homologous series.

Perform the experiments as outlined in Parts 1 and 2. The same concentrations of SDS and phosphate buffer are to be used. Again, the solutions you are required to make are the analyte solutions.

3. QUESTIONS

1) Calculate the retention factors (k') for all of the neutral analytes at the different SDS concentrations. Do the retention factors increase with analyte hydrophobicity, and if so why? How does k' change with an increase in surfactant (SDS) concentration? Why?
2) Construct log k' vs. Nc (carbon number) plots for the homologous series and calculate methylene selectivities at the two SDS concentrations. Is there a linear relationship? Show your correlation coefficients. If the plot is not linear, where do the deviations occur and why? Does methylene selectivity change with a change in SDS concentration? Why or why not?
3) Calculate the RI values for each of the neutral analytes. Some of the RI values can be obtained through extrapolation (see required reading for help). How does the RI value change with a change in SDS concentration? Why?
4) What is the t_{mc}/t_o value (measure of elution window) at the two SDS concentrations? Remember to measure from the start of the negative peak generated using acetonitrile. Does the elution range increase with an increase in SDS concentration? Why? (Hint: calculate the electroosmotic velocity and the electrophoretic mobility of the micelle.)
5) Calculate the retention factors for charged analytes using the k' equation that utilizes the μ_r value.
6) Calculate the RI values for the charged analytes at the two different SDS (25 and 50 mM) concentrations. Are they significantly different? Why or why not?

7) What effect could organic modifiers such as acetonitrile and methanol have on the retention indices generated for the neutral analytes?

8) Why does the operating current change with a change in SDS concentration? Explain.

9) Calculate the power per unit length (W/m) for each of the three running buffer systems (2, 25, 50 mM SDS systems). Remember to use the total length of capillary in your calculations.

3.1. Questions for Optional Section (if done)

1) Calculate the RI values for the neutral and charged analytes. Since there are only six commercially available nitroalkanes (Cl-C6), extrapolation for the RI value of benzophenone had to be done. Is this a disadvantage of the nitroalkane system in comparison to the alkylphenone system. Why or why not?

2) What advantage(s) does the use of the 1-nitroalkane series have over the alkylphenones as RI standards in MEKC?

3) Make a plot of log k vs. Nc (carbon number) for the l-nitroalkanes. Contrast this with the one generated for the alkylphenone homologous series.

4) Suggest other homologous series that could be used as RI standards in MEKC. List any advantages or disadvantages.

Experiment 5:
Chiral Separations Using MEKC

1. AIM

To separate enantiomers by MEKC with a commercially available chiral surfactant, N-dodecoxycarbonylvaline (DDCV).

Suggested reading:

A.G. Peterson, E.S. Ahuja, and J.P. Foley, *J. Chromatogr. B,* 683: 15–28 (1996), and references therein.

A.G. Peterson and J.P. Foley, *J. Microcol. Sep.*, 8, 427, 1996.

A.G. Peterson and J.P. Foley, *J. Chromatogr. B.*, 695, 131, 1997.

2. EXPERIMENTAL

2.1. Materials

Fused-silica capillary tubing of 75 μm i.d. × 363 μm o.d. and cut to a length of 40 to 45 cm, with an optical detection window installed if necessary.

2.2. Chemicals

Sulconazole (t_{mc} marker), CHES (zwitterionic buffer), triethylamine, (R) and (S) N-dodecoxycarbonylvaline (Waters Inc.), and the following racemic mixtures and/or pairs of pure enantiomers: atenolol, N-methylpseudoephedrine, norephedrine, pseudoephedrine, and pindolol.

2.3. Procedure

The fused-silica capillary should be activated by successive 15-min rinses with 1.0 *M* NaOH, 0.1 *M* NaOH, and running buffer. The applied voltage should be the highest voltage that gives a power per unit length of less than 1.5 W/m. The running buffer consists of a solution of 25 m*M* (R or S) N-dodecoxycarbonylvaline (DDCV, MW 325.9), 50 m*M* CHES 2-[N-cyclohexylamine] ethane sulfonic acid, MW 207.3), and 10 m*M* triethylamine. The buffer components should be placed in a beaker containing HPLC-grade water and sonicated. The pH should be monitored and 1 *M* NaOH added to keep the pH above 7.0 in order to avoid precipitation of the surfactant. The final solution in the beaker should then be adjusted to pH 8.8 and transferred to an appropriate volumetric flask and diluted to the mark. Stock solutions of all pairs of enantiomers should be prepared to a concentration of about 1 mg/mL in methanol. These stock solutions are then diluted 1:5 with running buffer to give a final concentration of 0.2 mg/mL. A t_{mc} marker such as sulconazole should be added to give about the same concentration in the final solution that is injected. Since the analytes are charged species, each analyte must also be run individually in the same buffer system without the presence of the DDCV micelles. The CZE is necessary to calculate the k' and the enantiose-lectivities.

3. QUESTIONS

1) Are the mean k' values of the racemic mixtures related to any physicochemical properties of these molecules? Compare the order of migration obtained using (1) free zone conditions (pH ~ 2.5) and (2) a nonchiral surfactant, such as sodium dodecyl sulfate; explain differences in migration in terms of charge density and hydrophobicity.
2) Can you accurately measure a low concentration of the slower-migrating enantiomer in the presence of an excess (>1000 times) of the faster-migrating isomer? Suggest and carry out an experiment to improve this accuracy, estimating limits of detection for both enantiomers in the presence of an excess concentration of the opposing antipode.

Experiment 6:
The Direct Separation of Amino Acid Enantiomer Derivatives Using Cyclodextrin-Modified Electrokinetic Chromatography (MEKC)

1. AIM

To resolve the enantiomers of D,L Leucine (DL-Leu) and D,L-phenylalanine (DL-Phe) as derivatives of naphthalene-2,3-dicarboxaldehyde (NDA) using cyclodextrin-modified electrokinetic chromatography (MEKC).

2. EXPERIMENTAL

2.1. Materials

All high-purity chemicals used in this experiment are commercially available from suppliers such as Aldrich or Sigma.

2.2. Derivatization of Amino Acids Using NDA

(where R=CH$_2$CH(CH$_3$)$_2$, for Leu, and R=CH$_2$Ph, for Phe)

The sample is dissolved in a volume (typically 100 µl) of 100 mM borate buffer adjusted to pH 9.0. To this solution, add an aliquot (10 µl) of an aqueous solution of 10 mM NaCN. Mix the solution well, then add a volume (80 µl) of a methanolic solution of 8 mM NDA. Leave the reaction in the dark at room temperature for 15 min, then analyze.

2.3. CE Separation Conditions

Capillary (uncoated fused-silica): i.d., 50 µm; total length, 57 cm; length from injection to detection, 50 cm. The capillary is conditioned by rinsing with 0. 1 M NaOH for 2 min, followed by 2 min with the running buffer.

Buffer: 61.8 mg boric acid, 259.4 mg γ-cyclodextrin, and 288.4 mg SDS are first dissolved in 20 ml of double distilled deionized water, and then pH is adjusted to 9.0 using 0.1 M NaOH.

513

Detection: UV absorbance at 254 nm or laser-induced fluorescence (LIF) detection set at excitation and emission wavelengths of 442 and 525 nm, respectively.

Voltage: A voltage of 25 kV is applied during the analysis, generating a current of about 60 μA.

Sample: A 5-sec high-pressure injection of water followed by 1-sec injection of the NDA reaction solution.

Temperature: 25°C

3. QUESTIONS

- Identify the derivatized amino acids and their enantiomers by coinjection with standards.
- Observe the migration order of individual amino acids and their respective enantiomers. Which enantiomer migrates first, D or L? Why?
- Comment on the differences between the peak profiles obtained using UV absorbance and LIF detection? Can you see a signal due to excess NDA when using either UV or fluorescence detection?
- Discuss the most likely mechanism for the formation of *transient diastereoisomers* in the chiral discrimination of NDA-derivatized amino acids.

Experiment 7:
The Resolution of Enantiomers of R,S-Naphthylethylamine (R,S-NEA) as Diastereoisomeric Derivatives of 2,3,4,6-tetra-O-acetyl-β-D-glucopyranosyl Isothiocyanate (GITC)

1. AIM

To resolve the enantiomers of an optically active amine such as R,S-naphthylethylamine (R,S-NEA) by first derivatizing with 2,3,4,6-tetra-O-acetyl-β-D-glucopyranosyl isothiocyanate (GITC) followed by separation of the diastereoisomeric derivative using cyclodextrin-modified electrokinetic chromatography (CD-MEKC).

2. EXPERIMENTAL

2.1. Materials

Buffers, the racemic mixture and the individual enantiomers of NEA, and the derivatizing agents GITC and the related isomer 2,3,4-tri-O-acetyl-α-D-arabinopyranosyl isothiocyanate (AITC) can be obtained from the usual chemical suppliers.

2.2. Derivatization of R,S-NEA with GITC or AITC

GITC   GITC-L-(+)-NH2-R   GITC-D-(-)-NH2-R

An aliquot (1 mg, 7×10^{-3} moles) of R,S-NEA (or one of the individual isomers) is dissolved in DMF (100 μl). To this is added a solution of GITC (2.72 mg, 7×10^{-3} moles) in DMF. The reaction solution is then left at room temperature for 30 min. Prior to analysis by CD-MEKC, the DMF reaction solution was diluted with water. Derivatization with AITC is carried out using the same procedure.

2.3. CE Separation Conditions

Capillary (uncoated fused-silica): i.d., 50 μm; total length, 57 cm; length from injection to detection, 50 cm. The capillary is conditioned by rinsing first with 0.1 M NaOH for 2-min, followed by a 2-min rinse with the running buffer.

Buffer: 93.6 mg NaH_2PO_4 (0.4 mmoles), 12.4 mg boric acid (0.2 mmoles), 454.0 mg β-cyclodextrin (0.4 mmoles) and 521.7 mg taurodeoxycholic acid (1 mmole) are dissolved in 20 ml of double distilled and deionized water. The pH is adjusted to 7 using 0.1 M NaOH.

Detection: UV detection at 254 nm.

Applied voltage: A voltage of 20 kV is applied during the analysis, generating a current of approximately 60 μA.

Sample: A 1-sec high-pressure injection of the GITC reaction solution.

Temperature: 25°C

3. QUESTIONS

- Confirm the identities of the GITC derivatives of the enantiomers of NEA by co-injecting the individual antipodes.
- Compare the resolution of the derivatized NEA with the underderivatized material by injecting a dilute solution of R,S-NEA using the same buffer system.
- Why does conversion of the enantiomers of NEA to their diastereoisomeric derivatives improve resolution?
- After derivatization of R,S-NEA with AITC, compare the migration order of the isomers with that obtained with GITC.

Experiment 8:
Selective Labeling and CE Analysis
of Acidic Monosaccharides

1. AIMS

- Labeling acidic monosaccharides to enable their detection and CE analysis
- Evaluation of labeled monosaccharide–borate complexation as a means for their CE separation
- Understanding the effect of the structure of monosaccharides on the stability of their borate complexes
- Calculation of effective electrophoretic mobilities

2. EXPERIMENTAL

2.1. Labeling Procedure

- Add separately 50 µl of 100 mM aqueous stock solution of 1-ethyl-3-(3-dimethylaminopropyl) carbodiimide hydrochloride (EDAC) to 1.0 mg each of D-glucuronic acid and D-galacturonic acid. To ensure a quantitative yield, the EDAC solution should be prepared just before starting the derivatization reaction, and used immediately.
- Next, add 100 µl of 100 mM aqueous stock solution of 7-aminonaphthalene-1,3-disulfonic acid (ANDSA) to the mixture, and stir for 3 hours at room temperature.

2.2. Separation Conditions

Capillary (bare fused-silica capillary):
Internal diameter, 50 µm; total length, 80 cm; effective length (length to detection point), 50 cm.

Prior to its first use, the capillary is flushed with 1.0 M NaOH for 5 min, followed by 10 min rinsing with 0.1 M NaOH and another 10 min with deionized water. It is then conditioned with the separation buffer for 10 min.

Detection:	247 nm
Separation voltage:	20 kV (inlet at anode, and outlet at cathode)
Analytes:	30 µl each of ANDSA-D-glucuronic acid and ANDSA-D-galacturonic acid, and 10 µl dimethyl sulfoxide (DMSO), which is a neutral maker, are added and mixed with 930 µl of deionized water.

0-8493-9127-X/98/$0.00+$.50
© 1998 by CRC Press LLC

Injection volume: Approximately 4.0 nl
Separation buffers: Buffer 1: 50 mM sodium borate, pH 10 adjusted with NaOH
                         Buffer 2: 100 mM sodium borate, pH 10, adjusted with NaOH
                         Buffer 3: 150 mM sodium borate, pH 10, adjusted with NaOH
Temperature: Ambient

2.3. CE of Monosaccharides

Run 1:   Rinse capillary with Buffer 1 for 2 min; inject the analytes and apply 20 kV
         separation voltage. Run time = 30 min.
Run 2:   Rinse capillary with Buffer 2 for 2 min; inject the analytes and apply 20 kV
         separation voltage. Run time = 30 min.
Run 3:   Rinse capillary with Buffer 3 for 2 min; inject the analytes and apply 20 kV
         separation voltage. Run time = 30 min.

3. RESULTS, CALCULATIONS, AND DISCUSSION

3.1. Effective Electrophoretic Mobilities

For the ANDSA-D-glucuronic and ANDSA-D-galacturonic peaks (ANDSA-D-galacturonic
will migrate first, followed by ANDSA-D-glucuronic), calculate the effective electrophoretic
mobility using the following equation:

$$\mu_{ep} = \frac{Ll}{V\,60}\left(\frac{1}{t_m} - \frac{1}{t_{EOF}}\right)$$

where L is the total length of the capillary in centimeters, l is the effective length (i.e., to the
detection point) of the capillary in centimeters, V is the separation voltage in volts, t_m is the
migration time of the analytes in minutes, and t_{EOF} is the migration time of DMSO in minutes
(a negative peak). Note, μ_{ep} will have a negative sign indicating that it is in the opposite
direction relative to EOF.

Buffer	t_{EOF} (min)	Glucuronic t_m (min)	Glucuronic μ_{ep} (cm² V⁻¹ S⁻¹)	Galacturonic t_m (min)	Galacturonic μ_{ep} (cm² V⁻¹ S⁻¹)
Buffer 1 50 mM Borate					
Buffer 2 100 mM Borate					
Buffer 3 150 mM Borate					

Plot μ_{ep} versus borate concentration.

3.2. Questions

1) How did the variation in the borate buffer concentration affect μ_{ep} of the labeled acidic
   monosaccharides?
2) What causes this variation?
3) Was the variation in the μ_{ep} the same for both monosaccharides? Explain.
4) Although both analytes have the same molecular weight, their μ_{ep} were affected differently
   by the variation of the borate concentration. Why?

4. LEARNING OBJECTIVES

After performing this experiment you should have demonstrated:

1) The ability to label acidic monosaccharides *via* their acidic groups
2) The effect of monosaccharides–borate complexation on the CE separation
3) The effect of the orientation of the hydroxyl groups in the analyte on the stability of the monosaccharide–borate complex.

Experiment 9:
Avoiding Protein Adsorption by the "Extreme pH" Approach

1. AIMS

- Understanding the phenomenon of proteins being adsorbed/desorbed on the capillary surface
- Calculation of theoretical plate numbers

2. EXPERIMENTAL

2.1. CE Conditions

Capillary (uncoated fused-silica): i.d., 50 μm; total length, 45 cm; length from injection to detection, 37 cm. The capillary is conditioned by rinsing 10 min with 0.1 M NaOH, followed by 10 min with deionized water and 10 min with the run buffer.

Detection: 214 nm.

Voltage applied during electrophoresis: a voltage which results in a current of approximately 30 μA is applied.

Sample (in deionized water): 2 mg lysozyme/ml. The isoelectric point (pI) of lysozyme is 11.0

Injection volume: 4 nl

Buffers: pH 2.0: 25 mM phosphoric acid, pH adjusted with NaOH. **pH 7.0:** 25 mM phosphoric acid, pH adjusted with NaOH. **pH 12.0:** 25 mM phosphoric acid, pH adjusted with NaOH.

Temperature: Ambient

2.2. CE of Lysozyme

- Run a capillary wash cycle with the pH 2.0 buffer
- Run 1: inject lysozyme and apply a voltage corresponding to a current of approximately 30 μA (~13 kV). Run time = 40 min
- Run 2: run a wash cycle with the pH 7.0 buffer, inject lysozyme, and apply a voltage corresponding to a current of approximately 30 μA (~16 kV). Run time = 40 min
- Run 3: run a wash cycle with the pH 12.0 buffer, inject lysozyme and apply a voltage corresponding to a current of approximately 30 μA (~5 kV). Run time = 40 min

3. RESULTS, CALCULATION, AND DISCUSSION

3.1. Theoretical Plate Numbers

For the lysozyme peak, calculate the number of theoretical plates in each electropherogram. The plate number can be calculated as follows:

$$N = 5.54 * \left(\frac{t_m}{w_{0.5}} \right)^2$$

where t_m is the migration time and $w_{0.5}$ is the peak width at half peak height.

Buffer:	t_m	$w_{0.5}$	N
pH 2.0 buffer			
pH 7.0 buffer			
pH 12.0 buffer			

3.2. Comment on the Following Points

a) In how many of the electropherograms did lysozyme elute as a peak?
b) In which electropherogram was the highest plate number obtained?
c) Explain the mechanisms causing lysozyme to elute as a peak at certain pH values, whereas it is adsorbed to the capillary surface at other conditions.
d) Discuss the stability of proteins at extreme pH values.
e) Suggest other methods to avoid adsorption of basic proteins when analysis is performed at neutral pH.

4. CONCLUSION

This experiment should have demonstrated

1) That by manipulation of the buffer pH, one can avoid the protein sticking to the capillary surface.
2) How theoretical plate numbers are calculated and that low plate numbers often indicate tailing peaks.

Note: It is recommended that this experiment be followed up with Experiment 10.

Experiment 10:
CE of a Basic Protein Using Phytic Acid
as an Ion-Pairing Agent

1. AIMS

- Understanding the phenomenon of proteins being adsorbed/desorbed to the capillary surface
- Understanding the principles of ion-pairing CE

2. EXPERIMENTAL

2.1. CE Conditions

Capillary (uncoated fused-silica):

- Internal diameter, 50 μm; total length, 45 cm
- Length from injection to detection, 37 cm

The capillary is conditioned by rinsing 10 min with 0.1 M NaOH, followed by 10 min with deionized water and 10 min with the run buffer.

Detection: 214 nm

Sample (in deionized water): 2 mg lysozyme/ml. Isoelectric point of lysozyme is 11.0.

Injection volume: 4 nl

Buffers: 0 mM phytic acid: 25 mM sodium phosphate buffer pH 7.0 (25 mM phosphoric acid, pH adjusted with NaOH). **5 mM phytic acid:** 25 mM sodium phosphate buffer pH 7.0 (25 mM phosphoric acid, pH adjusted with NaOH), 5 mM phytic acid added as sodium salt. **20 mM phytic acid:** 25 mM sodium phosphate buffer pH 7.0 (25 mM phosphoric acid, pH adjusted with NaOH), 20 mM phytic acid added as sodium salt. **60 mM phytic acid:** 25 mM sodium phosphate buffer pH 7.0 (25 mM phosphoric acid, pH adjusted with NaOH), 60 mM phytic acid added as sodium salt.

Temperature: Ambient

2.2. CE of Lysozyme

- Run a capillary wash cycle with the 0 mM phytic acid buffer.
- Run 1: inject lysozyme and apply a voltage corresponding to a current of approximately 30 μA (~16 kV). Run time = 80 min.
- Run 2: run a wash cycle with the 5 mM phytic acid buffer, inject lysozyme and apply a voltage corresponding to a current of approximately 30 μA (~11 kV). Run time = 80 min.

- Run 3: run a wash cycle with the 20 mM phytic acid buffer, inject lysozyme, and apply a voltage corresponding to a current of approximately 30 µA (~5 kV). Run time = 80 min.
- Run 4: run a wash cycle with the 60 mM phytic acid buffer, inject lysozyme, and apply a voltage corresponding to a current of approximately 60 µA (~4 kV). Run time = 80 min.

3. RESULTS, CALCULATION, AND DISCUSSION

3.1. Theoretical Plate Numbers

For the lysozyme peak calculate the number of theoretical plates in each electropherogram. The plate number can be calculated as follows:

$$N = 5.54 * \left(\frac{t_m}{w_{0.5}} \right)^2$$

where t_m is the migration time and $w_{0.5}$ is the peak width at half peak height.

Buffer:	t_m	$w_{0.5}$	N
0 mM phytic acid			
5 mM phytic acid			
20 mM phytic acid			
60 mM phytic acid			

3.2. Comment on the Following Points

a) In how many of the electropherograms did lysozyme migrate as a peak?
b) In which electropherogram was the highest plate number obtained?

4. CONCLUSION

This experiment should have demonstrated

1) That phytic acid is a powerful ion-pairing reagent which can be used to avoid adsorption of a basic protein to the negatively charged capillary surface at neutral pH.
2) How theoretical plate numbers are calculated and that low plate numbers often indicate tailing peaks.

Note: Before attempting this experiment, it is useful to complete Experiment 9.

Experiment 11:
Monitoring Protein Glycoforms Using Coated Capillaries and Acidic Buffers

1. AIMS

- Familiarize the experimentalist with the use of amine-coated capillaries
- Optimize electrophoretic conditions for the separation of protein glycoforms
- Calculate electrophoretic mobilities, electroosmotic flow, and theoretical plate numbers

2. EXPERIMENTAL

2.1. Material and Methods

Samples: Ribonuclease B (1 mg/mL in deionized water). Ribonuclease B is a glycoprotein of 124 amino acids with one N-linked glycosylation site at asparagine 38 which comprises a mixture of high mannose saccharides of the type $GlcNAc_2$ Man_x where x varies from 5 to 9 (GlcNAc: N-acetyl glucosamine and Man: mannose).

Capillary: Uncoated fused-silica i.d., 50 µm; total length, 97 cm; length to detector, 90 cm.

The capillary is first conditioned by rinsing with the following solutions: 1 M NaOH (15 min), deionized water (15 min), an aqueous solution of 5% (w:v) hexadimethrine bromide and 2% (v/v) of ethylene glycol (20 min), and 0.1 M formic acid (15 min). Between runs, the capillary is rinsed with 1 M NaOH (2 min), deionized water (3 min), an aqueous solution of 5% (w/v) hexadimethrine bromide and 2% (v/v) of ethylene glycol (3 min), and the running buffer (5 min).

Detection: 200 nm

Separation voltage: During the electrophoresis, a voltage of −30kV is applied at the injection end of the capillary. If a bipolar power supply is not available, a voltage of +30 kV can be applied at the collector buffer.

Injection volume: 20 nl

Separation buffers: 0.1, 0.5, 1.0, and 2.0 M HCOOH

Temperature: Ambient

2.2. CE-UV Analysis of Ribonuclease B

- Condition the capillary as described above, and conduct the separation using first a buffer of 0.1 M HCOOH. Replicate analyses (n = 3) might be required to ensure good separation conditions. The current should be approximately 10 to 15 µA. Note that shorter rinsing times are required between runs.

- Repeat the analysis using 0.5, 1.0, and 2.0 M HCOOH. The current should progressively increase (up to 60 µA) for higher acid strength buffers.

3. RESULTS, CALCULATION, AND DISCUSSION

3.1. Mobility Associated with the Electroosmotic Flow

Upon variation of the acidity of the separation buffer, the electroosmotic flow will change as reflected by a negative deflection of the baseline. If this deflection cannot be observed easily, mesityl oxide can be added to the sample to detect the electroosmotic flow. For each buffer system, calculate the mobility of the electroosmotic flow (μ_{eof}) according to the following equation:

$$\mu_{eof} = (L_t \cdot L_d)/V \cdot t_{eof})$$

where L_t is the total length of the capillary (m), L_d is the length of the capillary to the detector, V is the separation voltage (V), and t_{eof} is the migration time associated with the electroosmotic flow.

3.2. Electrophoretic Mobility

For the most abundant glycoform of ribonuclease B, calculate its electrophoretic mobility (μ_{ep}) as follows:

$$\mu_{ep} = [(L_t \cdot L_d)/V \cdot t_m)] - \mu_{eof}$$

where t_m is the migration time of the analyte.

3.3. Theoretical Plate Numbers

For each separation buffer, calculate the number of theoretical plates for the most abundant glycoform using the following equation:

$$N = 5.54[t_m /w_{0.5}]^2$$

where $w_{0.5}$ is the peak width at half height.

3.4. Resolution

For each separation buffer, calculate the resolution (R_s) between the two most abundant peaks (peak 1 and 2) as follows:

$$R_s = 2(t_1 - t_2)/(w_1 + w_2)$$

where w_1 and w_2 represent the peak width at the base of the peak.

Buffer	μ_{eof}	μ_{ep}	N	R_s
0.1 M HCOOH				
0.5 M HCOOH				
1.0 M HCOOH				
2.0 M HCOOH				

3.4. Discussion

a) Comment on the magnitude of the electrophoretic flow when buffers of high acid strength are used.

b) How do you explain the direction of the electrophoretic mobility of the glycoprotein?

c) Under which conditions do you obtain higher resolution and theoretical plate numbers? Can you provide an explanation for this observation?

d) The electrophoretic mobility of analyte in zone electrophoresis is inversely proportional to their molecular weight. Based on this, and on the fact that ribonuclease B has a variable number of neutral Man residues ranging from 5 to 9, can you identify each of the glycoforms on the electropherogram?

e) In certain batches of ribonuclease B the aglycone form of this protein can be observed. Based on previous observations, at which position in the electropherogram would this protein be expected?

4. CONCLUSION

The use of coated capillaries and acidic buffers should have demonstrated the following advantages for the separation of closely related glycoproteins:

1) The magnitude of the electroosmotic flow can be controlled by varying the acidity of the separation buffer, and improvement of separation efficiencies are generally obtained with higher acid concentrations.

2) Resolution of closely related glycoproteins can be obtained for simple oligosaccharide extension, and in some cases assignment of glycoproteins can be made using the relationship between the electrophoretic mobility and protein molecular weight.

3) Capillaries coated with basic polymers offer a good means to prevent protein adsorption to the surface when acidic buffers are used.

Experiment 12:
Separation of DNA Fragments

1. AIMS

- To understand the effects of the sample matrix components (e.g., salt) on the efficiency of DNA fragment peaks in capillary gel electrophoresis (CGE)
- To understand migration time shifts in CGE due to sample matrix effects
- To improve peak resolution when performing pressure injection in CGE by using a double injection technique (buffer injection followed by sample injection)

2. EXPERIMENTAL

The experiment requires a coated capillary and a suitable sieving gel. Various CE equipment manufacturers (e.g., Beckman, Dionex, Perkin-Elmer) offer DNA kits which contain pre-coated capillaries, DNA standards, sieving buffers, and recipe-type start-up protocols. This experiment has been performed with the Beckman eCAP ds DNA 1000 kit, in which a 100 µm i.d. capillary with a hydrophilic (polyacrylamide-type) coating is used to suppress the EOF. A gel buffer (linear polyacrylamide in TBE) is supplied, which must be rehydrated prior to the CE runs. However, this experiment can also be performed using suitable capillaries and buffers from other commercial or home-made sources. For example, the experiments described below should yield similar results when gel buffers based on polyethylene oxide, alkylcellulose derivatives, polyvinyl alcohol, or similar polymers are used, or when DNA standards other than ΦX 174 Hae III restriction fragments are used.

Capillary: Recommended capillary length is 30 cm to the detection window (37 cm total length). Install the capillary in the CE cartridge according to the instructions provided by the manufacturer. Set temperature on the instrument at 20°C.

Test mixtures; sample introduction: The test mixture consisting of ΦX 174 Hae III DNA fragments (72 to 1353 bp) is supplied at a concentration of 10 µg in 10 µl. This mixture also contains the reference marker Orange G. The test mixture should be diluted with filtered, deionized water to yield the final DNA concentrations which are introduced into the capillary. Pressure injections are performed using the low-pressure injection mode of the instrument (at 0.5 psi).

Gel buffer: As supplied in the eCAP ds DNA 1000 kit. Rehydrate according to the instructions, filter using a 0.45 µm filter before equilibrating the capillary with gel buffer. Before each run, rinse the capillary with gel buffer for 3 min at high pressure (20 psi).

Detection: UV absorbance, 254 nm. For experiments with "real" PCR samples, laser-induced fluorescence (LIF) detection is preferable because it offers much greater detectability; hence smaller DNA amounts can be injected without compromising sample detectability. With LIF detection, a DNA intercalating dye (e.g., EnhanCE used in the eCAP Ll Fluor dsDNA 1000 kit from Beckman) must be used.

Voltage: Use the reverse-polarity mode of the instrument (cathode on injection side) with a run voltage of 7.4 kV (electric field: 200 Wcm).

2.1. Effect of Injection Plug Length on Peak Efficiency

First, do a 10-sec pressure injection of the DNA test mixture (DNA concentration: 100 µg/mL), generating Electropherogram #1. Next, perform a 30-sec pressure injection (Electropherogram #2). Compare the migration times of the DNA peaks. Is there any difference between the two runs?

2.2. Effect of Salt Present in the Injection Plug. Double Injection Technique to Enhance Resolution

Dissolve the DNA standard in 20 mM NaCl at a DNA concentration of 50 µg/ml. Do a 10-sec pressure injection (Electropherogram #3). Compare this electropherogram with that of #1 and #2. Why are the peaks asymmetrical ("fronted")? Next, do a double injection (Electropherogram #4): first a 10-sec pressure injection of 0.1 M Tris-acetate buffer (pH 8.3), immediately followed by a 20-sec pressure injection of 50 µg/mL DNA standard in 20 mM NaCl. Try to rationalize the peak sharpening which should be visible in Electropherogram #4.

2.3. Effect of Salt Present in the Sample on Migration Times

Perform a 10-sec pressure injection of 100 µg/ml DNA standard in water (Electropherogram #5). The next CGE run (Electropherogram #6) is that of the DNA standard dissolved in 20 mM NaCl. Note the migration time shift, especially for the small DNA fragments migration in the early part of the electropherogram.

3. DISCUSSION

A detailed discussion of the phenomena observed in this experiment can be found in a publication by van der Schans et al., *J. Chromatogr. A,* 680 (1994), 511–516. Effects relating to electrokinetic and pressure injection are also discussed in Chapter 9.

With the replaceable gels used in this experiment, both pressure and electrokinetic injection are feasible. In CE, the pressure injection mode is generally recommended for quantitative work: the composition of the sample plug introduced into the capillary is exactly that of the sample vial from which the injection took place. It should be noted, however, that electrokinetic injection — when performed appropriately in conjunction with suitable background electrolytes — often yields more efficient peaks than pressure injection. The experiment discussed below addresses the pressure injection only.

The effect of the injection plug length is demonstrated in Electropherograms #1 and #2. With the present CE system, a 10-sec pressure injection should yield a plug length of about 1.3 mm, typical of many CE injection plug lengths. Much longer plug length will distort peak shape, as is visible in Electropherogram #2 where a threefold larger plug was injected. The "fronting" shoulders are due to high field conditions existing in the sample zone (relative to the rest of the capillary), leading to a stacking of DNA molecules at the gel-injection plug interface. When comparing Electropherograms #1 and #2, note that the migration times of the DNA standards stay approximately constant.

In the practice of CE, the separation performance is often strongly dependent on the composition of the sample solution. Desalting the sample by ultrafiltration may often enhance sample detectability, especially when used in conjunction with electrokinetic injection. The ultrafiltration procedure removes low MW sample constituents, resulting in efficient DNA peaks.

The effect of salt (20 mM NaCl) present in the sample on peak efficiency should be visible in Electropherogram #3. A combined isotachophoresis/stacking effect may be responsible for the fronting peaks seen in Electropherogram #3. By performing a double injection (first Tris-acetate buffer, then DNA standard), considerable peak sharpening should be obtainable. As shown in the diagram of Figure 1, this type of injection produces local electric field differences in the capillary (i.e., a lower field in the Tris-acetate zone compared to the sample) permitting relatively higher sample loads (note 30-sec injection time in Electropherogram #4 vs. 10-sec in Electropherogram #3).

The effect of salt on the migration times is visible in Electropherograms #5 and #6. This phenomenon is important when unknown samples are assessed for DNA size (bp's) through calibration with DNA size standards. If the standards have a different salt concentration compared to the sample, an erroneous bp assignment is possible. By comparing the DNA

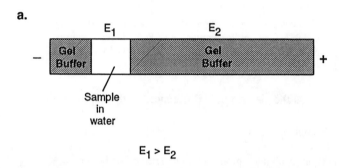

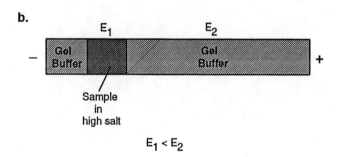

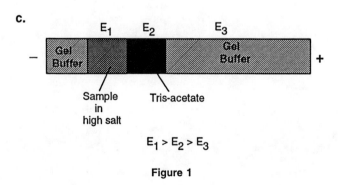

Figure 1

standard runs in #5 and #6, it can be seen that the smaller DNA fragments are relatively more effected by the presence of salt than the larger ones. This is so because the smaller DNA fragments are exposed longer to the lower field induced by the salt matrix compared to the larger DNA fragments.

The effects of co-injecting a DNA standard with PCR samples (using the double injection technique) was discussed earlier (see also Chapter 9).

4. CONCLUSION

The experiment described in this section demonstrates the following points:

- The sample matrix plays a significant role with respect to peak efficiency in the CGE of DNA. By performing a double injection in which a zone of Tris-acetate buffer is injected prior to the sample, considerable peak sharpening may be obtained for DNA-containing samples.
- The asymmetric peaks observed are due to localized field and stacking effects.
- Salt present in DNA-containing samples (e.g., PCR-derived) causes small migration time shifts which must be accounted for when the size of DNA is to be determined from comparison of electropherograms of samples with standards.

The diagram shows the electric field differences in the capillary for:

a) Injection of sample in water
b) Injection of sample in salt
c) Double injection involving a pre-injection of Tris-acetate buffer

Experiment 13:
Capillary Ion Electrophoresis of Nitrate and Nitrite in Urine

1. AIM

With the discovery of the biological importance of nitric oxide in human physiology, many assays for its metabolites nitrate and nitrite in biological fluids have been published. This experiment involves a simple CE assay which uses submicellar concentrations of the EOF reversal additive, tetradecyltrimethylammonium bromide (TTAB), with direct UV detection of the separated ions and reversed potential to rapidly separate the anions.

2. EXPERIMENTAL

2.1. Sample Preparation

Urine is collected in a clean container. Cool at 4°C to precipitate any salts and centrifuge to remove any precipitate. If required, the urine can be frozen at –20°C until analysis. Prior to analysis, pass through a 0.45 μm membrane filter.

Prepare a series of standard solutions of sodium nitrite and nitrate over the range 0 to 1 mM. In order to reduce background, use freshly collected distilled/deionized water.

2.2. CE Conditions

Mode: Free solution CE
Capillary: 44 cm (37 to window) × 50 μm i.d. bare fused-silica
Buffer: 150 mM sodium chloride 5 mM Tris pH 7.4 containing 1 mM TTAB
Detection: UV 212 nm
Sensitivity: Approximately 50 maufs (milliabsorbence units full scale)
Load: 4 sec hydrodynamic (approximately 5 ni) or –5 kV for 2 sec. Exact times will depend on the urine concentration
Run: –15 kV
Temperature: 25°C
Prior to commencing the day's runs the capillary should be conditioned with 0.1 M NaOH for 10 min followed by fresh running buffer for 10 min. Wash only with running buffer between runs.

2.3. Analyses

- Inject standard solutions of nitrite and nitrate (range 0 to 1 mM)

- Inject urine sample(s).
- Investigate difference in sensitivity between the electrokinetic and hydrodynamic methods of loading for both standards and urines by varying loading timings.

3. RESULTS

Depending on the instrumental configuration used interpretation of the results may vary.

1) Why does nitrite migrate before nitrate?
2) By comparison with the standard curve, calculate the concentration of nitrate in your urine sample.
3) By examination of the urine electropherogram, which method of injecting samples is more selective?

Experiment 14:
Determining Paracetamol in Urine

1. AIM

About 70 UV-absorbing peaks can be resolved by CE from a normal urine specimen in about 10 min. CE of urine can be useful in screening for changes in this pattern of peaks with disease and for determining drugs and their metabolites in urine following their ingestion.

Paracetamol (acetaminophen, mol weight 151.2) is a widely used nonprescription pain-killer. It is metabolized in the body by phase II processes to form a number of glucuronide and sulfate conjugates which are excreted in the urine. The metabolism of paracetamol can be readily followed by CE. The absorption maximum of paracetamol is 272 nm. The pK_a of paracetamol is about 9.4, whereas the conjugates will be negatively charged. Since parace-tamol is relatively neutral under conditions of high EOF, MEKC is to be preferred for the analysis.

The principal objective of this experiment is to follow the metabolic fate of paracetamol by analyzing urine samples taken at different time intervals after ingestion.

2. EXPERIMENTAL

2.1. Sample Preparation

Urine is collected in a clean container. Cool at 4°C to precipitate any salts and centrifuge to remove any precipitate. Prior to analysis, pass through a 0.45 μm membrane filter to remove any particulates.

If the paracetamol study is to be done, collect one urine sample from a normal, healthy individual immediately before administration of 500 mg of paracetamol. Then take urine samples at 1 hour, 2 hours, and 4 hours following the dose. (**Note. As with all drug-related tests, the drug should only be taken by healthy individuals with no indication of liver or kidney disease.**)

2.2. CE Conditions

Mode: Micellar electrokinetic chromatography
Capillary: 44 cm (37 to window) × 75 μm i.d. bare fused-silica
Buffer: 20 mM sodium tetraborate pH 10 containing 75 mM sodium dodecyl sulfate
Detection: UV detection (200 nm or variable wavelength or diode array)
Sensitivity: Approximately 50 maufs
Load: 4 sec hydrodynamic (approximately 50); this will depend on the urine concentration.

Run: 20 kV, for 15 min
Temperature: 25°C
The capillary is washed between runs with running buffer for 3 min. Prior to commencing the day's runs, the capillary is conditioned with 0.1 *M* NaOH for 10 min followed by fresh running buffer.

2.3. Analyses

Inject a 1 mg/ml solution of paracetamol in water.
Analyze each urine sample as above. If only fixed wavelength detection is available, run at both 200 nm and 272 nm.
Identify the peak for paracetamol in the urine by spiking a portion of the post-dose urine sample with an equal volume of paracetamol standard prior to injection.

3. RESULTS

Depending on the detector sensitivity and the wavelength used, a pattern of large and small peaks will be observed in normal urine when monitored at 200 nm. Relatively few peaks will be observed at 272 nm. The major peaks are creatinine, hippurate, urea, and uric acid,

1) Why are more peaks observed at 200 nm than 272 nm?
2) Using their structures, predict the migration order of the four major peaks in urine.

Following ingestion of paracetamol, a number of new peaks related to paracetamol and its metabolites will appear at both 200 nm and 272 nm. The ratio of these peaks will change with time until there is no paracetamol, and only signals from metabolites of this drug are seen in the electropherogram.

3) Calculate the concentration of paracetamol in urine samples taken at different time intervals after ingestion of the drug.
4) How, other than spiking, can the metabolites of paracetamol be determined? If diode array or scanning detection is available, peak identification can be performed by matching peak spectra to the paracetamol standard.

Index

biologically important, 459
carbohydrates, 456, 458
nucleosides, 453–455
nucleotides, 453–455
porphyrins, 458–459
thiols, 458
small exogenous molecules, separation of,
459–462
whole cells, separation of, 470–471
biological samples
description of, 443
detectability, 444
resolution, 443
selectivity, 444–446
sensitivity, 446–449
quantitative aspects, 451–452
sample preparation, 449–451
Borate-based electrolytes systems, for analyzing
carbohydrates
description of, 275–276
monosaccharides
derivatized, 306–312
underivatized, 303
oligosaccharides, underivatized, glycoaminoglycan-
derived, 314–318
Buffers
acidic, for separating protein glycoforms,
525–527
description of, 251
for enantiomer separation, 200
manufacturers and suppliers, 489
physical properties, 107
solutions, 491–492
types of, 251

C

Capillaries
coating of, 33–34
description of, 25
volume of, 26
Capillary electrophoresis
description of, 2, 4–5, 24–25
effluent, collection methods, 75–76
high-performance, *see* High-performance capillary
electrophoresis
history of, 5–6
home-built, 7
improvements in, 15–16
instrumentation, 6–8, 25–26, *see also specific mode*
mass spectrometry use with, 7
modes of
capillary gel electrophoresis, *see* Capillary gel
electrophoresis
capillary isoelectric focusing, *see* Capillary
isoelectric focusing
capillary isotachophoresis, *see* Capillary
isotachophoresis
capillary zone electrophoresis, *see* Capillary zone
electrophoresis
micellar electrokinetic chromatography, *see*
Micellar electrokinetic chromatography

modes of operation, 8
sensitivity improvements, 16
Capillary electrophoresis-mass spectrometry
advantages of, 55
deoxyribonucleic acid detection using, 414–415
description of, 7, 55, 59–61, 68
instrument developments
coupling to other ionization techniques, 72
description of, 67
mass analyzers, 73–74
sample loading enhancement, 69–72
separation formats, 67–69
interfaces for
coaxial configuration
continuous flow fast atom bombardment, 61
description of, 57
electrospray ionization, 58–61
design considerations, 57
liquid junction
continuous flow fast atom bombardment,
63–64
electrospray ionization, 61–63
sheathless, with microelectrospray ionization
application of, 66
comparison with coaxial interfaces, 65–66
description of, 64–65
ionization techniques
continuous flow fast atom bombardment, 57
description of, 56
electrospray, 56
ionspray, 56
mass analyzers, 73–74
methods of, 55
recent advancements in, 55
sample loading enhancements, 70
Capillary gel electrophoresis
fragment separation, experiment for, 529–532
for peptide and protein separation
applications, 381
classical *vs.* capillary techniques
description of, 375–376
Ferguson method, 377–381
mechanism, 381
sodium dodecyl sulfate mediated polyacrylamide,
374–376
Capillary isoelectric focusing
peptide and protein separation using
concomitant use with mass spectrometry, 392
description of, 382–384
without zone mobilization, 385–38390–392
zone mobilization, methods of
chemical, 385–386
electroosmotic, 387–390
hydrodynamic, 386–387
vs. slab gel, 382–384
Capillary isotachophoresis
description of, 95
for DNA analysis, 399
for DNA sample detectability, 407
for increasing sample load, 31–32, 373
manufacturers and suppliers, 489
of pharmaceutical drugs, 121

T

Tagging agents, *see also specific carbohydrate*
  for precolumn derivatization of carbohydrates,
      294–299
  for separation of derivatized homologous
      oligosaccharides, 322–324
Teicoplanin, chiral separation using, 230–231
Tetradecyltrimethyl ammonium bromide, 30–31
2,3,4,6-tetra-O-acetyl-β-D-glucopyranosyl
      isothiocyanate, for resolving R,S-
      naphthylethylamine, experiments,
      515–516
Thalidomide, 184–185
Thermo-optical absorbance
  deoxyribonucleic acid, 414
  principles of, 45
  procedure, 45–46
Tiselius, Arne, 1–2
Tricyclic antidepressants, capillary electrophoresis
      separation of, 121
Tryptophan, for direct UV detection of underivatized
      carbohydrates, 51, 284
TTAB, *see* Tetradecyltrimethyl ammonium bromide

U

Ubiquitous injection, of sample, 28–29
Ultraviolet absorbance detection
  direct
      detection limits of, 35
      for DNA analysis, 410–411
      oligosaccharides, 281
      tryptophan use, 51, 284
  indirect, 281–286
  of proteins, methods for enhancing sample
      detectability, 372–374
2-Undecyl-4-thiazolidine carboxylic acid, 216
Underivatized carbohydrates
  detection methods
      direct ultraviolet, 280–281
      electrochemical
          amperometric detection at constant potential,
              286, 289–290
          limits of, 288–289
          pulsed amperometric detection, *see* Pulsed
              amperometric detection
      indirect fluorescence
          description of, 281–286
          drawbacks, 281–286
      indirect ultraviolet
          description of, 281–286
          drawbacks, 286
  separation
      monosaccharides, 303–304

oligosaccharides
  glycoaminoglycan-derived, 313–318
  glycoprotein-derived, 318–320
  simple, 312–313
polysaccharides, 336–337
Urea, effect on chiral resolution using cyclodextrins, 199
Urine
  experimental assays
      nitrate and nitrite separation, 533–534
      paracetamol detection, 535–536
  screening of, for small molecules, 462–464

V

Vancomycin, chiral separation using, 227–229

W

Weak acids, dissociation constants of, capillary
      electrophoresis for, 16

Z

Zone electrophoresis, capillary
  applications, 67–68
  description of, 8, 136–137, 365
  for DNA analysis, 398–399
  free-solution, for separation of small organic
      molecules
      electroosmotic flow
          alterations in, 104–105
          description of, 100–101
          direct control of, 105
          illustration of, 100
          manipulation of, 104–105
          pH effects on, at constant ionic strength,
              103–104
          silica surface effects, 104–105
      principles of, 97–98
      sample injection, 110
          electrokinetic, 108
          hydrodynamic, 108–109
          stacking, 109
      separation efficiency
          description of, 99–101
          factors that limit, 99–100
          optimization
              complexation, 113–114
              description of, 110
              modifiers, 110–111
          peptide separations for, 112–113
          power dissipation effects on, 98–99
  and mass spectrometry
      description of, 59–61, 68
      sample loading enhancements, 70